Improving gut health in poultry

It is widely recognised that agriculture is a significant contributor to global warming and climate change. Agriculture needs to reduce its environmental impact and adapt to current climate change whilst still feeding a growing population, i.e. become more 'climate-smart'. Burleigh Dodds Science Publishing is playing its part in achieving this by bringing together key research on making the production of the world's most important crops and livestock products more sustainable.

Based on extensive research, our publications specifically target the challenge of climate-smart agriculture. In this way we are using 'smart publishing' to help achieve climate-smart agriculture.

Burleigh Dodds Science Publishing is an independent and innovative publisher delivering high quality customer-focused agricultural science content in both print and online formats for the academic and research communities. Our aim is to build a foundation of knowledge on which researchers can build to meet the challenge of climate-smart agriculture.

For more information about Burleigh Dodds Science Publishing simply call us on +44 (0) 1223 839365, email info@bdspublishing.com or alternatively please visit our website at www.bdspublishing.com.

Related titles:

Achieving sustainable production of eggs Volume 1: Safety and quality
Print (ISBN 978-1-78676-076-0); Online (ISBN 978-1-78676-078-4, 978-1-78676-079-1)

Achieving sustainable production eggs Volume 2: Animal welfare and sustainability
Print (ISBN 978-1-78676-080-7); Online (ISBN 978-1-78676-082-1, 978-1-78676-083-8)

Achieving sustainable production of poultry meat Volume 1: Safety, quality and sustainability
Print (ISBN 978-1-78676-064-7); Online (ISBN 978-1-78676-066-1, 978-1-78676-067-8)

Achieving sustainable production of poultry meat Volume 2: Breeding and nutrition
Print (ISBN 978-1-78676-068-5); Online (ISBN 978-1-78676-070-8, 978-1-78676-071-5)

Chapters are available individually from our online bookshop:
https://shop.bdspublishing.com

climate•SMART•publishing
IN AGRICULTURAL SCIENCE

BURLEIGH DODDS SERIES IN AGRICULTURAL SCIENCE
NUMBER 73

Improving gut health in poultry

Edited by Professor Steven C. Ricke,
University of Arkansas, USA

burleigh dodds
SCIENCE PUBLISHING

Published by Burleigh Dodds Science Publishing Limited
82 High Street, Sawston, Cambridge CB22 3HJ, UK
www.bdspublishing.com

Burleigh Dodds Science Publishing, 1518 Walnut Street, Suite 900, Philadelphia, PA 19102-3406, USA

First published 2020 by Burleigh Dodds Science Publishing Limited

Library of Congress Control Number: 2019952041

British Library Cataloguing in Publication Data
A catalogue record for this book is available from the British Library

ISBN 978-1-78676-304-4 (Print)
ISBN 978-1-78676-307-5 (PDF)
ISBN 978-1-78676-306-8 (ePub)
ISSN 2059-6936 (print)
ISSN 2059-6944 (online)

DOI 10.19103/AS.2019.0059

Typeset by Deanta Global Publishing Services, Dublin, Ireland

Contents

Series list

Title	Series number
Achieving sustainable cultivation of maize - Vol 1 From improved varieties to local applications Edited by: Dr Dave Watson, CGIAR Maize Research Program Manager, CIMMYT, Mexico	001
Achieving sustainable cultivation of maize - Vol 2 Cultivation techniques, pest and disease control Edited by: Dr Dave Watson, CGIAR Maize Research Program Manager, CIMMYT, Mexico	002
Achieving sustainable cultivation of rice - Vol 1 Breeding for higher yield and quality Edited by: Prof. Takuji Sasaki, Tokyo University of Agriculture, Japan	003
Achieving sustainable cultivation of rice - Vol 2 Cultivation, pest and disease management Edited by: Prof. Takuji Sasaki, Tokyo University of Agriculture, Japan	004
Achieving sustainable cultivation of wheat - Vol 1 Breeding, quality traits, pests and diseases Edited by: Prof. Peter Langridge, The University of Adelaide, Australia	005
Achieving sustainable cultivation of wheat - Vol 2 Cultivation techniques Edited by: Prof. Peter Langridge, The University of Adelaide, Australia	006
Achieving sustainable cultivation of tomatoes Edited by: Dr Autar Mattoo, USDA-ARS, USA & Prof. Avtar Handa, Purdue University, USA	007
Achieving sustainable production of milk - Vol 1 Milk composition, genetics and breeding Edited by: Dr Nico van Belzen, International Dairy Federation (IDF), Belgium	008
Achieving sustainable production of milk - Vol 2 Safety, quality and sustainability Edited by: Dr Nico van Belzen, International Dairy Federation (IDF), Belgium	009
Achieving sustainable production of milk - Vol 3 Dairy herd management and welfare Edited by: Prof. John Webster, University of Bristol, UK	010
Ensuring safety and quality in the production of beef - Vol 1 Safety Edited by: Prof. Gary Acuff, Texas A&M University, USA & Prof. James Dickson, Iowa State University, USA	011
Ensuring safety and quality in the production of beef - Vol 2 Quality Edited by: Prof. Michael Dikeman, Kansas State University, USA	012
Achieving sustainable production of poultry meat - Vol 1 Safety, quality and sustainability Edited by: Prof. Steven C. Ricke, University of Arkansas, USA	013
Achieving sustainable production of poultry meat - Vol 2 Breeding and nutrition Edited by: Prof. Todd Applegate, University of Georgia, USA	014

Acknowledgements

We wish to acknowledge the following for their help in reviewing particular chapters:

- Chapter 1: Emeritus Prof. Gerald Havenstein, North Carolina State University, USA
- Chapter 2: Prof. Jana Seifert, University of Hohenheim, Germany
- Chapter 10: Dr Doug Korver, University of Alberta, Canada
- Chapter 12: Dr Richard K. Gast, U. S. National Poultry Research Center - USDA-ARS, USA
- Chapter 13: Dr Steve Foley, FDA, USA
- Chapter 14: Dr Christi Swaggerty, USDA-ARS, USA
- Chapter 15: Prof Glenn Gibson, University of Reading, UK
- Chapter 16: Dr Mike Bedford, Research Director – AB Vista, UK; and Dr Maria Siwek, University of Science and Technology – Bydoszcz, Poland
- Chapter 17: Prof. Richard Ducatelle, University of Ghent, Belgium

Introduction

This collection summarises current research on the composition and function of the gastrointestinal tract in poultry, the factors that affect its function, and nutritional strategies to optimise poultry nutrition, health and environmental impact. Part 1 begins by summarising advances in sequencing and omics technologies to understand gut function. It then reviews our current understanding of the gut microbiota, the development of the gut microbiome over the life of the bird, and gut function in nutrient processing and immune response.

The second part of the book reviews what we know about factors affecting gut function and health. Chapters cover gastrointestinal diseases, the interaction between pathogens and the gut as well the impact of antibiotics. The final group of chapters discuss current research on the effectiveness of feed additives in optimising gut health, including probiotics, prebiotics, synbiotics, antimicrobials, essential oils and other botanicals as well as cereal grains.

Part 1 Understanding the gastrointestinal tract

Chapter 1 provides an overview of the development of the commercial poultry industry, and the impact of changes in production practices on poultry gut health. Beginning with an introduction to the origins of the broiler chicken, the chapter discusses developments in nutrition, genetic selection, poultry housing and veterinary care. All of these areas have created challenges in optimizing gut health which are explored in the rest of the book.

Chapter 2 summarizes the range of molecular tools used to analyze the gut microbiome in poultry such as T-RFLP, DGGE, and TGGE and clone library sequencing using the Sanger method. The chapter goes on to survey the development of next-generation sequencing techniques such as Roche 454, Illumina, and Ion Torrent. The chapter also discusses third-generation sequencing techniques including Pacbio SMRT sequencing and Oxford nanopore. Finally, the chapter introduces other 'omics' approaches such as metagenomics, metatranscriptomics and metaproteomics which are help advance our understanding of microbiome functions in poultry.

Building on Chapter 2, Chapter 3 summarizes the use of different 'omics' technologies (e.g. genomics, transcriptomics, proteomics, metabolomics) used to identify the response of chicken intestinal cells to various effectors. The chapter begins by introducing the functions, physiology and microbiota of the gastrointestinal tract in chickens. It then explores what omics technologies can tell researchers about gut function, using the example of the chicken intestine.

The chapter concludes with a case study on the proteomic analysis of the mucosal layer of the chicken gut.

Poultry gut microbiota include bacteria, archaea, protozoa, fungi and viruses. Their role in the gastrointestinal tract has a profound effect on the health and productivity of poultry. Chapter 4 outlines what we know about the establishment and development of the gut microbiota, some of the mechanisms by which the microbiota can affect poultry, and the ways that the microbiota can be manipulated to enhance poultry health and productivity. The chapter reviews the manipulation of microbiota for chicken health and productivity, covering the use of microbes to manipulate gut microbiota and also the use of feed additives and antibiotics. The chapter concludes with an overview of future trends in research.

Chapter 5 discusses the *in ovo* development of the chicken gut microbiome and its impact on later gut function. The gut of hatchlings contains both beneficial as well as pathogenic microorganisms derived from external and maternal sources. Research has focused on ways to optimize the enteric development of chicks and to assist in the establishment of intestinal bacteria populations that promote health and provide protection against invading pathogens. The chapter examines competitive exclusion cultures, specifically probiotics, prebiotics, synbiotics, and nutrients. The chapter looks at the prospects for the commercial *in ovo* use of these biologics and looks ahead to future research trends in this area.

In poultry production a tradeoff has traditionally been made between growth and efficiency, on the one hand, and immune potential and disease resistance, on the other. The emerging field of immunometabolism is an opportunity to eliminate this tradeoff and achieve both production efficiency and immune robustness. Chapter 6 provides an overview of metabolism and immunometabolism, including the most important links between metabolic pathways and immune pathways. The chapter discusses the absorption and metabolism of carbohydrates, amino acids and lipids in the poultry gut.

Chapter 6 then discusses metabolism and immune responses within the gut tissue, influenced by the feed and microbiota located in the lumen. It reviews the components of feed that lead to inflammation and how to mitigate this effect. The chapter concludes by looking at examples of how feeding the immune system with pre- and probiotics can both enhance growth and immune response in poultry.

Chapter 7 examines intestinal immunity and microbiota interactions with the immune system. The chapter considers the role of gut microbiota as an epigenetic regulator of gut function as well as the causes of dysregulation of gut functionality. As the chapter points out, the host immune response is important to maintaining microbial balance but can also be the cause of a disrupted microbiota (dysbiosis) which can contribute to disease. The chapter discusses microbiota interactions with the immune system, including microbiota-based metabolites and immunity as well as colonization resistance. It then focuses on dysregulation of gut functionality, looking at the causes of chronic, low grade inflammation, sterile and metabolic inflammation as well as pathobiont expansion, with an emphasis on nutritional strategies to avoid these conditions.

Part 2 Factors that impact the gastrointestinal tract and different types of birds

A comprehensive understanding of how various factors shape the intestinal microbiota in poultry can help develop new dietary and managerial interventions to enhance bird growth, maximize feed utilization efficiency, and lower enteric diseases caused by pathogens. Chapter 8 reviews the current understanding of how different factors (except diet and growth promoters that are covered in other chapters of this book) can affect the intestinal microbiota. These factors include genetics and breeds, hatchery conditions and environment, bedding and litter, climate and geographic regions, gender and diseases. This understanding provides the foundation for developing the nutritional and other management practices needed to optimize gut function.

As antibiotic resistance continues to evolve, finding alternatives to these chemical compounds that increases poultry performance has become imperative. Some of the most promising alternatives that been investigated include: bacteriocins, bacteriophage therapy, plant-derived phytochemicals, competitive exclusion of pathogens, and predatory bacteria. Chapter 9 places the use of antibiotics in poultry production in its historical context to understand the benefits that antibiotics have conferred on animal production to date. The chapter focuses in particular on the potential use of bacteriocins and plant-derived phytochemicals to replace the growth promoting and health benefits of the sub-therapeutic levels of antibiotics.

Gastrointestinal health plays a critical role in ensuring the overall health and productivity of livestock, including poultry. Antimicrobial growth promoters (AGP) have been used to maintain and promote gastrointestinal health. The move to phase out AGP in poultry production threatens to increase the incidence of enteric diseases such as necrotic enteritis. Chapter 10 reviews what we know about important enteric diseases and disorders, highlighting their etiology followed by possible nutritional interventions. These include feed additives such as plant-derived extracts, prebiotics, probiotics and organic acids, as possible alternatives to AGPs for disease control.

Although the majority of the microbiota in the poultry gut are commensal bacteria, pathogens are also present. Commensal and pathogenic microbes interact with each other, either positively or negatively, profoundly affecting host nutrition and incidence of infection. A better understanding of the gut pathogen-microbiota interaction is essential to address the current challenges in poultry production. As Chapter 11 highlights, recent studies using metagenomics have provided new insights into the interactions between the gut pathogens and commensal microbes in poultry. Chapter 11 reviews current understanding of the interaction between gut microbiota and pathogens in poultry. The intestinal pathogens discussed in this chapter include *Escherichia, Salmonella, Clostridium, Campylobacter, Eimeria* and viruses. In each case, the chapter summarizes what we know about these pathogens and their associated diseases, interactions with gut microbiota and what this means for health and nutrition.

Chapter 12 reviews current knowledge about the function and microbial ecology of the layer hen gastrointestinal tract. As the chapter points out, with the introduction of next generation sequencing, a more comprehensive identification of the laying hen gastrointestinal tract microbial population has emerged. This research has shown there are several factors that can influence the composition and function of the layer hen gastrointestinal tract, including age of the bird, diet, and type of feed amendment. Studies have identified the microbial communities in each compartment of the layer hen gastrointestinal tract and their impact on the host. Some compartments such as the ceca harbor a highly complex microbial population of fermentative microorganisms that produce short chain fatty acids. The ceca can also be colonized by foodborne *Salmonella* and some serovars such as *S.* Enteritidis can become invasive, infecting the reproductive tissues. The chapter shows how a variety of feed additives have been used to limit *Salmonella* colonization in laying hens and improve laying hen performance.

Part 3 Feed additives and gut health modulation

Foodborne pathogenic bacteria are all too often found as commensal or transient organisms in the gastrointestinal tract of poultry. Many of these organisms do not reveal themselves through illness in the bird, although some do. This means it is important find ways to apply treatment to all members of a flock, rather than simply treating 'sick' birds. After summarizing what we know about the gastrointestinal microbiota of poultry, Chapter 13 reviews alternatives to the use of antibiotics, discussing the use and effectiveness of organic acids, bacteriophages, sodium chlorate, and pro- and prebiotics.

Building on the overview in Chapter 13, the next group of chapters look at key feed additives, starting with Chapter 14 on probiotics. The chapter reviews current research on the safety and efficacy of individual monocultures for prophylactic and/or therapeutic use against *Salmonella* infections in poultry, under both laboratory and field conditions. There is a particular focus on the role of probiotics in preventing inflammation. The chapter discusses key issues and advances in the development of novel, cost-effective, feed-stable direct-fed microbials with potential for widespread application in poultry production. The chapter concludes with a review of the use of direct-fed microbials in commercial poultry diets.

As Chapter 15 highlights, prebiotics have been established as a series of feed compounds that serve as specific substrates for gastrointestinal tract (GIT) bacteria. Such compounds support those GIT bacteria that benefit the host and, in addition, can be antagonistic to foodborne pathogens and prevent their colonization in the GIT. As the chapter shows, prebiotics have been used primarily to prevent establishment of foodborne pathogens but have also received attention regarding their impact on overall GIT health. The chapter reviews the impact of prebiotics on bird health, GIT function, and prevention of foodborne pathogen GIT colonization. There is a particular focus on the impact of prebiotics on avian upper GIT health and optimizing function of the avian caecum.

As previous chapters have highlighted, the drive to ban the use of antibiotics in animal feed due to the current concern over the spread of antibiotic resistance genes makes the development of alternative prophylactics imperative. Chapter 16 reviews the combination of probiotics and prebiotics in synbiotics. It focusses particularly on the use of short chain fatty acids (SCFA) (especially butyrate). The chapter reviews the beneficial effects of SFCA on digestive physiology, blood flow and muscular activity, enterocyte proliferation and mucin production

As Chapter 17 points out, short chain organic acids have been employed as feed additives for a number of years. They have been primarily used for their antimicrobial properties, particularly in limiting *Salmonella* in feed and in the GIT. Short chain organic acids are also produced by indigenous gastrointestinal bacteria during fermentation. These are primarily generated in the cecum which is the site where most GIT microbial fermentation occurs. The chapter reviews current research on poultry GIT responses to short chain organic acids generated by GIT fermentative microorganisms, and how this can be optimized through feed interventions.

Chapter 18 discusses the role of essential oils and botanicals in improving gut function in poultry. The chapter focusses on four major functions of phytobiotics that could potentially contribute to gut health. These relate to digestive conditioning, antimicrobial properties, immunomodulation and gut microbiota modulation The chapter discusses the impact of these functions on performance as well as carcass and egg quality. Finally, the chapter looks ahead to future research trends in this area.

As Chapter 19 points out, cereal grains constitute the greatest proportion of most poultry diets. As dietary components, cereal grains provide most of the dietary energy and help to support the development of the structural and functional integrity of the digestive tract. However, many cereal grains contain one or more deleterious factors which may negatively affect the structural and functional development of the gut. Some of the key factors are carbohydrate in nature but there are non-carbohydrate fractions which function mainly as anti-nutrients. This chapter reviews what we know about the key components of cereal grains, particularly the components that may influence the development of intestinal structure and function. The chapter examines the role of cereal grain components on poultry gut function and the possible mechanisms by which these interactions take place and can be optimized.

Part 1

Understanding the gastrointestinal tract

Chapter 1

Commercial poultry production and gut function: a historical perspective

Dana Dittoe and Steven C. Ricke, University of Arkansas, USA; and Aaron Kiess, Mississippi State University, USA

1 Introduction

In 2016, it was estimated that the United States poultry industry provided 1 682 269 jobs with a total economic impact of $441.15 billion (NCC, 2016). The US industry also provided $68 billion in wages and $34 billion in government revenue (NCC, 2016). The broiler industry alone was estimated to provide 1 195 745 jobs with $68 billion in wages and contributing $313.12 billion in economic output (NCC, 2016). As of 2017, the US industry produced $42.7 billion in sales from the production of broilers, eggs, and turkeys (USDA, 2018). Broilers alone contributed 71% of annual sales amassing to $30.2 billion in 2017 (USDA, 2018), and in that year 8.91 billion broilers were produced. Thus, the US poultry industry contributes substantially to the economic welfare of this country and will continue to do so as it expands its markets, both domestically and internationally.

Since the beginnings of the modern industry in the early 1900s, the commercial industry has continued to advance and expand substantially. From

http://dx.doi.org/10.19103/AS.2019.0059.01

2016 to 2017, the industry grew 10% in sales (USDA, 2018), and this growth will continue because the demand for poultry and poultry products in the United States and around the world continues to rise. In 1960, the total amount of poultry consumed in the United States (chickens and turkeys) was 34.2 pounds (15.5 kg) per capita (NCC, 2019a). In 2018, the per capita consumption of total poultry was 110.0 pounds (49.9 kg), and it is forecasted to be 110.1 pounds (50.21 kg) in 2020 (NCC, 2019a).

The poultry industry will need to continue to grow to meet the demand related to the projected increases in our human population. In 2018, the US population was 326 766 748, but by 2050 it is projected to be 390 million (UN, 2017). Furthermore, the world's population will be 9.6 billion by 2050 (UN, 2017). To meet the rising demand for poultry products, both domestic and internationally, integrators will have to continue to develop and improve sustainable practices that will allow for more and more efficient production. Consumer demands for alternative production practices, however, including the demand for the removal of antibiotic from poultry feeds, have led to a shift in industry response. Some integrators have removed antibiotics from their feeds or they have removed antibiotics completely from the rearing process. Unfortunately, that removal has presented a major challenge to the industry due to increased disease levels, increased mortality, and losses in growth performance related to the resulting changes and imbalances in gut microbiota or 'dysbacteriosis' (Huyghebaert et al., 2011).

Due to the vast impact that the commercial poultry industry has on our human society, as well as its impacts on the economy of the United States, it is important to grasp the beginnings, the advancements, and the issues the industry has had to face in order to become the vertically integrated power house that it is today. The most significant of those advancements revolve around the incorporation of scientific advances made in nutrition, genetics, housing, and veterinary care. Therefore, it is the objective of the current chapter to provide a brief description of the history and development of the commercial poultry industry, and the impact of changes in production practices that are having impacts on poultry gut health.

2 Origins of the broiler chicken

In order to understand the current state of the poultry industry, it is important to understand the history of chicken domestication. The modern chicken, *Gallus gallus domestica*, is believed to have originated from the red jungle fowl of Asia (Sawyer, 1971). The red jungle fowl is one of four species within the jungle fowl genus, *Gallus*. That genus encompasses *Gallus gallus* (red jungle fowl), *Gallus varius* (green jungle fowl), *Gallus sonneratii* (grey jungle fowl), and *Gallus lafayetii* (ceylon jungle fowl) (Al-Nasser et al., 2007). *Gallus gallus*, the red jungle fowl, can still be found in regions of India, China, Java, Malaysia, Indonesia, and the Philippines (Al-Nasser et al., 2007). With the development of human societies, domestication of the red jungle fowl is thought to have occurred primarily due to its use for cultural and entertainment purposes (Crawford, 1990b). The earliest domestication is thought to have started around 5400 BC; however, it is also believed that most of the modern breeds were domesticated and developed primarily between 2500 and 2100 BC in the Harappan culture of the Indus Valley (Crawford, 1990a,b). Today's *Gallus gallus domesticus* is believed to have its origins about 3000 years ago (Crawford, 1990a).

Even before the domestication of the red jungle fowl, poultry had been utilized for both meat and egg production. Although it was not an uncommon occurrence to eat poultry in historic times, poultry was not the primary source of protein for humans. Chickens were used mainly for cockfighting until the mid-1800s when cockfighting was deemed illegal. After that occurred, chickens were sought after for exhibition purposes (Moreng and Avens, 1985; Crawford, 1990b). Even so, the consumption of *Gallus gallus* by humans was considered a luxury, and it generally occurred only at special events.

Domesticated chickens were brought to North America in 1607 by the initial settlers of Jamestown (Sawyer, 1971). The early breeds of domestic fowl were so numerous in the United States that an 'American Standard of Excellence' was created in 1873 to standardize the characteristics and requirements of the various so-called pure breeds of poultry in America (Sawyer, 1971). By the nineteenth century, chickens were an essential part of American agriculture, and by the onset of the twentieth century, chickens were commonly found on almost all farms. During the early twentieth century, chickens were mainly kept for their ability to produce eggs and the birds were used as a source of protein only on rare occasions (Sawyer, 1971).

It was not until 1923 that chickens were raised solely for meat consumption. In that year, Cecile Steele of Ocean View, Delaware, was mistakenly delivered 500 chicks instead of the 50 she had ordered (Williams, 1998). She decided to raise the extra birds as small broilers, and subsequently sold 387 chickens for 62 cents per pound to local families (Sawyer, 1971). Shortly thereafter, other surrounding families, hearing of her good fortune, sought to also profit from raising broiler chickens; and by 1925, Delaware was producing approximately 50000 broiler chickens per year (Sawyer, 1971). At about the same time, a disease that was referred to as 'range paralysis' hit the southern portion of Delaware and affected Leghorns of 12 weeks of age or older that were being used for egg production. This helped make younger birds of heavier breeds as the best alternative for poultry meat production, which increased the growth of the broiler industry (Sawyer, 1971). Although Cecile Steele has been noted by some as the woman who started the poultry meat industry, the first commercial broiler operation was actually established in 1880 in Hammonton, New Jersey

(Strausberg, 1995). Soon thereafter, the raising of chickens for meat production became common place in the northern and middle eastern parts of the United States and southern farmers soon began following the lead of their northern neighbors in producing chickens for meat consumption.

During the beginning of the twentieth century, cotton was the main cash crop for the southeastern United States. However, as times changed and government agencies became involved, cash incentives were provided for farmers who would stop producing cotton through the Agricultural Adjustment Administration (1933) and later through the 1936 Soil Conservation and Domestic Allotment Act (Gisolfi, 2006). This drastically changed the agricultural landscape of the south and spurred major changes in agricultural production. Georgia was especially affected by the diminishing harvest of cotton, and many of the cotton growers who remained no longer had a market in which to sell their products. This led individuals such as Jesse Jewell to enter the poultry raising industry (Sawyer, 1971; Gisolfi, 2006). Jesse Jewell and many others began selling chicks and feed on credit to farmers who eventually became known as poultry growers. When the birds reached their market weights, generally at about 12 to 16 weeks of age, they were hauled to a processing plant, sold, and the creditors and the growers were paid off (Gisolfi, 2006). Later, this method would eventually be coined vertical integration (Gisolfi, 2006). The decrease in cotton production, and the shift toward broiler production in Georgia would increase that state's poultry production to about 500 000 broilers by 1935 (Gisolfi, 2006). Although Georgia played an important role in the beginnings of the US broiler industry, a number of other southern states also made major contributions to the onset of the broiler industry.

Arkansas has not been widely accepted as the first state to raise poultry for meat, but it is well documented that the Arkansas broiler industry also had its beginning very early in the twentieth century. In fact, Arkansas' beginning was very similar to that of the Delmarva area. Arkansas found its start in 1916 when J. J. Glover and his daughter Edith raised 20 White Wyandottes purchased by mail, sold them for $1 a bird, and coined them as 'Arkansas Broilers' (Strausburg, 1995). However, poultry was not widely produced during this time as apples were the main commodity produced in northwest Arkansas in the 1920s, with there being over four and a half million apple trees in the state at that time (Strausberg, 1995). Around the same time, the northwestern Arkansas apple industry experienced drought and blight brought on by coddling moths, forcing the area to pursue other alternative commodities (Strausberg, 1995). One of those commodities was poultry. In 1921, the first Arkansas poultry hatchery was built by Jeff D. Brown, and in 1922, J. J. Glover raised 324 broilers with the promise of 55 cents of profit per bird (Strausberg, 1995). Although northwest Arkansas did have existing poultry farms, it was not until 1927, when a severe freeze ravished the apple industry, that poultry production became a highly sought-after activity (Strausberg, 1995). In the same year, Lloyd Peterson of Decatur, Arkansas, raised 500 Rhode Island Red chickens and made a $107 profit (Strausberg, 1995). As a result, Jeff Brown's hatchery group began lending credit to farmers in the form of chicks, feed, or both to entice others to invest in commercial broiler production (Strausberg, 1995). With assistance from county Extension agents, and scientists at the University of Arkansas and the Arkansas Experiment Station, Arkansas would eventually become one of the major leaders in the poultry industry, producing 6.9 billion pounds of broiler meat in 2017 (The Poultry Federation, 2017). Furthermore, a number of poultry companies such as George's Inc., Simmons Foods Inc., Cobb-Vantress Inc., Tyson Foods Inc., OK Foods Inc., and Twin River Foods, Inc. all originated in the state of Arkansas (The Poultry Federation, 2017).

As the broiler industry expanded after its humble beginnings, it did so with the support of the assistance provided by the US government and universities. Basic support for the modern poultry industry was really initiated during the late 1800s and early 1900s with the passage by the US government of the federal Hatch Act. That Act established the Land Grant University system for each state, and the Smith Lever Act which followed provided recurring funding for Agricultural Experiment Stations and extension programs in each state. A large number of the Land Grant universities developed poultry programs during the above time frame and hired professionals for doing teaching, research, and extension programming for poultry producers. At one time, 44 university-based Poultry Science programs were in existence in North America, but during the 1970s through to the 1990s, as poultry operations became larger and the industry became more and more concentrated in certain geographical areas of the country, many of those departments were either closed or merged with Animal Science Departments. So, by the early part of the twentieth century, only six Poultry Science Departments continued to exist, and all of those units are currently located in the Southeastern United States. A few other universities, however, have continued some poultry research, extension, and teaching as part of their integrated Animal Science Departments, and, most Veterinary Colleges also included poultry-related teaching, research, and extension activities. Contributions of these university-based programs were extensive, and probably the best example of how extensive those university-based poultry programs were can be seen via the history that Havenstein (2012) developed related to the poultry programs that were carried on at North Carolina State University from the year 1881 through 2010. All of the current and former university-based programs around the nation made similar contributions to the growth of the industry.

Of course, the US Department of Agriculture for many years has also led extensive research and development programs related to poultry. Numerous state- and/or regional university-sponsored extension conferences on nutrition, poultry management, poultry waste, and by-product management, and so forth, have also contributed greatly to the industry's growth and development; as have many annual state and national industry conventions and technical programs such as Fact Finding, the Southeastern Egg and Poultry Association, the Pacific Egg and Poultry Association, the National Poultry Breeder's Roundtables, and through meetings sponsored by a number of state-based industry associations. All of these university-, federal-, and industry-based programs have provided and are continuing to provide valuable inputs to the development of today's industry.

Even though the US broiler industry was started in the early 1900s, broiler production did not become a fully developed agricultural industry with major economic impact until the onset of the Second World War (Williams, 1998). Since then, the poultry industry has steadily grown and made drastic changes in its breeding, its production management (i.e. in hatchery operations; grow-out housing; ventilation and light control systems; feed milling systems; transportation systems; and litter source and management systems, etc.), its nutrition, and its animal health veterinary care practices (Hunton, 1990). Even with all of the developments that the industry has significantly benefitted from, it still faces many challenges for the future.

3 Vertical integration

Poultry production began as family backyard operations, and subsequently grew into a multibillion-dollar industry with approximately 35 major poultry businesses that control its operations from start to finish (NCC, 2018b). The rapid and successful development of the industry can largely be attributed to its willingness to incorporate scientific advancements, and to vertically integrate the various aspects of its overall production and marketing processes. The term 'vertical integration' was coined by Jesse Jewell of Gainesville, Georgia who in the mid-1930s was a visionary who had the foresight to begin furnishing farmers with chicks and feed on credit until the birds were heavy enough to sell back and settle the debt (Gisolfi, 2006). To begin this concept, Jewell contacted feed companies such as Ralston-Purina and Quaker Oats and local banks to receive credit for the feed and chicks (Gisolfi, 2006). He received credit easily and in turn, extended credit to farmers who were given baby chicks and feed in advance of their growouts. Balances were settled 12 to 16 weeks later when their broilers reached market weight, or when another processing merchant or distributor purchased them (Gisolfi, 2006). In the mid-1930s, farmers were not required to sell the broilers back to the dealer who had provided them as baby chicks, because farmers were not contractually bound to a specific processor.

By 1940, poultry businesses purchased and owned hatcheries, distribution facilities, and processing plants (Gisolfi, 2006). These early integrators limited their reliance on outside sources to maximize profits. They subsequently began to create contracts in the mid-1950s with individual farmers or growers and eventually implemented a contract payment system that was based on body weight and a 'feed-conversion plan' (Gisolfi, 2006). The contracts defined expectations for the farmers who supplied their own growing facilities, equipment, labor, heat, and litter; and, it indicated that they were to sell the grown broilers solely to the supplier they were in contract with. The 'feed-conversion plan' represented a method to determine the income the farmers would receive based on the pounds of chicken the merchants received from the pounds of feed the farmers used to raise the birds (Gisolfi, 2006). The contracts and the 'feed-conversion plan' may have taken away the independence of the farmer involved; however, it did establish uniform practices for contractual farming as a part of this vertically integrated business.

The business style of merchants such as Jesse Jewell and many others revolutionized the industry of today where vertical integration is still heavily relied upon. It is also apparent that the advancements made, especially in genetics (Havenstein et al., 2007), nutrition, housing, veterinary care, transportation, processing, product development, and marketing, have jointly allowed the broiler industry to become a much more efficient business, and one that has expanded to become the leading supplier of meat for human consumption worldwide. As such, the industry is an integral part of the US economy and the United States possesses the largest broiler chicken and turkey industries in the world. The US public consumes more chicken per capita now than any other country in the world, having consumed more than 93.5 pounds per capita in 2018 (NCC, 2019b). The wholesale value of industry shipments and consumer retail expenditures for chicken was 65 and 95 billion dollars, respectively, in 2018 (NCC, 2019b). Since the broiler industry is such an important aspect of the US economy, and of our citizens' diets and lifestyles, it is evident that the industry must continue to be vigilant in order to meet consumer demands in the future.

4 Nutrition

As the poultry industry on the eastern shore began to flourish in the Delmarva area in 1925, and then in the southern United States about 10 years later, so did the development of poultry feed systems and diets. Although Delmarva alone had ten major feed manufacturers present during the onset of the broiler industry, poultry feeds were primarily supplied by local feed mills, who held monopoly over the market (Sawyer, 1971). Large feed milling companies were more competitive, however, because of their greater resources which allowed them to develop substantial research on improving broiler diets. In 1925, the first complete broiler feed was developed by The Beacon Milling Company

(Sawyer, 1971). And as early as 1929, that company developed and introduced a coccidiosis control mash that drastically changed the developing field of poultry nutrition (Sawyer, 1971).

At the conclusion of the Second World War, the first high-energy broiler diet was introduced into the broiler industry (Sawyer, 1971). However, that diet was inadequate in its energy-protein ratio, and because of that the concept of 'calorie-protein ratio' was first introduced in 1955 (Sawyer, 1971). In an attempt to describe the energy-protein content in feed, in 1940, scientists at the Texas Agricultural Experiment Station investigated the metabolizable energy (ME) content in poultry feedstuffs (Fraps et al., 1940). Although the same group proposed a system for determining productive energy (PE), PE was shortly replaced with ME determination (Elwinger et al., 2016). In the second half of the twentieth century, research determining ME was conducted to evaluate apparent ME when corrected for zero nitrogen retention (AMEn) (Elwinger et al., 2016). Not long after, Ian Sibbald published an alternative method for determining ME that became known as the true metabolizable energy (TME) (Sibbald, 1976). Scientific advancements in the understanding of energy composition from poultry diets helped create our modern poultry feeding programs. However, that was only one of the components that contributed over the years to the development of improved nutritional practices for the poultry industry.

The primary components of poultry diets include water, carbohydrates, fats, proteins, minerals, and vitamins. The coining of the term 'protein' was first described in 1834, so throughout the second half of the nineteenth century biochemists and nutritionists began determining and elucidating the basic substructures of various proteins, the amino acids. At the onset of the first World Poultry Congress held in 1921 in The Hague, the Netherlands, there were no technical papers related to the evaluation of protein content in poultry diets. However, during the second Congress (Barcelona, Spain, 1924), there was one paper that described the protein structure of 30 different feedstuffs, as well as the digestibility coefficients of those feedstuffs. During the third Congress (Ottawa, Canada, 1927), a joint discussion of the importance of the theoretical and experimental foundation of protein requirements was held (Elwinger et al., 2016). Although the National Research Council's (NRC) *Nutrient Requirements of Poultry* was first published in 1944 (NRC, 1944), it was not until 1954 that the NRC elaborated on the actual requirements for crude protein, essential amino acids, and glycine for chicks, poults, and laying hens (NRC, 1954). Within the same 1954 publication, the NRC also described the arginine, lysine, methionine, cysteine, tryptophan, and glycine contents of various poultry feedstuffs (NRC, 1954).

After the 1950s, there were numerous elaborate attempts to better describe the requirements for protein and essential amino acid requirements of different poultry species, breeds, ages, and sex (Elwinger et al., 2016). The rapid expansion of knowledge through research led the NRC to compile the results from around 440 experiments and to publish those compilations in the NRC guidelines in 1994 (NRC, 1994). More recently, a meta-analysis was published by Sauvant et al. (2008) describing the various requirements affected by the previously mentioned fixed factors. However, there continues to be a need for updated requirements, so in 2016, the NRC announced that there would be a renewed effort to update the NRC *Nutrient Requirements of Poultry* (Elwinger et al., 2016) in the near future.

One of the greatest advances in poultry nutrition was initiated through the discovery of the animal protein factor (APF), B_{12} from the liver in 1948 (Rickes et al., 1948; Smith et al., 1952). Long before that, vitamins were known to be necessary components of poultry diets. As early as 1913, vitamin A (retinol) was detected by McCollum (Semba, 2012) and by the 1940s, all of today's vitamins had been elucidated with the last being vitamin B_{12} (Elwinger et al., 2016). Before the identification of vitamins was made, the poultry industry fed various microbial, plant, and animal origin ingredients in attempts to benefit production. After the identification of all of the various vitamins, it was realized that the supplementation of poultry diets with grain germs and maize germ oil, dried beet pulp, yeast, fish oil, and dried fish soluble were due to the presence of ß-carotene, vitamin E, vitamins of B-complex except B_{12}; vitamin D_2 after UV exposure; vitamins A, D_3, B_{12}; and vitamins of the B-complex, respectively (Elwinger et al., 2016).

In addition to the discovery of vitamins, improved knowledge of minerals and other feed additives and their requirements has also increased the success of the poultry industry. Minerals are required by all animal species for the formation and replacement of skeletal structure. Minerals also contribute to poultry health by being activators of enzymes and hormones, and through the maintenance of osmotic and acid-base homeostasis (Elwinger et al., 2016). For laying hens specifically, calcium (Ca) and phosphorus (P) and carbonate ions are required for metabolism to promote eggshell formation (Elwinger et al., 2016). Previously, it was believed that 1% Ca and 0.5% of available P would meet the requirements of growing chicks; however, this was costly and led to increased excreted P from poultry (Elwinger et al., 2016). The elucidation of the available phytase activity in cereal grains was determined in the early 1940s and promoted the decreased use of inorganic phosphates in poultry diets. The available phosphorus from wheat, triticale, and barley was high (50–70%), while the available P from corn, leguminous grains, soya meal, and rapeseed meal was less than 25% (Hoshi and Yoshida, 1977; Sauveur, 1983). Another advancement in meeting the optimal P level in poultry diets was the development of exogenous microbial phytases in the 1970s by Nelson et al. (1971). After the first phytase feed enzyme became commercially available in the Netherlands (Simons et al., 1990), the use of phytase became widespread and led to the innovative discovery that bone mineralization should be utilized to adequately determine and reduce the supplementation of C and P in poultry diets (Letourneau-Montminy et al., 2010). With phytase becoming a valuable resource to properly meet P requirements in poultry diets, other feed additives have been developed to meet requirement demands.

Although exogenous enzymes have been utilized in poultry diets as early as 1925 (Clickner and Follwell, 1925), their widespread distribution was not available until the 1980s (Elwinger et al., 2016). Many of the enzymes developed in the 1980s were created to degrade the non-soluble polysaccharides (NSP) that are present in various feedstuffs, thereby allowing for better nutrient absorption. Exogenous enzymes such as β-glucanase, xylanase, and protease have been effective in improving the nutritive value of barley (Hesselman and Aman, 1986), wheat and rye (Petterson and Aman, 1988), and protein sources, respectively. Currently, additional exogenous enzymes are being developed to further enhance nutrient absorption and reduce the overuse of feedstuffs in poultry diets, thus enabling poultry nutritionists to make considerable advancements in feed formulation.

Improvements in feed formulation were substantial from 1929 to 1969 and contributed to increased average bird weight from 2.82 pounds (1.28 kg) to 3.81 pounds (1.73 kg) (Sawyer, 1971). To evaluate the relative contribution of genetics and nutrition over time on the performance of commercial broilers, a study conducted by Havenstein et al. (2003) looked at changes in growth, livability, and feed conversion of 1957 and 2001 broilers when they were fed diets that were typical of diets used in those 2 years. The data showed that the average body weight of a Ross 308 fed a diet from 1957 reached a body weight of 2.126 g compared to 2.672 g when the same strain was fed the typical 2001 diet (Havenstein et al., 2003). Feed formulation and the ingredients utilized have continued to advance and improve the production and profitability of broilers, but the primary change in the growth performance of broilers has been brought about by the application of quantitative genetics by commercial poultry breeding companies. Genetic improvement has also been shown to be a major player in changes in the performance of turkeys (Havenstein et al., 2007) and commercial egg production stocks (Anderson et al., 2013).

5 Genetic selection

Although nutrition is an important factor that has led to the increase of broiler growth and size, genetic improvements have also played a major role in the change in development and growth of broilers. Improved nutrition was shown to have accounted for 10–15% of the change seen in broiler performance over the 46-year time span from 1957 to 2003, and genetics has contributed the rest

(Havenstein et al., 2003). During the twentieth and now well into the twenty-first century, the broiler and turkey industries have constantly demanded faster and faster growth and more and more efficient broilers and turkeys. Poultry breeders have responded by selecting each generation for improved growth rate and better feed efficiency. As a result, the days to market have continued to decrease, and the amount of feed required per unit of weight has also decreased dramatically. This decrease in efficiency has partially been brought about by shorter and shorter growing times being required to reach heavier and heavier market weights. The industry has also recognized that overhead costs per unit of meat processed also decline dramatically as heavier birds are processed. Therefore, the marketing of broilers has also changed from the sale of whole birds to the sale of further processed and cut-up products. Of course, in turn, the development of further processing has also spurred tremendous growth development in fast food restaurant chains, a number of which specialize in chicken products.

Previous to the concept of raising chickens solely for meat production, chickens were predominantly kept for egg production. Over the last 70–80 years, the breeders of egg production stocks have developed both white and brown strains that are smaller and more feed efficient. Those strains produce far more eggs, with better shell quality, with a specific average egg weight, and with better general livability over a given period of time, than did their ancestors. Egg production stocks have shown dramatic changes in performance (Anderson et al., 2013). Breeding for egg production has gone almost in the opposite direction (i.e. down in body weight and up in egg number and egg quality) from broiler and turkey breeding, but all three industries have been dramatically improved by the breeding process, thereby contributing greatly to the overall poultry industry's success.

Egg laying and broiler production are diametrically opposed, thus a breed designed solely for meat production would need to become a reality (Roberson et al., 1993). By the onset of the Second War World, there were numerous breeds of poultry that were used primarily for meat production; however, there was not a breed with the desired characteristics of a broad breast, such as the broad breasted turkey possessed (Gordy, 1974). The Chicken-of-Tomorrow Contest was first held in 1946 to address the need for a larger bird for poultry meat production, and in 1948 the first national contest was held at the University of Delaware (Gordy, 1974). In 1974, the winning bird, a White Cross, reached a weight of 5.7 pounds (2.59 kg) at approximately 7 weeks and 5 days, whereas the winning bird back in 1949, a New Hampshire-Rock Red Cross, required 13 weeks and 2 days to achieve the same weight (Gordy, 1974). One major advancement in genetics following the Chicken-of-Tomorrow Contest was the innovative development of a cross of a White Plymouth Rock male with a Cornish Gamebird female, that cross produced a broader breasted bird with more potential for breast meat development (Skinner, 1974). The Cornish Game breed originated in England, where it had been developed from Asiatic fighting stock that had traditionally been used for cockfighting, whereas the White Plymouth Rock had its origin in the New England area of North America (Crawford, 1990b). This innovative cross eventually became the primary foundation stock for most, if not all, of the male lines eventually used in modern broiler production. The Vantress organization was one of the major developers of such a male line, and eventually Vantress joined forces with the Cobb Breeding Company to form the current Cobb-Vantress Breeding organization. Several other breeding companies were also deeply involved in utilizing the descendants of the game bird cross as foundation stocks for their broiler male lines, including Ross, Hubbard, Cobb, Arbor Acres, Pilch, Ledbreast, and several others.

In the early 1950s, the description and application of quantitative genetics (initially led by Dr. Jay Lush and his students at Iowa State University, and subsequently followed by many others at a number of US Land Grant Universities) spurred the development of numerous commercial poultry breeding operations. Those commercial breeders hired quantitative geneticists to select and improve their stocks on an annual basis. The Chicken-of-Tomorrow Contest had demonstrated that some strains grew faster than others, but the application of the science of genetics and the development of specialized breeding companies for improving performance was the real driving force that led the broiler industry to point where flocks grown in 2016 reached market weights of 6.16 pounds (2.79 kg) at 47 days of age with a 1.86 feed-conversion ratio (NCC, 2017). It is apparent that breeding has contributed greatly to the overwhelming changes in broiler performance over the past 70–80 years. Of course, nutrition, housing, veterinary care, and many other factors have also played important roles in the broiler industry's success.

Over 25 years ago, Havenstein et al. (1992) predicted that the implementation of developing biotechnology methods and tools would benefit the industry. Specifically, they predicted that those tools would not only benefit the manufacturing of feed amendments such as modified feed grade enzymes and improved vaccines, but also for the direct genetic improvement of poultry breeding stock. Since then, the emergence of next-generation sequencing (NGS) technology in the Human Genome project has accelerated the development of sequencing instrumentation such that whole genome sequencing (WGS) has now become relatively commonplace (Heather and Chain, 2016; Hamdoun and Ehsan, 2017). These advancements led to a virtual explosion of applications ranging from WGS of specific foodborne pathogens for identifying and tracking specific causative organisms in a foodborne illness to characterizing gastrointestinal microbial communities in host animals such as chickens using 16S rRNA gene sequencing to generate taxonomic profiles (Ricke et al., 2017; Sekse et al., 2017; Pightling et al., 2018).

In poultry, the application of WGS has also advanced the understanding of the evolution of avian genome and the emergence of modern birds and progenitors of domestic chickens (Jarvis et al., 2014; Lawal et al., 2018). Application of RNA sequencing and differential transcriptomic analyses has led to the identification of genetic markers for chick resistance to *Salmonella* Enteritidis colonization (Li et al., 2018) while genotyping based on single nucleotide polymorphism (SNP) and microsatellite markers for quantitative trait loci mapping in commercial laying hens have been used to identify resistance to fowl typhoid caused by S. Gallinarum (Psifidi et al., 2018). It is likely that further elucidation of host resistance mechanisms should be useful for pathogen vaccine construction to enhance host immune response. It may also be possible to delineate host factors that while antagonistic to pathogen colonization in the gut are supportive of colonization by beneficial bacteria such as probiotic cultures.

Progress in the understanding of production traits is also becoming a reality. Liu et al. (2018) used a high-density SNP array to screen a population of laying hens from day of initial egg lay to 80 days of egg production to align specific SNPs with egg weight phenotypes and identify specific genes that could account for variation in age-dependent egg weights. Yuan et al. (2018) conducted a genome-wide association study of indigenous Chinese chicken breeds at the single marker and haplotype level loci mapping to establish that body weight should be considered polygenic with sufficient variability to suggest different genetic mechanisms accounting for the observed phenotypic variability in poultry breeds. As more fundamental genetic understanding is gained on commercial production, it is anticipated that detailed genetic information will be further linked to phenotypic traits and a more focused selection process can be implemented to optimize disease resistance, host gut health, and bird performance.

6 Housing

Improvements in broiler performance can be directly connected to improvements in the welfare of the birds involved, whether this includes housing, water, or veterinary care to treat and prevent diseases and mortalities. Before the onset of the modern poultry industry, many farms that had chickens were not concerned with their housing and let them roam free to roost in trees and be susceptible to predators and other external dangers (Skinner, 1974).

As the development of the broiler industry began, birds were raised in small sheds that had access to the outdoors. The sheds may have contained small heaters for the winter, but little else as the sheds were not controlled environments (Sawyer, 1971). One of first automated houses was developed in the late 1940s by the DeWitt brothers of Zeeland, Michigan. The brothers started with the production of an automated chain feeder (Sawyer, 1971). The automated feeder later became known as the 'Big Dutchman' and led to the DeWitt brothers to invent automated waterers, ventilation equipment, egg coolers, chick sorters, feed cleaners, and brooders (Sawyer, 1971). The Big Dutchman organization became one of the founding industries in automated housing.

The housing of the twenty-first century is vastly different than what was seen in the early 1900s. Broiler housing in the industry currently employs complete climate control facilities with the ability to manipulate temperature, relative humidity, air composition, air speed, air movement, and lighting. Because of the advancements in housing technology and machinery, poultry producers are able to maintain optimum control of the climate and environmental conditions in which their poultry are housed, which is also directly linked to the flock's growth rate, feed efficiency, and livability (Liang et al., 2013).

In more recent times, there has been some public movement for preference of organic, free-range, or pasture flock production of market birds. There are a multitude of reasons for this increase in popularity such as perceived benefits in animal welfare, bird health, and food safety (Berg, 2001; Van Loo et al., 2012). Under these circumstances, particularly for free-range birds, housing is minimal and environmental exposure is much greater than what birds would encounter in conventional commercial industry housing (Fanatico et al., 2007; Jacob et al., 2008). Consequently, challenges such as predation on birds and feed costs are primary concerns of pasture flock farmers (Hilimire, 2012). In addition, slow growing breeds may be used in these small operations resulting in a longer grow-out period (Fanatico et al., 2007). Given these differences in breed of birds, dietary modifications, and environmental challenges, it remains to be determined how these inputs would influence GIT health and function but one would anticipate there could be some detectable differences between birds raised in this type of operation compared to conventionally raised birds.

7 Veterinary care

Improvements in growth rate, feed efficiency, and livability of poultry have also been a direct effect of developments in veterinary care. From the beginning of the industry, medications and drugs were a considerable area of concern for the industry as growers sought to improve efficiency in the production of their market birds. Drug companies were thus very interested in developing markets for their products for the industry. Companies such as Hess and Clark, American Cyanamid, National Remedy Products Company, Whitmoyer, Western Condensing, Sterwin, Monsanto, Elanco, Vinland, Pfizer, Wyeth, Consolidated

Products, Merck, Commercial Solvents, Abbott, and numerous others were involved (Sawyer, 1971).

Veterinary care, however, was not a common practice for the early poultry industry, because servicing poultry was not seen as needed or beneficial. However, 'Doc' Salsbury introduced the industry to the idea of providing poultry with veterinary care and providing growers with a means for identifying and treating diseases and afflictions found in poultry to improve their performance (Sawyer, 1971). 'Doc' Salsbury introduced the industry to numerous innovative products and ideas such as his annual poultry school that was started in 1931, an experimental farm for conducting research on proposed products (1935), a product called Ren-O-Sal that was a leader in the growth promotion movement, feed medications (1950), and vaccines that could be administered through the drinking water (Sawyer, 1971).

Currently, it is common practice in the broiler industry to not only employ preventative measures such as 'Doc' Salsbury had recommended, but to utilize service technicians who monitor broiler farms and advise the growers who manage the facilities. Through innovative measures taken by the broiler industry, the industry has seen a dramatic decrease in mortality among its flocks from 1925 to 2016 as mortality decreased from 18 to 4.4 percent (NCC, 2017). Of course, general overall management and housing, nutrition, and genetics have also contributed greatly to the considerable improvement in livability of the industry's broilers.

8 Poultry industry challenges in gut health

Even with the drastic improvements that have been made over the past 70 to 80 years in the poultry industry, some challenges still exist. Challenges such as the increase in the incidence of foodborne illnesses, and in the increase in poultry diseases originating from pathogenic bacteria and viruses. The gastrointestinal tract (GIT) of the modern broiler contains a complex mixture of hundreds of different microbiota that consists of both commensal and pathogenic bacteria. In addition, commensal bacteria in poultry such as *Campylobacter jejuni* are also pathogenic in humans (Ayllón et al., 2017). Due to the complexity of the GIT in both humans and broilers, health of both the poultry produced and of the humans that consume poultry products can be significantly affected. It is therefore important to understand each specific challenge of this type in order to elucidate effective solutions and the preventative measures associated with them.

8.1 Foodborne illness

Foodborne illnesses originating from poultry products are primarily related to the presence of *Salmonella*, *Escherichia coli*, and *Campylobacter*. In 2015, foodborne illnesses were identified by FoodNet as being responsible for 20 098 cases of human infections, 4598 hospitalizations, and 77 deaths (CDC, 2017). Poultry is estimated to have been the source of 10% of all foodborne illnesses, 12% of all hospitalizations, and 19% of foodborne-related deaths from 1998 to 2008 (Painter et al., 2013). Poultry diseases are also said to have raised poultry production costs by 10 to 20% in developed countries (FOA, 2016). Even though these estimates include other foodborne diseases and illnesses, it does not diminish the detrimental effects of pathogenic bacteria on both poultry flocks and on the human consumers of poultry products.

Salmonella spp. are Gram-negative, facultative anaerobic bacteria that flourish at an optimal temperature of 37°C and they can cause typhoidal fever, enteric fevers, gastroenteritis, and septicemia (Duffy, 2009). Each year in the United States, the Center for Disease Control and Prevention (CDC) estimates that there are 1.2 million cases of salmonellosis (non-typhoidal) that result in approximately 450 deaths (Scallan et al., 2011). In 2012, 106 of the 831 foodborne outbreaks were directly related to *Salmonella,* and from the reported outbreaks that year 64% of the hospitalizations were confirmed to be caused by *Salmonella* (CDC, 2014c). In 2013, *Salmonella* accounted for the highest incidence of foodborne illness in the United States, with Enteritidis, Typhimurium, and Newport as the top serotypes involved (CDC, 2014a). *Salmonella* (nontyphoidal) is estimated to be the top cause of pathogen-related hospitalizations (19 336) and death (378) and the second leading cause of reported foodborne illnesses (1 027 561) (CDC, 2011). Consequently, incidences of *Salmonella* infections, especially Enteritidis, can be directly linked to poultry and eggs (CDC, 2014a; Finstad et al., 2012; Ricke, 2017). Although, S. Pullorum and Gallinarum were the primary *Salmonella enterica* serovars colonizing poultry in the 1900s, the implementation of vaccine procedures eliminated both Pullorum disease and fowl typhoid from commercial broiler flocks (Foley et al., 2011). With the eradication of those diseases, S. Enteritidis rose to become the dominant serovar for poultry, but its presence has been on the decline since the 1990s (Foley et al., 2011). More recently, S. Heidelberg and Kentucky have been identified as the two most identified serovars within commercial broiler flocks (Foley et al., 2011).

Campylobacter spp. are Gram-negative, microaerophilic bacteria that are the responsible agents for causing campylobacteriosis, the most identified source of bacterial gastroenteritis in the world (Duffy, 2009; Miller and Mandrell, 2005; Buzby et al., 1997; Adak et al., 2005; Horrocks et al., 2009). If the infection is not controlled, it can lead to severe complications including Guillain-Barré and reactive arthritis (Zia et al., 2003). Of the pathogenic strains of *Campylobacter, C. jejuni* (93%) and *C. coli* (7%) have been reported to be the leading causes of human campylobacteriosis (Gillespie et al., 2002). Due to their preferential environment of 37 to 42°C, *Campylobacter* spp. are commonly isolated within the GIT of poultry where the internal temperature is normally 41 to 42°C (CDC, 2014b; Duffy, 2009). In addition, *Campylobacter* has been isolated in the oral cavity, GIT, and reproductive organs of humans and other animals (Duffy, 2009). Campylobacteriosis has been identified as one of the leading causes of foodborne illness in the United States and is estimated to affect more than 1.3 million persons every year and result in approximately 76 deaths per year (CDC, 2006, 2014b). A 14% increase has been reported in *Campylobacter* illnesses in humans from 2006 to 2008 to 2012 (CDC, 2013). Since poultry are thought to be responsible for roughly half of *Campylobacter*-associated illnesses, it is imperative to mitigate and control this pathogen (Harris et al., 1986).

Although less prevalent, the contamination of poultry products with *Listeria* ssp. has also been noted (Rothrock et al., 2017). Human infection with *Listeria* ssp., listeriosis, can cause gastroenteritis similar to other foodborne pathogens but can also be life threatening in pregnant women and in those who are immunocompromised leading to miscarriages, stillbirths, bacteremia, and meningitis (Roberts and Wiedmann, 2003). Though the incidence of listeriosis is less than that of other foodborne pathogens, the severity and mortality is much higher (Scallan et al., 2011). The main *Listeria* spp. commonly associated with the contamination of poultry meat has been identified as *Listeria innocua* followed by *Listeria monocytogenes* and several other *Listeria* ssp. including *L. welshimeria, grayi*, and *ivanovii* (Rouger et al., 2017). Although contamination of poultry meat by *Listeria* is generally thought to occur in processing facilities, it is also a zoonotic disease capable of infecting poultry and residing within the GIT (Dhama et al., 2013). The *Listeria* pathogen commonly infects poultry when poultry are already colonized with other diseases such as coccidiosis, infectious coryza, salmonellosis, campylobacteriosis, and parasitic infections (Dhama et al., 2013). It is apparent that the industry, even with its drastic developments and improvements, still has a challenge with pathogenic bacteria on poultry products.

8.2 Poultry disease

Not only have pathogenic-related foodborne illnesses become an issue, but increased concern for disease outbreaks among poultry caused by other pathogenic bacteria has also occurred. Diseases such as necrotic enteritis, Fowl Typhoid, Paratyphoid, and colibacillosis, are all caused by pathogenic bacteria in the poultry industry (Aiello and Moses, 2016). Pathogens such as *Salmonella*, *Clostridium perfringens*, and *Escherichia coli* are main concerns for the industry and are commonly present in the poultry GIT. The presence of pathogens is a great concern, as it can cost the industry not only time and money, but the welfare of birds during growth.

Necrotic enteritis is an infection of the GIT caused by *Clostridium perfringens*, Type A or C (Porter, 1998). *Clostridium* spp. are anaerobic, Gram-positive bacilli that are spore forming and found in the soil and fresh water (Porter, 1998). Pathogenic Clostridia, such as *Clostridium perfringens*, can also naturally be found in the GIT of animals (Shapiro and Sarles, 1949). Necrotic enteritis occurs primarily in broilers between 2 and 6 weeks of age, producing enteric lesions in the small intestine (jejunum and ileum) (Porter, 1998; Cooper and Songer, 2010). Annually, necrotic enteritis outbreaks are estimated to cost over $2 billion globally (Van der Sluis, 2000). Furthermore, necrotic enteritis causes the intestines to become distended and filled with dark, brown fluid (Porter, 1998). It typically occurs simultaneously or as a secondary infection from a coccidia infection (Shane et al., 1985).

Coccidiosis is an enteric disease caused by parasites of the *Eimeria* species. Eimeria spp. (family Eimeriidae) belong to the phylum Apicomplexa and are obligate intracellular parasites further identified by specialized organelles unique to their species (Morrison, 2009). *Eimeria* have a short (4 to 6 days) life cycle consisting of two stages: an exogenous stage characterized by sporogony and an endogenous stage characterized by schizogony and gametogony (Blake and Tomley, 2014; McDougald, 2013). Several species of *Eimeria* are involved in inducing coccidiosis in poultry, infecting different segments of the GIT (Morgan et al., 2009; Haug et al., 2008), with *E. acervuline* colonizing the duodenum, *E. maxima* and *mitis* infecting the middle part of the small intestine, *E. tenella* colonizing the ceca, *E. brunetti* colonizing the ceca and the rectum, and *E. necatrix* colonizing the small intestine (Raman et al., 2011). The colonization of the GIT of chickens by *Eimeria* spp. is followed by the destruction of the intestinal epithelium and subsequent loss of performance, reduced feed intake, bloody diarrhea, and reduced weight gain (Gilbert et al., 2011; Dalloul and Lillehoj, 2005). Furthermore, *E. maxima, acervulina*, and *necatrix* are reported to induce necrotic enteritis in broilers (Williams, 2005). With *Eimeria* spp. being ubiquitous throughout poultry production facilities, it is extremely difficult to eradicate the disease from poultry (Quiroz-Castañeda and Dantán-González, 2015). Currently, the industry utilizes prophylactic measures in the form of anticoccidial drugs, ionophores and synthetic drugs, and live-attenuated vaccines in order to control or diminish the large economic impact coccidiosis has on the poultry industry (Chapman et al., 2010; Williams, 2002).

Fowl typhoid, paratyphoid, and Pullorum are associated with *Salmonella*-generated enteric lesions (Porter, 1998). As stated previously, *Salmonella* is a Gram-negative, facultative anaerobic bacterium that commonly resides in poultry. The GIT of poultry are common sites of colonization and thus, the fecal-oral route is a common mode of transmission (Porter, 1998). The aforementioned lesions predominately occur in the cecum of infected birds (Porter, 1998). *Salmonella* Pullorum and Gallinarum are responsible for Pullorum and fowl typhoid respectively (Porter, 1998). Both Pullorum and fowl typhoid affect young birds up to 3 weeks; however, fowl typhoid persists into adulthood. Because fowl typhoid can persist into adulthood, it can be transmitted from the hen through the egg by vertical transmission (Porter, 1998). White, raised, caseous lesions or gray, necrotic foci occur in the lungs, heart, spleen, liver, gizzard, and kidney of poultry infected with both Pullorum and fowl typhoid (Porter, 1998). Though S. Pullorum and Gallinarum were widespread in the United States in the early 1900s (USDA, 1997), the initiation of the National Poultry Improvement Plan (NPIP) in 1935 completely eradicated these diseases in commercial poultry flocks by the mid-1960s (Bäumler et al., 2000).

Paratyphoid encompasses most other serotypes, including S. Typhimurium, Enteritidis, Montevideo, and Heidelberg (Porter, 1998). Because these serotypes are commonly associated with foodborne illnesses, they are of high concern to the industry. Paratyphoid infections affect young birds with high mortality; however, older birds are non-symptomatic shedders of the disease (Porter, 1998). In the United States, poultry-related *Salmonella* costs the economy an estimated $966 million in direct and indirect costs each year (Callaway et al., 2008).

Colibacillosis in chickens is an infectious disease caused by *Escherichia coli,* a common pathogen found in the GIT of poultry (Porter, 1998; Gross, 1994). More recently, *E. coli* strains causing disease in poultry have been identified as Avian Pathogenic *E. coli* (APEC) (Dziva and Stevens, 2008). It is estimated that 10 to 15% of coliforms isolated from the avian alimentary tract belong to APEC serotypes (Harry and Hemsley, 1965). Yolk sac infection, respiratory disease complex (airsacculitis, perihepatitis, pericarditis), acute septicemia, salpingitis, peritonitis, synovitis, osteomyelitis, cellulitis, and enteric coligranuloma are all common syndromes due to the disease (Porter, 1998). APEC can be either a primary or secondary pathogen; however, due to its common presence in healthy birds, it is presumed to be more of an 'opportunistic' secondary pathogen (Porter, 1998). This is demonstrated by the presence of both virulent and avirulent *E. coli* colonizing and remaining in the GIT without systemic infection, until stressors induce extraintestinal translocation (Dominick and Jensen, 1984; Leitner and Heller, 1992). Therefore, stress is the main predisposing factor for inducing systemic APEC (Barnes and Gross, 1997). As stress plays an influential role in inducing infection and disease in birds, it is necessary to mitigate the stress poultry may endure to maintain a healthy flock and to reduce subsequent contamination of processed poultry.

It is well known that the GIT of poultry is the main source of contamination of poultry carcasses within processing facilities (Rigby et al., 1980). However, the feathers are also a major concern for contamination, as *Salmonella* spp. contamination normally occurs from external contaminants, whereas *Campylobacter* spp. occur internally (Bailey and Cox, 1991; Jones et al., 1991).

More recently, *Salmonella* outbreaks have been linked to animal contact, both indirect and direct (Steinmuller et al., 2006). Hale et al. (2012) estimate animal exposure to be responsible for 11% of *Salmonella* infections. As the colonization of the GIT occurs before processing, pathogenic bacteria must be controlled on the farm to mitigate the bacterial load sent to processing as well as to farmers who handle the birds.

9 Conclusions

As described previously, over the last century the poultry industry has grown into a multibillion-dollar business through advancements in nutrition, genetic selection, housing, veterinary care, and many other factors. A vast proportion of the success is due to the increased knowledge obtained through science and innovation. Although the industry has been very successful, the industry is faced with the challenge of maintaining GIT health. Since the GIT is a complex organ associated with the immune system and it is colonized by an array of microbiota, it is pertinent to fully understanding the GIT in order to maintain health and production. Furthermore, the industry must protect their flock's GIT by reducing or eliminating any and all factors that may negatively impact it, while at the same time utilizing those procedures that may benefit the host's GIT. These beneficial factors may not be consistent across different types of poultry, and thus the same preventative measures may not be compatible with all poultry types. Therefore, the industry must face these challenges by introducing novel feed additives in order to modulate gut health and maintain poultry performance.

10 Acknowledgements

The author DKD would like to acknowledge the Graduate College at the University of Arkansas for its support of the research and the stipend provided through the Distinguished Academy Fellowship and the continued support from the Center for Advanced Surface Engineering, under the National Science Foundation Grant No. OIA-1457888 and the Arkansas EPSCoR Program, ASSET III.

11 Where to look for further information

For further information please refer to the following resources:

Elwinger, K., Fisher, C., Jeroch, H., Sauveur, B., Tiller, H. and Whitehead, C. C. 2016. A brief history of poultry nutrition over the last hundred years. *Worlds Poult. Sci. J.* 72, 701–20. doi:10.1017/S004393391600074X.

Gordy, J. F. 1974. Broilers. In: *American Poultry History: 1823–1973*. American Printing and Publishing, Inc., Madison, WI, pp. 370–433

Sawyer, G. 1971. *The Agribusiness Poultry Industry: A History of Its Development.* Exposition Press Inc., Jericho, NY.
Strausberg, S. F. 1995. From hills and hollers: rise of the poultry industry in Arkansas. Arkansas Experimental Station Special Report 170. University of Arkansas, Fayetteville, AR.
Williams, W. H. 1998. *Delmarva's Chicken Industry: 75 Years of Progress.* Delmarva Poultry Industry, Inc., Georgetown, DE.

12 References

Adak, G. K., Meakins, S. M., Yip, H., Lopman, B. A. and O'Brien, S. J. 2005. Disease risks from foods, England and Wales, 1996–2000. *Emerg. Infect. Dis.* 11(3), 365–72. doi:10.3201/eid1103.040191.
Aiello, S. E. and Mose, M. A. 2016. *The Merck Veterinary Manual* (11th edn.). Merck & Co., Whitehouse Station, NJ.
Al-Nasser, A., Al-Khalaifa, H., Al-Saffar, A., Khail, F., Albahoun, M., Ragheb, G., Al-Haddad, A. and Mashaly, M. 2007. Overview of chicken taxonomy and domestication. *Worlds Poult. Sci. J.* 63(2), 285–300. doi:10.1017/S004393390700147X.
Anderson, K. E., Havenstein, G. B., Jenkins, P. K. and Osborne, J. 2013. Changes in commercial laying stock performance, 1958–2011: thirty-seven flocks of the North Carolina random sample and subsequent layer performance and management tests. *Worlds Poult. Sci. J.* 69(3), 489–514. doi:10.1017/S0043933913000536.
Ayllón, N., Jiménez-Marín, Á., Argüello, H., Zaldívar-López, S., Villar, M., Aguilar, C., Moreno, A., De La Fuente, J. and Garrido, J. J. 2017. Comparative proteomics reveals differences in host-pathogen interaction between infectious and commensal relationship with *Campylobacter jejuni*. *Front. Cell. Infect. Microbiol.* 7, 145. doi:10.3389/fcimb.2017.00145.
Bailey, J. S. and Cox, N. A. 1991. Internal colonization and external carriage of artificially inoculated *Salmonella typhimurium* from floor pens and cage reared chickens. *Poult. Sci.* 70(Suppl. 1), 142. (Abstr.).
Barnes, H. J. and Gross, W. B. 1997. Colibacillosis. In: Calnek, B. W., Barnes, H. J., Beard, C. W., McDougald, L. R. and Saif, Y. M. (Eds), *Diseases of Poultry* (10th edn.). Iowa State University Press, Ames, IA, pp. 131–41.
Bäumler, A. J., Hargis, B. M. and Tsolis, R. M. 2000. Tracing the origins of *Salmonella* outbreaks. *Science* 287(5450), 50–2. doi:10.1126/science.287.5450.50.
Berg, C. 2001. Health and welfare in organic poultry production. *Acta Vet. Scand. Suppl.* 95, 37–45. doi:10.1186/1751-0147-43-S1-S37.
Blake, D. P. and Tomley, F. M. 2014. Securing poultry production from the ever-present *Eimeria* challenge. *Trends Parasitol.* 30(1), 12–9. doi:10.1016/j.pt.2013.10.003.
Buzby, J. C., Allos, B. M. and Roberts, T. 1997. The economic burden of *Campylobacter*-associated Guillain-Barre' syndrome. *J. Infect. Dis.* 176, S192–7. doi:10.1086/513785.
Callaway, T. R., Edrington, T. S., Anderson, R. C., Byrd, J. A. and Nisbet, D. J. 2008. Gastrointestinal microbial ecology and the safety of our food supply as related to *Salmonella*. *J. Anim. Sci.* 86(14 Suppl.), E163–72. doi:10.2527/jas.2007-0457.
CDC. 2006. Preliminary FoodNet data on the incidence of infection with pathogens transmitted commonly through food–10 states, United States, 2005. *Morb. Mort. Wkly. Rep.* 55, 393–5.

CDC. 2011. CDC estimates of foodborne illness in the United States. National Center for Emerging & Zoonotic Infectious Diseases: division of foodborne, waterborne, and environmental diseases. Available at: www.cdc.gov/foodborneburden.

CDC. 2013. Surveillance for foodborne disease outbreaks - United States, 1998-2008. *Morb. Mort. Wkly. Rep.* 62(2), 1-34.

CDC. 2014a. Incidence and trends of infection with pathogens transmitted commonly through food – Foodborne Diseases Active Surveillance Network, 10 U.S. Sites, 2006-2013. *Morb. Mort. Wkly. Rep.* 63(15), 328-32.

CDC. 2014b. Campylobacter. Center for Disease Control and Prevention. Available at: www.cdc.gov/foodsafety/diseases.

CDC. 2014c. *Surveillance for Foodborne Disease Outbreaks, United States, 2012, Annual Report*. United States Department of Health and Human Services, CDC, Atlanta, GA.

CDC. 2017. *Foodborne Diseases Active Surveillance Network (FoodNet): FoodNet 2015 Surveillance Report (Final Data)*. United States Department of Health and Human Services, CDC, Atlanta, GA.

Chapman, H. D., Jeffers, T. K. and Williams, R. B. 2010. Forty years of monensin for the control of coccidiosis in poultry. *Poult. Sci.* 89(9), 1788-801. doi:10.3382/ps.2010-00931.

Clickner, F. H. and Follwell, E. H. 1925. Application of 'protozyme' by *Aspergillus orizae* to poultry feeding. *Poult. Sci.* 5, 241-7.

Cooper, K. K. and Songer, J. G. 2010. Virulence of *Clostridium perfringens* in an experimental model of poultry necrotic enteritis. *Vet. Microbiol.* 142(3-4), 323-8. doi:10.1016/j.vetmic.2009.09.065.

Crawford, R. D. 1990a. Poultry biology: origin and history of poultry species. In: Crawford, R. D. (Ed.), *Poultry Breeding and Genetics*. Elsevier Science Publishing Company, Amsterdam and New York, pp. 1-42.

Crawford, R. D. 1990b. Poultry genetic resources: evolution, diversity, and conservation. In: Crawford, R. D. (Ed.), *Poultry Breeding and Genetics*. Elsevier Science Publishing Company, Amsterdam and New York, pp. 43-60.

Dalloul, R. A. and Lillehoj, H. S. 2005. Recent advances in immunomodulation and vaccination strategies against coccidiosis. *Avian Dis.* 49(1), 1-8. doi:10.1637/7306-11150R.

Dhama, K., Verma, A. K., Rajagunalan, S., Kumar, A., Tiwari, R., Chakraborty, S. and Kumar, R. 2013. *Listeria monocytogenes* infection in poultry and its public health importance with special reference to food borne zoonoses. *Pak. J. Biol. Sci.* 16(7), 301-8. doi:10.3923/pjbs.2013.301.308.

Dominick, M. A. and Jensen, A. E. 1984. Colonisation and persistence of *Escherichia coli* in axenic and monoaxenic turkeys. *Am. J. Vet. Res.* 45(11), 2331-5.

Duffy, G. 2009. Pathogen control in primary production: meat, dairy and eggs. In Blackburn, C. de. W. and McClure, P. J. (eds), *Foodborne pathogens: hazards, risk analysis and control*, Second edition, Woodhead Publishing Limited, Cambridge, UK.

Dziva, F. and Stevens, M. P. 2008. Colibacillosis in poultry: unravelling the molecular basis of virulence of avian pathogenic *Escherichia coli* in their natural hosts. *Avian Pathol.* 37(4), 355-66. doi:10.1080/03079450802216652.

Elwinger, K., Fisher, C., Jeroch, H., Sauveur, B., Tiller, H. and Whitehead, C. C. 2016. A brief history of poultry nutrition over the last hundred years. *Worlds Poult. Sci. J.* 72, 701-20. doi:10.1017/S004393391600074X.

Fanatico, A. C., Pillai, P. B., Emmert, J. L. and Owens, C. M. 2007. Meat quality of slow- and fast-growing chicken genotypes fed low-nutrient or standard diets and raised indoors or with outdoor access. *Poult. Sci.* 86(10), 2245–55. doi:10.1093/ps/86.10.2245.

FAO. 2016. Poultry and animal health. Food and Agriculture Organization of the United Nations.

Finstad, S., O'Bryan, C. A., Marcy, J. A., Crandall, P. G. and Ricke, S. C. 2012. *Salmonella* and broiler production in the United States: relationship to foodborne salmonellosis. *Food Res. Int.* 45(2), 789–94. doi:10.1016/j.foodres.2011.03.057.

Foley, S. L., Nayak, R., Hanning, I. B., Johnson, T. J., Han, J. and Ricke, S. C. 2011. Population dynamics of *Salmonella enterica* serotypes in commercial egg and poultry production. *Appl. Environ. Microbiol.* 77(13), 4273–9. doi:10.1128/AEM.00598-11.

Fraps, G. S., Carlyle, E. C. and Fudge, J. F. 1940. Metabolisable energy of some chicken feeds. Texas Agricultural Experiment Station Bulletin No. 589.

Gilbert, E. R., Cox, C. M., Williams, P. M., McElroy, A. P., Dalloul, R. A., Ray, W. K., Barri, A., Emmerson, D. A., Wong, E. A. and Webb, K. E. 2011. *Eimeria* species and genetic background influence the serum protein profile of broilers with coccidiosis. *PLoS ONE* 6(1), e14636. doi:10.1371/journal.pone.0014636.

Gillespie, I. A., O'Brien, S. J., Frost, J. A., Adak, G. K., Horby, P., Swan, A. V., Painter, M. J., Neal, K. R. and Campylobacter Sentinel Surveillance Scheme Collaborators. 2002. A case-case comparison of *Campylobacter coli* and *Campylobacter jejuni* infection: a tool for generating hypotheses. *Emerg. Infect. Dis.* 8(9), 937–42. doi:10.3201/eid0809.010817.

Gisolfi, M. R. 2006. From crop to contract farming: the roots of agribusiness in the American south, 1929–1939. *Agr. Hist.* 80(2), 167–89. doi:10.1525/ah.2006.80.2.167.

Gordy, J. F. 1974. Broilers. In: *American Poultry History 1823-1973*. American Printing and publishing, Inc., Madison, WI, pp. 370–433.

Gross, W. G. 1994. Diseases due to *Escherichia coli* in poultry. In: Gayles, C. L. (Ed.), *Escherichia coli in Domestic Animals and Humans*. CAB International, Tuscon, AZ, pp. 237–59.

Hale, C. R., Scallan, E., Cronquist, A. B., Dunn, J., Smith, K., Robinson, T., Lathrop, S., Tobin-D'Angelo, M. and Clogher, P. 2012. Estimates of enteric illness attributable to contact with animals and their environments in the United States. *Clin. Infect. Dis.* 54(5), S472–9. doi:10.1093/cid/cis051.

Hamdoun, Z. and Ehsan, H. 2017. Aftermath of the human genome project: an era of struggle and discovery. *Turk. J. Biol.* 41, 403–18. doi:10.3906/biy-1609-77.

Harris, N. V., Weiss, N. S. and Nolan, C. M. 1986. The role of poultry and meats in the etiology of *Campylobacter jejuni* serogroups. *Epidemiol. Infect.* 76(407), 477.

Harry, E. G. and Hemsley, L. A. 1965. The association between the presence septicaemia strains of *Escherichia coli* in the respiratory tract and intestinal tracts of chickens and the occurrence of colisepticaemia. *Vet. Rec.* 77, 35 -40.

Haug, A., Gjevre, A. G., Thebo, P., Mattsson, J. G. and Kaldhusdal, M. 2008. Coccidial infections in commercial broilers: epidemiological aspects and comparison of *Eimeria* species identification by morphometric and polymerase chain reaction techniques. *Avian Pathol.* 37(2), 161-70. doi:10.1080/03079450801915130.

Havenstein, G. B. 2012. *History of the Department of Poultry Science and Other Poultry Related Programs at North Carolina State University 1881-2010*. North Carolina Agricultural Foundation, Inc., College of Agriculture and Life Sciences, North

Carolina State University, Raleigh NC, pp. 1-474. Plus appendix of all NCSU poultry related publs.

Havenstein, G. B., Crittenden, L. B., Pettite, J. N., Qureshi, H. A. and Foster, D. N. 1992. Application of biotechnology in the poultry industry. *Anim. Biotechnol.* 3(1), 15-36. doi:10.1080/10495399209525760.

Havenstein, G. B., Ferket, P. R. and Qureshi, M. A. 2003. Growth, livability, and feed conversion of 1957 versus 2001 broilers when fed representative 1957 and 2001 broiler diets. *Poult. Sci.* 82(10), 1500-8. doi:10.1093/ps/82.10.1500.

Havenstein, G. B., Ferket, P. R., Grimes, J. L., Qureshi, M. A. and Nestor, K. E. 2007. Comparison of the performance of 1966 versus 2003-type turkeys, when fed representative 1966 and 2003 turkey diets: growth rate, livability, and feed conversion. *Poult. Sci.* 86(2), 232-40. doi:10.1093/ps/86.2.232.

Heather, J. M. and Chain, B. 2016. The sequence of sequencers: the history of sequencing DNA. *Genomics* 107(1), 1-8. doi:10.1016/j.ygeno.2015.11.003.

Hesselman, K. and Aman, P. 1986. The effect of β-glucanase on the utilization of starch and nitrogen by broiler chicks fed on barley of low or high viscosity. *Anim. Feed Sci. Technol.* 15(2), 83-93. doi:10.1016/0377-8401(86)90015-5.

Hilimire, K. 2012. The grass is greener: farmers' experiences with pastured poultry. *Renew. Agric. Food Syst.* 27(3), 173-9. doi:10.1017/S1742170511000287.

Horrocks, S. M., Anderson, R. C., Nisbet, D. J. and Ricke, S. C. 2009. Incidence and ecology of *Campylobacter jejuni* and *coli* in animals. *Food Microbiol.* 15(1-2), 18-25. doi:10.1016/j.anaerobe.2008.09.001.

Hoshi, H. and Yoshida, M. 1977. Phosphorus availability of 25 feed ingredients determined by bioassay on toe ash content. *Jpn. Poult. Sci.* 14(6), 274-8. doi:10.2141/jpsa.14.274.

Hu, K., Johnson, J., Florens, L., Fraunholz, M., Suravajjala, S., DiLullo, C., Yates, J., Roos, D. S. and Murray, J. M. 2006. Cytoskeletal components of an invasion machine–the apical complex of *Toxoplasma gondii*. *PLoS Pathog.* 2(2), e13. doi:10.1371/journal.ppat.0020013.

Hunton, P. 1990. Industrial breeding and selection. In: Crawford, R. D. (Ed.), *Poultry Breeding and Genetics*. Elsevier Science Publishing Company, Amsterdam and New York, pp. 985-1028.

Huyghebaert, G., Ducatelle, R. and Van Immerseel, F. 2011. An update on alternative to antimicrobial growth promoter for broilers. *Vet. J.* 187, 182-8. doi:10.1016/j.tvjl.2010.03.003.

Jacob, J. P., Griggs, J. P. and Bender, J. B. 2008. Characterization of small-scale antibiotic-free broiler production in Minnesota. *J. Appl. Poult. Res.* 17(3), 412-20. doi:10.3382/japr.2007-00057.

Jarvis, E. D., Mirarab, S., Aberer, A. J., Li, B., Houde, P., Li, C., Ho, S. Y. W., Faircloth, B. C., Nabholz, B., Howard, J. T., Suh, A., Weber, C. C., da Fonseca, R. R., Li, J., Zhang, F., Li, H., Zhou, L., Narula, N., Liu, L., Ganapathy, G., Boussau, B., Bayzid, M. S., Zavidovych, V., Subramanian, S., Gabaldón, T., Capella-Gutiérrez, S., Huerta-Cepas, J., Rekepalli, B., Munch, K., Schierup, M., Lindow, B., Warren, W. C., Ray, D., Green, R. E., Bruford, M. W., Zhan, X., Dixon, A., Li, S., Li, N., Huang, Y., Derryberry, E. P., Bertelsen, M. F., Sheldon, F. H., Brumfield, R. T., Mello, C. V., Lovell, P. V., Wirthlin, M., Schneider, M. P., Prosdocimi, F., Samaniego, J. A., Vargas Velazquez, A. M., Alfaro-Núñez, A., Campos, P. F., Petersen, B., Sicheritz-Ponten, T., Pas, A., Bailey, T., Scofield, P., Bunce, M., Lambert, D. M., Zhou, Q., Perelman, P., Driskell, A. C., Shapiro, B., Xiong, Z., Zeng, Y., Liu, S., Li,

Z., Liu, B., Wu, K., Xiao, J., Yinqi, X., Zheng, Q., Zhang, Y., Yang, H., Wang, J., Smeds, L., Rheindt, F. E., Braun, M., Fjeldsa, J., Orlando, L., Barker, F. K., Jønsson, K. A., Johnson, W., Koepfli, K. P., O'Brien, S., Haussler, D., Ryder, O. A., Rahbek, C., Willerslev, E., Graves, G. R., Glenn, T. C., McCormack, J., Burt, D., Ellegren, H., Alström, P., Edwards, S. V., Stamatakis, A., Mindell, D. P., Cracraft, J., Braun, E. L., Warnow, T., Jun, W., Gilbert, M. T. and Zhang, G. 2014. Whole-genome analyses resolve early branches in the tree of life of modern birds. *Science* 346(6215), 1320–31. doi:10.1126/science.1253451.

Jones, F. T., Axtell, R. C., Rives, D. V., Scheideler, S. E., Tarver, F. R., Walker, R. L. and Wineland, M. J. 1991. A survey of *Campylobacter jejuni* contamination in modern broiler production and processing systems. *J. Food Prot.* 54(4), 259–62. doi:10.4315/0362-028X-54.4.259.

Lawal, R. A., Al-Atiyat, R. M., Aljumaah, R. S., Silva, P., Mwacharo, J. M. and Hanotte, O. 2018. Whole-genome resequencing of red junglefowl and indigenous village chicken reveal new insights on the genome dynamics of the species. *Front. Genet.* 9, 264. doi:10.3389/fgene.2018.00264.

Leitner, G. and Heller, E. D. 1992. Colonisation of *Escherichia coli* in young turkeys and chickens. *Avian Dis.* 36(2), 211–20. doi:10.2307/1591493.

Letourneau-Montminy, M. P., Narcy, A., Lescoat, P., Bernier, J. F., Magnin, M., Pomar, C., Nys, Y., Sauvant, D. and Jondreville, C. 2010. Meta-analysis of phosphorus utilisation by broilers receiving corn-soybean meal diets: influence of dietary calcium and microbial phytase. *Animal* 4(11), 1844–53. doi:10.1017/S1751731110001060.

Li, P., Fan, W., Everaert, N., Liu, R., Li, Q., Zheng, M., Cui, H., Zhao, G. and Wen, J. 2018. Messenger RNA sequencing and pathway analysis provide novel insights into the susceptibility to *Salmonella* Enteritidis infection in chickens. *Front. Genet.* 9, 256. doi:10.3389/fgene.2018.00256.

Liang, Y., Kidd, M., Watkins, S. and Tabler, G. 2013. Effect of commercial broiler house retrofit: a 4-year study of live performance, *J. Appl. Poult. Res.*, 22: 211–16.

Liu, Z., Sun, C., Yan, Y., Li, G., Wu, G., Liu, A. and Yang, N. 2018. Genome-wide association analysis of age-dependent egg weights in chickens. *Front. Genet.* 9, 128. doi:10.3389/fgene.2018.00128.

McDougald, L. R. 2013. Coccidiosis. In: Swayne, D. E. (Ed.), *Diseases of Poultry* (13th edn.). Wiley-Blackwell, Hoboken, NJ, p. 1408.

Miller, W. G. and Mandrell, R. E. 2005. Prevalence of *Campylobacter* in the food and water supply: incidence, outbreaks, isolation and detection. In: Ketley, J. M. and Konkel, M. E. (Eds), *Campylobacter: Molecular and Cellular Biology*. Horizon Bioscience, Wymondham, UK, pp. 101–63.

Moreng, R. and Avens, J. S. 1985. Classification, nomenclature, and showing of poultry. In: *Poultry Science and Production*. Reston Publishing Co., Inc. Prentice-Hall Company, Reston, VA, p. 1645.

Morgan, J. A., Morris, G. M., Wlodek, B. M., Byrnes, R., Jenner, M., Constantinoiu, C. C., Anderson, G. R., Lew-Tabor, A. E., Molloy, J. B., Gasser, R. B. and Jorgensen, W. K. 2009. Realtime polymerase chain reaction (PCR) assays for the specific detection and quantification of seven *Eimeria* species that cause coccidiosis in chickens. *Mol. Cell. Probes* 23(2), 83–9. doi:10.1016/j.mcp.2008.12.005.

Morrison, D. A. 2009. Evolution of the *Apicomplexa*: where are we now? *Trends Parasitol.* 25(8), 375–82. doi:10.1016/j.pt.2009.05.010.

NCC. 2016. *U.S. Poultry Industry Provides 1.6 Million Jobs; Economic Output of $441 Billion*. National Chicken Council. Available at: https://www.nationalchickencouncil.

org/u-s-poultry-industry-provides-1-6-million-jobs-economic-output-441-billion/ (accessed on 26 June 2019).

NCC. 2017. *U.S. Broiler Performance*. National Chicken Council. Available at: https://www.nationalchickencouncil.org/about-the-industry/statistics/u-s-broiler-performance/ (accessed on 26 June 2019).

NCC. 2019a. *Per Capita Consumption of Poultry and Livestock, 1965 to Forecast 2020, in Pounds*. National Chicken Council. Available at: https://www.nationalchickencouncil.org/about-the-industry/statistics/per-capita-consumption-of-poultry-and-livestock-1965-to-estimated-2012-in-pounds/ (accessed 26 on June 2019).

NCC. 2019b. *Broiler Chicken Industry Key facts 2019*. National Chicken Council. Available at: https://www.nationalchickencouncil.org/about-the-industry/statistics/broiler-chicken-industry-key-facts/ (accessed 26 June 2019).

Nelson, T. S., Shieh, T. R., Wodzinski, R. J. and Ware, J. H. 1971. Effect of supplemental phytase on utilization of phytate phosphorus by chicks. *J. Nutr.* 101(10), 1289–93. doi:10.1093/jn/101.10.1289.

NRC. 1944. *Recommended Nutrient Allowances for Poultry*. (1st edn.). National Academies Press, Washington DC.

NRC. 1954. *Nutrient Requirements of Poultry, Number 1*. National Academies Press, Washington DC, p. 37.

NRC. 1994. *Nutrient Requirements of Poultry* (9th rev. edn.). National Academies Press, Washington DC, p. 155.

Painter, J. A., Hoekstra, R. M., Ayers, T., Tauxe, R. V., Braden, C. R., Angulo, F. J. and Griffin, P. M. 2013. Attribution of foodborne illnesses, hospitalizations, and deaths to food commodities by using outbreak data, United States, 1998–2008. *Emerg. Infect. Dis.* 19(3), 407–15. doi:10.3201/eid1903.111866.

Petterson, D. and Aman, P. 1988. Effects of enzyme supplementation of diets based on wheat, rye or triticale on their productive value for broiler chickens. *Anim. Feed Sci. Technol.* 20(4), 313–24. doi:10.1016/0377-8401(88)90005-3.

Pightling, A. W., Pettengill, J. B., Luo, Y., Baugher, J. D., Rand, H. and Strain, E. 2018. Interpreting whole-genome sequence analyses of foodborne bacteria for regulatory applications and outbreak investigations. *Front. Microbiol.* 9, 1482. doi:10.3389/fmicb.2018.01482.

Porter, R. E. 1998. Bacterial enteritidis of poultry. *Poult. Sci.* 77(8), 1159–65. doi:10.1093/ps/77.8.1159.

Psifidi, A., Russell, K. M., Matika, O., Sánchez-Molano, E., Wigley, P., Fulton, J. E., Stevens, M. P. and Fife, M. S. 2018. The genomic architecture of fowl typhoid resistance in commercial layers. *Front. Genet.* 9, 519. doi:10.3389/fgene.2018.00519.

Quiroz-Castañeda, R. E. and Dantán-González, E. 2015. Control of avian coccidiosis: future and present natural alternatives. *BioMed. Res. Int.* 2015, 430610 (Article ID). doi:10.1155/2015/430610.

Raman, M., Banu, S., Gomathinayagam, S. and Raj, G. 2011. Lesion scoring techniques for assessing the virulence and pathogenicity of Indian field isolates of avian *Eimeria* species. *Vet. Archiv.* 81(2), 259–71.

Ricke, S. C. 2017. Insights and challenges of *Salmonella* infections in laying hens. *Curr. Opin. Food Sci.* 18, 43–9. doi:10.1016/j.cofs.2017.10.012.

Ricke, S. C., Hacker, J., Yearkey, K., Shi, Z., Park, S. H. and Rainwater, C. 2017. Unravelling food production microbiomes: concepts and future directions. In: Ricke, S. C.,

Atungulu, G. G., Park, S. H. and Rainwater, C. E. (Eds), *Food and Feed Safety Systems and Analysis*. Elsevier Inc., San Diego, CA, pp. 347–74. Chapter 19.
Rickes, E. L., Brink, N. G., Koniuszy, F. R., Wood, T. R. and Folkers, K. 1948. Crystalline vitamin B_{12}. *Science* 107(2781), 396–7. doi:10.1126/science.107.2781.396.
Rigby, C. E., Pettiet, J. R., Baker, M. F., Bentley, A. H., Salomons, M. O. and Lior, H. 1980. Sources of *Salmonellae* in an uninfected commercially-processed broiler flock. *Can. J. Comp. Med.* 44(3), 267–74.
Roberts, A. J. and Wiedmann, M. 2003. Pathogen, host and environmental factors contributing to the pathogenesis of listeriosis. *Cell. Mol. Life Sci.* 60(5), 904–18. doi:10.1007/s00018-003-2225-6.
Robinson, F. E., Wilson, J. L., Yu, M. W., Fasenko, G. M. and Hardin, R. T. 1993. The relationship between body weight and reproductive efficiency in meat-type chickens. *Poult. Sci.* 72(5), 912–22. doi:10.3382/ps.0720912.
Rothrock, M. J., Davis, M. L., Locatelli, A., Bodie, A., McIntosh, T. G., Donaldson, J. R. and Ricke, S. C. 2017. *Listeria* occurrence in poultry flocks: detection and potential implications. *Front. Vet. Sci.* 4, 125. doi:10.3389/fvets.2017.00125.
Rouger, A., Tresse, O. and Zagorec, M. 2017. Bacterial contaminants of poultry meat: sources, species and dynamics. *Microorganisms* 5(3), 50–66. doi:10.3390/microorganisms5030050.
Sauvant, B., Scmidely, P., Daudin, J. J. and St-Pierre, N. R. 2008. Meta-analyses of experimental data in animal nutrition. *Animal* 2, 1203–14. doi:10.1017/S1751731108002280.
Sauveur, B. 1983. Bio-availability for poultry of plant-origin phosphorus. Methodological criticisms and results. In: *Proceedings of 4th European Symposium on Poultry Nutrition*, Tours, France, pp. 103–13.
Sawyer, G. 1971. *The Agribusiness Poultry Industry: A History of Its Development*. Exposition Press Inc., Jericho, NY.
Scallan, E., Hoekstra, R. M., Angulo, F. J., Tauxe, R. V., Widdowson, M. A., Roy, S. L., Jones, J. L. and Griffin, P. M. 2011. Foodborne illness acquired in the United States-major pathogens. *Emerg. Infect. Dis.* 17(1), 7–15. doi:10.3201/eid1701.p11101.
Sekse, C., Holst-Jensen, A., Dobrindt, U., Johannessen, G. S., Li, W., Spilsberg, B. and Shi, J. 2017. High throughput sequencing for detection of foodborne pathogens. *Front. Microbiol.* 8, 2029. doi:10.3389/fmicb.2017.02029.
Semba, R. D. 2012. On the 'discovery' of vitamin A. *Ann. Nutr. Metab.* 61(3), 192–8. doi:10.1159/000343124.
Shane, S. M., Gyimah, J. E., Harrington, K. S. and Snider III, T. G. 1985. Etiology and pathogenesis of necrotic enteritis. *Vet. Res. Commun.* 9(4), 269–87. doi:10.1007/BF02215151.
Shapiro, S. K. and Sarles, W. B. 1949. Microorganisms in the intestinal tract of normal chickens. *J. Bacteriol.* 58(4), 531–44.
Sibbald, I. R. 1976. A bioassay for true metabolizable energy in feeding stuffs. *Poult. Sci.* 55(1), 303–8. doi:10.3382/ps.0550303.
Simons, P. C., Versteegh, H. A., Jongbloed, A. W., Kemme, P. A., Slump, P., Bos, K. D., Wolters, M. G., Beudeker, R. F. and Verschoor, G. J. 1990. Improvement of phosphorus availability by microbial phytase in broilers and pigs. *Br. J. Nutr.* 64(2), 525–40. doi:10.1079/bjn19900052.
Skinner, J. 1974. *American Poultry History: 1823–1973*. The American Poultry Historical Society, Inc., Madison, WI.

Smith, E. L., Fantes, K. H., Ball, S., Waller, J. G., Emery, W. B., Anslow, W. K. and Walker, A. D. 1952. B_{12} vitamins (cobalamins). 1. Vitamins B_{12c} and B_{12d}. *Biochem. J.* 52(3), 389-95. doi:10.1042/bj0520389.

Steinmuller, N., Demma, L., Bender, J. B., Eidson, M. and Angulo, F. J. 2006. Outbreaks of enteric disease associated with animal contact: not just a foodborne problem anymore. *Clin. Infect. Dis.* 43(12), 1596-602. doi:10.1086/509576.

Strausberg, S. F. 1995. From hills and hollers: rise of the poultry industry in Arkansas. Arkansas Experimental Station Special Report 170. University of Arkansas, Fayetteville, AR.

The Poultry Federation. 2017. *Resources*. Available at: https://www.thepoultryfederation.com/resources (accessed on 3 December 2018).

United Nations, Department of Economic and Social Affairs, Population Division. 2017. *World Population Prospects: The 2017 Revision*. United Nations.

U.S. Department of Agriculture. 1997. *The National Poultry Improvement Plan and Auxiliary Provisions*. United States Department of Agriculture, Animal and Plant Health Inspection Service, Washington DC.

U.S. Department of Agriculture, National Agricultural Statistics Service. 2018. *Poultry-Production and Value 2017 Summary*. United States Department of Agriculture.

Williams, W. H. 1998. *Delmarva's Chicken Industry: 75 Years of Progress*. Delmarva Poultry Industry, Inc., Georgetown, DE.

Williams, R. B. 2002. Anticoccidial vaccines for broiler chickens: pathways to success. *Avian Pathol.* 31(4), 317-53. doi:10.1080/03079450220148988.

Williams, R. B. 2005. Intercurrent coccidiosis and necrotic enteritis of chickens: rational, integrated disease management by maintenance of gut integrity. *Avian Pathol.* 34(3), 159-80. doi:10.1080/03079450500112195.

Van der Sluis, W. 2000. *Clostridial enteritis* is an often underestimated problem. *World Poult.* 16, 42-3.

Van Loo, E. J., Alali, W. and Ricke, S. C. 2012. Food safety and organic meats. *Annu. Rev. Food Sci. Technol.* 3, 203-25. doi:10.1146/annurev-food-022811-101158.

Yuan, Y., Peng, D., Gu, X., Gong, Y., Sheng, Z. and Hu, X. 2018. Polygenic basis and variable genetic architectures contribute to the complex nature of body weight – a genome-wide study in four Chinese indigenous chicken breeds. *Front. Genet.* 9, 229. doi:10.3389/fgene.2018.00229.

Zia, S., Wareing, D., Sutton, C., Bolton, E., Mitchell, D. and Goodacre, J. A. 2003. Health problems following *Campylobacter jejuni* Enteritis in a Lancashire population. *Rheumatology* 42(9), 1083-8. doi:10.1093/rheumatology/keg303.

Chapter 2

Advances in sequence technologies for generating poultry gut microbiome data

Xiaofan Wang and Jiangchao Zhao, University of Arkansas, USA

1 Introduction

Poultry (chicken, duck, and turkey) production possesses a high feed conversion rate, making it one of the most efficient meat production systems in the world. Early colonization of an exogenous microbiome in the gastro-intestinal (GI) tract occurs after hatching and affects the digestion process, immune system, and reproduction (Vispo and Karasov, 1997; Stanley et al., 2014; Clavijo and Flórez, 2018). Different organs harbor distinct microbial communities, which have a combined impact on the overall health of the host. The GI tract of poultry is composed of the esophagus, crop, proventriculus, gizzard, duodenum, jejunum, ileum, cecum, colon, and cloaca (Pan and Yu, 2014). Crops and gizzards store short-term pulverized grains that are high in starch. *Lactobacillus* and *Clostridiaceae* are dominant in these two organs and are responsible for starch digestion and lactate fermentation (Clavijo and Flórez, 2018). The cecum, which contains high cellulose digesta, serves as the main site for water

http://dx.doi.org/10.19103/AS.2019.0059.02

reabsorption and is the ideal niche for bacterial fermentation and growth. The bacteria associated with the cecum are mainly Firmicutes (*Clostridiaceae*), Bacteroides, and Proteobacteria (Clench and Mathias, 1995; Mesa et al., 2017). These bacteria generate energies from fermentation and provide nonfood-sourced essential amino acids and fatty acids for host physiological maintenance (LeBlanc et al., 2013; Pan and Yu, 2014).

Gut microbiota perturbations, that is, dysbiosis, are important etiologic factors in aviculture. This dysbiosis is not due to a single factor but is caused by a combination of other gut syndromes. Dysbiosis could result in various long-term effects such as gradually damaged mucosal membranes and retention of non-efficiently digested non-starch soluble polysaccharides (Immerseel et al., 2004). Some low abundance commensal bacteria – *Campylobacter coli*, *Salmonella enterica*, *Escherichia coli*, and *Clostridium perfringens* – have a pathogenic impact when their numbers increase to a critical level (Immerseel et al., 2004). For instance, *Campylobacter coli* induces severe enterocolitis in birds and also in humans when undercooked contaminated meat is consumed (Oakley et al., 2014). High levels of certain *E.coli* strains are associated with intestinal infections and respiratory diseases (Castellanos et al., 2017). High numbers of *C. perfringens* can cause necrotic enteritis (Keyburn et al., 2008). These diseases in the breeding industry lead to enormous economic losses each year and indirectly threaten human health. The current poultry industry widely accepts the significant role that a healthy and balanced gut microbiome plays in growth performance and reproductive traits. As a result, microbiome analysis has become more critical than ever before.

The availability of cultivable strains such as *Lactococcus lactis*, *Lactobacillus acidophilus*, *Bacillus acidipropionici*, and *Bifidobacterium* isolated from milk, cheese, and infant feces (Teuber, 1995; Shulman et al., 2007; Schultz, 2008) have greatly improved our understanding of their biological functions, metabolic preferences, and intrinsic relationships with their host (Medvecky et al., 2018). However, the majority of the avian gut microbiome is not cultivable yet. Important breakthroughs such as the discovery of the DNA double-helix, 16S rRNA sequence technique, and fluorescence in situ hybridization (FISH) in the late nineteenth century greatly expanded our knowledge of the microbiome world which now includes the unculturable groups (Brosius et al., 1978; Dahm, 2005; Hiergeist et al., 2015; Cui et al., 2016). Early twentieth-century inventions such as denaturing gradient gel electrophoresis (DGGE), temperature gradient gel electrophoresis (TGGE), and terminal restriction fragment length polymorphism (T-RFLP) allowed us to generally define microbiome community structures under different biological statuses. Another milestone for microbiome analysis was the establishment of the next-generation sequencing technique, which generates a large volume of high-throughput data used to disclose in-depth microbial community profiles. The use of these sequencing-based techniques together with advanced animal models such as germ-free mice and pigs significantly improved our understanding of the microbiome/host relationship in human and animals with respect to multiple physiological aspects such as digestive function, the immune system, reproduction, and disease diagnosis (Sochocka et al., 2019; Williams, 2014).

Research on the interaction between the microbiome and host is very important, but equally important is the research regarding the internal organization and communication between microbiome community members. Traditional microbiome analysis measures changes in community diversity (alpha and beta diversity) and link bacterial strains or operational taxonomic units (OTUs) to certain biological meanings through advanced correlation analysis and function annotations. Bioinformatics and statistical pipelines have been developed to reveal the bacterial-bacterial interactions including both antagonistic and synergistic relationships (De Angelis et al., 2006; Torok et al., 2008).

Microbiome-driven studies focusing on the relationship between specific bacterial taxa and the improvements in growth performance are relatively new but have potential future applications in livestock production. The emergence of various advanced high-throughput sequencing techniques at a reduced cost has made it both possible and affordable for producers to incorporate microbiome studies into different aspects of animal production such as nutrition, digestibility, animal health, and well-being. In the next sections, we will illustrate the development of microbiome analytical methods and their advantages and disadvantages in poultry microbiome analysis.

2 Culture-dependent microbiome analysis

Prior to the advent of high-throughput sequencing techniques, culture-dependent methods were the principal practice in historical microbiome analysis for over a century. In the past, because of limited knowledge and equipment, microbes were generally classified based on physical or biochemical properties (Hiergeist et al., 2015). These culture techniques are used for active cell quantification and physiological function studies. The colony-forming units (CFU) disclosed from agar plates represented an important quantification method for both pure cultured isolates and microbiome mixtures. Isolated bacterial strains provide great opportunities for discovering their physiological and metabolic features, which cannot be achieved in mixed culture (Bedbury and Duke, 1983).

Direct plating is widely used in specific bacterial quantification in food safety inspection and disease diagnosis (Butzler and Oosterom, 1991; Kim et al., 2017). Selective agars for dominant pathogens have been extensively developed and have been made commercially available. For example, *Salmonella spp.* and *Escherichia coli*, the top two worldwide food-borne pathogens, can be enumerated by a broad range of selective agars such as xylose lysine deoxycholate agar and *E. coli* Petrifilm, respectively (Panisello et al., 2000; Altekruse et al., 2009). The United States ranks as the top country for *Campylobacter ssp.* contamination in poultry meat products, the main source for human campylobacteriosis (Oyarzabal et al., 2004). Routine inspection of *Campylobacter ssp.* can be achieved with direct plating of carcass rinses on CampyCefex and campy-Line agar (Oyarzabal et al., 2005). Colony count serves as an easy and inexpensive method for viable cells visualization, but different bacteria species are not distinguishable by colony morphology. Hence, 16S rRNA sequencing on randomly picked colonies is required to validate the bacterial strain of interest.

In the 1980s, early poultry culturomic studies recovered approximately 60% of microscope cell count (3×10^{10} cells/g) in cecal microflora from 5-week-old chickens using multiple nonselective media (rumen fluid, blood agar, etc.) anaerobically. The resulting isolates mainly belong to anaerobic cocci and *Streptococci*, *Peptostreptococcus*, *Propionibacterium*, *Eubacterium*, *Bacteroides*, and *Clostridium* (Barnes et al., 1972; Salanitro et al., 1974; Gaskins et al., 2002). Most rare species such as *Cyanobacteria*, *Spirochaetes*, *Synergisteles*, *Fusobacteria*, *Tenericutes*, and *Verrucomicrobia* were not cultivable. In one of the recent culturomic works of 2018, Medvecky et al. (2018) isolated a total of 133 novel strains from chicken cecum using Wilkins-Chalgren anaerobe agar under different growth conditions. Some of these strains belonged to the minor phyla members such as *Verrucomicrobia*, *Elusimicrobia,* and *Synergistetes* (Medvecky et al., 2018). Establishment of culturing methods is also critical to novel strain cultivation. Culturing strategies such as supplementation of liver and fecal extracts and the dilution of cecal materials in the media promote the growth of some fastidious strains. Despite this progress, much is still unknown about the portion of the culturable bacteria in the poultry gut. The low recovery rate using old culture techniques can easily mask both the real microbial structural differences between bacterial communities and biological findings because the uncultured groups were excluded from the analysis.

3 Terminal restriction fragment length polymorphism (T-RFLP)

The terminal restriction fragment length polymorphism (T-RFLP) is a popular fingerprinting technique that has been widely used in distinguishing community structure diversity in biological samples. Fluorescent dyes such as 6-carboxyfluorescein (6-FAM), ROX, TAMRA, and hexachlorofluorescein are labeled on the 5′ end of both primers to generate the fluorescent fragment terminal. Amplicons contain various nucleotide sequences and can be cleaved by a restriction enzyme into different length fragments with a 4-base pair recognition site. These fragments are separated in a capillary or polyacrylamide electrophoresis to obtain a stacked band profile incorporating both the molecular size and fluorescent intensity information. Appropriate fluorescent platforms are recruited to generate a fingerprint for each sample and allow direct comparisons between the sample groups. The band patterns or peak height changes indicate the dynamics of the microbial community structure (Osborn et al., 2000; Sakamoto et al., 2003).

Crhanova et al. (2011) successfully evaluated the complex changes of the chicken cecal microbiome during the first 19 days after hatching by using the T-RFLP technique based on the 16S rRNA gene. As indicators of bacterial species, a gradual increase of T-RFLP peaks was discovered in cecal microflora on days 1 (n = 5), 4 (n = 14), and 19 (n = 42) (Crhanova et al., 2011). In another study, Gong et al. reported on the differences in bacterial population between different geographic locations along the GI tract such as ileum, cecum, mucosa, and lumen of the ileum with T-RFLP. They profiled the microbial communities by using two restriction enzymes for comparison (AluI and MspI) and found that the ileum has lower diversity than the cecum (Gong et al., 2002).

Albeit proficient and easily accessible, fragment-based profiling has unavoidable drawbacks such as low resolution and false peaks. Some bacterial strains share one single brief piece of nucleotide sequence on the same 16S rDNA region that can commonly be recognized by restriction enzymes. This led to the same length terminal fragments from different bacteria to merge together and become undistinguishable on the electropherogram. To avoid this coincidence, three or more restriction enzymes are recommended for use in order to provide more fingerprint references. Even so, the resolution of T-RFLP is still low compared with current sequencing techniques, especially when performing the richness and evenness measurement. Background noises generated from DNA extraction and PCR reaction are usually unpredictable. These DNA 'noises' incorporated into PCR reactions are difficult to exclude in subsequent peak readings. Some research groups attempt to diminish this bias with repetitions and threshold adjustments (Dunbar et al., 2001). However, such strategies will enlarge the variations on low abundance bacterial members and reduce the community richness and diversity (Blackwood et al., 2007).

4 Denaturing gradient gel electrophoresis (DGGE) and temperature gradient gel electrophoresis (TGGE)

Denaturing gradient gel electrophoresis (DGGE) and temperature gradient gel electrophoresis (TGGE) are two length-independent molecular methods aimed at separating DNA fragments based on chemical and physical characters. The double-helix structure of the DNA molecule is maintained by the Watson-Crick base pairs (A-T and G-C). The G-C is connected with three hydrogen bonds, while A-T is connected with two hydrogen bonds. DNA fragments containing higher GC pairs are more stable and have a higher melting point. The sensitivity to temperature serves as the principle for TGGE to separate the DNAs with a variant base composition. For DGGE, a gradient of the chemical denaturant is used to drive the denaturing of DNA as opposed to temperature for TGGE. Both techniques possess high resolutions and have been widely used in biomedical and ecological research such as mutations and SNPs detections (Fodde and Losekoot, 1994; Enwall and Hallin, 2009; Knapp, 2009).

DGGE was introduced to microbiome profiling by Muyzer in 1993. The bacterial 16S rDNA V3 regions from microbial mats, Wadden Sea sediment, and wastewater were amplified by PCR and then separated by DGGE. The community profiles showed up to 10 distinguishable bands in each lane, and each of them represented a bacterial group from original specimens. Commercial software is available to interpret the DGGE and TGGE-generated band patterns, and bioinformatics analysis including both quantitative and phylogenetical microbiome diversities can be performed using the software (Ahn et al., 2009; Lalande et al., 2013). DGGE and TGGE were widely used to study gut microbial composition dynamics in poultry in response to diet changes such as antibiotic administrations and probiotics/prebiotics supplementation. Dietary antibiotics such as virginiamycin and bacitracin methylene di-salicylate possess a dose-dependent impact on avian gut flora compositions, indicated by certain faded or appeared DGGE DNA bands (Zhou et al., 2007). A study utilizing the PCR/DGGE method for the first time in broiler chickens disclosed the beneficial effect of dietary supplementation of fermented cottonseed meal, which contains high beneficial bacteria and prebiotics (organic acids), on the gut microbiome structure as indicated by reducing *E.coli* and increasing *Lactobacillus* populations in both ileum and cecum (Sun et al., 2013). However, the relative abundance of microbiome compositions based on 16S rDNA lacks adequate accuracy due to inconsistent 16S rRNA gene copy numbers in the various species. For that purpose, single-copy housekeeping genes such as rpoB can be used for bacterial quantification (Piterina and Pembroke, 2013).

5 16S ribosomal RNA clone library sequencing

In 1990, the 16S rDNA sequencing technique was first used to reveal novel community members from Octopus Spring mat: eight distinct members shared up to 94.7% similarity with previously isolated strain libraries. Some of those discovered included *Synechococcus lividus*, *Chloroflexus aurantiacus*, *Isophaera pallid*, *Methanobacterium thermoautotrophicum* (Ward et al., 1990). Since its initial introduction into microbiome identification, the 16S rRNA clone library was used as a gold standard in most microbiome analytical methods including the aforementioned T-RFLP, DGGE, and TGGE as well as next-generation sequencing (NGS), which rapidly enlarged the population of ribosomal RNA databases (Yarza et al., 2008).

As an important component of the ribosomes, 16S rRNA has the major role of protein translation in bacteria and archaea. Due to the slow rate of evolution, the 16S rRNA gene is highly conserved in all prokaryotes, thus phylogenetics can be comprehensively reconstructed in culture-independent conditions (Dowling and Wald, 1960). The 16S rRNA gene is composed of approximately 1500 base pairs including nine hypervariable and ten conserved regions. Variable regions determine the secondary structures of the ribosome subunits as well as taxonomic classifications such as phylum and genus (Yang et al., 2016). The alternate arrangement of hypervariable and conserved regions allows flexible primer design for targeting interested variable regions. The 16S rRNA sequence-based phylogenetic reconstructions have different depth strategies such as single or multiple successive variable regions or the whole 16S rRNA sequences. The long fragments V3, V4, and V5 are the most hypervariable regions and are the most widely used. Presently, many public assessable retrieval systems such as SILVA, RDP, GreenGenes, or NCBI provide numerous taxonomic databases.

6 Next-generation sequencing (*Roche 454*, Illumina, and Ion Torrent)

The launch of NGS techniques, also called second-generation sequencing, marks the arrival of the big data era. These techniques obey the 'sequencing-by-synthesis' principle allowing the generation of hundreds of megabase pairs of nucleotide sequences within a short period of time. The large data output and low cost provide a robust technique in studying subjects-wide microbiome structure, bacterial communications, mutation mapping, and so on. There are several NGS platforms such as *Roche 454*, Illumina, and Ion Torrent sequencing.

Roche 454. This method begins from a library construction: genomic DNA is nebulized into 300–600 base pairs and short adapters A and B are ligated onto both ends of the fragments. These adapters carry complementary binding sites that bind to the oligomers located on the bead surface in one unique sequence per bead, and then emulsion PCR occurs to generate DNA strand libraries. The sequence-ready beads are transferred onto a PicoTiterPlate (PTP) at one bead per well with the required enzyme mix. The pyrosequencing in all PTP wells occurs simultaneously. When synthesizing the new complementary strand upon the bead-bond DNA template, the ligation of each nucleotide emits a signal from luciferase activity and is computed into specific nucleotide information for each well (Mardis, 2008). Extracted data after passing a systematic quality check is then exported for the downstream analysis. The Roche 454 was widely used in gut microbiota studies, such as the dietary phosphorus induced spatial community variations in chicken (Witzig et al., 2015). However, the popularity of this technique continued for a relatively short period, as other, more robust sequencing technologies emerged, such as Illumina genome analyzer.

Illumina MiSeq and HiSeq. Different from the 454, Illumina MiSeq and HiSeq library amplicons are generated from PCR reactions with specifically designed primers. Besides flow cell adapters, linkers, forward and reverse primers, and index barcodes are also incorporated into the primers to enable multiplex sequencing of a large number of samples on the same lane. The indexed amplicons from various samples are purified through normalization and then equally combined, followed by a DNA integrity and quality check (QC). Before loading the library into a sequencer cartridge, 5–20% of Phix control is added to increase the diversity of the library. Loaded amplicon libraries are presented in a low density on the flow cells, but after the bridge amplification cycle, up to 1,000 identical copies will be created as a 'Cluster'. Each cluster, representing a single unique nucleotide sequence, emits the same fluorescence and is captured by an image detector after each cycle (Diaz-Sanchez et al., 2013). The sequencing undergoes 3 or 4 steps depending on single or dual indexing on the primers: Read 1 (forward direction), Index 1 (i7) Read, Index 2 (i5) Read (only for dual-indexed workflows), and Read 2 (reversal direction). When all sequencing steps are finished, the Illumina program performs a QC on all reads. The QC-passed sequences (forward and reverse reads) are traced back to the Sample ID depending on the Index information. Finally, all reads are available as FASTQ types for download from the BaseSpace sequence hub. Currently, the Illumina Hiseq and Miseq techniques are two very important sequencing systems due to the generation of big data with high quality and at a low cost. The high precise sequencing platform with limited background noises also paves a wider road for linking gut microbiome to both phenotype and genetics, allowing for improvements in breeding. Pandit and colleagues measured the caecal microbiome variations between chicken breeds (indigenous Indian breeds Kadaknath and Aseel) and two global commercial broiler lines – Cobb400 (AAU) and Ross 308 (TANUVAS) – using Miseq targeting on V3–V4 region and disclosed that breeds and lines present a major driver for microbiome structures (Pandit et al., 2018). For instance, genera *Geobacillus*, *Cyclobacterium*, *Caldicellulosiruptor*, *Thermobaculum*, and so on are Kadaknath birds specific (over 0.1% abundance in all birds) while *Slackia*, *Cronobacter*, *Phascolarctobacterium*, unclassified *Alphaproteobacteria*, *Oceanimonas*, and so on are Aseel specific.

Ion Torrent. Ion Torrent sequencing, first released in 2010 by Torrent Systems Inc., is another method of 'sequencing by synthesis' that provides an alternative platform for specific research purposes. It shares the majority of sequencing procedures as *Roche 454*, but instead of fluorescence detection, Ion Torrent sequencing captures the pH changes generated from released hydrogen bonds and translates this signal into corresponding nucleotides. Specifically, incorporation of deoxyribonucleoside triphosphate (dNTP) into the DNA polymerization can lead to the release of pyrophosphate and an H+. The released H+ changes the solution pH which then can be detected by an ion-sensitive field-effect transistor (Rusk, 2011). This pH detector technique depends directly on real-time natural biochemistry features instead of recruiting complex fluorescence labeling, luciferase tags, lasers, and cameras used in other platforms (Diaz-Sanchez et al., 2013). Ion semiconductor sequencing technology was once the leading and most rapid sequencing machine in the world, finishing a set of genes within 90 minutes. With its rapid, high accuracy, low cost properties, longer testing fragments (400–500), and simple handling, Ion Torrent sequencing has broad applications in microbiome disease etiology, drug resistance testing, cancer diagnosis, viral identification, and gene mutation investigations (Powers et al., 2012; Hsiao et al., 2016; Bhogoju et al., 2018; Sajnani et al., 2018). In poultry science, the Ion Personal Genome Machine (PGM) system was used to detect microbiome community differences within different poultry types, such as chicken and Guinea fowl, based on the 16S rRNA gene sequences. The phylogenetic mapping between two bird types disclosed detailed compositions at the phylum, family and species levels, which presented basic references in gut microbiome studies. The Ion PGM system was also used to assist investigations in understanding the viral metagenome in respiratory infected poultry, the complete genome of high abundant viruses such as avian gyrovirus 2, gyrovirus 4, anemia virus, mongoose feces-associated gemycircularvirus b strain 160b, human gyrovirus type 1, and Gyrovirus Tu789 (Sajnani et al., 2018).

The *Roche 454* GS FLX+ system allows up to 800 bp libraries sequenced, longer than Illumina and Ion Torrent platforms, and thus it benefits long-reads purposes such as genome or transcriptome assembly. Sequencing errors such as homopolymer, chimeric reads, and systematic errors contribute to the partial misrepresentation of microbial populations or quantitative bias, which may lead to masking the real biological significance in diversity analysis and disease diagnosis (Huse et al., 2007; Schloss et al., 2011). Illumina MiSeq and HiSeq are frequently reported to possess the lowest error rates compared to the *Roche 454*, Ion Torrent, or other NGS sequencing platforms (Metzker, 2010; Manley et al., 2016). Illumina and Ion Torrent platforms generate larger throughputs than *Roche 454*, which is able to obtain deeper community structures (Loman et al., 2012). Ion Torrent has the fastest sequencing speed with acceptable error rates. More machine systems and reagent kits with specific research goals are available from Illumina. Illumina NextSeq series is designed for whole-genome and transcriptome sequencing, while Ion PGM™ Chips v2 (314™, 316™, and 318™) generate various-size throughputs.

7 Third-generation sequencing (Pacbio SMRT and Oxford Nanopore MinION)

The third-generation sequencing technique (TGS) is the beginning of a new revolution for genetic studies with a promising future due to the longer read length, higher speed, and epigenetic detection properties. TGS is based on a single-molecule real-time sequencing technology that generates unprecedented lengths, which largely alleviate the alignment challenges, especially for genomes containing numerous repetitive regions. Of note, the Oxford Nanopore technique is much smaller in size and is portable (MinION) because it does not require complex library preparation and sample processing. Because of its portability and rapid output, genetic sequencing can be performed anytime and anywhere. However, TGS has its drawbacks such as lower accuracy which have slowed down the spread of the application. Two recently delivered TGS platforms Pacbio Single-Molecule Realtime (SMRT) Sequencing, and Oxford Nanopore Technologies' MinION will be discussed in the next section.

PacBio SMRT Sequencing. The central part of the PacBio SMRT device is the SMRT cell. It is composed of 150000 single sequencing units called zero-mode waveguide (ZMW), a highly advanced light detector which is the smallest in the world. On the bottom of each ZMW, a single polymerase is located, which manages the replication of a prior-prepared DNA template called the SMRTbell. This template is a single-stranded circular DNA generated by double end-hairpin adapter ligation of the target DNA. The replication begins when the hairpin adapter binds to the polymerase. Each nucleotide incorporation generates two signals: a pulse and an emission spectrum from fluorescent-labeled nucleotides. The fluorescence output color is translated into bases while the synchronous pulse rate and rhythm disclose the presence of methylation (Koren and Phillippy, 2015). The entire replication of SMRTbell generates two reads; however, this can double or triple when the template is short or if time permits. Generation of multiple reads from the same ZMW can improve its accuracy (Rhoads and Au, 2015). However, the low efficiency of the ZMW reduces the overall success rate and produces low throughput with a high error rate. Because of this, PacBio SMRT cannot compete with NGS such as Illumina Miseq and Hiseq in regard to large-scale microbiome analyses at this moment.

Oxford Nanopore Technologies' MinION. Oxford Nanopore MinION became available on the market in 2014. It identifies DNA bases by measuring electrical conductivity generated when passing through a protein nanopore (Lu et al., 2016). Once available, its portability caught the attention of investigators in many fields such as clinical research, fieldwork, and disease surveillance. The MinION weighs less than 100 g and thus can be taken to just about any place for immediate sequencing. Without complex sample handling, microbiome identification via 16S rRNA takes less than 40 minutes. Also, the built-in NCBI databases directly blast the generated reads and report the taxonomic information automatically. Although the nanopore technique has not been widely applied in agriculture yet, its application regarding fast clinical diagnosis will fulfill the sample-to-answer requirement (Russell et al., 2018).

8 Microbiome, metagenomics, and metatranscriptomics

Since scientists have noticed a critical relationship between the microbiome and environmental health, efforts probing into the microbial world will never end. Early microbiome studies relied on a limited group of culturable bacterial strains for observing their morphology, physiology, and growth. With the advent of molecular methods such as PCR, Sanger sequencing, T-RFLP, DGGE, TGGE, and NGS, the focus on whole bacterial cells moved toward the field of genetic materials, where investigators no longer struggled with cell cultivability. Metagenomics, which captures all of the genes present in a target sample, is most suitable for a comprehensive analysis of microbial community compositions. It also provides glimpses into biological functions through annotation of dominantly presented genes. However, not all genes are translated into proteins so metagenomics may generate biased function interpretations (Aguiar-Pulido et al., 2016). In response to environmental changes, a microbiome community performs physiological and metabolism adjustments when challenged by a new environment. These physiological changes are accompanied by changes in macromolecular compositions (Konopka and Wilkins, 2012). Hence, omics techniques such as metatranscriptomics, the gathering of all expressed genetic products (mainly mRNA), and metaproteomics, the collection of proteins from an entire microbial community, are necessary to study the microbial functions in a specific niche.

Metagenomic analysis of chicken feces revealed considerable glycolytic and non-starch polysaccharide digestions occurring in the cecum by the presented genes of glycosyl hydrolases, polysaccharides (e.g. Xylane), and degrading enzymes (Sergeant et al., 2014). Another chicken fecal sourced metaproteomic analysis indicated the existences of pyruvate kinases, and a high abundance of stress proteins in response to cold, toxic oxygen molecules, and oxidative stresses (Tang et al., 2014). Metatranscriptomics was widely recruited for analyzing physiological principles in host tissues in the avian world. For instance, intestinal and muscular sites RNA were sequenced for interpreting the phenotype-related feed efficiencies, while gustatory tissues were targeted for RNAseq to study taste sensation functions (Cui et al., 2017; Reyer et al., 2018). In addition, the cecum is one of the main locations encountering frequent pathogen infections such as *Eimeria tenella*, and RNA sequencing of pathogen-affected intestinal tissues could provide insights into the mechanisms of pathogenesis at the gene expression level (Wang et al., 2019).

In a recent paper, Parreira et al. (2016), for the first time, characterized the transcriptome of the poultry-sourced pathogen *netB*-positive type A *Clostridium perfringens* in early adaptation and colonization in the chicken gut and depicted how nutrition status and environment affect the gene expression of the pathogen. For a better understanding of how pathogens react to the host and initiate diseases, combinations of multiple -omics from both host and pathogen sides are highly recommended.

9 Conclusion and future trends

Despite the great promise that NGS holds in the poultry industry, many concerns remain for future microbiome analysis including experimental design, statistical power, and analytical pipeline. Proposing a meaningful assumption ahead of the experiment is critical for future interpretation of results. The assumptions need to obey logical scientific knowledge and possess biological meaning, or they will be pointless even when supported by statistical data. When comparing microbiome structures in different environments or biological groups, clarifying host background such as sex, age, and breed is necessary, especially for medication and disease studies. Questionable subjects should be excluded from the group. Retaining adequate statistical power is critical to generating convincing results. For high-throughput genetic studies, statistical power can be improved by the correct sample size, which can be calculated by an online-sourced calculator or commercial tools (Sham and Purcell, 2014). Of note, recruiting an oversized group may not bring extra power but more burdens on sample handling and experimental cost. Generally used pipelines for data computation include quality control, sequence clustering, assigning taxonomy, diversity measurement, and data visualization (Plummer et al., 2015). Publicly shared computational platforms such as QIIME2 and mothur provide professional and easy-to-follow pipelines for researchers to use. However, different algorithms in different platforms vary in species classification and running speed (Plummer et al., 2015).

10 Where to look for further information

There are a number of journals that record most recent gut microbiome projects:

Microbiome: https://microbiomejournal.biomedcentral.com/
Frontiers in microbiology: https://www.frontiersin.org/journals/microbiology

Veterinary Microbiology: https://www.journals.elsevier.com/veterinary-microbiology.

Some recently published journal papers that demonstrate the significances of poultry gut microbiome and detection technology:

http://www.fbae.org/2009/FBAE/website/images/PDF%20files/Imporatant%20Publication/Next%20generation%20of%20DNA%20sequencing.pdf (Shang et al., 2018)

https://microbiomejournal.biomedcentral.com/articles/10.1186/s40168-018-0590-5 (Huang et al., 2018).

11 References

Aguiar-Pulido, V., Huang, W., Suarez-Ulloa, V., Cickovski, T., Mathee, K. and Narasimhan, G. 2016. Metagenomics, metatranscriptomics, and metabolomics approaches for microbiome analysis: supplementary issue: bioinformatics methods and applications for big metagenomics data. *Evolutionary Bioinformatics* 12. doi:10.4137/EBO.S36436.

Ahn, J. H., Kim, Y. J., Kim, T., Song, H. G., Kang, C. and Ka, J. O. 2009. Quantitative improvement of 16S rDNA DGGE analysis for soil bacterial community using real-time PCR. *Journal of Microbiological Methods* 78(2), 216–22. doi:10.1016/j.mimet.2009.06.001.

Altekruse, S. F., Berrang, M. E., Marks, H., Patel, B., Shaw, W. K., Saini, P., Bennett, P. A. and Bailey, J. S. 2009. Enumeration of *Escherichia coli* cells on chicken carcasses as a potential measure of microbial process control in a random selection of slaughter establishments in the United States. *Applied and Environmental Microbiology* 75(11), 3522–7. doi:10.1128/AEM.02685-08.

Barnes, E. M., Mead, G. C., Barnuml, D. A. and Harry, E. G. 1972. The intestinal flora of the chicken in the period 2 to 6 weeks of age, with particular reference to the anaerobic bacteria. *British Poultry Science* 13(3), 311–26. doi:10.1080/00071667208415953.

Bedbury, H. P. and Duke, G. E. 1983. Cecal microflora of turkeys fed low or high fiber diets: enumeration, identification, and determination of cellulolytic activity. *Poultry Science* 62(4), 675–82. doi:10.3382/ps.0620675.

Bhogoju, S., Nahashon, S., Wang, X., Darris, C. and Kilonzo-Nthenge, A. 2018. A comparative analysis of microbial profile of Guinea fowl and chicken using metagenomic approach. *PLoS ONE* 13(3), e0191029. doi:10.1371/journal.pone.0191029.

Blackwood, C. B., Hudleston, D., Zak, D. R. and Buyer, J. S. 2007. Interpreting ecological diversity indices applied to terminal restriction fragment length polymorphism data: insights from simulated microbial communities. *Applied and Environmental Microbiology* 73(16), 5276–83. doi:10.1128/AEM.00514-07.

Brosius, J., Palmer, M. L., Kennedy, P. J. and Noller, H. F. 1978. Complete nucleotide sequence of a 16S ribosomal RNA gene from *Escherichia coli*. *Proceedings of the National Academy of Sciences of the United States of America* 75(10), 4801–5. doi:10.1073/pnas.75.10.4801.

Butzler, J. P. and Oosterom, J. 1991. Campylobacter: pathogenicity and significance in foods. *International Journal of Food Microbiology* 12(1), 1-8. doi:10.1016/0168-1605(91)90043-o.

Castellanos, L. R., Donado-Godoy, P., León, M., Clavijo, V., Arevalo, A., Bernal, J. F., Timmerman, A. J., Mevius, D. J., Wagenaar, J. A. and Hordijk, J. 2017. High heterogeneity of *Escherichia coli* sequence types harbouring ESBL/AmpC genes on IncI1 plasmids in the Colombian poultry chain. *PLoS ONE* 12(1), e0170777. doi:10.1371/journal.pone.0170777.

Clavijo, V. and Flórez, M. J. V. 2018. The gastrointestinal microbiome and its association with the control of pathogens in broiler chicken production: a review. *Poultry Science* 97(3), 1006–21. doi:10.3382/ps/pex359.

Clench, M. H. and Mathias, J. R. 1995. The avian cecum: a review. *The Wilson Bulletin* 107(1), 93–121.

Crhanova, M., Hradecka, H., Faldynova, M., Matulova, M., Havlickova, H., Sisak, F. and Rychlik, I. 2011. Immune response of chicken gut to natural colonisation by gut microflora and to *Salmonella enterica* serovar Enteritidis infection. *Infection and immunity:IAI* 79(7), 2755–63.

Cui, C., Shu, W. and Li, P. 2016. Fluorescence *in situ* hybridization: cell-based genetic diagnostic and research applications. *Frontiers in Cell and Developmental Biology* 4, 89. doi:10.3389/fcell.2016.00089.

Cui, X., Marshall, B., Shi, N., Chen, S. Y., Rekaya, R. and Liu, H. X. 2017. RNA-Seq analysis on chicken taste sensory organs: an ideal system to study organogenesis. *Scientific Reports* 7(1), 9131. doi:10.1038/s41598-017-09299-7.

Dahm, R. 2005. Friedrich Miescher and the discovery of DNA. *Developmental Biology* 278(2), 274–88. doi:10.1016/j.ydbio.2004.11.028.

De Angelis, M., Siragusa, S., Berloco, M., Caputo, L., Settanni, L., Alfonsi, G., Amerio, M., Grandi, A., Ragni, A. and Gobbetti, M. 2006. Selection of potential probiotic lactobacilli from pig feces to be used as additives in pelleted feeding. *Research in Microbiology* 157(8), 792–801. doi:10.1016/j.resmic.2006.05.003.

Diaz-Sanchez, S., Hanning, I., Pendleton, S. and D'Souza, D. 2013. Next-generation sequencing: the future of molecular genetics in poultry production and food safety. *Poultry Science* 92(2), 562–72. doi:10.3382/ps.2012-02741.

Dowling, J. E. and Wald, G. 1960. The biological function of vitamin a acid. *Proceedings of the National Academy of Sciences of the United States of America* 46(5), 587–608.

Dunbar, J., Ticknor, L. O. and Kuske, C. R. 2001. Phylogenetic specificity and reproducibility and new method for analysis of terminal restriction fragment profiles of 16S rRNA genes from bacterial communities. *Applied and Environmental Microbiology* 67(1), 190–7. doi:10.1128/AEM.67.1.190-197.2001.

Enwall, K. and Hallin, S. 2009. Comparison of T-RFLP and DGGE techniques to assess denitrifier community composition in soil. *Letters in Applied Microbiology* 48(1), 145–8. doi:10.1111/j.1472-765X.2008.02498.x.

Fodde, R. and Losekoot, M. 1994. Mutation detection by denaturing gradient gel electrophoresis (DGGE). *Human Mutation* 3(2), 83–94. doi:10.1002/humu.1380030202.

Gaskins, H. R., Collier, C. T. and Anderson, D. B. 2002. Antibiotics as growth promotants: mode of action. *Animal Biotechnology* 13(1), 29–42. doi:10.1081/ABIO-120005768.

Gong, J., Forster, R. J., Yu, H., Chambers, J. R., Wheatcroft, R., Sabour, P. M. and Chen, S. 2002. Molecular analysis of bacterial populations in the ileum of broiler chickens and comparison with bacteria in the cecum. *FEMS Microbiology Ecology* 41(3), 171-9. doi:10.1111/j.1574-6941.2002.tb00978.x.

Hiergeist, A., Gläsner, J., Reischl, U. and Gessner, A. 2015. Analyses of intestinal microbiota: culture versus sequencing. *ILAR Journal* 56(2), 228–40. doi:10.1093/ilar/ilv017.

Hsiao, Y. P., Lu, C. T., Chang-Chien, J., Chao, W. R. and Yang, J. J. 2016. Advances and applications of Ion Torrent personal genome machine in cutaneous squamous cell carcinoma reveal novel gene mutations. *Materials* 9(6), 464. doi:10.3390/ma9060464.

Huang, P., Zhang, Y., Xiao, K., Jiang, F., Wang, H., Tang, D., Liu, D., Liu, B., Liu, Y. and He, X. 2018. The chicken gut metagenome and the modulatory effects of plant-derived benzylisoquinoline alkaloids. *Microbiome* 6(1), 211.

Huse, S. M., Huber, J. A., Morrison, H. G., Sogin, M. L. and Welch, D. M. 2007. Accuracy and quality of massively parallel DNA Pyrosequencing. *Genome Biology* 8(7), R143. doi:10.1186/gb-2007-8-7-r143.

Immerseel, F. V., De Buck, J., Pasmans, F., Huyghebaert, G., Haesebrouck, F. and Ducatelle, R. 2004. Clostridium perfringens in poultry: an emerging threat for animal and public health. *Avian Pathology: Journal of the W.V.P.A* 33(6), 537–49. doi:10.1080/03079450400013162.

Keyburn, A. L., Boyce, J. D., Vaz, P., Bannam, T. L., Ford, M. E., Parker, D., Di Rubbo, A., Rood, J. I. and Moore, R. J. 2008. NetB, a new toxin that is associated with avian necrotic enteritis caused by *Clostridium perfringens*. *PLoS Pathogens* 4(2), e26. doi:10.1371/journal.ppat.0040026.

Kim, Y. G., Sakamoto, K., Seo, S. U., Pickard, J. M., Gillilland, M. G., Pudlo, N. A., Hoostal, M., Li, X., Wang, T. D., Feehley, T., Stefka, A. T., Schmidt, T. M., Martens, E. C., Fukuda, S., Inohara, N., Nagler, C. R. and Núñez, G. 2017. Neonatal acquisition of *Clostridia* species protects against colonization by bacterial pathogens. *Science* 356(6335), 315–9. doi:10.1126/science.aag2029.

Knapp, L. A. 2009. Single nucleotide polymorphism screening with denaturing gradient gel electrophoresis. In: Komar, A. (Ed.), *Single Nucleotide Polymorphisms. Methods in Molecular Biology™ (Methods and Protocols)* (vol. 578). Humana Press, Totowa, NJ, pp. 137–51.

Konopka, A. and Wilkins, M. J. 2012. Application of meta-transcriptomics and-proteomics to analysis of in situ physiological state. *Frontiers in Microbiology* 3, 184. doi:10.3389/fmicb.2012.00184.

Koren, S. and Phillippy, A. M. 2015. One chromosome, one contig: complete microbial genomes from long-read sequencing and assembly. *Current Opinion in Microbiology* 23, 110–20. doi:10.1016/j.mib.2014.11.014.

Lalande, J., Villemur, R. and Deschênes, L. 2013. A new framework to accurately quantify soil bacterial community diversity from DGGE. *Microbial Ecology* 66(3), 647–58. doi:10.1007/s00248-013-0230-3.

LeBlanc, J. G., Milani, C., de Giori, G. S., Sesma, F., Van Sinderen, D. and Ventura, M. 2013. Bacteria as vitamin suppliers to their host: a gut microbiota perspective. *Current Opinion in Biotechnology* 24(2), 160–8. doi:10.1016/j.copbio.2012.08.005.

Loman, N. J., Misra, R. V., Dallman, T. J., Constantinidou, C., Gharbia, S. E., Wain, J. and Pallen, M. J. 2012. Performance comparison of benchtop high-throughput sequencing platforms. *Nature Biotechnology* 30(5), 434–9. doi:10.1038/nbt.2198.

Lu, H., Giordano, F. and Ning, Z. 2016. Oxford nanopore MinION sequencing and genome assembly. *Genomics, Proteomics and Bioinformatics* 14(5), 265–79. doi:10.1016/j.gpb.2016.05.004.

Manley, L. J., Ma, D. and Levine, S. S. 2016. Monitoring error rates in Illumina sequencing. *Journal of Biomolecular Techniques: JBT* 27(4), 125–8. doi:10.7171/jbt.16-2704-002.

Mardis, E. R. 2008. The impact of next-generation sequencing technology on genetics. *Trends in Genetics: TIG* 24(3), 133–41. doi:10.1016/j.tig.2007.12.007.

Medvecky, M., Cejkova, D., Polansky, O., Karasova, D., Kubasova, T., Cizek, A. and Rychlik, I. 2018. Whole genome sequencing and function prediction of 133 gut anaerobes isolated from chicken caecum in pure cultures. *BMC Genomics* 19(1), 561. doi:10.1186/s12864-018-4959-4.

Mesa, D., Lammel, D. R., Balsanelli, E., Sena, C., Noseda, M. D., Caron, L. F., Cruz, L. M., Pedrosa, F. O. and Souza, E. M. 2017. Cecal microbiota in broilers fed with prebiotics. *Frontiers in Genetics* 8, 153. doi:10.3389/fgene.2017.00153.

Metzker, M. L. 2010. Sequencing technologies–the next generation. *Nature Reviews. Genetics* 11(1), 31–46. doi:10.1038/nrg2626.

Oakley, B. B., Lillehoj, H. S., Kogut, M. H., Kim, W. K., Maurer, J. J., Pedroso, A., Lee, M. D., Collett, S. R., Johnson, T. J. and Cox, N. A. 2014. The chicken gastrointestinal microbiome. *FEMS Microbiology Letters* 360(2), 100–12. doi:10.1111/1574-6968.12608.

Osborn, A. M., Moore, E. R. and Timmis, K. N. 2000. An evaluation of terminal-restriction fragment length polymorphism (T-RFLP) analysis for the study of microbial community structure and dynamics. *Environmental Microbiology* 2(1), 39–50. doi:10.1046/j.1462-2920.2000.00081.x.

Oyarzabal, O. A., Hawk, C., Bilgili, S. F., Warf, C. C. and Kemp, G. K. 2004. Effects of postchill application of acidified sodium chlorite to control *Campylobacter* spp. and *Escherichia coli* on commercial broiler carcasses. *Journal of Food Protection* 67(10), 2288–91. doi:10.4315/0362-028x-67.10.2288.

Oyarzabal, O. A., Macklin, K. S., Barbaree, J. M. and Miller, R. S. 2005. Evaluation of agar plates for direct enumeration of *Campylobacter* spp. from poultry carcass rinses. *Applied and Environmental Microbiology* 71(6), 3351–4. doi:10.1128/AEM.71.6.3351-3354.2005.

Pan, D. and Yu, Z. 2014. Intestinal microbiome of poultry and its interaction with host and diet. *Gut Microbes* 5(1), 108–19. doi:10.4161/gmic.26945.

Pandit, R. J., Hinsu, A. T., Patel, N. V., Koringa, P. G., Jakhesara, S. J., Thakkar, J. R., Shah, T. M., Limon, G., Psifidi, A., Guitian, J. J. M., Hume, D. A., Tomley, F. M., Rank, D. N., Raman, M., Tirumurugaan, K. G., Blake, D. P. and Joshi, C. G. 2018. Microbial diversity and community composition of caecal microbiota in commercial and indigenous Indian chickens determined using 16S rDNA amplicon sequencing. *Microbiome* 6(1), 115. doi:10.1186/s40168-018-0501-9.

Panisello, P. J., Rooney, R., Quantick, P. C. and Stanwell-Smith, R. 2000. Application of foodborne disease outbreak data in the development and maintenance of HACCP systems. *International Journal of Food Microbiology* 59(3), 221–34. doi:10.1016/s0168-1605(00)00376-7.

Parreira, V. R., Russell, K., Athanasiadou, S. and Prescott, J. F. 2016. Comparative transcriptome analysis by RNAseq of necrotic enteritis *Clostridium perfringens* during *in vivo* colonization and *in vitro* conditions. *BMC Microbiology* 16(1), 186. doi:10.1186/s12866-016-0792-6.

Piterina, A. V. and Pembroke, J. T. 2013. Use of PCR-DGGE based molecular methods to analyse microbial community diversity and stability during the thermophilic stages of an ATAD wastewater sludge treatment process as an aid to performance monitoring. *ISRN Biotechnology* 2013, 162645. doi:10.5402/2013/162645.

Plummer, E., Twin, J., Bulach, D. M., Garland, S. M. and Tabrizi, S. N. 2015. A comparison of three bioinformatics pipelines for the analysis of preterm gut microbiota using 16S rRNA gene sequencing data. *Journal of Proteomics and Bioinformatics* 8(12), 283. doi:10.4172/jpb.1000381.

Powers, M., Watkins, W., Potucek, Y. and Warner, D. 2012. Whole genome mtDNA sequencing on the Ion Torrent PGM. *Journal of Biomolecular Techniques: JBT* 23 (Suppl.), S37.

Reyer, H., Metzler-Zebeli, B. U., Trakooljul, N., Oster, M., Muráni, E., Ponsuksili, S., Hadlich, F. and Wimmers, K. 2018. Transcriptional shifts account for divergent resource allocation in feed efficient broiler chickens. *Scientific Reports* 8(1), 12903. doi:10.1038/s41598-018-31072-7.

Rhoads, A. and Au, K. F. 2015. PacBio sequencing and its applications. *Genomics, Proteomics and Bioinformatics* 13(5), 278–89. doi:10.1016/j.gpb.2015.08.002.

Rusk, N. 2011. Torrents of sequence. *Nature Methods* 8(1), 44. doi:10.1038/nmeth.f.330.

Russell, J. A., Campos, B., Stone, J., Blosser, E. M., Burkett-Cadena, N. and Jacobs, J. L. 2018. Unbiased strain-typing of arbovirus directly from mosquitoes using nanopore sequencing: a field-forward biosurveillance protocol. *Scientific Reports* 8(1), 5417. doi:10.1038/s41598-018-23641-7.

Sajnani, M. R., Sudarsanam, D., Pandit, R. J., Oza, T., Hinsu, A. T., Jakhesara, S. J., Solosanc, S., Joshi, C. G. and Bhatt, V. D. 2018. Metagenomic data of DNA viruses of poultry affected with respiratory tract infection. *Data in Brief* 16, 157–60. doi:10.1016/j.dib.2017.11.033.

Sakamoto, M., Takeuchi, Y., Umeda, M., Ishikawa, I. and Benno, Y. 2003. Application of terminal RFLP analysis to characterize oral bacterial flora in saliva of healthy subjects and patients with periodontitis. *Journal of Medical Microbiology* 52(1), 79–89. doi:10.1099/jmm.0.04991-0.

Salanitro, J. P., Fairchilds, I. G. and Zgornicki, Y. D. 1974. Isolation, culture characteristics, and identification of anaerobic bacteria from the chicken cecum. *Applied Microbiology* 27(4), 678–87.

Schloss, P. D., Gevers, D. and Westcott, S. L. 2011. Reducing the effects of PCR amplification and sequencing artifacts on 16S rRNA-based studies. *PLoS ONE* 6(12), e27310. doi:10.1371/journal.pone.0027310.

Schultz, M. 2008. Clinical use of *E. coli* Nissle 1917 in inflammatory bowel disease. *Inflammatory Bowel Diseases* 14(7), 1012–8. doi:10.1002/ibd.20377.

Sergeant, M. J., Constantinidou, C., Cogan, T. A., Bedford, M. R., Penn, C. W. and Pallen, M. J. 2014. Extensive microbial and functional diversity within the chicken cecal microbiome. *PLoS ONE* 9(3), e91941. doi:10.1371/journal.pone.0091941.

Sham, P. C. and Purcell, S. M. 2014. Statistical power and significance testing in large-scale genetic studies. *Nature Reviews. Genetics* 15(5), 335–46. doi:10.1038/nrg3706.

Shang, Y., Kumar, S., Oakley, B. and Kim, W. K. 2018. Chicken gut microbiota: importance and detection technology. *Frontiers in Veterinary Science* 5, 254.

Shulman, S. T., Friedmann, H. C. and Sims, R. H. 2007. Theodor Escherich: the first pediatric infectious diseases physician? *Clinical Infectious Diseases: an Official Publication of the Infectious Diseases Society of America* 45(8), 1025–9. doi:10.1086/521946.

Sochocka, M., Donskow-Łysoniewska, K., Diniz, B. S., Kurpas, D., Brzozowska, E. and Leszek, J. 2019. The gut microbiome alterations and inflammation-driven pathogenesis of Alzheimer's disease–a critical review. *Molecular Neurobiology* 56(3), 1841–51. doi:10.1007/s12035-018-1188-4.

Stanley, D., Hughes, R. J. and Moore, R. J. 2014. Microbiota of the chicken gastrointestinal tract: influence on health, productivity and disease. *Applied Microbiology and Biotechnology* 98(10), 4301-10. doi:10.1007/s00253-014-5646-2.

Sun, H., Tang, J. W., Fang, C. L., Yao, X. H., Wu, Y. F., Wang, X. and Feng, J. 2013. Molecular analysis of intestinal bacterial microbiota of broiler chickens fed diets containing fermented cottonseed meal. *Poultry Science* 92(2), 392-401. doi:10.3382/ps.2012-02533.

Tang, Y., Underwood, A., Gielbert, A., Woodward, M. J. and Petrovska, L. 2014. Metaproteomics analysis reveals the adaptation process for the chicken gut microbiota. *Applied and Environmental Microbiology* 80(2), 478-85. doi:10.1128/AEM.02472-13.

Teuber, M. 1995. The genus Lactococcus. In: Wood, B. J. B. and Holzapfel, W. H. (Eds), *The Genera of Lactic Acid Bacteria. The Lactic Acid Bacteria* (vol. 2). Springer, Boston, MA, pp. 173-234.

Torok, V. A., Ophel-Keller, K., Loo, M. and Hughes, R. J. 2008. Application of methods for identifying broiler chicken gut bacterial species linked with increased energy metabolism. *Applied and Environmental Microbiology* 74(3), 783-91. doi:10.1128/AEM.01384-07.

Vispo, C. and Karasov, W. H. 1997. The interaction of avian gut microbes and their host: an elusive symbiosis. In: Mackie, R. I. and White, B. A. (Eds), *Gastrointestinal Microbiology*. Chapman & Hall Microbiology Series. Springer, Boston, MA, pp. 116-55.

Wang, X., Zou, W., Yu, H., Lin, Y., Dai, G., Zhang, T., Zhang, G., Xie, K., Wang, J. and Shi, H. 2019. RNA sequencing analysis of chicken cecum tissues following *Eimeria tenella* infection *in vivo*. *Genes* 10(6), 420. doi:10.3390/genes10060420.

Ward, D. M., Weller, R. and Bateson, M. M. 1990. 16S rRNA sequences reveal numerous uncultured microorganisms in a natural community. *Nature* 345(6270), 63-5. doi:10.1038/345063a0.

Williams, S. C. P. 2014. Gnotobiotics. *Proceedings of the National Academy of Sciences of the United States of America* 111(5), 1661. doi:10.1073/pnas.1324049111.

Witzig, M., da Silva, A. C., Green-Engert, R., Hoelzle, K., Zeller, E., Seifert, J., Hoelzle, L. E. and Rodehutscord, M. 2015. Spatial variation of the gut microbiota in broiler chickens as affected by dietary available phosphorus and assessed by T-RFLP analysis and 454 pyrosequencing. *PLoS ONE* 10(11), e0143442. doi:10.1371/journal.pone.0143442.

Yang, B., Wang, Y. and Qian, P. Y. 2016. Sensitivity and correlation of hypervariable regions in 16S rRNA genes in phylogenetic analysis. *BMC Bioinformatics* 17(1), 135. doi:10.1186/s12859-016-0992-y.

Yarza, P., Richter, M., Peplies, J., Euzeby, J., Amann, R., Schleifer, K. H., Ludwig, W., Glöckner, F. O. and Rosselló-Móra, R. 2008. The All-Species Living Tree project: a 16S rRNA-based phylogenetic tree of all sequenced type strains. *Systematic and Applied Microbiology* 31(4), 241-50. doi:10.1016/j.syapm.2008.07.001.

Zhou, H., Gong, J., Brisbin, J. T., Yu, H., Sanei, B., Sabour, P. and Sharif, S. J. P. S. 2007. Appropriate chicken sample size for identifying the composition of broiler intestinal microbiota affected by dietary antibiotics, using the polymerase chain reaction-denaturing gradient gel electrophoresis technique. *Poultry Science* 86(12), 2541-9. doi:10.3382/ps.2007-00267.

Chapter 3

Omics technologies for connecting host responses with poultry gut function

Jana Seifert and Bruno Tilocca, University of Hohenheim, Germany

1 Introduction

Poultry production is increasing worldwide, providing meat and eggs for growing customer demands. Sustainable production that saves feed resources and considers animal welfare has to be achieved in order to fulfil these requirements in the future. For both aspects, optimal gut function is needed to maintain animal health and optimize feed efficiency. Animal gut health is still not clearly defined and the mere existence of a disease and an imbalanced gut microbiome can alter gut function (Celi et al., 2017). In accordance with increasing demands, breeders are focussing on chickens with high growth and reproductive traits, thus body organs including the intestinal tract have to evolve. The intestinal tract is important for feed conversion into energy for the host and the maintenance of immune functions. Different external factors such as pathogens or hazardous compounds can cause damage to the digestive system, which leads to inflammation, lesions and the loss of gut integrity. To avoid this, comprehensive knowledge of gut physiology is necessary and is of ongoing interest to animal scientists. Access to genomic information of the chicken in 2004 (International Chicken Genome Sequencing, 2004) was one of the starting points for the use of state-of-the-art 'omics' technologies, that is transcriptomics, proteomics and metabolomics, to study gene transcripts

http://dx.doi.org/10.19103/AS.2019.0059.03

and products in chickens. Since then, the intestinal tract of chickens has been investigated at the molecular level to identify transcripts that are involved in gut development, how response mechanisms are expressed after pathogenic infection and how the metabolome of the chicken is affected by different dietary treatments. The overall aim is to identify biomarkers for gut functions, which remains challenging (Celi et al., 2018).

This chapter will only briefly summarize the gut sections and the available 'omics' technologies, as these aspects have been excellently reviewed by other authors (Scanes, 2015). A comprehensive overview of the recent studies using 'omics' techniques focussing on the intestinal tract of chickens is discussed. These publications highlight the variability of objectives and targets that can be studied at the molecular level. Although the use of 'omics' is not trivial and expertise is essential, this chapter aims to detail the key benefits of its use to gain comprehensive insight into physiologic mechanisms, which are essential in order to improve gut function.

2 Gastrointestinal tract - functions, physiology and microbiota

The gastrointestinal tract of chickens is shorter in comparison to mammals and is adapted to enable the bird to fly (Denbow, 2015). Breeding strategies to optimize egg production and muscle growth have demanded an adaptation of the gut to enhance feed efficiency by a higher feed uptake (Svihus, 2014). This drastic change in nutrition can cause gut dysfunctions, as it is still not fully adapted to recent production strategies and feed formulations. As an optimized use of feed is needed to fulfil animal requirements and save resources, the functionality of the gut requires detailed investigation, in order to understand problems occurring in animal production associated with gut dysfunction and to support animal welfare. The chicken gut is divided into several sections as is known from other animal species. The main function is to digest feed compounds into smaller molecules that are absorbed by epithelial cells. In addition, the gut has important immunological functions in the protection against pathogenic invasion (Svihus, 2014; Denbow, 2015). Upon hatching, the chicken gut is essentially sterile but is quickly colonized by a myriad of microorganisms, including bacteria, archaea, fungi and viruses. The onset of the gut microbiota is a dynamic phenomenon, where composition richness and diversity increases with animal growth and is strongly influenced by both host (e.g. genetics, gender etc.) and environmental parameters (e.g. oxygen concentration, pH, diet etc.) (Del Chierico et al., 2012; Pan and Yu, 2014). The commensal microbiota of chickens changes along and across the digestive tract. Members of Lactobacillaceae are the predominant bacteria in the upper part of the digestive tract, whereas in distal sections, the bacterial diversity increases drastically concomitant with high fermentation activity (Borda-Molina et al., 2018).

2.1 Crop

The main function of the crop is storage and feed moistening, especially if large amounts of feed are available for the animal and the storage capacity of the gizzard is limited. Depending on the retention time of the feed in the crop, digestive processes may begin accompanied by the activation of feed supplements such as enzymes, which result in the formation of metabolites such as organic acids. Starch hydrolyses and lactate formation are catalysed by bacterial enzymes from *Lactobacillus* sp., *Bifidobacterium* sp. and Enterobacteriaceae which are found in the crop (Borda-Molina et al., 2016; Tilocca et al., 2016; Stanley et al., 2014; Oakley et al., 2014).

2.2 Proventriculus and gizzard

The proventriculus and gizzard are part of the two-chambered stomach in chickens. Feed transported from the crop to the proventriculus is further chemically and enzymatically digested here via the secretion of hydrochloric acid and pepsinogen. Mechanical treatment of the feed and gastric proteolysis occurs in the gizzard (Svihus, 2014; Denbow, 2015). The microbiota of the gizzard exhibits significant similarities to the crop with a predominance of *Lactobacillus* sp. (Choi et al., 2014; Sekelja et al., 2012).

2.3 Small intestine

The duodenal loop is the initial segment of the small intestine followed by the jejunum and ileum. The digestive process starts in the duodenum as a result of the reflux of bile acids and pancreatic juices and rapid balancing to more neutral pH conditions. The jejunum is the largest segment of the small intestine, and the digestion and absorption of fat, starch and proteins are almost complete at the end of this section (Svihus, 2014). The combined retention time of the duodenum and jejunum was found to be shorter than that of the ileum (Rougiére and Carré, 2010), which indicates an efficient physiological function by host and microbial enzymes in these sections and transport capacities for carbohydrates, amino acids, peptides, fatty acids and electrolytes. As most of the food components are almost completely digested in the previous gut sections, the ileum is described as being the major site for water and mineral absorption (Svihus, 2014). Thus, digestibility measurements assessing nutrient absorption and turnover in chickens are usually performed using ileum digesta samples. The microbiota of the small intestine is still characterized by the dominance of Lactobacillaceae, which varies in species distribution depending on the feeding regime (Witzig et al., 2015). In addition, Clostridiaceae and *Enterococcus* have also been identified here (Xiao et al., 2017).

2.4 Caeca

One pair of caeca is arranged as a two-blinded sac at the junction between the ileum and the colon. The size of the caeca is diet dependent as it functions as an important fermentation chamber in the chicken intestine. The content of the caeca originates from ileal and renal material. Water and electrolytes are primarily absorbed here along with short-chain fatty acids (Svihus et al., 2013). The considerable fermentation activity is caused by a variety of bacterial species that suppress the dominance of Lactobacillaceae, which are under-represented here, whereas bacteroidia, clostridia, bacilli and betaproteobacteria are the dominant bacterial classes in the caeca (Xiao et al., 2017; Nordentoft et al., 2011; Borda-Molina et al., 2018).

3 Omics technologies – how to use and what can they tell us?

Technological advances achieved over the past decade have marked a profound change in the methodologies adopted for the investigation of animal biology and microbial communities colonizing diverse anatomical sites, moving from the traditional approaches to the state-of-the-art 'omics' sciences, which employ a holistic view of the diverse molecular levels of organization of a given organism. Thus, the focus is on the detection and quantitation of the major biomolecules such as genomic DNA (genomics), ribosomal RNA (transcriptomics), proteins (proteomics) and metabolites (metabolomics), enabling a comprehensive picture of the biology and physiology of the investigated living system (Zoetendal et al., 2008).

3.1 Genomics

The first part of the chicken genome sequence was available in 2000, as a large amount of expressed sequence tags had been published (Abdrakhmanov et al., 2000). Four years later, the complete chicken genome and corresponding polymorphisms were available (International Chicken Genome Sequencing, 2004, International Chicken Polymorphism Map, 2004). These genomic explorations allowed the development of DNA microarrays (see below) and other downstream tools such as proteomics where the annotated genome sequence is needed to identify gene products. In addition to chickens, other poultry species including turkey, duck and quail (Kawahara-Miki et al., 2013; Dalloul et al., 2010; Huang et al., 2013) have also been sequenced and access to their genetic code allows future 'omics' applications.

3.2 Transcriptomics

The next level of biological information is stored in RNA, where mRNA and miRNA are usually used to conduct the transcriptional profiling of a sample. Transcriptomics, or functional genomics analysis, is performed by either using DNA microarrays or massively parallel next-generation sequencing (RNAseq) (Porter, 2015). DNA microarrays were designed just after the first draft genome of the chicken was sequenced and several chicken-specific gene expression microarrays are now available to target expressed sequence tags (Gheyas and Burt, 2013). This methodology needs *a priori* knowledge of the genes of interest, thus sequence-specific probes need to be designed that are also of limited applicability for some regions of the genome. Due to the technological progress in sequencing technologies, the robust and routine use of RNAseq will replace microarrays for transcriptional profiling of different physiological states in chickens. However, microarrays are still of value in studies with hundreds of samples and defined genomic sites.

3.3 Proteomics

In general, proteomics includes various qualitative and quantitative methods which are based on mass spectrometric measurements of peptides and intact proteins, respectively. The first challenge is the sample preparation and protein extraction protocol, especially if complex organismal samples such as tissues or intestinal content/faeces are of interest (Lippolis and Nally, 2018). This should be adjusted to each matrix as the co-extraction of feed particles and litter should be avoided. The subsequent steps are dependent upon the research question and available technological platform. In general, if peptides are analysed, proteins have to be digested with a single protease or mixture of proteases either in gel or in solution, and purified/desalted prior to chromatographic separation and mass spectrometric analysis (Kunec and Burgess, 2015; Bilic et al., 2018).

Gel-based proteomics: two-dimensional gel electrophoresis (2DE) was the starting point of proteomics. This method is based on the separation of protein extracts using first, the isoelectric point and second, the molecular weight of the proteins. The resulting gel image shows the separation of the proteins in a sample and can be used to compare gel images of different samples and to visually identify the differential presence and abundance of proteins. The method was enhanced by using fluorescent dyes to stain single protein extracts, which enables the simultaneous separation of different samples in one gel and improves quantification via the fluorescent intensity of the protein spot (2DE-DIGE). This increases the reproducibility of the method and allows the use of an internal standard. Protein identification is achieved by picking protein spots from the gel, either manually or using robotics, protease digestion and mass spectrometric measurement. Gel-based proteomics also includes the first separation via a 1D SDS polyacrylamide gel (PAGE), which achieves a first separation and purification of the protein extract if longer chromatographic runs or 2D chromatography are not possible to perform.

Gel-free proteomics: this includes the proteolytic digestion of the proteins in solution and the shotgun analysis of the complex peptide solution in one run. The peptides are subjected to an extended 2D chromatographic run coupled to a high-resolution mass spectrometer. Chemical or *in vivo* labelling of the peptides can be coupled to obtain an absolute quantification result.

The protein identification described above is either based on a previous reference gel image, indicating a differential protein abundance by the intensity in the gel, or by shotgun analysis, which gives the benefit of a complete list of proteins in the sample of interest. Quantification can be achieved either by using label-free or label-based methods but targeted quantification is not possible as peptide fragmentation is more or less random and low abundant proteins could be missed. If proteins or protein groups are of special interest and the expected masses of the respective peptides are known, multiple/selected reaction monitoring (Picotti and Aebersold, 2012) can be used, which requires a different mass spectrometer from the ones used above. This technique increases the sensitivity and precision of protein quantification, which is of particular value for biomarker detection.

3.4 Metabolomics

A huge number of small molecules can be identified in the intestinal tract of chickens that are produced by host and microbial metabolism as well as the turnover of feed. The inventory of these metabolites can be measured with mass spectrometry coupled to liquid or gas chromatography (LC-MS, GC-MS), or spectroscopy such as nuclear magnetic resonance (NMR) (Shao et al., 2018). The methods will either allow an untargeted shotgun analysis of all metabolites, without a complete identification of the compounds but with the possibility of comparing different samples based on the metabolomics profile, or a targeted analysis. The latter requires the availability of reference compounds and an adequate metabolite database.

The combination of some of the 'omics' methodologies described above, either genomics-proteomics, transcriptomics-proteomics, transcriptomics-metabolomics or proteomics-metabolomics, would achieve a more precise understanding of biological processes and would enable the identification of why and what is changing, for example at periods of intestinal dysbiosis or pathogenic infection (Fig. 1). Currently, numerous studies are using single 'omics' tools, see text below, which does not provide as much value as adding

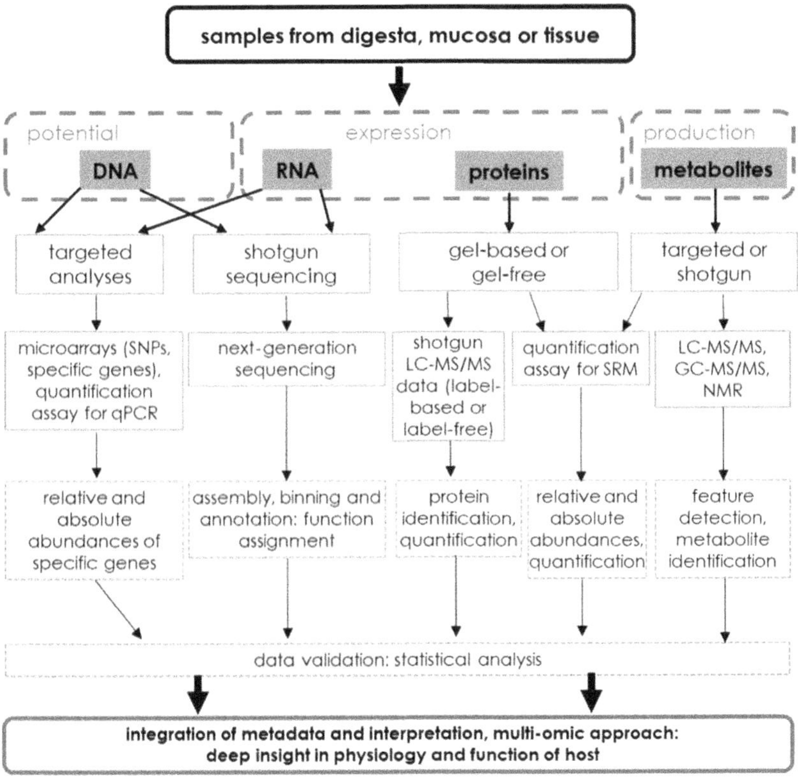

Figure 1 Omics - four pillars to study the chicken intestine on molecular levels.

the analysis of another biomolecule. Such multi-omics studies are of great value and should be considered with more emphasis in future studies.

4 Application of omics to study the chicken intestine

The 'omics' tools mentioned above are beginning to be used more frequently to describe and understand the growth and development of the chicken's gastrointestinal tract, to investigate the influence of nutrition and to show the effect of infection and diseases on the epithelial lining (Table 1).

4.1 Chicken growth and development

Bird development begins in an egg where the embryo matures until hatching. During this embryonic phase, the gut anatomy and function are also formed to enable the chicks to digest the first solid feed immediately after hatching. This embryonic development is important to acquire in robust chicks and

Table 1 Selection of publications using transcriptomics, proteomics and metabolomics to study the chicken gastrointestinal tract in respect of development, nutrition and diseases

	Sample	Method	Reference
Transcriptomics			
Comparison of gene expression profiles in the chicken intestine of two genetically different chicken lines, 24 h after a *Salmonella enteritidis* inoculation	Jejunum	Microarray	van Hemert et al. (2006)
Embryonic development of epithelial cells in turkeys	Duodenum	Microarray	de Oliveira et al. (2009)
Development of epithelial cells 3 weeks after hatch	Jejunum	Microarray	Schokker et al. (2009)
Intestinal expression of genes differing between control and *Salmonella*-infected chicken in a time-dependent manner	Jejunum, caecum	Microarray	Schokker et al. (2011)
Expression of immunogenes after *C. jejuni* infection	Caecum	RNASeq	Connell et al. (2012)
RNA expression of two chicken lines depending on NE susceptibility	Mucosa, small intestine	RNASeq	Dinh et al. (2014)
Differential regulation of microRNA transcriptome in chicken lines resistant and susceptible to NE disease	Intestinal intraepithelial lymphocytes and spleen	RNASeq	Hong et al. (2014)
miRNA profiles in NE-induced White Leghorn chickens	Jejunum mucosa	RNAseq	Rengaraj et al. (2016)
miRNA profiles during NE disease (Fayoumi chicken lines)	Intestinal mucosa	RNAseq	Rengaraj et al. (2017)
Comparative transcriptome analysis by RNAseq of NE *Clostridium perfringens* during *in vivo* colonization and *in vitro* conditions	Duodenum	RNAseq	Parreira et al. (2016)
miRNA profiles to study effects of probiotics in *Salmonella*-infected newly hatched chickens	Caecum	RNAseq	Chen et al. (2017)
miRNA profiles to study effects of *Salmonella* infections in White Leghorn chickens	Caecum	RNAseq	Wu et al. (2017)
Proteomics			
Changes during hatching and early post-hatch period in small intestine	Small intestine	2DE PAGE, MALDI-TOF-TOF	Gilbert et al. (2010)
Proteome changes in the intestinal mucosa of broiler activated by probiotic *E. faecium*	Small intestine	2DE-DIGE, LC-Chip ESI-QTOF-MS	Luo et al. (2013)
Quantitative proteome changes in plasma after challenge with *E. coli* LPS	Plasma	TMT-labelled peptides, LC-MS/MS	Horvatic et al. (2019)
Structures/metabolome			
Analyses of *o*-glucans from intestinal mucosa infected with *C. jejuni*	Mucosa	LC-MS	Struwe et al. (2015)
Metabolomic changes during treatment with antibiotic growth promoters (broilers)	Ileum digesta	UHPLC/MS/MS^2	Gadde et al. (2018)
Relationships between digestive efficiency and metabolomic profiles of serum and intestinal contents in chickens	Serum, ileum and caecal digesta	NMR	Beauclercq et al. (2018)

was studied using RNA microarrays by de Oliveira et al. (2009). They sampled duodenal tissue from turkey embryos at three time points before and once directly after hatching. Total RNA was extracted and cDNA was hybridized on the NCSU_Chicken_JS1 microarray (96 oligos). The data showed a time-dependent effect on cell proliferation where the expression of genes related to animal development became more abundant during embryonic development. Functional development was demonstrated as on the day of hatching, disaccharide and oligopeptide digestion should be possible, which was proved by the higher abundance of monosaccharide and peptide transporter gene expression. This study showed the functional capacity of the duodenum for digestion processes and provided initial evidence for possible *in ovo* feeding strategies (de Oliveira et al., 2009). The development of the jejunum in the first 21 days of a chicken was studied using RNA extracts and the Agilent 4 x 44K chicken array (Schokker et al., 2009). A high expression of genes involved in morphological and functional development was identified directly after hatching and subsequently declined. The immunological response of the chicks could be clustered into three phases based on gene expression patterns during the first days of life. The proteins of the brush border membrane of the small intestine were analysed by Gilbert et al. (2010). This study focussed first on age-dependent protein expression and second, on the differences between two genetic lines of broilers. Samples were taken on the day of hatching and 1, 3, 7 and 14 days post-hatch. Extracted proteins were separated using 2DE and mass spectrometry (MALDI-TOF/TOF). Approximately 10% of the protein spots were differentially expressed among different biological functions depending on the breed and age of the animals. This is one of the first studies to identify some of the target proteins that can be used to distinguish the correlation between feed conversion and host physiology and metabolism.

4.2 Feeding strategies for growth promotion and growth efficiency

Intestinal health, which is related to animal health, is strongly affected by diet. Feed supplements such as pre- and probiotics are commonly used currently to improve feed conversion, growth or egg production. This means that there is a need to understand the biological processes stimulated by feed supplements. In the past, several studies have focussed on growth performance and changes related to gut microbiota but only a few studied the response of the intestinal tissue in depth. The mucosal layer, sampled from the intestinum tenue, of broilers fed with *Enterococcus faecium* was analysed using proteomics by Luo et al. (2013). The extracted proteins were separated using 2D fluorescence difference gel electrophoresis and differentially expressed proteins were analysed by mass spectrometry. The daily growth and feed intake of the animals were unaffected by probiotic supplementation; however, a change of the immune organs, gut morphology and microbiota composition was observed. Protein analysis revealed the differential abundance of 42 mucosal proteins related to the immune and antioxidant systems and mucosal structure, which appear to have key roles in nutrient and energy consumption and nutrient absorption (Luo et al., 2013).

Alteration of the intestinal metabolome, which is a product of host and microbial activity, caused by the use of antibiotic growth promoters was studied recently in ileum samples from broiler chickens (Gadde et al., 2018). Bacitracin and virginiamycin were used separately and alteration of the metabolite profiles, detected with mass spectrometry, were found to be antibiotic specific. For example, an increased level of polyunsaturated fatty acids (PUFAs) was only found in bacitracin-supplemented chickens. PUFAs are synthesized by the host and the increased level indicates a decreased intestinal absorption capacity in these chickens. The metabolite profile of both antibiotic treatments showed an increased level of kynurenine, kynurenate and quinolinate. These metabolites are involved in inflammatory and immune responses and neurological pathways. Thus, the antibiotic treatments seemed to influence the passage rate of the intestinal content which may result in enhanced weight gain of the animals. These studies impressively showed the importance of investigating the biology of these treatments to better understand the mode of action and to support a more efficient use of such antibiotics. A correlation between digestive efficiency and the metabolite profile of the intestine and serum was recently attempted by Beauclercq et al. (2018). They aimed to identify potential predictive biomarkers for digestive efficiency by using ^{1}H-NMR metabolomics. In general, they could distinguish the metabolite profile of ileum, caecum and serum samples. A predictive model for the correlation between digestive efficiency and metabolite profiles was tested. For serum samples, amino acids in particular could be used for prediction, whereas fumarate in the ileum and glucose in the caeca showed the best matches. As the measurement of digestive efficiency is time-consuming and difficult to conduct for a large number of animals, the analyses of serum samples by ^{1}H-NMR metabolomics appeared to be an appropriate strategy to predict this trait.

4.3 Infection and diseases affecting intestinal health

The negative impact caused by pathogenic microorganisms towards animal health is widely known and potential pathogens are well described. However, the biological mechanisms behind these infections and other intestinal diseases are unclear. The most prominent pathogens in poultry are *Salmonella* sp., *Campylobacter* sp., *Clostridium* sp., *Escherichia coli* and *Eimeria* sp.

Salmonella is a typical pathogenic bacterium which colonizes the chicken's intestine and penetrates the mucosal layer. The consequences are diarrhoea and intestinal lesions, which are especially problematic for young chickens as well as humans if crossover of contaminated meat or eggs occurs. Thus, the defence and response mechanisms of chickens against a *Salmonella* infection have been investigated by several authors. Van Hemert et al. (2006) compared the changes of gene expression profiles in two different chicken lines after a one-day *Salmonella* infection. Jejunal tissue was used to extract RNA which was further analysed using a whole-genome oligonucleotide array and a cDNA microarray. They detected an upregulation of genes involved in the innate immune system and wound healing in both chicken lines, whereas some other upregulated genes were uniquely expressed only in one of the two lines. Thus, the response after a *Salmonella* infection is specific to the chicken breed and has to be studied individually. In line with this result, several other studies have shown the effect of *Salmonella* infection, for example Cheeseman et al., 2007, Schokker et al., 2010. Resolution of the complexity of the different gene expression patterns was attempted via gene association network analyses (Schokker et al., 2011). This should lead to the identification of highly interacting (*hub*) genes which indicate a host-specific response towards *Salmonella* infection. RNA extracted from jejunal samples of control and infected chickens was analysed using a 4 × 44k chicken array. The results indicate a higher expression of genes involved in defence and pathogen response as well as cellular communication. The role of miRNA in the control of gene expression was recently studied for *Salmonella* infection (Wu et al., 2017; Chen et al., 2017). The two studies analysed the differential expression of miRNAs and their target genes in *Salmonella*-infected animals and identified some key miRNAs which require further validation.

Campylobacter jejuni is a common bacterium that colonizes the chicken intestine and can cause severe human bacterial gastroenteritis. Chickens are not ubiquitously colonized with *C. jejuni* and it looks like gut-associated immune mechanisms control colonization. Connell et al. (2012) analysed the caecal tissue of *C. jejuni*-susceptible birds and *C. jejuni*-resistant birds by RNAseq. The mRNA sequence analyses showed 219 differentially expressed genes suggesting an early active immune response to *C. jejuni* in resistant animals and an altered host-microbe interaction. *C. jejuni* colonization is also controlled by mucin composition, as shown by Alemka et al. (2010). The results indicate that the strongest inhibition of *C. jejuni* binding was found in mucins from the large intestine > small intestine > caecum. Thus, the chemical structure of the mucins along the chicken gastrointestinal tract was characterized to gain further insight into colonization potential (Struwe et al., 2015).

Clostridium perfringens causes necrotic enteritis, which is an acute disease that induces weight loss due to decreased appetite and sudden death. It is becoming of more interest as the ban of antibiotic growth promoters in several countries has triggered the increase of *C. perfringens* infections in chickens and, by carryover effects, also in humans. Hence, several authors have been interested in studying the pathogenicity as well as host response. Most of the host-related publications used RNAseq to identify differently expressed mRNA genes or miRNA (Dinh et al., 2014; Hong et al., 2014; Rengaraj et al., 2016; Rengaraj et al., 2017; Parreira et al., 2016). The published data yielded considerable information on the regulation of host-pathogen interactions and, together with future studies, will lead to the discovery of genetic markers for future breeding strategies.

Another type of pathogenic invasion in chickens is caused by the secretion of lipopolysaccharides (LPS), an endotoxin from *E. coli*. This usually occurs in the intestine where LPS are released and enter the bloodstream by passing through the enterocytes. Proteome changes in plasma samples after an LPS subcutaneous injection were recently studied by Horvatic et al. (2019) to determine if certain proteins can be used as predictive markers of an *E. coli* infection. They used tandem mass tag labelling and LC-MS/MS to quantify the differential abundance of proteins 12 h after LPS injection. The results showed 87 proteins that were differentially abundant in LPS and control groups, demonstrating the value of the methodology as well as possible biomarkers for future diagnostic tools.

The summary of applications listed above show the value of 'omics' technologies to enhance knowledge of host responses to different internal and external factors. Over the last few years, RNAseq has become more common as the methodology is established in numerous laboratories and sequencing facilities are widely distributed. With respect to the organ of interest here, that is the intestine, proteomics and metabolomics have been less used to date, which is probably due to limited and expensive measurements and less experience of the techniques.

5 Case study: proteomic analysis of the mucosal layer along the chicken gut - host and microbiome

A metaproteomics approach was performed to describe the functional peculiarities of the diverse gastrointestinal sections and investigate the complex interconnecting network existing among the chicken intestinal mucosal layer and its associated bacterial community. Section inter-individual variation was profiled by referring to samples from four animals of the same age and kept on the same dietary regimen.

Mucosal tissue from the crop, ileum and caeca were independently sampled from four 25-day-old broiler chickens (strain Ross 308) and subjected to the sample preparation protocols of UHPLC-MS/MS measurement, as previous described (Tilocca et al., 2016). Briefly, each of these samples was initially treated with multiple centrifugation/resuspension steps in order to enrich the fraction of the target cells (i.e. host and bacterial cells) and reduce the influence of the digesta and feed particles. Recovered cells were lysed using both mechanical/physical and chemical methods. Extracted proteins were then trapped in the first centimetre of an SDS-PAGE for further clean-up of the proteins from the myriad of interfering compounds of the mucosal environment (e.g. polysaccharides, lipids etc.). In-gel digestion yielded tryptic peptides that were purified and measured using a UHPLC-MS/MS (Q Exactive plus, Thermo Scientific) system (Tilocca et al., in preparation).

With regard to the bioinformatics data analysis, obtained raw data were subjected to a two-step database search approach, in an attempt to maximize the protein identification rate and reduce the false discovery rate. Specifically, raw data were searched against a customized sample-specific database obtained from a previous database-dependent search against a publicly available database (Tilocca et al., 2016). Inference of the protein identities was performed via open-source proteomic software implementing algorithms for the label-free quantification of the identified protein repertoire (Cox et al., 2014). Identified proteins were functionally classified into Cluster of Orthologous Groups (COG) and Kyoto Encyclopedia of Genes and Genomes (KEGG) data repositories, providing a snapshot of the main ongoing activities in the mucosal layer and its associated microbiota, on a section basis.

The identified mucosa-related dataset accounted for a total of 4526 proteins, whereas 496 proteins were attributed to the mucosa-associated bacterial community. With regard to the distribution of proteins across the investigated gastrointestinal sections, a higher number of both host and bacterial proteins were identified in the ileum, followed by the crop and caeca.

In the host mucosal proteome, the trend of shared proteins between the sections reflected the gastrointestinal anatomic structure, with a higher number of proteins shared among crop–ileum and ileum–caeca than crop–caeca. This is also supported by statistically significant differences scored by the analysis of variance analysis ($p<0.001$). Nevertheless, no inter-individual variability was highlighted by the same test ($p>0.05$).

Functional analysis of the mucosal proteome has been performed via its classification into diverse data repositories such as KEGG and COG. A visual depiction of these data is provided by proteomaps (Fig. 2) (Liebermeister et al., 2014). The mucosal proteome of the four investigated chickens showed that the crop section exhibited major concern for 'cellular processes' (Fig. 2, red polygons), such as vesicular transport, cytoskeleton and cell growth, when compared to the other sections. The opposite trend was observed for the 'metabolism' class (Fig. 2, yellow polygons), with a higher expression in the ileum and caeca sections. Here, energy metabolism and central carbon metabolism of this section shows a homogeneous distribution of the affiliated proteins among sections of the investigated animals.

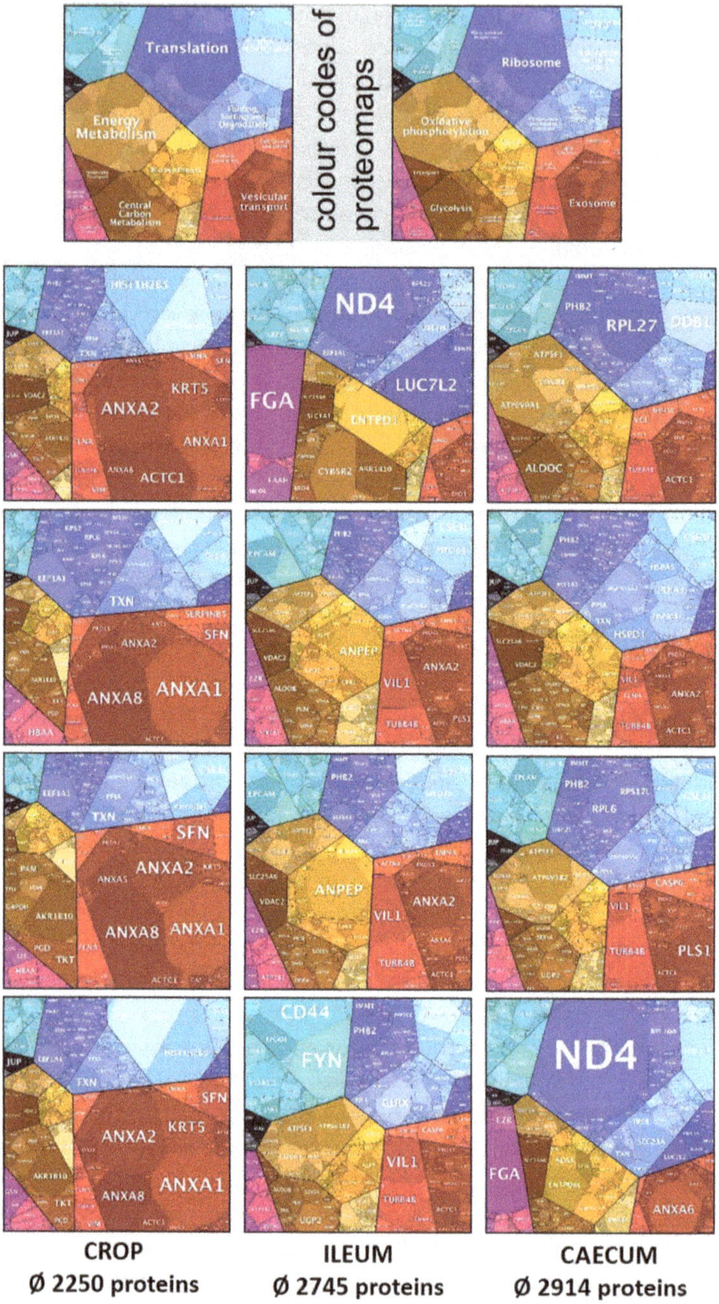

Figure 2 Functional distribution of mucosal chicken proteins identified in four broiler chickens (vertical order) and the average number of identified proteins in the crop, ileum and caecum samples.

The genetic information processing class (Fig. 2, blue polygons) showed a general section-conserved distribution. Nevertheless, a deeper investigation of the affiliated proteins revealed a diverse expression set both at the section and inter-individual levels, suggesting strong evidence for how the surrounding environment (i.e. the gastrointestinal sections) and the individual features influence the physiology of each animal.

Results concerning the mucosa-associated microbiota relied on a relatively low number of identified proteins. Indeed, only 137 bacterial proteins were identified for the crop section, 351 in the ileum and 155 in the caeca, which is most likely due to the adopted sample preparation protocols. In studies based on intestinal content or faeces, these protocols enabled the identification of a higher metaproteome coverage (Tilocca et al., 2016; Tang et al., 2014). Thus, adjustments for future mucosa-based studies are desirable. However, the phylogenetic assessment on a protein basis identified six different phyla over the investigated sections and animals. Proteobacteria appear the most abundant phyla regardless of the section, followed by Firmicutes and Actinobacteria. Bacteroidetes, Aquificae and Cyanobacteria were identified at a low-relative abundance in all sections (Fig. 3). Although major bacterial phyla were identified in all animals and sections, a strong variability is shown in Fig. 3 both between the sections and at the inter-individual level. Even though an analysis of variance with permutation showed a statistical difference for only the crop-ileum (p = 0.034), no statistical significance was detected for the ileum-caeca and crop-caeca datasets. In addition, no inter-individual variability was described by the statistical tests, which is probably due to the 'reduced' number of identified proteins leading to misinterpretation of the gastrointestinal-specific metaproteomes.

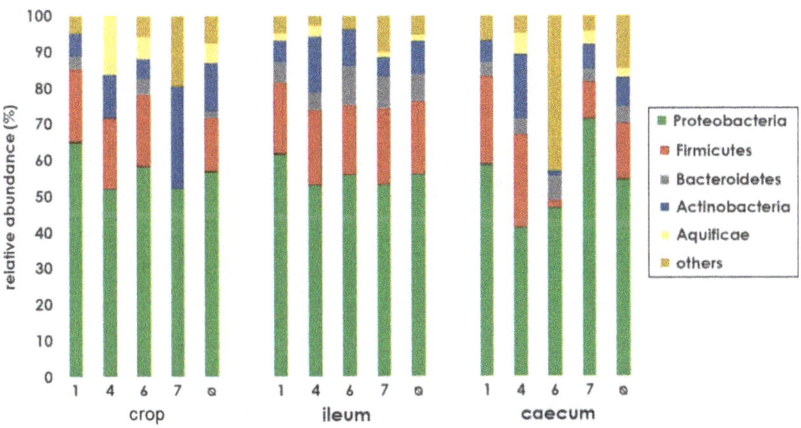

Figure 3 Relative abundance of bacterial phyla in mucosal samples of four broiler chickens. Identification and quantification based on protein label-free quantification.

Proteins affiliated to Proteobacteria in the ileal section were mostly involved in cell-to-cell communication (*quorum sensing*) and coordination of a variety of activities such as energy metabolism, virulence, biofilm formation and antibiotic biosynthesis. Similarly, proteins belonging to Firmicutes were involved in energy metabolism, but proteins involved in chemotaxis and signal transduction were also identified. Proteins affiliated to Actinobacteria were mainly represented by chaperones involved in cell protection against diverse environmental stressors, followed by other entries involved in carbohydrate and amino acid metabolism. Bacteroidetes were featured by the expression of proteins involved in a dynamic exchange between the bacterial cell and the surrounding host environment. Several processes are implicated such as cell structure maintenance, cell-to-cell adhesion and regulation of bactericidal agents suggesting Bacteroidetes are important modulators of the mucosal microbiota structure, especially in response to environmental stress.

A similar trend was observed by the functional categorization of the bacterial proteins into COG and KEGG classes, which revealed the ileum showed the highest functional variability, followed by the caeca and crop. The energy metabolism biochemical pathway is the most abundant in all sections, with abundance values in the ileum and caeca almost double that of the crop. This is in line with the physiologic role of these gastrointestinal sections, where the highest energy production is expected in the caudal intestinal sections. Section distribution of other biochemical pathways was considered rather homogeneous, except for the stress response pathway, which was more abundant in the caeca. In contrast, the two-component system biochemical pathway, mainly represented by proteins involved in nutrient exchange and environmental stimuli responses, exhibited higher expression in the ileum.

To conclude, although this study provided evidence of a section-specific physiology and microbiota, methodological improvements are desirable in order to gain further and deeper knowledge of host-microbiota interactions. Here, a complementary approach among 'omics' science is of crucial importance for better and more accurate data acquisition and interpretation.

6 Summary and future trends

The 'omics' tools, transcriptomics, proteomics and metabolomics, enable more comprehensive understanding of the biological processes of the chicken intestine on a molecular basis. In comparison with the classical biochemical approaches, such as Western blots of transporter proteins or tight junctions, or digestibility measurements, these recent technologies allow holistic insight of host physiology and function. Currently, most studies use single 'omics' tools to study a defined treatment or infection. This can be improved by using combinations of 'omics' methods and developing a systems biology approach.

This knowledge can be transferred into practice by performing correlation analyses between metadata with data input from daily weight gain, feed efficiency, egg production and so forth, and 'omics' data. The integration of these multivariate data is a considerable challenge in the field. First, bioinformatics software is needed to analyse data which are derived from different technology platforms that provide an abundance of data of transcripts, proteins or metabolites. Second, these datasets have to be combined by overcoming mismatches between the different biological levels (DNA, RNA, proteins and metabolites). A third challenge is the correlation between 'omics' data with the metadata to identify biological patterns. At the moment, limited knowledge is available regarding prediction analyses based on these combinations. However, such a strategy would be of great value as the sole record of large biological datasets without any additional information regarding the animals and treatments are useless. Although these strategies are still in their infancy in the field of animal science, much can be learned and derived from human and medical studies. Biomarkers, genes, proteins or metabolites, can be identified and used to develop rapid detection strategies for upcoming diseases or infections in a flock. Microarrays are of particular value as they allow a relatively fast screening of hundreds of samples in a short time if the target genes are known, which will support animal welfare and reduce subsequent costs for the farmers. Thus, animal scientists need to learn more about 'omics' technologies and bioinformatics to benefit from these technologies and to better understand animal biology at the molecular level.

7 Where to look for further information

The topic of intestinal health in chickens is receiving increasing attention, resulting in an increased number of review and research articles as well as book chapters. This book chapter cites the most important current articles on 'omics' methodologies as well as the chicken intestine.

8 References

Abdrakhmanov, I., Lodygin, D., Geroth, P., Arakawa, H., Law, A., Plachy, J., Korn, B. and Buerstedde, J.-M. 2000. A large database of chicken bursal ESTs as a resource for the analysis of vertebrate gene function. *Genome Research* 10(12), 2062–9. doi:10.1101/gr.10.12.2062.

Alemka, A., Whelan, S., Gough, R., Clyne, M., Gallagher, M. E., Carrington, S. D. and Bourke, B. 2010. Purified chicken intestinal mucin attenuates *Campylobacter jejuni* pathogenicity *in vitro*. *Journal of Medical Microbiology* 59(8), 898–903. doi:10.1099/jmm.0.019315-0.

Beauclercq, S., Nadal-Desbarats, L., Hennequet-Antier, C., Gabriel, I., Tesseraud, S., Calenge, F., Le Bihan-Duval, E. and Mignon-Grasteau, S. 2018. Relationships between digestive efficiency and metabolomic profiles of serum and intestinal contents in chickens. *Scientific Reports* 8(1), 6678. doi:10.1038/s41598-018-24978-9.

Bilic, P., Kules, J., Galan, A., Gomes De Pontes, L., Guillemin, N., Horvatic, A., Festa Sabes, A., Mrljak, V. and Eckersall, P. D. 2018. Proteomics in veterinary medicine and animal science: neglected scientific opportunities with immediate impact. *Proteomics* 18(14), e1800047. doi:10.1002/pmic.201800047.

Borda-Molina, D., Vital, M., Sommerfeld, V., Rodehutscord, M. and Camarinha-Silva, A. 2016. Insights into broilers' gut microbiota fed with phosphorus, calcium and phytase supplemented diets. *Frontiers in Microbiology* 7, 2033. doi:10.3389/fmicb.2016.02033.

Borda-Molina, D., Seifert, J. and Camarinha-Silva, A. 2018. Current perspectives of the chicken gastrointestinal tract and its microbiome. *Computational and Structural Biotechnology Journal* 16, 131–9. doi:10.1016/j.csbj.2018.03.002.

Celi, P., Cowieson, A. J., Fru-Nji, F., Steinert, R. E., Kluenter, A.-M. and Verlhac, V. 2017. Gastrointestinal functionality in animal nutrition and health: new opportunities for sustainable animal production. *Animal Feed Science and Technology* 234, 88–100. doi:10.1016/j.anifeedsci.2017.09.012.

Celi, P., Verlhac, V., Pérez Calvo, E., Schmeisser, J. and Kluenter, A.-M. 2018. Biomarkers of gastrointestinal functionality in animal nutrition and health. *Animal Feed Science and Technology*. doi:10.1016/j.anifeedsci.2018.07.012.

Cheeseman, J. H., Kaiser, M. G., Ciraci, C., Kaiser, P. and Lamont, S. J. 2007. Breed effect on early cytokine mRNA expression in spleen and cecum of chickens with and without *Salmonella enteritidis* infection. *Developmental and Comparative Immunology* 31(1), 52–60. doi:10.1016/j.dci.2006.04.001.

Chen, Q., Tong, C., Ma, S., Zhou, L., Zhao, L. and Zhao, X. 2017. Involvement of microRNAs in probiotics-induced reduction of the cecal inflammation by *Salmonella* Typhimurium. *Frontiers in Immunology* 8, 704. doi:10.3389/fimmu.2017.00704.

Choi, J. H., Kim, G. B. and Cha, C. J. 2014. Spatial heterogeneity and stability of bacterial community in the gastrointestinal tracts of broiler chickens. *Poultry Science* 93(8), 1942–50. doi:10.3382/ps.2014-03974.

Connell, S., Meade, K. G., Allan, B., Lloyd, A. T., Kenny, E., Cormican, P., Morris, D. W., Bradley, D. G. and O'Farrelly, C. 2012. Avian resistance to *Campylobacter jejuni* colonization is associated with an intestinal immunogene expression signature identified by mRNA sequencing. *PLoS ONE* 7(8), e40409. doi:10.1371/journal.pone.0040409.

Cox, J., Hein, M. Y., Luber, C. A., Paron, I., Nagaraj, N. and Mann, M. 2014. Accurate proteome-wide label-free quantification by delayed normalization and maximal peptide ratio extraction, termed MaxLFQ. *Molecular and Cellular Proteomics: MCP* 13(9), 2513–26. doi:10.1074/mcp.M113.031591.

Dalloul, R. A., Long, J. A., Zimin, A. V., Aslam, L., Beal, K., Blomberg, Le Ann, Bouffard, P., Burt, D. W., Crasta, O., Crooijmans, R. P., et al. 2010. Multi-platform next-generation sequencing of the domestic turkey (*Meleagris gallopavo*): genome assembly and analysis. *PLoS Biology* 8(9), e1000475. doi:10.1371/journal.pbio.1000475.

De Oliveira, J. E., Druyan, S., Uni, Z., Ashwell, C. M. and Ferket, P. R. 2009. Prehatch intestinal maturation of turkey embryos demonstrated through gene expression patterns. *Poultry Science* 88(12), 2600–9. doi:10.3382/ps.2008-00548.

Del Chierico, F., Vernocchi, P., Bonizzi, L., Carsetti, R., Castellazzi, A. M., Dallapiccola, B., De Vos, W., Guerzoni, M. E., Manco, M., Marseglia, G. L., et al. 2012. Early-life gut microbiota under physiological and pathological conditions: the central role of combined meta-omics-based approaches. *Journal of Proteomics* 75(15), 4580–7. doi:10.1016/j.jprot.2012.02.018.

Denbow, D. M. 2015. Gastrointestinal anatomy and physiology. In: Scanes, C. G. (Ed.), *Sturkie's Avian Physiology* (6th edn.). Elsevier Academic Press, New York, NY, 337–66.

Dinh, H., Hong, Y. H. and Lillehoj, H. S. 2014. Modulation of microRNAs in two genetically disparate chicken lines showing different necrotic enteritis disease susceptibility. *Veterinary Immunology and Immunopathology* 159(1–2), 74–82. doi:10.1016/j.vetimm.2014.02.003.

Gadde, U. D., Oh, S., Lillehoj, H. S. and Lillehoj, E. P. 2018. Antibiotic growth promoters virginiamycin and bacitracin methylene disalicylate alter the chicken intestinal metabolome. *Scientific Reports* 8(1), 3592. doi:10.1038/s41598-018-22004-6.

Gheyas, A. A. and Burt, D. W. 2013. Microarray resources for genetic and genomic studies in chicken: a review. *Genesis* 51(5), 337–56. doi:10.1002/dvg.22387.

Gilbert, E. R., Williams, P. M., Ray, W. K., Li, H., Emmerson, D. A., Wong, E. A. and Webb, K. E. 2010. Proteomic evaluation of chicken brush-border membrane during the early posthatch period. *Journal of Proteome Research* 9(9), 4628–39. doi:10.1021/pr1003533.

Hong, Y. H., Dinh, H., Lillehoj, H. S., Song, K. D. and Oh, J. D. 2014. Differential regulation of microRNA transcriptome in chicken lines resistant and susceptible to necrotic enteritis disease. *Poultry Science* 93(6), 1383–95. doi:10.3382/ps.2013-03666.

Horvatic, A., Guillemin, N., Kaab, H., McKeegan, D., O'Reilly, E., Bain, M., Kules, J. and Eckersall, P. D. 2019. Quantitative proteomics using tandem mass tags in relation to the acute phase protein response in chicken challenged with *Escherichia coli* lipopolysaccharide endotoxin. *Journal of Proteomics* 192, 64–77. pii:S1874-3919(18)30313-0. doi:10.1016/j.jprot.2018.08.009.

Huang, Y., Li, Y., Burt, D. W., Chen, H., Zhang, Y., Qian, W., Kim, H., Gan, S., Zhao, Y., Li, J., et al. 2013. The duck genome and transcriptome provide insight into an avian influenza virus reservoir species. *Nature Genetics* 45(7), 776–83. doi:10.1038/ng.2657.

International Chicken Genome Sequencing Consortium. 2004. Sequence and comparative analysis of the chicken genome provide unique perspectives on vertebrate evolution. *Nature* 432(7018), 695–716. doi:10.1038/nature03154.

International Chicken Polymorphism Map Consortium. 2004. A genetic variation map for chicken with 2.8 million single-nucleotide polymorphisms. *Nature* 432(7018), 717–22. doi:10.1038/nature03156.

Kawahara-Miki, R., Sano, S., Nunome, M., Shimmura, T., Kuwayama, T., Takahashi, S., Kawashima, T., Matsuda, Y., Yoshimura, T. and Kono, T. 2013. Next-generation sequencing reveals genomic features in the Japanese quail. *Genomics* 101(6), 345–53. doi:10.1016/j.ygeno.2013.03.006.

Kunec, D. and Burgess, S. C. 2015.Avian proteomics. In: Scanes, C. G. (Ed.), *Sturkie's Avian Physiology* (6th edn.). Academic Press, San Diego, CA. Chapter 3, 25–37.

Liebermeister, W., Noor, E., Flamholz, A., Davidi, D., Bernhardt, J. and Milo, R. 2014. Visual account of protein investment in cellular functions. *Proceedings of the National Academy of Sciences of the United States of America* 111(23), 8488–93. doi:10.1073/pnas.1314810111.

Lippolis, J. D. and Nally, J. E. 2018. Considerations for farm animal proteomic experiments: an introductory view gel-based versus non-gel-based approaches. In: De Almeida,

A. M., Eckersall, D. and Miller, I. (Eds), *Proteomics in Domestic Animals: from Farm to Systems Biology*. Springer International Publishing, Cham.

Luo, J., Zheng, A., Meng, K., Chang, W., Bai, Y., Li, K., Cai, H., Liu, G. and Yao, B. 2013. Proteome changes in the intestinal mucosa of broiler (*Gallus gallus*) activated by probiotic *Enterococcus faecium*. *Journal of Proteomics* 91, 226–41. doi:10.1016/j.jprot.2013.07.017.

Nordentoft, S., Molbak, L., Bjerrum, L., De Vylder, J., Van Immerseel, F. and Pedersen, K. 2011. The influence of the cage system and colonisation of *Salmonella* enteritidis on the microbial gut flora of laying hens studied by T-RFLP and 454 Pyrosequencing. *BMC Microbiology* 11, 187. doi:10.1186/1471-2180-11-187.

Oakley, B. B., Lillehoj, H. S., Kogut, M. H., Kim, W. K., Maurer, J. J., Pedroso, A., Lee, M. D., Collett, S. R., Johnson, T. J. and Cox, N. A. 2014. The chicken gastrointestinal microbiome. *FEMS Microbiology Letters* 360(2), 100–12. doi:10.1111/1574-6968.12608.

Pan, D. and Yu, Z. 2014. Intestinal microbiome of poultry and its interaction with host and diet. *Gut Microbes* 5(1), 108–19. doi:10.4161/gmic.26945.

Parreira, V. R., Russell, K., Athanasiadou, S. and Prescott, J. F. 2016. Comparative transcriptome analysis by RNAseq of necrotic enteritis *Clostridium perfringens* during in vivo colonization and *in vitro* conditions. *BMC Microbiology* 16(1), 186. doi:10.1186/s12866-016-0792-6.

Picotti, P. and Aebersold, R. 2012. Selected reaction monitoring-based proteomics: workflows, potential, pitfalls and future directions. *Nature Methods* 9(6), 555–66. doi:10.1038/nmeth.2015.

Porter, T. E. 2015. Transcriptomics of physiological systems. In: Scanes, C. G. (Ed.), *Sturkie's Avian Physiology* (6th edn.). Academic Press, San Diego, CA. Chapter 2, 15–23.

Rengaraj, D., Truong, A. D., Lee, S. H., Lillehoj, H. S. and Hong, Y. H. 2016. Expression analysis of cytosolic DNA-sensing pathway genes in the intestinal mucosal layer of necrotic enteritis-induced chicken. *Veterinary Immunology and Immunopathology* 170, 1–12. doi:10.1016/j.vetimm.2015.12.010.

Rengaraj, D., Truong, A. D., Ban, J., Lillehoj, H. S. and Hong, Y. H. 2017. Distribution and differential expression of microRNAs in the intestinal mucosal layer of necrotic enteritis induced Fayoumi chickens. *Asian-Australasian Journal of Animal Sciences* 30(7), 1037–47. doi:10.5713/ajas.16.0685.

Rougière, N. and Carrè, B. 2010. Comparison of gastrointestinal transit times between chickens from D+ and D- genetic lines selected for divergent digestion efficiency. *Animal: an International Journal of Animal Bioscience* 4(11), 1861–72. doi:10.1017/S1751731110001266.

Scanes, C. G. 2015. *Sturkie's Avian Physiology*. Elsevier Academic Press, New York, NY.

Schokker, D., Hoekman, A. J. W., Smits, M. A. and Rebel, J. M. J. 2009. Gene expression patterns associated with chicken jejunal development. *Developmental and Comparative Immunology* 33(11), 1156–64. doi:10.1016/j.dci.2009.06.002.

Schokker, D., Smits, M. A., Hoekman, A. J., Parmentier, H. K. and Rebel, J. M. 2010. Effects of *Salmonella* on spatial-temporal processes of jejunal development in chickens. *Developmental and Comparative Immunology* 34(10), 1090–100. doi:10.1016/j.dci.2010.05.013.

Schokker, D., De Koning, D. J., Rebel, J. M. J. and Smits, M. A. 2011. Shift in chicken intestinal gene association networks after infection with *Salmonella*. *Comparative Biochemistry and Physiology. Part D, Genomics and Proteomics* 6(4), 339–47. doi:10.1016/j.cbd.2011.07.004.

Sekelja, M., Rud, I., Knutsen, S. H., Denstadli, V., Westereng, B., Næs, T. and Rudi, K. 2012. Abrupt temporal fluctuations in the chicken fecal microbiota are explained by its gastrointestinal origin. *Applied and Environmental Microbiology* 78(8), 2941-8. doi: 10.1128/AEM.05391-11.

Shao, F. J., Ying, Y. T., Tan, X., Zhang, Q. Y. and Liao, W. T. 2018. Metabonomics profiling reveals biochemical pathways associated with pulmonary arterial hypertension in broiler chickens. *Journal of Proteome Research* 17(10), 3445-53. doi:10.1021/acs.jproteome.8b00316.

Stanley, D., Hughes, R. J. and Moore, R. J. 2014. Microbiota of the chicken gastrointestinal tract: influence on health, productivity and disease. *Applied Microbiology and Biotechnology* 98(10), 4301-10. doi:10.1007/s00253-014-5646-2.

Struwe, W. B., Gough, R., Gallagher, M. E., Kenny, D. T., Carrington, S. D., Karlsson, N. G. and Rudd, P. M. 2015. Identification of *O*-glycan structures from chicken intestinal mucins provides insight into *Campylobactor jejuni* pathogenicity. *Molecular and Cellular Proteomics: MCP* 14(6), 1464-77. doi:10.1074/mcp.M114.044867.

Svihus, B. 2014. Function of the digestive system. *Journal of Applied Poultry Research* 23(2), 306-14. doi:10.3382/japr.2014-00937.

Svihus, B., Choct, M. and Classen, H. L. 2013. Function and nutritional roles of the avian caeca: a review. *World's Poultry Science Journal* 69(2), 249-64. doi:10.1017/S0043933913000287.

Tang, Y., Underwood, A., Gielbert, A., Woodward, M. J. and Petrovska, L. 2014. Metaproteomics analysis reveals the adaptation process for the chicken gut microbiota. *Applied and Environmental Microbiology* 80(2), 478-85. doi:10.1128/AEM.02472-13.

Tilocca, B., Witzig, M., Rodehutscord, M. and Seifert, J. 2016. Variations of phosphorous accessibility causing changes in microbiome functions in the gastrointestinal tract of chickens. *PLoS ONE* 11(10), e0164735. doi:10.1371/journal.pone.0164735.

Tilocca, B., Ghilardi, A. and Seifert, J. in preparation. Analysis of the bacterial and host proteins in the gastrointestinal tract of chickens using a metaproteomics approach.

van Hemert, S., Hoekman, A. J. W., Smits, M. A. and Rebel, J. M. J. 2006. Early host gene expression responses to a *Salmonella* infection in the intestine of chickens with different genetic background examined with cDNA and oligonucleotide microarrays. *Comparative Biochemistry and Physiology. Part D, Genomics and Proteomics* 1(3), 292-9. doi:10.1016/j.cbd.2006.05.001.

Witzig, M., Camarinha-Silva, A., Green-Engert, R., Hoelzle, K., Zeller, E., Seifert, J., Hoelzle, L. E. and Rodehutscord, M. 2015. Spatial variation of the gut microbiota in broiler chickens as affected by dietary available phosphorus and assessed by T-RFLP analysis and 454 pyrosequencing. *PLoS ONE* 10(12), e0145588. doi:10.1371/journal.pone.0145588.

Wu, G., Qi, Y., Liu, X., Yang, N., Xu, G., Liu, L. and Li, X. 2017. Cecal MicroRNAome response to *Salmonella enterica* serovar Enteritidis infection in White Leghorn Layer. *BMC Genomics* 18(1), 77. doi:10.1186/s12864-016-3413-8.

Xiao, Y., Xiang, Y., Zhou, W., Chen, J., Li, K. and Yang, H. 2017. Microbial community mapping in intestinal tract of broiler chicken. *Poultry Science* 96(5), 1387-93. doi:10.3382/ps/pew372.

Zoetendal, E. G., Rajilic-Stojanovic, M. and De Vos, W. M. 2008. High-throughput diversity and functionality analysis of the gastrointestinal tract microbiota. *Gut* 57(11), 1605-15. doi:10.1136/gut.2007.133603.

Chapter 4

Understanding gut microbiota in poultry

Robert Moore, RMIT University, Australia

1 Introduction

The gut microbiota is comprised of diverse populations of microbes, including bacteria, archaea, protozoa, fungi and viruses, that reside within the gastrointestinal tract (GIT). Studies on various animals, including chickens, have shown that the gut microbiota has profound effects on the host and, indeed, the microbiota-host interaction is so important that the combined host-microbiota pairing has been called a superorganism (Sleator, 2010). Such terminology emphasises the integrated, co-evolved, nature of the gut microbiota's association with the host. The host-microbiota interaction is highly complex. The gut microbiota is made up of many hundreds of different types of microbes which interact not only with the host but with other members of the microbiota. The host (chicken) gut is made up of different cell types and within each compartment of the GIT different environments are established with different pH levels, oxidation states and nutrient compositions. Each of these environments supports microbiotas with different phylogenetic and function compositions. This highly complex host-microbiota system must maintain functional homeostasis in order to efficiently extract nutrients and energy from feed to produce the highly productive chickens that the poultry industry relies on to deliver a safe, high-quality and efficiently produced protein supply. If we are to fully understand the biology of poultry and how to most efficiently produce birds and eggs, then a detailed functional understanding of the gut microbiota is important. The key issues are (1) how the microbiota

http://dx.doi.org/10.19103/AS.2019.0059.04

becomes established and develops over time, (2) how it interacts with the host and (3) how it can be nurtured or manipulated to ensure the health and productivity of poultry.

2 The microbiota of chickens

2.1 Establishment, maturation and phylogenetic structure

The composition of the gut microbiota develops over time, with rapid changes occurring early in life followed by more subtle changes as the bird ages. Until recently it was thought that the digestive tract was essentially sterile at hatch; a blank space ready to be populated with the microbes that the bird first encounters post-hatch. However, some researchers have reported that developing embryos can carry fairly complex microbial populations, albeit at low densities (Ding et al., 2017; Ilina et al., 2016; Kizerwetter-Świda and Binek, 2008). There are technical challenges that need to be considered when studying such low-density samples and so the results and conclusions are somewhat controversial and need careful consideration and replication before the full picture is clearly revealed (Cuperus et al., 2018; Klaschik et al., 2002; Salter et al., 2014; Silkie et al., 2008). Most studies that have investigated the development of chicken gut microbiota have started with the post-hatch chick. The microbiota in the first few days post-hatch is usually dominated by Enterobacteriaceae. During the first week, post-hatch, the microbiota increases in complexity and there is a shift from Gram-negative to Gram-positive bacteria, and Firmicutes and Bacteroidetes start to dominate and a number of other phyla are also generally represented, including Actinobacteria, Proteobacteria and a variable range of other phyla at low abundance (Ballou et al., 2016). The microbiota composition continues to develop over the subsequent week or two, attaining a fairly stable 'adult' configuration by week two or three (Ijaz et al., 2018; Oakley et al., 2014a; Ranjitkar et al., 2016).

The structure of the gut microbiota varies across the different gastrointestinal compartments (Ranjitkar et al., 2016). The upper GIT, including the crop, gizzard and small intestine, is usually dominated by *Lactobacillus* whereas the microbiota of the caeca and colon are more complex, usually dominated by Firmicutes, and show greater variability between birds (Choi et al., 2014; Yan et al., 2017). Within each gastrointestinal compartment there are also various microenvironments which are differentially populated by different members of the gut microbiota. Studies on other animals, likely to also be indicative of the situation in chickens, have shown that at a gross level there are microbes that favour colonisation of the mucosa, within the mucous layer and in close proximity to the epithelium, and other microbes which are principally found in the lumen, in digesta and have only transient association, if any, with the mucosa (Stanley et al., 2018; Zoetendal et al., 2002). The microenvironment within which effects can be exerted, and the permanence or transience of colonisation, will affect the degree of influence that the microbes are likely to have on the host and other members of the microbiota.

There is no single standard gut microbiota composition shared by all chickens. It has been found that microbiota composition can show significant variations between flocks and even birds within a flock can exhibit variation in microbiota composition (Cuperus et al., 2018; Stanley et al., 2013b). The differences in composition can be substantial, even to the level of different phyla dominating; for example, caecal microbiotas of birds in some flocks are dominated by Firmicutes whereas other flocks are dominated by Bacteroidetes. There has been some effort to use the variation in microbiota composition to define a variety of enterotypes, each representing a particular microbiota structure. Kaakoush et al. (2014) characterised the faecal microbiota of broiler birds into four enterotypes and suggested that carriage of *Campylobacter* and *Helicobacter* species varied across the enterotypes. However, it is not clear that there really are biologically meaningful distinct enterotypes; there may instead be a continuum of microbiota structures and the arbitrary division into enterotypes may not be a particularly useful way to think about variable microbiota structures. The enterotype concept was first raised in the context of human microbiota investigations (Arumugam et al., 2011) but has since been reconsidered and remains somewhat controversial (Costea et al., 2018).

If the potential of gut microbiota to influence health and productivity is to be understood and harnessed it is important to consider the root causes and consequences of variation in microbiota structures as this may give insights into which microbiota structures are the most resilient and how desired changes can be produced. Bird genetics, environment and feed can all influence the gut microbiota but considerable variations in microbiota composition can be seen in flocks in which all these potential variables are held constant (Stanley et al., 2013b). Therefore, it is clear that some drivers of microbiota variation are independent of these known influences. Difference in gut microbiota composition across birds within a flock can be seen early in life and so at least some of the variability is established during the initial microbial colonisation of the gut. We have hypothesised that the unnatural conditions under which commercial chickens are hatched is one of the main factors influencing subsequent microbial colonisation (Stanley et al., 2013b). In their natural state chicks would be brooded and hatched in a nest and have contact with adult birds. The nest and the adult birds provide a source of 'natural' bird microbiota which would be expected to seed the gut of the newly hatched chick. However, under modern hatchery conditions there is an attempt to minimise microbial exposures so that the chances of colonisation with foodborne pathogens such as *Salmonella* and *Campylobacter* are reduced. The consequence of this is that newly hatched chicks do not have access to normal chicken microbiota and hence are colonised by a largely random collection of microbes derived from various environmental sources, including from transport boxes and other equipment, human handlers, the first feed and water that they are exposed to and aerosolised microbes suspended in the air. Chick exposure to each of these sources is variable and random. It is further hypothesised that the earliest microbial colonisers can influence colonisation by subsequent microbes. Early colonisers establish a microenvironment that can be conducive or resistant to subsequent colonisation and so the type of microbes involved in the very earliest colonisation of the gut can influence the structure of the gut microbiota throughout the life of chickens.

Although hatching chicks can be exposed to a wide variety of microbes from different sources, there are obviously constraints on the types of microbiota structures that can develop within the gut of chickens. Although the gut microbiota can be quite variable, this variability is still within clear limits - there are only certain types of bacteria that can colonise and survive in the GIT. The environmental conditions within the gut dictate which microbes can colonise and persist and which are killed or are only transiently present. The constraints are obviously very tight as there are vast arrays of environmental microbes that are never found in the guts of chickens. So, although birds hatched in a modern hatchery are not exposed to 'natural' chicken microbiota, they do in most cases end up being colonised by a functional microbiota that supports good health and high levels of productivity.

2.2 The technology used for microbiota characterisation - advantages and limitations

The study of chicken gut microbiota has been enhanced over the last decade by the advent of molecular methods that have allowed microbiota characterisation without culturing. Prior to this our knowledge of the composition of gut microbiota was largely dependent on the direct culturing of the microbes present. Culturing gave a very limited view of complex microbiota because many bacterial species are currently unculturable and hence invisible to these traditional methods. The study of microbiota really took off with the advent and use of next-generation sequencing (NGS) systems that facilitated the identification and quantitation of microbes within complex populations, whether they were culturable or not. It should still be understood that the NGS methods do not capture an absolutely complete picture of the gut microbiota as there are still limitations on the depth to which most studies can go. Typically, recent microbiota studies use the Illumina MiSeq platform to sequence PCR produced amplicons of the 16S ribosomal RNA genes that are ubiquitous in all bacteria (or 18S rRNA genes for eukaryotes). Good studies usually generate approximately 30 000–50 000 sequence reads from each microbiota sample tested. At that depth of coverage the method can reliably detect bacteria down to an abundance level of about 0.01%. That is certainly an impressive achievement compared to what could previously be achieved with culturing methods but in the context of the density and complexity of microbial populations in the caecum the resolution is limited when it is considered that the caeca contain about 10^{12} bacteria/g. Therefore, a bacterium that was present at an abundance of 0.01% would be represented by approximately 10^{8} bacteria/g. Bacteria colonising the caeca at lower abundance than that can still be of very great biological significance. For example, foodborne pathogens such as *Campylobacter jejuni* are typically present within the caeca at levels of 10^{6}–10^{8}/g and so may be only marginally and unreliably detected by our current microbiota characterisation methods. Sometimes, when specific bacteria within the microbiota are of interest, it is still necessary to use culturing or specific molecular methods such as quantitative PCR to assess population levels.

In addition to the issue of resolution there are also technical issues regarding the reproducibility and accuracy of the 16S rRNA gene amplicon sequencing methods; there have been many commentaries on such issues (Brooks et al., 2015; Ibarbalz et al., 2014; Pollock et al., 2018; Rintala et al., 2017). The main conclusion has been that although the methods may not give a perfect picture of the exact composition of the gut microbiota they are very capable of indicating the general composition and relative changes in microbial populations that can occur over time or for comparison of different treatment groups, as long as all within-study analysis is carried out using the same methods and analysis pipelines.

Whole metagenomics, in which total microbiota DNA is sequenced instead of 16S and 18S amplicons, can also be used to characterise the phylogenetic and functional structure of microbiota; however, the amount of sequencing required means that far fewer samples can be investigated for each dollar invested and hence the amplicon approach is still the most widely used technique.

Experimental design is a further important consideration when performing microbiota studies. Because of the known intrinsic variations in microbiota composition between flocks and even within flocks, and the strong influence of feed and environment on microbiota structure, legitimate conclusions can only be drawn when experiments are properly designed, controlling these known variables, considering pen and flock effects, and sampling sufficient birds to give adequate statistical power.

The take home message is that the advantages and limitations of the methods used to assess microbiota need to be understood and considered when interpreting the information generated and reaching conclusions.

3 Functional interaction of microbiota and host

The functional interaction of gut microbiota with the chicken is critical to health and productivity. At its simplest the interaction includes a complementarity of functional capabilities, with the host providing feedstuffs and an environment in which microbes can thrive and the microbes in turn provide essential nutritional components that the host cannot make itself (e.g. vitamin K) and release nutrients from structurally complex carbohydrate molecules that the host could not access without microbial fermentation. However, the host-microbiota interaction is much more complex than just this ecological niche provision and nutrient release. There is a long and growing list of functional ways in which gut microbiota is known to interact with the host to alter and augment host physiological processes. In model systems, microbiota has been shown to influence endocrine function and hence affect glucose homeostasis and appetite regulation (Cani and Knauf, 2016). It can influence lipid metabolism and bile recycling and can modulate the physical characteristics and function of the gut via changes in gut integrity and morphological characteristics such as villus height, and can affect musculature, vasculature and peristalsis (Furuse and Okumura, 1994; Sagar et al., 2015; Sen et al., 2012). A more surprising finding in recent years has been the importance of the gut-brain axis and the role of gut microbiota in influencing that connection. There are aspects of both neural and endocrine communication between the gut and brain that have been shown, in model systems, to be affected by gut microbiota composition (Carabotti et al., 2015). The importance of these connections is just beginning to be studied in chickens and it is likely to have relevance as it may affect behaviour and responses to stress (Calefi et al., 2016).

3.1 The role of gut microbiota in nutritional functions

It has long been recognised that gut microbiota plays a key role in nutrition and energy capture. The gut microbiota expands the metabolic potential of the chicken by providing micronutrients and enzymes that the chicken cannot synthesise. Of particular note is the importance of gut microbiota in breaking down complex carbohydrates to release sugars that would otherwise be inaccessible to the chicken, as the chicken lacks the hydrolytic enzymes to degrade the complex carbohydrate molecules. The gut microbiota has the potential to encode a much broader range of anabolic and catabolic pathways than the chicken. The latest analysis of the chicken genome has identified 19 119 protein-coding genes (Warren et al., 2017). In contrast, the gut microbiota is made up of many different species of microbes, each with its own complement of genes, in total contributing far more unique genes and metabolic capabilities than the chicken. Estimates of the number of bacterial species present in the mature chicken gut microbiota have varied but it is probably conservative to suggest that there would be at least 1000 different species. The genomes of bacteria range in size but, again conservatively, they would on average be approximately three megabases in size and encode approximately 3000 genes. This suggests that within the gut microbiota there would be of the order of three million protein-coding genes. Many of these will be functionally similar but even with a high level of redundancy it can be seen that the gut microbiota makes a very substantial contribution to the coding potential and hence metabolic capabilities of the host-microbiota superorganism.

The metabolic potential of the gut microbiota can be estimated from various omics approaches including proteomics and metabolomics but currently the most widely used approaches to defining microbiota metabolic potential are to undertake whole metagenome analysis or use a bioinformatics approach to predict likely function from phylogenetic information (e.g. PICRUSt (Langille et al., 2013)). The genomic approaches have shown that variation in the functional potential of the gut microbiota is not as pronounced as the variation in phylogenetic composition of the microbiota (Oakley et al., 2014b). This suggests that the functional properties of the chicken gut microbiota can be equally fulfilled by a phylogenetically diverse collection of bacteria.

The caecum of the chicken is a major site of fermentation in which complex carbohydrates that the chicken itself is incapable of digesting can be broken down into subunits, which the chicken can capture, by the action of the caecal microbiota. Many of the bacteria that are resident in the caeca and play a role in these critical fermentation processes produce short chain fatty acids (SCFA) such as acetate, propionate and butyrate. Butyrate, in particular, is an important by-product of fermentation as it is a key energy supply for the host colonocytes and has anti-inflammatory properties. Therefore, the bacteria present in the caeca have a critical role in the productivity of birds, as an optimal microbiota will maximise the energy and nutrition that is extracted from the diet.

3.2 Immune functions

The gut microbiota is involved in defence against pathogens by both direct interactions and via induction and training of appropriate immune responses (Kabat et al., 2014). Much of the research about the complex interactions between microbiota and the host immune system has been carried out in model systems, such as the mouse, and has not yet been fully translated and confirmed in chicken studies, although there are a few studies in chickens which have linked disruption of the gut microbiota in young chicks to changes in immune development (Schokker et al., 2017; Simon et al., 2016). However, it is reasonable to expect that most of the findings can be extrapolated to interactions between the chicken immune system and its microbiota (Kogut, 2013). Innate and acquired immune functions must be able to differentiate self from non-self and direct immune attack against potentially pathogenic invaders whilst avoiding inappropriate immune responses against self that could result in autoimmunity and against commensal bacteria or food antigens that could result in inappropriate inflammatory responses. The passage of material through the digestive system exposes the host to a vast array of foreign antigens from the ingested food and the resident gut microbiota; the host immune system must be trained to not overreact to food antigens and antigens in the normal gut microbiota whilst mounting an effective immune response against potentially threatening pathogens. Microbes are ubiquitous in the environment, so a host is continually sampling a huge variety of microbes and the antigens they carry. There are estimated to be millions of bacterial species in the environment but of these there are only about 1400 species that have been shown to have any pathogenic potential in animals (Editorial, 2011). It can therefore be seen that the large majority of microbial antigens that a host is exposed to do not signal a threat; it is only the relatively rare pathogenic organisms which need to be actively excluded by an immune response. Immune responses are energy intensive and so inappropriate responses are an unnecessary drain on an animal's energy reserves; in the case of chickens, excessive immune activity would suppress productivity. Early and appropriate exposure to food antigens and antigens from harmless commensal bacteria appears to be important in training the immune system to develop appropriately (Klipper et al., 2001).

3.3 Bacteria involved in functional interactions with the host

A series of studies have shown correlations between gut microbiota composition and productivity outcomes in broiler birds (Apajalahti and Vienola, 2016; Han et al., 2016; Stanley et al., 2012a, 2013a, 2016; Yan et al., 2017). It has been more challenging to demonstrate causal relationships so it is not completely clear that particular microbiota structures actually induce peak performance, but that is certainly the working hypothesis that is driving much of the interest in studying gut microbiota composition. There are some general themes that can be drawn from the above studies. High performance, which can be measured in various ways, including feed conversion ratio, body weight gain or apparent metabolisable energy, is often associated with higher phylogenetic diversity of the caecal microbiota, a reduction in potentially pathogenic bacterial genera and an increase in SCFA-producing bacterial genera. There are a wide variety of bacteria that can produce SCFAs, including various clostridia, *Bacteroides*, *Bifidobacterium*, *Ruminococcus*, *Peptostreptococcus*, *Fusobacterium*, *Lactobacillus*, *Streptococcus*, *Ruminococcus*, *Eubacterium* and *Roseburia* species.

Other bacteria commonly found in chicken microbiota are likely to have important functional roles. *Faecalibacterium prausnitzii* is a common and often abundant member of the gut microbiota and is not only a notable butyrate-producing bacterium but also has a strong influence on immune development (Miquel et al., 2013). Similarly the segmented filamentous bacteria appear to play an important role in immune development (Hedblom et al., 2018). Although the roles of these bacteria in development of the immune system have only been demonstrated in mouse and human studies, it is likely that they function in a similar manner in chickens. *Akkermansia muciniphila* degrades mucin, induces a thickening of the mucus layer and an increase in gut barrier function, has anti-inflammatory activity and appears to be a key influence on energy retention and use by the host (Ottman et al., 2017). Despite these apparently positive effects induced by *A. muciniphila* the limited studies reported in chickens indicate that it may be negatively correlated with weight gain (Han et al., 2016).

Lactobacilli generally make up a large proportion of the microbiota in the upper GIT. Besides producing SCFAs the lactobacilli also have an important role in excluding pathogens, by both passive (niche occupation) and active (antimicrobial protein expression) means, and are involved in bile deconjugation and recycling (Cole and Fuller, 1984). There are many different species of *Lactobacillus*, and although they are generally regarded as beneficial bacteria, and many have been used as probiotics, it is also apparent that some lactobacilli are correlated with poor performance in chickens (Stanley et al., 2016).

3.4 Microbiota and disease

Much of the interest in microbiota-host interactions in humans is because of the connections between gut microbiota and inflammatory diseases, such as diabetes, and intestinal diseases, such as inflammatory bowel disease. In poultry the focus has been more on the direct impacts of microbiota on productivity and the efficient capture and use of nutrients and energy. Of course, good gut health is an essential element of productivity so researchers have also addressed the involvement of microbiota in poultry intestinal health, for example, in the context of important diseases such as necrotic enteritis and salmonellosis (Antonissen et al., 2016; Lacey et al., 2018; Stanley et al., 2012b; Videnska et al., 2013) and in dysbiosis (Ducatelle et al., 2018; Kogut, 2013). To date the major observations have been that the gut microbiota is clearly modified following these diseases, but it remains to be seen whether changes in the microbiota play a significant role in actually precipitating disease. It is certainly true that many of the factors that predispose birds to develop necrotic enteritis do cause perturbations in gut microbiota (Moore, 2016; Wu et al., 2014). As with much microbiota research, the cause-and-effect relationship can be difficult to establish and many of our current conclusions are drawn from correlative findings that require further investigation to establish causal linkages.

4 Microbiota manipulation for chicken health and productivity

Different approaches can be taken to the manipulation of gut microbiota; at the most fundamental level treatments can be designed to either increase the abundance of beneficial bacteria or reduce the abundance of detrimental bacteria (e.g. pathogens). Manipulation can be highly focused, with the introduction of specific beneficial microbes, that is probiotics, or can be broad and holistic, such as the use of feed additives aimed at suppressing or enhancing particular microbial populations. Historically, the use of in-feed antibiotics, as antibiotic growth promoters, has been the most widespread method used to modify the gut microbiota, reduce pathogen loads and increase productivity (Dibner and Richards, 2005). However, in recent decades there have been strong moves to reduce and restrict such antibiotic use, starting first with regulatory restrictions in Europe and recently spreading to widespread reduction in usage in North America, driven largely by retailer and consumer demand (Castanon, 2007). The motivation to reduce antibiotic use in animal production has been driven by a perceived link between such use and the increasing levels of antibiotic resistance seen in pathogenic bacteria that infect humans. Although evidence for such linkage is sparse the precautionary principle has driven the changes that are occurring. With the reduced reliance on in-feed antibiotics there has been a strong industry push to develop and assess alternative means to enhance productivity and suppress pathogens and many of these approaches result in modifications to the gut microbiota.

Linkage of the need to reduce antibiotic use and the promise of productivity and health improvements facilitated by the use of beneficial food additives has resulted in intensive efforts to look for ways to nurture and manipulate the microbiota using a wide range of feed additives.

4.1 Use of microbes to manipulate the gut microbiota

Direct application of microbial products can be used to modify the gut microbiota and environment. Probiotics are defined as live bacterial cultures that have beneficial effects on the treated host. They are sometimes referred to as direct feed antimicrobials. They are widely used, and many products are available commercially (Moore, 2017). Complex, undefined microbial preparations, called competitive exclusion products, have also been used to encourage development of the gut microbiota with a view to excluding colonisation by pathogens such as *Salmonella* and *Clostridium perfringens* (Wagner, 2006). Because of their undefined nature quality control is problematic and, therefore, the use of such products is not permitted in many regions.

There are many probiotics on the market making it difficult for end users to decide which may be worth using. The end users need information on how a probiotic may help their flock, when the product could be expected to be useful and what it might do. The mechanism of action of particular probiotics is not always clearly defined but there are many ways in which probiotics can theoretically work. Perhaps the simplest and clearest way in which a probiotic may have a beneficial effect is via the expression of antimicrobial compounds such as bacteriocins. Bacteriocins are small proteins that can directly kill certain target bacteria. Probiotics with antimicrobial activities have been used to target specific pathogens within the microbiota, such as *C. perfringens*, the cause of necrotic enteritis. Studies have demonstrated the effectiveness of some probiotic strains in reducing *C. perfringens* colonisation and the severity of necrotic enteritis infections and hence the potential value of the probiotics (Caly et al., 2015). However, there needs to be some caution in adopting such strains as the antimicrobial activity is rarely so specific that it only affects *C. perfringens* - the antimicrobial activity is also likely to affect other related bacteria, some of which, rather than being pathogens, may be quite beneficial, for example the SCFA-producing clostridia. Other probiotic products have less well-defined modes of action and may influence the composition of the gut microbiota in less direct ways, for example, by modulating the host immune system or altering the local microenvironments within the gut by producing acid, enzymes (e.g. bile salt hydrolases) or metabolites that can cross-feed other bacteria within the gut microbiota.

When attempting to manipulate the microbiota it is important to take into account that the composition of the gut microbiota varies in different compartments of the GIT, reflecting the different functions and requirements of different sections of the gut. These different environments must be considered when assessing existing probiotics or considering where to sample from and what sort of tests would be informative when isolating, developing and assessing new probiotics. The proventriculus and gizzard are acidic environments and are exposed to moderate levels of oxygen whereas the caecum is anaerobic and closer to neutral pH; the resident microbiota in each section must be adapted to cope with these very different environments and must deliver functions that are appropriate to the gut section in which they are most predominant.

4.2 Feed additives

A wide variety of feed additives are used commercially and have the potential to function by altering the composition of the gut microbiota, whether that impact is advertised or not in the promotion of the products. Although feed additives can have significant effects on gut microbiota composition and function, the basic formulation of the feed is of more profound significance. However, there is often less flexibility in its composition as it is usually driven largely by the local price and availability of raw materials (Crisol-Martínez et al., 2017a; Munyaka et al., 2016).

Prebiotic additives, which are complex carbohydrates that are indigestible by host enzymes, are specifically designed to feed particular classes of bacteria, such as lactobacilli and bifidobacteria, to increase their abundance and the level of SCFAs that they produce (Van Immerseel et al., 2017). Induction of higher levels of butyrate can be particularly useful as it has multiple beneficial effects, including providing an important source of energy for gut epithelial cells, supporting intestinal integrity, stimulating the immune system and cross-feeding other beneficial microbial populations. Both refined and crude feed additives containing complex indigestible carbohydrates can be effective in modulating gut microbiota (Pourabedin and Zhao, 2015; Ricke, 2018). Refined complex carbohydrates that have proven to be effective in some experimental work include inulin/fructo-oligosaccharide, galacto-oligosaccharides, mannan-oligosaccharides, isomalto-oligosaccharide, and xylo-oligosaccharide (XOS) (Bucław, 2016; Chacher et al., 2017; De Maesschalck et al., 2015; Hughes et al., 2017; Ricke, 2015; Zhang et al., 2003). Other more complex products, such as yeast cell wall products that contain complex oligosaccharides as well as other bioactive compounds, have been shown to modulate gut microbiota (Roto et al., 2015). These prebiotics have several other potential mechanisms of action that can modulate the gut microbiota. Some can directly inhibit the colonisation of Gram-negative bacteria (e.g. *Escherichia coli*, *Salmonella*, *Campylobacter*) by binding to bacterial flagella and interfering with mucosal association (Ganner and Schatzmayr, 2012; Ramirez-Hernandez et al., 2015; Xu et al., 2017). The probiotic complex carbohydrates have also been reported to stimulate immune responses, acting in an adjuvant-like fashion (Janardhana et al., 2009). In this and other studies it has generally been interpreted that this is a direct effect, but some elements of the effect may also be influenced by altered gut microbiota composition – this remains to be investigated.

The addition of exogenous enzymes such as amylases, xylanases, glucanases, proteases, lysozyme and phytase can modify both the biochemical and physical properties of the digesta and improve nutrient availability. These changes also modify the gut microbiota which in turn can alter fermentation efficiency and metabolite production, thus modulating gut health (Abdel-Latif et al., 2017; Józefiak et al., 2010; Kiarie et al., 2013; Wu et al., 2017; Yin et al., 2018). The modified environment and microbiota produced by some in-feed enzymes have also been shown to reduce the caecal load of pathogens such as *C. jejuni* and *Brachyspira intermedia* in poultry (Wealleans et al., 2017).

Organic acids are produced by the gut microbiota but can also be used as useful feed additives. They have various modes of action and can modify the gut microbiota and increase the abundance of potentially beneficial bacteria such as lactobacilli and bifidobacteria while reducing *Salmonella* and *Campylobacter* colonisation (Dittoe et al., 2018; Liu et al., 2017; Nava et al., 2009; Van Deun et al., 2008).

Phytogenic products such as essential oils, herbs and spices are becoming popular additives to use in the face of the widespread withdrawal of antibiotics from many poultry production systems around the world. As with so many of the alternative feed supplements that are available, they are likely to exert positive effects on the gut health of birds by a variety of mechanisms, but an important action appears to be the modification of the gut microbiota (Murugesan et al., 2015; Wati et al., 2015).

4.3 Antibiotics

Although the use of antibiotics as in-feed growth promoters is now being phased out in many regions, they are still widely used for therapeutic purposes. It is interesting to note that the gut microbiota of chickens is surprisingly resilient to low-dose transient use of antibiotics. A number of studies have found that various antibiotic treatment regimens produce fairly minor changes in microbiota composition and those changes disappear soon after the removal of antibiotics (Crisol-Martínez et al., 2017b; Wisselink et al., 2017). Therapeutic treatments that use higher antibiotic dosages may result in more damage to the gut microbiota and in those cases, it might be useful to help repopulate the gut by using a multi-strain probiotic.

4.4 Novel approaches to microbiota manipulation

Nanoparticulate metals have been investigated as feed supplements in poultry because nanoparticle delivery alters the bioavailability and tissue uptake of the metals (Joshua et al., 2016). There are very few reports on the effects of metal nanoparticles on gut microbiota but one recent study showed that selenium nanoparticles induced changes in the microbiota, increasing the abundance of beneficial bacteria such as *F. prausnitzii* and lactobacilli (Gangadoo et al., 2018). Another study suggested that delivery of a variety of metal nanoparticles altered microbiota composition and could be of use in correcting gut dysbiosis (Yausheva et al., 2018). The above studies used metals that are required as micronutrients; in a different approach to microbiota manipulation, one study has investigated whether the antibacterial activity of silver nanoparticles could be usefully deployed in poultry feed. They found that at the dosage and nanoparticle form used the silver nanoparticles had no effect on the gut microbiota (Vadalasetty et al., 2018).

An examination of materials that can absorb organic compounds has demonstrated that zeolite, bentonite and biochar could reduce the GIT carriage of certain pathogenic bacterial species (e.g. *C. jejuni*, *Helicobacter pullorum* and *Gallibacterium anatis*) while having no significant effect on overall microbiota diversity (Prasai et al., 2016).

Currently, probiotics and prebiotics are delivered in the feed but *in ovo* delivery of prebiotics and probiotics has been investigated in a few studies (Miśta et al., 2017; Pedroso et al., 2016). It is not yet clear whether this may be a useful approach to microbiota manipulation, but it certainly has the advantage of trying to intervene very early on in the establishment of the gut microbiota. However, further work is probably required to identify suitable products designed for this purpose (Dunislawska et al., 2017; Roto et al., 2016) and more advanced *in ovo* injection systems, that can reliably target particular compartments within the developing embryo, may be required to ensure efficacy and minimise reductions in hatchability. An alternative approach to ensuring early influences on the developing chick microbiota is to use probiotics and competitive exclusion products in the hatchery with delivery on the egg surface or hatching chick by egg dipping/painting, spraying or gel applications (Abdelrahman, 2015; Donaldson et al., 2017).

4.5 Translating the research on microbiota manipulation to industry use

A wide variety of prebiotics, probiotics and other feed additives have been demonstrated to modify gut microbiota composition and improve bird performance under experimental conditions. These findings do not always translate to better bird performance in the field when the products are used on commercial flocks. It may be informative to consider why the experimental findings are not always found to hold in field settings. Before the role of gut microbiota was understood and considered such variation in results was difficult to explain based just on the consideration of feed, environments and bird genetics – if rations, housing and genetics were similar then why were variable results found? Our newly acquired knowledge regarding the role and variability of the gut microbiota provides at least a partial explanation for variability in outcomes when the microbiota manipulating additives are used. The products will only be effective if the pre-existing microbiota in the treated birds is suboptimal and capable of being modified to provide better health and productivity outcomes. If the pre-existing microbiota is already in an optimal configuration then no improvement would be possible. Alternatively, a suboptimal microbiota may not contain members that can respond to the applied products. In this instance synbiotic products, that contain both a prebiotic and a probiotic that can benefit from the prebiotic component, may be more reliably effective. Also, of potential importance is the types of flocks and feed used in research and product development; a product developed for corn-fed broiler birds may not be effective on wheat-fed layer birds as the underlying gut microbiota may be substantially different. Further research needs to be done to validate treatments under different management constraints and practices.

5 Future trends

Much of our knowledge regarding host-microbiota interactions in poultry has been generated from studies of broiler birds housed in experimental facilities. As we know that the life history of a bird including such factors as housing, environment, feed and handling all influence gut microbiota composition, we must be cautious when extrapolating results from experimental settings to production settings. Similarly, birds with different genetics (e.g. broilers versus layers or Cobb versus Ross), housed in different systems (e.g. barn versus free-range) or fed different diets (e.g. corn-based versus wheat-based) are likely to develop gut microbiotas of different compositions. Hence, it is important that the developing body of knowledge generated by experimental work is eventually translated, adapted and tested by field studies in real production facilities. Of course, validating field study findings can sometimes be difficult as it is not always possible to include the control groups that assist in the interpretation of results. However, the real test of the value of the knowledge of host-microbiota interactions is whether it can be practically used to improve the industry. Currently probiotics, prebiotics and other feed supplements are addressing some production needs, at least partially by influencing gut microbiota composition and function. In the future, with our increasing understanding of the mechanistic basis of action of many of these products, more refined and targeted products for microbiota manipulation should become available.

There is considerable scope to identify and develop new strains of bacteria for use as probiotics. Areas of probiotic research that could be of value include (1) the development of probiotic strains that are more effective at colonising and persisting within the gut microbiota and thus do not need to be continually dosed; (2) the development of at-hatch delivered probiotics that 'programme' the gut microbiota to mature in a consistent and beneficial way; (3) the development of probiotic products from bacteria species that are currently difficult to commercialise because of their inherent physical and biological properties, for example, the strict anaerobe *F. prausnitzii*, but which clearly have the potential to beneficially influence the host; and (4) the development of consortia of co-functional bacterial strains that work together (e.g. by cross-feeding) to establish desirable environmental conditions for microbiota establishment.

6 Where to look for further information

Gut microbiota studies are a booming area of research because the host-microbiota interaction is now recognised to be of fundamental importance to a wide range of health and productivity outcomes. Hence, there is a rapidly expanding body of scientific literature. The research field is now so extensive that it is difficult for even closely involved researchers to keep up with all the developments; for the less involved reader it is probably best to start with broad review articles (Borda-Molina et al., 2018; Clavijo and Flórez, 2018; Pan and Yu, 2014; Stanley et al., 2014) and then only drill down into original papers that may be of relevance to particular interests.

Because of the wide and growing interest in gut microbiota, particularly with respect to the influence on human health, there are now many popular science books that deal with the subject, although all focus on microbiota in humans. Relevant books include (1) *Follow Your Gut: How the Ecosystem in Your Gut Determines Your Health, Mood, and More* by Rob Knight; (2) *10% Human: How Your Body's Microbes Hold the Key to Health and Happiness* by Alanna Collen; (3) *The Good Gut: Taking Control of Your Weight, Your Mood, and Your Long-Term Health* by Justin Sonnenburg; (4) *Understanding the Gut Microbiota* by Gerald W. Tannock; (5) *The Human Superorganism: How the Microbiome Is Revolutionizing the Pursuit of a Healthy Life* by Rodney Dietert; and (6) *Gut: The Inside Story of Our Body's Most Underrated Organ* by Giulia Enders.

A third important source of information is the manufacturers and marketers of products such as prebiotics and probiotics. These companies often have dossiers of information to support their products. The quality and accessibility of such information is quite variable but much of it is informative. It is important that the end users of such products continue to encourage the suppliers to improve their information offering and in particular provide evidence of the mechanisms of action and reproducibility of the claimed outcomes of product use.

7 References

Abdel-Latif, M. A., El-Far, A. H., Elbestawy, A. R., Ghanem, R., Mousa, S. A. and Abd El-Hamid, H. S. 2017. Exogenous dietary lysozyme improves the growth performance and gut microbiota in broiler chickens targeting the antioxidant and non-specific immunity mRNA expression. *PLoS One* 12(10), e0185153. doi:10.1371/journal.pone.0185153.

Abdelrahman, W. 2015. The application of probiotics in the hatchery. *Int. Hatchery Practice* 29, 11.

Antonissen, G., Eeckhaut, V., Van Driessche, K., Onrust, L., Haesebrouck, F., Ducatelle, R., Moore, R. J. and Van Immerseel, F. 2016. Microbial shifts associated with necrotic enteritis. *Avian Pathol.* 45(3), 308–12. doi:10.1080/03079457.2016.1152625.

Apajalahti, J. and Vienola, K. 2016. Interaction between chicken intestinal microbiota and protein digestion. *Anim. Feed Sci. Tech.* 221, 323–30. doi:10.1016/j.anifeedsci.2016.05.004.

Arumugam, M., Raes, J., Pelletier, E., Paslier, D. L., Yamada, T., Mende, D. R., Fernandes, G. R., Tap, J., Bruls, T., Batto, J.-M., et al. 2011. Enterotypes of the human gut microbiome. *Nature* 473, 174–80. doi:10.1038/nature09944.

Ballou, A. L., Ali, R. A., Mendoza, M. A., Ellis, J. C., Hassan, H. M., Croom, W. J. and Koci, M. D. 2016. Development of the chick microbiome: how early exposure influences future microbial diversity. *Front. Vet. Sci.* 3, 2. doi:10.3389/fvets.2016.00002.

Borda-Molina, D., Seifert, J. and Camarinha-Silva, A. 2018. Current perspectives of the chicken gastrointestinal tract and its microbiome. *Comput. Struc. Biotech. J.* 16, 131–9. doi:10.1016/j.csbj.2018.03.002.

Brooks, J. P., Edwards, D. J., Harwich, M. D., Rivera, M. C., Fettweis, J. M., Serrano, M. G., Reris, R. A., Sheth, N. U., Huang, B., Girerd, P., et al. 2015. The truth about metagenomics: quantifying and counteracting bias in 16S rRNA studies. *BMC Microbiol.* 15, 66. doi:10.1186/s12866-015-0351-6.

Bucław, M. 2016. The use of inulin in poultry feeding: a review. *J. Anim. Physiol. Anim. Nutr. (Berl)* 100(6), 1015–22. doi:10.1111/jpn.12484.

Calefi, A. S., da Silva Fonseca, J. G., Cohn, D. W. H., Honda, B. T. B., Costola-de-Souza, C., Tsugiyama, L. E., Quinteiro-Filho, W. M., Piantino Ferreira, A. J. and Palermo-Neto, J. 2016. The gut-brain axis interactions during heat stress and avian necrotic enteritis. *Poult. Sci.* 95, 1005–14. doi:10.3382/ps/pew021.

Caly, D. L., D'Inca, R., Auclair, E. and Drider, D. 2015. Alternatives to antibiotics to prevent necrotic enteritis in broiler chickens: a microbiologist's perspective. *Front. Microbiol.* 6, 1336. doi:10.3389/fmicb.2015.01336.

Cani, P. D. and Knauf, C. 2016. How gut microbes talk to organs: the role of endocrine and nervous routes. *Mol. Metab.* 5(9), 743–52. doi:10.1016/j.molmet.2016.05.011.

Carabotti, M., Scirocco, A., Maselli, M. A. and Severi, C. 2015. The gut-brain axis: interactions between enteric microbiota, central and enteric nervous systems. *Ann. Gastroenterol.* 28(2), 203–9.

Castanon, J. I. R. 2007. History of the use of antibiotic as growth promoters in European poultry feeds. *Poult. Sci.* 86(11), 2466–71. doi:10.3382/ps.2007-00249.

Chacher, M. F. A., Kamran, Z., Ahsan, U., Ahmad, S., Koutoulis, K. C., Din, H. G. Q. U. and Cengiz, Ö. 2017. Use of mannan oligosaccharide in broiler diets: an overview of underlying mechanisms. *Worlds Poult. Sci. J.* 73(4), 831–44. doi:10.1017/S0043933917000757.

Choi, J. H., Kim, G. B. and Cha, C. J. 2014. Spatial heterogeneity and stability of bacterial community in the gastrointestinal tracts of broiler chickens. *Poult. Sci.* 93(8), 1942–50. doi:10.3382/ps.2014-03974.

Clavijo, V. and Flórez, M. J. V. 2018. The gastrointestinal microbiome and its association with the control of pathogens in broiler chicken production: a review. *Poult. Sci.* 97(3), 1006–21. doi:10.3382/ps/pex359.

Cole, C. B. and Fuller, R. 1984. Bile acid deconjugation and attachment of chicken gut bacteria: their possible role in growth depression. *Br. Poult. Sci.* 25(2), 227–31. doi:10.1080/00071668408454861.

Costea, P. I., Hildebrand, F., Arumugam, M., Bäckhed, F., Blaser, M. J., Bushman, F. D., de Vos, W. M., Ehrlich, S. D., Fraser, C. M., Hattori, M., et al. 2018. Enterotypes in the landscape of gut microbial community composition. *Nat. Microbiol.* 3(1), 8–16. doi:10.1038/s41564-017-0072-8.

Crisol-Martínez, E., Stanley, D., Geier, M. S., Hughes, R. J. and Moore, R. J. 2017a. Sorghum and wheat differentially affect caecal microbiota and associated performance characteristics of meat chickens. *Peer J.* 5, e3071. doi:10.7717/peerj.3071.

Crisol-Martínez, E., Stanley, D., Geier, M. S., Hughes, R. J. and Moore, R. J. 2017b. Understanding the mechanisms of zinc bacitracin and avilamycin on animal production: linking gut microbiota and growth performance in chickens. *Appl. Microbiol. Biotechnol.* 101(11), 4547–59. doi:10.1007/s00253-017-8193-9.

Cuperus, T., Kraaij, M. D., Zomer, A. L., Dijk, A. van and Haagsman, H. P. 2018. Immunomodulation and effects on microbiota after *in ovo* administration of chicken cathelicidin-2. *PLoS One* 13(6), e0198188. doi:10.1371/journal.pone.0198188.

De Maesschalck, C., Eeckhaut, V., Maertens, L., De Lange, L., Marchal, L., Nezer, C., De Baere, S., Croubels, S., Daube, G., Dewulf, J., et al. 2015. Effects of xylo-oligosaccharides on broiler chicken performance and microbiota. *Appl. Environ. Microbiol.* 81(17), 5880–8. doi:10.1128/AEM.01616-15.

Dibner, J. J. and Richards, J. D. 2005. Antibiotic growth promoters in agriculture: history and mode of action. *Poult. Sci.* 84(4), 634–43. doi:10.1093/ps/84.4.634.

Ding, J., Dai, R., Yang, L., He, C., Xu, K., Liu, S., Zhao, W., Xiao, L., Luo, L., Zhang, Y., et al. 2017. Inheritance and establishment of gut microbiota in chickens. *Front. Microbiol.* 8, 1967. doi:10.3389/fmicb.2017.01967.

Dittoe, D. K., Ricke, S. C. and Kiess, A. S. 2018. Organic acids and potential for modifying the avian gastrointestinal tract and reducing pathogens and disease. *Front. Vet. Sci.* 5, 216. doi:10.3389/fvets.2018.00216.

Donaldson, E. E., Stanley, D., Hughes, R. J. and Moore, R. J. 2017. The time-course of broiler intestinal microbiota development after administration of cecal contents to incubating eggs. *Peer J.* 5, e3587. doi:10.7717/peerj.3587.

Ducatelle, R., Goossens, E., De Meyer, F., Eeckhaut, V., Antonissen, G., Haesebrouck, F. and Van Immerseel, F. 2018. Biomarkers for monitoring intestinal health in poultry: Present status and future perspectives. *Vet. Res.* 49(1), 43. doi:10.1186/s13567-018-0538-6.

Dunislawska, A., Slawinska, A., Stadnicka, K., Bednarczyk, M., Gulewicz, P., Jozefiak, D. and Siwek, M. 2017. Synbiotics for broiler chickens – *in vitro* design and evaluation of the influence on host and selected microbiota populations following *in ovo* delivery. *PLoS One* 12(1), e0168587. doi:10.1371/journal.pone.0168587.

Editorial. 2011. Microbiology by numbers. *Nat. Rev. Microbiol.* 9(9), 628. doi:10.1038/nrmicro2644.

Furuse, M. and Okumura, J. 1994. Nutritional and physiological characteristics in germ-free chickens. *Comp. Biochem. Physiol. A Physiol.* 109(3), 547–56. doi:10.1016/0300-9629(94)90193-7.

Gangadoo, S., Dinev, I., Chapman, J., Hughes, R. J., Van, T. T. H., Moore, R. J. and Stanley, D. 2018. Selenium nanoparticles in poultry feed modify gut microbiota and increase abundance of *Faecalibacterium prausnitzii*. *Appl. Microbiol. Biotechnol.* 102(3), 1455–66. doi:10.1007/s00253-017-8688-4.

Ganner, A. and Schatzmayr, G. 2012. Capability of yeast derivatives to adhere enteropathogenic bacteria and to modulate cells of the innate immune system. *Appl. Microbiol. Biotechnol.* 95(2), 289–97. doi:10.1007/s00253-012-4140-y.

Han, G. G., Kim, E. B., Lee, J., Lee, J. Y., Jin, G., Park, J., Huh, C. S., Kwon, I. K., Kil, D. Y., Choi, Y. J., et al. 2016. Relationship between the microbiota in different sections of the gastrointestinal tract, and the body weight of broiler chickens. *SpringerPlus* 5(1), 911. doi:10.1186/s40064-016-2604-8.

Hedblom, G. A., Reiland, H. A., Sylte, M. J., Johnson, T. J. and Baumler, D. J. 2018. Segmented filamentous bacteria – metabolism meets immunity. *Front. Microbiol.* 9, 1991. doi:10.3389/fmicb.2018.01991.

Hughes, R. A., Ali, R. A., Mendoza, M. A., Hassan, H. M. and Koci, M. D. 2017. Impact of dietary galacto-oligosaccharide (GOS) on chicken's gut microbiota, mucosal gene expression, and *Salmonella* colonization. *Front. Vet. Sci.* 4, 192. doi:10.3389/fvets.2017.00192.

Ibarbalz, F. M., Pérez, M. V., Figuerola, E. L. M. and Erijman, L. 2014. The bias associated with amplicon sequencing does not affect the quantitative assessment of bacterial community dynamics. *PLoS One* 9(6), e99722. doi:10.1371/journal.pone.0099722.

Ijaz, U. Z., Sivaloganathan, L., McKenna, A., Richmond, A., Kelly, C., Linton, M., Stratakos, A. C., Lavery, U., Elmi, A., Wren, B. W., et al. 2018. Comprehensive longitudinal microbiome analysis of the chicken cecum reveals a shift from competitive to environmental drivers and a window of opportunity for *Campylobacter*. *Front. Microbiol.* 9, 2452. doi:10.3389/fmicb.2018.02452.

Ilina, L. A., Yildirim, E. A., Nikonov, I. N., Filippova, V. A., Laptev, G. Y., Novikova, N. I., Grozina, A. A., Lenkova, T. N., Manukyan, V. A., Egorov, I. A., et al. 2016. Metagenomic bacterial community profiles of chicken embryo gastrointestinal tract by using T-RFLP analysis. *Dokl. Biochem. Biophys.* 466, 47–51. doi:10.1134/S1607672916010130.

Janardhana, V., Broadway, M. M., Bruce, M. P., Lowenthal, J. W., Geier, M. S., Hughes, R. J. and Bean, A. G. D. 2009. Prebiotics modulate immune responses in the gut-associated lymphoid tissue of chickens. *J. Nutr.* 139(7), 1404-9. doi:10.3945/jn.109.105007.

Joshua, P. P., Valli, C. and Balakrishnan, V. 2016. Effect of *in ovo* supplementation of nano forms of zinc, copper, and selenium on post-hatch performance of broiler chicken. *Vet. World* 9(3), 287–94. doi:10.14202/vetworld.2016.287-294.

Józefiak, D., Rutkowski, A., Kaczmarek, S., Jensen, B. B., Engberg, R. M. and Højberg, O. 2010. Effect of β -glucanase and xylanase supplementation of barley- and rye-based diets on caecal microbiota of broiler chickens. *Br. Poult. Sci.* 51(4), 546–57. doi:10.1080/00071668.2010.507243.

Kaakoush, N. O., Sodhi, N., Chenu, J. W., Cox, J. M., Riordan, S. M. and Mitchell, H. M. 2014. The interplay between *Campylobacter* and *Helicobacter* species and other gastrointestinal microbiota of commercial broiler chickens. *Gut Pathog.* 6, 18. doi:10.1186/1757-4749-6-18.

Kabat, A. M., Srinivasan, N. and Maloy, K. J. 2014. Modulation of immune development and function by intestinal microbiota. *Trends Immunol.* 35(11), 507–17. doi:10.1016/j.it.2014.07.010.

Kiarie, E., Romero, L. F. and Nyachoti, C. M. 2013. The role of added feed enzymes in promoting gut health in swine and poultry. *Nutr. Res. Rev.* 26(1), 71–88. doi:10.1017/S0954422413000048.

Kizerwetter-Świda, M. and Binek, M. 2008. Bacterial microflora of the chicken embryos and newly hatched chicken. *J. Anim. Feed Sci.* 17(2), 224–32. doi:10.22358/jafs/66602/2008.

Klaschik, S., Lehmann, L. E., Raadts, A., Hoeft, A. and Stuber, F. 2002. Comparison of different decontamination methods for reagents to detect low concentrations of bacterial 16S DNA by real-time-PCR. *Mol. Biotechnol.* 22(3), 231–42. doi:10.1385/MB:22:3:231.

Klipper, E., Sklan, D. and Friedman, A. 2001. Response, tolerance and ignorance following oral exposure to a single dietary protein antigen in *Gallus domesticus*. *Vaccine* 19(20–22), 2890–7. doi:10.1016/S0264-410X(00)00557-0.

Kogut, M. H. 2013. The gut microbiota and host innate immunity: regulators of host metabolism and metabolic diseases in poultry? *J. Appl. Poult. Res.* 22(3), 637–46. doi:10.3382/japr.2013-00741.

Lacey, J. A., Stanley, D., Keyburn, A. L., Ford, M., Chen, H., Johanesen, P., Lyras, D. and Moore, R. J. 2018. *Clostridium perfringens*-mediated necrotic enteritis is not influenced by the pre-existing microbiota but is promoted by large changes in the post-challenge microbiota. *Vet. Microbiol.* 227, 119–26. doi:10.1016/j.vetmic.2018.10.022.

Langille, M. G. I., Zaneveld, J., Caporaso, J. G., McDonald, D., Knights, D., Reyes, J. A., Clemente, J. C., Burkepile, D. E., Vega Thurber, R. L., Knight, R., et al. 2013. Predictive functional profiling of microbial communities using 16S rRNA marker gene sequences. *Nat. Biotechnol.* 31(9), 814–21. doi:10.1038/nbt.2676.

Liu, Y., Yang, X., Xin, H., Chen, S., Yang, C., Duan, Y. and Yang, X. 2017. Effects of a protected inclusion of organic acids and essential oils as antibiotic growth promoter alternative on growth performance, Intestinal morphology and gut microflora in broilers. *Anim. Sci. J.* 88(9), 1414–24. doi:10.1111/asj.12782.

Miquel, S., Martín, R., Rossi, O., Bermúdez-Humarán, L. G., Chatel, J. M., Sokol, H., Thomas, M., Wells, J. M. and Langella, P. 2013. *Faecalibacterium prausnitzii* and human intestinal health. *Curr. Opin. Microbiol.* 16(3), 255–61. doi:10.1016/j.mib.2013.06.003.

Miśta, D., Króliczewska, B., Pecka-Kiełb, E., Kapuśniak, V., Zawadzki, W., Graczyk, S., Kowalczyk, A., Łukaszewicz, E. and Bednarczyk, M. 2017. Effect of *in ovo* injected prebiotics and synbiotics on the caecal fermentation and intestinal morphology of broiler chickens. *Anim. Prod. Sci.* 57(9), 1884–92. doi:10.1071/AN16257.

Moore, R. J. 2016. Necrotic enteritis predisposing factors in broiler chickens. *Avian Pathol.* 45(3), 275–81. doi:10.1080/03079457.2016.1150587.

Moore, R. J. 2017. Probiotics, prebiotics and other feed additives to improve gut function and immunity in poultry. In: Applegate, T. (Ed.), *Achieving Sustainable Production of Poultry Meat Volume 2*. Burleigh Dodds Science Publishing Limited, Cambridge, UK, pp. 181-206. Chapter 10.

Munyaka, P. M., Nandha, N. K., Kiarie, E., Nyachoti, C. M. and Khafipour, E. 2016. Impact of combined β-glucanase and xylanase enzymes on growth performance, nutrients utilization and gut microbiota in broiler chickens fed corn or wheat-based diets. *Poult. Sci.* 95(3), 528–40. doi:10.3382/ps/pev333.

Murugesan, G. R., Syed, B., Haldar, S. and Pender, C. 2015. Phytogenic feed additives as an alternative to antibiotic growth promoters in broiler chickens. *Front. Vet. Sci.* 2, 21. doi:10.3389/fvets.2015.00021.

Nava, G. M., Attene-Ramos, M. S., Gaskins, H. R. and Richards, J. D. 2009. Molecular analysis of microbial community structure in the chicken ileum following organic acid supplementation. *Vet. Microbiol.* 137(3–4), 345–53. doi:10.1016/j.vetmic.2009.01.037.

Oakley, B. B., Buhr, R. J., Ritz, C. W., Kiepper, B. H., Berrang, M. E., Seal, B. S. and Cox, N. A. 2014a. Successional changes in the chicken cecal microbiome during 42 days of growth are independent of organic acid feed additives. *BMC Vet. Res.* 10, 282. doi:10.1186/s12917-014-0282-8.

Oakley, B. B., Lillehoj, H. S., Kogut, M. H., Kim, W. K., Maurer, J. J., Pedroso, A., Lee, M. D., Collett, S. R., Johnson, T. J. and Cox, N. A. 2014b. The chicken gastrointestinal microbiome. *FEMS Microbiol. Lett.* 360(2), 100–12. doi:10.1111/1574-6968.12608.

Ottman, N., Geerlings, S. Y., Aalvink, S., de Vos, W. M. and Belzer, C. 2017. Action and function of *Akkermansia muciniphila* in microbiome ecology, health and disease. *Best Prac. Res. Clin. Gastroent.* 31(6), 637–42. doi:10.1016/j.bpg.2017.10.001.

Pan, D. and Yu, Z. 2014. Intestinal microbiome of poultry and its interaction with host and diet. *Gut Microbes* 5(1), 108–19. doi:10.4161/gmic.26945.

Pedroso, A. A., Batal, A. B. and Lee, M. D. 2016. Effect of *in ovo* administration of an adult-derived microbiota on establishment of the intestinal microbiome in chickens. *Am. J. Vet. Res.* 77(5), 514-26. doi:10.2460/ajvr.77.5.514.

Pollock, J., Glendinning, L., Wisedchanwet, T. and Watson, M. 2018. The madness of microbiome: attempting to find consensus 'best practice' for 16S microbiome studies. *Appl. Environ. Microbiol.* 84(7). doi:10.1128/AEM.02627-17.

Pourabedin, M. and Zhao, X. 2015. Prebiotics and gut microbiota in chickens. *FEMS Microbiol. Lett.* 362(15), fnv122. doi:10.1093/femsle/fnv122.

Prasai, T. P., Walsh, K. B., Bhattarai, S. P., Midmore, D. J., Van, T. T. H., Moore, R. J. and Stanley, D. 2016. Biochar, bentonite and zeolite supplemented feeding of layer chickens alters intestinal microbiota and reduces *Campylobacter* load. *PLoS One* 11(4), e0154061. doi:10.1371/journal.pone.0154061.

Ramirez-Hernandez, A., Rupnow, J. and Hutkins, R. W. 2015. Adherence reduction of *Campylobacter jejuni* and *Campylobacter coli* strains to HEp-2 cells by mannan oligosaccharides and a high-molecular-weight component of cranberry extract. *J. Food Prot.* 78(8), 1496-505. doi:10.4315/0362-028X.JFP-15-087.

Ranjitkar, S., Lawley, B., Tannock, G. and Engberg, R. M. 2016. Bacterial succession in the broiler gastrointestinal tract. *Appl. Environ. Microbiol.* 82(8), 2399-410. doi:10.1128/AEM.02549-15.

Ricke, S. C. 2015. Potential of fructooligosaccharide prebiotics in alternative and nonconventional poultry production systems. *Poult. Sci.* 94(6), 1411-8. doi:10.3382/ps/pev049.

Ricke, S. C. 2018. Impact of prebiotics on poultry production and food safety. *Yale J. Biol. Med.* 91(2), 151-9.

Rintala, A., Pietilä, S., Munukka, E., Eerola, E., Pursiheimo, J. P., Laiho, A., Pekkala, S. and Huovinen, P. 2017. Gut microbiota analysis results are highly dependent on the 16S rRNA gene target region, whereas the impact of DNA extraction is minor. *J. Biomol. Tech.* 28(1), 19-30. doi:10.7171/jbt.17-2801-003.

Roto, S. M., Rubinelli, P. M. and Ricke, S. C. 2015. An introduction to the avian gut microbiota and the effects of yeast-based prebiotic-type compounds as potential feed additives. *Front. Vet. Sci.* 2, 28. doi:10.3389/fvets.2015.00028.

Roto, S. M., Kwon, Y. M. and Ricke, S. C. 2016. Applications of *in ovo* technique for the optimal development of the gastrointestinal tract and the potential influence on the establishment of its microbiome in poultry. *Front. Vet. Sci.* 3, 63. doi:10.3389/fvets.2016.00063.

Sagar, N. M., Cree, I. A., Covington, J. A. and Arasaradnam, R. P. 2015. The interplay of the gut microbiome, bile acids, and volatile organic compounds. *Gastroent. Res. Prac.* 2015, 398585. doi:10.1155/2015/398585.

Salter, S. J., Cox, M. J., Turek, E. M., Calus, S. T., Cookson, W. O., Moffatt, M. F., Turner, P., Parkhill, J., Loman, N. J. and Walker, A. W. 2014. Reagent and laboratory contamination can critically impact sequence-based microbiome analyses. *BMC Biol.* 12, 87. doi:10.1186/s12915-014-0087-z.

Schokker, D., Jansman, A. J. M., Veninga, G., de Bruin, N., Vastenhouw, S. A., de Bree, F. M., Bossers, A., Rebel, J. M. J. and Smits, M. A. 2017. Perturbation of microbiota in one-day old broiler chickens with antibiotic for 24 hours negatively affects intestinal immune development. *BMC Genomics* 18(1), 241. doi:10.1186/s12864-017-3625-6.

Sen, S., Ingale, S. L., Kim, Y. W., Kim, J. S., Kim, K. H., Lohakare, J. D., Kim, E. K., Kim, H. S., Ryu, M. H., Kwon, I. K., et al. 2012. Effect of supplementation of *Bacillus subtilis* LS

1-2 to broiler diets on growth performance, nutrient retention, caecal microbiology and small intestinal morphology. *Res. Vet. Sci.* 93(1), 264–8. doi:10.1016/j.rvsc.2011.05.021.

Silkie, S. S., Tolcher, M. P. and Nelson, K. L. 2008. Reagent decontamination to eliminate false-positives in *Escherichia coli* qPCR. *J. Microbiol. Methods* 72(3), 275–82. doi:10.1016/j.mimet.2007.12.011.

Simon, K., Verwoolde, M. B., Zhang, J., Smidt, H., de Vries Reilingh, G., Kemp, B. and Lammers, A. 2016. Long-term effects of early life microbiota disturbance on adaptive immunity in laying hens. *Poult. Sci.* 95(7), 1543–54. doi:10.3382/ps/pew088.

Sleator, R. D. 2010. The human superorganism - of microbes and men. *Med. Hypotheses* 74(2), 214–5. doi:10.1016/j.mehy.2009.08.047.

Stanley, D., Denman, S. E., Hughes, R. J., Geier, M. S., Crowley, T. M., Chen, H., Haring, V. R. and Moore, R. J. 2012a. Intestinal microbiota associated with differential feed conversion efficiency in chickens. *Appl. Microbiol. Biotechnol.* 96(5), 1361-9. doi:10.1007/s00253-011-3847-5.

Stanley, D., Keyburn, A. L., Denman, S. E. and Moore, R. J. 2012b. Changes in the caecal microflora of chickens following *Clostridium perfringens* challenge to induce necrotic enteritis. *Vet. Microbiol.* 159(1-2), 155–62. doi:10.1016/j.vetmic.2012.03.032.

Stanley, D., Geier, M. S., Denman, S. E., Haring, V. R., Crowley, T. M., Hughes, R. J. and Moore, R. J. 2013a. Identification of chicken intestinal microbiota correlated with the efficiency of energy extraction from feed. *Vet. Microbiol.* 164(1-2), 85–92. doi:10.1016/j.vetmic.2013.01.030.

Stanley, D., Geier, M. S., Hughes, R. J., Denman, S. E. and Moore, R. J. 2013b. Highly variable microbiota development in the chicken gastrointestinal tract. *PLoS One* 8(12), e84290. doi:10.1371/journal.pone.0084290.

Stanley, D., Hughes, R. J. and Moore, R. J. 2014. Microbiota of the chicken gastrointestinal tract: influence on health, productivity and disease. *Appl. Microbiol. Biotechnol.* 98(10), 4301–10. doi:10.1007/s00253-014-5646-2.

Stanley, D., Hughes, R. J., Geier, M. S. and Moore, R. J. 2016. Bacteria within the gastrointestinal tract microbiota correlated with improved growth and feed conversion: challenges presented for the identification of performance enhancing probiotic bacteria. *Front. Microbiol.* 7, 187. doi:10.3389/fmicb.2016.00187.

Stanley, D., Moore, R. J. and Wong, C. H. Y. 2018. An insight into intestinal mucosal microbiota disruption after stroke. *Sci. Rep.* 8(1), 568. doi:10.1038/s41598-017-18904-8.

Vadalasetty, K. P., Lauridsen, C., Engberg, R. M., Vadalasetty, R., Kutwin, M., Chwalibog, A. and Sawosz, E. 2018. Influence of silver nanoparticles on growth and health of broiler chickens after infection with *Campylobacter jejuni*. *BMC Vet. Res.* 14(1), 1. doi:10.1186/s12917-017-1323-x.

Van Deun, K., Haesebrouck, F., Van Immerseel, F., Ducatelle, R. and Pasmans, F. 2008. Short-chain fatty acids and L-lactate as feed additives to control *Campylobacter jejuni* infections in broilers. *Avian Pathol.* 37(4), 379–83. doi:10.1080/03079450802216603.

Van Immerseel, F., Eeckhaut, V., Moore, R. J., Choct, M. and Ducatelle, R. 2017. Beneficial microbial signals from alternative feed ingredients: a way to improve sustainability of broiler production? *Microb. Biotechnol.* 10(5), 1008-11. doi:10.1111/1751-7915.12794.

Videnska, P., Sisak, F., Havlickova, H., Faldynova, M. and Rychlik, I. 2013. Influence of *Salmonella enterica* serovar Enteritidis infection on the composition of chicken cecal microbiota. *BMC Vet. Res.* 9, 140. doi:10.1186/1746-6148-9-140.

Wagner, R. D. 2006. Efficacy and food safety considerations of poultry competitive exclusion products. *Mol. Nutr. Food Res.* 50(11), 1061–71. doi:10.1002/mnfr.200600058.

Warren, W. C., Hillier, L. W., Tomlinson, C., Minx, P., Kremitzki, M., Graves, T., Markovic, C., Bouk, N., Pruitt, K. D., Thibaud-Nissen, F., et al. 2017. A new chicken genome assembly provides insight into avian genome structure. *G3-Genes Genom. Genet.* 7(1), 109–17. doi:10.1534/g3.116.035923.

Wati, T., Ghosh, T. K., Syed, B. and Haldar, S. 2015. Comparative efficacy of a phytogenic feed additive and an antibiotic growth promoter on production performance, caecal microbial population and humoral immune response of broiler chickens inoculated with enteric pathogens. *Anim. Nutr.* 1(3), 213–9. doi:10.1016/j.aninu.2015.08.003.

Wealleans, A. L., Walsh, M. C., Romero, L. F. and Ravindran, V. 2017. Comparative effects of two multi-enzyme combinations and a *Bacillus* probiotic on growth performance, digestibility of energy and nutrients, disappearance of non-starch polysaccharides, and gut microflora in broiler chickens. *Poult. Sci.* 96(12), 4287–97. doi:10.3382/ps/pex226.

Wisselink, H. J., Cornelissen, J. B. W. J., Mevius, D. J., Smits, M. A., Smidt, H. and Rebel, J. M. J. 2017. Antibiotics in 16-day-old broilers temporarily affect microbial and immune parameters in the gut. *Poult. Sci.* 96(9), 3068–78. doi:10.3382/ps/pex133.

Wu, S. B., Stanley, D., Rodgers, N., Swick, R. A. and Moore, R. J. 2014. Two necrotic enteritis predisposing factors, dietary fishmeal and *Eimeria* infection, induce large changes in the caecal microbiota of broiler chickens. *Vet. Microbiol.* 169(3–4), 188–97. doi:10.1016/j.vetmic.2014.01.007.

Wu, D., Wu, S. B., Choct, M. and Swick, R. A. 2017. Performance, intestinal microflora, and amino acid digestibility altered by exogenous enzymes in broilers fed wheat- or sorghum-based diets. *J. Anim. Sci.* 95(2), 740–51. doi:10.2527/jas.2016.0411.

Xu, X., Qiao, Y., Peng, Q., Gao, L. and Shi, B. 2017. Inhibitory effects of YCW and MOS from *Saccharomyces cerevisiae* on *Escherichia coli* and *Salmonella pullorum* adhesion to Caco-2 cells. *Front. Biol.* 12(5), 370–5. doi:10.1007/s11515-017-1464-0.

Yan, W., Sun, C., Yuan, J. and Yang, N. 2017. Gut metagenomic analysis reveals prominent roles of *Lactobacillus* and cecal microbiota in chicken feed efficiency. *Sci. Rep.* 7, 45308. doi:10.1038/srep45308.

Yausheva, E., Miroshnikov, S. and Sizova, E. 2018. Intestinal microbiome of broiler chickens after use of nanoparticles and metal salts. *Environ. Sci. Pollut. Res. Int.* 25(18), 18109–20. doi:10.1007/s11356-018-1991-5.

Yin, D., Yin, X., Wang, X., Lei, Z., Wang, M., Guo, Y., Aggrey, S. E., Nie, W. and Yuan, J. 2018. Supplementation of amylase combined with glucoamylase or protease changes intestinal microbiota diversity and benefits for broilers fed a diet of newly harvested corn. *J. Anim. Sci. Biotechnol.* 9, 24. doi:10.1186/s40104-018-0238-0.

Zhang, W. F., Li, D. F., Lu, W. Q. and Yi, G. F. 2003. Effects of isomalto-oligosaccharides on broiler performance and intestinal microflora. *Poult. Sci.* 82(4), 657–63. doi:10.1093/ps/82.4.657.

Zoetendal, E. G., von Wright, A., Vilpponen-Salmela, T., Ben-Amor, K., Akkermans, A. D. L. and de Vos, W. M. 2002. Mucosa-associated bacteria in the human gastrointestinal tract are uniformly distributed along the colon and differ from the community recovered from feces. *Appl. Environ. Microbiol.* 68(7), 3401–7. doi:10.1128/AEM.68.7.3401-3407.2002.

Chapter 5

In ovo development of the chicken gut microbiome and its impact on later gut function

E. David Peebles, Mississippi State University, USA

1 Introduction

Villi enterocytes in the intestines of 14-day-old chicken embryos lack active transport and carbohydrase activity (Moran, 1985), and the intestines of chicks are normally not fully capable of nutrient digestion and absorption until 14 days posthatch (Vieira and Moran, 1999). Therefore, the early nutritional demands of the chicken embryo are met by absorption of essential nutrients from the yolk by the yolk sac membrane (Yadgary et al., 2013). The yolk sac membrane encompasses the yolk and becomes vascularized during the first week of incubation, and subsequently acquires absorptive capabilities by Day 10 of incubation (Speake et al., 1998). Expansion of the villi surface area in the yolk sac is proportionate to that which occurs in the embryonic intestine (Holdsworth and Wilson, 1967). By Day 14 of incubation, the intestinal villi enterocytes of embryos are capable of immunoglobulin uptake (Moran, 1985). Furthermore, intestine weight as a proportion of the total body weight of Ross × Ross broiler embryos has been shown to increase 3.5-fold during the last three days of incubation, with the development of villi leading to associated

http://dx.doi.org/10.19103/AS.2019.0059.05

increases in maltase, sodium-glucose transporter, aminopeptidase and ATPase activities by Day 19 of incubation (Uni et al., 2003).

Changes in the morphological and physiological characteristics of the small intestine epithelium that occur as the bird matures (Konarzewski et al., 1990; Karcher and Applegate, 2008) allows for its adaptation to dietary changes (Noy and Sklan, 2001) as well as its functionality as a barrier to various external environmental factors, including microorganisms (Farhadi et al., 2003). The tight junctions in the intercellular spaces between individual enterocytes of the intestinal epithelium are established by day of hatch (Karcher and Applegate, 2008). Microbiome colonization of the intestinal tract of the chicken embryo begins as early as Day 16 of incubation (Pedroso, 2009). However, the initiation of gastric hydrochloric acid secretion by the embryo on Day 18 of incubation can influence gut microbiome inhabitation (Kabir, 2009).

The gut of a young posthatch chick can acquire a wide variety of externally (Pan and Yu, 2014) and maternally (Keller et al., 1995; Miyamoto et al., 1999) derived bacteria that may include those that are beneficial as well as pathogenic. Ding et al. (2017) have provided evidence to support the contention that part of the microorganisms contained in early chicken embryos are maternally inherited and that the amount and diversity of gut microorganisms involved in embryo and chick nutrient metabolism and adaptive responses to infection and other external factors are subsequently influenced by their environment and genetic constitution. For instance, the trans-ovarian transmission of pathogens such as *Salmonella* Enteritidis may occur (Thiagarajan et al., 1994). Furthermore, microbial contamination of the eggshell can influence the microbiota composition in the hatchling's gut (Kabir, 2009), as bacteria are capable of invading the interior of an egg by penetration through the pores of the eggshell (Berrang et al., 1999; Cook et al., 2003; Kizerwetter-Świda and Binek, 2008). Kizerwetter-Świda and Binek (2008) have established the primary presence of *Enterococcus* cocci in the liver, yolk sac and caecal contents of chicken embryos at 18 and 20 days of incubation, and in newly hatched chicks. The presence of other bacteria that were observed in some of those tissue samples included *Staphylococcus* sp., *Escherichia coli*, *Clostridium tertium*, *Klebsiella* sp. and *Enterobacter* sp. The authors surmised that pathogenic bacterial infections can occur during the embryonic period, with distinct infections in chicks at hatch. Also, more specifically, as embryonic age progresses, gut microbiota profiles can become more differentiated and the numbers of certain groups of bacteria can become more predominant.

McReynolds et al. (2000) have observed that feed restriction for 6 h improved the establishment of competitive exclusion culture-associated bacteria in neonatal chicks after they were administered by oral gavage. The dense population of microbiota that eventually inhabit and interact within the gastrointestinal tract of poultry are further affected by the diet of the host, and in turn can influence the maturation and function of the immune and digestive systems of their host. Cisek and Binek (2014) have stated that microorganisms can interact directly with the lining of the gut, which may then affect the physiology and immunological condition of the bird. The bird's immune response to its microbiome is strictly regulated, with anti-inflammatory mechanisms in place that allow it to tolerate normal commensal bacterial populations (Brisbin et al., 2008). Nevertheless, the function of the digestive system can more specifically be affected by the influence of the gut microbiome on intestinal morphology, including villus height and crypt depth (Pan and Yu, 2014). Sears (2000) has shown that enteric pathogens and their virulence factors can alter the function of intestinal epithelial tight junctions as a barrier to pathogenic microorganisms.

Although the rate of intestinal development in the chicken embryo varies with incubation temperature (Southwell, 2006), the most rapid development of the intestine commonly occurs within several days before and after hatch (Iji et al., 2001), with prime adjustments in the positioning and expression of brush-border proteins occurring in the small intestine that help prepare the emerging chick in its transition from a primarily lipid- to a carbohydrate- and protein-based diet (Uni et al., 2003). However, chicks commonly experience a delay in being provided external nutrition from feed through the first few days after hatch, which occurs in conjunction with the exhaustion of their source of nutrients from internal egg reserves (Batal and Parsons, 2002; Shinde Tamboli et al., 2017). Collectively, these factors that impact the period during which the bird undergoes a nutritional and subsequent metabolic transition have made the *in ovo* delivery of various vaccines, supplements and various other biologics more attractive to poultry producers (Peebles, 2018).

2 *In ovo* use of biologics to shape the gut microbiome

Automated multi-egg *in ovo* injection machines are currently used in over 90% of broiler hatcheries in the United States, and have the capability of injecting between 25 000 and 62 000 eggs per hour (Peebles, 2018). The *in ovo* injection of 50 µL solution volumes is routinely practiced commercially during egg transfer from setter to hatcher between 17.5 and 19.2 days of incubation (Ricks et al., 1999; Williams, 2011; Peebles, 2018). The most effective site of delivery is in the amnion or body of the embryo, excluding the eyes and head (Wakenell et al., 2002), which can normally be achieved by injecting at an approximate 2.49 cm depth through the air cell and underlying membranes at the large end of the broiler hatching egg (Keralapurath et al., 2010). In this manner, compounds can be injected without traumatizing the embryo and chick handling and labour costs can be reduced (Johnston et al., 1997). Upon the *in ovo* delivery of biologics into the amnion, they are afterward imbibed by the embryo, which results in their subsequent absorption on the mucosae of the digestive and respiratory tracts (Williams, 2011). A partial absorption of proteins from an albumen and amniotic fluid mixture transpires by enterocyte activity in the duodenum and jejunum (Moran, 2007).

Using immunocytochemical methods, Jochemsen and Jeurissen (2002) determined the localization of *in ovo*-injected fluorescent microspheres (40 nm) and the live infectious bursal disease virus in various organs, including the lungs and intestines, of chicken embryos. Upon manual injection of 0.1 mL solution volumes into the amnion at 18 days of incubation using a 23-gauge needle, fluorescent microspheres were detected in the lungs and intestines of embryos 24 h and 48 h after having been administered. Live virus was also observed in the lungs 24 h and 48 h after injection and in the intestines 48 h after injection. The results confirmed that the introduction of substances into the amnion at 18 days of incubation will subsequently enter the mouth of the embryo for eventual aspiration into the respiratory tract and ingestion into the intestinal tract. Therefore, it can be concluded that the localization of biologics into the respiratory and digestive tracts of chicken embryos can be effectively achieved by their *in ovo* injection into the amnion on Day 18 of incubation.

The early provision of feed to the growing chick is known to enhance growth, and it is proposed that that this effect is due to nutrient stimulation of intestinal growth, with an associated increase in intestinal absorptive surface area. Nutrients may likewise stimulate intestinal peristalsis and the subsequent increased utilization of yolk (Noy and Sklan, 1998). Therefore, the very early provision of biologics to developing embryos should advance their gut development or immune status, allowing for an enhanced maturation and functionality of their intestine, as well as their overall health, prior to and after hatch.

It is critical to establish a gut microbiome in growing chicks that provides them with protection from various invasive pathogens that can disrupt their viability and growth. This has become a challenge of increasing proportions resulting from the progressive elimination of antibiotics in poultry diets. Although not currently banned in the United States, as of 1 January 2006, approval for the use of antibiotics as growth promoters was withdrawn in the European Union (Castanon, 2007). Antibiotics have been used to protect birds from disease by altering the gut microbiome, including a reduction in pathogenic bacteria, without negatively affecting growth (Dibner and Richards, 2005; Gao et al., 2017). Nevertheless, a minimization of changes in intestinal microbiota in response to antibiotics can be associated with a delay in intestinal maturation (Bedford, 2000; Gao et al., 2017). Therefore, efforts have been made to establish substitutive measures to protect poultry against pathogens in antibiotic-free operations (Peebles, 2018). In this endeavour, increasing numbers of studies have been conducted to test the effectiveness of the *in ovo* administration of competitive exclusion cultures (Seuna, 1979; Bielke et al., 2003), probiotics (Edens et al., 1997; Madej and Bednarczyk, 2016; Triplett et al., 2018), prebiotics (Rastall and Gibson, 2015; Sobolewska et al., 2017), synbiotics (combination of probiotics and prebiotics) (Maiorano et al., 2012; Pruszynska-Oszmalek et al., 2015; Madej and Bednarczyk, 2016) or immune lymphokines (from culture supernatants of concanavalin A-stimulated T cells derived from *S.* Enteritidis-immune hens) (McGruder et al., 1995), in providing the bird a defensive mechanism against pathogenic bacterial infections.

Xu et al. (2003) have suggested that the morphological alterations witnessed in the intestines of birds are not a direct effect of these supplements, but rather an indirect effect through changes in the profile of the gut microbiome. It has also been suggested that providing chicks with the earliest possible exposure to competitive exclusion cultures or probiotics before their subjection to pathogenic bacteria would afford them the most effective protection (Seuna, 1979; Edens et al., 1997). It has been further stressed that in addition to exerting beneficial effects in their host, probiotics must also remain viable under usual storage conditions and be acceptable for industrial application (Kabir, 2009).

Bacteria that have thus far proven to be satisfactory candidates as probiotics for *in ovo* administration include *Lactococcus reuteri*, *Lactococcus lactis*, *Enterococcus faecium*, *Bifidobacterium animalis* and *Lactobacillus acidophilus* (Edens et al., 1997; de Oliveira et al., 2014; Madej and Bednarczyk, 2016; Triplett et al., 2018). These beneficial bacteria can alter the gut microbiota of birds (Schleifer, 1985; Edens et al., 1997). In addition, the components of the immune system of birds that have been shown to be stimulated by these beneficial bacteria upon *in ovo* delivery include lymphocyte infiltration of the gut-associated lymphoid tissue (Madej and Bednarczyk, 2016), the production and the development of various immune organs (Sławińska et al., 2014) and immunoglobulin synthesis (Madej et al., 2015; Roto et al., 2016). Nevertheless, the effectiveness of probiotics and prebiotics in altering the immune system and intestinal microbiota is influenced by the environmental conditions to which the bird is exposed to and the level of stress it experiences (Patterson and Burkholder, 2003).

Specific information from various research studies regarding the administration of various biologics employed for microbiome development in the chicken gut using *in ovo* technology and their impact on later gut function will be discussed in more detail in the following sections.

3 Competitive exclusion cultures

Competitive exclusion cultures may contain multiple species and strains of defined or undefined non-pathogenic bacteria that are derived from a natural source, such as the caeca or intestines of other birds. In the poultry intestines, the establishment of competitive exclusion cultures has been improved and subsequent pathogen colonization inhibited by the administration of competitive exclusion cultures to hatchlings via oral gavage (Lloyd et al., 1977; Pivnick et al., 1981; McReynolds et al., 2000). Upon including a multispecies probiotic culture of bacteria isolated from chicken intestines (*E. faecium*, *B. animalis*, *Pediococcus acidilactici*, *Lactobacillus salivarius* and *L. reuteri*) in the diets of broilers, Mountzouris et al. (2010) observed a decrease in the population of coliform bacteria in their caeca. Ghareeb et al. (2012) likewise decreased the caecal population of *Campylobacter jejuni* in broilers with the dietary use of a multispecies probiotic culture containing *E. faecium*, *P. acidilactici*, *L. salivarius* and *L. reuteri*.

Cox et al. (1992) tested the potential protective effects of the *in ovo* injection of an anaerobically grown competitive exclusion culture against *Salmonella* infections in broilers. The undefined culture, derived from the caeca of healthy adult chickens, was manually injected into the air cells of broiler hatching eggs using a 22-gauge hypodermic needle at 18 days of incubation. After the injection of 100 μL volumes of 1:1000 and 1:1000000 dilutions of the culture, the broiler hatchlings from those eggs were subsequently challenged by oral gavage with 10^3, 10^5 or 10^7 colony forming units (cfu) of a nalidixic-acid-resistant strain of *Salmonella* Typhimurium. Hatchability of the experimental birds approximated commercially acceptable levels, indicating that the culture was not detrimental to their late embryological development. Furthermore, improved resistance to the *S.* Typhimurium challenge in chicks that received the culture treatment suggested that the *in ovo* application of competitive exclusion cultures may be used to pragmatically provide broiler chicks protection against *Salmonella* infections.

Meijerhof and Hulet (1997) also tested for the potential protective effects of *in ovo* applied competitive exclusion cultures against *Salmonella* infections in broilers. The dry cultures used contained an assortment of several species of selected bacteria. Culture suspensions (1 mg dry culture/0.2 mL fluid) were manually injected into the air cell of the egg or body of the broiler embryo at 18 days of incubation using a 20-gauge needle. Only air cell injections resulted in acceptable hatchability levels. However, it was suggested that the reductions in hatchability that were observed were related to increased bacterial penetration through the membrane underlying the air cell. Chicks that hatched from eggs that received culture injections in their air cells likewise experienced increased bacterial contamination, which resulted in increased mortality and associated reduced yolk uptake, especially during the first week of posthatch life. Furthermore, those chicks also had higher levels of *Salmonella* contamination at four weeks of posthatch age in comparison to those that received culture by oral gavage at hatch. In additional testing for the culture's ability to provide ultimate protection against the pathogen, broilers were also subsequently challenged by oral gavage with 10^4 cfu of *Salmonella* Panama at hatch. In comparison to birds that received culture by oral gavage at hatch, *Salmonella*-positive incidence at four weeks of age was greater in birds that were given culture by *in ovo* injection, indicating that oral gavage provided better protection than did *in ovo* injection against a subsequent *Salmonella* infection.

With the objective of determining the effects of the *in ovo* administration of a probiotic competitive exclusion culture on the intestinal microbiota of chicks from two different genetic lineages of broilers, Pedroso et al. (2016) manually injected a 0.1 mL volume of a solution, containing from 3.3×10^5 to 2.6×10^8 viable bacteria derived from the microbiota of adult birds, into the amnion of Cobb 500 (modern) and Athens Canadian Random Bred (heritage) broiler hatching eggs at 18 days of incubation. The development and diversity of microbiota in the caecal contents of chicks from both lineages at 12 h after hatch were enhanced by the *in ovo* injected culture; however, a decrease in the prevalence of the undesirable species of bacteria was only achieved in chicks belonging to the heritage line.

4 Probiotics

Probiotics are specifically identified live microbials that have commonly been used as feed supplements to protect poultry from intestinal infections from pathogenic bacteria (Dahiya et al., 2005; Kabir, 2009; Pan and Yu, 2014). In a previous review article, Kabir (2009) has included *Lactobacillus*, *Streptococcus*, *Bacillus*, *Bifidobacterium*, *Enterococcus*, *Aspergillus*, *Candida* and *Saccharomyces* as probiotic species that have been used in poultry nutrition. Cisek and Binek (2014) have also added *E. coli* and *Lactococcus* to that list as microbial species that have been used as probiotics. In addition to enhancing their host's immune function (Ng et al., 2009; Yang et al., 2012), their activities against pathogens can involve the production of bacteriocins (Murry et al., 2004), neutralization of enterotoxins (Knap et al., 2010), counteraction of dysbiosis (Cisek and Binek, 2014), maintenance of intestinal homeostasis (Brisbin et al., 2008), alteration of host and bacterial digestive enzyme activity and subsequent host metabolism (Jin et al., 2000), and competition for nutrients and mucosal attachment sites in the gut (Lan et al., 2005; Lawley and Walker, 2013). Probiotics have been found to increase antibody production in association with altered spleen and bursa weights in broilers subjected to a sheep red blood cell challenge (Kabir et al., 2004).

Probiotic bacteria may interact with intestinal tissue to result in a rearrangement of epithelial cells and to enhance the bacteria's ability to populate the mucosal lining (Sansonetti, 2001; Patterson and Burkholder,

2003). When used as dietary supplements, probiotics have been observed to increase duodenal villi height and the ileal villus height:crypt depth ratio of broilers (Chae et al., 2012). The probiotic species that led to those effects included *L. acidophilus*, *Bacillus subtilis* and *Saccharomyces cerevisiae*. The dietary inclusion of *L. salivarius* (Pascual et al., 1999), *L. reuteri* (Nakphaichit et al., 2011) and *B. subtilis* (Molnár et al., 2011) have been shown to decrease populations of unfavourable bacteria such as *Salmonella* and *Campylobacter* in the chicken intestines. However, it has been suggested that a mixture of *L. acidophilus* and *Lactobacillus fermentum* may better allow for the exclusion of *S.* Typhimurium and *E. coli* from the gastrointestinal tracts of chickens than a single species of *Lactobacillus* (Cisek and Binek, 2014). In contrast to the aforementioned reports, Ribeiro et al. (2007) failed to observe an influence of the dietary use of the probiotic Lac XCL 5X™ on the mortality, performance or intestinal micrometry of Ross 308 broilers. The production of antibodies against a subsequent *S.* Enteritidis challenge were also not influenced by the dietary use of the probiotic, although caecal *S.* Enteritidis colonization was diminished. The use of yeast cultures as a feed supplement has led to shallower intestine epithelium crypt depths in broilers with an associated lower allocation of energy needed for the support of epithelial tissue maintenance (Gao et al., 2008). In addition, Singh et al. (2009) used yeast cultures in combination with the probiotics *L. acidophilus* and *Streptococcus faecium* to enhance the growth performance of broilers, and Shareef and Al-Dabbagh (2009) used *S. cerevisiae* as a supplementary probiotic in the diets of broilers to increase their feed efficiency and body weight gain.

Results of further tests concerning the effects of various probiotics administered by *in ovo* injection have been reported in various articles. Although effects of the *in ovo* injection of probiotics on yolk sac membrane function in chickens are lacking in the literature, an appreciable number of studies have been conducted to investigate their effects on the gut microbiome and its function. In an early probiotic research article targeting the intestines of poultry, Edens et al. (1997) have advised caution concerning the *in ovo* use of competitive exclusion cultures. They have warned against their *in ovo* use because of the lack of specific identification of a particular protective strain or strains in the culture, and the possible inclusion of harmful bacteria having proteolytic activity and gas- and toxin-producing capabilities. These researchers have also asserted that of the potential competitive exclusion agents, only *L. reuteri* has been shown to not adversely affect hatchability and to display safe and effective posthatch effects by decreasing *Salmonella* and *E. coli* colonization in chicks and poults, while increasing the villus height and crypt depth of intestinal mucosa. A series of studies on chicks and poults were conducted by these investigators that involved the injection of 10^6 or 10^7 *L. reuteri* cells into the air cell, amnion or directly into the body proper of the embryo on Days 18 and 19 of incubation. Although not specified, it is assumed that the *in ovo* injections were administered manually, in a manner similar to that of Cox et al. (1992). The *in ovo* use of 0.2 mg of the antibiotic Gentamicin® and the response to a subsequent *S.* Typhimurium challenge (10^3 cfu per bird) by oral gavage at hatch were likewise tested. Edens et al. (1997) concluded that the *in ovo* use of *L. reuteri* alone or in combination with Gentamicin®, followed by supplementation of *L. reuteri* in the feed or drinking water, can potentially control enteric pathogen colonization in poultry.

Nevertheless, the effectiveness of the *in ovo* application of *L. acidophilus*, *L. fermentum* or *L. salivarius* in conferring protection against *S.* Enteritidis colonization in the caeca of chicks was evaluated by Yamawaki et al. (2013). A 0.1 mL volume of the probiotic bacteria were separately inoculated into the air cells of Cobb 500 broiler hatching eggs at 18 days of incubation. The hatchlings were later challenged by oral gavage on Day 2 posthatch with a 0.5 mL volume of inoculum containing 8.3×10^8 cfu/mL of *S.* Enteritidis. On Day 5 posthatch, there was no decrease in *S.* Enteritidis colonization in the caeca of the chicks in response to the *in ovo* administration of any of the three probiotics, suggesting that when provided by air cell injection, these probiotics do not offer any potential benefit in inhibiting *S.* Enteritidis colonization of the caeca in broiler chicks.

de Oliveira et al. (2014) used *B. subtilis* and *E. faecium* as probiotic bacteria to test for their effectiveness in protecting Ross 308 broilers against *S.* Enteritidis infection. A 500 µL volume of either *B. subtilis* (16×10^9 cfu/egg) or *E. faecium* (5×10^9 cfu/egg) was manually injected into the amnion of eggs with a 23-gauge needle at 17.5 days of incubation. The *in ovo* injections were administered alone or in combination with the dietary inclusion of the probiotics during an 18-day grow-out period. Chicks were later challenged by oral gavage on Day 4 posthatch with 0.4 mL of 10^6 cfu of *S.* Enteritidis. The *in ovo* or dietary provision of the probiotics only partially recovered chick performance when compared to chicks fed antibiotics. Nevertheless, upon evaluation at the end of the rearing period, it was observed that the *in ovo* injection and subsequent dietary inclusion of *E. faecium* reduced the number of *S.* Enteritidis-positive chicks, indicating that the *in ovo* colonization of probiotics has the potential to be effectively used as an integral countermeasure against pathogenic bacterial infections.

In more recent work, Teague et al. (2017) experimented with the use of a defined probiotic culture containing lactic acid bacteria (FloraMax®-B11), that has been shown to enhance the development of normal intestinal microbiota in poultry. On Day 18 of incubation, the amnions of embryonated eggs were manually injected with 10^4 cfu of FloraMax®-B11. Hatchlings subsequently received a 10^4 cfu oral gavage of *S.* Enteritidis. In the absence of any adverse effects on hatchability, the probiotic reduced lactose-positive Gram-negative bacterial levels in the gut and the incidence of caecal tonsil *S.* Enteritidis infection. In addition, it increased bird BW in association with a greater ileum villi surface area. It was concluded that FloraMax®-B11 not only allowed for a normal hatching process, but also improved broiler BW and decreased *S.* Enteritidis recovery during the first seven days of broiler posthatch life.

Also in a recent study, Pender et al. (2017) tested the effects of the *in ovo* supplementation of a commercial probiotic product (Primalac W/S), that contained *L. acidophilus*, *Lactobacillus casei*, *E. faecium* and *Bifidobacterium bifidum*, on the performance and immune-related gene expression of Cobb 500 broiler chicks. A 100 μL volume of probiotic product was manually injected into the amnion on Day 18 of incubation. On Day 6 posthatch, 1×10^7 bacteria decreased Toll-like receptor-4 and interferon-γ gene expression in the ileum. In the caecal tonsils, 1×10^6 bacteria decreased Toll-like receptor-2 gene expression, whereas 1×10^5 and 1×10^6 bacteria decreased Toll-like receptor-4 gene expression. At that same posthatch period, 1×10^5 bacteria reduced ileal interleukin-3 and interleukin-4 gene expression. Nevertheless, in association with the modulatory effects of Primalac W/S on gene expression in the ileum and caecal tonsils, it improved broiler performance during the first week of posthatch life. In a review article by Cox and Dalloul (2015), in which the results of the above work conducted in broilers using Primalac W/S were discussed, it was concluded that the main effect of probiotics in a non-challenge context is the downregulation of host immune-related gene expression in the intestine. Furthermore, when the broilers in that work were subjected to a posthatch *Eimeria* challenge, the *in ovo* administration of the probiotic product reduced mortality and gross intestinal lesion severity, suggesting that the *in ovo* use of the product may also help diminish the negative effects of coccidiosis.

Similar to de Oliveira et al. (2014), Triplett et al. (2018) also experimented with the use of the probiotic, *B. subtilis*, but also tested *L. acidophilus* and *B. animalis* as other potential probiotic species for use in automated commercial *in ovo* injection. A 50 μL volume of commercial diluent containing 10^3, 10^4, 10^5 or 10^6 cfu of one of the three bacterial species was injected at 18 days of incubation into the amnion of Ross 708 broiler hatching eggs. The hatchability of fertile eggs was significantly reduced by *B. subtilis*, with the occurrence of progressive decreases as the concentration of *B. subtilis* increased. Hatchability reached a nadir of 1.67% in response to *B. subtilis* at 10^5 cfu, but conversely was not affected by the injection of up to 10^6 cfu of *L. acidophilus* or *B. animalis*. Likewise, 10^6 cfu of *L. acidophilus* had no adverse effect on hatch residue analysis results, whereas 10^5 and 10^6 cfu of *B. animalis* increased dead pipped chick and contaminated egg percentages. Triplett et al. (2018), therefore, concluded that *L. acidophilus* and *B. animalis,* but not *B. subtilis*, may be suitable probiotics for *in ovo* use in broiler hatching eggs. Further research will be necessary in order to ascertain the impact of *L. acidophilus* and *B. animalis* on the microbiome profile and function of the gut in broilers subsequent to their administration in accordance with the procedures described by Triplett et al. (2018).

5 Prebiotics

Prebiotics are nutritive substrates, primarily polysaccharides, that are indigestible to the host and are used to provide a substrate for select beneficial bacteria, which would provide them a proliferative advantage over harmful bacteria in the intestine of their host (Pan and Yu, 2014; Roto et al., 2015). Considerable research has been conducted on the potential efficacy of yeast metabolites as prebiotic-like compounds as supplements in animal feeds. However, additional work is needed to establish the efficacy of the use of these compounds in poultry (Roto et al., 2015). Patterson and Burkholder (2003) have provided a listing of the various prebiotics that have been investigated in the literature for use in poultry. These include fructo-oligosaccharides, inulin-type fructo-oligosaccharides, transgalactooligosaccharides, lactulose, lactitol, maltooligosaccharides, xylo-oligosaccharides, mannan-oligosaccharides, stachyose, raffinose and sucrose thermal oligosaccharides. In the intestines of broilers, they have been observed to favour the growth of strains of the beneficial probiotic bacteria *Bifidobacterium* and *Lactobacillus* (Xu et al., 2003; Jung et al., 2008), and to decrease populations of the pathogenic bacteria *Clostridium perfringens* and *E. coli* (Xu et al., 2003; Kim et al., 2011). They have also decreased caecal populations of *S.* Enteritidis in layers (Donalson et al., 2008) and *S. Typhimurium* in broilers (Spring et al., 2000). Mannan-oligosaccharides are also included in the list of prebiotics, but unlike the others above, do not enhance beneficial bacterial numbers.

Upon their dietary inclusion, prebiotics have been reported to increase the height of villi and the villus height:crypt depth ratios in the intestines of chickens (Sonmez and Eren, 1999; Xu et al., 2003). It has been noted that while inducing changes in intestinal microbiome composition in favour of beneficial over pathogenic bacteria, diets containing prebiotics, such as fructo-oligosaccharides and galactooligosaccharides, can also promote the activity of various digestive enzymes in association with increased histomorphological development (Xu et al., 2003; Jung et al., 2008; Pan and Yu, 2014). Conversely, dietary use of the prebiotic Bio Mos™ had no effect on Ross 308 broiler livability, performance or intestinal morphometry (Ribeiro et al., 2007). Furthermore, although *S.* Enteritidis counts were reduced by the prebiotic after an *S.* Enteritidis challenge, the antibody response to the challenge was not affected by the use of the prebiotic. However, like probiotics, it has not been until more recently that prebiotics have been employed through *in ovo* application. The following are various types of these studies that have been conducted to test for their efficacy when administered in different locations within the egg.

Bednarczyk et al. (2016) compared the efficacies of the prebiotics DiNovo® [a commercial prebiotic that is an extract of beta-glucans from the *Laminaria* species of seaweed, and contains laminarin (exhibits immunomodulatory activities) and fucoidan (possesses antiviral and antibacterial properties)], Bi²tos (non-digestible transgalactooligosaccharide from milk lactose) and RFO (raffinose family oligosaccharides; an extract of the seeds of *Lupinus luteus*), when provided by means of *in ovo* inoculation, drinking water or by the combination of both. The *in ovo* injected prebiotics were dissolved in 0.2 mL of physiological saline in graded doses, and were delivered on Day 12 of incubation. Birds in the in-water treatment received 12 mL of the prebiotics that were dissolved in water (20 mg of prebiotic/mL) during the first seven days of posthatch life. The optimal doses of the DiNovo® and Bi²tos prebiotics were determined to be 0.88 mg and 3.5 mg per embryo, respectively. Furthermore, all three prebiotics increased body weight gain, feed intake and feed conversion ratio during the first 21 days posthatch, regardless of mode of administration. However, because there was no synergistic interaction between in-water and *in ovo* supplementation, it was concluded that *in ovo* administration of the prebiotics may be used to effectively replace posthatch in-water supplementation.

When inoculated *in ovo* by manual injection on Day 12 of incubation, the RFO prebiotic α-galactoside has been used to define the microbiome profile in broiler hatchlings (Villaluenga et al., 2004; Pilarski et al., 2005). Villaluenga et al. (2004) injected 0.2 mL of Ringer's solution containing the oligosaccharide into the air cells of eggs, and subsequently found higher levels of bifidobacteria in the colons of those broilers at 2 days posthatch, indicating that oligosaccharides have potential for inclusion in the development of substitutes for antibiotics in poultry. Using the same methods employed by Villaluenga et al. (2004), Pilarski et al. (2005) established that the *in ovo* administration of oligosaccharide preparations can further lead to long-term (through six weeks posthatch) perpetuation of high caecal levels of bifidobacteria in broilers, and further suggested that the *in ovo* injection of very low doses of these prebiotics can be used as an effective substitute for their use as a supplement in broiler diets.

The same RFO used by Villaluenga et al. (2004) and Pilarski et al. (2005) were tested by Bednarczyk et al. (2011), Maiorano et al. (2012) and Sławińska et al. (2014). Bednarczyk et al. (2011) established a positive *in vitro* effect on *B. bifidum* numbers and the benefits of their *in ovo* application on Day 12 of incubation for chicken posthatch development. Maiorano et al. (2012) showed that the *in ovo*-delivered prebiotic had no effect on the histopathology, pH or abdominal fat and cholesterol contents of broiler pectoral muscle. Sławińska et al. (2014) later more specifically demonstrated that *in ovo* use of the prebiotic alone increased relative bursal weight and the bursa to spleen index of Ross 308 broilers in comparison to a synbiotic containing lactose, *L. acidophilus* and *S. faecium*.

More recently, the *in ovo* use of the prebiotics, inulin (1.76 mg) and Bi^{2}tos (0.528 mg), were tested for their effects on the posthatch morphology of the central and peripheral lymphoid organs (Madej et al., 2015), the presence of immune cells in the gut-associated lymphoid tissue (Madej and Bednarczyk, 2016), the digestive potency of the pancreas (Pruszynska-Oszmalek et al., 2015) and the long-term transcriptomic effects in the spleen, caecal tonsils and large intestines (Sławińska et al., 2016) of Ross 308 broilers. The prebiotics (0.2 mL solutions) were injected into the air cells of the eggs containing live embryos on Day 12 of incubation using a dedicated automatic system. The *in ovo* administration of both prebiotics had no adverse effects on the development of the immune system, including the relative weights and histological structures and morphologies of the bursa of Fabricius, thymus and spleen through Day 35 posthatch. The prebiotics also exerted no adverse effects on the gut-associated lymphoid tissue or the health statuses of the liver and pancreas of the birds, but rather resulted in an increased activity of pancreatic trypsin one and two weeks after hatch, and an increased rate of B-cell colonization in their peripheral lymphoid organs and by T-cells in their caecal tonsils one week after hatch. Gut-associated lymphoid tissue is responsible for creating mucosal immune responses (Brisbin et al., 2008), thereby advocating that inulin and Bi^{2}tos were effective in stimulating specific immune cell function in the gut-associated lymphoid tissue of broilers. Furthermore, *in ovo* application of Bi^{2}tos was observed to be the most potent stimulator of the transcriptome of gut-associated lymphoid tissue in the broilers. It has been deduced that the basis for the stimulation of specific immune cell function, particularly in the caecal tonsils, may be due to its precipitation of native gut microbiota in the embryo and an associated indirect regulation of gene expression.

Effects of the *in ovo* injection of DiNovo® (0.88 mg dissolved in 0.2 mL of physiological saline) on broiler performance and duodenal histomorphology were determined by Sobolewska et al. (2017). Using a dedicated automatic system on Day 12 of incubation, the prebiotic was injected into the air cells of Ross 308 broiler hatching eggs and the holes in their shells were subsequently sealed. Without having any significant effect on bird body weight, feed conversion ratio, European broiler index or mortality on Day 42 posthatch, differential treatment effects were observed for several duodenal histomorphological variables on Days 21 and 42. Duodenum weight, length, diameter, cross-sectional area and muscular layer thickness were not significantly affected by treatment at either rearing age. However, in comparison to saline-injected controls, DiNovo® increased villi width and crypt depth on Day 21, and decreased villus height, width, surface area and crypt depth on Day 42. It was concluded that air cell injection of the DiNovo® prebiotic influenced various histomorphological features of the duodenum in Ross 308 broilers on Days 21 and 42 posthatch without influencing their growth performance. In addition, the influence of the *in ovo* injection of DiNovo® on villus width and crypt depth observed on Day 42 were opposite to those observed on Day 21.

6 Synbiotics

Synbiotics are products that contain both probiotics and prebiotics (Roto et al., 2016). However, these products also more specifically involve a synergism between the two, where the prebiotics selectively favour the probiotics. Ribeiro et al. (2007) have reported that the dietary use of a synbiotic containing a combination of the prebiotic Bio Mos™ and the probiotic Lac XCL 5X™ did not produce any caecal and liver colonization or immune response results in Ross 308 broilers that were significantly different from those resulting from the individual use of each of the additives. However, research concerning the dietary inclusion of a synbiotic (Biomin IMBO) containing *E. faecium* and a prebiotic derived from chicory showed that it could be used to promote growth and improve gut health in broilers (Awad et al., 2009). The efficacy of synbiotics administered by *in ovo* injection has also been tested in subsequent studies.

Influences of the *in ovo* injection of synbiotics on the meat quality (Maiorano et al., 2012) and development and structure of the immune organs (Sławińska et al., 2014) in Ross 308 broilers have been explored. Effects of solutions (0.2 mL) containing 10 μL of a suspension (10^3 cfu) of the specific lactic acid bacterium, *L. lactis*, and 190 μL of an RFO prebiotic were tested. The synbiotic was manually injected into the air cells of the eggs on Day 12 of incubation. *In ovo* administration of the synbiotic was shown to increase feed conversion ratio without affecting pectoral muscle quality traits or histopathology (Maiorano et al., 2012), but was shown to stimulate immune organ development (Sławińska et al., 2014). Specifically, thymocyte presence in the cortex of the thymus on Days 21 and 42 posthatch were increased by the *in ovo* use of the synbiotic. Conversely, in comparison to a different synbiotic containing lactose, *L. acidophilus* and *S. faecium*, the synbiotic increased carcass yield (Maiorano et al., 2012), relative weight of the bursa of Fabricius (Sławińska et al., 2014) and bursa/spleen ratio (Sławińska et al., 2014) on Day 42 posthatch.

Using the same procedures described earlier by Madej et al. (2015) and Madej and Bednarczyk (2016), further tests were conducted in Ross 308 broilers using synbiotics (0.2 mL) that consisted of *L. lactis* subsp. *lactis* IBB SL1 in combination with inulin or *L. lactis* subsp. *cremoris* IBB SC1 in combination with Bi²tos. The prebiotic (inulin or Bi²tos) solution volumes were 180 μL and there were 1000 cfu of either bacteria suspended in 20 μL volumes of prebiotic solution. The specific posthatch effects of the synbiotics delivered by *in ovo* injection on the central and peripheral lymphoid organ morphologies (Madej et al., 2015), immune cell compositions of the gut-associated lymphoid tissues (Madej and Bednarczyk, 2016), the degree of influence of the pancreas on digestion (Pruszynska-Oszmalek et al., 2015) and the body weight, caecal fermentation and intestinal morphology (Miśta et al., 2016) of the birds were examined. Both combinations injected into the air cells of the eggs on Day 12 of incubation exhibited no detrimental effects on the development of the immune systems or the absolute and relative pancreatic weights of the broilers. Increases in the body weight, short-chain fatty acid caecal profile and villus length:crypt depth ratio in the mucosa of the jejunum of the birds in response to both synbiotic combinations were most pronounced in the inulin and *L. lactis* subsp. *lactis* IBB SL1 combination treatment group. However, the Bi^{2}tos and *L. lactis* subsp. *cremoris* IBB SC1 combination displayed modulatory effects on the central and peripheral lymphatic organs that included decreases in the cortex/medulla ratio in the thymus and the development of the cortex region in the follicles of the bursa of Fabricius on Day 21 posthatch. That combination also promoted germinal centre formation in the spleen on Days 21 and 35, as well as an age-dependent increase in the spleen/bursa of Fabricius ratio. The two synbiotic combinations demonstrated no unfavourable effects on gut-associated lymphoid tissue development, but rather stimulated T- and B-cell colonization in caecal tonsils on Day 7 posthatch. Those combinations also stimulated pancreatic amylase and trypsin activities on Days 1, 7, 14, 21 and 34 posthatch, and increased T-cell numbers in the ileum and the rate of B-cell colonization of the caecal tonsils on Day 21 posthatch. The Bi^{2}tos and *L. lactis* subsp. *cremoris* IBB SC1 combination also stimulated pancreatic lipase activity on Days 3, 7, 21 and 34 posthatch, and increased the rate of establishment of B cells in the peripheral lymphoid organs. Besides modulating the central and peripheral lymphatic organ development of broilers, it was surmised that the *in ovo* administration of these synbiotics was effective in stimulating the development of their gut-associated lymphoid tissue. It was also suggested that *in ovo* use of the synbiotics were more effective than the singular use of either prebiotic. This contention is supported by the finding of Sławińska et al. (2016), that a combination of *L. lactis* subsp. *lactis* IBB2955 and inulin were involved in regulating the expression of two gut-associated lymphoid tissue genes.

7 Other biologics

Using an immune lymphokine prepared from culture supernatants of concanavalin A-stimulated T cells derived from *S.* Enteritidis-immune White Leghorn hens (McGruder et al., 1993), McGruder et al. (1995) experimented with its *in ovo* use as a potential protective measure for neonatal White Leghorn chicks against *S.* Enteritidis infection. The immune lymphokine preparation was manually injected into the amnion on Day 18 of incubation using a tuberculin syringe and a 25-gauge needle. Chicks were subsequently challenged by oral gavage with 5×10^4 cfu of *S.* Enteritidis bacteria. Although gut morphology was not investigated, it was noted that peripheral blood heterophil bactericidal activity was increased and *S.* Enteritidis invasion of the liver and spleen was reduced by the immune lymphokine.

Although their specific effects on the gut microbiome were not determined, Uni and Ferket (2004) have established that the *in ovo* feeding of solutions containing carbohydrates and protein markedly augmented the enteric development of broiler chicks. This was demonstrated by findings that the gastrointestinal tracts of *in ovo*-fed embryos exhibited enhanced intestinal development by 48 h after *in ovo* feeding, and that of hatchlings was comparable in development to that of conventionally fed chicks at two days posthatch. Furthermore, jejunum villi height in embryos at 19, 20 and 21 (hatch) days of incubation was significantly increased in response to *in ovo* feeding. These results would suggest that the *in ovo* feeding of these nutrients was likewise favourable to the colonization of beneficial bacteria as well as the overall microbiome environment in the gut. Kadam et al. (2013) have advocated that *in ovo* feeding may provide the bird with an excellent start towards an appropriate response to the introduction of enteric antigens. Enhanced intestinal development and function in poultry have also been reported in response to the *in ovo* administration of peptide YY (Coles et al., 1999); beta-hydroxy-beta-methylbutyrate (Tako et al., 2004); amino acids (Bhanja et al., 2004); threonine (Kadam et al., 2008); arginine, isoleucine and serine (Bakyaraj et al., 2011); and zinc, iodine or iron (Bakyaraj et al., 2011).

8 Conclusion and future trends

The *in ovo* administration of various biologics has been shown to promote the intestinal development of chicks and provide them protection against pathogenic bacterial infections. These biologics have included competitive exclusion cultures, defined probiotics, prebiotics, synbiotics and various nutrients. In the literature, certain probiotic bacteria have been reported to increase posthatch body weight, reduce pathogenic bacterial infections and increase the villi surface area in the intestines of chicks. Probiotics have likewise been observed to stimulate chick immune organ development and pancreatic enzyme activity, and to modulate immune-related gene expression. Table 1 provides a compilation of probiotic bacteria regarded as being beneficial to the health of poultry when administered by *in ovo* injection alone or in combination with a prebiotic (synbiotic).

The benefits obtained through the singular *in ovo* administration of assorted prebiotics have resulted in increased body weight gain and carcass yield, immune organ maturation, establishment of lymphatic organ T- and B-cell populations, pancreatic enzyme activity and overall gut health. However, the use of synbiotics rather than the singular use of prebiotics may be more effective. Table 2 provides a compilation of prebiotics regarded as being beneficial to the health of poultry when administered by *in ovo* injection alone or in combination with a probiotic (synbiotic).

Because the effectiveness of the *in ovo* application of these biologics is dependent upon the environmental and physiological conditions of the bird during both the embryonic and posthatch phases of its life, further research is needed to refine the treatment regimens that will optimize their effects for pragmatic commercial use.

Table 1 Probiotic bacteria regarded as being beneficial to the health of poultry when administered by *in ovo* injection and the specific physiological variables noted as being beneficially affected

Probiotic bacteria	Variable
Bifidobacterium animalis	Did not impact hatchability of fertile eggs at concentrations up to 10^6 cfu/50 µL (Triplett et al., 2018).
Enterococcus faecium	*In ovo* injection and dietary inclusion reduced *S.* Enteritidis infection (de Oliveira et al., 2014).
FloraMax®-B11[a]	Decreased gut lactose positive Gram-negative bacterial levels and caecal tonsil *S.* Enteritidis infection. Increased body weight and ileum villi surface area (Teague et al., 2017).
Lactobacillus acidophilus	Did not impact hatchability of fertile eggs at concentrations up to 10^6 cfu/50 µL (Triplett et al., 2018).
Lactococcus lactis[b]	Stimulated central and peripheral lymphoid organ development (Madej et al., 2015), and immune cell colonization in gut-associated lymphoid tissue (Madej and Bednarczyk, 2016). Increased carcass yield (Maiorano et al., 2012); body weight and jejunum villus length:crypt depth ratio (Miśta et al., 2016); and pancreatic enzyme activity (Pruszynska-Oszmalek et al., 2015). Stimulated immune organ development (Sławińska et al., 2014).
Lactococcus reuteri	Decreased *Salmonella* and *E. coli* colonization and increased villus height and crypt depth in the gut (Edens et al., 1997).
Primalac W/S[c]	Increased body weight and modulated gene expression in the ileum and caecal tonsils (Pender et al., 2017). Reduced mortality and gross intestinal lesion severity in broilers subjected to posthatch *Eimeria* challenge (Cox and Dalloul, 2015).

[a] Contains lactic acid bacteria.
[b] Used in combination with a prebiotic to form a synbiotic.
[c] Contains *Lactobacillus acidophilus*, *Lactobacillus casei*, *Enterococcus faecium* and *Bifidobacterium bifidum*.

Table 2 Prebiotics regarded as being beneficial to the health of poultry when administered by *in ovo* injection and the specific physiological variables noted as being beneficially affected

Prebiotic	Variable
DiNovo®[a]	Increased body weight gain, feed intake and feed conversion ratio (Bednarczyk et al., 2016). Had variable age-related influences on duodenum histomorphology (Sobolewska et al., 2017).
Bi²tos[b]	Increased posthatch activity of pancreatic trypsin (Pruszynska-Oszmalek et al., 2015); increased B-cell proliferation in secondary lymphatic organs (Madej et al., 2015); increased T-cell colonization in caecal tonsils (Madej and Bednarczyk, 2016); and body weight gain, feed intake and feed conversion ratio (Bednarczyk et al., 2016). Potent stimulator of gut-associated lymphoid tissue transcriptome (Sławińska et al., 2016).
Bi²tos[b,d]	Increased development of cortex region of follicles of bursa of Fabricius and germinal centre formation in the spleen, and rate of establishment of B cells in peripheral lymphatic organs (Madej et al., 2015). Stimulated pancreatic amylase, lipase and trypsin activities (Pruszynska-Oszmalek et al., 2015); stimulated T- and B-cell colonization in caecal tonsils (Madej and Bednarczyk, 2016).
RFO[c]	Increased colonic (Villaluenga et al., 2004) and caecal (Pilarski et al., 2005) *Bifidobacteria* levels. Positive *in vitro* effect on *Bifidobacterium bifidum* numbers (Bednarczyk et al., 2011); increased bursal weight and bursa to spleen ratio (Sławińska et al., 2014); increased body weight gain, feed intake and feed conversion ratio (Bednarczyk et al., 2011, 2016).
RFO[c,d]	Increased carcass yield (Maiorano et al., 2012); stimulated immune organ development (Sławińska et al., 2014).
Inulin	Increased posthatch activity of pancreatic trypsin (Pruszynska-Oszmalek et al., 2015); increased B-cell proliferation in secondary lymphatic organs (Madej et al., 2015); and T-cell colonization in caecal tonsils (Madej and Bednarczyk, 2016).
Inulin[d]	Stimulated pancreatic amylase and trypsin activities (Pruszynska-Oszmalek et al., 2015); increased body weight, short-chain fatty acid caecal profile and jejunum villus length:crypt depth ratio (Miśta et al., 2016); stimulated T- and B-cell colonization in caecal tonsils (Madej and Bednarczyk, 2016); regulated expression of two gut-associated lymphoid tissue genes (Sławińska et al., 2016).
Derivative of chicory[d]	Promoted growth and gut health (Awad et al., 2009).

[a] Extract of beta-glucans that contains laminarin and fucoidan.
[b] Non-digestible transgalactooligosaccharide from milk lactose.
[c] Raffinose family of oligosaccharides.
[d] Used in combination with a probiotic to form a synbiotic.

9 Where to look for further information

It is suggested that future research be directed towards determining the optimal use of synbiotics during the various developmental phases of commercial poultry. More specifically, the effects of the *in ovo* administration of synbiotics in neonatal poultry should be more thoroughly explored. It is recommended that potential investigators refer to the Guidelines for the Evaluation of

Probiotics in Food published by the Food and Agriculture Organization of the United Nations (FAO) and the World Health Organization for the safe use of probiotics. Furthermore, a review of the various methods employed for the *in ovo* administration of different substances in poultry is provided by Peebles (2018), and additional perspectives on the use of various materials for modifying the microbiome of the gastrointestinal tract in poultry are provided by Roto et al. (2016). Organizations such as the Poultry Science Association and the World's Poultry Science Association are involved with the advancement and dissemination of information on these topics.

10 References

Awad, W. A., Ghareeb, K., Abdel-Raheem, S. and Böhm, J. (2009), Effects of dietary inclusion of probiotic and symbiotic on growth performance, organ weights, and intestinal histomorphology of broiler chickens, *Poult. Sci.*, 88, 49-55.

Bakyaraj, S., Bhanja, S. K., Majumdar, S. and Dash, B. (2011), Modulation of posthatch growth and immunity through *in ovo* supplemented nutrients in broiler chickens, *J. Sci. Food Agric.*, 92, 313-20.

Batal, A. B. and Parsons, C. M. (2002), Effect of fasting versus feeding oasis after hatching on nutrient utilization in chicks, *Poult. Sci.*, 81, 853-9.

Bedford, M. (2000), Removal of antibiotic growth promoters from poultry diets: Implications and strategies to minimize subsequent problems, *World Poult. Sci. J.*, 56(4), 347-65.

Bednarczyk, M., Urbanowski, M., Gulewicz, P., Kasperczyk, K., Maiorano, G., et al. (2011), Field and *in vitro* study on prebiotic effect of raffinose family of oligosaccharides in chickens, *B. Vet. I. Pulawy*, 55, 465-9.

Bednarczyk, M., Stadnicka, K., Kozłowska, I., Abiuso, C., Tavaniello, S., et al. (2016), Influence of different prebiotics and mode of their administration on broiler chicken performance, *Animal*, 10(8), 1271-9.

Berrang, M. E., Frank, J. F., Buhr, R. J., Bailey, J. S. and Cox, N. A. (1999), Eggshell membrane structure and penetration by *Salmonella* Typhimurium, *J. Food Prot.*, 62, 73-6.

Bhanja, S. K., Mandal, A. B. and Johri, T. S. (2004), Standardization of injection site, needle length, embryonic age and concentration of amino acids for *in ovo* injection in broiler breeder eggs, *Indian J. Poult. Sci.*, 39, 105-11.

Bielke, L. R., Elwood, A. L., Donoghue, D. J., Donoghue, D. M., Newberry, L. A., et al. (2003), Approach for selection of individual enteric bacteria for competitive exclusion in turkey poults, *Poult. Sci.*, 82, 1378-82.

Brisbin, J. T., Gong, J. and Sharif, S. (2008), Interactions between commensal bacteria and the gut-associated immune system of the chicken, *Anim. Health Res. Rev.*, 9(1), 101-10.

Castanon, J. I. (2007), History of the use of antibiotic as growth promoters in European poultry feeds, *Poult. Sci.*, 86(11), 2466-71.

Chae, B., Ingale, S., Kim, J., Kim, K., Sen, S., et al. (2012), Effect of dietary supplementation of probiotics on performance, caecal microbiology and small intestinal morphology of broiler chickens, *Anim. Nutr. Feed Technol.*, 12, 1-12.

Cisek, A. A. and Binek, M. (2014), Chicken intestinal microbiota function with a special emphasis on the role of probiotic bacteria, *Pol. J. Vet. Sci.*, 17(2), 385-94.

Coles, B. A., Croom, W. J., Brake, J., Daniel, L. R., Christensen, V.L., et al. (1999), *In ovo* peptide YY administration improves growth and feed conversion ratios in week-old broiler chicks, *Poult. Sci.*, 78, 1320-2.

Cook, M. I., Beissinger, S. R., Toranzos, G. A., Rodriguez, R. A. and Arendt, W. J. (2003), Trans-shell infection by pathogenic micro-organisms reduces the shelf life of non-incubated bird's egg: A constraint on the onset of incubation? *Proc. R. Soc. London Ser. B*, 270, 2233-40.

Cox, C. M. and Dalloul, R. A. (2015), Immunomodulatory role of probiotics in poultry and potential *in ovo* application, *Benef. Microbes*, 68(1), 45-52.

Cox, N. A., Bailey, J. S., Blankenship, L. C. and Gildersleeve, R. P. (1992), *In ovo* administration of a competitive exclusion culture treatment to broiler embryos, *Poult. Sci.*, 71, 1781-4.

Dahiya J., Wilkie, D., van Kessel, A. and Drew, M. (2005), Potential strategies for controlling necrotic enteritis in broiler chickens in post-antibiotic era, *Anim. Feed Sci. Technol.*, 129, 60-88.

de Oliveira, J. E., van der Hoeven-Hangoor, E., van de Linde, I. B., Montijn, R. C. and van der Vossen, J. M. B. M. (2014), *In ovo* inoculation of chicken embryos with probiotic bacteria and its effect on posthatch *Salmonella* susceptibility, *Poult. Sci.*, 93, 818-29.

Dibner, J. J. and Richards, J. D. (2005), Antibiotic growth promoters in agriculture: History and mode of action, *Poult. Sci.*, 84(4), 634-43.

Ding, J., Dai, R., Yang, L., He, C., Xu, K., et al. (2017), Inheritance and establishment of gut microbiota in chickens, *Front. Microbiol.*, 8, 1967.

Donalson, L. M., McReynolds, J. L., Kim, W. K., Chalova, V. I., Woodward, C. L., et al. (2008), The influence of fructooligosaccharide prebiotic combined with alfalfa molt diets on the gastrointestinal tract fermentation, *Salmonella* Enteritidis infection, and intestinal shedding in laying hens, *Poult. Sci.*, 87, 1253-62.

Edens, F. W., Parkhurst, C. R., Casas, I. A. and Dobrogosz, W. J. (1997), Principles of *ex ovo* competitive exclusion and *in ovo* administration of *Lactobacillus reuteri*, *Poult. Sci.*, 76, 179-96.

Farhadi, A., Banan, A., Fields, J. and Keshavarzian, A. (2003), Intestinal barrier: An interface between health and disease, *J. Gastroenterol. Hepatol.*, 18, 479-97.

Gao, J., Zhang, H. J., Yu, S. H., Wu, S. G., Yoon, I., et al. (2008), Effects of yeast culture in broiler diets on performance and immunomodulatory functions, *Poult. Sci.*, 87, 1377-84.

Gao, P., Ma, C., Sun, Z., Wang, L., Huang, S., et al. (2017), Feed-additive probiotics accelerate yet antibiotics delay intestinal microbiota maturation in broiler chicken, *Microbiome*, 5(1), 91.

Ghareeb, K., Awad, W. A., Mohnl, M., Porta, R., Biarnés, M., et al. (2012), Evaluating the efficacy of an avian-specific probiotic to reduce the colonization of *Campylobacter jejuni* in broiler chickens, *Poult. Sci.*, 91, 1825-32.

Holdsworth, C. D. and Wilson, T. H. (1967), Development of active sugar and amino acid transport in the yolk sac and intestine of the chicken, *Am. J. Physiol.*, 212, 233-40.

Iji, P. A., Saki, A. and Tivey, D. R. (2001), Body and intestinal growth of broiler chicks on a commercial starter diet. I. Intestinal weight and mucosal development, *Br. Poult. Sci.*, 42, 505-13.

Jin, L. Z., Ho, Y. W., Abdullah, N. and Jalaludin, S. (2000), Digestive and bacterial enzyme activities in broilers fed diets supplemented with *Lactobacillus* cultures, *Poult. Sci.*, 79, 886-91.

Jochemsen, P. and Jeurissen, S. H. M. (2002), The localization and uptake of *in ovo* injected soluble and particulate substances in the chicken, *Poult. Sci.*, 81, 1811-17.

Johnston, P. A., Liu, H., O'Connell, T., Phelps, P., Bland, M., et al. (1997), Applications of *in ovo* technology, *Poult. Sci.*, 76, 165-78.

Jung, S. J., Houde, R., Baurhoo, B., Zhao, X. and Lee, B. H. (2008), Effects of galacto-oligosaccharides and a bifidobacteria lactis-based probiotic strain on the growth performance and fecal microflora of broiler chickens, *Poult. Sci.*, 87, 1694-9.

Kabir, S. M. L. (2009), The role of probiotics in the poultry industry, *Int. J. Mol. Sci.*, 10(8), 3531-46.

Kabir, S. M. L., Rahman, M. M., Rahman, M. B., Rahman, M. M. and Ahmed, S. U. (2004), The dynamics of probiotics on growth performance and immune response in broilers, *Int. J. Poult. Sci.*, 3, 361-4.

Kadam, D. M. M., Bhanja, S. K., Mandal, A., Thakur, R., Vasan, P., et al. (2008), Effect of *in ovo* threonine supplementation on early growth, immunological responses and digestive enzyme activities in broiler chickens, *Br. Poult. Sci.*, 49, 736-41.

Kadam, M. M., Barekatain, M. R., Bhanja, S. K. and Iji, P. A. (2013), Prospects of *in ovo* feeding and nutrient supplementation for poultry: The science and commercial applications-a review, *J. Sci. Food Agric.*, 93(15), 3654-61.

Karcher, D. M. and Applegate, T. (2008), Survey of enterocyte morphology and tight junction formation in the small intestine of avian embryos, *Poult. Sci.*, 87, 339-50.

Keller, L. H., Benson, C. E., Krotec, K. and Eckroade, R. J. (1995), *Salmonella* enteritidis colonization of the reproductive tract and forming and freshly laid eggs of chickens, *Infect. Immun.*, 63, 2443-9.

Keralapurath, M. M., Corzo, A., Pulikanti, R., Zhai, W. and Peebles, E. D. (2010), Effects of *in ovo* injection of L-carnitine on hatchability and subsequent broiler performance and slaughter yield, *Poult. Sci.*, 89, 1497-501.

Kim, G. B., Seo, Y. M., Kim, C. H. and Paik, I. K. (2011), Effect of dietary prebiotic supplementation on the performance, intestinal microflora, and immune response of broilers, *Poult. Sci.*, 90, 75-82.

Kizerwetter-Świda, M. and Binek, M. (2008), Bacterial microflora of the chicken embryos and newly hatched chicken, *J. Anim. Feed Sci.*, 17, 224-32.

Knap, I., Lund, B., Kehlet, A. B., Hofacre, C. and Mathis, G. (2010), *Bacillus licheniformis* prevents necrotic enteritis in broiler chickens, *Avian Dis.*, 54, 931-5.

Konarzewski, M., Lilja, C., Kozlowski, J. and Lewonczuk, B. (1990), On the optimal growth of the alimentary tract in avian postembryonic development, *J. Zool.*, 222, 89-101.

Lan, Y., Verstegen, M., Tamminga, S. and Williams, B. (2005), The role of the commensal gut microbial community in broiler chickens, *World Poult. Sci. J.*, 61, 95-104.

Lawley, T. D. and Walker, A. W. (2013), Intestinal colonization resistance, *Immunology*, 138, 1-11.

Lloyd, A. B., Cumming, R. B. and Kent, R. D. (1977), Prevention of *Salmonella typhimurium* infection in poultry by pretreatment of chickens and poults with intestinal extract, *Aust. Vet. J.*, 53, 82-7.

Madej, J. P. and Bednarczyk, M. (2016), Effect of *in ovo*-delivered prebiotics and synbiotics on the morphology and specific immune cell composition in the gut-associated lymphoid tissue, *Poult. Sci.*, 95, 19-29.

Madej, J. P., Stefaniak, T. and Bednarczyk, M. (2015), Effect of *in ovo*-delivered prebiotics and synbiotics on lymphoid-organs morphology in chickens, *Poult. Sci.*, 94, 1209-19.

Maiorano, G., Sobolewska, A., Cianciullo, D., Walasik, K., Elminowska-Wenda, G., et al. (2012), Influence of *in ovo* prebiotic and synbiotic administration on meat quality of broiler chickens, *Poult. Sci.*, 91, 2963–9.

McGruder, E. D., Ray, P. M., Tellez, G. I., Kogut, M. H., Corrier, D. E., et al. (1993), *Salmonella enteritidis* immune leukocyte-stimulated factors: Effects on increased resistance to *Salmonella* organ invasion in day-old Leghorn chicks, *Poult. Sci.*, 72, 2264–71.

McGruder, E. D., Ramirez, G. A., Kogut, M. H., Moore, R. W., Corrier, D. E., et al. (1995), *In ovo* administration of *Salmonella enteritidis*-immune lymphokines confers protection to neonatal chicks against *Salmonella enteritidis* organ infectivity, *Poult. Sci.*, 74(1), 18–25.

McReynolds, J. L., Caldwell, D. Y., Barnhart, E. T., Deloach, J. R., McElroy, A. P., et al. (2000), The effect of *in ovo* or day-of hatch subcutaneous antibiotic administration on competitive exclusion culture (PREEMPT™) establishment in neonatal chickens, *Poult. Sci.*, 79, 1524–30.

Meijerhof, R. and Hulet, R. M. (1997), *In ovo* injection of competitive exclusion culture in broiler hatching eggs, *J. Appl. Poult. Res.*, 6, 260–6.

Miśta, D., Króliczewska, B., Pecka-Kiełb, E., Kapuśniak, V., Zawadzki, W., et al. (2016), Effect of *in ovo* injected prebiotics and synbiotics on the caecal fermentation and intestinal morphology of broiler chickens, *Anim. Prod. Sci.*, 57(9), 1884–92.

Miyamoto, T., Kitaoka, D., Withanage, G. S. K., Fukata, T., Sasai, K., et al. (1999), Evaluation of the efficacy of *Salmonella enteritidis* oil-immersion bacterin in an intravaginal challenge model in hens, *Avian Dis.*, 43, 497–505.

Molnár, A. K., Podmaniczky, B., Kürti, P., Tenk, I., Glávits, R., et al. (2011), Effect of different concentrations of *Bacillus subtilis* on growth performance, carcase quality, gut microflora and immune response of broiler chickens, *Br. Poult. Sci.*, 52, 658–65.

Moran Jr., E. T. (1985), Digestion and absorption of carbohydrates in fowl through perinatal development, *J. Nutr.*, 115, 665–74.

Moran Jr., E. T. (2007), Nutrition of the developing embryo and hatchling, *Poult. Sci.*, 86, 1043–9.

Mountzouris K. C., Tsitrsikos, P., Palamidi, I., Arvaniti, A., Mohnl, M., et al. (2010), Effects of probiotic inclusion levels in broiler nutrition on growth performance, nutrient digestibility, plasma immunoglobulins, and cecal microflora composition, *Poult. Sci.*, 89, 58–67.

Murry Jr., A. C., Hinton Jr., A. and Morrison, H. (2004), Inhibition of growth of *Escherichia coli*, *Salmonella typhimurium*, and *Clostridia perfringens* on chicken feed media by *Lactobacillus salivarius* and *Lactobacillus plantarum*, *Int. J. Poult. Sci.*, 3, 603–7.

Nakphaichit M., Thanomwongwattana, S., Phraephaisarn, C., Sakamoto, N., Keawsompong, S., et al. (2011), The effect of including *Lactobacillus reuteri* KUB-AC5 during post-hatch feeding on the growth and ileum microbiota of broiler chickens, *Poult. Sci.*, 90, 2753–65.

Ng, S. C., Hart, A. L., Kamm, M. A., Stagg, A. J. and Knight, S. C. (2009), Mechanisms of action of probiotics: Recent advances, *Inflamm. Bowel Dis.*, 15, 300–10.

Noy, Y. and Sklan, D. (1998), Metabolic responses to early nutrition, *J. Appl. Poul. Res.*, 7, 437–51.

Noy, Y. and Sklan, D. (2001), Yolk and exogenous feed utilization in the posthatch chick, *Poult. Sci.*, 80, 1490–5.

Pan, D. and Yu, Z. (2014), Intestinal microbiome of poultry and its interaction with host and diet, *Gut Microbes*, 5(1), 108–19.

Pascual, M., Hugas, M., Badiola, J. I., Monfort, J. M. and Garriga, M. (1999), *Lactobacillus salivarius* CTC2197 prevents *Salmonella Enteritidis* colonization in chickens, *Appl. Environ. Microbiol.*, 65, 4981-6.

Patterson, J. A. and Burkholder, K. M. (2003), Application of prebiotics and probiotics in poultry production, *Poult. Sci.*, 82, 627-31.

Pedroso, A. A. (2009), Which came first: The egg or its microbiota? *P. I. P.*, 109, 1-5.

Pedroso, A. A., Batal, A. B. and Lee, M. D. (2016), Effect of *in ovo* administration of an adult-derived microbiota on establishment of the intestinal microbiome in chickens, *Am. J. Vet. Res.*, 77(5), 514-26.

Peebles, E. D. (2018), *In ovo* applications in poultry: A review, *Poult. Sci.*, 97, 2322-38.

Pender, C. M., Kim, S., Potter, T. D., Ritzi, M., M., Young, M., et al. (2017), *In ovo* supplementation of probiotics and its effects on performance and immune-related gene expression in broiler chicks, *Poult. Sci.*, 96, 1052-62.

Pilarski, R., Bednarczyk, M., Lisowski, M., Rutkowski, A., Bernacki, Z., et al. (2005), Assessment of the effect of β-galactosides injected during embryogenesis on selected chicken traits, *Folia Biol. (Kraków)*, 53, 13-20.

Pivnick, H. B., Blanchfield, B. and D'Aoust, J. Y. (1981), Prevention of *Salmonella* infection in chicks by treatment with fecal culture from mature chickens (Nurmi cultures), *J. Food Prot.*, 44, 909-13.

Pruszynska-Oszmalek, E., Kolodziejski, P. A., Stadnicka, K., Sassek, M., Chalupka, D., et al. (2015), *In ovo* injection of prebiotics and synbiotics affects the digestive potency of the pancreas in growing chickens, *Poult. Sci.*, 94, 1909-16.

Rastall, R. A. and Gibson, G. R. (2015), Recent developments in prebiotics to selectively impact beneficial microbes and promote intestinal health, *Curr. Opin. Biotechnol.*, 32, 42-6.

Ribeiro, A. M. L., Vogt, L. K., Canal, C. W., Cardoso, M. R. I., Labres, R. V., et al. (2007), Effects of prebiotics and probiotics on the colonization and immune response of broiler chickens challenged with *Salmonella Enteritidis*, *Braz. J. Poult. Sci.*, 9(3), 193-200.

Ricks, C. A., Avakian, A., Bryan, T., Gildersleeve, R., Haddad, E., et al. (1999), *In ovo* vaccination technology, *Adv. Vet. Med.*, 41, 495-515.

Roto, S. M., Rubinelli, P. M. and Ricke, S. C. (2015), An introduction to the avian gut microbiota and the effects of yeast-based prebiotic-type compounds as potential feed additives, *Front. Vet. Sci.*, 2(Article 28), 1-35.

Roto, S. M., Kwon, Y. M. and Ricke, S. C. (2016), Applications of *in ovo* technique for the optimal development of the gastrointestinal tract and the potential influence on the establishment of its microbiome in poultry, *Front. Vet. Sci.*, 3, (Article 63), 1-25.

Sansonetti, P. J. (2001), Rupture, invasion and inflammatory destruction of the intestinal barrier by *Shigella*, making sense of prokaryote-eukaryote cross-talks, *FEMS Microbiol. Rev.*, 25, 3-14.

Schleifer, J. H. (1985), A review of the efficacy and mechanism of competitive exclusion for the control of *Salmonella* in poultry, *World Poult. Sci. J.*, 41, 72-83.

Sears, C. L. (2000), Molecular physiology and pathophysiology of tight junctions. V. Assault of the tight junction by enteric pathogens, *Am. J. Physiol. Gastrointest. Liver Physiol.*, 279, G1129-G1134.

Seuna, E. (1979), Sensitivity of young chickens to *Salmonella typhimurium* var. *Copenhagen* and *S. infantis* infection and the preventative effect of cultured intestinal microflora, *Avian Dis.*, 23, 392-400.

Shareef, A. M. and Al-Dabbagh, A. S. A. (2009), Effect of probiotic (*Saccharomyces cerevisiae*) on performance of broiler chicks, *Iraqi J. Vet. Sci.*, 23, 23–9.

Shinde Tamboli, A. S., Goel, A., Mehra, M., Rokade, J. J., Bhadauria, P., et al. (2017), Delayed post-hatch feeding affects the performance and immunocompetence differently in male and female broiler chickens, *J. Appl. Anim. Res.*, 46 (1), 306–13.

Singh, S. K., Niranjan, P. S., Singh, U. B., Koley, S. and Verma, D. N. (2009), Effects of dietary supplementation of probiotics on broiler chicken, *Anim. Nutr. Feed Technol.*, 9, 85–90.

Sławińska, A., Siwek, M., Żylińska, J., Bardowski, J., Brzezińska, J., et al. (2014), Influence of synbiotics delivered *in ovo* on immune organs development and structure, *Folia Biol. (Kraków)*, 62(3), 277–85.

Sławińska, A., Plowiec, A., Siwek, M., Jaroszewski, M. and Bednarczyk, M. (2016), Long-term transcriptomic effects of prebiotics and synbiotics delivered *in ovo* in broiler chickens, *PLoS ONE*, 11(12), e0168899.

Sobolewska, A., Elminowska-Wenda, G., Bogucka, J., Dankowiakowska, A., Kulakowska, A., et al. (2017), The influence of *in ovo* injection with the prebiotic DiNovo® on the development of histomorphological parameters of the duodenum, body mass and productivity in large-scale poultry production conditions, *J. Anim. Sci. Biotechnol.*, 8, 45–62.

Sonmez, G. and Eren, M. (1999), Effects of supplementation of zinc bacitracin, mannan oligosaccharide, and probiotic into the broiler feeds on morphology of the small intestine, *Vet. Fac. Dergisi Uludag Univ.*, 18, 125–38.

Southwell, B. R. (2006), Staging of intestinal development in the chick embryo, *Anat. Rec. A Discov. Mol. Cell Evol. Biol.*, 288(8), 909–20.

Speake, B. K., Murray, A. M. and Noble, R. C. (1998), Transport and transformations of yolk lipids during development of the avian embryo, *Prog. Lipid Res.*, 37, 1–32.

Spring, P., Wenk, C., Dawson, K. A. and Newman, K. E. (2000), The effects of dietary mannaoligosaccharides on cecal parameters and the concentrations of enteric bacteria in the ceca of *Salmonella*-challenged broiler chicks, *Poult. Sci.*, 79, 205–11.

Tako, E., Ferket, P. R. and Uni, Z. (2004), Effects of *in ovo* feeding of carbohydrates and beta-hydroxy-beta-methylbutyrate on the development of chicken intestine, *Poult. Sci.*, 83, 2023–8.

Teague, K. D., Graham, L. E., Dunn, J. R., Chang, H. H., Anthony, N., et al. (2017), *In ovo* evaluation of FloraMax®-B11 on Marek's disease HVT vaccine protective efficacy, hatchability, microbiota composition, morphometric analysis, and *Salmonella Enteritidis* infection in broiler chickens, *Poult. Sci.*, 96, 2074–82.

Thiagarajan, D., Saeed, A. M. and Asem, E. K. (1994), Mechanism of transovarian transmission of *Salmonella enteritidis* in laying hens, *Poult. Sci.*, 73(1), 89–98.

Triplett, M. D., Zhai, W., Peebles, E. D., McDaniel, C. D. and Kiess, A. S. (2018), Investigating commercial *in ovo* technology as a strategy for introducing probiotic bacteria to broiler embryos, *Poult. Sci.*, 97, 658–66.

Uni, Z. and Ferket, P. R. (2004), Methods for early nutrition and their potential, *World Poult. Sci. J.*, 60, 101–11.

Uni, Z., Tako, E., Gal-Garber, O. and Sklan, D. (2003), Morphological, molecular, and functional changes in the chicken small intestine of the late-term embryo, *Poult. Sci.*, 82, 1747–54.

Vieira, S. L. and Moran Jr., E. T. (1999), Effects of egg of origin and chick post-hatch nutrition on broiler live performance and meat yields, *World Poult. Sci. J.*, 55, 125–42.

Villaluenga C. M., Wardeńska, M., Pilarski, R., Bednarczyk, M. and Gulewicz, K. (2004), Utilization of the chicken embryo model for assessment of biological activity of different oligosaccharides, *Folia Biol. (Kraków)*, 52, 135–42.

Wakenell, P. S., Bryan, T., Schaeffer, J., Avakian, A., Williams, C., et al. (2002), Effects of *in ovo* vaccine delivery route on herpesvirus of turkeys/SB-1 efficacy and viremia, *Avian Dis.*, 46, 274–80.

Williams, C. J. (2011), *In ovo* vaccination and chick quality, *Int. Hatch. Prac.*, 19, 7–13.

Xu, Z. R., Hu, C. H., Xia, M. S., Zhan, X. A. and Wang, M. Q. (2003), Effects of dietary fructooligosaccharide on digestive enzyme activities, intestinal microflora and morphology of male broilers, *Poult. Sci.*, 82, 1030–6.

Yadgary, L., Kedar, O., Adepeju, O. and Uni, Z. (2013), Changes in yolk sac membrane absorptive area and fat digestion during chick embryonic development, *Poult. Sci.*, 92, 1634–40.

Yamawaki, R. A., Milbradt, E. L., Coppola, M. P., Rodrigues, J. C. Z., Andreatti Filho, R. L., et al. (2013), Effect of immersion and inoculation *in ovo* of *Lactobacillus* spp. in embryonated chicken eggs in the prevention of *Salmonella Enteritidis* after hatch, *Poult. Sci.*, 92, 1560–3.

Yang, C. M., Cao, G. T., Ferket, P. R., Liu, T. T., Zhou, L., et al. (2012), Effects of probiotic, *Clostridium butyricum*, on growth performance, immune function, and cecal microflora in broiler chickens, *Poult. Sci.*, 91, 2121–9.

Chapter 6

Understanding gut function in poultry: immunometabolism at the gut level

Ryan J. Arsenault, University of Delaware, USA

1 Introduction

Metabolism is the entire network of chemical reactions carried out by living cells (Horton et al., 2002). Metabolism is most often separated into reactions that degrade molecules, called catabolism, or reactions that create molecules, called anabolism. The intermediates of these two reactions are referred to as metabolites. The reactions involving low molecular weight molecules in catabolism or anabolism is called intermediary metabolism.

1.1 Nutrient absorption

The lumen of the gastrointestinal tract can be considered outside of the body from a metabolic perspective. Only when the nutrients are absorbed within or through the epithelial barrier of the gut are the nutrients considered 'inside' the bird and are able to be utilized. Intestinal uptake capacity of macronutrients in chickens matches the individual's nutrient needs; this is unlike mammals where uptake capacity far exceeds immediate host needs (Denbow, 2015). This upper limit on uptake or absorption is a possible limit on growth potential. In the modern commercial bird under high production pressure, rapid growth requires significant feeding time to ensure adequate absorption of required macronutrients. In the following subsections, the absorption and metabolism

http://dx.doi.org/10.19103/AS.2019.0059.07

of carbohydrates, amino acids and lipids in poultry are described, this is the first step in studying immunometabolism.

1.1.1 Carbohydrates

Carbohydrates are absorbed in the gut through both passive and active transport, though active transport, which requires the expenditure of energy, represents approximately 80% of glucose absorption (Riesenfeld et al., 1980). Glucose absorption occurs more rapidly in the small intestine than in the caeca, and it also occurs in the colon (Denbow, 2015). Absorption of nearly all of the glucose from starch occurs in the small intestine. Within the small intestine the greatest proportion of glucose is absorbed by the duodenum (Riesenfeld et al., 1980). On the luminal surface of the epithelial barrier glucose is absorbed by the sodium-glucose linked transporter (SGLT)-1 system, found in the small intestine and colon (Denbow, 2015). Fructose is transported by the glucose transporter (GLUT) 5-type system (Garriga et al., 2004). Once inside the epithelial cells glucose molecules are sent to the basolateral side of the epithelial barrier via the GLUT-2 transporter. As mentioned, some glucose absorption occurs in the caecum, which appears to have a greater capacity to absorb low concentration sugars (Vinardell and Lopera, 1987).

1.1.2 Amino acids and peptides

Amino acid transport across the gut barrier occurs by active transport, as with sugars. This process is coupled to Na^{+}, requires ATP for every unit of energy and is saturable (Denbow, 2015). The majority of amino acid absorption occurs in the small intestine, though some does occur in the crop, gizzard and proventriculus. Peptide absorption in the poultry gut usually occurs in the form of di- and tripeptides. Peptide absorption occurs through the peptide transporter 1 (PepT1) system. Peptides are absorbed more readily than amino acids. The preference for peptides may be due to the efficiency gained by allowing more amino acids to be absorbed per unit energy; three amino acids are transported within one tripeptide, and also to maintain a mixed pool of amino acids rather than favouring some over others in absorption. The caeca has a relatively greater capacity to transport amino acids than sugars (Moreto et al., 1991).

1.1.3 Fatty acids

Fatty acid (FA) absorption occurs mainly in the distal half of the jejunum and to some extent in the ileum (Denbow, 2015). In birds the bile duct enters the duodenum more distally, thus absorption occurs in the jejunum rather than the duodenum. Lipids from feed are hydrolysed in the gut lumen and the released free FAs are absorbed by the epithelial cells of the gut barrier (Buyse and Decuypere, 2015). Mammals rely on the lymphatic system to collect and transport FAs away from the gut, since poultry lack a lymphatic system lipids are packaged into portomicrons that pass directly into the hepatic portal blood supply from the endothelial cells of the gut (Bensadoun and Rothfeld, 1972). Volatile FAs (VFAs), acetate, propionate and butyrate are absorbed along the gastrointestinal tract (Denbow, 2015). VFAs are generated by bacteria of the microbiota and are an important energy source for the host, especially the cells of the gut. VFA levels in the caeca are high (Annison et al., 1968), likely due to the extensive bacterial fermentation occurring there. VFAs are absorbed by the small intestine and caeca by passive transport, as the concentration in the lumen is far greater than within epithelial cells of the gut and on the basolateral side of the gut barrier (Sudo and Duke, 1980).

1.2 Carbohydrate metabolism

Glucose absorbed in the intestine can be broken down to produce energy via the metabolic pathways, glycolysis, the tricarboxylic acid (TCA) cycle and the pentose phosphate pathway. Glucose can be stored as glycogen in the liver, muscles and other tissues or can be converted to FAs for longer-term storage. Despite the aggressive breeding for growth and efficiency in modern commercial chickens, there is no difference in circulating concentrations of glucose between red jungle fowl and commercial chickens (Scanes, 2015a). Circulating concentrations of glucose decline post-hatch and during grow out (Sinsigalli et al., 1987). There is a measurable decrease in circulating glucose concentrations associated with pathogenic infections (Davis et al., 1995), likely due to the mobilization of this energy source for use in the immune response. The highest intake of glucose occurs in the brain and heart. There is high GLUT2 expression in the liver and kidneys but low expression in the intestine (Duarte et al., 2011). Though there have been claims that chickens do not respond to insulin, they do produce insulin which stimulates glucose uptake and GLUT1 expression in avian myoblasts (Zhao et al., 2012). Chickens also generate a signal transduction response up and down the insulin receptor signalling pathway when exposed to *Salmonella* (Arsenault et al., 2013). Finally, the glucose transporter GLUT8 is expressed in chicken and its activity is insulin dependent (Seki et al., 2003).

1.3 Lipid metabolism

The portomicrons generated in the gut are transported to extrahepatic tissues via the portal vein where they are partially hydrolysed, and then they are sent to the liver for processing (Denbow, 2015). *De novo* lipids generated from carbohydrates and amino acids as well as those absorbed from the gut are packaged in the liver with protein and cholesterol and sent through the bloodstream to the tissues (Buyse and Decuypere, 2015). FAs are taken up by cells and either oxidized for energy production or stored as fat in adipose tissue.

1.4 Protein metabolism

The production of protein in the chicken is carried out using the pool of amino acids absorbed from feed in the gut as well as non-essential amino acids synthesized within the bird. Transporters on the surface of cells present throughout the body translocate amino acids into the cell. Protein synthesis is high in the muscle, gastrointestinal tract and immune cells of the bird (Scanes, 2015b), sites of either rapid growth or high cellular turnover. Amino acids can also be catabolized for energy, a major amino acid used as an energy source is glutamine (Mathew et al., 1993), which is metabolized to α-ketoglutarate and enters the TCA cycle to produce energy. During the first week post-hatch proline uptake in the small intestine is high relative to glucose (Soriano and Planas, 1998). Proline uptake in the caecum is also high early in chick development (Moreto et al., 1991), possibly because proline is a preferred amino acid in early growth stages of chicks and the importance of proline in chick growth has been documented for many decades (Roy and Bird, 1959). Growth rate is rapid at this early stage of development and amino acid uptake is high to match (Scanes, 2015b).

2 Immunometabolism

This section will focus on the metabolism and immune responses within the gut tissue, influenced by the feed and microbiota located in the lumen.

The gut is the critical nutritional organ and its role in nutrient absorption feeds every cell and tissue in the body. An immense research effort has been ongoing since the advent of the commercial broiler and layer industries to optimize the uptake of macronutrients into the bird through improved gut absorption. Effective and efficient nutrient absorption is the first and major step in cost-effective animal production. As a result, feed efficiency is of critical importance to poultry producers, as higher feed efficiency both increases the amount of commodity produced and reduces costs. Any increase in feed efficiency from a whole animal perspective must occur in the gut. Any defect in the gut's ability to extract nutrients from feed can have a profound impact on both bird growth and disease susceptibility. Perhaps overlooked in the quest for optimum feed efficiency is the fact that the gut is the largest immunological organ in the body (Holmgren et al., 1992). A wide variety of immune cells, of both innate and adaptive immune system, defence peptides and immunoglobulins are produced or matured in the gut tissue. The gastrointestinal system is the largest lymphoid organ in the body in both humans (Mayer, 2000) and chickens (Jones, 1972) and secretes the most antibody in humans (Mayer, 2000).

Besides being a nutritional organ and an immunological organ the third role of this organ, intimately related to the first two, is to house and foster the resident gut microbiota. The microbiota/microbiome, a central component of gut physiology, must be considered in any discussion of the gut and immunometabolism. In the context of nutrition, the bacteria that make up the microbiota provide a crucial role breaking down otherwise indigestible feed components that then become available to the host providing digestible and absorbable metabolites (Pflughoeft and Versalovic, 2012). The importance of the metabolism carried out by bacteria in the gut cannot be overstated. A significant amount of the biology of bacteria is devoted to intermediary metabolism, 900 genes encode enzymes used in intermediary metabolism in *Escherichia coli* representing approximately 130 pathways, these genes account for 21% of the genome (Horton et al., 2002). This percentage holds roughly true for fungi such as *Saccharomyces cerevisiae* (20% of genes are metabolic) and nematodes such as *Caenorhabditis elegans* (28% of genes are metabolic).

In the context of immunity, the microbial population matures (Flint et al., 2012) and stimulates the gut immune system for optimum immune performance and potential (Hooper et al., 2012). The microbiota aids in the development of the mucosal immune system and the resident microorganisms are competitors for pathogens that enter the gut. A large proportion of disease-causing microorganisms in agricultural animals enter the host via the gut (Barrington et al., 2002). A healthy and homeostatic microbiota represents a complete ecological environment. If every niche of this ecosystem is filled by a commensal microorganism there is no opportunity for a pathogen to enter and take up residence, this competitive exclusion of pathogens helps maintain disease-free poultry (Hooper et al., 2012; Pflughoeft and Versalovic, 2012). Commensal microorganisms are required for proper stimulation and development of the gut immune system to maintain a balance between tolerance and active immune response (Hooper et al., 2012; Taschuk and Griebel, 2012).

After decades of studying the immune system and the wide variety of tissues and cells that make it up, researchers have identified core cellular signal transduction pathways that constitute the two branches of immunity: the innate and adaptive immune responses. Microbial-associated molecular patterns (MAMPs) are common motifs found in microorganisms that are recognized by receptors of the innate immune system. MAMPs were previously referred to as pathogen-associated molecular patterns (PAMPs) before it was fully realized how important these patterns are in commensal microbe's stimulation of a properly functioning immune system. These MAMPs are recognized by receptors including the Toll-like receptors (TLRs), nucleotide-binding oligomerization domain (NOD)-like receptors (NLRs), retinoic acid-inducible gene (RIG)-I-like receptors (RLRs) and C-type lectin receptors (CLRs). The adaptive immune system has its own repertoire of receptors, which usually interact with other immune cells rather than MAMPs, activating unique signal transduction pathways, these include the B-cell and T-cell receptors, among a plethora of others.

The catabolic and anabolic metabolism of the cell was often considered in parallel with the other physiological functions of immune cells, the biochemical processes that provided energy and building blocks for cell activities rather than a part of the immune response. Early research into metabolic pathways and their role in immunity began with the understanding that obesity resulted in a generalized inflammatory response (Xu et al., 2003) which could lead to metabolic syndrome and diabetes. The field of research integrating metabolism and immunity has been styled immunometabolism (Mathis and Shoelson, 2011). Generally immunometabolism is considered from two perspectives; (1) the role immune cells play in the metabolism of tissues and the resulting effect on the host, for example M1 macrophages and the metabolism of adipose tissue and (2) the role of metabolic pathways in immune cells and the effect on the immune system, for example the activation of macrophages by the induction of aerobic glycolysis (Pearce and Pearce, 2013).

Studying immunometabolism involves characterizing the direct intracellular pathway links between metabolism and immunity (Kelly and O'Neill, 2015; Pearce and Pearce, 2013). Central regulators of immunometabolic responses are the energy-sensing signalling proteins that integrate both cellular energy levels and immune response signals; examples include mammalian target of rapamycin (mTOR), adenosine monophosphate-activated protein kinase (AMPK) and sirtuins. The protein synthesis pathway is regulated by mTOR; protein synthesis allows general cell growth and differentiation or the production of cytokines and other immune proteins. mTOR is also involved in T-cell maturation (Pollizzi and Powell, 2015), determining whether the cell becomes an effector T cell or a regulatory T cell (Chi, 2012). AMPK is an energy sensor that monitors the ratio of AMP:ATP, activating either anabolic or catabolic processes. It is also involved in innate immune response and has a direct link to mTOR (O'Neill and Hardie, 2013). Metabolism-induced reprogramming of immune pathways occurs through sirtuin enzymatic activity (Preyat and Leo, 2013). The perspective that separated immunity and metabolism resulted in a focus on targeting immune pathways for infectious disease and metabolic pathways for growth/metabolic disorders, without considering the knock-on effects on the other. With an integrated approach, we can broaden our potential targets for disease intervention without compromising growth, and encourage more efficient production and feed conversion without limiting the immune potential of the bird.

Profiles of the immune cells' metabolism have shown that the type of metabolism carried out by the cells determine the immune function. In dendritic cells and macrophages, LPS and other immune ligands trigger the switch from oxidative phosphorylation to aerobic glycolysis, and leads to profound immune activity changes (Kelly and O'Neill, 2015). These changes include a release of proinflammatory cytokines, an increase in cell migration and the utilization of oxidative phosphorylation machinery (Krawczyk et al., 2010; Rodríguez-Prados et al., 2010) for the production of immune effectors such as nitric oxide (NO) and reactive oxygen species (ROS) (Everts et al., 2012; West et al., 2011). This metabolic change is so critical to immune activity that without the change to aerobic glycolysis and away from oxidative phosphorylation the electron transport chain proteins in the mitochondria would not be available to generate these potent antimicrobial molecules. The change in metabolism when immune cells are activated has been described as akin to the Warburg effect in cancerous cells, leading to aerobic glycolysis fuelled proliferation and activity (Palsson-McDermott and O'Neill, 2013).

As mentioned above, what metabolic processes are activated can help determine the ultimate function of immune cells (Fig. 1). Effector T cells undergo glycolysis, often aerobic glycolysis, utilizing this rapid energy production to carry out immediate immune activities and proliferate. Quiescent T cells, such as T regulatory cells, rely on glycolysis, the TCA cycle and oxidative phosphorylation (Pearce and Pearce, 2013). Memory T-cell metabolism is biased towards free FA metabolism, and the cell proliferative

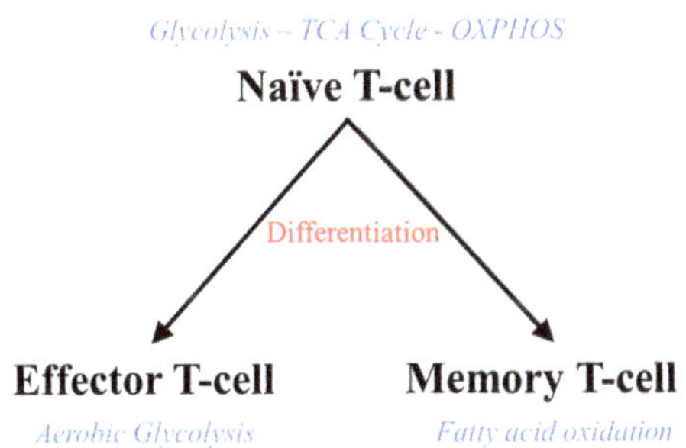

Figure 1 T-cell differentiation and metabolism. T cells (black) display a different metabolic phenotype (blue) depending on their differentiation status. Naïve T cells carry out a balanced glycolysis, TCA cycle, oxidative phosphorylation (OXPHOS) metabolism. Effector T cells carry out aerobic glycolysis to generate rapid energy and active immune pathways. Memory T cells are long lived and carry out slow, steady fatty acid oxidation for energy. Figure based on Pearce and Pearce (2013) review.

machinery is turned off. This metabolic profile gives memory T cells a lower metabolic rate and a longer lifespan, allowing them to survive and circulate longer than the shorter-lived and glycolytic effector T cells (van der Windt et al., 2012).

Gene association networks and protein-protein interaction databases provide an amalgamation of data on how genes and proteins connect in function and response to stimuli. Classic textbook descriptions of immune and metabolic pathways often show them as completely distinct linear pathways that lead directly from metabolic uptake or receptor binding through a series of intermediate steps to changes in gene expression or cell function. However, within a cell (or tissue) these responses are extremely complex networks of interactions with extensive overlap in pathway members and responses. Different cell types and tissues 'recycle' pathway members, resulting in different cellular phenotypes and the ultimate effect depends on the cell type and its host context.

An example of this is the interplay of metabolic pathways and immune pathways that can be illustrated using protein-protein interaction databases. Considering the most common metabolic pathways, such as glycolysis, TCA cycle, FA oxidation/synthesis and amino anabolism/catabolism (Fig. 2), and common immune pathways, such as TLR, NLR, CLR, RLR, TCR and BCR signalling (Fig. 3), one can readily observe an extensive interaction network between the two with no means to distinguish where the metabolic processes end and the immune pathways begin (Fig. 4).

2.1 Immunometabolism and poultry production

In animal production research feeding for growth and encouraging metabolism in the direction of muscle growth or egg production has been the goal. In poultry, the immune system, especially the innate immune system, was seen as an energy sink that reduced the growth rate of broilers or the production capacity of layers and reduced feed conversion (Klasing, 1988b; Klasing and Korver, 1997). Quantifying the amount of energy required to mount an immune response showed how innate immunity was costly to growth (Klasing, 2007). Data shows that the growth-promoting ability of antibiotics is due to their anti-inflammatory effects on the host tissue, perhaps more than any effect on the microbiome or the prevention of infectious diseases (Niewold, 2007). Even in humans, infection early in life has been linked to a reduction in adult size; the incidence of disease in childhood has been shown to affect the ultimate height reached in adulthood (Bozzoli et al., 2009). Pushing metabolism towards production goals and treating the immune system as an energy cost to production has resulted in breeding programmes for both broilers and layers that have de-emphasized immunological robustness. Studies have shown that wild fowl or even heritage

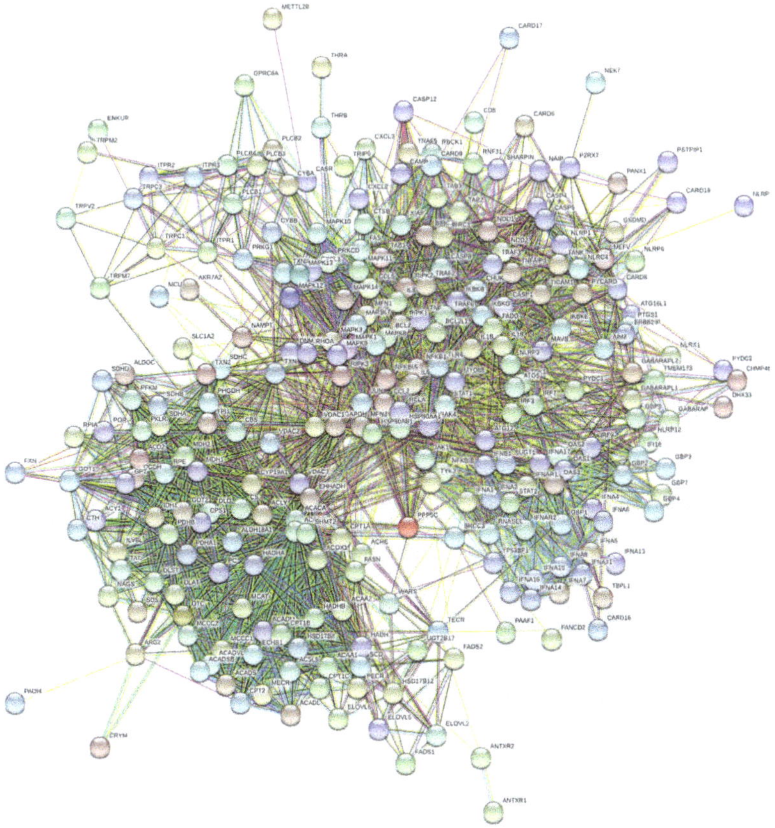

Figure 2 Immunometabolic interactions. Protein-protein interaction networks of metabolic pathways glycolysis, TCA cycle, fatty acid oxidation/synthesis, amino acid anabolism/catabolism.

breeds of commercial chicken are more appropriately immunologically active and better able to resist pathogenic infections than modern commercial broilers (Cheema et al., 2003; Qureshi and Havenstein, 1994) and turkeys (Genovese et al., 2006). On the other hand, when a pathogenic infection does become established in a modern commercial chicken, often the response, usually the innate immune inflammatory response, is out of proportion resulting in a loss of homeostasis and runaway inflammation. Heritage birds are less likely to develop gastrointestinal diseases such as coccidiosis or necrotic enteritis than modern birds. Heritage breeds also do not show an age-dependent susceptibility to infection during grow out like modern birds; the heritage breeds maintain a better homeostatic immune potential throughout grow out while the ramp up in growth after two weeks in modern birds appears to reduce the potential to deal with infectious pathogens (Johnson et al., 2018).

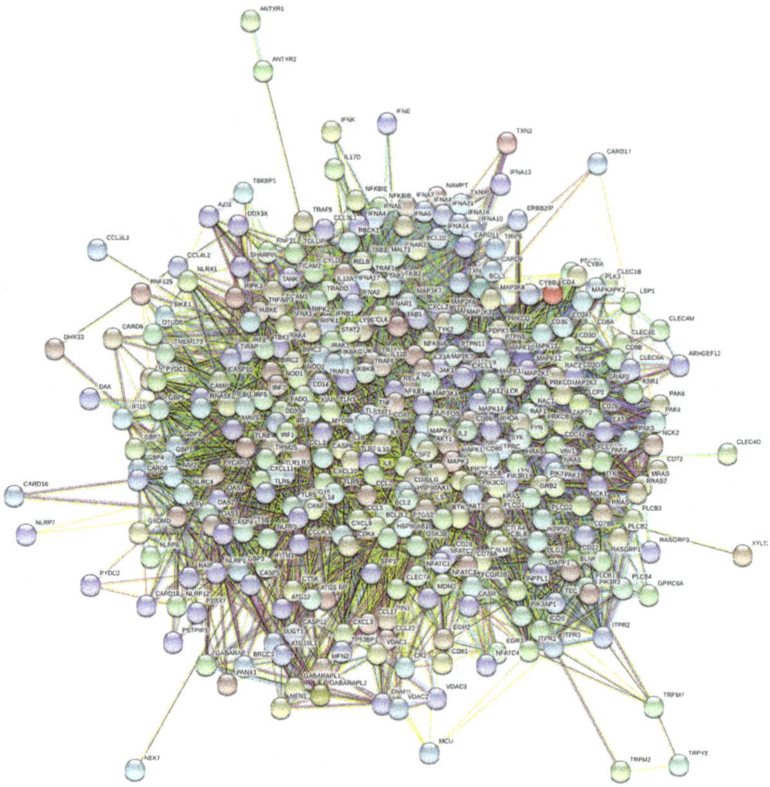

Figure 3 Immunometabolic interactions. Protein-protein interaction networks of immune pathways TLR, NLR, CLR, RLR, TCR and BCR signalling.

This more complete understanding of how infectious diseases alter metabolism indicates that striving for growth at the expense of the immune system may have more production consequences than simply increased susceptibility to diseases. Even subclinical *Salmonella* infections have resulted in muscle growth issues and the deposition of fat rather than muscle in broiler breast muscle, and metabolic pathways related to FA synthesis are induced in the breast muscle of birds given a subclinical *Salmonella* dose (Arsenault et al., 2013).

As mentioned previously, the goal in feeding the metabolism of broiler birds is to encourage the deposition of muscle, and in layer hens it is egg production. Feed formulations have been effective in encouraging these productive uses of nutrients. In chickens, fat deposition is not as significant an issue given excess nutrient intake as it is in humans. However, fatty muscle and abdominal fat production in chickens is a production concern (Leenstra, 1986). It has been reported that a subclinical infection with *Salmonella*

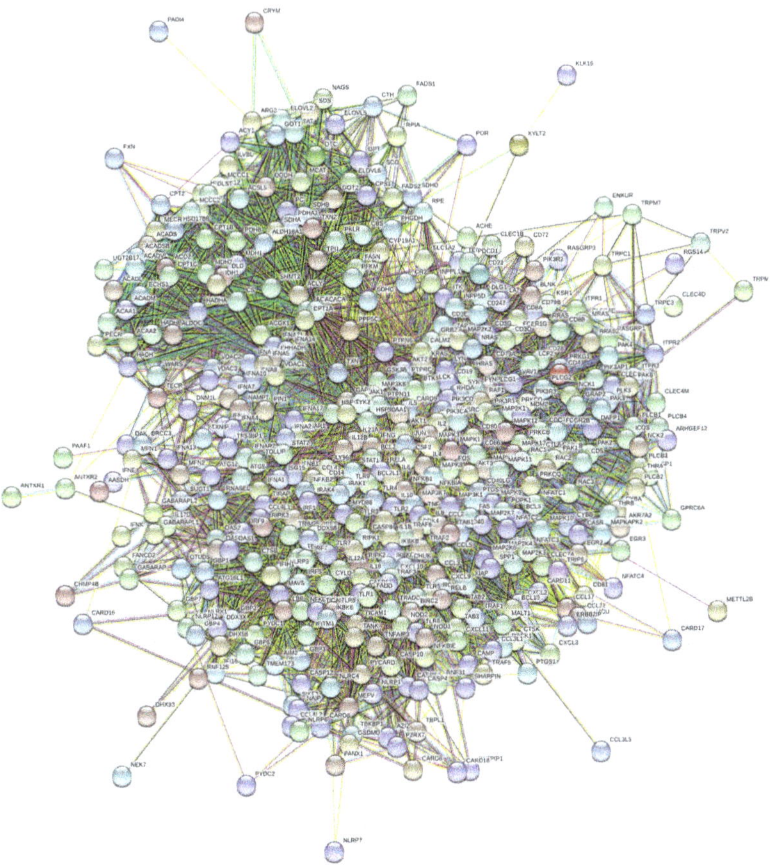

Figure 4 Immunometabolic interactions. Protein-protein interaction networks of the combined pathways from Fig. 2 and 3. The extensive interaction network observed shows how integrated immune and metabolic responses are within a cell.

Typhimurium (Arsenault et al., 2013) results in the activation of anabolic FA signalling in the breast muscle of broiler birds. The connection between a gut infection and muscle metabolism is unknown but may have to do with the link between inflammation of the gut and a generalized inflammation leading to lipid production. Unproductive abdominal fat accumulation and its inevitable inflammatory consequences are likely the result of excess energy intake and the need to store this energy in a bioavailable form for further use once the protein production machinery has been maxed out (Griffin et al., 1992).

A significant production issue facing the modern broiler industry is white striping and woody breast disease. These diseases appear to be immunometabolic in nature (Mutryn et al., 2015). The extreme and rapid growth of the breast muscle bred into modern broiler chickens, due to the desire for this white meat in the U.S. consumer market, generates considerable metabolic stress on the myocytes. This stress initiates pathways that mimic an inflammatory response, recruiting immune cells to the site, initiating an unregulated FA synthesis response (white striping) (Petracci et al., 2014) and an inflammatory macrophage response leading to muscle myopathy (woody breast) (Mutryn et al., 2015).

3 Assessing metabolic gut function

Numerous techniques and research tools are currently employed to characterize the immune state of the poultry gut. However, the amount of tools and reagents at the disposal of poultry scientists is significantly less than what is available for researchers working with humans, mice or rats. Some techniques often used in poultry research include, but are not limited to, quantitative polymerase chain reaction (qPCR), enzyme-linked immunosorbent assay (ELISA), western blot, immunohistochemistry and fluorescence-activated cell sorting (FACS). Measuring metabolism has been a significantly more difficult task than measuring immune response, and has generally relied on the characterization of enzyme activities in metabolic pathways or the identification of metabolites and metabolic intermediates as a measurement of metabolic state. Metabolites are often identified using mass spectroscopy (MS), though other techniques have been used and new methods are currently being employed and developed. In the following subsections, three methods of measuring metabolism are discussed.

3.1 Mass spectroscopy

MS is a common and fundamental technique in the study of metabolites (Dettmer et al., 2007). Modern mass spectrometers contain three major components: an ion source, a mass analyser and a detector; this is preceded by a molecular separation apparatus. A sample for metabolic analysis is usually separated based on size or other physical characteristic using either gas chromatography (GC), liquid chromatography (LC) or capillary electrophoresis (CE). This separation aids in the input of a distinct, clean, more easily interpreted sample that is also more efficiently ionized by the ion source. Once the sample leaves the separation apparatus it is ionized, this step fragments the sample components and gives them a charge. The mass analyser and detector generate a mass spectrum that is represented as a mass-to-charge ratio. The mass spectrum is then analysed based on a library of known spectrum to identify the metabolites found in the test sample. In this way a metabolic profile or metabolome can be deciphered from the sample and provide an abundance of information on the metabolic state of the animal, tissue or cell undergoing analysis.

3.2 Radiolabelling

A common technique to study the metabolism of a specific metabolic substrate is to incorporate a radiolabelled element, such as ^{3}H or ^{14}C. The radiolabelled substrate can be tracked through its metabolic transformation in a cell, tissue or *in vivo*. This technique can be applied to carbohydrates, proteins or lipids. Often this technique is coupled with nuclear magnetic resonance (NMR) spectroscopy that can detect certain isotopes within larger compounds (Horton et al., 2002).

3.3 Post-translational protein modifications

As mentioned in the previous subsections, the identification and quantification of the metabolites generated by metabolic pathways has been the standard way of studying metabolism. This is in contrast to work in immunology or cancer biology that relies on the study of signal transduction pathways that are mediated by the post-translational modification of proteins catalysed by enzymes, predominantly kinases. Kinases are enzymes that phosphorylate proteins, altering their function in some way. Both metabolic and signal transduction pathways rely on enzymes, the difference lies in the enzymatic action generating a metabolite or the propagation of a signal via sequential post-translational modification, respectively. Turning to the biochemical techniques and tools developed to study enzyme activity we can study the states of metabolic pathways in tissues and cells. The major druggable targets in cancer biology are the kinase enzymes (Chen et al., 2007); these are often the oncogenes that, when mutated, result in uncontrolled growth and proliferation leading to cancer. Kinases regulate many cellular processes and response to stimuli. Kinase enzymes are also part of critical metabolic pathways including glycolysis, the breakdown of glucose to produce energy for the cell. Figure 5 shows the enzymatic steps in glycolysis that lead from glucose to pyruvate; pyruvate can then be used for aerobic (or anaerobic) glycolysis and the production of lactate, or be converted into acetyl-coA and processed through the TCA cycle and oxidative phosphorylation in the mitochondria. At four points along the glycolysis pathway, kinases carry out phosphorylation that can be measured. In addition, many of the enzymes of glycolysis are regulated by phosphorylation post-translational modifications themselves to alter their enzymatic activities, including hexokinase (Zhang et al., 2017b), phosphofructokinase (Sale et al., 1987) and pyruvate kinase (Kawaguchi et al., 2001). Studying the phosphorylation states of these enzymes, or the phosphorylation states of their metabolic substrates, can be significantly easier than conducting a global MS study of cellular or tissue metabolites. In the following subsections are described two techniques to measure phosphorylation that can be applied to the study of metabolism.

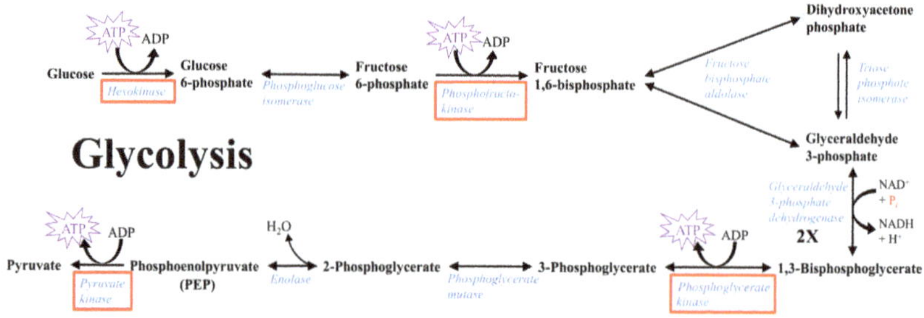

Figure 5 Glycolysis. The steps of glycolysis with the four kinases of the metabolic pathway highlighted. By studying phosphorylation one can measure several steps of the glycolysis process. In addition, many of these enzymes' (in blue) activities are regulated by phosphorylation themselves.

3.3.1 Phospho-specific antibodies

Antibodies can be generated to recognize only specific epitopes that contain a phosphorylated residue, amino acid sequences containing these phosphorylated epitopes are referred to as phosphopeptides or phosphoproteins (Mandell, 2003). If a sample, from a cell or tissue, contains the phosphopeptide or phosphoprotein the phospho-specific antibody will bind to it and this binding can be visualized in a variety of ways. This technique can be applied to standard lab techniques including western blot or ELISA. Phospho-specific antibodies can also be immobilized on an array format and an experimental sample exposed to hundreds of antibodies simultaneously.

3.3.2 Kinome peptide arrays

Kinome peptide arrays capitalize on recent advances in bioinformatics providing an alternative to MS in identifying and measuring metabolism. The kinome is classically defined as the complete set of kinases found in an organism's genome; it can also refer to the set of active kinases in an organism, tissue or cell (Arsenault et al., 2011). Peptide arrays are becoming an increasingly common, high-throughput experimental technique to study protein biochemistry, including the activity of kinases, in a cell or tissue. Kinome peptide arrays incorporate peptides, representing kinase recognition target sequences, immobilized in an array format on a glass slide. When exposed to a sample (cell or tissue lysate) containing active kinases and ATP (as a phosphate group donor), the kinases phosphorylate their respective target peptide sequence on the array. This phosphorylation can be visualized by either using γP^{32} ATP (and capturing the radioactive decay on a film screen) or using phospho-specific antibodies or using a phospho-specific fluorescence dye that binds to phosphorylated peptides (Arsenault et al., 2011).

New '-omics' techniques are often designed for use on standard biomedical species, mice, rats or human tissue/cells. Using novel bioinformatics techniques (Trost et al., 2013) custom kinome peptide arrays have been designed for use in poultry species, including chicken (Arsenault and Kogut, 2012), turkey (Arsenault et al., 2014) and duck (Pagano et al., 2018).

4 Inflammatory feed components

In the animal science field, feed-induced inflammation is a concern. Something as simple as an excess of feed can lead to changes in immune response (Klasing, 1988a). Excess carbohydrates engage metabolic pathways and feed inflammatory responses. There is evidence that simple sugars have a direct effect on MAPK signalling (Peeters et al., 2017), leading to proliferative and proinflammatory responses. This response to excess glucose looks metabolically like the Warburg effect seen first in cancer cells but also in highly proliferative immune cells (Palsson-McDermott and O'Neill, 2013). The simplistic decision to increase production by feeding more carbohydrate energy into the bird's system may be counterproductive if it generates inflammation, siphoning off energy. In addition, an inflammatory gut is more open to certain pathogenic infections, a classic example being *Salmonella* (Zhang et al., 2018). *Salmonella* is dependent on an inflammatory response at the epithelial barrier of the gut in order to invade the gut tissue. Other gut diseases, perhaps including necrotic enteritis, may be inflammatory in nature.

Certain feed ingredients can lead to an inflammatory gut response, examples include non-digestible components of wheat and rye in chickens (Teirlynck et al., 2009). These feed components often contain immune ligands, or compounds that mimic the structure of immune ligands that engage pathogen receptors on the host cell surface. One current animal feed strategy involves adding enzymes to break down certain indigestible and/or inflammatory feed components in the gut, with the aim to reduce immune response and redirect this energy to growth (Choct, 2006). β-galactomannans found in soybean meal, soybean being a significant component of poultry feed, engage certain CLRs within the gut generating an innate immune signalling cascade and an inflammatory response (Arsenault et al., 2017). Feed enzymes that break up these immune ligand-like feed components have been marketed as 'energy-sparing' enzymes as they prevent the diversion of energy into an inflammatory response from growth/production.

It is not only immune ligand-like feed components or carbohydrates that can generate inflammatory immune responses in the gut, but amino acids are also used as an energy sensor and energy source for cells. Dietary amino acids form the building blocks of host protein, and are critical for laying down muscle or egg production in poultry. However, the direct feeding of high doses of protein, pure amino acids or enzymes that generate more absorbable dietary protein can overwhelm the metabolic pathways and generate spillover into immune pathways. One mechanism for this is excess amino acids are converted to glucose by the gluconeogenic pathway resulting in the same excess energy state as high carbohydrate feeding (Azzout-Marniche et al., 2007).

5 Feeding immunometabolism

Manipulation of the gut, both immunologically and metabolically, is a multi-billion dollar animal health research and development enterprise. This includes prebiotics, probiotics, antibiotics, anti-parasitics, feed enzymes and feed additives, among others. There has been a significant amount of research into nutrition's effects on immunity (Korver, 2012) and the use of pre- and probiotic feed ingredients to improve growth and disease resistance (Griggs and Jacob, 2005; Hume, 2011; Patterson and Burkholder, 2003). A more complete understanding of how the absorbed metabolic components of feed and microbiota fermentates interact with the metabolic pathways of the immune system will allow us to manipulate the immune response to produce more immunologically robust birds and counteract some of the immune deficiencies that have been bred into modern poultry. Two examples are described in this section.

Depending on their dietary source and structure, FAs can have different impacts on inflammation and the immune system more generally. Fats from corn or soybean oil have been shown to reduce humoral immune response and growth, while polyunsaturated fats from fish oil show the opposite effect (Fritsche et al., 1991). A feed-induced reduction in humoral response could have a significant negative effect on resistance to pathogens that require an antibody response, including important production diseases that are vaccinated against, such as Marek's disease virus and coccidiosis. On the other hand, supplementation of feed with fish oil may enhance the protective effect of these vaccines, as well as the bird's natural adaptive response to infections encountered in the field.

The supplementation of the broiler diet with L-arginine has been shown to be beneficial to gut health during a *Clostridium perfringens* (Cp) infection (Zhang et al., 2017a). L-arginine increased tight junction gene expression and the expression of IFNγ, IL-10 and NOD1 reducing Cp and improving gut barrier integrity and function. An improvement in gut barrier function is expected to be protective against a variety of pathogens that must cross the epithelial cell lining in order to establish infection, this includes various species of *Salmonella* (Zhang et al., 2018) and *Campylobacter jejuni* (Hermans et al., 2012).

6 Conclusion and future trends

These examples of feeding the 'immunometabolic system' are only a few of the numerous avenues under active research right now. This is a research trend with significant potential, as it does not rely on the trade-offs of growth versus immunity that have been made to date. Some claim we have reached or exceeded the growth potential of poultry, as evidenced by woody breast and other diseases, and are advocating production of slower growing poultry (Tallentire et al., 2018). A nuanced understanding of the mechanisms of action of prebiotics, probiotics, antibiotics, anti-parasitics, feed enzymes, feed additives and feed formulations can allow for enhanced immune responses, generate more resistant birds and maintain the impressive and increasing feed conversion efficiency and growth rate achieved so far. Such a perspective is imperative in order to feed the world in a sustainable and profitable manner, and we have not scratched the surface of understanding of gut health from an immunometabolic perspective.

7 Where to look for further information

Immunometabolism is an emerging field and its application to animal agriculture is evolving. Those interested in further information on the topic will need to consult a variety of sources.

- For a general overview of avian physiology, including gut health and function, *Sturkie's Avian Physiology* (6th edn.). 2014. Scanes, C. G. (Ed.), San Diego: Academic Press is an excellent resource.
- Frontiers in Veterinary Science has published an ebook on the topic of gut health and animal production, *Gut Health: The New Paradigm in Food Animal Production*. 2016. Arsenault, R.J. and Kogut, M.H. (Eds), Lausanne, Switzerland: Frontiers Media, SA.
- Symposium on Gut Health in Production of Food Animals (https://guthealthsymposium.com) is an annual meeting that includes the latest research in immunometabolism of agriculturally important animals.
- A well-written introduction to immunometabolism from an immunology perspective has been written by Dr. Luke O'Neill, a foundational researcher in the field, and collaborators: O'Neill, L.A., Kishton, R.J. and Rathmell, J. 2016. A guide to immunometabolism for immunologists. *Nature Reviews Immunology* 16(9), 553.

8 References

Annison, E. F., Hill, K. J. and Kenworthy, R. 1968. Volatile fatty acids in the digestive tract of the fowl. *Br. J. Nutr.* 22(2), 207–16. doi:10.1079/BJN19680026.

Arsenault, R. J. and Kogut, M. H. 2012. Chicken-specific peptide arrays for kinome analysis: Flight for the flightless. *Biotechnology* 7, 79–89.

Arsenault, R., Griebel, P. and Napper, S. 2011. Peptide arrays for kinome analysis: New opportunities and remaining challenges. *Proteomics* 11(24), 4595–609. doi:10.1002/pmic.201100296.

Arsenault, R. J., Napper, S. and Kogut, M. H. 2013. *Salmonella enterica* Typhimurium infection causes metabolic changes in chicken muscle involving AMPK, fatty acid and insulin/mTOR signaling. *Vet. Res.* 44(1), 35. doi:10.1186/1297-9716-44-35.

Arsenault, R. J., Trost, B. and Kogut, M. H. 2014. A comparison of the chicken and turkey proteomes and phosphoproteomes in the development of poultry-specific immuno-metabolism kinome peptide arrays. *Front. Vet. Sci.* 1(22), 22. doi:10.3389/fvets.2014.00022.

Arsenault, R. J., Lee, J. T., Latham, R., Carter, B. and Kogut, M. H. 2017. Changes in immune and metabolic gut response in broilers fed β-mannanase in β-mannan-containing diets. *Poult. Sci.* 96(12), 4307–16. doi:10.3382/ps/pex246.

Azzout-Marniche, D., Gaudichon, C., Blouet, C., Bos, C., Mathé, V., Huneau, J. F. and Tomé, D. 2007. Liver glyconeogenesis: A pathway to cope with postprandial amino acid excess in high-protein fed rats? *Am. J. Physiol. Regul. Integr. Comp. Physiol.* 292(4), R1400–7. doi:10.1152/ajpregu.00566.2006.

Barrington, G. M., Gay, J. M. and Evermann, J. F. 2002. Biosecurity for neonatal gastrointestinal diseases. *Vet. Clin. North Am. Food Anim. Pract.* 18(1), 7–34. doi:10.1016/S0749-0720(02)00005-1.

Bensadoun, A. and Rothfeld, A. 1972. The form of absorption of lipids in the chicken, *Gallus domesticus*. *Proc. Soc. Exp. Biol. Med.* 141(3), 814–7. doi:10.3181/00379727-141-36878.

Bozzoli, C., Deaton, A. and Quintana-Domeque, C. 2009. Adult height and childhood disease. *Demography* 46(4), 647–69. doi:10.1353/dem.0.0079.

Buyse, J. and Decuypere, E. 2015. Adipose tissue and lipid metabolism. In: Scanes, C. G. (Ed.), *Sturkie's Avian Physiology* (6th edn.). San Diego: Academic Press, pp. 443–53. Chapter 19. doi:10.1016/B978-0-12-407160-5.00019-1.

Cheema, M. A., Qureshi, M. A. and Havenstein, G. B. 2003. A comparison of the immune response of a 2001 commercial broiler with a 1957 randombred broiler strain when fed representative 1957 and 2001 broiler diets. *Poult. Sci.* 82(10), 1519–29. doi:10.1093/ps/82.10.1519.

Chen, J., Zhang, X. and Fernández, A. 2007. Molecular basis for specificity in the druggable kinome: Sequence-based analysis. *Bioinformatics* 23(5), 563–72. doi:10.1093/bioinformatics/btl666.

Chi, H. 2012. Regulation and function of mTOR signalling in T cell fate decisions. *Nat. Rev. Immunol.* 12(5), 325–38. doi:10.1038/nri3198.

Choct, M. 2006. Enzymes for the feed industry: past, present and future. *Worlds Poult. Sci. J.* 62(1), 5–16. doi:10.1079/WPS200480.

Davis, J. F., Castro, A. E., de la Torre, J. C., Scanes, C. G., Radecki, S. V., Vasillatos-Younken, R., Doman, J. T. and Teng, M. 1995. Hypoglycemia, enteritis, and spiking mortality in Georgia broiler chickens: Experimental reproduction in broiler breeder chicks. *Avian Dis.* 39(1), 162–74. doi:10.2307/1591998.

Denbow, D. M. 2015. Gastrointestinal anatomy and physiology. In: Scanes, C. G. (Ed.), *Sturkie's Avian Physiology* (6th edn.). San Diego: Academic Press, pp. 337–66. Chapter 14. doi:10.1016/B978-0-12-407160-5.00014-2.

Dettmer, K., Aronov, P. A. and Hammock, B. D. 2007. Mass spectrometry-based metabolomics. *Mass Spectrom. Rev.* 26(1), 51–78. doi:10.1002/mas.20108.

Duarte, C. R. A., Vicentini-Paulino, M. L. M., Buratini, J., Castilho, A. C. S. and Pinheiro, D. F. 2011. Messenger ribonucleic acid abundance of intestinal enzymes and transporters in feed-restricted and refed chickens at different ages. *Poult. Sci.* 90(4), 863–8. doi:10.3382/ps.2010-01015.

Everts, B., Amiel, E., Windt, G. J. W. van der, Freitas, T. C., Chott, R., Yarasheski, K. E., Pearce, E. L. and Pearce, E. J. 2012. Commitment to glycolysis sustains survival of NO-producing inflammatory dendritic cells. *Blood* 120(7), 1422–31. doi:10.1182/blood-2012-03-419747.

Flint, H. J., Scott, K. P., Louis, P. and Duncan, S. H. 2012. The role of the gut microbiota in nutrition and health. *Nat. Rev. Gastroenterol. Hepatol.* 9(10), 577–89. doi:10.1038/nrgastro.2012.156.

Fritsche, K. L., Cassity, N. A. and Huang, S. C. 1991. Effect of dietary fat source on antibody production and lymphocyte proliferation in chickens. *Poult. Sci.* 70(3), 611-7. doi:10.3382/ps.0700611.

Garriga, C., Barfull, A. and Planas, J. M. 2004. Kinetic characterization of apical D-fructose transport in chicken jejunum. *J. Membr. Biol.* 197(1), 71-6. doi:10.1007/s00232-003-0640-0.

Genovese, K. J., He, H., Lowry, V. K., Swaggerty, C. L. and Kogut, M. H. 2006. Comparison of heterophil functions of modern commercial and wild-type Rio Grande turkeys. *Avian Pathol.* 35(3), 217–23. doi:10.1080/03079450600711029.

Griffin, H. D., Guo, K., Windsor, D. and Butterwith, S. C. 1992. Adipose tissue lipogenesis and fat deposition in leaner broiler chickens. *J. Nutr.* 122(2), 363–8. doi:10.1093/jn/122.2.363.

Griggs, J. P. and Jacob, J. P. 2005. Alternatives to antibiotics for organic poultry production. *J. Appl. Poult. Res.* 14(4), 750–6. doi:10.1093/japr/14.4.750.

Hermans, D., Pasmans, F., Heyndrickx, M., Van Immerseel, F., Martel, A., Van Deun, K. and Haesebrouck, F. 2012. A tolerogenic mucosal immune response leads to persistent *Campylobacter jejuni* colonization in the chicken gut. *Crit. Rev. Microbiol.* 38(1), 17–29. doi:10.3109/1040841X.2011.615298.

Holmgren, J., Czerkinsky, C., Lycke, N. and Svennerholm, A. M. 1992. Mucosal immunity: Implications for vaccine development. *Immunobiology* 184(2-3), 157–79. doi:10.1016/S0171-2985(11)80473-0.

Hooper, L. V., Littman, D. R. and Macpherson, A. J. 2012. Interactions between the microbiota and the immune system. *Science* 336(6086), 1268–73. doi:10.1126/science.1223490.

Horton, H. R., Moran, L. A., Ochs, R. S., Rawn, D. J. and Scrimgeour, K. G. 2002. *Principles of Biochemistry* (3rd edn.). Upper Saddle River, NJ: Prentice Hall.

Hume, M. E. 2011. Historic perspective: Prebiotics, probiotics, and other alternatives to antibiotics. *Poult. Sci.* 90(11), 2663–9. doi:10.3382/ps.2010-01030.

Johnson, C., Aylward, B., Whelan, R. and Arsenault, R. J. 2018. Comparison of modern and heritage broiler strain's responses to CpG using immunometabolic analysis of gut tissue. *XV Avian Immunology Research Group Meeting*, Oxford, UK.

Jones, E. A. 1972. Immunoglobulins and the gut. *Gut* 13(10), 825–35. doi:10.1136/gut.13.10.825.

Kawaguchi, T., Takenoshita, M., Kabashima, T. and Uyeda, K. 2001. Glucose and cAMP regulate the L-type pyruvate kinase gene by phosphorylation/dephosphorylation of the carbohydrate response element binding protein. *Proc. Natl. Acad. Sci. USA* 98(24), 13710–5. doi:10.1073/pnas.231370798.

Kelly, B. and O'Neill, L. A. 2015. Metabolic reprogramming in macrophages and dendritic cells in innate immunity. *Cell Res.* 25(7), 771–84. doi:10.1038/cr.2015.68.

Klasing, K. C. 1988a. Influence of acute feed deprivation or excess feed intake on immunocompetence of broiler chicks. *Poult. Sci.* 67(4), 626–34. doi:10.3382/ps.0670626.

Klasing, K. C. 1988b. Nutritional aspects of leukocytic cytokines. *J. Nutr.* 118(12), 1436–46. doi:10.1093/jn/118.12.1436.

Klasing, K. C. 2007. Nutrition and the immune system. *Br. Poult. Sci.* 48(5), 525–37. doi:10.1080/00071660701671336.

Klasing, K. C. and Korver, D. R. 1997. Leukocytic cytokines regulate growth rate and composition following activation of the immune system. *J. Anim. Sci.* 75, 58–67. doi:10.2134/animalsci1997.75Supplement_258x.

Korver, D. R. 2012. Implications of changing immune function through nutrition in poultry. *Anim. Feed Sci. Technol.* 173(1–2), 54–64. doi:10.1016/j.anifeedsci.2011.12.019.

Krawczyk, C. M., Holowka, T., Sun, J., Blagih, J., Amiel, E., DeBerardinis, R. J., Cross, J. R., Jung, E., Thompson, C. B., Jones, R. G., et al. 2010. Toll-like receptor-induced changes in glycolytic metabolism regulate dendritic cell activation. *Blood* 115(23), 4742–9. doi:10.1182/blood-2009-10-249540.

Leenstra, F. R. 1986. Effect of age, sex, genotype and environment on fat deposition in broiler chickens – A review. *Worlds Poult. Sci. J.* 42(1), 12–25. doi:10.1079/WPS19860002.

Mandell, J. W. 2003. Phosphorylation state-specific antibodies: Applications in investigative and diagnostic pathology. *Am. J. Pathol.* 163(5), 1687–98. doi:10.1016/S0002-9440(10)63525-0.

Mathew, A., Grdisa, M. and Johnstone, R. M. 1993. Nucleosides and glutamine are primary energy substrates for embryonic and adult chicken red cells. *Biochem. Cell Biol.* 71(5–6), 288–95. doi:10.1139/o93-043.

Mathis, D. and Shoelson, S. E. 2011. Immunometabolism: An emerging frontier. *Nat. Rev. Immunol.* 11(2), 81–3. doi:10.1038/nri2922.

Mayer, L. 2000. Mucosal immunity and gastrointestinal antigen processing. *J. Pediatr. Gastroenterol. Nutr.* 30, S4–12. doi:10.1097/00005176-200001001-00002.

Moreto, M., Amat, C., Puchal, A., Buddington, R. K. and Planas, J. M. 1991. Transport of L-proline and alpha-methyl-D-glucoside by chicken proximal cecum during development. *Am. J. Physiol. Gastrointest. Liver Physiol.* 260(3), G457–63. doi:10.1152/ajpgi.1991.260.3.G457.

Mutryn, M. F., Brannick, E. M., Fu, W., Lee, W. R. and Abasht, B. 2015. Characterization of a novel chicken muscle disorder through differential gene expression and pathway analysis using RNA-sequencing. *BMC Genomics* 16, 399. doi:10.1186/s12864-015-1623-0.

Niewold, T. A. 2007. The nonantibiotic anti-inflammatory effect of antimicrobial growth promoters, the real mode of action? A hypothesis. *Poult. Sci.* 86(4), 605–9. doi:10.1093/ps/86.4.605.

O'Neill, L. A. J. and Hardie, D. G. 2013. Metabolism of inflammation limited by AMPK and pseudo-starvation. *Nature* 493(7432), 346–55. doi:10.1038/nature11862.

Pagano, G., Johnson, C., Hahn, D. C. and Arsenault, R. J. 2018. A new tool for studying waterfowl immune and metabolic responses: Molecular level analysis using kinome profiling. *Ecol. Evol.* 8(16), 8537–46. doi:10.1002/ece3.4370.

Palsson-McDermott, E. M. and O'Neill, L. A. J. 2013. The Warburg effect then and now: From cancer to inflammatory diseases. *BioEssays* 35(11), 965–73. doi:10.1002/bies.201300084.

Patterson, J. A. and Burkholder, K. M. 2003. Application of prebiotics and probiotics in poultry production. *Poult. Sci.* 82(4), 627–31. doi:10.1093/ps/82.4.627.

Pearce, E. L. and Pearce, E. J. 2013. Metabolic pathways in immune cell activation and quiescence. *Immunity* 38(4), 633–43. doi:10.1016/j.immuni.2013.04.005.

Peeters, K., Leemputte, F. V., Fischer, B., Bonini, B. M., Quezada, H., Tsytlonok, M., Haesen, D., Vanthienen, W., Bernardes, N., Gonzalez-Blas, C. B., et al. 2017. Fructose-1,6-bisphosphate couples glycolytic flux to activation of Ras. *Nat. Commun.* 8(1), 922. doi:10.1038/s41467-017-01019-z.

Petracci, M., Mudalal, S., Babini, E. and Cavani, C. 2014. Effect of white striping on chemical composition and nutritional value of chicken breast meat. *Ital. J. Anim. Sci.* 13(1), 3138. doi:10.4081/ijas.2014.3138.

Pflughoeft, K. J. and Versalovic, J. 2012. Human microbiome in health and disease. *Annu. Rev. Pathol.* 7, 99–122. doi:10.1146/annurev-pathol-011811-132421.

Pollizzi, K. N. and Powell, J. D. 2015. Regulation of T cells by mTOR: The known knowns and the known unknowns. *Trends Immunol.* 36(1), 13–20. doi:10.1016/j.it.2014.11.005.

Preyat, N. and Leo, O. 2013. Sirtuin deacylases: A molecular link between metabolism and immunity. *J. Leukoc. Biol.* 93(5), 669–80. doi:10.1189/jlb.1112557.

Qureshi, M. A. and Havenstein, G. B. 1994. A comparison of the immune performance of a 1991 commercial broiler with a 1957 randombred strain when fed 'typical' 1957 and 1991 broiler diets. *Poult. Sci.* 73(12), 1805–12. doi:10.3382/ps.0731805.

Riesenfeld, G., Sklan, D., Bar, A., Eisner, U. and Hurwitz, S. 1980. Glucose absorption and starch digestion in the intestine of the chicken. *J. Nutr.* 110(1), 117–21. doi:10.1093/jn/110.1.117.

Rodríguez-Prados, J. C., Través, P. G., Cuenca, J., Rico, D., Aragonés, J., Martín-Sanz, P., Cascante, M. and Boscá, L. 2010. Substrate fate in activated macrophages: A comparison between innate, classic, and alternative activation. *J. Immunol.* 185(1), 605–14. doi:10.4049/jimmunol.0901698.

Roy, D. N. and Bird, H. R. 1959. Stimulation of chick growth by proline. *Poult. Sci.* 38(1), 192–6. doi:10.3382/ps.0380192.

Sale, E. M., White, M. F. and Kahn, C. R. 1987. Phosphorylation of glycolytic and gluconeogenic enzymes by the insulin receptor kinase. *J. Cell. Biochem.* 33(1), 15–26. doi:10.1002/jcb.240330103.

Scanes, C. G. 2015a. Carbohydrate metabolism. In: Scanes, C. G. (Ed.), *Sturkie's Avian Physiology* (6th edn.). San Diego: Academic Press, pp. 421–41. Chapter 18. doi:10.1016/B978-0-12-407160-5.00018-X.

Scanes, C. G. 2015b. Protein metabolism. In: Scanes, C. G. (Ed.), *Sturkie's Avian Physiology* (6th edn.). San Diego: Academic Press, pp. 455–67. Chapter 20. doi:10.1016/B978-0-12-407160-5.00020-8.

Seki, Y., Sato, K., Kono, T., Abe, H. and Akiba, Y. 2003. Broiler chickens (Ross strain) lack insulin-responsive glucose transporter GLUT4 and have GLUT8 cDNA. *Gen. Comp. Endocrinol.* 133(1), 80–7. doi:10.1016/S0016-6480(03)00145-X.

Sinsigalli, N. A., McMurtry, J. P., Cherry, J. A. and Siegel, P. B. 1987. Glucose tolerance, plasma insulin and immunoreactive glucagon in chickens selected for high and low body weight. *J. Nutr.* 117(5), 941–7. doi:10.1093/jn/117.5.941.

Soriano, M. E. and Planas, J. M. 1998. Developmental study of alpha-methyl-D-glucoside and L-proline uptake in the small intestine of the White Leghorn chicken. *Poult. Sci.* 77(9), 1347–53. doi:10.1093/ps/77.9.1347.

Sudo, S. Z. and Duke, G. E. 1980. Kinetics of absorption of volatile fatty acids from the ceca of domestic turkeys. *Comp. Biochem. Physiol. A Physiol.* 67(2), 231–7. doi:10.1016/0300-9629(80)90268-6.

Tallentire, C. W., Leinonen, I. and Kyriazakis, I. 2018. Artificial selection for improved energy efficiency is reaching its limits in broiler chickens. *Sci. Rep.* 8(1), 1168. doi:10.1038/s41598-018-19231-2.

Taschuk, R. and Griebel, P. J. 2012. Commensal microbiome effects on mucosal immune system development in the ruminant gastrointestinal tract. *Anim. Health Res. Rev.* 13(1), 129–41. doi:10.1017/S1466252312000096.

Teirlynck, E., Bjerrum, L., Eeckhaut, V., Huygebaert, G., Pasmans, F., Haesebrouck, F., Dewulf, J., Ducatelle, R. and Van Immerseel, F. 2009. The cereal type in feed influences gut wall morphology and intestinal immune cell infiltration in broiler chickens. *Br. J. Nutr.* 102(10), 1453–61. doi:10.1017/S0007114509990407.

Trost, B., Kindrachuk, J., Määttänen, P., Napper, S. and Kusalik, A. 2013. PIIKA 2: An expanded, web-based platform for analysis of kinome microarray data. *PLoS ONE* 8(11), e80837. doi:10.1371/journal.pone.0080837.

van der Windt, G. J. W., Everts, B., Chang, C. H., Curtis, J. D., Freitas, T. C., Amiel, E., Pearce, E. J. and Pearce, E. L. 2012. Mitochondrial respiratory capacity is a critical regulator of CD8+ T cell memory development. *Immunity* 36(1), 68–78. doi:10.1016/j.immuni.2011.12.007.

Vinardell, M. P. and Lopera, M. T. 1987. Jejunal and cecal 3-oxy-methyl-D-glucose absorption in chicken using a perfusion system in vivo. *Comp. Biochem. Physiol. A Comp. Physiol.* 86(4), 625–7. doi:10.1016/0300-9629(87)90612-8.

West, A. P., Brodsky, I. E., Rahner, C., Woo, D. K., Erdjument-Bromage, H., Tempst, P., Walsh, M. C., Choi, Y., Shadel, G. S. and Ghosh, S. 2011. TLR signalling augments macrophage bactericidal activity through mitochondrial ROS. *Nature* 472(7344), 476–80. doi:10.1038/nature09973.

Xu, H., Barnes, G. T., Yang, Q., Tan, G., Yang, D., Chou, C. J., Sole, J., Nichols, A., Ross, J. S., Tartaglia, L. A., et al. 2003. Chronic inflammation in fat plays a crucial role in the development of obesity-related insulin resistance. *J. Clin. Invest.* 112(12), 1821–30. doi:10.1172/JCI19451.

Zhang, B., Lv, Z., Li, H., Guo, S., Liu, D. and Guo, Y. 2017a. Dietary l-arginine inhibits intestinal *Clostridium perfringens* colonisation and attenuates intestinal mucosal injury in broiler chickens. *Br. J. Nutr.* 118(5), 321–32. doi:10.1017/S0007114517002094.

Zhang, J., Wang, S., Jiang, B., Huang, L., Ji, Z., Li, X., Zhou, H., Han, A., Chen, A., Wu, Y., et al. 2017b. c-Src phosphorylation and activation of hexokinase promotes tumorigenesis and metastasis. *Nat. Commun.* 8, 13732. doi:10.1038/ncomms13732.

Zhang, K., Griffiths, G., Repnik, U. and Hornef, M. 2018. Seeing is understanding: *Salmonella*'s way to penetrate the intestinal epithelium. *Int. J. Med. Microbiol.* 308, 97–106. doi:10.1016/j.ijmm.2017.09.011.

Zhao, J. P., Bao, J., Wang, X. J., Jiao, H. C., Song, Z. G. and Lin, H. 2012. Altered gene and protein expression of glucose transporter 1 underlies dexamethasone inhibition of insulin-stimulated glucose uptake in chicken muscles. *J. Anim. Sci.* 90(12), 4337–45. doi:10.2527/jas.2012-5100.

Chapter 7

Understanding gut function in poultry: the role of commensals, metabolites, inflammation and dysbiosis in intestinal immune function and dysfunction

Michael H. Kogut, USDA-ARS, USA

1 Introduction

The gastrointestinal tract, or 'gut', regulates homeostasis of the microbiological, physiological and physical functions that allows the host to endure infectious and non-infectious stressors that it encounters (Crhanova et al., 2011; Sansonetti, 2004; Maslowski and Mackay, 2011; Quinteiro-Filho et al., 2012). Because the gut has the greatest surface area separating the environmentally exposed lumen and the internal subepithelial tissue, it is constantly exposed to infectious and non-infectious stressors making it an active immune organ containing more resident immune cells than any other organ in the host. The gut mucosal immune system, a highly regulated network of innate and acquired elements, provides a remarkable ability to respond and modify to these extremely diverse encounters (Thaiss et al., 2014; Honda and Littman, 2016). The development of the different divisions of the immune response has corresponded with the acquisition and maintenance of a symbiotic microbiota. The microbiota trains, stimulates and functionally adjusts the different features of the immune system (Hooper and MacPherson, 2010; Hooper et al., 2012).

http://dx.doi.org/10.19103/AS.2019.0059.08
Published by Burleigh Dodds Science Publishing Limited, 2020.

2 Intestinal immunity

Like the systemic immune system, the mucosal immune system is made up of a network of innate and acquired elements. However, unlike the systemic immune system, the intestinal immune system has two distinct functions: the ability to respond to pathobionts (potential pathogenic microbes), invasive pathogens and microbial products while also maintaining a state of tolerance to the diverse and beneficial commensal intestinal microbes (Broom and Kogut, 2018). Both systems working together through innate immune sensing using pattern recognition receptors (PRRs) on epithelial cells and professional immune cells in the lamina propria (dendritic cells and macrophages) trigger immune pathways resulting in microbial killing and the activation of various acquired immune effector T cells (Th1, Th2, Th17, Treg), all the while keeping the resident microbiota in check without generating an overt inflammatory response.

The intestinal innate defences are characterized by a 'mucosal firewall', a system of barriers that separates the lumenal side of the intestine from the subepithelial tissues (Macpherson et al., 2009; Belkaid and Hand, 2014). The reliability of the mucosal firewall is vital for the interactions between the immune system components and the intestinal contents. The first component of the mucosal firewall is the microbiological barrier where the microbiota live in or at the upper mucus layer. These commensal bacteria function to provide colonization resistance against pathogen colonization, produce metabolites/components that modulate immune signalling and promote immune homeostasis (Garrett et al., 2010; Belkaid and Hand, 2014; Belkaid and Harrison, 2017). The second firewall is the chemical barrier consisting of the mucus overlaying the gut epithelium. The mucus regulates contact between the commensal bacteria and the epithelial cells. This division between the epithelium and commensals is achieved by the activity of the mucus produced by goblet cells in the epithelium, antimicrobial peptides released by the epithelial cells and mucosal IgA produced by dendritic cells in the intestine (Vaishnava et al., 2011; Belkaid and Harrison, 2017). The third component of the firewall is the physical barrier provided by the single cell epithelial cell layer. The intestinal epithelium is a single cell layer that assists the absorption of nutrients while providing a physical barrier that prevents both pathogen invasion and extra-intestinal translocation of commensal microbes. Besides being the primary barrier preventing a microbial breach of the intestine, the epithelial cells should also be considered a part of the cellular component of the innate immune response possessing PRRs for sensing microbial-associated molecular patterns (MAMPS), but also capable of producing cytokines and chemokines to drive an inflammatory response against pathogen infection. The final component of the mucosal firewall is the immunological barrier where the professional

Published by Burleigh Dodds Science Publishing Limited, 2020.

immune cells (macrophages, dendritic cells, lymphocytes) reside in the lamina propria (Abraham and Medzhitov, 2011). Further innate sensing of microbes is conducted by the macrophages and dendritic cells which can present antigens to T cells resulting in the differentiation and activation of various T-cell subsets (Th1, Th2, Th17 or Treg) (Abraham and Medzhitov, 2011). Specialized epithelial cells of the gastrointestinal tract function together with lymphoid, myeloid and stromal cells to secrete mucus, antimicrobial peptides, IgA and chemokines that limit direct contact between the epithelium and infectious agents and activate target cells that mediate innate defences (Medzhitov, 2001; Akira et al., 2006; Abreu et al., 2005; Kawai and Akira, 2009; Mantis and Forbes, 2010). The importance of these epithelial defence mechanisms is highlighted by the ability of enteric pathogens to target these mechanisms to achieve invasion and dissemination (Awad et al., 2017; Lu and Walker, 2001; Fischbach et al., 2006; Goto et al., 2017; Alemka et al., 2012).

This infiltration of immune cell in lamina propria is inversely correlated with weight gain (Belote et al., 2018), showing that this final component of the mucosal firewall has a metabolic cost for the host that affects animal performance (Kogut and Klasing, 2009).

3 Microbiota interactions with the immune system

The host-microbiota interaction that affects the host metabolism, immunity and health is exceedingly complex (Marchesi et al., 2016). This crosstalk is mediated by dietary nutrients, host and microbiota metabolites, microbial structural components, as well as antimicrobial compounds. Microbiota growth and anatomical location are regulated by the host through production of non-specific antimicrobial peptides such as defensins (Xiao et al., 2006; Bommineni et al., 2014), IgA (Lammers et al., 2010; Den Hartog et al., 2016) and miRNAs that regulate bacterial transcripts and bacterial growth (Liu et al., 2016).

The commensal microbes in the intestinal tract sense the local environment to induce biochemical pathways to activate bacterial metabolism that allows them to avoid, alter and/or survive host innate immune killing. Furthermore, some microbial-based molecules can promote specific commensal processes that are beneficial to both host and microbe. Similarly, the host detects the microbes through their production of specific molecules or components with unique molecular patterns that leads to the activation of innate and acquired immune responses. Thus, the adaptation of the commensal bacteria (as well as viruses and fungi) living in the intestine of a host has resulted in a mutually beneficial coexistence for both microbiota and host during homeostasis (Kogut, 2013; Kogut et al., 2017; Broom and Kogut, 2018). The interdependent relationship between the host and microbiota pointedly influences the host immune response to induce an immune tolerance to commensal microbes while also maintaining responsiveness to invading pathogens (Bene et al., 2017; Guo et al., 2017; Shi et al., 2017). Altering the intestinal microbial communities disturbs this immune balance and leads to immune dysregulation and susceptibility to diseases.

Sensing of the microbiota by PRRs generates a number of mechanisms that promote the host-microbiota relationship while preventing infection by pathogenic organisms. Microbial signals induce pro-inflammatory cytokines such as IL-23 and IL-1β from macrophages and DCs that then activate IL-17 and IL-22 production by T cells, leading to the production of steady-state physiological inflammation (Kogut et al., 2018). DCs can carry microbiota antigens to the Peyer's patches and/or small lymphoid follicles in the avian intestine, where they drive the differentiation of regulatory T cells (Tregs) and Th17 T cells that, in turn, induce the differentiation of IgA-producing plasma B cells that secrete further amounts of IgA.

3.1 Microbiota-based metabolites and immunity

The microbiota is directly engaged in maintaining the functional innate immunity of the host. The host immune system consistently senses the intestinal microenvironment to determine the metabolic state and colonization status (Levy et al., 2016). In the steady state, the metabolites and/or components of the commensal microbiota are recognized by various PRRs, including toll-like receptors (TLRs) and NOD-like receptors (NLRs), to regulate intestinal epithelial barrier function and cellular lifespan of phagocytes, and to induce secretion of antimicrobial peptides and IgA (Levy et al., 2016; Blacher et al., 2017). Furthermore, beneficial bacteria ferment dietary fibres to produce small chain fatty acids (SCFA) which stimulate the production of anti-inflammatory cytokines (Levy et al., 2016; Blacher et al., 2017) that drives the production of regulatory T cells (Tregs). In addition, the microbiota influences the priming signal of the inflammasome activation that leads to the transcription of IL-6, as well as pro-IL-1β and pro-IL-18. The gut microbiota is involved in maintaining intestinal immune homeostasis by stimulating different arms of the T-cell response. Segmented filamentous bacteria (SFB) are potent promoters of Th17 cells in the intestine, whereas polysaccharide A from the commensal *Bacteroides fragilis* stimulates the generation of Tregs (Levy et al., 2017). Alternatively, pattern recognition by TLRs and NLRs can also induce the maintenance of tolerance (Levy et al., 2017).

Lastly, it has become readily apparent that the intestinal immune system can also detect the metabolic state of the microbiota by recognition of microbial metabolites via their PRRs (Levy et al., 2017; Blacher et al., 2017). The microbiota, using a number of biochemical pathways, metabolizes both diet- and host-derived metabolites that then influence various components of the intestinal immune system. For example, the microbiota converts non-digestible fibres to SCFA that have a number of anti-inflammatory activities (Postler and Ghosh, 2017). Dietary tryptophan can be degraded by the microbiota into indoles which promote epithelial cell barrier function (Postler and Ghosh, 2017). Likewise, the microbiota can metabolize dietary arginine to polyamines that inhibit the production of pro-inflammatory cytokines by macrophages (Postler and Ghosh, 2017). The microbiota converts primary host-derived hepatic bile acids to secondary bile acids that inhibit pro-inflammatory cytokine secretion by DCs and macrophages (Thaiss et al., 2014). Besides having a repertoire of metabolite-sensing receptors, the host has developed immune signalling pathways (inflammasomes) expressed in various intestinal cell subsets (macrophages, DCs, epithelial cells, T cells) that recognize microbial-mediated metabolic activity that can stimulate anti-microbial activity involved in stable colonization of the intestine (Levy et al., 2015; Wang et al., 2015; Birchenough et al., 2016).

Therefore, there is intimate crosstalk between the microbiota and the host that is steered by metabolite secretion and immune signalling that has critical influence in animal health and disease through multiple physiological functions of the host.

3.2 Colonization resistance

The commensal bacteria also provide protection to the host from colonization by exogenous pathogens by a process known as colonization resistance (Van der Waaij et al., 1971; Buffie and Pamer, 2013; Rangan and Hang, 2017). Two primary mechanisms of colonization resistance have been identified: direct, where the microbiota are in direct competition with pathogen colonization, and indirect, where the commensal microbiota stimulate the innate and acquired immune systems as described in the previous sections.

The direct competition of colonization resistance involves multiple processes that include the following: (1) *occupying microbial niches* - specific commensal microbes can prevent pathogen colonization of the intestinal mucosa by occupying the niche where a pathogen would normally establish (Belkaid and Hand, 2014; Sassone-Corsi and Raffatellu, 2015); (2) *limiting carbon sources* - individual commensals, such as *Bacteroides thetaiotaomicron* can metabolize fucose (sugar) molecules, thereby preventing the availability of this sugar moiety for certain pathogens in the intestine (Belkaid and Hand, 2014; Sassone-Corsi and Raffatellu, 2015); (3) *siderophore production* - some commensals possess the genes for the production and acquisition of the metal ion (iron) via iron chelators (siderophores) that can uptake iron limiting its availability to pathogens, especially during gut inflammation (Belkaid and Hand, 2014; Deriu et al., 2013); (4) *production of antimicrobial compounds* - some Enterobacteriaceae commensals produce antimicrobial compounds, such as bacteriocins, that target competitor pathogens (Chassaing and Cascales, 2018); (5) *contact-dependent delivery of toxins* - some commensals can express a type VI secretion system (T6SS), a needle-like injection system that injects toxic proteins into close competitors in a contact-dependent manner (Pezoa et al., 2013; Sana et al., 2016; Chassaing and Cascales, 2018).

The indirect mechanisms of colonization resistance against enteric pathogens are mediated by microbiota-stimulated activation of both host innate and acquired immunity (Belkaid and Hand, 2014; Sassone-Corsi and Raffatellu, 2015; Rangan and Hang, 2017). Commensal bacteria can indirectly control pathogen colonization by stimulating the intestinal barrier function and innate immunity as described above. In this case, the commensal bacteria, through the production of metabolites or release of surface components (LPS, peptidoglycans, DNA etc.), are recognized by the PRRs on the intestinal epithelial and professional immune cells that result in the production and secretion of mucins, secretory IgA (sIgA) and antimicrobial peptides, all of which either increase the barrier function of the mucosal firewall or are lethal to pathogens (Belkaid and Hand, 2014; Sassone-Corsi and Raffatellu, 2015). Furthermore, the commensal microbiota can enhance epithelial barrier function by producing SCFA, such as butyrate, from dietary fibres (Guilloteau et al., 2010). T-cell subsets in the intestinal lamina propria are involved in the establishment and maintenance of colonization resistance. Diverse populations of commensal bacteria in the intestine generate a balanced T-helper/T-regulatory status. For example, segmented filamentous bacteria promote acquired immunity by T cells by stimulating Th17 cells whereas other commensals, such as *Clostridium* and *Bacteroides fragilis*, induce the expansion of T-regulatory cells that can regulate inflammatory responses through the production of IL-10 (Lee and Mazmanian, 2010; Round and Mazmanian, 2009; O'Mahony et al., 2008; Ivanov et al., 2009).

4 Gut microbiota as an epigenetic regulator of gut function

Epigenetics involves genomic modifications through post-translational and post-transcriptional modifications induced by environmental factors, but without modifying the nucleotide sequence of the host cell (Shenderov, 2012). Epigenetic mechanisms regulate transcriptional control by external environmental cues such as diet, stress events, disease, infections and host-microbe crosstalk (Shenderov, 2012; Chen et al., 2017; Woo and Alenghat, 2017; Grabiec and Potempa, 2018). Since epigenetic events do not alter the DNA, the epigenomic effects are associated with the attachment of different chemical groups to DNA, histones and chromatin post-translationally and the epigenetic alterations can persist for several generations (Furrow et al., 2011; Shenderov, 2012). These epigenetic alterations both affect the chromatin structure and serve as recognition elements for proteins with motifs dedicated to binding particular modifications.

Since the gut microbiota plays such a pivotal role in poultry metabolism, microbiota-induced epigenetic alterations by dietary nutrients could be a significant environmental factor affecting poultry performance and health. Based on studies in mammals, microbiota-generated metabolites of dietary components can be epigenetic activators of gene expression that modify or inhibit enzymes involved in epigenetic pathways (Alenghat et al., 2013; Alenghat and Artis, 2014; Hullar and Fu, 2014; Woo and Alenghat, 2017). This can best be exemplified by the production of SCFA (acetate, propionate, butyrate) produced by intestinal microbiota by bacterial fermentation of non-digestible carbohydrates (Hu and Guo, 2007; Guilloteau et al., 2010; Jiang et al., 2015). Butyrate is best known for its beneficial effects on intestinal barrier function, anti-inflammatory activity and as the primary source of energy to intestinal epithelial cells (Hamer et al., 2008; Guilloteau et al., 2010; Leonel and Alvarez-Leite, 2012; Huang et al., 2015). Butyrate regulates these biological activities of host gut health by functioning as a histone deacetylase inhibitor (HDAC) (Canani et al., 2012; Liu et al., 2018). Butyrate anti-inflammatory activity is mediated by HDAC suppression of NF-κB in phagocytic cells and dendritic cells, increased production of anti-inflammatory cytokines and increased differentiation of naive T cells into T-regulatory cells (Usami et al., 2008; Arpaia et al., 2013; Furusawa et al., 2013; Smith et al., 2013; Chang et al., 2014). Other microbial metabolites derived from dietary components, such as nicotinamide adenine dinucleotide (NAD)-dependent deacetylases called surtuins, have been shown to mediate the regulation of epigenetic modifications, including DNA methylation noncoding RNAs and histone modification, in the host intestinal immune barrier function of mammals (Kobayashi et al., 2012; Ganal et al., 2012; Singh et al., 2012). Further research is needed to determine whether such gut microbiota metabolite-mediated epigenetic modifications of the immuno-barrier function occur in the poultry intestine.

5 Dysregulation of gut functionality

Gut function is regulated by a number of factors including mucosal immunity, the gut microbiota/microbiome and extrinsic environmental factors, such as diet and stress (e.g. overcrowding or heat stress). Specifically, the microbiota can be altered by the ingestion of antibiotics, infection by pathogens, diet and other host- and environmental-dependent events. The plasticity of the microbiome has been implicated in numerous disease conditions, and an unfavourable alteration of the commensal structure of gut microbiota is referred to as 'dysbiosis', which includes a reduction in the number of tolerogenic bacteria and an overgrowth of potentially pathogenic bacteria (pathobionts) that can penetrate the intestinal epithelium and induce disease in certain genetic or environmental contexts (Chow and Mazmanian, 2010; Oakley et al., 2014; Ducatelle et al., 2018).

Dysbiosis (dysbacteriosis) is considered a perturbation in the microbiota composition and function resulting in an imbalance between beneficial and harmful bacteria that result in an aberrant immune response against the commensal bacteria. In a healthy gut, bacteria work synergistically, with each bacteria being an integral link in the production of metabolites that are used by the host. However, during dysbiosis an aberrant host-inflammatory response targets commensal bacteria resulting in a loss of bacterial diversity and essential functions with subsequent reduced metabolic activity depriving the host of multiple end products that reduces gut health and bird performance (Cisek and Binek, 2014; Ducatelle et al., 2015). At this time we do not have an understanding of the exact drivers of dysbiosis in poultry, although intestinal inflammation is probably involved (Ducatelle et al., 2018; Kogut et al., 2018), but the question to be answered is whether this is the cause or a consequence of the disruption of the host-microbe gut homeostasis (Zeng et al., 2017; Singh et al., 2016; Sommer et al., 2017; Weiss and Hennet, 2017). The natural differences in the diversity of the gut microbiota of the phyla Firmicutes and Bacteroides (Oakley et al., 2014) deteriorate into dysbiosis when conditions decrease the abundance of these obligate anaerobes and promote the growth of facultative bacteria of the phylum Proteobacteria, specifically members of the order Enterobacteriaceae including *Salmonella* and *Clostridium perfringens*.

5.1 Chronic, low-grade inflammation

5.1.1 Sterile inflammation

A low-grade chronic inflammation, in the absence of an infection, in response to a chemical (oxidative stress), physical (microbiota-derived components) and metabolic (non-digestible compounds from diet) stimuli is known as sterile inflammation (Rubartelli et al., 2013). With a sterile inflammatory response, the stimulus persists without being eliminated suggesting that collateral damage is the cause of the disease (Rock et al., 2010). In sterile inflammation, PRRs not only recognize microbiome-derived components like LPS, peptidoglycan (PGN) and bacterial DNA (CpG) following microbiome bacterial cell wall turnover or bacterial lysis, but also DAMPs released from host cells after oxidative stress-induced cell death (Rock et al., 2010). Furthermore, certain ingredients used in commercial broiler diets contain dietary non-starch polysaccharides (NSP) that are indigestible by poultry, but represent a potential energy source which can be utilized with the proper addition of exogenous enzymes (Meng et al., 2005). For example, soybean meal is a primary source of vegetable protein that contains around 3% soluble NSP and 16% insoluble NSP, consisting mainly of mannans and galactomannans (Slominski, 2011). β-mannan is similar in structure to surface components of multiple pathogenic microbes and is recognized by a specific PRR, the mannose receptor, that stimulates the intestinal innate immune system leading to a purposeless energy-draining immune response called feed-induced immune response (FIIR) (Hsiao et al., 2006; Martinez-Cummer, 2015). In addition, gliadin, a component of wheat gluten, has been shown to alter expression of tight junction proteins and increase intestinal permeability, which has been implicated to contribute to the induction of inflammation (De Punder and Pruimboom, 2015).

The largest potential source of endogenous triggers of sterile inflammation in the intestine is the 10^7–10^{11} bacteria/g in the microbiota. For example, PGN is the foremost component of gram-positive bacterial cell walls as well as being present in gram-negative bacterial cells walls (McDonald et al., 2005). Likewise, LPS (gram-negative bacterial cell wall) and unmethylated CpG oligodeoxynucleotide of all prokaryotic DNA are all ligands of host TLRs and NLRs (Keestra et al., 2013; He et al., 2012). Both PGN and LPS are shed when bacteria in the microbiota divide and all components are released upon bacterial lysis (Myhre et al., 2006). Under homeostatic conditions, these components play a role in physiological inflammation. However, under stress conditions, such as feed-induced inflammation, oxidative and heat stress, or feeding diets with high levels of NSPs that increased feed passage time and/or decrease intestinal motility, as well as the frequency of reverse peristalsis, results in prolonging the exposure to these triggers resulting in the induction of a low-grade sterile inflammatory response with resultant increase of dysbiosis and dysfunction of a healthy gut (Sacranie et al., 2007).

5.1.2 Metabolic inflammation

A second type of chronic, low-grade inflammation generated by excessive nutrient intake and the metabolic surplus fosters intestinal dysfunction by activating the same signalling transduction molecules and pathways as immune responses to infections (Hotamisligil, 2006; Kogut, 2013). Metabolic dysfunction appears to take centre stage by integrating signals from both the immune and metabolic systems (Kogut, 2013). Nutrient excess from diets can result in overloads of metabolites that act as DAMPs that can be sensed by PRRs of the intestinal innate immune system (Assamann and Finlay, 2016; Land, 2015; Lackey and Olefsky, 2016), thereby simulating a chronic inflammation. PRRs, such as TLR4 and NLRP3, act as metabolite sensors activated by different metabolite DAMPs including free fatty acids, carbohydrates and lipids (Gregor and Hotamisligil, 2011). Continuous activation of these PRRs through excessive nutrient intake leads to the production of pro-inflammatory cytokines, such as IL-1β and IL-6, which maintain the low-grade inflammatory response.

When nutrient excess leads to activation of PRR-mediated inflammation, there is a concurrent increase in metabolic pathways in the mitochondria of activated immune cells (Tannahill et al., 2013; McGettrick and O'Neill, 2013) that results in the overproduction of various metabolites via the Krebs cycle (succinate, citrate, NAD^+) and glycolysis (lactate) (Tannahill et al., 2013; McGettrick and O'Neill, 2013; Tretter et al., 2016). All four metabolites have been shown to be signals for multiple inflammatory mediators that result in a low-grade chronic inflammation (Tannahill et al., 2013; Loftus and Finlay, 2016). Increased succinate production results in an increase in stabilization of the transcription factor, hypoxia-inducible factor (HIF-1α) (Tannahill et al., 2013; Pucino et al., 2017). The stabilization of HIF-α results in increased glycolysis and the persistent production of the pro-inflammatory cytokine IL-1β (Tannahill et al., 2013; Loftus and Finlay, 2016). Likewise, citrate is also increased via the Krebs cycle and is a signal for the increased production of inflammatory mediators such as reactive oxygen species, nitric oxide and prostaglandins (Infantino et al., 2011). Lactate is a waste product produced in the immune cell cytoplasm at the end of glycolysis that affects local T-cell immunity by inhibiting T-cell motility, inducing the change of CD4+ cells to Th17 pro-inflammatory T-cell subset that leads to the production of IL-17, which is a hallmark of chronic inflammation (Haas et al., 2015).

5.1.3 Pathobiont expansion

Regardless of the source of inflammation, the inflamed gut provides a physiological, chemical and nutritive opportunity for pathobionts in the microbiota to exploit for their own benefit. Although dietary changes can lead to alterations in microbiota composition, these alterations are confined to the species level (Sonnenburg and Backhed, 2016). On the other hand, gut inflammation in poultry provides both an environmental condition that can be exploited and/or sustained by the pathobionts for growth over the competing commensal microbiota (Zeng et al., 2017) and stimulates the activation of genes that code for pathways of physiologic and metabolic properties that allow them to survive and 'bloom' in the inflamed gut (Stecher, 2015). For example, during inflammation thiosulphate from damaged intestinal epithelium is oxidized to tetrathionate by reactive oxygen species which is then utilized by *Salmonella* as an electron acceptor for anaerobic respiration so that the pathogen can outcompete the microbiota (Winter et al., 2010). Furthermore, tetrathionate is an electron acceptor that allows *Salmonella* to utilize ethanolamine released from the host tissue for fermentation while the majority of gut microbiota are unable to use ethanolamine (Thiennimitr et al., 2011). In addition, in the inflamed gut, *Salmonella* is resistant to a variety of host-derived defensins and antimicrobial peptides (Zeng et al., 2017; Stecher, 2015) including lipocalin-2

Published by Burleigh Dodds Science Publishing Limited, 2020.

which is an antimicrobial protein that prevents iron acquisition by commensal bacteria by binding to the siderophore, enterobactin. However, *Salmonella* produces salmochelin, a glycosylated variant of enterobactin, which is not bound by lipocalin-2 and uses the iron transporter, thus giving it a competitive advantage over the iron-negative commensals (Raffatellu et al., 2009). The dysbiotic depletion of Firmicutes, that is clostridia, results in the reduction in butyrate production, which induces a switch in the intestinal epithelial cells from oxidative metabolism to lactate fermentation. *Salmonella* is able to use the lactate as a nutrient further allowing it to outcompete the commensal bacteria (Gillis et al., 2018). Lastly, under homeostatic conditions, the commensal microbiota are extremely efficient at consuming sugars leaving limited sugars available to pathobionts. During inflammation-induced dysbiosis, the reduction of commensal competitors for the sugars results in *Salmonella* and *C. perfringens* that are able to exploit the carbohydrate pool for growth (Ng et al., 2013).

6 Future trends and conclusion

The maintenance of a healthy status is complex and relies on a delicate balance between the immune system and the normal endogenous microbiota (Brisbane et al., 2005; Mwangi et al., 2010; Oakley et al., 2014). The normal microbiota confers many benefits to the intestinal physiology of the host. However, when this balance is upset (dysbiosis), pathogens that arrive or that have already been present, but in numbers too small to cause disease, take the opportunity to multiply. The intestinal microbiota is a positive health asset to poultry health that influences the normal structural and functional development of the mucosal immune response. Future studies building on the gene and organism catalogues established thus far will need to include increasingly detailed investigations of meta-transcriptomes and meta-proteomes. These studies will help to answer questions concerning the role host immunity or genetics play in shaping patterns of diversity, elucidating the functional changes that dictate the microbiome functions in given contexts, its interactions with the host and functional alterations that accompany the conversion of a healthy microbiome towards a disease-driving configuration, and allow us to more fully understand the links between the chicken microbiome, health and disease.

7 References

Abraham, C. and Medzhitov, R. 2011. Interaction between the host innate immune system and microbes in inflammatory bowel disease. *Gastroenterology* 140(6), 1729-37. doi:10.1053/j.gastro.2011.02.012.

Abreu, M. T., Fukata, M. and Arditi, M. 2005. TLR signaling in the gut in health and disease. *J. Immunol.* 174(8), 4453-60. doi:10.4049/jimmunol.174.8.4453.

Akira, S., Uematsu, S. and Takeuchi, O. 2006. Pathogen recognition and innate immunity. *Cell* 124(4), 783–801. doi:10.1016/j.cell.2006.02.015.

Alemka, A., Corcionivoschi, N. and Bourke, B. 2012. Defense and adaptation: The complex inter-relationship between *Campylobacter jejuni* and mucus. *Front. Cell. Infect. Microbiol.* 2, Article 15. doi:10.3389/fcimb.2012.00015.

Alenghat, T. and Artis, D. 2014. Epigenomic regulation of host-microbe interactons. *Trends Immunol.* 35(11), 518–25. doi:10.1016/j.it.2014.09.007.

Alenghat, T., Osborne, L. C., Saenz, S. A., Kubuley, D., Ziegler, C. G. K., Mullican, S. E., Choi, I., Grunberg, S., Sinha, R., Wynosky-Dolfi, M., et al. 2013. Histone deacetylase 3 coordinates commensal-bacteria-dependent intestinal homeostasis. *Nature* 504(7478), 153–7. doi:10.1038/nature12687.

Arpaia, N., Campbell, C., Fan, X., Dikly, S., van der Veeken, J., deRoos, P., Liu, H., Cross, J. R., Pfeffer, K., Coffer, P. J., et al. 2013. Metabolites produced by commensal bacteria promote peripheral regulatory T-cell generation. *Nature* 504(7480), 451–5. doi:10.1038/nature12726.

Assamann, N. and Finlay, D. K. 2016. Metabolic regulation of immune responses: Therapeutic opportunities. *J. Clin. Invest.* 126(6), 2031–9. doi:10.1172/JCI83005.

Awad, W. A., Hess, C. and Hess, M. 2017. Enteric pathogens and their toxin-induced disruption of the intestinal barrier through alteration of tight junctions in chickens. *Toxins* 9(2), 60. doi:10.3390/toxins9020060.

Belkaid, Y. and Hand, T. W. 2014. Role of the microbiota in immunity and inflammation. *Cell* 157(1), 121–41. doi:10.1016/j.cell.2014.03.011.

Belkaid, Y. and Harrison, O. J. 2017. Homeostatic immunity and the microbiota. *Immunity* 46(4), 562–76. doi:10.1016/j.immuni.2017.04.008.

Belote, B. L., Tujimoto-Silva, A., Hummelgen, P. H., Sanches, A. W. D., Wammes, J. C. S., Hayashi, R. M. and Santin, E. 2018. Histological parameters to evaluate intestinal health on broilers challenged with *Eimeria* and *Clostridium perfringens* with or without enramycin as growth promoter. *Poult. Sci.* 97(7), 2287–94. doi:10.3382/ps/pey064.

Bene, K., Varga, Z., Petrov, V. O., Boyko, N. and Rahnavolgyi, E. 2017. Gut microbiota species can invoke both inflammatory and tolerogenic immune responses in human dendritic cells mediated by retinoic acid receptor alpha ligation. *Front. Immunol.* 8, Article 427. doi:10.3389/fimmu.2017.00427.

Birchenough, G. M. H., Nystrom, E. E. L., Johansson, M. E. V. and Hansson, G. C. 2016. A sentinel goblet cell guards the colonic crypt by triggering NLRP6-dependent Muc2 secretion. *Science* 352(6293), 1535–42. doi:10.1126/science.aaf7419.

Blacher, E., Levy, M., Tatirovsky, E. and Elinov, E. 2017. Microbiome-modulated metabolites at the interface of host immunity. *J. Immunol.* 198(2), 572–80. doi:10.4049/jimmunol.1601247.

Bomminieni, Y. R., Pham, G. H., Sunkara, L. T., Achanta, M. and Zhang, G. 2014. Immune regulatory activities of fowlcidin-1, a cathelicidin host defense peptide. *Mol. Immunol.* 59(1), 55–63. doi:10.1016/j.molimm.2014.01.004.

Brisbane, J. T., Gong, J. and Sharif, S. 2005. Interactions between commercial bacteria and the gut-associated immune systems of the chicken. *Anim. Health Res. Rev.* 9, 101–10.

Broom, L. J. and Kogut, M. H. 2018. Inflammation: Friend or foe for animal production? *Poult. Sci.* 97(2), 510–4. doi:10.3382/ps/pex314.

Buffie, C. G. and Pamer, E. G. 2013. Microbiota-mediated colonization resistance against intestinal pathogens. *Nat. Rev. Immunol.* 13(11), 790–801. doi:10.1038/nri3535.

Canani, R. B., DiCostanzo, M. and Leone, L. 2012. The epigenetic effects of butyrate: Potential therapeutic implications for clinical practice. *Clin. Epigenetics* 4(1), 4. doi:10.1186/1868-7083-4-4.

Chang, P. V., Hao, L., Offermanns, S. and Medzhitov, R. 2014. The microbial metabolite butyrate regulates intestinal macrophage function via histone deacetylase inhibition. *Proc. Natl. Acad. Sci. U.S.A.* 111(6), 2247–52. doi:10.1073/pnas.1322269111.

Chassaing, B. and Cascales, E. 2018. Antibacterial weapons: Targeted destruction in the microbiota. *Trends Microbiol.* 26(4), 329–38. doi:10.1016/j.tim.2018.01.006.

Chen, B., Sun, L. and Zhang, X. 2017. Integration of microbiome and epigenome to decipher the pathogenesis of autoimmune diseases. *J. Autoimmun.* 83, 31–42. doi:10.1016/j.jaut.2017.03.009.

Chow, J. and Mazmanian, S. K. 2010. A pathobionts of the microbiota balances host colonization and intestinal inflammation. *Cell Host Microbe.* 7(4), 265–76. doi:10.1016/j.chom.2010.03.004.

Cisek, A. A. and Binek, M. 2014. Chicken intestinal microbiota function with a special emphasis on the role of probiotic bacteria. *Pol. J. Vet. Sci.* 17, 385–94.

Crhanova, M., Hradecka, H., Faldynova, M., Matulova, M., Havlickova, H., Sisak, F. and Rychlik, I. 2011. Immune response of chicken gut to natural colonization by gut microflora and to *Salmonella enterica* serovar Enteritidis infection. *Infect. Immun.* 79(7), 2755–63. doi:10.1128/IAI.01375-10.

De Punder, K. and Pruimboom, L. 2015. Stress induces endotoxemia and low-grade inflammation by increasing barrier permeability. *Front. Immunol.* 6, 223. doi:10.3389/fimmu.2015.00223.

Den Hartog, G., De Vries-Reilingh, G., Wehmaker, A. M., Savelkoul, H. F. J., Parmentier, H. K. and Lammers, A. 2016. Intestinal immune maturation is accompanied by temporal changes in the composition of the microbiota. *Benefic. Microbes* 7(5), 677–85. doi:10.3920/BM2016.0047.

Deriu, E., Liu, J. Z., Pezeshki, M., Edwards, R. A., Ochoa, R. J., Contreras, H., Libby, S. J., Fang, F. F. and Raffatellu, M. 2013. Probiotic bacteria reduce *Salmonella* Typhimurium intestinal colonization by competing for iron. *Cell Host Microbe.* 14(1), 26–37. doi:10.1016/j.chom.2013.06.007.

Ducatelle, R., Eeckhaut,V., Haesebrouke, F. and van Immerseel, F. 2015. A review on prebiotics and probiotics for the control of dysbiosis: Present status and future perspectives. *Animal* 9(1), 43–8. doi:10.1017/S1751731114002584.

Ducatelle, R., Goossens, E., De Meyer, F., Eeckhaut, V., Antonissen, G., Haesebrouck, F. and Van Immerseel, F. 2018. Biomarkers for monitoring intestinal health in poultry: Present status and future perspectives. *Vet. Res.* 49(1), 43. doi:10.1186/s13567-018-0538-6.

Fischbach, M. A., Lin, H., Zhou, L., Yu, Y., Abergel, R. J., Liu, D. R., Raymond, K. N., Wanner, B. L., Strong, R. K., Walsh, C. T., et al. 2006. The pathogen-associated *iroA* gene cluster mediates bacterial evasion of lipocalin-2. *Proc. Natl. Acad. Sci. U.S.A.* 103(44), 16502–7. doi:10.1073/pnas.0604636103.

Furrow, R. E., Christiansen, F. B. and Feldman, M. W. 2011. Environment-sensitive epigenetics and the heritability of complex diseases. *Genetics* 189(4), 1377–87. doi:10.1534/genetics.111.131912.

Furusawa, Y., Obata, Y., Fukuda, S., Endo, T. A., Nakato, G., Takahashi, D., Nakanishi, Y., Uetake, C., Kato, K., Kato, T., et al. 2013. Commensal microbe-derived butyrate induces the differentiation of colonic regulatory T cells. *Nature* 504(7480), 446-50. doi:10.1038/nature12721.

Ganal, S. C., Sanos, S. L., Kalifass, C., Oberle, K., Johner, C., Kirschning, C., Lienenklaus, S., Weiss, S., Staeheli, P., Aichele, P., et al. 2012. Priming of natural killer cells by nonmucosal mononuclear phagocytes requires instructive signals from commensal microbiota. *Immunity* 37(1), 171-86. doi:10.1016/j.immuni.2012.05.020.

Garrett, W. S., Gordon, J. I. and Glimcher, L. H. 2010. Homeostasis and inflammation in the intestine. *Cell* 140(6), 859-70. doi:10.1016/j.cell.2010.01.023.

Gillis, C. C., Hughes, E. R., Spiga, L., Winter, M. G., Zhu, W., Furtado de Carvalho, T., Chanin, R. B., Behrendt, C. L., Hooper, L. V., Santos, R. L., et al. 2018. Dysbiosis-associated change in host metabolism generates lactate to support *Salmonella* growth. *Cell Host Microbe*. 23(1), 54-64.e6. doi:10.1016/j.chom.2017.11.006.

Goto, R., Miki, T., Nakimura, N., Fujimoto, M. and Okada, N. 2017. *Salmonella* Typhimurium PagP- and Ugtl-dependent resistance to antimicrobial peptides contribute to gut colonization. *PLoS ONE* 12, e0190095. doi:10.1371/journal.pone.0190095.

Grabiec, A. M. and Potempa, J. 2018. Epigenetic regulation of bacterial infections: Targeting histone deacetylases. *Crit. Rev. Microbiol.* 44(3), 336-50. doi:10.1080/1040841X.2017.1373063.

Gregor, M. F. and Hotamisligil, G. S. 2011. Inflammatory mechanisms in obesity. *Annu. Rev. Immunol.* 29, 415-45. doi:10.1146/annurev-immunol-031210-101322.

Guilloteau, P., Martin, L., Eeckhaut, V., Ducatelle, R., Zabielski, R. and Van Immerseel, F. 2010. From the gut to the peripheral tissues: The multiple effects of butyrate. *Nutr. Res. Rev.* 23(2), 366-84. doi:10.1017/S0954422410000247.

Guo, C.-J., Chang, F.-Y., Wyche, T. P., Backus, K. M., Nayfach, S., Pollard, K. S., Craik, C. S., Cravett, B. F., Clardy, J., Voigt, C. A., et al. 2017. Discovery of reactive microbiota-derived metabolites that inhibit host proteases. *Cell* 168, 617-28.

Haas, R., Smith, J., Rocher-Ros, V., Nadkarni, S., Montero-Melendez, T., D'Acquisto, F., Bland, E. J., Bombardieri, M., Pitzalis, C., Perretti, M., et al. 2015. Lactate regulates metabolic and pro-inflammatory circuits in control of T cell migration and effector functions. *PLoS Biol.* 13(7), e1002202. doi:10.1371/journal.pbio.1002202.

Hamer, H. M., Jonkers, D., Venema, K., Vanhoutvin, S., Troost, F. J. and Brummer, R.-J. 2008. Review article: The role of butyrate on colonic function. *Aliment. Pharmacol. Ther.* 27, 104-19.

He, H., Genovese, K. J., Swaggerty, C. L., MacKinnon, K. M. and Kogut, M. H. 2012. Co-stimulation with TLR3 and TLR21 ligands synergistically up-regulates Th-1 cytokine, IFN-gamma and regulatory cytokine, IL10 expression in chicken monocytes. *Dev. Comp. Immunol.* 36(4), 756-60. doi:10.1016/j.dci.2011.11.006.

Honda, K. and Littman, D. R. 2016. The microbiota in adaptive immune homeostasis and disease. *Nature* 535(7610), 75-84. doi:10.1038/nature18848.

Hooper, L. V. and MacPherson, A. J. 2010. Immune adaptations that maintain homeostasis with the intestinal microbiota. *Nat. Rev. Immunol.* 10(3), 159-69. doi:10.1038/nri2710.

Hooper, L. V., Litman, D. R. and MacPherson, A. J. 2012. Interactions between the microbiota and the immune system. *Science* 336(6086), 1268-73. doi:10.1126/science.1223490.

Hotamisligil, G. S. 2006. Inflammation and metabolic disorders. *Nature* 444(7121), 860-7. doi:10.1038/nature05485.

Hsiao, H. Y., Anderson, D. M. and Dale, N. M. 2006. Levels of β-mannan in soybean meal. *Poult. Sci.* 85(8), 1430-2. doi:10.1093/ps/85.8.1430.

Hu, Z. and Guo, Y. 2007. Effects of dietary sodium butyrate supplementation on the intestinal morphological structure, absorptive function, and gut flora in chickens. *Anim. Feed Sci. Technol.* 132, 240-9.

Huang, C., Song, P., Fan, P., Hou, C., Thacker, P. and Ma, X. 2015. Dietary sodium butyrate decreases postweaning diarrhea by modulating intestinal permeability and changing the bacterial communities in weaned piglets. *J. Nutr.* 145(12), 2774-80. doi:10.3945/jn.115.217406.

Hullar, M. A. J. and Fu, B. C. 2014. Diet, the gut microbiome, and epigenetics. *Cancer J.* 20(3), 170-5. doi:10.1097/PPO.0000000000000053.

Infantino, V., Convertini, P., Cucci, L., Panaro, M. A., DiNoia, M. A., Calvello, R., Palmieri, F. and Iacobazzi, V. 2011. The mitochondrial citrate carrier: A new player in inflammation. *Biochem. J.* 438(3), 433-6. doi:10.1042/BJ20111275.

Ivanov, I. I., Atarashi, K., Manel, N., Brodie, E. L., Shima, T., Karaoz, U., Wei, D., Goldfarb, K. C., Santee, C. A., Lynch, S. V., et al. 2009. Induction of intestinal Th17 cell by segmented filamentous bacteria. *Cell* 139(3), 485-98. doi:10.1016/j.cell.2009.09.033.

Jiang, Y., Zhang, W., Gao, F. and Zhou, G. 2015. Effect of sodium butyrate on intestinal inflammatory response to lipopolysaccharide in broiler chickens. *Can. J. Anim. Sci.* 95(3), 389-95. doi:10.4141/cjas-2014-183.

Kawai, T. and Akira, S. 2009. The roles of TLRs, RLRs and NLRs in pathogen recognition. *Int. Immunol.* 21(4), 317-37. doi:10.1093/intimm/dxp017.

Keestra, A. M., de Zoete, M. R., Bowman, L. I., Vaezirad, M. M. and van Putten, J. P. 2013. Unique features of chicken Toll-like receptors. *Dev. Comp. Immunol.* 14, 316-23.

Kobayashi, T., Matsuoka, K., Shiekh, S. Z., Russo, S. M., Mishima, Y., Collins, C., deZoeten, E. F., Karp, C. L., Ting, J. P. Y., Sartor, R. B., et al. 2012. IL-10 regulates *Il12b* expression via histone deacetylation: Implications for intestinal macrophage homeostasis. *J. Immunol.* 189(4), 1792-9. doi:10.4049/jimmunol.1200042.

Kogut, M. H. 2013. The gut microbiome and host innate immunity: Regulators of host metabolism and metabolic diseases in poultry? *J. Appl. Poult. Res.* 22, 637-46.

Kogut, M. H. and Klasing, K. 2009. An immunologist's perspective on nutrition, immunity, and infectious diseases: Introduction and overview. *J. Appl. Poult. Res.* 18(1), 103-10. doi:10.3382/japr.2008-00080.

Kogut, M. H., Yin, X., Yuan, J. and Broom, L. 2017. Gut health in poultry. *CAB Rev.* 12(31). doi:10.1079/PAVSNNR201712031.

Kogut, M. H., Genoese, K. J., Swaggerty, C. L., He, H. and Broom, L. 2018. Inflammatory phenotypes in the intestine of poultry: Not all inflammation is created equal. *Poult. Sci.* 97(7), 2339-46. doi:10.3382/ps/pey087.

Lackey, D. E. and Olefsky, J. M. 2016. Regulation of metabolism by the innate immune system. *Nat. Rev. Endocrinol.* 12(1), 15-28. doi:10.1038/nrendo.2015.189.

Lammers, A., Wieland, W. H., Kruijt, L., Jansma, A., Straetemans, T., Schots, A., den Hartog, G. and Parmentier, H. K. 2010. Successive immunoglobulin and cytokine expression in the small intestine of juvenile chicken. *Devel. Comp. Immunol.* 34, 1252-62.

Land, W. G. 2015. The role of damage-associated molecular patterns (DAMPs) in human diseases: Part II: DAMPs as diagnostics, prognostics and therapeutics in clinical medicine. *Sultan Qaboos Univ. Med. J.* 15(2), e157-70.

Lee, Y. K. and Mazmanian, S. K. 2010. Has the microbiota played a critical role in the evolution of the adaptive immune system? *Science* 330(6012), 1768-73. doi:10.1126/science.1195568.

Leonel, A. J. and Alvarez-Leite, J. I. 2012. Butyrate: Implications for intestinal function. *Curr. Opin. Clin. Nutr. Metab. Care* 15(5), 474–9. doi:10.1097/MCO.0b013e32835665fa.

Levy, M., Thiass, C. A., Katz, M. N., Suez, J. and Elinav, E. 2015. Inflammasomes and the microbiota - Partners in the preservation of mucosal homeostasis. *Semin. Immunopathol.* 37(1), 39–46. doi:10.1007/s00281-014-0451-7.

Levy, M., Thiass, C. A. and Elinav, E. 2016. Metabolites: Messengers between the microbiota and the immune system. *Genes Dev.* 30(14), 1589-97. doi:10.1101/gad.284091.116.

Levy, M., Blacher, E. and Elinav, E. 2017. Microbiome, metabolites, and host immunity. *Curr. Opin. Microbiol.* 35, 8–15. doi:10.1016/j.mib.2016.10.003.

Liu, S., da Cunha, A. P., Rezende, R. M., Cialic, R., Wei, Z., Bry, L., Comstock, L. E., Gandhi, R. and Weiner, H. L. 2016. The host shapes the gut microbiota via fecal microRNA. *Cell Host Microbe.* 19(1), 32–43. doi:10.1016/j.chom.2015.12.005.

Liu, H., Wang, J., He, T., Becker, S., Zhang, G., Li, D. and Ma, X. 2018. Butyrate: A double-edged sword for health? *Adv. Nutr.* 9(1), 21–9. doi:10.1093/advances/nmx009.

Loftus, R. M. and Finlay, D. K. 2016. Immunometabolism: Cellular metabolism turns immune regulator. *J. Biol. Chem.* 291(1), 1–10. doi:10.1074/jbc.R115.693903.

Lu, L. and Walker, W. A. 2001. Pathologic and physiologic interactions of bacteria with the gastrointestinal epithelium. *Am. J. Clin. Nutr.* 73(6), 1124S–30S. doi:10.1093/ajcn/73.6.1124S.

Macpherson, A. J., Slack, E., Geuking, M. B. and McCoy, K. D. 2009. The mucosal firewalls against commensal intestinal microbes. *Semin. Immunopathol.* 31(2), 145–9. doi:10.1007/s00281-009-0174-3.

Mantis, N. J. and Forbes, S. J. 2010. Secretory IgA: Arresting microbial pathogens at epithelial borders. *Immunol. Invest.* 39(4–5), 383–406. doi:10.3109/08820131003622635.

Marchesi, J. R., Adams, D. H., Fava, F., Hermes, G. D. A., Hirschfield, G. M., Hold, G., Quraishi, M. N., Kinross, J., Smidt, H., Tuohy, K. M., et al. 2016. The gut microbiota and host health: A new clinical frontier. *Gut* 65(2), 330–9. doi:10.1136/gutjnl-2015-309990.

Martinez-Cummer, M. A. 2015. Immunogenic ingredients in poultry: Applications of innovative concepts leading to sustainable solutions for improved productivity. *Aust. Poult. Sci. Proc.* 2105, 209–12.

Maslowski, K. M. and Mackay, C. R. 2011. Diet, gut microbiota, and immune responses. *Nat. Immunol.* 12(1), 5–9. doi:10.1038/ni0111-5.

McDonald, C., Inohara, N. and Nunez, G. 2005. Peptidoglycan signaling in innate immunity and inflammatory disease. *J. Biol. Chem.* 280(21), 20177–80. doi:10.1074/jbc.R500001200.

McGettrick, A. F. and O'Neill, L. A. 2013. How metabolism generates signals during innate immunity and inflammation. *J. Biol. Chem.* 288(32), 22893–8. doi:10.1074/jbc.R113.486464.

Medzhitov, R. 2001. Toll-like receptors and innate immunity. *Nat. Rev. Immunol.* 1(2), 135–45. doi:10.1038/35100529.

Meng, X., Slominski, B. A., Nyachoti, C. M., Campbell, L. D. and Geunter, W. 2005. Degradation of cell wall polysaccharides by combinations of carboxylase enzymes and their effect on nutrient utilization and broiler chicken performance. *Poult. Sci.* 84(1), 37–47. doi:10.1093/ps/84.1.37.

Mwangi, W. N., Beal, R. K., Powers, C., Wu, X., Humphrey, T., Watson, M., Bailey, M., Friedman, A. and Smith, A. L. 2010. Regional and global changes in TCRαβ T cell repertoires in the gut are dependent upon the complexity of the enteric microflora. *Dev. Comp. Immunol.* 34(4), 406-17. doi:10.1016/j.dci.2009.11.009.

Myhre, A. E., Aasen, A. O., Thiermann, C. and Wang, J. E. 2006. Peptidogylcan - An endotoxin in its own right? *Shock* 25(3), 227-35. doi:10.1097/01.shk.0000191378.55274.37.

Ng, K. M., Ferryra, J. A., Higgenbottom, S. K., Lynch, J. B., Kashyap, P. C., Gopinath, S., Naidu, N., Choudhury, B., Weimer, B. C., Monack, D. M., et al. 2013. Microbiota-liberated host sugars facilitate post-antibiotic expansion of enteric pathogens. *Nature* 502(7469), 96-9. doi:10.1038/nature12503.

Oakley, B. B., Lillhoj, H. S., Kogut, M. H., Kim, W. K., Maurer, J. J., Pedroso, A., Lee, M. D., Collett, S. R., Johnson, T. J. and Cox, N. A. 2014. The chicken gastrointestinal microbiome. *FEMS Microbiol. Lett.* 360(2), 100-12. doi:10.1111/1574-6968.12608.

O'Mahony, C., Scully, P., O'Mahony, D., Murphy, S., O'Brien, F., Lyons, A., Sherlock, G., MacSharry, J., Kiely, B., Shanhan, F., et al. 2008. Commensal-induced regulatory T cells mediate protection against pathogen-stimulated NF-κB activation. *PLoS Pathog.* 11, e100112. doi:10.1371/journal.pone.0100112.

Pezoa, D., Yang, H. J., Blundel, C. J., Santiviago, C. A., Andrews-Polymenis, H. L. and Contreras, I. 2013. The type VI secretion system encoded in SPI-6 plays a role in gastrointestinal colonization and systemic spread of *Salmonella enterica* serovars Typhimurium in the chicken. *PLoS ONE* 8(5), e63917. doi:10.1371/journal.pone.0063917.

Postler, T. S. and Ghosh, S. 2017. Understanding the holobiont: How microbial metabolites affect human health and shape the immune system. *Cell Metab.* 26(1), 110-30. doi:10.1016/j.cmet.2017.05.008.

Pucino, V., Bombardieri, M., Pitzalis, C. and Mauro, C. 2017. Lactate at the crossroads of metabolism, inflammation, and autoimmunity. *Eur. J. Immunol.* 47(1), 14-21. doi:10.1002/eji.201646477.

Quinteiro-Filho, W. M., Rodrigues, M. V., Ribeiro, A., Ferraz-de-Paula, V., Pinheiro, M. L., Sa, L. R. M., Ferreira, A. J. P. and Palermo-Neto, J. 2012. Acute heat stress impairs performance parameters and induces mild intestinal enteritis in broiler chickens: Role of acute hypothalamic-pituitary adrenal axis activation. *J. Anim. Sci.* 90(6), 1986-94. doi:10.2527/jas.2011-3949.

Raffatellu, M., George, M. D., Akiyama, Y., Hornsby, M. J., Nuccio, S. P., Paixao, T. A., Butler, B. P., Chu, H., Santos, R. L., Berger, T., et al. 2009. Lipocalin-2 resistance confers an advantage to *Salmonella enterica* serotype Typhimurium for growth and survival in the inflamed intestine. *Cell Host Microbe.* 5(5), 476-86. doi:10.1016/j.chom.2009.03.011.

Rangan, K. J. and Hang, H. C. 2017. Biochemical mechanisms of pathogens restriction by intestinal bacteria. *Trends Biochem. Sci.* 42(11), 887-98. doi:10.1016/j.tibs.2017.08.005.

Rock, K. L., Latz, E., Ontiveros, F. and Kono, H. 2010. The sterile inflammatory response. *Annu. Rev. Immunol.* 28, 321-42.

Round, J. L. and Mazmanian, S. K. 2009. The gut microbiota shapes intestinal immune responses during health and disease. *Nat. Rev. Immunol.* 9(5), 313-23. doi:10.1038/nri2515.

Rubartelli, A., Lotze, M. T., Latz, E. and Manfredi, A. 2013. Mechanisms of sterile inflammation. *Front. Immunol.* 4, Article 398. doi:10.3389/fimmu.2013.00398.

Sacranie, A., Iji, P. A., Mikkelson, L. L. and Choct, M. 2007. Occurrence of reverse peristalsis in broiler chickens. *Aust. Poult. Sci. Proc.* 2007, 161-4.

Sana, T. G., Flaugnatti, N., Lugo, K. A., Lam, L. H., Jacobson, A., Baylot, V., Durand, E., Journet, L., Cascales, E. and Monack, D. M. 2016. *Salmonella* Typhimurium utilizes a T6SS-mediated antibacterial weapon to establish in the host gut. *Proc. Natl. Acad. Sci. U.S.A.* 113, E5044-51.

Sansonetti, P. J. 2004. War and peace at mucosal surfaces. *Nat. Rev. Immunol.* 4(12), 953-64. doi:10.1038/nri1499.

Sassone-Corsi, M. and Raffatellu, M. 2015. No vacancy: How beneficial microbes cooperate with immunity to provide colonization resistance to pathogens. *J. Immunol.* 194, 4089-93.

Shenderov, B. A. 2012. Gut indigenous microbiota nd epigenetics. *Microb. Ecol. Health Dis.* 2012, 23.

Shi, N., Xi, N., Duan, X. and Niu, H. 2017. Interaction between the gut microbiome and mucosal immune system. *Mil. Med. Res.* 4, 14. doi:10.1186/s40779-017-0122-9.

Singh, N., Shirdel, E. A., Waldron, L., Zhang, R. H., Jurisca, I. and Cornelli, E. M. 2012. The murine caecal microRNA signature depends on the presence of the endogenous microbiota. *Int. J. Biol. Sci.* 8(2), 171-86. doi:10.7150/ijbs.8.171.

Singh, V. P., Proctor, S. D. and Willing, B. P. 2016. Koch's postulates: Microbial dysbiosis and inflammatory bowel disease. *Clin. Microbiol. Infect.* 22(7), 594-9. doi:10.1016/j.cmi.2016.04.018.

Slominski, B. A. 2011. Review: Recent advances in research on enzymes for poultry diets. *Poult. Sci.* 90, 2013-23.

Smith, P. M., Howitt, M. R., Panikov, N., Michaud, M., Gallini, C. A., Bohlooly-Y, M., Glickman, J. N. and Garrett, W. S. 2013. The microbial metabolites, short-chain fatty acids, regulate colonic T reg cell homeostasis. *Science* 341(6145), 569-73. doi:10.1126/science.1241165.

Sommer, F., Anderson, J. M., Bharti, R., Raes, J. and Rosenstiel, P. 2017. The resilence of the intestinal microbiota influences health and disease. *Nat. Rev. Microbiol.* 15(10), 630-8. doi:10.1038/nrmicro.2017.58.

Sonnenburg, J. L. and Backhed, F. 2016. Diet-microbiota interactions as moderators of human metabolism. *Nature* 535(7610), 56-64. doi:10.1038/nature18846.

Stecher, B. 2015. The roles of inflammation, nutrient availability and the commensal microbiota in enteric pathogen infection. *Microbiol. Spectr.* 3(3). doi:10.1128/microbiolspec.MBP-0008-2014.

Tannahill, G. M., Curtis, A. M., Adamik, J., Palsson-McDermott, E. M., McGettrick, A. F., Goel, G., Frezza, C., Bernard, N. J., Kelly, B., Foley, N. H., et al. 2013. Succinate is an inflammatory signal that induces IL-1β through HIF1-α. *Nature* 496(7444), 238-42. doi:10.1038/nature11986.

Thaiss, C. A., Levy, M., Suez, J. and Elinav, E. 2014. The interplay between the innate immune system and the microbiota. *Curr. Opin. Immunol.* 26, 41-8. doi:10.1016/j.coi.2013.10.016.

Thiennimitr, P., Winter, S. E., Winter, M. G., Xavier, M. N., Tolstikov, V., Huseby, D. L., Sterzenbach, T., Tsolis, R. M., Roth, J. R. and Bäumler, A. J. 2011. Intestinal inflammation allows *Salmonella* to use ethanolamine to compete with the microbiota. *Proc. Natl. Acad. Sci. U.S.A.* 108(42), 17480-5. doi:10.1073/pnas.1107857108.

Tretter, L., Patocs, A.and Chinopoulas, C. 2016. Succinate, an intermediate in metabolism, signal transduction, ROS, hypoxia, and tumorigenesis. *Biochim. Biophys. Acta* 1857(8), 1086–101.

Usami, M., Kishimoto, K., Ohata, A., Miyoshi, M., Aoyama, M., Fueda, Y. and Kotani, J. 2008. Butyrate and trichostatin A attenuate nuclear factor κB activation and tumor necrosis factor α secretion and increase prostaglandin E_2 secretion in human peripheral blood mononuclear cells. *Nutr. Res.* 28(5), 321-8. doi:10.1016/j.nutres.2008.02.012.

Vaishnava, S., Yamamoto, M., Severson, K. M., Ruhn, K. A., Yu, X., Koren, O., Ley, R., Wakeland, E. K. and Hooper, L. V. 2011. The antibacterial lectin RegIIIγ promotes the spatial segregation of microbiota and host in the intestine. *Science* 334(6053), 255–8. doi:10.1126/science.1209791.

Van der Waaij, D., Berguis-de-Vries, J. M. and Lekkerkerk-van der Wees, J. E. C. 1971. Colonization resistance of the digestive tract in conventional and antibiotic-treated mice. *J. Hyg.* 69(3), 405–11. doi:10.1017/S0022172400021653.

Wang, P., Zhu, S., Yang, L., Cui, S., Pan, W., Jackson, R., Zheng, Y., Rongvaux, A., Sun, Q., Yang, G.,, et al. 2015. Nlrp6 regulates intestinal antiviral innate immunity. *Science* 350(6262), 826–30. doi:10.1126/science.aab3145.

Weiss, G. A. and Hennet, T. 2017. Mechanisms and consequences of intestinal dysbiosis. *Cell. Mol. Life Sci.* 74, 2959–77.

Winter, S. E., Thienimitr, P., Winter, M. G., Butler, B. P., Huseby, D. L., Crawford, R. W., Russell, J. M., Bevins, C. L., Adams, L. G., Tsolis, R. M., et al. 2010. Gut inflammation provides a respiratory electron acceptor for *Salmonella*. *Nature* 467(7314), 426–9. doi:10.1038/nature09415.

Woo, V. and Alenghat, T. 2017. Host-microbiota interactions: Epigenomic regulation. *Curr. Opin. Immunol.* 44, 52–60. doi:10.1016/j.coi.2016.12.001.

Xiao, Y., Cai,Y., Bommineni, Y. R., Fernando, S. C., Prakash, O., Gilliland, S. E. and Zhang, G. 2006. Identification and functional characterization of three chicken cathelicidins with potent antimicrobial activity. *J. Biol. Chem.* 281(5), 2858–67. doi:10.1074/jbc.M507180200.

Zeng, M. Y., Inohara, N. and Nunez, G. 2017. Mechanisms of inflammation-driven bacterial dysbiosis in the gut. *Mucosal Immunol.* 10(1), 18–26. doi:10.1038/mi.2016.75.

Part 2

Factors that impact the gastrointestinal tract and different types of birds

Chapter 8

Genetics and other factors affecting intestinal microbiota and function in poultry

Michael D. Cressman, The Ohio State University, USA; Jannigje G. Kers, Utrecht University, The Netherlands; and Lingling Wang and Zhongtang Yu, The Ohio State University, USA

1 Introduction

The intestines of poultry, like that of other animals, harbor a diverse, complex, and dynamic community of microbes. This microbiota contains diverse species of microbes, dominated by bacteria, with most of them being commensal bacteria and remaining to be cultured and characterized, taxonomically, metabolically, physiologically, and ecologically. Although once controversial, the preponderance of evidence supports that this intestinal microbiota, as the intestinal microbiota in other animals and humans, is a required organ that contributes to many aspects of poultry nutrition, productivity, and disease risk and resistance. Such recognition promoted many studies to explore various strategies to modulate the intestinal microbiota in poultry with the ultimate goal to enhance feed efficiency and decrease diseases. To develop rational and knowledge-based intervention strategies, a basic understanding of how the intestinal microbiota can be affected and shaped is required. To that end, many studies have investigated the intestinal microbiota of poultry, especially chickens and turkeys, with a goal to understand its diversity and composition (Wei et al., 2013a; Oakley et al., 2014) and how the intestinal microbiota, its dynamics, and functionality are affected by different factors, which include host genetics, age, feed (and thus nutrient availability), feed additives (biotic and

http://dx.doi.org/10.19103/AS.2019.0059.09

abiotic), health and disease status, and environment (e.g. litter, geography, and season) (Pan and Yu, 2014; Stanley et al., 2014; Kers et al., 2018).

The intestines of newly hatched birds are sterile (several studies showed presence of bacteria in newly hatched birds), but bacterial colonization starts immediately after hatching (van der Wielen et al., 2002). The intestinal microbiota undergoes rapid temporal successions, from simple to complex and diverse. 'Everything is everywhere, but, environment selects (Baas Becking, 1934).' This principle of microbial ecology applies to the intestinal microbiota of poultry. The intestinal microbiota of mature bids, especially the cecal microbiota, is vastly diverse, complex, and dynamic. The vast diversity is reflected by the large number of species, while the complexity is manifested as several orders of difference in abundance (Wei et al., 2013a; Ferrario et al., 2017). The intestinal microbiota is also dynamic, especially during the first week post hatch. Furthermore, most of the intestinal microbes are not culturable or difficult to culture in laboratory media (Ferrario et al., 2017). These attributes have challenged researchers who attempted to comprehensively characterize the intestinal microbiota of poultry. The recent advancement of DNA sequencing technologies and bioinformatics, and concurrently affordable cost to sequence a large number of metagenomes have created unprecedented opportunities to understand many important aspects of the intestinal microbiota, including diversity, composition, structure, and interaction with the host, diets, and other factors that can profoundly affect the intestinal microbiota. The objective of this chapter is to review the current state of understanding of the factors that can affect the intestinal microbiota and its functions of poultry, with a focus on poultry genetics and breeds, gender, diseases, and environments including hatchery conditions, litter conditions, climate, and geography.

2 Characteristics of poultry intestines as an environmental for microbiota

The intestinal tract of poultry has several structural differences compared to that of mammalians, and these differences can affect the intestinal microbiota and should be taken into consideration when investigating poultry intestinal microbiota and how it can be affected by various factors. First, the intestinal tract of poultry is shorter (relative to the body length) than that of most mammalians. As such, the digesta passes the intestinal tract faster, resulting in shorter retention time (5–6 h in total), in poultry than in mammalians (Shires et al., 1987). The short retention time will select fast-growth microbes. Second, one pair of ceca are the major large intestine in poultry, and they harbor most of the intestinal microbes. The interdigitating villi meshwork at the cecal entrance functions as a filter, excluding large digesta particles and only allowing fine particles and fluid to enter the ceca. More importantly, the ceca do not have plug flow, diminishing the longitudinal gradients in nutrients and fermentation products found in the colons. In the case of broiler chickens, the birds grow and mature fast, within 5–6 weeks; thus, their gut microbiota develops and matures rapidly, causing rapid successions and dynamic responses. Therefore, the responses of the intestinal microbiota to changing factors can be more rapid in poultry than in other animals.

3 Factors that affect the development and function of intestinal microbiota in poultry

3.1 Genetics and breeds

The genetic background of a host is an important factor that can influence its intestinal microbiota diversity and composition (Benson et al., 2010; Org et al., 2015; Han et al., 2016). Considerable physiological differences exist between layer-type and broiler-type chickens as a result of breeding programs that have selected laying hens with the best egg production versus broilers with the best meat production. After many decades, these chicken breeding programs seem to have affected intestinal physiology (Uni et al., 1996) and immune function (Simon et al., 2014). Morphological differences in the intestinal tissue, such as villus height, villus width, and crypt depth, between laying hens and broilers affect the intestinal absorptive area and have been associated with a higher body weight of broilers (Uni et al., 1996). Moreover, the production of IgA, IgM, and IgY in the ileum has been shown to be higher in broilers compared to laying hens measured at 20 weeks of age (Simon et al., 2014). These and other differences in intestinal physiology and immune system development between laying hens and broilers can influence the microbiota diversity and composition and vice versa. Indeed, one study showed clear differences in microbiota composition between 3-week-old broilers and 62-week-old laying hens (Videnska et al., 2014). However, the large age difference, differences in potential exposure to microbes in the housing environment, and substantial differences in the composition of the feed between the young broilers and the old laying hens probably had influenced the microbiota composition. Another study found a higher number of cecal CD4+ and CD8+ cells in layer-type chickens compared to broiler-type chickens (Han et al., 2016). This study also found that broiler-type chickens had an increased interleukin-8 (IL-8) expression at 7 days post challenge with *Campylobacter jejuni* compared to layer-type chickens. These authors concluded that broiler-type chickens had a stronger immune response compared to layer-type chickens following *C. jejuni* challenge (Han et al., 2016). Besides the differences in the immune response, the colonization pattern of commensal bacteria and the response to *C. jejuni* challenge also appeared to be different between layer-type chickens and broilers (Han et al., 2016). The cecal microbiota of the former changed to a greater extent after the *C. jejuni* challenge than that of the latter (Han et al., 2016). This indicates that differences in chicken breeds or genetic lines can have an impact on the immune system and the microbes in their intestine. However, in future studies investigating host genetic effect on intestinal microbiota, potential confounding factors, such as feed, age, housing and environmental conditions, should be eliminated or taken into consideration.

In addition to genetic differences between layer-type and broiler-type chickens, there are also genetic differences among different lines within a chicken breeds of the same chicken type (Ponsuksili et al., 1998; Xu et al., 2018), and such genetic difference can also affect the intestinal microbiota in poultry. In one study (Persoons et al., 2011), the breed of broilers was found to be a risk factor associated with the colonization by antibiotic-resistant strains of *E. coli*. In another experimental study, different broiler breeds were shown to differ significantly in disease susceptibility to necrotic enteritis (Jang et al., 2013). Using different broiler breeds that were hatched in the same hatchery, Kim et al. (2015) demonstrated that each breed also had its own unique ileal microbiota when analyzed at the age of 20 days. In that study, the phylum *Bacteroidetes* was found in the ileal content of 20-day-old Cobb broilers, but it was not found in the same intestinal region of Ross broilers of the same age. On the other hand, the phylum *Actinobacteria* was found in the ileal content of Ross broilers, but not in the ileal content of Cobb broilers. The finding of Kim et al. (2015) was corroborated by other studies, that is, absence of *Bacteroidetes* and presence of *Actinobacteria* in Ross broilers at 21 days of age (Nakphaichit et al., 2011) and 25 days of age (Pourabedin et al., 2015), but absence of *Actinobacteria* and presence of *Bacteroidetes* in Cobb broilers at 23 days of age (Mohd Shaufi et al., 2015). In one study (Han et al., 2016), however, *Bacteroidetes* was found to account for 22% of the total ileal microbiota in the ileum of 18-day-old Ross broilers. The above discrepancy with respect to the occurrence of these bacterial phyla may be attributable to inevitable differences in diet or other experimental conditions, sampling age, sequencing and analysis methodologies, and/or reference databases used in their microbiota analyses. Therefore, it is difficult to compare microbiota results obtained from different studies that differed in many factors that affect the intestinal microbiota and the analysis results. Considering these confounding factors, Kers et al. (2018) compiled published intestinal microbiota data (sequence data of 16S rRNA gene amplicons generated using 454 pyrosequencing and MiSeq sequencing) from multiple studies and then compared the intestinal microbiota of Ross versus that of Cobb broilers. They showed that the phylum *Actinobacteria* was present in cecal samples in all four Cobb studies (100%) and in three of the eight Ross studies (38%), and that the phylum *Bacteroidetes* was present in all four Cobb studies (100%) and in six of eight Ross studies (75%) (Kers et al., 2018). The above differences in these two phyla might suggest effects of poultry breeds on the intestinal microbiota due to variations in the immune responses caused by differences in the genetic background (Emam et al., 2014). However, it is more likely that Cobb and Ross broilers have different feed and early exposure to microbiota due to differences in parent flocks (Schokker et al., 2015). It should be noted that diversity and comparison of intestinal microbiota at low taxonomic levels (e.g. genus and operational taxonomic units (OTUs)) will almost certainly reveal greater differences, but few studies have made such comparisons.

Different lines within certain broiler breeds exhibit different growth potential and feed conversion ratio (FCR). Broiler lines with low FCR, indicating a more efficient use of feed for growth, showed a higher abundance of *Lactobacillus* in fecal samples compared to the broilers lines with a high FCR (Zhao et al., 2013; Meng et al., 2014; Mignon-Grasteau et al., 2015). In another study (Schokker et al., 2015), the composition of the intestinal microbiota between broiler lines was found to differ, while the microbial diversity did not, which might suggest that different chicken lines can harbor different microorganisms for the same intestinal function. The mechanisms behind the variation in intestinal microbiota between different broiler lines remain obtuse, but it has been suggested that genetic background and the immune system influence the establishment of intestinal microbiota after hatch (Schokker et al., 2015). Breeding programs for high production may result in co-microevolution of the immune system and the microbiota of chickens (Yang et al., 2017), though other factors, such as differences in exposure to microbial communities, cannot be excluded. Certain rumen bacteria were found to impact feed efficiency in beef cattle (Paz et al., 2018) and so were some intestinal bacteria in broiler chickens (Stanley et al., 2013). It will be intriguing if certain intestinal bacteria can serve as 'microbial markers' in poultry breeding for improved feed conversion and resistance to pathogen infections.

3.2 Hatchery conditions and environment

Different from newborn mammalians, newly hatched chicks have no direct contact with their mothers (hens) in commercial hatcheries. Thus, parental influence on the development of intestinal microbiota is greatly reduced. However, bacteria were detected in poultry embryos though the source, diversity and succession of the embryo bacteria remain unexplored (Kers et al., 2018). Eggs can acquire parental microbes, especially bacteria, during fertilization and egg formation in the oviduct (Cox et al., 2012; Ding et al., 2017).

Bacteria can penetrate the eggshell, shell membrane, and albumin, especially within the first several minutes after oviposition because the cuticle outside of the shell is immature and some pores may be open (Gantois et al., 2009). The penetration of bacteria is driven by the temperature differential between the warm body temperature of the hens and the cooler environment. Moisture can facilitate this process by forming 'sweating' eggs (Gantois et al., 2009; Cox et al., 2012). Although the mechanism of the 'inheritance' of maternal intestinal microbes and establishment of both maternal and environmental bacteria in eggs remain elusive, various bacteria have been found to be capable of penetrating the eggshell, shell membrane, and albumin of fresh laid eggs (Gantois et al., 2009). Some of these bacteria can be the primary colonizers of the intestines of poultry.

In modern commercial hatcheries, several hygiene measures, such as egg washing and fumigation, are routinely used to reduce contamination of the eggs and minimize the spread of pathogens. However, quite a few studies revealed the existence and persistence of pathogenic bacteria, including *Salmonella*, *Campylobacter*, and *Clostridium* in hatcheries, on bedding paper tower, egg fragments, fluff, walls, and workers' boots (Cox et al., 1990, 2012; Craven et al., 2001a,b). During hatching, bacteria can be transmitted within the hatchery and colonize the intestines in newly hatched chicks (Bailey et al., 1994; Cason et al., 1994). When young chicks emerge from the egg at the end of hatching, they can ingest bacteria on the shell, which can be contaminated with environmental bacteria. A recent study showed that inoculating fecal microbiota on the eggshells could reduce variations in the intestinal microbiota among young chicks (Donaldson et al., 2017). Moreover, although newly hatched chicks spend no more than a few days of their life at hatcheries before being transported to chicken houses, the environmental bacteria at hatcheries can affect the early intestinal colonization process because their intestines are very receptive to bacterial colonization. However, no research can be found in the literature that examined how the environmental microbes in the hatchery colonize or affect the bacterial colonization of the intestines of newly hatched chicks. It is desirable to control infection by pathogens (Corrier et al., 1995; Walker and Sander, 2004), but facilitated early colonization using competitive exclusion cultures can reduce the risk of pathogen infection in young chicks. Future research is warranted to investigate how and to what extent the microbes at hatcheries can affect the early intestinal colonization and the potential long-term impact on the eventual intestinal microbiota of chickens and turkeys. One technical difficulty that needs to be overcome is to recover the DNA from day 0 chicks for sequencing analysis of 16S rRNA gene amplicons or metagenomes (Donaldson et al., 2017).

3.3 Bedding and litter

Intensive production of poultry involves the rearing of birds indoors. Egg-laying type chickens can be reared in caged or cage-free environments, whereas virtually all broiler-type chickens and turkeys are raised in cage-free environments. Birds raised cage-free spend the vast majority of their lives in contact with the ground, and in order to provide them with some protection and comfort a regionally available bedding material (e.g. wood shavings or dust, cereal grass or hulls, compost, peat, sand, etc.) is spread across the floor surface at a depth of approximately 5–15 cm. The bedding material absorbs moisture, dilutes manure, and provides a clean, comfortable, and safe environment for the bird. During the grow-out period, the bedding materials accumulate moisture, manure, and other materials (e.g. spilt feed, feathers, soil, etc.) at which point it is generally referred to as litter. In an effort to better understand the impact that litter has on the commercial broiler house environment and bird performance, the production rates and physiochemical properties of poultry litter (i.e. temperature, moisture, pH, ammonia, and nutrients) have been studied extensively (Toghyani et al., 2010). Variability in litter conditions (physicochemical and microbiological) is attributed to a number of factors, including but not limited to bedding material, geography, season, stocking density, and single- versus multiuse application (Bolan et al., 2010).

Both fresh bedding materials and poultry litter have been shown to possess unique and diverse microbiotas (Cressman et al., 2010; Wang et al., 2016). Early studies relied on cultivation-based techniques to identify and enumerate broad classifications of microorganisms, such as aerobes and anaerobes linked to the decomposition of organic and nitrogenous compounds found within poultry litter, as well as coliforms, molds, and yeasts (Halbrook et al., 1951; Schefferle, 1965; Lovett et al., 1971). Subsequent studies have identified pathogens within poultry litter, such as species of *Escherichia, Staphylococcus, Clostridium,* and *Salmonella* (Schefferle, 1965; Alexander et al., 1968; McCann et al., 1998; Terzich et al., 2000; McCrea et al., 2005, 2008; Williams and Macklin, 2013), and helped explain the variability in microbial composition and abundance observed in litter samples originating from different bedding materials (Fries et al., 2005; Macklin et al., 2005), litter depths (Barker et al., 2010; Williams and Macklin, 2013), geography (Terzich et al., 2000), bird types, and production systems (Omeira et al., 2006). More recent studies have employed molecular techniques to elucidate a more detailed view and understanding of the litter microbiota. Lu et al. (2003) analyzed 16S rRNA gene clone libraries constructed from litter samples at various stages of reuse and bird exposure from commercial broiler farms located in the state of Georgia in the United States. From a total of 340 clones analyzed, four major phyla were identified: *Firmicutes* (62.05%), *Actinobacteria* (24.70%), *Proteobacteria* (12.94%), and

Bacteroidetes (0.29%). In total, 31 genera were identified, of which *Salinicoccus* and members of *Lactobacillaceae* were most abundant. Cressman et al. (2010) analyzed 16S rRNA gene clone libraries constructed from fresh pine shavings litter and reused litter (approximately 2 years old) that had both been exposed to commercial broiler chicks for 7 days. The majority of the bacteria from both fresh pine shavings litter (77.3% of clones) and reused litter (98.7% of clones) belonged to the phylum *Firmicutes*; however, the fresh pine shavings litter also contained bacteria belonging to the phylum *Proteobacteria*, which was not represented by the clones derived from reused litter. Such a contrast in the occurrence of litter bacteria of intestinal origin versus litter bacteria of environmental origin has been observed in another study (Wang et al., 2016).

Recently, most studies used next-generation sequencing (NGS) technologies to comprehensively characterize the microbiota in the litter and the gut of poultry birds. We used pyrosequencing to characterize the microbiotas between fresh pine shavings litter and reused litter (approximately 1 year old) that had been exposed to broiler chicks for 35 days (Wang et al., 2016). Compared to the previous study, greater bacterial diversity was revealed, allowing us to conclude that litter management regimes (fresh versus reused) can impact the litter microbiota at the species level. Fresh pine shaving litter contained bacteria not commonly found in the GI tract (e.g. species of *Acinetobacter*, *Devosia*, *Luteimonas*, *Trichococcus*, and *Yaniella*), while the reused litter contained bacteria of both intestinal origin and environmental bacteria that became adapted to the reused litter environment, such as halo- or bile-tolerant bacteria (e.g., species of *Salinicoccus*, *Oceanobacillus*, and *Dialister*). Other studies have demonstrated both temporal and spatial shifts of the microbiota structure occur within poultry litter. Using PCR-DGGE, Cressman et al. (2010) showed that the microbiotas found both within fresh pine shavings litter and within reused litter (approximately 2 years old) underwent temporal community succession as broilers grew to 42 days of age. This succession was observed to a greater extent in the fresh pine shavings litter. As a result, the authors concluded that the manure droppings from the birds had a more profound impact on the biotic and abiotic conditions of the fresh litter, as compared to the reused litter. Using similar qualitative techniques, Lovanh et al. (2007) were able to show that spatial variations in the microbiota structure exist within poultry litter from commercial broiler houses. Principal component analysis (PCA) revealed that litter moisture had the greatest impact on litter microbial diversity, while litter temperature had the second greatest impact on diversity. Litter from around the feeders and drinkers, as well as litter located in the back of the poultry house (closest to the exhaust fans), exhibited greater bacterial diversity than litter sampled elsewhere in the house, an observation explained by the similar physical conditions (i.e. moisture and compaction) of those litters. Spatial variations in the litter microbiota from litter collected in commercial broiler houses, as well as the correlation between litter moisture and litter microbial diversity, have been observed in other studies (Dumas et al., 2011).

Virtually all commercially hatched meat-type birds, as well as some commercially egg-type birds, are placed in direct contact with either bedding or litter. In such an environment, young birds will almost immediately begin to peck and ingest some litter materials. Sufficient evidence exists to suggest that litter is a microbial reservoir that has the potential to inoculate and thus influence the development and diversity of the gut microbiotas of birds reared in contact with litter (Cressman et al., 2010; Wei et al., 2013b; Mancabelli et al., 2016; Wang et al., 2016; De Cesare et al., 2019). This has the potential to be advantageous, assuming that a 'healthy' litter microbiota promotes the development of a healthy gut microbiota. Numerous pioneering studies have demonstrated how fecal inocula from healthy mature birds (chickens and turkeys) given to hatchlings provide some form of probiotic or protective effect within the gut (Nurmi and Rantala, 1973; Lloyd et al., 1977; Snoeyenbos et al., 1978; Rigby and Pettit, 1980; Pivnick et al., 1981). *Salmonella* colonization within the broiler gut has been reported to decrease in broilers reared on reused litter (Fanelli et al., 1970; Duff et al., 1973; Gustafson and Kobland, 1984), while Corrier et al. (1993) demonstrated a reduction in the *Salmonella* colonization in the ceca, liver, and spleen of Leghorn chicks fed broiler litter at 5% inclusion of the diet. Conversely, several studies suggested the litter microbiota as a potential source of enteric pathogens. Indeed, poultry litter has been shown to harbor pathogens (Terzich et al., 2000; Lu et al., 2003; Wei et al., 2013b), as well as antibiotic-resistant strains of bacteria (Kelley et al., 1998; Dhanarani et al., 2009; You et al., 2013). These unwanted bacteria have the potential to amplify and evolve by means of horizontal gene transfer throughout a commercial flock (Pan and Yu, 2014). However, reuse of litter was not shown to increase the occurrence or level of *Campylobacter* across sequential farming cycles (Chinivasagam et al., 2016).

While some studies have investigated the effect of poultry litter materials, conditions, or management strategies on the host gut microbiota (Torok et al., 2009; Samli et al., 2010; Wei et al., 2013b; Wang et al., 2016; De Cesare et al., 2019), concurrent analysis of the litter and gut microbiotas is necessary to truly elucidate the reciprocal relationship between the two. To date, only a small number of studies have looked at the microbiota of both poultry litter and the gut (Cressman et al., 2010; Oakley et al., 2013; Wei et al., 2013b; Mancabelli et al., 2016; Wang et al., 2016; De Cesare et al., 2019), while only one study has looked at both microbiotas and their reciprocal effects in a time-dependent manner (Cressman et al., 2010). Taken together, the biotic and abiotic conditions within poultry litter have been studied rather extensively; however, the effect that those conditions have on the poultry gut microbiota, bird health, and efficient meat and egg production warrants more attention.

The reciprocal relationship between the litter and gut microbiotas has been studied somewhat in broiler chickens, while a significant knowledge gap exists in the literature on this relationship as it pertains to egg-laying chickens, turkeys, or ducks. A greater understanding in this area is especially urgent, as more meat-type birds are raised without antibiotics to meet the growing demand for No Antibiotic Ever (NAE) poultry meat. In addition, as many food retail and food service providers fulfill their commitments to source cage-free eggs in the coming years, a greater proportion of egg-laying chickens will be reared in environments that provide them with daily contact of floor litter. The increasingly powerful metagenomics would greatly facilitate the undertaking to comprehensively investigate the reciprocal relationship between litter and the gut microbiota of poultry.

3.4 Climate and geographic regions

Local climatic conditions of poultry houses can have an impact on the intestinal microbiota of chickens and turkeys because climate can affect the microbes present in the environment, including the bedding materials and airborne bacteria inside of the poultry houses. Studies examining the effects of climate on microbiota using 16S rRNA gene amplicon sequencing are, however, scarce. Although it is likely, at least in some parts of the world, that seasons may impact the intestinal microbiota, only two studies were found in the literature that specifically reported seasonal differences in bacterial genera detected in the cecal samples. In one study, the cecal content samples collected from the northeastern part of the state of Georgia in the United States were found to have fewer bacterial genera in winter compared to spring or summer (Oakley et al., 2018). In contrast, in the other study in Austria where fecal samples were collected in all four seasons in one flock between 2003 and 2006 and another flock in 2013, no seasonal effect was noted (Sofka et al., 2015). The discrepancy between the above two studies could be attributable to differences in climate (humid subtropical Georgia at 33.7490°N versus temperate Austria at 48.2082°N), geography (Europe versus the United States), flock management, and other factors. Because intestinal microbiota can affect the susceptibility to pathogen infections in association with climate, future research is needed to investigate the effects of climate on intestinal microbiota and risk of infectious diseases.

Heat stress has been the focus of many studies, especially in broilers, due to the negative effects on production performance (Zhang et al., 2017; Sohail et al., 2015). Several studies investigated the effect of heat stress on intestinal microbiota, but all of them only analyzed the bacteria that are perceived important because their focus was to evaluate prebiotics or probiotics for their efficacy in abating the adverse effect of heat stress (Lan et al., 2004; Sohail et al., 2015; Varmuzova et al., 2015; Zhang et al., 2017). Compared with the broilers (both Ross and Cobb) in the control group, the heat-stressed birds exhibited a decreased abundance of *Lactobacillus* and *Bifidobacterium*, but increased populations of *Clostridium* in the small intestines (Song et al., 2014) and cecum (Zhang et al., 2017). In contrast, one study showed increased *Lactobacillus* in the cecum of heat-stressed broilers (Sohail et al., 2015). In many studies, heat stress increased intestinal colonization by *Salmonella* (Burkholder et al., 2008; Soliman et al., 2009; Zhang et al., 2017) and *E. coli* (Laudadio et al., 2012; Zhang et al., 2017). It was postulated that alterations in the composition of the intestinal microbiota, and thus colonization resistance, caused by heat stress might be a major contributing factor to the increased colonization by *E. coli* (Laudadio et al., 2012) and *Salmonella* (Burkholder et al., 2008; Soliman et al., 2009) in heat-stressed birds. The impact of the heat stress on the cecal microbiota was considerably less profound compared to that of the microbiota of small intestines, suggesting that the microbiota in the ileum may be more sensitive to stress or temperature changes than the cecal microbiota (Burkholder et al., 2008). It should be noted that all the above studies did not pair-feed the birds. Because heat stress lowers feed intake (Zhang et al., 2017) and feed intake can have a considerable impact on the intestinal microbiota, the above-mentioned effects of heat stress on the intestinal microbiota may be attributed to heat stress as well as lowered feed intake. Future studies shall use pair feeding and expand the scope of microbial analysis, beyond the few known groups of bacteria.

Only a few studies found in the literature have addressed the geographical impact on the intestinal microbiota of chickens (Videnska et al., 2014; Zhou et al., 2016; Siegerstetter et al., 2017). Despite the controlled environments established in commercial poultry houses worldwide to reduce the influence of climatic conditions, it is likely that geographical regions can differ in local poultry house climate and management, feed composition and processing, and medication, and these differences may also affect intestinal microbiota. In one study comparing intestinal microbiota between birds reared in Austria and Northern Ireland, between-sample β-diversity analysis based on principal coordinates analysis (PCoA) did not show clear separation among the two geographical locations, but geographical effects were noted in total bacterial abundance in the ileal and cecal content (but not feces), Shannon and Simpson within sample α-diversity, indices of the cecal content, relative abundance of several phyla (including *Firmicutes, Bacteroidetes, Actinobacteria,* and *Tenericutes*) and genera (including *Lactobacillus, Ruminococcus, Streptococcus,* and *Faecalibacterium*) (Siegerstetter et al., 2017). Interestingly, the geographical differences varied between different intestinal regions (ileum versus cecum) and gender (female versus male). Unfortunately, information on the geographical location and the climate to which the birds were exposed were not measured or described in literature, and therefore it is difficult to evaluate to what extent these factors may influence the intestinal microbiota.

3.5 Gender

The segregation of commercial poultry production systems has resulted in layer-type chicken flocks predominantly consisting of female chickens, whereas males and females are often raised together in broiler flocks. In studies using other animal models, males are usually preferred to avoid the effects of cyclical reproductive hormone levels (Zucker and Beery, 2010). Many studies on broilers and intestinal microbiota only generated data from males or the sex of the broilers is unknown. This gender bias in literature might influence our understanding of microbiota development in chickens.

Broiler males generally have a higher growth rate and lower FCR than broiler females. However, differences in intestinal microbiota between male and female broilers are not influenced only by growth rate. Lumpkins et al. (2008) showed that although no difference in growth rate was observed between male and female broilers until day 21, differences in intestinal microbiota composition were already detected from day 3 onwards. In that study, the intestinal microbiota, as examined using denaturing gradient gel electrophoresis (DGGE) of PCR-amplified 16S rRNA gene fragments, showed less than 30% similarity in DGGE profiles between male and female birds (Lumpkins et al., 2008). Another study, in which female and male broilers (age 22 and 42 days) were compared using quantitative PCR (qPCR), showed differences in the abundance of *Lactobacillus salivarius*, *L. crispatus*, *L. aviaries*, and *E. coli* in cecal content (Torok et al., 2013). In a study using chickens of 245 days of age and two body weight lines (high versus low), the relative abundance of 48 microbial species was significantly different between males and females (Zhao et al., 2013). Additionally, an interaction between probiotic treatment (dietary yeast) and gender was found with respect to the abundance of *Bifidobacterium* in 42-day-old broilers (Mountzouris et al., 2015). Furthermore, divergence by sex and body weight was shown for the cecal microbiota (Lee et al., 2017), but male and female broilers were housed separately, which might have created confounding factors contributing to the observed gender differences in the intestinal microbiota. These results reinforce that gender may affect the intestinal microbiota in poultry, but studies that eliminate potential confounding factors are needed to reveal the actual impact of gender on the intestinal microbiota in poultry.

3.6 Diseases

The poultry intestinal microbiota contains pathogens, particularly pathogenic bacteria, viruses, and parasites as coccidia. These pathogens can cause clinical diseases when their abundance reaches a threshold, particularly under predisposing or stressful conditions (Williams, 2005; Moore, 2016; Prescott et al., 2016a,b). Interested readers are directed to Chapter 11 of this book for a detailed discussion of enteric diseases in chickens. These pathogens can directly affect the intestinal microbiota of chickens through microbe-microbe interactions or indirectly through modifications of the intestinal environment, including the increased supply of host-derived nutrients and increased redox potential due to leaky gut caused by some toxins (e.g. NetB) and inflammation, and change in immune responses. Because such alterations of the intestinal environment are concomitantly accompanied with the infectious disease caused by these pathogens, it is difficult to determine the direct and indirect effect of disease on the intestinal microbiota. Readers interested in the impact of enteric pathogens on intestinal microbiota are referred to Chapter 12 of this book.

4 Future trends and conclusion

A growing body of evidence has documented the importance of the intestinal microbiota to the nutrient utilization, FCR, and health and diseases in poultry. To improve feed conversion and enhance host health, recent research interests have gravitated toward reprograming the development of the intestinal microbiota in young birds. Much has been learned about the effect of diet and antimicrobial growth promoters (AGP), but much remains to be learned about how other factors modulate the development and eventual assemblage of the intestinal microbiota. Below are several of the important research areas that can help understand the interactions between chickens and the intestinal microbiota as affected by some of the factors discussed in this chapter.

4.1 Links among host genotypes, intestinal microbiota, and production traits

Studies on different breeds and lines of broiler have shown that the host genotypes can affect some of the important production traits of chickens, such as feed conversion efficiency and health. The genetic control of the intestinal microbiota and linkage to feed utilization was also demonstrated in other animals (Lindholm-Perry et al., 2014; Sasson et al., 2017). Some intestinal bacteria of mammalians are considered inheritable. Future research to establish the inheritability of intestinal bacteria and the linkage among chicken genotypes, intestinal microbiota, and feed conversion and disease resistance can help select and breed chickens with desirable production traits.

4.2 Litter microbiota and intestinal microbiota

Litter has a profound effect on the intestinal microbiota of chickens, especially young chicks when they first arrive in a chicken house. Reused litter was shown to accelerate the establishment of the intestinal microbiota. Future research to determine the effect of the litter microbiota in the reused litter on the intestinal microbiota of young chicks and mature chickens, especially the occurrence and load of pathogens, will help the management of litter while reducing the risk of vertical transmission of pathogens.

4.3 Modulation of intestinal microbiota development in young chicks and poults

The nearly sterile intestines of newly hatched chicks are receptive to bacterial colonization and provide an opportunity to program the development of their intestinal microbiota. Several studies have demonstrated that *in ovo* inoculation with probiotic strains can reduce the colonization of chicks with *Salmonella enteritidis* (de Oliveira et al., 2014) or *E. coli* (Majidi-Mosleh et al., 2017). *In ovo* inoculation with Aviguard®, a competitive exclusion culture, was shown to expedite the bacterial colonization in the intestines of young chicks and decrease the occurrence of undesirable bacteria (Pedroso et al., 2016). Future research is needed to evaluate if *in ovo* inoculation can be used as an effective strategy to jumpstart and program the development of intestinal microbiota in chicks and poults. Fecal microbiota transplantation (FMT) has also been used to treat and reduce susceptibility to disease in humans (Paramsothy et al., 2017) and animals (Niederwerder, 2018). One study has tested FMT in broiler chickens, and little effects were noted on fecal microbiota or residual feed intake (Siegerstetter et al., 2018). More studies are needed to evaluate how FMT affect feed efficiency under different conditions and in different efficiency measure (e.g. FCR).

4.4 Impact on the functionality of the intestinal microbiota

From an ecological perspective, the intestines of poultry are a rather selective habitat, primarily due to the availability of a narrow spectrum of substrates (feed ingredients) at large quantities. As a result, the intestinal microbiota manifests high functional redundancy with different taxa of microbes carrying out the same or similar functions, including digesting the same substrates and producing the same fermentation products and other metabolites (Sollinger et al., 2018). Functional redundancy is shown in the human fecal microbiota (Turnbaugh et al., 2009), the rumen of ruminants (Weimer, 2015; Sollinger et al., 2018), and the chicken intestinal microbiota (Oakley et al., 2014). As such, factors that affect the taxonomic diversity and composition of the intestinal microbiota may not necessarily have functional consequence in poultry. Holistic research of various factors on the intestinal microbiota shall include investigation of the taxonomic diversity and composition as well as the functional diversity, the latter of which can be achieved using metagenomics, metatranscriptomics, and metabolomics of the intestinal microbiota in chickens.

5 Where to look for further information

Intestinal microbiota research is a rapidly developing area and the most up-to-date scientific information and new learning are published in many scientific journals including microbiology journals and poultry science journals and scientific conferences. The following journals are among the journals that publish the research in the area discussed in this chapter. For technical information, interested readers are referred to Chapters 2 and 3.

5.1 Scientific journals

- *Animal Microbiome* (BMC, ISSN: 2524-4671)
- *Applied and Environmental Microbiology* (ASM, ISSN: 0099-2240)
- *Beneficial Microbes* (Wageningen Academic Publishers, 18762883)
- *British Poultry Science* (Taylor & Francis, ISSN: 14661799)
- *European Poultry Science* (Verlag Eugen Ulmer, ISSN: 16129199)
- *Frontiers in Microbiology* (Electronic ISSN: 1664-302X)
- *Frontiers in Veterinary Science* (Electronic ISSN: 2297-1769)
- *Gut Microbes* (Landes BioScience, ISSN: 19)
- *International Journal of Poultry Science* (Asian Network for Scientific Information, ISSN: 16828356)
- *Journal of Animal Physiology and Animal Nutrition* (Wiley, ISSN: 1439-0396)
- *Journal of Applied Poultry Research* (Oxford University Press, ISSN: 15370437)
- *Journal of Poultry Science* (Japan Poultry Science Association, ISSN: 13467395)
- *Microbiome* (BioMed Central, ISSN: 20492618)
- *PLoS ONE* (PLoS ONE, eISSN: 1932-6203)
- *Poultry Science* (Oxford University Press, ISSN: 0032-5791)
- *World's Poultry Science Journal* (Cambridge University Press, ISSN: 00439339)

5.2 Major scientific conferences

- International Production & Processing Expo
- International Conference on Poultry Intestinal Health
- Poultry Science Association Annual Meeting, USA
- World Veterinary Poultry Association Congress
- World's Poultry Congress

6 References

Alexander, D. C., Carriere, J. A. and McKay, K. A. 1968. Bacteriological studies of poultry litter fed to livestock. *Can. Vet. J.* 9(6), 127-31.

Baas Becking, L. G. M. 1934. *Geobiologie of inleiding tot de milieukunde*. W.P. Van Stockum & Zoon, The Hague, the Netherlands (in Dutch).

Bailey, J. S., Cox, N. A. and Berrang, M. E. 1994. Hatchery-acquired salmonellae in broiler chicks. *Poult. Sci.* 73(7), 1153-7. doi:10.3382/ps.0731153.

Barker, K. J., Purswell, J. L., Davis, J. D., Parker, H. M., Kidd, M. T., McDaniel, C. D. and Kiess, A. S. 2010. Distribution of bacteria at different poultry litter depths. *Int. J. Poult. Sci.* 9(1), 10-3. doi:10.3923/ijps.2010.10.13.

Benson, A. K., Kelly, S. A., Legge, R., Ma, F., Low, S. J., Kim, J., Zhang, M., Oh, P. L., Nehrenberg, D., Hua, K., Kachman, S. D., Moriyama, E. N., Walter, J., Peterson, D. A. and Pomp, D. 2010. Individuality in gut microbiota composition is a complex polygenic trait shaped by multiple environmental and host genetic factors. *Proc. Natl Acad. Sci. U. S. A.* 107(44), 18933-8. doi:10.1073/pnas.1007028107.

Bolan, N. S., Szogi, A. A., Chuasavathi, T., Seshadri, B., Rothrock, M. J. and Panneerselvam, P. 2010. Uses and management of poultry litter. *Worlds Poult. Sci. J.* 66(4), 673-98. doi:10.1017/S0043933910000656.

Burkholder, K. M., Thompson, K. L., Einstein, M. E., Applegate, T. J. and Patterson, J. A. 2008. Influence of stressors on normal intestinal microbiota, intestinal morphology, and susceptibility to *Salmonella* Enteritidis colonization in broilers. *Poult. Sci.* 87(9), 1734-41. doi:10.3382/ps.2008-00107.

Cason, J. A., Cox, N. A. and Bailey, J. S. 1994. Transmission of *Salmonella* Typhimurium during hatching of broiler chicks. *Avian Dis.* 38(3), 583-8. doi:10.2307/1592082.

Chinivasagam, H. N., Estella, W., Rodrigues, H., Mayer, D. G., Weyand, C., Tran, T., Onysk, A. and Diallo, I. 2016. On-farm *Campylobacter* and *Escherichia coli* in commercial broiler chickens: re-used bedding does not influence *Campylobacter* emergence and levels across sequential farming cycles. *Poult. Sci.* 95(5), 1105-15. doi:10.3382/ps/pew003.

Corrier, D. E., Hargis, B. M., Hinton Jr., A. and DeLoach, J. R. 1993. Protective effect of used poultry litter and lactose in the feed ration on *Salmonella enteritidis* colonization of Leghorn chicks and hens. *Avian Dis.* 37(1), 47-52. doi:10.2307/1591456.

Corrier, D. E., Nisbet, D. J., Scanlan, C. M., Hollister, A. G., Caldwell, D. J., Thomas, L. A., Hargis, B. M., Tomkins, T. and Deloach, J. R. 1995. Treatment of commercial broiler chickens with a characterized culture of cecal bacteria to reduce salmonellae colonization. *Poult. Sci.* 74(7), 1093-101. doi:10.3382/ps.0741093.

Cox, N. A., Bailey, J. S., Mauldin, J. M. and Blankenship, L. C. 1990. Presence and impact of *Salmonella* contamination in commercial broiler hatcheries. *Poult. Sci.* 69(9), 1606-9. doi:10.3382/ps.0691606.

Cox, N. A., Richardson, L. J., Maurer, J. J., Berrang, M. E., Fedorka-Cray, P. J., Buhr, R. J., Byrd, J. A., Lee, M. D., Hofacre, C. L., O'Kane, P. M., Lammerding, A. M., Clark, A. G., Thayer, S. G. and Doyle, M. P. 2012. Evidence for horizontal and vertical transmission in *Campylobacter* passage from hen to her progeny. *J. Food Prot.* 75(10), 1896-902. doi:10.4315/0362-028.JFP-11-322.

Craven, S. E., Stern, N. J., Bailey, J. S. and Cox, N. A. 2001a. Incidence of *Clostridium perfringens* in broiler chickens and their environment during production and processing. *Avian Dis.* 45(4), 887-96. doi:10.2307/1592868.

Craven, S. E., Cox, N. A., Stern, N. J. and Mauldin, J. M. 2001b. Prevalence of *Clostridium perfringens* in commercial broiler hatcheries. *Avian Dis.* 45(4), 1050–3. doi:10.2307/1592887.

Cressman, M. D., Yu, Z., Nelson, M. C., Moeller, S. J., Lilburn, M. S. and Zerby, H. N. 2010. Interrelations between the microbiotas in the litter and in the intestines of commercial broiler chickens. *Appl. Environ. Microbiol.* 76(19), 6572–82. doi:10.1128/AEM.00180-10.

De Cesare, A., Caselli, E., Lucchi, A., Sala, C., Parisi, A., Manfreda, G. and Mazzacane, S. 2019. Impact of a probiotic-based cleaning product on the microbiological profile of broiler litters and chicken caeca microbiota. *Poult. Sci.* doi:10.3382/ps/pez148.

de Oliveira, J. E., van der Hoeven-Hangoor, E., van de Linde, I. B., Montijn, R. C. and van der Vossen, J. M. 2014. *In ovo* inoculation of chicken embryos with probiotic bacteria and its effect on posthatch *Salmonella* susceptibility. *Poult. Sci.* 93(4), 818–29. doi:10.3382/ps.2013-03409.

Dhanarani, T. S., Shankar, C., Park, J., Dexilin, M., Kumar, R. R. and Thamaraiselvi, K. 2009. Study on acquisition of bacterial antibiotic resistance determinants in poultry litter. *Poult. Sci.* 88(7), 1381–7. doi:10.3382/ps.2008-00327.

Ding, J., Dai, R., Yang, L., He, C., Xu, K., Liu, S., Zhao, W., Xiao, L., Luo, L., Zhang, Y. and Meng, H. 2017. Inheritance and establishment of gut microbiota in chickens. *Front. Microbiol.* 8, 1967. doi:10.3389/fmicb.2017.01967.

Donaldson, E. E., Stanley, D., Hughes, R. J. and Moore, R. J. 2017. The time-course of broiler intestinal microbiota development after administration of cecal contents to incubating eggs. *Peer J.* 5, e3587. doi:10.7717/peerj.3587.

Duff, R. H., Ross, J. G. and Brown, D. D. 1973. The influence of litter on *Salmonella typhimurium* infection in poultry. *Avian Pathol.* 2(4), 263–8. doi:10.1080/03079457309353802.

Dumas, M. D., Polson, S. W., Ritter, D., Ravel, J., Gelb, J., Jr., Morgan, R. and Wommack, K. E. 2011. Impacts of poultry house environment on poultry litter bacterial community composition. *PLoS ONE* 6(9), e24785. doi:10.1371/journal.pone.0024785.

Emam, M., Mehrabani-Yeganeh, H., Barjesteh, N., Nikbakht, G., Thompson-Crispi, K., Charkhkar, S. and Mallard, B. 2014. The influence of genetic background versus commercial breeding programs on chicken immunocompetence. *Poult. Sci.* 93(1), 77–84. doi:10.3382/ps.2013-03475.

Fanelli, M. J., Sadler, W. W. and Brownell, J. R. 1970. Preliminary studies of persistence of salmonellae in poultry litter. *Avian Dis.* 14(1), 131–41. doi:10.2307/1588564.

Ferrario, C., Alessandri, G., Mancabelli, L., Gering, E., Mangifesta, M., Milani, C., Lugli, G. A., Viappiani, A., Duranti, S., Turroni, F., Ossiprandi, M. C., Hiyashi, R., Mackie, R., van Sinderen, D. and Ventura, M. 2017. Untangling the cecal microbiota of feral chickens by culturomic and metagenomic analyses. *Environ. Microbiol.* 19(11), 4771–83. doi:10.1111/1462-2920.13943.

Fries, R., Akcan, M., Bandick, N. and Kobe, A. 2005. Microflora of two different types of poultry litter. *Br. Poult. Sci.* 46(6), 668–72. doi:10.1080/00071660500395483.

Gantois, I., Ducatelle, R., Pasmans, F., Haesebrouck, F., Gast, R., Humphrey, T. J. and Van Immerseel, F. 2009. Mechanisms of egg contamination by *Salmonella* Enteritidis. *FEMS Microbiol. Rev.* 33(4), 718–38. doi:10.1111/j.1574-6976.2008.00161.x.

Gustafson, R. H. and Kobland, J. D. 1984. Factors influencing *Salmonella* shedding in broiler chickens. *J. Hyg. (Lond)* 92(3), 385–94. doi:10.1017/s0022172400064603.

Halbrook, E. R., Winter, A. R. and Sutton, T. S. 1951. The microflora of poultry house litter and droppings. *Poult. Sci.* 30(3), 381–8. doi:10.3382/ps.0300381.

Han, Z., Willer, T., Pielsticker, C., Gerzova, L., Rychlik, I. and Rautenschlein, S. 2016. Differences in host breed and diet influence colonization by *Campylobacter jejuni* and induction of local immune responses in chicken. *Gut Pathog.* 8, 56. doi:10.1186/s13099-016-0133-1.

Jang, S. I., Lillehoj, H. S., Lee, S. H., Lee, K. W., Lillehoj, E. P., Hong, Y. H., An, D. J., Jeoung, D. H. and Chun, J. E. 2013. Relative disease susceptibility and clostridial toxin antibody responses in three commercial broiler lines coinfected with *Clostridium perfringens* and *Eimeria maxima* using an experimental model of necrotic enteritis. *Avian Dis.* 57(3), 684–7. doi:10.1637/10496-011813-ResNote.1.

Kelley, T. R., Pancorbo, O. C., Merka, W. C. and Barnhart, H. M. 1998. Antibiotic resistance of bacterial litter isolates. *Poult. Sci.* 77(2), 243–7. doi:10.1093/ps/77.2.243.

Kers, J. G., Velkers, F. C., Fischer, E. A. J., Hermes, G. D. A., Stegeman, J. A. and Smidt, H. 2018. Host and environmental factors affecting the intestinal microbiota in chickens. *Front. Microbiol.* 9, 235. doi:10.3389/fmicb.2018.00235.

Kim, J. E., Lillehoj, H. S., Hong, Y. H., Kim, G. B., Lee, S. H., Lillehoj, E. P. and Bravo, D. M. 2015. Dietary *Capsicum* and *Curcuma longa* oleoresins increase intestinal microbiome and necrotic enteritis in three commercial broiler breeds. *Res. Vet. Sci.* 102, 150–8. doi:10.1016/j.rvsc.2015.07.022.

Lan, Y., Xun, S., Tamminga, S., Williams, B. A., Verstegen, M. W. and Erdi, G. 2004. Real-time PCR detection of lactic acid bacteria in cecal contents of *Eimeria tenella*-infected broilers fed soybean oligosaccharides and soluble soybean polysaccharides. *Poult. Sci.* 83(10), 1696–702. doi:10.1093/ps/83.10.1696.

Laudadio, V., Dambrosio, A., Normanno, G., Khan, R. U., Naz, S., Rowghani, E. and Tufarelli, V. 2012. Effect of reducing dietary protein level on performance responses and some microbiological aspects of broiler chickens under summer environmental conditions. *Avian Biol. Res.* 5(2), 88–92. doi:10.3184/175815512X13350180713553.

Lee, K. C., Kil, D. Y. and Sul, W. J. 2017. Cecal microbiome divergence of broiler chickens by sex and body weight. *J. Microbiol.* 55(12), 939–45. doi:10.1007/s12275-017-7202-0.

Lindholm-Perry, A. K., Kuehn, L. A., Oliver, W. T., Kern, R. J., Cushman, R. A., Miles, J. R., McNeel, A. K. and Freetly, H. C. 2014. DNA polymorphisms and transcript abundance of PRKAG2 and phosphorylated AMP-activated protein kinase in the rumen are associated with gain and feed intake in beef steers. *Anim. Genet.* 45(4), 461–72. doi:10.1111/age.12151.

Lloyd, A. B., Cumming, R. B. and Kent, R. D. 1977. Prevention of *Salmonella typhimurium* infection in poultry by pretreatment of chickens and poults with intestinal extracts. *Aust. Vet. J.* 53(2), 82–7. doi:10.1111/j.1751-0813.1977.tb14891.x.

Lovanh, N., Cook, K. L., Rothrock, M. J., Miles, D. M. and Sistani, K. 2007. Spatial shifts in microbial population structure within poultry litter associated with physicochemical properties. *Poult. Sci.* 86(9), 1840–9. doi:10.1093/ps/86.9.1840.

Lovett, J., Messer, J. W. and Read, R. B., Jr. 1971. The microflora of southern Ohio poultry litter. *Poult. Sci.* 50(3), 746–51. doi:10.3382/ps.0500746.

Lu, J., Sanchez, S., Hofacre, C., Maurer, J. J., Harmon, B. G. and Lee, M. D. 2003. Evaluation of broiler litter with reference to the microbial composition as assessed by using 16S rRNA and functional gene markers. *Appl. Environ. Microbiol.* 69(2), 901–8. doi:10.1128/aem.69.2.901-908.2003.

Lumpkins, B. S., Batal, A. B. and Lee, M. 2008. The effect of gender on the bacterial community in the gastrointestinal tract of broilers. *Poult. Sci.* 87(5), 964–7. doi:10.3382/ps.2007-00287.

Macklin, K. S., Hess, J. B., Bilgili, S. F. and Norton, R. A. 2005. Bacterial levels of pine shavings and sand used as poultry litter. *J. Appl. Poult. Res.* 14(2), 238–45. doi:10.1093/japr/14.2.238.

Majidi-Mosleh, A., Sadeghi, A. A., Mousavi, S. N., Chamani, M. and Zarei, A. 2017. Ileal MUC2 gene expression and microbial population, but not growth performance and immune response, are influenced by in ovo injection of probiotics in broiler chickens. *Br. Poult. Sci.* 58(1), 40–5. doi:10.1080/00071668.2016.1237766.

Mancabelli, L., Ferrario, C., Milani, C., Mangifesta, M., Turroni, F., Duranti, S., Lugli, G. A., Viappiani, A., Ossiprandi, M. C., van Sinderen, D. and Ventura, M. 2016. Insights into the biodiversity of the gut microbiota of broiler chickens. *Environ. Microbiol.* 18(12), 4727–38. doi:10.1111/1462-2920.13363.

McCann, M. A., Martin, S. A. and Waltman II, W. D. 1998. Microbiological survey of Georgia poultry litter. *J. Appl. Poult. Res.* 7(1), 90–8. doi:10.1093/japr/7.1.90.

McCrea, B. A., Norton, R. A., Macklin, K. S., Hess, J. B. and Bilgili, S. F. 2005. Recovery and genetic similarity of *Salmonella* from broiler house drag swabs versus surgical shoe covers. *J. Appl. Poult. Res.* 14(4), 694–9. doi:10.1093/japr/14.4.694.

McCrea, B. A., Macklin, K. S., Norton, R. A., Hess, J. B. and Bilgili, S. F. 2008. Recovery and genetic diversity of *Escherichia coli* isolates from deep litter, shallow litter, and surgical shoe covers. *J. Appl. Poult. Res.* 17(2), 237–42. doi:10.3382/japr.2007-00067.

Meng, H., Zhang, Y., Zhao, L., Zhao, W., He, C., Honaker, C. F., Zhai, Z., Sun, Z. and Siegel, P. B. 2014. Body weight selection affects quantitative genetic correlated responses in gut microbiota. *PLoS ONE* 9(3), e89862. doi:10.1371/journal.pone.0089862.

Mignon-Grasteau, S., Narcy, A., Rideau, N., Chantry-Darmon, C., Boscher, M. Y., Sellier, N., Chabault, M., Konsak-Ilievski, B., Le Bihan-Duval, E. and Gabriel, I. 2015. Impact of selection for digestive efficiency on microbiota composition in the chicken. *PLoS ONE* 10(8), e0135488. doi:10.1371/journal.pone.0135488.

Mohd Shaufi, M. A., Sieo, C. C., Chong, C. W., Gan, H. M. and Ho, Y. W. 2015. Deciphering chicken gut microbial dynamics based on high-throughput 16S rRNA metagenomics analyses. *Gut Pathog.* 7, 4. doi:10.1186/s13099-015-0051-7.

Moore, R. J. 2016. Necrotic enteritis predisposing factors in broiler chickens. *Avian Pathol.* 45(3), 275–81. doi:10.1080/03079457.2016.1150587.

Mountzouris, K. C., Dalaka, E., Palamidi, I., Paraskeuas, V., Demey, V., Theodoropoulos, G. and Fegeros, K. 2015. Evaluation of yeast dietary supplementation in broilers challenged or not with *Salmonella* on growth performance, cecal microbiota composition and *Salmonella* in ceca, cloacae and carcass skin. *Poult. Sci.* 94(10), 2445–55. doi:10.3382/ps/pev243.

Nakphaichit, M., Thanomwongwattana, S., Phraephaisarn, C., Sakamoto, N., Keawsompong, S., Nakayama, J. and Nitisinprasert, S. 2011. The effect of including *Lactobacillus reuteri* KUB-AC5 during post-hatch feeding on the growth and ileum microbiota of broiler chickens. *Poult. Sci.* 90(12), 2753–65. doi:10.3382/ps.2011-01637.

Niederwerder, M. C. 2018. Fecal microbiota transplantation as a tool to treat and reduce susceptibility to disease in animals. *Vet. Immunol. Immunopathol.* 206, 65–72. doi:10.1016/j.vetimm.2018.11.002.

Nurmi, E. and Rantala, M. 1973. New aspects of *Salmonella* infection in broiler production. *Nature* 241(5386), 210–1. doi:10.1038/241210a0.

Oakley, B. B., Morales, C. A., Line, J., Berrang, M. E., Meinersmann, R. J., Tillman, G. E., Wise, M. G., Siragusa, G. R., Hiett, K. L. and Seal, B. S. 2013. The poultry-associated microbiome: network analysis and farm-to-fork characterizations. *PLoS ONE* 8(2), e57190. doi:10.1371/journal.pone.0057190.

Oakley, B. B., Lillehoj, H. S., Kogut, M. H., Kim, W. K., Maurer, J. J., Pedroso, A., Lee, M. D., Collett, S. R., Johnson, T. J. and Cox, N. A. 2014. The chicken gastrointestinal microbiome. *FEMS Microbiol. Lett.* 360(2), 100–12. doi:10.1111/1574-6968.12608.

Oakley, B. B., Vasconcelos, E. J. R., Diniz, P. P. V. P., Calloway, K. N., Richardson, E., Meinersmann, R. J., Cox, N. A. and Berrang, M. E. 2018. The cecal microbiome of commercial broiler chickens varies significantly by season. *Poult. Sci.* 97(10), 3635–44. doi:10.3382/ps/pey214.

Omeira, N., Barbour, E. K., Nehme, P. A., Hamadeh, S. K., Zurayk, R. and Bashour, I. 2006. Microbiological and chemical properties of litter from different chicken types and production systems. *Sci. Total Environ.* 367(1), 156–62. doi:10.1016/j.scitotenv.2006.02.019.

Org, E., Parks, B. W., Joo, J. W., Emert, B., Schwartzman, W., Kang, E. Y., Mehrabian, M., Pan, C., Knight, R., Gunsalus, R., Drake, T. A., Eskin, E. and Lusis, A. J. 2015. Genetic and environmental control of host-gut microbiota interactions. *Genome Res.* 25(10), 1558–69. doi:10.1101/gr.194118.115.

Pan, D. and Yu, Z. 2014. Intestinal microbiome of poultry and its interaction with host and diet. *Gut Microbes* 5(1), 108–19. doi:10.4161/gmic.26945.

Paramsothy, S., Paramsothy, R., Rubin, D. T., Kamm, M. A., Kaakoush, N. O., Mitchell, H. M. and Castano-Rodriguez, N. 2017. Faecal microbiota transplantation for inflammatory bowel disease: a systematic review and meta-analysis. *J. Crohns Colitis* 11(10), 1180–99. doi:10.1093/ecco-jcc/jjx063.

Paz, H. A., Hales, K. E., Wells, J. E., Kuehn, L. A., Freetly, H. C., Berry, E. D., Flythe, M. D., Spangler, M. L. and Fernando, S. C. 2018. Rumen bacterial community structure impacts feed efficiency in beef cattle. *J. Anim. Sci.* 96(3), 1045–58. doi:10.1093/jas/skx081.

Pedroso, A. A., Batal, A. B. and Lee, M. D. 2016. Effect of *in ovo* administration of an adult-derived microbiota on establishment of the intestinal microbiome in chickens. *Am. J. Vet. Res.* 77(5), 514–26. doi:10.2460/ajvr.77.5.514.

Persoons, D., Haesebrouck, F., Smet, A., Herman, L., Heyndrickx, M., Martel, A., Catry, B., Berge, A. C., Butaye, P. and Dewulf, J. 2011. Risk factors for ceftiofur resistance in Escherichia coli from Belgian broilers. *Epidemiol. Infect.* 139(5), 765–71. doi:10.1017/S0950268810001524.

Pivnick, H., Blanchfield, B. and D'Aoust, J. Y. 1981. Prevention of *Salmonella* infection in chicks by treatment with fecal cultures from mature chickens (Nurmi Cultures). *J. Food Prot.* 44(12), 909–16. doi:10.4315/0362-028X-44.12.909.

Ponsuksili, S., Wimmers, K. and Horst, P. 1998. Evaluation of genetic variation within and between different chicken lines by DNA fingerprinting. *J. Hered.* 89(1), 17–23. doi:10.1093/jhered/89.1.17.

Pourabedin, M., Guan, L. and Zhao, X. 2015. Xylo-oligosaccharides and virginiamycin differentially modulate gut microbial composition in chickens. *Microbiome* 3, 15. doi:10.1186/s40168-015-0079-4.

Prescott, J. F., Smyth, J. A., Shojadoost, B. and Vince, A. 2016a. Experimental reproduction of necrotic enteritis in chickens: a review. *Avian Pathol.* 45(3), 317–22. doi:10.1080/03079457.2016.1141345.

Prescott, J. F., Parreira, V. R., Mehdizadeh Gohari, I., Lepp, D. and Gong, J. 2016b. The pathogenesis of necrotic enteritis in chickens: what we know and what we need to know: a review. *Avian Pathol.* 45(3), 288–94. doi:10.1080/03079457.2016.1139688.

Rigby, C. E. and Pettit, J. R. 1980. Observations on competitive exclusion for preventing *Salmonella typhimurium* infection of broiler chickens. *Avian Dis.* 24(3), 604–15. doi:10.2307/1589796.

Samli, H. E., Dezcan, S., Koc, F., Ozduven, M. L., Okur, A. A. and Senkoylu, N. 2010. Effects of *Enterococcus faecium* supplementation and floor type on performance, morphology of erythrocytes and intestinal microbiota in broiler chickens. *Br. Poult. Sci.* 51(4), 564–8. doi:10.1080/00071668.2010.507241.

Sasson, G., Kruger Ben-Shabat, S., Seroussi, E., Doron-Faigenboim, A., Shterzer, N., Yaacoby, S., Berg Miller, M. E., White, B. A., Halperin, E. and Mizrahi, I. 2017. Heritable bovine rumen bacteria are phylogenetically related and correlated with the cow's capacity to harvest energy from its feed. *MBio* 8(4). doi:10.1128/mBio.00703-17.

Schefferle, H. E. 1965. The microbiology of built up poultry litter. *J. Appl. Bacteriol.* 28(3), 403–11. doi:10.1111/j.1365-2672.1965.tb02170.x.

Schokker, D., Veninga, G., Vastenhouw, S. A., Bossers, A., de Bree, F. M., Kaal-Lansbergen, L. M., Rebel, J. M. and Smits, M. A. 2015. Early life microbial colonization of the gut and intestinal development differ between genetically divergent broiler lines. *BMC Genomics* 16, 418. doi:10.1186/s12864-015-1646-6.

Shires, A., Thompson, J. R., Turner, B. V., Kennedy, P. M. and Goh, Y. K. 1987. Rate of passage of corn-canola meal and corn-soybean meal diets through the gastrointestinal tract of broiler and White Leghorn chickens. *Poult. Sci.* 66(2), 289–98. doi:10.3382/ps.0660289.

Siegerstetter, S. C., Schmitz-Esser, S., Magowan, E., Wetzels, S. U., Zebeli, Q., Lawlor, P. G., O'Connell, N. E. and Metzler-Zebeli, B. U. 2017. Intestinal microbiota profiles associated with low and high residual feed intake in chickens across two geographical locations. *PLoS ONE* 12(11), e0187766. doi:10.1371/journal.pone.0187766.

Siegerstetter, S. C., Petri, R. M., Magowan, E., Lawlor, P. G., Zebeli, Q., O'Connell, N. E. and Metzler-Zebeli, B. U. 2018. Fecal microbiota transplant from highly feed-efficient donors shows little effect on age-related changes in feed-efficiency-associated fecal microbiota from chickens. *Appl. Environ. Microbiol.* 84(2). doi:10.1128/AEM.02330-17.

Simon, K., de Vries Reilingh, G., Kemp, B. and Lammers, A. 2014. Development of ileal cytokine and immunoglobulin expression levels in response to early feeding in broilers and layers. *Poult. Sci.* 93(12), 3017–27. doi:10.3382/ps.2014-04225.

Snoeyenbos, G. H., Weinack, O. M. and Smyser, C. F. 1978. Protecting chicks and poults from *Salmonellae* by oral administration of "normal" gut microflora. *Avian Dis.* 22(2), 273–87. doi:10.2307/1589539.

Sofka, D., Pfeifer, A., Gleiß, B., Paulsen, P. and Hilbert, F. 2015. Changes within the intestinal flora of broilers by colonisation with *Campylobacter jejuni*. *Berl. Münch. Tierärztl. Wochenschr.* 128(3-4), 104–10.

Sohail, M. U., Hume, M. E., Byrd, J. A., Nisbet, D. J., Shabbir, M. Z., Ijaz, A. and Rehman, H. 2015. Molecular analysis of the caecal and tracheal microbiome of heat-stressed broilers supplemented with prebiotic and probiotic. *Avian Pathol.* 44(2), 67-74. doi: 10.1080/03079457.2015.1004622.

Soliman, E. S., Taha, E., Infante, K. D., Laboy, K., Sobieh, M. A. and Reddy, P. G. 2009. Stressors influence on *Salmonella enterica* serovar Enteritidis colonization in broilers. *Am. J. Anim. Vet. Sci.* 4(3), 42–8. doi:10.3844/ajavsp.2009.42.48.

Sollinger, A., Tveit, A. T., Poulsen, M., Noel, S. J., Bengtsson, M., Bernhardt, J., Frydendahl Hellwing, A. L., Lund, P., Riedel, K., Schleper, C., Højberg, O. and Urich, T. 2018. Holistic assessment of rumen microbiome dynamics through quantitative metatranscriptomics reveals multifunctional redundancy during key steps of anaerobic feed degradation. *mSystems* 3(4). doi:10.1128/mSystems.00038-18.

Song, J., Xiao, K., Ke, Y. L., Jiao, L. F., Hu, C. H., Diao, Q. Y., Shi, B. and Zou, X. T. 2014. Effect of a probiotic mixture on intestinal microflora, morphology, and barrier integrity of broilers subjected to heat stress. *Poult. Sci.* 93(3), 581–8. doi:10.3382/ps.2013-03455.

Stanley, D., Geier, M. S., Denman, S. E., Haring, V. R., Crowley, T. M., Hughes, R. J. and Moore, R. J. 2013. Identification of chicken intestinal microbiota correlated with the efficiency of energy extraction from feed. *Vet. Microbiol.* 164(1–2), 85–92. doi:10.1016/j.vetmic.2013.01.030.

Stanley, D., Hughes, R. J. and Moore, R. J. 2014. Microbiota of the chicken gastrointestinal tract: influence on health, productivity and disease. *Appl. Microbiol. Biotechnol.* 98(10), 4301–10. doi:10.1007/s00253-014-5646-2.

Terzich, M., Pope, M. J., Cherry, T. E. and Hollinger, J. 2000. Survey of pathogens in poultry litter in the United States. *J. Appl. Poult. Res.* 9(3), 287–91. doi:10.1093/japr/9.3.287.

Toghyani, M., Gheisari, A., Modaresi, M., Tabeidian, S. A. and Toghyani, M. 2010. Effect of different litter material on performance and behavior of broiler chickens. *Appl. Anim. Behav. Sci.* 122(1), 48–52. doi:10.1016/j.applanim.2009.11.008.

Torok, V. A., Hughes, R. J., Ophel-Keller, K., Ali, M. and Macalpine, R. 2009. Influence of different litter materials on cecal microbiota colonization in broiler chickens. *Poult. Sci.* 88(12), 2474–81. doi:10.3382/ps.2008-00381.

Torok, V. A., Dyson, C., McKay, A. and Ophel-Keller, K. 2013. Quantitative molecular assays for evaluating changes in broiler gut microbiota linked with diet and performance. *Anim. Prod. Sci.* 53(12), 1260–8. doi:10.1071/AN12272.

Turnbaugh, P. J., Hamady, M., Yatsunenko, T., Cantarel, B. L., Duncan, A., Ley, R. E., Sogin, M. L., Jones, W. J., Roe, B. A., Affourtit, J. P., Egholm, M., Henrissat, B., Heath, A. C., Knight, R. and Gordon, J. I. 2009. A core gut microbiome in obese and lean twins. *Nature* 457(7228), 480–4. doi:10.1038/nature07540.

Uni, Z., Noy, Y. and Sklan, D. 1996. Development of the small intestine in heavy and light strain chicks before and after hatching. *Br. Poult. Sci.* 37(1), 63–71. doi:10.1080/00071669608417837.

van der Wielen, P. W., Keuzenkamp, D. A., Lipman, L. J., van Knapen, F. and Biesterveld, S. 2002. Spatial and temporal variation of the intestinal bacterial community in commercially raised broiler chickens during growth. *Microb. Ecol.* 44(3), 286–93. doi:10.1007/s00248-002-2015-y.

Varmuzova, K., Matulova, M. E., Gerzova, L., Cejkova, D., Gardan-Salmon, D., Panheleux, M., Robert, F., Sisak, F., Havlickova, H. and Rychlik, I. 2015. *Curcuma* and *Scutellaria* plant extracts protect chickens against inflammation and *Salmonella* Enteritidis infection. *Poult. Sci.* 94(9), 2049–58. doi:10.3382/ps/pev190.

Videnska, P., Rahman, M. M., Faldynova, M., Babak, V., Matulova, M. E., Prukner-Radovcic, E., Krizek, I., Smole-Mozina, S., Kovac, J., Szmolka, A., Nagy, B., Sedlar, K., Cejkova, D. and Rychlik, I. 2014. Characterization of egg laying hen and broiler fecal microbiota in poultry farms in Croatia, Czech Republic, Hungary and Slovenia. *PLoS ONE* 9(10), e110076. doi:10.1371/journal.pone.0110076.

Walker, S. E. and Sander, J. E. 2004. Effect of BioSentry 904 and ethylenediaminetetraacetic acid-tris disinfecting during incubation of chicken eggs on microbial levels and productivity of poultry. *Avian Dis.* 48(2), 238–43. doi:10.1637/7049.

Wang, L., Lilburn, M. and Yu, Z. 2016. Intestinal microbiota of broiler chickens as affected by litter management regimens. *Front. Microbiol.* 7, 593. doi:10.3389/fmicb.2016.00593.

Wei, S., Morrison, M. and Yu, Z. 2013a. Bacterial census of poultry intestinal microbiome. *Poult. Sci.* 92(3), 671–83. doi:10.3382/ps.2012-02822.

Wei, S., Gutek, A., Lilburn, M. and Yu, Z. 2013b. Abundance of pathogens in the gut and litter of broiler chickens as affected by bacitracin and litter management. *Vet. Microbiol.* 166(3–4), 595–601. doi:10.1016/j.vetmic.2013.06.006.

Weimer, P. J. 2015. Redundancy, resilience, and host specificity of the ruminal microbiota: implications for engineering improved ruminal fermentations. *Front. Microbiol.* 6, 296. doi:10.3389/fmicb.2015.00296.

Williams, R. B. 2005. Intercurrent coccidiosis and necrotic enteritis of chickens: rational, integrated disease management by maintenance of gut integrity. *Avian Pathol.* 34(3), 159–80. doi:10.1080/03079450500112195.

Williams, Z. T. and Macklin, K. S. 2013. Stratification of bacterial concentrations, from upper to lower, in broiler litter. *J. Appl. Poult. Res.* 22(3), 492–8. doi:10.3382/japr.2012-00705.

Xu, L., He, Y., Ding, Y., Liu, G. E., Zhang, H., Cheng, H. H., Taylor, R. L., Jr. and Song, J. 2018. Genetic assessment of inbred chicken lines indicates genomic signatures of resistance to Marek's disease. *J. Anim. Sci. Biotechnol.* 9, 65. doi:10.1186/s40104-018-0281-x.

Yang, G., Yao, J., Yang, W., Jiang, Y., Du, J., Huang, H., Gu, W., Hu, J., Ye, L., Shi, C., Shan, B. and Wang, C. 2017. Construction and immunological evaluation of recombinant *Lactobacillus plantarum* expressing SO7 of *Eimeria tenella* fusion DC-targeting peptide. *Vet. Parasitol.* 236, 7–13. doi:10.1016/j.vetpar.2017.01.023.

You, Y., Hilpert, M. and Ward, M. J. 2013. Identification of Tet45, a tetracycline efflux pump, from a poultry-litter-exposed soil isolate and persistence of tet(45) in the soil. *J. Antimicrob. Chemother.* 68(9), 1962–9. doi:10.1093/jac/dkt127.

Zhang, C., Zhao, X. H., Yang, L., Chen, X. Y., Jiang, R. S., Jin, S. H. and Geng, Z. Y. 2017. Resveratrol alleviates heat stress-induced impairment of intestinal morphology, microflora, and barrier integrity in broilers. *Poult. Sci.* 96(12), 4325–32. doi:10.3382/ps/pex266.

Zhao, L., Wang, G., Siegel, P., He, C., Wang, H., Zhao, W., Zhai, Z., Tian, F., Zhao, J., Zhang, H., Sun, Z., Chen, W., Zhang, Y. and Meng, H. 2013. Quantitative genetic background of the host influences gut microbiomes in chickens. *Sci. Rep.* 3, 1163. doi:10.1038/srep01163.

Zhou, X., Jiang, X., Yang, C., Ma, B., Lei, C., Xu, C., Zhang, A., Yang, X., Xiong, Q., Zhang, P., Men, S., Xiang, R. and Wang, H. 2016. Cecal microbiota of Tibetan chickens from five geographic regions were determined by 16S rRNA sequencing. *Microbiologyopen* 5(5), 753–62. doi:10.1002/mbo3.367.

Zucker, I. and Beery, A. K. 2010. Males still dominate animal studies. *Nature* 465(7299), 690. doi:10.1038/465690a.

Chapter 9

Antibiotics and gut function: historical and current perspectives

Jeferson M. Lourenço, Darren S. Seidel and Todd R. Callaway, University of Georgia, USA

1 Introduction

Undeniably, antibiotics are one of the most successful forms of chemotherapy in the history of medicine as they have contributed to the control of infectious diseases and saved many lives (Bhargava and Srivastava, 2017). The term 'antibiotic' was first used in 1890 as an antonymic approach to the word 'symbiosis', to describe the antagonistic action between different microorganisms. Not long after, in 1893, mycophenolic acid became the first antibiotic isolated from nature (from the mould *Penicillium glaucum*) and scientists demonstrated that this compound could inhibit the growth of *Bacillus anthracis* (Nicolaou and Rigol, 2018).

Antibiotics have played an essential role in the human food production chain for more than 50 years as they have been used to treat animal diseases, for prophylactic/metaphylactic purposes, as well as growth promotors (Greer, 2016; Aminov, 2010). In fact, it is estimated that the global consumption of antimicrobials in animals is twice that of humans, and in the United States it is estimated that approximately 80% of all antimicrobials consumed by the nation are used in food animal production, to treat and prevent diseases and improve growth (Van Boeckel et al., 2015). Moreover, with an ever-increasing population, the global need for antimicrobials in the food animal production segment is projected to increase by 67% between 2010 and 2030, with the top three consuming countries remaining unchanged: China, the United States and Brazil, respectively (Van Boeckel et al., 2015).

http://dx.doi.org/10.19103/AS.2019.0059.10

Despite their successful use for several decades, bacterial resistance to antibiotics has always existed and was noted shortly after the introduction of penicillin; however, the emergence of more resistant (and consequently more dangerous) bacterial strains have occurred more frequently in the last 20 years (Salyers and Shoemaker, 2006; Fair and Tor, 2014). A recent publication by Gelband et al. illustrates the extension of this problem and its implications on human health, as they found that antibiotic resistance is responsible for more than 2 million infections and 23 000 deaths each year in the United States alone, at a cost of US$35 billion to the US economy. Unfortunately, these noteworthy numbers are not restricted to the United States, as the number of deaths due to antibiotic-resistant infections in Europe is estimated at 25 000 per year (Gelband et al., 2015). Although the exact mechanism has not been completely elucidated, some researchers have linked the development of antimicrobial drug-resistant bacteria to the inclusion of subtherapeutic levels of antibiotics (STLAB) in livestock feeds (Mathews, 2016; Mathew et al., 2007). This suggestion stems from the fact that many of the antimicrobials used in livestock production are the same as (or at least closely related to) the drugs used in human health care, which may cause the transmission of resistant organisms from animals to humans through the direct handling of animals, or even from the edible products derived from them.

The gastrointestinal tract of chickens harbours trillions of microorganisms such as viruses, fungi, archaea and chiefly bacteria. In commercial chickens, the gut colonization process starts immediately after hatching by contact with their housing environment, and one determinant factor in this process is their litter (Oakley et al., 2014). Most of the biological interactions between the chickens and their gut microorganisms are beneficial to both of them. In fact, the gut microbes play an important role in modulating the development of birds' digestive system (Pan and Yu, 2014) and immune system (Torok et al., 2011). In addition, the microorganisms are benefited from the nearly constant conditions provided by their host (e.g. temperature), as well as the pool of nutrients coming from the host's diet, to thrive. On the other hand, certain portions of the diet which are otherwise undigestible to chickens - such as fibre - are converted by the microbes into useful products like water-soluble B vitamins and short-chain fatty acids, which are then absorbed and used by the birds. The normal microbial diversity in the gastrointestinal tract of chickens can be disrupted by numerous agents, including parasitic infections and administration of antibiotics.

In poultry nutrition, the use of STLAB to increase growth rates has been in use since the late 1940s (Lorian, 1980), a practice that has led to a greater number of genes associated with antibiotic resistance in the chicken microbiome, as well as a higher risk of transmitting these genes to humans, since many pathogens can colonize both the human and chicken gastrointestinal tracts (Yeoman et al., 2012). Supporting such claims, some studies have found association between the use of STLAB and the emergence of pathogens resistant to fluoroquinolones, vancomycin and fourth-generation cephalosporins (Danzeisen et al., 2011). As a precautionary measure, the EU has banned the use of antimicrobial drugs for growth-promoting purposes in livestock production (Mathews, 2016). Similarly, in 2005, the US Food and Drug Administration (FDA) banned the use of enrofloxacin in poultry due to an increase in fluoroquinolone-resistant *Campylobacter* (Danzeisen et al., 2011). However, a higher incidence of necrotic enteritis has been associated with antimicrobial removal from poultry feeds (Cross et al., 2004). Moreover, results have shown that the removal of antibiotics from diets does not result in the disappearance of the antimicrobial resistance genes from a food animal population (Bager et al., 1999; DANMAP, 2006). Currently, the poultry industry is evaluating strategies to completely eliminate the use of antibiotics that have human analogs. This chapter discusses the historical and current perspectives on the use of antimicrobial drugs in poultry production.

2 Historical perspectives on antibiotics in poultry production

The history of antibiotics and antimicrobials alike, began in the early 1900s in the human health discipline to combat serious illnesses through controlling bacterial populations in the human gastrointestinal tract. After extensive research and scientific discoveries, animal scientists began investigating how these newly discovered chemicals could improve animal agriculture performance and prevent disease spread. Antibiotics may be used at one or more phases during the domesticated fowl's life cycle. These chemicals were first experimentally introduced into phases of poultry production in the mid-twentieth century and documented in publications as early as the 1940s (Berg et al., 1952; Couch et al., 1957; Moore et al., 1946). In 1951, the US FDA approved the use of antibiotics in animal additives without the need for a veterinary prescription (Castanon, 2007). The data in Table 1 illustrates the use of antibiotics for growth promotion and coccidiosis according to Jones and Ricke (2003).

Antibiotic use may begin as early as prior to hatching (or *in ovo*; Tavakkoli et al., 2014), and be continued throughout the birds' entire productive life. Figure 1 summarizes the opportunities for antibiotic usage in commercial broiler production systems.

Sulphonamides-, penicillin- and tetracycline-derived antibiotics and bacitracin were the first antibiotics utilized in studies to assess performance and health benefits of antibiotic use in poultry (Couch et al., 1957; Berg et al., 1952). At the dawn of antibiotic usage in poultry production, the intention was to improve growth performance by inhibiting the development of

Table 1 Antimicrobial compounds that were approved for use in broiler feed without a veterinary prescription that promote growth or treatment of coccidiosis, in 2001[a]

Drug name	Indication for use
Amprolium	Coccidiosis
Arsanilic acid	Growth promotion, feed efficiency
Bacitracin methylene disalicylate	Rate of weight gain, feed efficiency
Bacitracin zinc	Rate of weight gain, feed efficiency
Bambermycins	Rate of weight gain, feed efficiency
Chlortetracycline	Rate of weight gain, feed efficiency
Decoquinate	Coccidiosis
Diclazuril	Coccidiosis
Halofuginone hydrobromide	Coccidiosis
Lincomycin	Rate of weight gain, feed efficiency
Maduramicin ammonium	Coccidiosis
Monensin	Coccidiosis
Narasin	Coccidiosis
Narasin/nicarbazin	Coccidiosis
Nicarbazin	Coccidiosis
Oxytetracycline	Rate of weight gain, feed efficiency
Penicillin	Rate of weight gain, feed efficiency
Robenidine hydrochloride	Coccidiosis
Roxarsone	Rate of weight gain, feed efficiency
Salinomycin	Coccidiosis
Semduramicin	Coccidiosis
Sulphadimethoxine and ormetoprim 5:3	Coccidiosis
Tylosin	Rate of weight gain, feed efficiency
Virginiamycin	Rate of weight gain, feed efficiency
Zoalene	Coccidiosis

[a] Source: abstracted and modified from Jones and Ricke (2003).

specific microorganisms (e.g. *Clostridium* sp. (Watkins et al., 1997)) within the gastrointestinal tracts of chickens (Immerseel et al., 2004) and turkeys (Luangtongkum et al., 2006), which acted as 'drags' on production efficiency.

Historically speaking, the effectiveness of antibiotic usage in poultry production were rooted in 'trial-and-error' examples of empirical nature and the industry-adapted antibiotic utilization and use protocols to improve poultry health in modern production systems, and these procedures and the industry co-evolved together. Couch et al. (1957) conducted a study that examined penicillin (oral & injected), inactivated penicillin (oral and injected), aureomycin (injected) and bacitracin (injected) with New Hampshire chicks. From the results, the authors illustrate how antibiotic usage in poultry diets

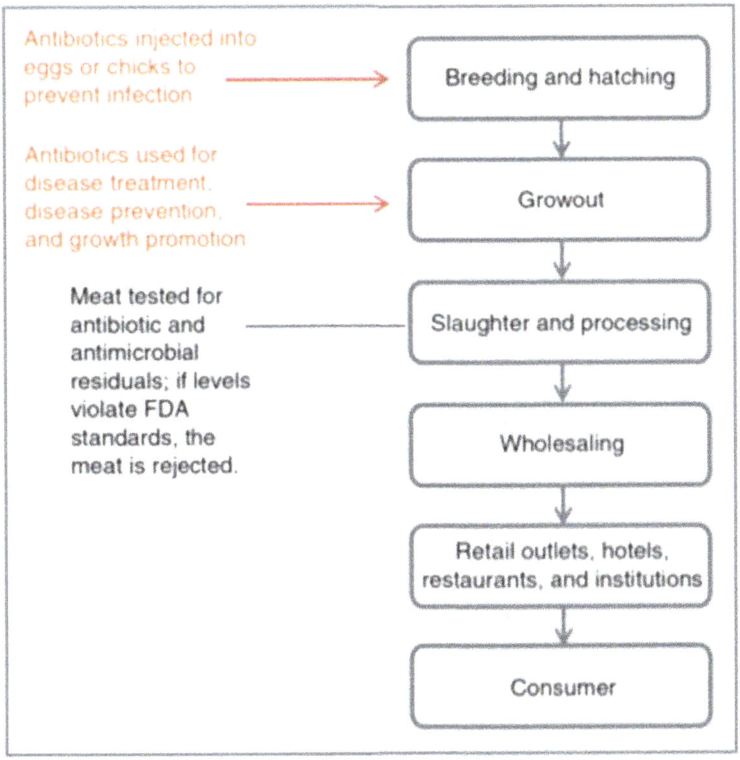

Figure 1 Phases of broiler production and antibiotic use. Source: adapted from Sneeringer et al. (2015) (USDA Economic Research Service).

increased average weight over antibiotic-free diets, with the exception of sodium benzylpenicilloate and beta-diethylaminoethanol. Despite these promising results, over a decade earlier, Moore et al. (1946) were actually the first researchers to describe the positive effects of antimicrobial usage on growth response in chicks. Studies such as these established the early results and knowledge around antibiotic utilization in poultry production, but as in most food animal producing industries, the poultry industry began to evolve from small and outdoor facilities to much large confined-feeding operations. Therefore, the early methods of administration of antibiotics were no longer practical through injection/oral methods, instead, researchers developed administration techniques that involved supplementing chickens and turkeys with antibiotics through water sources and/or within the diet of poultry feeds (Vranjes and Wenk, 1996; Rosen, 1995; Yang et al., 2009). Adaptation of the production setting and new science behind usage and antibiotic development have been the primary drivers of antibiotic use in the poultry industry from the 1950s until today. In modern production systems, simply walking through a large flock of chickens or turkeys to administer antibiotics to a single sick bird adds stress and have more adverse effects than treating the group as a whole. After the right dosage is calculated and approved for use, producers can administer the antibiotics through water nipple systems or through feed troughs. After the group shows signs of recovery from the disease, further antibiotic administration is continued through diet if antibiotics are needed. With the concerns of improper antibiotic stewardship and antibiotic resistance, the FDA made a public announcement in 2012 asking to phase out antibiotic usage in livestock and poultry production when used to promote animal growth, whereas European counties have established bans on certain antimicrobials as growth promoters because of resistance-related human health concerns (Castanon, 2007; Dibner and Richards, 2005). The implementation in the United States of the Veterinary Feed Directive (VFD) to regulate antibiotic use in animals is the latest expression of the need to reduce antibiotic usage, especially in growth-promoting use. Ionophores, bacitracin, bambermycin and tiamulin will remain over the counter, and injectables, boluses or other dosages are not affected by the VFD protocols.

As agricultural science and microbiology have progressed, we have discovered that the bacteria that survive the effects of antibiotics (resistant) often have the capabilities to reproduce, and the antibiotic resistance provides a competitive advantage to resistant organisms in the presence of antibiotics. This results in a generational competitive advantage which becomes difficult to control without the creation of new bacterial control methods. However, despite the public perception around antibiotic usage as growth promotants, antibiotics are commonly used to prevent disease spread, control illness outbreaks in poultry facilities, to treat specific animals or groups and must follow strict protocols administered by the FDA. Animal welfare and antimicrobial stewardship are current global issues and researchers have been investigating trends in antibiotic usage and long-term effects of misusage (Van Boeckel et al., 2015). Antibiotic usage in the poultry industry has been changing with science and with production methods as well. The history of the industry shows us that regardless of what type of poultry production facility is being utilized, antibiotics are one tool to help producers promote health and wellness of their flocks. In the mid-twentieth century, the poultry industry was predominantly small farms, whereas modern-day facilities house thousands of chickens or turkeys and require more intense management practices to maintain overall flock health. A majority of these facilities are contract facilities and an avian nutritionist/specialist has the final say in antibiotic usage based on contractual metrics. Currently, finding substitutes to traditionally used antibiotics in the poultry industry is crucial, and some compounds appear as promising alternatives which are discussed in the next section.

3 Future perspectives on antibiotics in poultry production

As antibiotic resistance continues to evolve, finding alternatives to these chemical compounds that increase poultry performance has become imperative. Some of the most promising alternatives that have been investigated include bacteriocins, bacteriophage therapy, plant-derived phytochemicals, competitive exclusion of pathogens and predatory bacteria. In general, these treatments have a more narrow target than the commonly used broad-spectrum antibiotics (in fact, it may actually be only one specific bacterium in some cases) which can be beneficial to the host animal, since other commensal members of its microbial community are not affected by the treatments, and there are no ripples of unintended consequences. However, to date, there is a lack of commercial products available in the market. In addition, further research on their impacts on animal performance and impacts on quality of the final product are still needed. The next section focuses on the potential use of bacteriocins and plant-derived phytochemicals to replace the growth-promoting and health benefits of the STLAB.

3.1 Bacteriocins

Bacteriocins are among the most examined antimicrobial substances in the food industry and veterinary sciences (Klaenhammer, 1988; Jack et al., 1995; Kierończyk et al., 2017). They consist of small antimicrobial peptides produced by certain bacteria which have activity against other bacteria, including some antibiotic-resistant strains. Bacteriocins' mode of action is based on cell-membrane permeabilization or inhibition of membrane formation (Kierończyk et al., 2017). Given their proteinaceous nature, bacteriocins can be artificially synthesized through gene-based peptide engineering (Cotter et al., 2013; Schulz et al., 2015). While some bacteriocins have high specific activity and are able to target certain pathogens without affecting the commensal population, others have a broad-spectrum and can be used to treat infections of unknown aetiology (Cotter et al., 2013). The use of bacteriocins to treat infections has had promising results. A particular bacteriocin (nisin) has been shown to target *Streptococcus pneumoniae* and to be up to 16 times more active than vancomycin in an intravenous regimen (Goldstein et al., 1998), while other bacteriocins effectively controlled infections by *Staphylococcus aureus* in the respiratory tract and intraperitoneally (De Kwaadsteniet et al., 2009; Mota-Meira et al., 2005). Similarly, it has been reported that another bacteriocin (planosporicin) had a good efficacy in *Streptococcus pyogenes*-induced septicaemia in mice (Castiglione et al., 2007).

Specifically to their use in poultry nutrition, bacteriocins produced by the microorganisms *Paenibacillus polymyxa* (bacteriocin B602) and *Lactobacillus salivarius* (bacteriocin OR7) have been successfully used to eliminate *Campylobacter*. Cole et al. (2006) tested a concentration of 250 mg of either one of those bacteriocins per kg of feed offered to turkey poults. Their results showed that both bacteriocins reduced the concentrations of *Campylobacter* in the caecal contents of poults to nondetectable levels. Moreover, they noticed that duodenal crypt depth and goblet cell numbers were also reduced in animals treated with either one of the bacteriocins, compared to controls. The authors hypothesized that this modification in intestinal morphology may have contributed to the observed reduction in *Campylobacter* counts, since this microorganism has the ability to sequester itself within the intestinal crypts as a survival strategy. Furthermore, because the goblet cells secrete mucin glycoproteins, which *Campylobacter* can use as a nutrient source, the reduction in goblet cell number by bacteriocins may also have limited *Campylobacter* colonization in poults' caeca.

Józefiak et al. (2012) investigated the effects of a bacteriocin isolated from *Carnobacterium divergens* AS7 (divercin AS7) at 200 mg/kg of diet on the performance of broiler chickens - challenged or not - with a mixture of *Clostridium perfringens* isolates. Overall, their findings relative to the inclusion of bacteriocin in the diet were positive. For instance, the overall feed conversion rate (from day 1 to day 42) was not affected by addition of bacteriocin to the diet in animals not challenged with *C. perfringens*; however, in the challenged group of birds, divercin AS7 significantly improved feed conversion rate. Indeed, it re-established it to the same levels of non-challenged animals (shifting it from 1.77 back to 1.68 g feed:g gain). Similarly, crude-fat digestibility and diet-apparent metabolizable energy were numerically improved by the inclusion of bacteriocin when birds were infected with *C. perfringens* (by approximately 8.1% and 5.2%, respectively). Finally, their histomorphology results showed mixed responses: a decrease in crypt depth due to the inclusion of bacteriocin was observed when birds were not challenged with *C. perfringens*, but the opposite effect was observed in birds that were challenged with *C. perfringens* isolates. Authors hypothesized that these divergent morphological results may be due to the long-term effects caused by a *C. perfringens* challenge, suggesting that bacteriocin effects on intestinal morphology may depend on the health status of the gastrointestinal tract. Despite of this unclear effect on intestinal morphology, the study concluded that divercin AS7 reduced the negative effects related to the *C. perfringens* challenge, since it improved important traits like feed conversion rate and apparent metabolizable energy.

In another study using broiler chickens, Kierończyk et al. (2017) evaluated the effects of dietary supplementation with the bacteriocin nisin both alone and in combination with the ionophores monensin or salinomycin. Overall, they found no additive effects between the two ionophores and nisin; however, crude protein digestibility was increased by the presence of nisin in the diet.

Moreover, body weight gain and feed conversion rate were both significantly improved by nisin, particularly when birds were between 1 and 14 days-old, when these two traits were improved by 13.0% and 10.7%, respectively. The study also found that nisin decreased the weight of jejunum as a percentage of birds' body weight (from 1.4% to 1.2% of total body weight), which the authors related to an intensified nutrient absorption and utilization, given that a reduced intestinal size results in less energy expended on maintenance, and consequently more energy available for growth. Finally, the study examined one important issue in modern poultry production - tibiotarsus health - and found no impact of nisin supplementation on tibiotarsus mineralization; therefore, this comprehensive study concluded that nisin may be used as a novel growth promoter in broiler nutrition.

3.2 Phytochemicals

Phytochemicals are a group of biologically active compounds found in plants. It has been shown that these compounds can significantly alter microbial populations, reduce pathogens, reduce antimicrobial resistant bacteria and improve growth performance in broilers (Cowan, 1999; Nascimento et al., 2000; Doughari et al., 2009; Grilli et al., 2011, 2013). Thus, the use of organic acids and other phytochemicals derived from plants have been advanced as potential replacers of STLAB (Mitsch et al., 2004; Patra and Saxena, 2009). Some important phytochemicals with antimicrobial activity are terpenoid compounds (which are the primary components of the essential oils). Antimicrobial activity has been reported in the essential oils of thyme, rosemary, marjoram and oregano (Cross et al., 2004).

The effect of two different blends of essential oil components on the proliferation of *Clostridium perfringens* in the intestines broilers was investigated by Mitsch et al. (2004). These authors tested one blend of essential oil composed mainly of thymol (the essential oil from thyme) and another blend composed mainly of thymol plus carvacrol (from oregano). On day 14 of their trial, both essential oil blends significantly reduced *Clostridium perfringens* concentration in the contents of the jejunum. However, on day 30, only the blend composed of thymol plus carvacrol significantly decreased *C. perfringens* concentration in the jejunum, compared to the control group. Authors concluded that those essential oil blends can efficiently control the proliferation of *C. perfringens* in broilers' intestine, which may in turn reduce the risk of necrotic enteritis; however, different compositions of essential oil blends may have different efficiencies in this respect (Mitsch et al., 2004).

Giannenas et al. (2004) investigated the use of ground oregano as a dietary supplement in broiler chickens for control of *Eimeria tenella*. They included different levels of oregano in their birds' diet (2.5, 5.0, 7.5 and

10.0 g/kg of feed) and compared their performance to a group of birds consuming the anticoccidial lasalocid (included at 75 mg/kg of feed). The ground oregano utilized contained 0.07% thymol and 1.22% carvacrol. The number of oocyst excreted by the broilers infected with *Eimeria tenella* was significantly reduced in birds consuming oregano. Even the group of birds consuming the lowest level of oregano inclusion (2.5 g/kg of feed) had their oocyst counts decreased by less than half of what was found in the control group, and such numbers were reduced even further by the other levels of inclusion. Authors also evaluated intestinal lesions and reported significant improvements by the inclusion of oregano, except for the highest level of inclusion (10.0 g/kg of feed). Lastly, their animal performance data also showed very positive results regarding the use of oregano: although feed intake measured at the end of their study (day 35) was severely reduced by the challenge with *E. tenella*, the inclusion of oregano offset this problem. Similarly, bird body weights on day 35 were improved by all levels of oregano inclusion, especially the rates of 5.0 g/kg and 7.5 g/kg of feed, as they yielded results comparable to birds receiving the anticoccidial lasalocid, and the exact same effect was observed for feed conversion ratio. Overall, they concluded that supplementation with ground oregano could reduce the adverse effects of *E. tenella* infection and improve bird performance, particularly when included at the rates of 5.0 g/kg or 7.5 g/kg of feed, since birds in those groups had the same performance as the ones fed lasalocid. This indicates that oregano can be administered to broiler chickens as an alternative to ionophores.

The efficacy of carvacrol in controlling *Salmonella* has also been demonstrated. Upadhyaya et al. (2016) investigated the efficacy of this phytochemical (and two others - eugenol and β-resorcylic acid) as coating for reducing contamination with *Salmonella enterica* serovar Enteritidis on shell eggs. Their results showed that all of the three tested phytochemicals reduced *Salmonella enterica* to undetectable levels on day 3 of egg storage, when present at 0.75% of coating. However, the coatings containing carvacrol significantly reduced *S. enterica* from day 0, even when included in a concentration as low as 0.25%. Furthermore, this beneficial effect remained unchanged until the last day of their experiment (day 7 of egg storage).

Extracts from blueberry and blackberry pomaces have also shown promise as alternative growth factors for broilers due to their bioactive effects: Salaheen et al. (2017) reported that these alternative products increased body weight gain by approximately 6% compared to a control group not receiving them. In addition, they found that the gastrointestinal microbiome of birds fed those phytochemicals had higher *Firmicutes:Bacteroidetes* ratios than did controls. Moreover, the use of these bioactive extracts did not increase the incidence of caecal antibiotic resistomes, indicating that the supplementation with those bioactive extracts may be safe in terms of development and transfer of antibiotic-resistant genes among microbial communities.

4 Conclusion

As highlighted, antibiotics have saved millions of lives around the world and are an invaluable and irreplaceable component of the human public health system. Antibiotics have also found a home in veterinary medicine, and have been crucial to increasing the efficiency of poultry production worldwide. The reduction in illnesses amongst poultry from antibiotic treatment, as well as the improvements in feed efficiency, have resulted in the feeding of untold millions of humans and is a true success story in agricultural science. Such successful history is well documented in the scientific literature with numerous examples of how the use of antibiotics have evolved in the past seven decades, as well as their great contribution to bringing the food-producing industry to where it is today. However, antibiotic resistance is a direct result of antibiotic use (Gelband et al., 2015). Thus, with the advent of increased antibiotic resistance worldwide, constraints on antibiotic usage have been placed to reduce the spread of resistance. Although the exact mechanism by which each microorganism develops resistance to antibiotics is not completely understood, it is known that antibiotic-resistant genes can be disseminated as a direct consequence of selective pressure of organisms (Mathew et al., 2007) through selection of microorganisms that possess resistant genes. Thus being, the extensive use of antimicrobials in agriculture has been regarded as one of the main culprits for increasing the prevalence of antibiotic-resistant bacteria of human significance (Mathew et al., 2007), reducing the effective lifespan of medicinal antimicrobials. Moreover, there is a shortage of new families of antibiotics that could potentially compensate for resistance to existing antibiotics (Cotter et al., 2013).

Given this scenario, an enormous amount of pressure is being placed on the food-producing chain to reduce or even ban the use of antibiotics. If such chemicals are to be withdrawn from use as feed additives, alternative strategies must be introduced to offset the adverse effects on production. Therefore, there is a growing interest in the identification of alternative feed additives that can support animal production while satisfying the contemporary requirements and consumer perceptions (Giannenas et al., 2004). In this regard, some compounds such as bacteriocins and phytochemicals appear as promising alternatives. They have now been studied for a few decades and their inclusion in poultry diets has shown encouraging results. The examples presented here and many others found in the literature demonstrate that, indeed, these compounds can embody the future of poultry production.

5 Where to look for further information

For additional information on the use of antibiotics in the livestock industry, bacterial resistance to antibiotics and how the restriction of antibiotics affects animal production, the interested reader is referred to:

- USDA Economic Research Service. Economic Research Report Number 200, November 2015: Economics of Antibiotic Use in U.S. Livestock Production (https://ageconsearch.umn.edu/bitstream/229202/2/err200.pdf).
- The State of the World's Antibiotics 2015 (https://journals.co.za/docserver/fulltext/mp_whsa/8/2/mp_whsa_v8_n2_a4.pdf?expires=1550772079&id=id&accname=guest&checksum=A8C13993F70E11676EB16C7A903DA903).
- Antibiotics and Bacterial Resistance in the 21st Century (https://journals.sagepub.com/doi/pdf/10.4137/PMC.S14459).
- A Brief History of the Antibiotic Era: Lessons Learned and Challenges for the Future (https://www.ncbi.nlm.nih.gov/pmc/articles/PMC3109405/pdf/fmicb-01-00134.pdf).
- Centers for Disease Control and Prevention (https://www.cdc.gov/drugresistance/about.html).

Additional information on alternatives to antibiotics can be found here:

- National Academy of Sciences (https://nam.edu/alternatives-to-antibiotics-why-and-how/).
- Bacteriocins to Control *Campylobacter* spp. in Poultry–A Review (https://academic.oup.com/ps/article/89/8/1763/1564658).
- Bacteriocins – A Viable Alternative to Antibiotics? (https://www.nature.com/articles/nrmicro2937.pdf).
- The Effect of Two Different Blends of Essential Oil Components on the Proliferation of *Clostridium perfringens* in the Intestines of Broiler Chickens (https://academic.oup.com/ps/article/83/4/669/1569143).

6 References

Aminov, R. I. 2010. A brief history of the antibiotic era: lessons learned and challenges for the future. *Frontiers in Microbiology* 1, 134. doi:10.3389/fmicb.2010.00134.

Bager, F., Aarestrup, F. M., Madsen, M. and Wegener, H. C. 1999. Glycopeptide resistance in *Enterococcus faecium* from broilers and pigs following discontinued use of avoparcin. *Microbial Drug Resistance* 5(1), 53–6. doi:10.1089/mdr.1999.5.53.

Berg, L. R., Carver, J. S., Bearse, G. E. and McGinnis, J. 1952. Antibiotics in the nutrition of laying hens. Washington Agr. Expt. Sta. Bull. 534.

Bhargava, A. and Srivastava, S. 2017. *Biotechnology: Recent Trends and Emerging Dimensions*. CRC Press, Boca Raton, FL.

Castanon, J. I. R. 2007. History of the use of antibiotic as growth promoters in European poultry feeds. *Poultry Science* 86(11), 2466–71. doi:10.3382/ps.2007-00249.

Castiglione, F., Cavaletti, L., Losi, D., Lazzarini, A., Carrano, L., Feroggio, M., Ciciliato, I., Corti, E., Candiani, G., Marinelli, F., et al. 2007. A novel lantibiotic acting on bacterial cell wall synthesis produced by the uncommon actinomycete *Planomonospora* sp. *Biochemistry* 46(20), 5884–95. doi:10.1021/bi700131x.

Cole, K., Farnell, M. B., Donoghue, A. M., Stern, N. J., Svetoch, E. A., Eruslanov, B. N., Volodina, L. I., Kovalev, Y. N., Perelygin, V. V., Mitsevich, E. V., et al. 2006. Bacteriocins reduce *Campylobacter* colonization and alter gut morphology in turkey poults. *Poultry Science* 85(9), 1570–5. doi:10.1093/ps/85.9.1570.

Cotter, P. D., Ross, R. P. and Hill, C. 2013. Bacteriocins - a viable alternative to antibiotics? *Nature Reviews Microbiology* 11(2), 95–105. doi:10.1038/nrmicro2937.

Couch, J. R., Quisenberry, J. H., Camp, A. A., Creech, B. G. and Reid, B. L. 1957. Antibiotics and aresenicals in poultry nutrition. Bulletin/Texas Agricultural Experiment Station no. 871.

Cowan, M. M. 1999. Plant products as antimicrobial agents. *Clinical Microbiology Reviews* 12(4), 564–82. doi:10.1128/CMR.12.4.564.

Cross, D. E., Hillman, K., Fenlon, D., Deans, S. G., McDevitt, R. M., Acamovic, T., Stewart, C. S. and Pennycott, T. W. 2004. Antibacterial properties of phytochemicals in aromatic plants in poultry diets. *Poisonous Plants and Related Toxins* 18, 175–80.

DANMAP. 2006. DANMAP 2005 - Use of antimicrobial agents and occurrence of antimicrobial resistance in bacteria from food animals, foods and humans in Denmark (Online). Statens Serum Institut, Danish Veterinary and Food Administration, Danish Medicines Agency and Danish Institute for Food and Veterinary Research. Available at: http://www.danmap.org/pdffiles/danmap_2005.pdf (accessed on 17 August 2007).

Danzeisen, J. L., Kim, H. B., Isaacson, R. E., Tu, Z. J. and Johnson, T. J. 2011. Modulations of the chicken cecal microbiome and metagenome in response to anticoccidial and growth promoter treatment. *PLoS One* 6(11), e27949. doi:10.1371/journal.pone.0027949.

De Kwaadsteniet, M., Doeschate, K. T. and Dicks, L. M. T. 2009. Nisin F in the treatment of respiratory tract infections caused by *Staphylococcus aureus*. *Letters in Applied Microbiology* 48(1), 65–70. doi:10.1111/j.1472-765X.2008.02488.x.

Dibner, J. J. and Richards, J. D. 2005. Antibiotic growth promoters in agriculture: history and mode of action. *Poultry Science* 84(4), 634–43. doi:10.1093/ps/84.4.634.

Doughari, J. H., Human, I. S., Bennade, S. and Ndakidemi, P. A. 2009. Phytochemicals as chemotherapeutic agents and antioxidants: possible solution to the control of antibiotic resistant verocytotoxin producing bacteria. *Journal of Medicinal Plants Research* 3(11), 839–48.

Fair, R. J. and Tor, Y. 2014. Antibiotics and bacterial resistance in the 21st century. *Perspectives in Medicinal Chemistry* 6, 25–64. doi:10.4137/PMC.S14459.

Gelband, H., Molly Miller, P., Pant, S., Gandra, S., Levinson, J., Barter, D., White, A. and Laxminarayan, R. 2015. The state of the world's antibiotics 2015. *Wound Healing Southern Africa* 8(2), 30–4.

Giannenas, I. A., Florou-Paneri, P., Papazahariadou, M., Botsoglou, N. A., Christaki, E. and Spais, A. B. 2004. Effect of diet supplementation with ground oregano on performance of broiler chickens challenged with *Eimeria tenella*. *Archiv fur Geflugelkunde* 68, 247–52.

Goldstein, B. P., Wei, J., Greenberg, K. and Novick, R. 1998. Activity of nisin against *Streptococcus pneumoniae*, *in vitro*, and in a mouse infection model. *Journal of Antimicrobial Chemotherapy* 42(2), 277–8. doi:10.1093/jac/42.2.277.

Greer, K. 2016. *Antibiotic Use in U.S Livestock Production*. Nova Science Publishers, Inc, New York, NY. eBook Collection (EBSCOhost).

Grilli, E., Tugnoli, B., Formigoni, A., Massi, P., Fantinati, P., Tosi, G. and Piva, A. 2011. Microencapsulated sorbic acid and nature-identical compounds reduced *Salmonella* Hadar and *Salmonella* enteritidis colonization in experimentally infected chickens. *Poultry Science* 90(8), 1676–82. doi:10.3382/ps.2011-01441.

Grilli, E., Vitari, F., Domeneghini, C., Palmonari, A., Tosi, G., Fantinati, P., Massi, P. and Piva, A. 2013. Development of a feed additive to reduce caecal *Campylobacter jejuni* in broilers at slaughter age: from *in vitro* to *in vivo*, a proof of concept. *Journal of Applied Microbiology* 114(2), 308–17. doi:10.1111/jam.12053.

Immerseel, F. V., Buck, J. D., Pasmans, F., Huyghebaert, G., Haesebrouck, F. and Ducatelle, R. 2004. *Clostridium perfringens* in poultry: an emerging threat for animal and public health. *Avian Pathology: Journal of the W.V.P.A.* 33(6), 537–49. doi:10.1080/03079450400013162.

Jack, R. W., Tagg, J. R. and Ray, B. 1995. Bacteriocins of gram-positive bacteria. *Microbiological Reviews* 59(2), 171–200.

Jones, F. T. and Ricke, S. C. 2003. Observations on the history of the development of antimicrobials and their use in poultry feeds. *Poultry Science* 82(4), 613–7. doi:10.1093/ps/82.4.613.

Józefiak, D., Sip, A., Rutkowski, A., Rawski, M., Kaczmarek, S., Wołuń-Cholewa, M., Engberg, R. M. and Højberg, O. 2012. Lyophilized *Carnobacterium divergens* AS7 bacteriocin preparation improves performance of broiler chickens challenged with *Clostridium perfringens*. *Poultry Science* 91(8), 1899–907. doi:10.3382/ps.2012-02151.

Kierończyk, B., Sassek, M., Pruszyńska-Oszmałek, E., Kołodziejski, P., Rawski, M., Świątkiewicz, S. and Józefiak, D. 2017. The physiological response of broiler chickens to the dietary supplementation of the bacteriocin nisin and ionophore coccidiostats. *Poultry Science* 96(11), 4026–37. doi:10.3382/ps/pex234.

Klaenhammer, T. R. 1988. Bacteriocins of lactic acid bacteria. *Biochimie* 70(3), 337–49. doi:10.1016/0300-9084(88)90206-4.

Lorian, V. 1980. Effects of subminimum inhibitory concentrations of antibiotics on bacteria. In: Lorian, V. (Ed.), *Antibiotics in Laboratory Medicine*. Williams & Wilkins, Baltimore, MD.

Luangtongkum, T., Morishita, T. Y., Ison, A. J., Huang, S., McDermott, P. F. and Zhang, Q. 2006. Effect of conventional and organic production practices on the prevalence and antimicrobial resistance of *Campylobacter* spp. in poultry. *Applied and Environmental Microbiology* 72(5), 3600–7. doi:10.1128/AEM.72.5.3600-3607.2006.

Mathew, A. G., Cissell, R. and Liamthong, S. 2007. Antibiotic resistance in bacteria associated with food animals: a United States perspective of livestock production. *Foodborne Pathogens and Disease* 4(2), 115–33. doi:10.1089/fpd.2006.0066.

Mathews, K. H. 2016. Antimicrobial drug use and veterinary costs in US livestock production. In: Greer, K. (Ed.), *Antibiotic Use in U.S. Livestock Production*. Nova Science Publishers, Inc, New York, NY. eBook Collection (EBSCOhost).

Mitsch, P., Zitterl-Eglseer, K., Köhler, B., Gabler, C., Losa, R. and Zimpernik, I. 2004. The effect of two different blends of essential oil components on the proliferation of *Clostridium perfringens* in the intestines of broiler chickens. *Poultry Science* 83(4), 669–75. doi:10.1093/ps/83.4.669.

Moore, P. R., Evenson, A., Luckey, T. D., McCoy, E., Elvehjem, C. A. and Hart, E. B. 1946. Use of sulfasuxidine, streptothricin, and streptomycin in nutritional studies with the chick. *Journal of Biological Chemistry* 165(2), 437–41.

Mota-Meira, M., Morency, H. and Lavoie, M. C. 2005. *In vivo* activity of mutacin B-Ny266. *Journal of Antimicrobial Chemotherapy* 56(5), 869–71. doi:10.1093/jac/dki295.

Nascimento, G. G. F., Locatelli, J., Freitas, P. C. and Silva, G. L. 2000. Antibacterial activity of plant extracts and phytochemicals on antibiotic-resistant bacteria. *Brazilian Journal of Microbiology* 31(4), 247–56. doi:10.1590/S1517-83822000000400003.

Nicolaou, K. C. and Rigol, S. 2018. A brief history of antibiotics and select advances in their synthesis. *Journal of Antibiotics* 71(2), 153–84. doi:10.1038/ja.2017.62.

Oakley, B. B., Lillehoj, H. S., Kogut, M. H., Kim, W. K., Maurer, J. J., Pedroso, A., Lee, M. D., Collett, S. R., Johnson, T. J. and Cox, N. A. 2014. The chicken gastrointestinal microbiome. *FEMS Microbiology Letters* 360(2), 100–12. doi:10.1111/1574-6968.12608.

Pan, D. and Yu, Z. 2014. Intestinal microbiome of poultry and its interaction with host and diet. *Gut Microbes* 5(1), 108–19. doi:10.4161/gmic.26945.

Patra, A. K. and Saxena, J. 2009. Dietary phytochemicals as rumen modifiers: a review of the effects on microbial populations. *Antonie van Leeuwenhoek* 96(4), 363–75. doi:10.1007/s10482-009-9364-1.

Rosen, G. D. 1995. Antibacterials in poultry and pig nutrition. In: Wallace, R. J. and Chesson, A. (Eds), *Biotechnology in Animal Feeds and Animal Feeding*. VCH Publishers Inc., New York, NY, p. 172.

Salaheen, S., Kim, S. W., Haley, B. J., Van Kessel, J. A. S. and Biswas, D. 2017. Alternative growth promoters modulate broiler gut microbiome and enhance body weight gain. *Frontiers in Microbiology* 8, 2088. doi:10.3389/fmicb.2017.02088.

Salyers, A. and Shoemaker, N. B. 2006. Reservoirs of antibiotic resistance genes. *Animal Biotechnology* 17(2), 137–46. doi:10.1080/10495390600957076.

Schulz, S., Stephan, A., Hahn, S., Bortesi, L., Jarczowski, F., Bettmann, U., Paschke, A. K., Tusé, D., Stahl, C. H., Giritch, A., et al. 2015. Broad and efficient control of major foodborne pathogenic strains of *Escherichia coli* by mixtures of plant-produced colicins. *Proceedings of the National Academy of Sciences* 112(40), E5454–60. doi:10.1073/pnas.1513311112.

Sneeringer, S., MacDonald, J. M., Key, N., McBride, W. D. and Mathews, K. 2015. Economics of antibiotic use in US livestock production. U.S. Department of Agriculture, Economic Research Service. ERR-200.

Tavakkoli, H., Derakhshanfar, A. and Noori Gooshki, S. 2014. The effect of florfenicol egg-injection on embryonated chicken egg. *International Journal of Advanced Biological and Biomedical Research* 2(2), 496–503.

Torok, V. A., Hughes, R. J., Mikkelsen, L. L., Perez-Maldonado, R., Balding, K., McAlpine, R., Percy, N. J. and Ophel-Keller, K. 2011. Identification and characterization of potential performance related gut microbiota in broiler chickens across various feeding trials. *Applied and Environmental Microbiology* 77, 5868–78. doi:10.1128/AEM.00165-11.

Upadhyaya, I., Yin, H. -B., Surendran Nair, M., Chen, C. -H., Lang, R., Darre, M. J. and Venkitanarayanan, K. 2016. Inactivation of *Salmonella* enteritidis on shell eggs by coating with phytochemicals. *Poultry Science* 95(9), 2106–11. doi:10.3382/ps/pew152.

Van Boeckel, T. P., Brower, C., Gilbert, M., Grenfell, B. T., Levin, S. A., Robinson, T. P., Teillant, A. and Laxminarayan, R. 2015. Global trends in antimicrobial use in food animals. *Proceedings of the National Academy of Sciences* 112(18), 5649–54. doi:10.1073/pnas.1503141112.

Vranjes, M. V. and Wenk, C. 1996. Influence of *Trichoderma viride* enzyme complex on nutrient utilization and performance of laying hens in diets with and without antibiotic supplementation. *Poultry Science* 75(4), 551–5. doi:10.3382/ps.0750551.

Watkins, K. L., Shryock, T. R., Dearth, R. N. and Saif, Y. M. 1997. *In-vitro* antimicrobial susceptibility of *Clostridium perfringens* from commercial turkey and broiler chicken origin. *Veterinary Microbiology* 54(2), 195–200. doi:10.1016/S0378-1135(96)01276-X.

Yang, Y., Iji, P. A. and Choct, M. 2009. Dietary modulation of gut microflora in broiler chickens: a review of the role of six kinds of alternatives to in-feed antibiotics. *World's Poultry Science Journal* 65(1), 97–114. doi:10.1017/S0043933909000087.

Yeoman, C. J., Chia, N., Jeraldo, P., Sipos, M., Goldenfeld, N. D. and White, B. A. 2012. The microbiome of the chicken gastrointestinal tract. *Animal Health Research Reviews* 13(1), 89–99. doi:10.1017/S1466252312000138.

Chapter 10

Gastrointestinal diseases of poultry: causes and nutritional strategies for prevention and control

Raveendra R. Kulkarni, North Carolina State University, USA; Khaled Taha-Abdelaziz, University of Guelph, Canada and Beni-Suef University, Egypt; and Bahram Shojadoost, Jake Astill and Shayan Sharif, University of Guelph, Canada

1 Introduction

The growing global population and a need to meet the current as well as the future demand for high-value animal protein have put immense pressure on the livestock industry, including poultry, to enhance food animal production. According to a recent WHO/FAO joint report, annual meat production is projected to increase from 218 million tonnes in 1997–1999 to 376 million tonnes by 2030 (Salter, 2017). In this context, more efficient and sustainable food production systems that provide economically viable feed conversion efficiencies while also reducing the ecological footprint of livestock production are needed. Successful animal production depends largely on efficient feed conversion. Considering that the intestine is the primary site for digestion and nutrient absorption and that it is also the primary point of contact between the host and the external environment, including infectious agents, it is important to recognize that 'gut health' forms an integral part of a sustainable food animal production system. Therefore, maintaining a healthy intestinal environment is a critical element in ensuring the overall health and productivity of poultry flocks.

The prophylactic use of antibiotics as growth promoters (AGPs) has been a management practice for several decades in poultry production. However, the use of antibiotics in livestock may lead to the emergence of antibiotic

http://dx.doi.org/10.19103/AS.2019.0059.11

resistance in bacteria. Therefore, restrictions on dietary antibiotics for poultry have been imposed in several countries, and several other jurisdictions have plans to phase out AGPs in the near future. Restrictions on antibiotic usage may have negative consequences for performance, animal welfare, and general health of poultry, particularly for gastrointestinal (GI) disorders (Suresh et al., 2018). AGPs have been effective at reducing the burden of enteric infections, including necrotic enteritis (NE), in chickens. As a result, removal of AGPs to offer an 'antibiotic-reduced/free' environment in broiler production has posed a threat to the industry as it has led to a spike in enteric infections, which has significant negative implications for the industry. In this chapter, important enteric diseases or disorders will be discussed, briefly highlighting their etiology followed by possible nutritional interventions, including feed additives, as possible alternatives to AGPs for disease control.

2 Gastrointestinal (GI) tract diseases

Gastrointestinal (GI) diseases affecting poultry can be of infectious or noninfectious origin (Table 1). The etiological factors can range from infectious agents, environmental factors, and management practices, including feed and water, each of which can adversely affect the growth rate and feed conversion efficiency (Szkotnicki, 2013).

2.1 Infectious diseases

As the GI tract provides a large mucosal surface area allowing for digestion and absorption of ingested feed, it also poses a greater risk of being exposed to a variety of infectious agents. Enteric diseases, in many instances, are complex due to the involvement of more than one infectious agent, including bacterial, viral, fungal, or parasitic microorganisms (Weber et al., 2016). Infectious agents can gain access to poultry via different routes, such as oral or aerosol, and spread within or between farms can occur through contamination of feed, water, litter, fomites, or air. It is of note that the vertical route of transmission is also a major threat for introducing certain infectious agents, such as *Salmonella* or *Escherichia coli* during the early days of chick life (Calnek, 2015). Some of the important enteric disease agents that affect poultry are discussed briefly here with an emphasis on two economically important diseases: NE and coccidiosis.

2.1.1 Bacteria

Low-grade damage to the intestinal tract by pathogenic bacteria may cause poor feed conversion and a decreased rate of body weight gain in poultry flocks. More severe enteric damage by bacterial infections results in overt

Table 1 Some of the important etiological agents/factors affecting GI health of poultry

Infectious origin				Noninfectious origin		
				Management		
Bacteria	Viruses	Parasites	Fungi	Environment	Feed	Water
Salmonella *E. coli* Clostridia	Rota Reo Entero Astro Corona Adeno Influenza Paramyxo	Coccidia Histomonas Worms	Candida	Stocking density Temperature Humidity, Air quality (ventilation and ammonia) Feeder/waterer placement	Type Quality Palatability Content Mycotoxins	Quality Palatability

disease and high mortality (Adedokun and Olojede, 2018). Of the bacterial pathogens, *Salmonella* and Clostridia are considered the most important bacteria that affect poultry. To a certain extent, Campylobacteriosis caused by *Campylobacter jejuni*, *C. coli*, and *C. lardis* also poses intestinal health problems such as distension of the intestinal tract and diarrhea. However, most chickens carry *Campylobacter* asymptomatically and the importance of *Campylobacter* in poultry is due to its foodborne zoonotic potential (Marotta et al., 2015).

2.1.1.1 Salmonellosis

Salmonellosis in poultry, also sometimes referred to as pullorum or bacillary white diarrhea, is a serious disease of chickens of all ages caused by *Salmonella enterica* serovar Pullorum. While the disease is acute in young chicks, consisting of severe clinical signs and mortality/morbidity that peaks at 100% in 7–10-day-old birds, older birds exhibit chronic infection and remain subclinical (Wigley et al., 2005). The pathogen can be vertically spread through the egg (transovarian) or on the egg surface (by fecal contamination), or by feed and water contamination, or by bird-to-bird (horizontal) transmission (Berchieri et al., 2001, Barrow et al., 2012). It is noteworthy that pullorum has been categorized as a notifiable disease and has been eradicated from most commercial flocks around the world; however, disease incidence is still common in backyard and commercial flocks in some developing countries (Barrow et al., 2012). Certain serovars of *Salmonella* such as *Salmonella enterica* serovar Typhimurium and *Salmonella enterica* serovar Enteritidis are carried by poultry asymptomatically but pose a great zoonotic threat to humans.

2.1.1.2 Necrotic Enteritis (NE)

Clostridial infections are a major problem in poultry, and *C. perfringens*, the most important *Clostridium* species, can cause several clinical manifestations and lesions, including NE, necrotic dermatitis, ulcerative enteritis, and cholangiohepatitis, as well as gizzard erosion (Lovland and Kaldhusdal, 1999). Of these, NE has been considered the most economically important enteric disease in recent years due to the increasing demand for restrictions on the use of AGPs in poultry production. NE is a multifactorial disease and is of utmost importance to broiler production, as economic losses in the global poultry industry due to NE, both clinical and subclinical, are estimated to be $6 billion/year according to a recent report (Goossens et al., 2017). The clinical form of the disease often results in necrosis of the small intestine and is associated with high mortality. In addition to mortality, other economic losses due to NE, particularly in the case of subclinical NE, are attributed to reduced weight gain, higher feed conversion, and overall poor performance. These problems are due to chronic intestinal mucosal damage resulting in diminished nutrient digestion and absorption.

Clostridium perfringens bacteria are found widespread in breeder farms, hatcheries, grow-out houses, and processing plants, and are considered to be part of the normal intestinal flora of chickens (Craven et al., 2003). Some of the predisposing factors of NE include dietary contents, such as high animal protein and cereal grains (wheat, barley, and rye), that induce high viscosity of intestinal contents, in addition to other factors such as infection with coccidia or other mucosal pathogens. NE lesions can be among the most severe of any disease that affects the chicken intestine (Al-Sheikhly and Truscott, 1977). The pathogenesis of NE is complex because of the involvement of many microbial factors such as enzymes, adhesion molecules, house-keeping molecules, and importantly, tissue degrading toxins such as NetB, alpha-toxin, and TpeL, all of which contribute to virulence of *C. perfringens* (Goossens et al., 2017; Prescott et al., 2016). Recent reports also suggest that there are NE-causing strains that possess certain signature NE-associated virulence gene(s) that are absent in commensal avirulent non-NE-causing isolates of *C. perfringens* (Van Immerseel et al., 2016; Lepp et al., 2010). Intestinal mucosal damage that occurs during coccidiosis in chickens is usually considered as one of the most important predisposing factors, as coccidiosis often occurs just before or concurrently with outbreaks of NE in the field (Mot et al., 2014). However, in turkeys, mucosal damage is usually caused by coccidiosis, in addition to ascaridiasis, and viral hemorrhagic enteritis (HE) (Gazdzinski and Julian, 1992; Palya et al., 2007).

Although NE is very common in broilers, NE outbreaks can also occur in laying hens, particularly near the onset of laying or during the peak of production (Dhillon et al., 2004). Affected layers are weak, depressed, and have reduced egg production. Moreover, these birds show diarrhea with occasional high mortality with necrotic lesions. Cases in pullets have also been reported to be on the rise, particularly in pullets predisposed to coccidial infections (Hofacre et al., 2018). Importantly, an increasing demand for egg producers to adopt cage-free rearing systems is likely to challenge birds with more GI health issues including NE (Elwinger et al., 1992; Fossum et al., 1988). Antibiotics are the current best choice of treatment; however, many of them leave residues in eggs, forcing a withdrawal period that leads to economic losses. It is also often difficult to control NE with antibiotics as the disease progresses very rapidly sometimes producing irreversible intestinal damage. Hence, it is best to prevent NE rather than treat it.

2.1.2 Viruses

Most viral enteric infections occur in the first 3 weeks of life, which may cause diarrhea. These infections cause high morbidity but low mortality. Viral enteric infections may facilitate bacterial replication and attachment to the gut membrane, which can make the condition more severe. Determining the causative agent of enteritis or enteric disease is often complicated by the presence of more than one infectious agent, including combinations of bacteria, viruses, and parasites. For example, reoviruses have been isolated from flocks exhibiting enteric problems (Benavente and Martinez-Costas, 2007); however, whether this virus is the primary agent is questionable. In this context, it is clear that reoviruses can interact with other infectious agents of chickens such as *E. coli*, infectious bursal disease virus, and *Eimeria* resulting in increased pathological effects and economic losses (Weber et al., 2016; Benavente and Martinez-Costas, 2007). Reoviruses may be one of the several possible causative agents associated with various malabsorption syndromes, such as poult enteritis complex (PEC) and poult enteritis mortality syndrome (PEMS) in young turkeys, and runting-stunting syndrome (RSS) in broiler chickens (Mettifogo et al., 2014). Reovirus infections in broiler chicks result in viral arthritis, and a general lack of performance including diminished weight gain, poor feed conversions, chronic feed passage problems, uneven growth rate, and reduced marketability and sometimes mortality (Clavijo and Florez, 2018). Investigations of enteric diseases in chickens and turkeys have also focused on turkey coronavirus (TCoV), turkey and chicken astrovirus, avian orthoreovirus, avian rotavirus, torovirus, parvovirus, and several unknown 'small round viruses' (Dhama et al., 2015). Of note, a subset of these viruses has been isolated from birds affected by viral arthritis-tenosynovitis, stunting syndrome, respiratory disease, enteric disease, immunosuppression, and malabsorption syndrome (Cortez et al., 2017). Generally, virus-induced enteritis results in reduced daily weight gain, impaired feed conversion, and decreased flock uniformity (Guy, 1998).

Although enteric viral infections are commonly seen in young birds, older age groups can also be affected. For example, HE caused by type II adenovirus is an acute viral enteric disease of older turkeys and is characterized by depression, bloody droppings, and death (Beach et al., 2009). Importance of

HE in turkeys is also related to the immunosuppressive nature of the virus which may exacerbate other diseases. However, the outcome of disease depends on many factors, including the age and immune status of affected birds, the virulence of the infecting virus, the presence of other infectious agents, nutrition, management practices, and environmental factors (Guy, 1998). Overall, diagnosing the causative agent of the malabsorption syndrome in poultry is challenging due to the possible involvement of more than one virus.

2.1.3 Parasites

Parasites that pose a serious challenge to poultry production are often internal parasites that include largely the protozoa. Commercial poultry are often infested with protozoan parasites, some of which can cause moderate-to-severe disease. Of utmost significance to the poultry industry, coccidiosis is one of the most common and economically important diseases of chickens worldwide, as described in detail in the next section. It is noteworthy that the intensive poultry production systems and high-density flocks increase bird susceptibility to protozoan parasites that have short and direct life cycles, such as coccidiosis (Williams, 2005). The other important protozoan disease that affects poultry is histomoniasis or blackhead, which is caused by *Histomonas meleagridis*. This pathogen causes severe lesions in the ceca and liver of many gallinaceous birds, of which turkeys are the most susceptible population (Abraham et al., 2014; McDougald et al., 2012; McDougald and Fuller, 2005). While histomoniasis in turkeys causes high mortality, sometimes approaching 100%, lower mortality rates of 10–20% are observed in chickens and many outbreaks may even go unnoticed. An interesting feature of this disease is that *H. meleagridis* is carried from host to host by eggs of the cecal worm *Heterakis gallinarum*. It is noteworthy that a previous observation indicated that the lesions of histomoniasis can be found more severe in turkeys when *C. perfringens* were present (Agunos et al., 2013).

Additionally, to a certain extent, helminths can also affect intestinal health in poultry, particularly those that are reared on used litter. Ascarids are the most common worms in poultry and *Ascaridia galli* is the most common ascarid, residing mostly in the intestinal lumen causing weight loss in birds (Jansson et al., 2010). *A. galli* has also been shown to transmit viruses and bacteria. Unlike cestodes and flatworms, the life cycle of Ascarids is direct. Economic losses due to Ascarid infections in chickens are largely attributed to parasite-induced anemia, retarded growth, and mortality.

2.1.3.1 Coccidiosis

Coccidiosis has been studied for decades and is a disease of the intestinal tract of almost all domestic and wild animals. For economic and disease control reasons, coccidiosis is one of the most important and challenging diseases of poultry, particularly in chickens (Chapman et al., 2013). The disease is caused by members of the *Eimeria* genus that are host-specific and do not require an intermediate host for their life cycle completion. Coccidia are tissue-trophic and multiply rapidly; their replication cycle involves different intermediate intracellular stages in different segments of the small and large intestine (Fatoba and Adeleke, 2018). These parasites undergo at least two asexual stages (schizogony) and one sexual stage (gametogony) during their replication cycle. The developmental stage of sporogony occurs outside of the host during which the oocysts undergo sporulation and the sporulated oocysts become infective to susceptible birds when they are ingested. The clinical disease results in severe lesions including erosive and hemorrhagic lesions in the intestinal segments leading to mortality, as well as impaired nutrient digestion and absorption, poor bird growth and performance, and mortality (Williams, 2005).

Many species of coccidia are widespread in countries where poultry are produced on a commercial basis. The spread of *Eimeria* from bird to bird and from flock to flock depends on the survival of oocysts of the parasite in the litter or soil. Chickens can be affected by nine species of *Eimeria*, of which six, namely *E. acervulina*, *E. maxima*, *E. brunetti*. *E. necatrix*, *E. mitis*, and *E. tenella*, are considered important. All avian *Eimeria* infect only one poultry species with the exception of *E. dispersa* which may infect and cause disease in turkeys, quails, and pheasants. Some of the coccidial members such as *E. maxima*, *E. necatrix* and *E. tenella* are deep tissue invaders and can cause severe necrosis, hemorrhage of the intestinal mucosa, and bloody diarrhea resulting in mortality (Barbour et al., 2015). Culled birds often appear pale due to anemia, and prior to death birds exhibit depression, poor weight gain and feed conversion, and a drop in egg production.

The presence of other microorganisms in chickens can impact coccidial infections, for example, it has been shown that certain indigenous bacterial species such as *Streptococcus faecalis*, *E. coli*, *Lactobacillus* species, and *Bacteroides* species may play a role in pathogenesis of cecal coccidiosis (Bradley and Radhakrishnan, 1973). It has also been shown that immunosuppressive diseases or conditions may act in concert with *Eimeria* to produce a more severe coccidiosis. Examples include Marek's disease and infectious bursal disease, both of which may interfere with the development of immunity to coccidiosis, and thus, may influence the severity of disease outcome (Williams, 2005). Coccidial infection in chickens is also thought to affect the outcome of other infectious diseases; for example, cecal coccidiosis may contribute to increased severity of histomoniasis in chickens (McDougald and Fuller, 2005). Additionally, in the context of NE, it is well known that subclinical coccidiosis acts as an important contributing factor in facilitating the growth and multiplication of *C.*

perfringens and thereby, development of NE in broiler chickens by inducing mucosal damage (Al-Sheikhly and Truscott, 1977; Williams, 2005).

2.2 Noninfectious factors

2.2.1 Management and environment

Good management practices are key to successful poultry production. These practices include elements such as the design and structure of poultry barns, environmental conditions (ventilation, temperature, and litter condition), stocking density, feed and water supply, as well as the knowledge and experience of the people who manage the poultry operations (Powers and Capelari, 2017; Kiarie and Mills, 2019). These factors affect each other and can promote or adversely influence the health of the flock.

2.2.2 Toxins

Mycotoxins, the toxic secondary metabolites of fungi are one of the major causes of enteric disease in poultry (Guerre, 2016), and their presence in poultry feed has been identified as a widespread cause of economic losses due to impaired health status and reduced performance (Sklan et al., 2003). Although the effects of mycotoxins vary depending on the age and type of the bird, some common effects are reduced feed efficiency, growth performance, immunity, egg production, and hatchability along with increased mortality and organ lesions (mainly liver and kidney). There are over 100 known mycotoxins, and when the moisture content of grains rises, fungal growth and toxin production can result, leading to consumption of toxin-containing feed and subsequent disease (Greco et al., 2014). Mycotoxins include aflatoxin produced by *Aspergillus* sp., trichothecene toxins (T-2), diacetoxyscirpenol (DAS), and deoxynivalenol (DON or Vomitoxin) produced by *Fusarium* sp., and ochratoxin produced by *Aspergillus* sp. While birds with aflatoxicosis typically demonstrate hemorrhages in the intestinal tract, muscles, and skin along with enlarged kidneys and liver, the effects of trichothecenes include hemorrhagic and necrotic lesions in the gizzard and proventriculus, and also atrophy of the bursa of Fabricius and thymus (Barnes et al., 2001). In combination with other factors, mycotoxins can predispose or exacerbate outbreaks of enteric diseases. For example, it has been shown that ochratoxin A and coccidial infections can interact to adversely affect broiler growth and performance (Koynarski et al., 2007).

2.2.3 Diet and nutrition

Not only is the GI tract responsible for physiological functions including digestion and absorption of macro- and micronutrients, it also acts as a physical barrier between the host and the environment, underscoring the importance of GI health to protect against invading pathogens (Kiarie and Mills, 2019) and nutrition can influence poultry GI health and bird susceptibility to enteric diseases. Factors associated with diet, such as feed palatability, quality, and type of feed as well as feeding strategies including restricted feeding, can negatively affect GI health by altering intestinal microbiota and/or enzymatic activity leading to enteric disorders (Gelli et al., 2017). Additionally, the amount of fiber and non-starch polysaccharides (NSP) in the feed, and the content and quality of the ingredients used in the feed, are also key factors in maintaining intestinal integrity. Dietary fibers from plant cell walls, NSP, and lignins have the capacity to absorb water in the intestine leading to higher amounts of bulky material in the intestine, thus increasing the viscosity and decreasing the digestibility of food (Bach Knudsen et al., 2017). For example, cereals used in poultry diets contain various levels of NSP such as β-glucans and arabinoxylans that resist digestion and increase the viscosity of the lumen contents, thus increasing the digesta retention time and facilitating bacterial colonization in the small intestine (Bach Knudsen et al., 2017). Diets that include barley, wheat, rye, and oat contain high amounts of NSP, which has been associated with an increased number of *C. perfringens* in the chicken intestine and higher mortality due to NE (Annett et al., 2002). To this end, it is of note that that in commercial poultry diets, where these cereals are used, NSP-degrading enzymes such as xylanase, beta-glucanase, mannanase, and galactosidase are used to decrease viscosity and eliminate the negative effects of NSP (Choct et al., 1995; Ferrandis Vila et al., 2018).

Dietary protein content is also an important factor that influences GI health in birds. Different protein sources of animal and plant origin are used to meet the protein demands of poultry, and some protein sources could play an important role in predisposing chickens to certain intestinal diseases. For example, chickens fed with diets containing high amounts of animal protein have been shown to have a higher number of *C. perfringens* in the intestine (Drew et al., 2004), and a higher amount of fish meal in feed may predispose broilers to NE (Shojadoost et al., 2012). Additionally, the physical texture of the feed can influence digestion and thereby affect bird susceptibility to enteric diseases. For example, broiler chickens fed with finely ground wheat showed elevated mortality due to NE and coccidiosis compared to chickens fed a coarsely ground wheat-based diet (Choct et al., 1996). Subsequent reports also suggested that feeding whole wheat to broiler chickens can reduce the intestinal burden of *Salmonella* and *C. perfringens* (Engberg et al., 2004). To this end, Korver et al. (2004) suggested that when the GI tract is in a healthy state, inclusion of whole wheat into the diet may help improve digestive tract function, but when the integrity of the intestinal tract is impaired, inclusion of whole wheat may decrease performance of the GI tract. Apart from the texture, factors such as excessive dust levels in the feed can reduce its palatability and improperly stored feed may allow fungal growth and rancidity, thus affecting its nutritive content and leading to impairment of GI health (Kiarie and Mills, 2019). Additionally, inadequate feeder or water space and distribution can result in a non-uniform feed/water intake and, thus, lead to precipitation of enteric disease (Powers and Capelari, 2017).

3 Nutritional interventions

Over the last decade or so, much research has focused on devising dietary strategies for controlling GI diseases or disorders in poultry. These include either manipulation of feed composition (ingredients and nutritional content) or finding feed additives to replace AGPs (Adedokun and Olojede, 2018; Morrissey et al., 2014). In this section, we have discussed both nutritional modulation and research on feed additives focused on enhancing GI health and resistance to enteric infections in poultry. Some of the nutritional components of poultry feed that have been extensively studied in the context of GI health issues are dietary fiber, starch, protein, and fat. Important effects of these constituents on GI health and disease are discussed.

3.1 Dietary modulation

3.1.1 Dietary fiber

Dietary fiber content of poultry feed and its effects on GI health and the resident microflora have been studied in detail (Leeson et al., 1997; Morrissey et al., 2014). Studies focused on water-insoluble fiber indicate positive effects on nutrient digestion despite the fact that insoluble fiber has only a limited nutritional value in poultry due to its low fermentability (Carre et al., 1995). A previous study showed that dietary supplementation with 10% oat hulls can significantly increase gizzard weight and pancreatic enzyme activity (Gonzalez-Alvarado et al., 2008). Furthermore, these authors also showed that the oat hulls can help overcome the adverse effects of poor intestinal conditions caused by soluble NSP by reducing digesta moisture content. It is noteworthy that the high-molecular-weight soluble NSP, which increase the viscosity of luminal content, can adversely affect GI tract development and nutrient digestion and thus can have a negative impact on production performance and intestinal health (Langhout et al., 2000; Maisonnier et al., 2003). As mentioned earlier, increased mucus production might exacerbate overgrowth of mucolytic bacteria such as *C. perfringens* in the small intestine; therefore, the mucus-increasing fiber components, such as NSP, and anti-nutritional factors, such as phytate, need to be taken into consideration when formulating feed (Paiva et al., 2013). In this context, it has been proposed that low levels of NSP along with optimal insoluble fiber levels may help modify the intestinal microbiota through mechanisms such as production of volatile butyrate fatty acids and thus, resist *C. perfringens* and coccidial infections (Zubair et al., 1996).

3.1.2 Dietary protein

With the growing trend of phasing out AGPs, dietary management of GI health has become critical in poultry production, and the inclusion of high-quality protein in poultry diets is an important dietary management strategy. From a nutritional standpoint, animal by-products (fish meal, meat, and bonemeal) are generally considered of much higher quality than plant protein sources because the amino acid balance of the animal products more closely reflects the nutritional requirements of the bird (Beski et al., 2015). However, it is also of note that overheating of animal by-products during the rendering process may reduce the digestibility/availability of amino acids. Low-quality protein in the diet is associated with poor protein digestibility and may result in the accumulation of large amounts of undigested protein in the hindgut, stimulating the growth of proteolytic and potentially pathogenic bacteria including *C. perfringens*. In this context, several studies have examined the relative effects of animal proteins (fish meal, meat, and bonemeal and feather meal) or vegetable proteins (potato, pea and soy protein concentrates, or corn gluten meal) on *C. perfringens* counts in the ileum and cecum (Dahiya et al., 2007; Palliyeguru et al., 2010). The conclusions from these studies indicated that vegetable protein sources, except for potato protein concentrate, could reduce *C. perfringens* counts and that the dietary glycine was very critical such that *C. perfringens* numbers were positively correlated with glycine content in the diet and in the intestinal contents (Dahiya et al., 2005). To this end, while the crude protein levels in the diet may be reduced to offer a better intestinal health, it is equally important to incorporate the required levels of important amino acids, such as glycine. Therefore, the use of feed ingredients that can reduce incidence of intestinal disorders would be helpful for diet optimization without in-feed antibiotics.

3.1.3 Dietary fats

Dietary fats and fatty acids are also important nutritional factors that affect GI health and their effects are highly dependent on quantity, type, and quality of fat and fatty acid composition. Studies have indicated that unsaturated lipids are better digested than saturated lipids and that certain components such as soya, coconut oils, and milk lipids tend to reduce the *C. perfringens*-induced NE in broiler chickens (Gilbert et al., 2018). A previous report showed that fat digestibility during coccidial and *C. perfringens* infections is generally reduced and that the use of coconut oil in diets could ameliorate the disease due to its superior digestion compared to other oils (Moore, 2016; Adams et al., 1996). Persia et al. (2006) observed that 15% fish meal in a broiler diet can ameliorate *E. acervulina* infection and increase amino acid digestion and absorption; this was likely related to the effect fish oil has on ameliorating infection-induced inflammatory responses (Persia et al., 2006). Therefore, it seems that impaired intestinal health requires a change in energy source from saturated fats to sources containing unsaturated fatty acids or medium-chain fatty acids and that unsaturated vegetable oils or oils that contain high levels of medium-chain fatty acids have antimicrobial effects.

3.2 Feed additives

Antibiotics have been used in the poultry industry for decades to prevent disease and to improve feed efficiency and growth. However, concerns about increasing antimicrobial resistance among bacteria and the threat of its spread through the food chain into the human population demands the need for effective alternatives to antibiotics (Mehde et al., 2018). Among a series of experimental alternatives, some have shown to be effective in enhancing GI health and helping with resistance against enteric infections, including plant-derived extracts (essential oils (EO)), prebiotics, probiotics, and organic acids (OA), and they are described and discussed in the next section.

3.2.1 Essential oils (EO)

Essential oils (EO) that constitute a major class of plant-derived extracts have been shown to offer beneficial effects for poultry health, immunity, and feed digestion (Farhadi et al., 2017; Hernandez et al., 2006). EO, which are secondary metabolites of plants, and can be extracted from different plant parts, are known to possess antimicrobial properties and have been suggested as potential replacements for antibiotics for use in livestock (Burt, 2004; Varel et al., 2007). Additionally, they are shown to exert antioxidant activity and, thus, help in boosting host immune functions (Kim et al., 2016; Shahid et al., 2018). Specific evidence in the context of EO-mediated antimicrobial activity for use as a feed additive in animal feeds also comes from studies showing antiviral (Garozzo et al., 2009), antifungal (Sakkas and Papadopoulou, 2017), and antiparasitic (Pessoa et al., 2002) activities. Furthermore, due to their aromatic properties, they have been shown to have a positive impact on reducing odors and ammonia levels in poultry houses, thus improving bird performance (Varel et al., 2007). Of around 3000 known EO, around 300 are available on the market, of which many are available for use in the livestock industry including poultry (Kalemba and Kunicka, 2003). The main components of EO are terpenoids (citronellol and menthol), aliphatic hydrocarbons (phenols and thymols), and aromatic organic compounds (cinnamaldehyde and phelandral) (Delaquis et al., 2002). The activity of EO has been mainly attributed to the major components that comprise 85% of the herbal extract, of which phenolic compounds are the key component that offers antibacterial activity (Cosentino et al., 1999). In addition to enhancing feed palatability, EO are able to assist in digestion through the induction of GI peristaltic movements and secretion of saliva and digestive enzymes (Zhai et al., 2018).

3.2.1.1 Essential oils and antimicrobial activity

Effects of EO on various bacterial, viral, and fungal pathogens have been widely reported (Burt, 2004; Kalemba and Kunicka, 2003; Mith et al., 2014; Sakkas and Papadopoulou, 2017). The EO extracted from bay, cinnamon, clove, and thyme were shown to reduce the growth of bacteria *in vitro* including those foodborne pathogens, namely *C. jejuni*, *Salmonella enterica* serovar Enteritidis, and *E. coli* (Smith-Palmer et al., 1998). Another study that used the non-water-soluble subfraction of EO (carvacrol and thymol) showed significant *in vitro* antimicrobial activity against *E. Coli*, *Enterococcus faecalis*, *Pseudomonas aeruginosa*, *Salmonella enterica* serovar Enteritidis, and *Streptococcus pyogenes* (Gulluce et al., 2003). Furthermore, there have been many reports that describe the effectiveness of EO against *C. perfringens*, the causative agent of NE in chickens. For example, EO combinations of thymol, eugenol, carvacrol, curcumin, and piperine showed a reduction in the number and intestinal colonization of *C. perfringens* in broiler chickens (Mitsch et al., 2004). Additionally, this study also showed that a combination of ginger oil and carvacrol EO can reduce intestinal lesions and improve broiler performance. In this context, the authors suggested that in addition to their direct effects on *C. perfringens*, EO could enhance the growth and colonization of intestinal microbiota that may inhibit proliferation of *C. perfringens*.

In the context of foodborne pathogens present in the poultry gut, there has been ample evidence to suggest that EO can reduce the intestinal burden of *E. coli* and *Salmonella*. To this end, a recent study showed a reduction in the number of *Salmonella*, *E.coli*, and Clostridia in the cecum of broiler chickens, when a mixture of plant extracts (fennel, melissa balm, peppermint, anise, oak, clove, and thyme) was used as a feed supplement (Wati et al., 2015). It has also been shown that supplementation of broiler chicken feed with cinnamon oil reduced the number of *E. coli* in the prececal digesta without altering the number of lactobacilli (Gomathi et al., 2018). Furthermore, an EO blend (carvacrol, thymol, eucalyptol, and lemon) administered via drinking water was shown to reduce *Salmonella enterica* serovar Heidelberg counts in the crop (Alali et al., 2013). In addition to their effects on foodborne enteric bacterial pathogens, EO have also been shown to exert anti-protozoan activity in poultry, particularly against coccidia. Previous studies have found that EO can modulate intestinal microbiota populations in chickens that were challenged with a combination of *Eimeria* oocysts containing *E. acervulina, E. maxima, and E. tenella* (Hume et al., 2006; Oviedo-Rondon et al., 2006). Supplementation of poultry feed with eucalyptus and peppermint demonstrated a reduction in lesion scores and number of intestinal oocysts as well as decreased weight loss in birds challenged with eight prevalent *Eimeria* spp (*E. acervulina, E. brunetti, E. hagani, E. maxima, E. mivati, E. necatrix, E. praecox, and E. tenella*) (Barbour et al., 2015). The authors in this study concluded that this blend of EO could be used as an alternative to anticoccidial drugs to control coccidial infections in chickens. In a recent comprehensive *in vitro* study, Jitviriyanon et al. (2016) demonstrated an *in vitro* oocysticidal effect of ten EO extracted from different indigenous plants, of which some, especially those extracted from *Boesenbergia pandurata* and *Ocimum basilicum*, were able to induce degenerative changes in oocysts of *E. tenella* and inhibited sporulation (Jitviriyanon et al., 2016). Another study, using the herb extract from *Aloe secundiflora* leaves, also showed a marked reduction in clinical signs and lesions as well as a dose-dependent reduction in fecal oocyst counts in *E. tenella*-challenged chickens (Kaingu et al., 2017).

In summary, different EO have been shown to have antimicrobial activities against many bacterial and protozoan pathogens, including *C. perfringens, Salmonella, E. coli, C. jejuni*, coccidia, and others in the intestine of poultry, while also not affecting the normal microbiota population. Indeed, the use of EO seems promising for the poultry industry as a reasonable alternative for AGP usage. However, attention also needs to be paid to the fact that certain bacteria such as *Salmonella* Typhimurium, *Salmonella* Enteritidis, *E. coli, Staphylococcus aureus*, and *E. faecalis* can gain adaptation ability to resist antimicrobial activities of EO (oregano, cinnamon, and other oils), when used as feed additives (Becerril et al., 2012; Melo et al., 2015).

3.2.2 Prebiotics

Prebiotics are nondigestible oligosaccharide carbohydrate compounds, and examples include fructooligosaccharides (FOS, derived from grains), galactooligosaccharides (GOS, derived from milk), and mannan-oligosaccharides (MOS, derived from the cell wall of the yeast *Saccharomyces cerevisiae*), each of which have potential to improve poultry health and reduce the burden of enteric pathogens (Hughes et al., 2017; Ricke, 2018). Prebiotics exert their beneficial effects on the host through improvement of intestinal function, modulation of host immune responses, and modification of the gut microbiota (M'Sadeq et al., 2015a). In the context of their immunomodulatory activities, *in ovo* administration or dietary inclusion of GOS and inulin has been shown to reduce the expression of proinflammatory cytokines and signaling molecules in lymphoid organs of chickens (Hughes et al., 2017; Smirnov et al., 2005; Slawinska et al., 2016). Emerging data also indicate that dietary supplementation of oligosaccharides can reduce colonization of *C. perfringens*, *E. coli*, and *Salmonella* by selectively promoting proliferation of beneficial bacteria such as bifidobacters and lactic acid-producing bacteria and/or by blocking the sites of bacterial attachment on the intestinal epithelium (Kim et al., 2011; Pourabedin and Zhao, 2015; Teng and Kim, 2018). Furthermore, fermentation of these oligosaccharides by resident microbiota produces short-chain fatty acids, mainly acetate, propionate, butyrate, and other by-products, which, in turn, can enhance host defense against infections (Pourabedin and Zhao, 2015; Teng and Kim, 2018). Prebiotics have also been shown to be effective against enteric protozoan parasites. For example, supplementation of a *Saccharomyces cerevisiae* fermentation product and galactoglucomannan oligosaccharide-arabinoxylan provided significant protection against *E. maxima* and *E. acervulina* infections in broiler chickens (Faber et al., 2012; Lensing et al., 2012). However, the effectiveness of prebiotics in controlling NE in chickens has been somewhat less conclusive. While some studies have shown that dietary inclusion of arabinoxylo-oligosaccharides, MOS, or yeast cell wall extract can reduce NE lesions and mortality (Keerqin et al., 2017; M'Sadeq et al., 2015b), one study did not observe any beneficial effects when the same or other oligosaccharides were used (Hofacre et al., 2018). The reason for these discrepant findings could be due to the different types and dosage regimens of oligosaccharides and/or the strain *C. perfringens* used in these studies. Nonetheless, prebiotics seem to offer some beneficial effects as feed additives in reducing enteric pathogen burdens in poultry.

3.2.3 Probiotics

Probiotics are beneficial microbes that confer various health benefits to the host (Taha-Abdelaziz et al., 2018). Several species of the bacterial genera, such as *Lactobacillus*, *Streptococcus*, *Enterococcus*, *Enterobacter*, and *Bifidobacteria*, and yeasts, such as *Saccharomyces*, *Torulopsis*, *Aspergillus*, and *Candida*, have been widely used as beneficial microbes. In the context of poultry, in addition to their roles in improving bird performance and GI health, supplementation with probiotics or their by-products have been shown to provide protection against various foodborne and enteric pathogens such as *E. coli*, *Salmonella enterica* serovar Enteritidis, *C. perfringens*, *C. jejuni*, and coccidia (Nakphaichit et al., 2011; Pan and Yu, 2014; Pascual et al., 1999; Shin et al., 2008; Strompfova et al., 2010).

Accumulating evidence also suggests that compared to single-strain probiotics, multi-strain probiotics exhibit greater efficacy for improving intestinal and immune functions and controlling bacterial pathogens, such as *Salmonella*, and protozoan parasites, such as *E. tenella* (Chen et al., 2012; Lensing et al., 2012). This is attributed to synergistic interactions between probiotic strains in mixtures (Chapman et al., 2013). Additionally, routes of administration of probiotics such as in feed, water, spray, and also experimentally via oral gavage seem to influence their beneficial effects. For example, administration of probiotics in drinking water can improve bird performance and enhance resistance against mixed *Eimeria* infection (*E. acervulina*, *E. maxima*, and *E. tenella*) in broiler chickens compared to in-feed supplementation (Ritzi et al., 2014). This observation may be attributed to the enhanced viability of probiotics in water during passage through the GI tract as it shortens the gastric transit time, thereby reducing the negative impacts of gastric acid and digestive secretions.

Probiotics can exert their protective effects against microbial pathogens either directly by inhibiting their growth, attenuating their virulence, and by competing with them for space and nutrients (referred to as competitive exclusion), or indirectly by modulating the host immune system and intestinal microbiome composition (Adedokun and Olojede, 2018; Brisbin et al., 2010). Some of the mechanisms of probiotics in the context of certain important enteric pathogens of poultry are briefly discussed in the following section.

3.2.3.1 Competitive exclusion and inhibition of pathogen growth

Mounting evidence indicates that mucosal adhesion is an important prerequisite for probiotics to establish colonization and is regarded as a key element for selection of probiotic candidates (Tuomola et al., 2001). In light of the fact that competitive coexistence is one of the mechanisms of pathogen exclusion, a recent study has demonstrated a relative dominance of *Lactobacillus* and *Bifidobacterium* sp. in the intestinal microbiota of chickens when they were given as dietary supplements. The competition capacity of probiotic bacteria, however, may vary according to the probiotic strain and the microbial competitor. For instance, oral administration of two doses of 10^9 colony forming units of *Lactobacillus johnsonii* strain FI9785 was not sufficient to confer protection against *C. jejuni* (Manes-Lazaro et al., 2017), while a single dose of this strain was sufficient to attain a competitive advantage over *C. perfringens* (La Ragione et al., 2004). Another study has also shown that administration of a lower dose of *L. johnsonii* R-17504 reduces *Salmonella enterica* serovar Enteritidis colonization (Van Coillie et al., 2007). Nevertheless, it is noteworthy that co-administration of other products such as prebiotics, OA, and EO with probiotics has been shown to enhance their protective efficacy against a wide range of enteric pathogens, such as *C. perfringens* and *Eimeria* spp (Pourabedin and Zhao, 2015; Alagawany et al., 2018; Giannenas et al., 2012). The addition of these agents likely facilitates growth and proliferation of probiotic bacteria in addition to inducing immunomodulatory effects on the host immune system (Caly et al., 2017; Namkung et al., 2004).

Numerous studies have also demonstrated that probiotics and their metabolites display broad-spectrum bactericidal activity against a wide range of Gram-negative and Gram-positive bacteria when tested *in vitro* (Guo et al., 2017). For example, Teo and Tan (2005) demonstrated that a single strain of *Bacillus subtilis* exhibits broad-spectrum inhibitory activity against various strains of *C. perfringens*, *C. jejuni*, and *Campylobacter coli*. To this end, attempts have been made to identify the antimicrobial compounds produced by probiotic bacteria. Among these compounds, bacteriocins have been investigated and several bacteriocins have been shown to possess antagonistic activity against various strains of *C. perfringens*, including pediocin A, divercin, nisin, subtilin, and bacteriocin-like inhibitory substance (Caly et al., 2017; Dabard et al., 2001; Jozefiak et al., 2012; Sharma et al., 2014; Udompijitkul et al., 2012). The production of these active metabolites is thought to provide probiotics with a competitive growth advantage over pathogenic microorganisms.

3.2.3.2 Attenuation of virulence factors

Existing evidence indicates that motility and adhesion are essential factors for microbial colonization of the GI tract, and the alteration of the expression of genes responsible for these attributes could lead to a reduction in the ability of the pathogen to adhere to and colonize mucosal surfaces (Haiko and Westerlund-Wikstrom, 2013; Ribet and Cossart, 2015). In addition to effects on motility and adhesion, *in vitro* studies have shown that probiotic exposure of pathogenic bacteria, such as *C. perfringens*, *C. jejuni*, and pathogenic *Salmonella*, results in downregulation of genes responsible for invasion, biofilm formation, and toxin and auto-inducer production (Guo et al., 2017; Li et al., 2011; Muyyarikkandy and Amalaradjou, 2017; Najarian et al., 2019).

3.2.3.3 Improvement of intestinal morphology

Maintenance of intestinal mucosal barrier integrity is essential for the prevention of enteric infections. In addition to their role in the induction of innate responses in epithelial cells, probiotic supplementation has been shown to improve intestinal morphology and mucus barrier function (Aliakbarpour et al., 2012; Smirnov et al., 2005), thereby enhancing bird performance as well as providing resistance against enteric pathogens, including *C. perfringens* and coccidia. For example, when used without anticoccidials, different probiotic bacteria (*Enterococcus faecium*, *Bifidobacterium*, *L reuteri*, *L salivarius*, and *Bacillus subtilis*) have been shown to mitigate the negative effects of *Eimeria* in broiler chickens (Giannenas et al., 2012, 2014). Furthermore, combination of a probiotic mixture (*Enterococcus*, *Bifidobacterium*, *Pediococcus*, and *Lactobacillus*) with an anticoccidial vaccine was shown to provide enhanced protection against *Eimeria* infection in broilers compared to those that received vaccine-only controls (Ritzi et al., 2016; Lin et al., 2017). A recent study has also shown that *B. licheniformis* supplementation was not only able to restore intestinal integrity but also the ileum and cecal microbial balance in chickens challenged with *C. perfringens* (Lin et al., 2017).

3.2.3.4 Modulation of intestinal immune responses

In the context of mucosal immunity, it has been shown that lactobacillus probiotic supplementation enhances antigen-specific and natural antibodies and also alters the expression of cytokines, antimicrobial peptides, and T cell surface markers in gut-associated lymphoid tissue (GALT), which, in turn, may enhance resistance to bacterial and parasitic pathogens in chickens (Akbari et al., 2008; Brisbin et al., 2010; Haghighi et al., 2008). Administration of a mixture of probiotic bacteria to chickens resulted in a significant increase in antibody-mediated immune responses to antigens, such as sheep red blood cells (Haghighi et al., 2005). Further, Haghighi et al. (2006) observed an enhanced production of natural antibodies in probiotic-treated chickens (Haghighi et al., 2006). Additionally, oral administration of *Lactobacillus* in chickens has been shown to affect antibody- and cell-mediated immune responses (Brisbin et al., 2011, 2012). Some of these immune-enhancing effects may be attributed to the structural constituents of probiotic bacteria, including their DNA and cell wall components (Brisbin et al., 2008). It is possible that interactions between bacterial components, such as DNA and peptidoglycan, and cells of the immune system, various innate or adaptive immune pathways are triggered. For example, administration of probiotics to chickens results in alteration of cytokine and antimicrobial peptide gene expression in cecal tonsils after infection with *Salmonella* (Akbari et al., 2008). The *in vitro* effects of *Lactobacillus* species on chicken immune system cells have been demonstrated, marked by regulation of gene expression in subsets of cecal tonsil cells (Brisbin et al., 2011, 2012) and splenocytes (Brisbin et al., 2010), in addition to enhancing the function of macrophages (Brisbin et al., 2015). In the context of *Eimeria* infection, probiotic administration has been shown to increase the levels of T helper 1-type cytokines (interferon-gamma) in both serum and intestinal secretions and the number of IELs expressing certain cell surface markers (CD3, CD4, CD8, and αβ T cell receptor). These immune-enhancing activities are associated with attenuation of the virulence and reproductive capacity of *Eimeria* spp (Dalloul et al., 2003). Similarly, dietary supplementation of *Lactobacillus johnsonii* elicits intestinal mucosal immunity in the ileum and cecal tonsils associated with enhanced resistance against subclinical NE in broiler chickens (Wang et al., 2017). Although there is evidence that individual or combinations of probiotic bacteria can modulate immune system gene expression and, thus, the immune functions, it is still not very clear as to whether probiotics can completely replace AGP in commercial practice and that they can effectively mitigate severe infectious challenges.

Taken together, it is evident that probiotics may serve as an important component of strategies aimed at developing alternatives to AGP in poultry. Although a vast amount of research has been devoted to understanding the dynamic microbial ecosystem in the intestine and the host-microbe or microbe-microbe interactions, there is still considerable work required to dissect out the precise mechanisms by which probiotics can provide general or GI health benefits in poultry.

3.2.4 Organic acids (OA)

Organic acids (OA) belong to a broad class of compounds that have important roles in various fundamental metabolic processes of host physiological machinery. These compounds, including formic, fumaric, propionic, citric, and lactic acid, and their salts (e.g. calcium formate, calcium propionate), have been traditionally used in animal feeds for reducing bacterial and fungal growth (Dittoe et al., 2018). Considering their safety and antimicrobial activity, OA are classified as 'feed preservatives' in Europe (Adil et al., 2010). As their use in animal production has proven to be beneficial, increasing evidence accumulated over the last two decades has indicated that the use of OA could also contribute to increased weight gain, higher feed conversion rates, and reduced incidence of GI-related health issues in livestock (Mikkelsen et al., 2009). In poultry, OA have been used either in feed or in drinking water with the objective of reducing intestinal pathogen burden and the associated toxic microbial metabolites. This practice has also been shown to improve nutrient digestibility, thereby enhancing bird performance and immune health of the avian intestine (Diarra and Malouin, 2014). The antimicrobial effects of OA, including short-chain and medium-chain fatty acids, also seem to depend on both the concentration of the acid and the microbial pathogen exposed to the acid (Adil et al., 2010). These acidifiers are used to benefit production in three ways in poultry operations: (1) OA added to the feed facilitates prevention of bacterial or mold growth in feed and also reduces the pH in the crop, (2) OA given via drinking water inhibits microbial growth in the water and reduces the pH of the crop and intestinal contents to facilitate pathogen control, and (3) OA sprayed onto the poultry litter can affect the bacteria that facilitate the breakdown of uric acid and, thus, limit the amount of ammonia release (Rodjan et al., 2018).

OA treatments composed of individual acids and blends of several acids have been found to exert antimicrobial activities (Gadde et al., 2017). Important enteric bacterial pathogens that affect poultry including *Salmonella*, *Campylobacter*, *C. perfringens*, and *E. coli* have been shown to be controlled by supplementation of OA in feed or in water (Van Immerseel et al., 2006). For example, a study by Koyuncu et al. (2013) showed that treatment of pelleted and compound mash feeds with formic acid and different blends of formic acid, propionic acid, and sodium formate could significantly reduce *Salmonella* counts in the feed (Koyuncu et al., 2013). The current practice of drinking water acidification in the broiler industry has improved bird performance as well as reduced pathogen load in the water and in the crop and proventriculus, coupled with an optimal regulation of intestinal microflora and adequate digestion of feed (Dittoe et al., 2018). In support of this, Bourassa et al. (2018) showed that incorporation of OA (lactic acid, acetic acid, or formic acid) in the drinking water during pre-transport feed withdrawal can reduce *Salmonella*, *E. coli*, and *Campylobacter* contamination of crops and broiler carcasses at processing (Bourassa et al., 2018).

Similarly, mixtures of OA (fumaric acid, calcium format, calcium propionate, potassium sorbate, calcium butyrate, calcium lactate, and hydrogenated vegetable oil) were found to be more efficacious than an AGP (Enramycin) at decreasing intestinal *C. perfringens*, *E. coli*, and *Salmonella* spp. (Manafi et al., 2019). Fernandez-Rubio et al. (2009) also suggested that supplementation of 0.2% encapsulated OA to the diet might improve the proliferation of useful commensal microbiota (*Lactobacillus* spp.) and diminish the population of pathogenic bacteria in poultry intestinal contents (Fernandez-Rubio et al., 2009).

Collectively, the aforementioned evidence shows that OA supplementation of poultry feed or water can help reduce the load of enteric pathogens, including *E. coli*, *C. perfringens*, *Salmonella*, and *Campylobacter* and, thus, improve the health and performance of poultry. However, to what extent OA can effectively replace AGP in commercial production remains to be yet further investigated.

4 Conclusion and future trends

Careful and well-thought-out nutritional intervention strategies involving the use of feed ingredients and feed additives may be exploited to promote gut health and development, and to reduce GI disease burden in poultry. These strategies may exert their effects via three different pathways: (1) Enhancement of intestinal integrity and functions, including establishment of beneficial microbial population, (2) reduction of enteric pathogen burden, and (3) modulation of host immune responses. Clearly, strategies aimed at reducing the incidence of enteric infections are critical for the productivity, sustainability, and profitability of the poultry industry. Vast research in the last two decades has put forth several candidate feed additives that have potential as alternatives to AGP in poultry production. These include prebiotics, probiotics, phytogenic feed additives (EO), antimicrobial peptides, bacteriophages, antibodies, enzymes, and acids, each of which can impact the incidence and severity of GI diseases, including NE and coccidiosis. Although many approaches have been proposed, combination of more than one approach seems necessary to enhance the performance and GI health of poultry. Some examples include choosing feed ingredients that have higher digestibility and nutritional value, a combination of prebiotics, probiotics, and EO that are known to benefit gut health and, importantly, proper management practices. However, a challenging question that still remains to be answered is, whether these AGP replacements can actually protect the birds in the face of a serious challenge, particularly in a commercial poultry setting? Many studies have shown the efficacy of AGP alternatives under a low-challenge environment and that such testing under experimental conditions in the absence of an active infectious challenge somewhat limits their potential applications. Furthermore, these AGP alternative strategies or approaches will require suitable adaptation to existing feeding programs and poultry production practices considering the wide variation in global climate and in housing and management practices. Nevertheless, given the increasing global demand for high-quality protein for humans from poultry sources, combined with the looming potential for decreases in production as AGP are removed from the industry, employing new nutritional intervention strategies is essential to enhance the GI and overall health of farmed poultry while also reducing the impact of enteric diseases to a considerable degree.

5 Where to look for further information

An introduction to the topic of GI diseases as well as a thorough understanding of the disease pathophysiology and microbiology aspects can be found in the book titled 'Diseases of Poultry', 14th edition by David E Swayne (Editor-in-chief) published by Wiley Blackwell publishers. Considerable insight and detailed information about the role of nutrition and enteric microbiota in the maintenance of optimal gut-health and the future roadmap for the use of various types and forms of feed additives, including probiotics, as antimicrobial alternatives can be obtained from two review articles authored by Adedokun and Olojede (2018) and Clavijo and Florez (2018), which are cited here in this chapter.

Additionally, the United States Department of Agriculture (USDA), which is the largest funding agency for animal researchers, provides up-to-date information about various topical aspects of animal health and disease surveillance and prevention, including those affecting poultry (https://www.usda.gov/topics/animals). In particular, the USDA's National Institute of Food and Agriculture (NIFA) that provides ample information highlighting important animal research areas, including the area of 'reduced use of antibiotics' has been an encouraging source for many poultry researchers to develop high impact research programs (https://www.nifa.usda.gov/topic/animal-health). Furthermore, an online U. S. Federal Science database resource (https://www.science.gov/) provides a very helpful access to various scientific reports and publications, including those in animal research and development.

6 References

Abraham, M., McDougald, L. R. and Beckstead, R. B. 2014. Blackhead disease: reduced sensitivity of *Histomonas meleagridis* to nitarsone *in vitro* and *in vivo*. *Avian. Dis.*, 58, 60-3.

Adams, C., Vahl, H. A. and Veldman, A. 1996. Interaction between nutrition and *Eimeria acervulina* infection in broiler chickens: development of an experimental infection model. *Br. J. Nutr.*, 75, 867-73.

Adedokun, S. A. and Olojede, O. C. 2018. Optimizing gastrointestinal integrity in poultry: the role of nutrients and feed additives. *Front. Vet. Sci.*, 5, 348.

Adil, S., Banday, T., Bhat, G. A., Mir, M. S. and Rehman, M. 2010. Effect of dietary supplementation of organic acids on performance, intestinal histomorphology, and serum biochemistry of broiler chicken. *Vet. Med. Int.*, 2010, 479485.

Agunos, A., Carson, C. and Leger, D. 2013. Antimicrobial therapy of selected diseases in turkeys, laying hens, and minor poultry species in Canada. *Can. Vet. J.*, 54, 1041-52.

Akbari, M. R., Haghighi, H. R., Chambers, J. R., Brisbin, J., Read, L. R. and Sharif, S. 2008. Expression of antimicrobial peptides in cecal tonsils of chickens treated with probiotics and infected with *Salmonella enterica* serovar typhimurium. *Clin. Vaccine Immunol.*, 15, 1689-93.

Al-Sheikhly, F. and Truscott, R. B. 1977. The pathology of necrotic enteritis of chickens following infusion of crude toxins of *Clostridium perfringens* into the duodenum. *Avian Dis.*, 21, 241-55.

Alagawany, M., Abd El-Hack, M. E., Farag, M. R., Sachan, S., Karthik, K. and Dhama, K. 2018. The use of probiotics as eco-friendly alternatives for antibiotics in poultry nutrition. *Environ. Sci. Pollut. Res. Int.*, 25, 10611-18.

Alali, W. Q., Hofacre, C. L., Mathis, G. F. and Faltys, G. 2013. Effect of essential oil compound on shedding and colonization of *Salmonella enterica* serovar Heidelberg in broilers. *Poult. Sci.*, 92, 836-41.

Aliakbarpour, H. R., Chamani, M., Rahimi, G., Sadeghi, A. A. and Qujeq, D. 2012. The *Bacillus subtilis* and lactic acid bacteria probiotics influences intestinal mucin gene expression, histomorphology and growth performance in broilers. *Asian-Australas. J. Anim. Sci.*, 25, 1285-93.

Annett, C. B., Viste, J. R., Chirino-Trejo, M., Classen, H. L., Middleton, D. M. and Simko, E. 2002. Necrotic enteritis: effect of barley, wheat and corn diets on proliferation of *Clostridium perfringens* type A. *Avian Pathol.*, 31, 598-601.

Bach Knudsen, K. E., Norskov, N. P., Bolvig, A. K., Hedemann, M. S. and Laerke, H. N. 2017. Dietary fibers and associated phytochemicals in cereals. *Mol. Nutr. Food Res.*, 61.

Barbour, E. K., Bragg, R. R., Karrouf, G., Iyer, A., Azhar, E., Harakeh, S. and Kumosani, T. 2015. Control of eight predominant *Eimeria* spp. involved in economic coccidiosis of broiler chicken by a chemically characterized essential oil. *J. Appl. Microbiol.*, 118, 583-91.

Barnes, D. M., Kirby, Y. K. and Oliver, K. G. 2001. Effects of biogenic amines on growth and the incidence of proventricular lesions in broiler chickens. *Poult. Sci.*, 80, 906-11.

Barrow, P. A., Jones, M. A., Smith, A. L. and Wigley, P. 2012. The long view: *Salmonella*–the last forty years. *Avian Pathol.*, 41, 413-20.

Beach, N. M., Duncan, R. B., Larsen, C. T., Meng, X. J., Sriranganathan, N. and Pierson, F. W. 2009. Persistent infection of turkeys with an avirulent strain of turkey hemorrhagic enteritis virus. *Avian Dis.*, 53, 370-5.

Becerril, R., Nerin, C. and Gomez-Lus, R. 2012. Evaluation of bacterial resistance to essential oils and antibiotics after exposure to oregano and cinnamon essential oils. *Foodborne Pathog. Dis.*, 9, 699-705.

Benavente, J. and Martinez-Costas, J. 2007. Avian reovirus: structure and biology. *Virus Res.*, 123, 105-19.

Berchieri Jr., A., Murphy, C. K., Marston, K. and Barrow, P. A. 2001. Observations on the persistence and vertical transmission of *Salmonella enterica* serovars Pullorum and Gallinarum in chickens: effect of bacterial and host genetic background. *Avian Pathol.*, 30, 221-31.

Beski, S. S. M., Swick, R. A. and Iji, P. A. 2015. Specialized protein products in broiler chicken nutrition: a review. *Anim. Nutr.*, 1, 47-53.

Bourassa, D. V., Wilson, K. M., Ritz, C. R., Kiepper, B. K. and Buhr, R. J. 2018. Evaluation of the addition of organic acids in the feed and/or water for broilers and the subsequent recovery of *Salmonella* Typhimurium from litter and ceca. *Poult. Sci.*, 97, 64-73.

Bradley, R. E. and Radhakrishnan, C. V. 1973. Coccidiosis in chickens: obligate relationship between *Eimeria tenella* and certain species of cecal microflora in the pathogenesis of the disease. *Avian Dis.*, 17, 461-76.

Brisbin, J. T., Gong, J. and Sharif, S. 2008. Interactions between commensal bacteria and the gut-associated immune system of the chicken. *Anim. Health Res. Rev.*, 9, 101-10.

Brisbin, J. T., Gong, J., Parvizi, P. and Sharif, S. 2010. Effects of lactobacilli on cytokine expression by chicken spleen and cecal tonsil cells. *Clin. Vaccine Immunol.*, 17, 1337-43.

Brisbin, J. T., Gong, J., Orouji, S., Esufali, J., Mallick, A. I., Parvizi, P., Shewen, P. E. and Sharif, S. 2011. Oral treatment of chickens with lactobacilli influences elicitation of immune responses. *Clin. Vaccine Immunol.*, 18, 1447-55.

Brisbin, J. T., Parvizi, P. and Sharif, S. 2012. Differential cytokine expression in T-cell subsets of chicken caecal tonsils co-cultured with three species of *Lactobacillus*. *Benef. Microbes*, 3, 205-10.

Brisbin, J. T., Davidge, L., Roshdieh, A. and Sharif, S. 2015. Characterization of the effects of three Lactobacillus species on the function of chicken macrophages. *Res. Vet. Sci.*, 100, 39-44.

Burt, S. 2004. Essential oils: their antibacterial properties and potential applications in foods–a review. *Int. J. Food Microbiol.*, 94, 223-53.

Calnek, B. W. 2015. Avian diseases: the creation and evolution of P. Philip Levine's enduring gift. *Avian Dis.*, 59, 1-6.

Caly, D. L., Chevalier, M., Flahaut, C., Cudennec, B., Al Atya, A. K., Chataigne, G., D'inca, R., Auclair, E. and Drider, D. 2017. The safe enterocin Dd14 is a leaderless two-peptide bacteriocin with anti-*Clostridium perfringens* activity. *Int. J. Antimicrob. Agents*, 49, 282-9.

Carre, B., Gomez, J. and Chagneau, A. M. 1995. Contribution of oligosaccharide and polysaccharide digestion, and excreta losses of lactic acid and short chain fatty acids, to dietary metabolisable energy values in broiler chickens and adult cockerels. *Br. Poult. Sci.*, 36, 611-29.

Chapman, H. D., Barta, J. R., Blake, D., Gruber, A., Jenkins, M., Smith, N. C., Suo, X. and Tomley, F. M. 2013. A selective review of advances in coccidiosis research. *Adv. Parasitol.*, 83, 93-171.

Chen, C. Y., Tsen, H. Y., Lin, C. L., Yu, B. and Chen, C. S. 2012. Oral administration of a combination of select lactic acid bacteria strains to reduce the *Salmonella* invasion and inflammation of broiler chicks. *Poult. Sci.*, 91, 2139-47.

Choct, M., Hughes, R. J., Trimble, R. P., Angkanaporn, K. and Annison, G. 1995. Non-starch polysaccharide-degrading enzymes increase the performance of broiler chickens fed wheat of low apparent metabolizable energy. *J. Nutr.*, 125, 485-92.

Choct, M., Hughes, R. J., Wang, J., Bedford, M. R., Morgan, A. J. and Annison, G. 1996. Increased small intestinal fermentation is partly responsible for the anti-nutritive activity of non-starch polysaccharides in chickens. *Br. Poult. Sci.*, 37, 609-21.

Clavijo, V. and Florez, M. J. V. 2018. The gastrointestinal microbiome and its association with the control of pathogens in broiler chicken production: a review. *Poult. Sci.*, 97, 1006-21.

Cortez, V., Meliopoulos, V. A., Karlsson, E. A., Hargest, V., Johnson, C. and Schultz-Cherry, S. 2017. Astrovirus biology and pathogenesis. *Annu. Rev. Virol.*, 4, 327-48.

Cosentino, S., Tuberoso, C. I., Pisano, B., Satta, M., Mascia, V., Arzedi, E. and Palmas, F. 1999. In-vitro antimicrobial activity and chemical composition of Sardinian *Thymus* essential oils. *Lett. Appl. Microbiol.*, 29, 130-5.

Craven, S. E., Cox, N. A., Bailey, J. S. and Cosby, D. E. 2003. Incidence and tracking of *Clostridium perfringens* through an integrated broiler chicken operation. *Avian. Dis.*, 47, 707-11.

Dabard, J., Bridonneau, C., Phillipe, C., Anglade, P., Molle, D., Nardi, M., Ladire, M., Girardin, H., Marcille, F., Gomez, A. and Fons, M. 2001. Ruminococcin A, a new lantibiotic produced by a *Ruminococcus gnavus* strain isolated from human feces. *Appl. Environ. Microbiol.*, 67, 4111-8.

Dahiya, J. P., Hoehler, D., Wilkie, D. C., Van Kessel, A. G. and Drew, M. D. 2005. Dietary glycine concentration affects intestinal *Clostridium perfringens* and lactobacilli populations in broiler chickens. *Poult. Sci.*, 84, 1875-85.

Dahiya, J. P., Hoehler, D., Van Kessel, A. G. and Drew, M. D. 2007. Effect of different dietary methionine sources on intestinal microbial populations in broiler chickens. *Poult. Sci.*, 86, 2358-66.

Dalloul, R. A., Lillehoj, H. S., Shellem, T. A. and Doerr, J. A. 2003. Intestinal immunomodulation by vitamin A deficiency and lactobacillus-based probiotic in *Eimeria acervulina*-infected broiler chickens. *Avian. Dis.*, 47, 1313-20.

Delaquis, P. J., Stanich, K., Girard, B. and Mazza, G. 2002. Antimicrobial activity of individual and mixed fractions of dill, cilantro, coriander and eucalyptus essential oils. *Int. J. Food. Microbiol.*, 74, 101-9.

Dhama, K., Saminathan, M., Karthik, K., Tiwari, R., Shabbir, M. Z., Kumar, N., Malik, Y. S. and Singh, R. K. 2015. Avian rotavirus enteritis - an updated review. *Vet. Q.*, 35, 142-58.

Dhillon, A. S., Roy, P., Lauerman, L., Schaberg, D., Weber, S., Bandli, D. and Wier, F. 2004. High mortality in egg layers as a result of necrotic enteritis. *Avian. Dis.*, 48, 675-80.

Diarra, M. S. and Malouin, F. 2014. Antibiotics in Canadian poultry productions and anticipated alternatives. *Front. Microbiol.*, 5, 282.

Dittoe, D. K., Ricke, S. C. and Kiess, A. S. 2018. Organic acids and potential for modifying the avian gastrointestinal tract and reducing pathogens and disease. *Front. Vet. Sci.*, 5, 216.

Drew, M. D., Syed, N. A., Goldade, B. G., Laarveld, B. and Van Kessel, A. G. 2004. Effects of dietary protein source and level on intestinal populations of *Clostridium perfringens* in broiler chickens. *Poult. Sci.*, 83, 414-20.

Elwinger, K., Schneitz, C., Berndtson, E., Fossum, O., Teglof, B. and Engstom, B. 1992. Factors affecting the incidence of necrotic enteritis, caecal carriage of *Clostridium perfringens* and bird performance in broiler chicks. *Acta. Vet. Scand.*, 33, 369-78.

Engberg, R. M., Hedemann, M. S., Steenfeldt, S. and Jensen, B. B. 2004. Influence of whole wheat and xylanase on broiler performance and microbial composition and activity in the digestive tract. *Poult. Sci.*, 83, 925-38.

Faber, T. A., Dilger, R. N., Hopkins, A. C., Price, N. P. and Fahey Jr., G. C. 2012. Effects of oligosaccharides in a soybean meal-based diet on fermentative and immune responses in broiler chicks challenged with *Eimeria acervulina*. *Poult. Sci.*, 91, 3132-40.

Farhadi, D., Karimi, A., Sadeghi, G., Sheikhahmadi, A., Habibian, M., Raei, A. and Sobhani, K. 2017. Effects of using eucalyptus (*Eucalyptusglobulus* L.) leaf powder and its essential oil on growth performance and immune response of broiler chickens. *Iran. J. Vet. Res.*, 18, 60-2.

Fatoba, A. J. and Adeleke, M. A. 2018. Diagnosis and control of chicken coccidiosis: a recent update. *J. Parasit. Dis.*, 42, 483-93.

Fernandez-Rubio, C., Ordonez, C., Abad-Gonzalez, J., Garcia-Gallego, A., Honrubia, M. P., Mallo, J. J. and Balana-Fouce, R. 2009. Butyric acid-based feed additives help protect broiler chickens from *Salmonella* Enteritidis infection. *Poult. Sci.*, 88, 943-8.

Ferrandis Vila, M., Trudeau, M. P., Hung, Y. T., Zeng, Z., Urriola, P. E., Shurson, G. C. and Saqui-Salces, M. 2018. Dietary fiber sources and non-starch polysaccharide-degrading enzymes modify mucin expression and the immune profile of the swine ileum. *PLoS One*, 13, e0207196.

Fossum, O., Sandstedt, K. and Engstrom, B. E. 1988. Gizzard erosions as a cause of mortality in White Leghorn chickens. *Avian Pathol.*, 17, 519-25.

Gadde, U., Kim, W. H., Oh, S. T. and Lillehoj, H. S. 2017. Alternatives to antibiotics for maximizing growth performance and feed efficiency in poultry: a review. *Anim. Health. Res. Rev.*, 18, 26-45.

Garozzo, A., Timpanaro, R., Bisignano, B., Furneri, P. M., Bisignano, G. and Castro, A. 2009. *In vitro* antiviral activity of *Melaleuca alternifolia* essential oil. *Lett. Appl. Microbiol.*, 49, 806-8.

Gazdzinski, P. and Julian, R. J. 1992. Necrotic enteritis in turkeys. *Avian. Dis.*, 36, 792-8.

Gelli, A., Becquey, E., Ganaba, R., Headey, D., Hidrobo, M., Huybregts, L., Verhoef, H., Kenfack, R., Zongouri, S. and Guedenet, H. 2017. Improving diets and nutrition through an integrated poultry value chain and nutrition intervention (Selever) in Burkina Faso: study protocol for a randomized trial. *Trials*, 18, 412.

Giannenas, I., Papadopoulos, E., Tsalie, E., Triantafillou, E., Henikl, S., Teichmann, K. and Tontis, D. 2012. Assessment of dietary supplementation with probiotics on performance, intestinal morphology and microflora of chickens infected with *Eimeria tenella*. *Vet. Parasitol.*, 188, 31-40.

Giannenas, I., Tsalie, E., Triantafillou, E., Hessenberger, S., Teichmann, K., Mohnl, M. and Tontis, D. 2014. Assessment of probiotics supplementation via feed or water on the growth performance, intestinal morphology and microflora of chickens after experimental infection with *Eimeria acervulina*, *Eimeria maxima* and *Eimeria tenella*. *Avian Pathol.*, 43, 209-16.

Gilbert, M. S., Ijssennagger, N., Kies, A. K. and Van Mil, S. W. C. 2018. Protein fermentation in the gut; implications for intestinal dysfunction in humans, pigs, and poultry. *Am. J. Physiol. Gastrointest. Liver Physiol.*, 315, G159-70.

Gomathi, G., Senthilkumar, S., Natarajan, A., Amutha, R. and Purushothaman, M. R. 2018. Effect of dietary supplementation of cinnamon oil and sodium butyrate on carcass characteristics and meat quality of broiler chicken. *Vet. World*, 11, 959-64.

Gonzalez-Alvarado, J. M., Jimenez-Moreno, E., Valencia, D. G., Lazaro, R. and Mateos, G. G. 2008. Effects of fiber source and heat processing of the cereal on the development and pH of the gastrointestinal tract of broilers fed diets based on corn or rice. *Poult. Sci.*, 87, 1779-95.

Goossens, E., Valgaeren, B. R., Pardon, B., Haesebrouck, F., Ducatelle, R., Deprez, P. R. and Van Immerseel, F. 2017. Rethinking the role of alpha toxin in *Clostridium perfringens*-associated enteric diseases: a review on bovine necro-haemorrhagic enteritis. *Vet. Res.*, 48, 9.

Greco, M. V., Franchi, M. L., Rico Golba, S. L., Pardo, A. G. and Pose, G. N. 2014. Mycotoxins and mycotoxigenic fungi in poultry feed for food-producing animals. *Sci World J.*, 2014, 968215.

Guerre, P. 2016. Worldwide mycotoxins exposure in pig and poultry feed formulations. *Toxins (Basel)*, 8.

Gulluce, M., Sokmen, M., Daferera, D., Agar, G., Ozkan, H., Kartal, N., Polissiou, M., Sokmen, A. and Sahin, F. 2003. *In vitro* antibacterial, antifungal, and antioxidant activities of the essential oil and methanol extracts of herbal parts and callus cultures of *Satureja hortensis* L. *J. Agric. Food. Chem.*, 51, 3958-65.

Guo, S., Liu, D., Zhang, B., Li, Z., Li, Y., Ding, B. and Guo, Y. 2017. Two *Lactobacillus* species inhibit the growth and alpha-toxin production of *Clostridium perfringens* and induced proinflammatory factors in chicken intestinal epithelial cells *in vitro*. *Front. Microbiol.*, 8, 2081.

Guy, J. S. 1998. Virus infections of the gastrointestinal tract of poultry. *Poult. Sci.*, 77, 1166-75.

Haghighi, H. R., Gong, J., Gyles, C. L., Hayes, M. A., Sanei, B., Parvizi, P., Gisavi, H., Chambers, J. R. and Sharif, S. 2005. Modulation of antibody-mediated immune response by probiotics in chickens. *Clin. Diagn. Lab. Immunol.*, 12, 1387-92.

Haghighi, H. R., Gong, J., Gyles, C. L., Hayes, M. A., Zhou, H., Sanei, B., Chambers, J. R. and Sharif, S. 2006. Probiotics stimulate production of natural antibodies in chickens. *Clin. Vaccine. Immunol.*, 13, 975-80.

Haghighi, H. R., Abdul-Careem, M. F., Dara, R. A., Chambers, J. R. and Sharif, S. 2008. Cytokine gene expression in chicken cecal tonsils following treatment with probiotics and *Salmonella* infection. *Vet. Microbiol.*, 126, 225-33.

Haiko, J. and Westerlund-Wikstrom, B. 2013. The role of the bacterial flagellum in adhesion and virulence. *Biology (Basel)*, 2, 1242–67.

Hernandez, F., Garcia, V., Madrid, J., Orengo, J., Catala, P. and Megias, M. D. 2006. Effect of formic acid on performance, digestibility, intestinal histomorphology and plasma metabolite levels of broiler chickens. *Br. Poult. Sci.*, 47, 50–6.

Hofacre, C. L., Smith, J. A. and Mathis, G. F. 2018. An optimist's view on limiting necrotic enteritis and maintaining broiler gut health and performance in today's marketing, food safety, and regulatory climate. *Poult. Sci.*, 97, 1929–33.

Hughes, R. A., Ali, R. A., Mendoza, M. A., Hassan, H. M. and Koci, M. D. 2017. Impact of dietary galacto-oligosaccharide (Gos) on chicken's gut microbiota, mucosal gene expression, and *Salmonella* colonization. *Front. Vet. Sci.*, 4, 192.

Hume, M. E., Clemente-Hernandez, S. and Oviedo-Rondon, E. O. 2006. Effects of feed additives and mixed *Eimeria* species infection on intestinal microbial ecology of broilers. *Poult. Sci.*, 85, 2106–11.

Jansson, D. S., Nyman, A., Vagsholm, I., Christensson, D., Goransson, M., Fossum, O. and Hoglund, J. 2010. Ascarid infections in laying hens kept in different housing systems. *Avian Pathol.*, 39, 525–32.

Jitviriyanon, S., Phanthong, P., Lomarat, P., Bunyapraphatsara, N., Porntrakulpipat, S. and Paraksa, N. 2016. *In vitro* study of anti-coccidial activity of essential oils from indigenous plants against *Eimeria tenella*. *Vet. Parasitol.*, 228, 96–102.

Jozefiak, D., Sip, A., Rutkowski, A., Rawski, M., Kaczmarek, S., Wolun-Cholewa, M., Engberg, R. M. and Hojberg, O. 2012. Lyophilized *Carnobacterium divergens* As7 bacteriocin preparation improves performance of broiler chickens challenged with *Clostridium perfringens*. *Poult. Sci.*, 91, 1899–907.

Kaingu, F., Liu, D., Wang, L., Tao, J., Waihenya, R. and Kutima, H. 2017. Anticoccidial effects of *Aloe secundiflora* leaf extract against *Eimeria tenella* in broiler chicken. *Trop. Anim. Health. Prod.*, 49, 823–8.

Kalemba, D. and Kunicka, A. 2003. Antibacterial and antifungal properties of essential oils. *Curr. Med. Chem.*, 10, 813–29.

Keerqin, C., Morgan, N. K., Wu, S. B., Swick, R. A. and Choct, M. 2017. Dietary inclusion of arabinoxylo-oligosaccharides in response to broilers challenged with subclinical necrotic enteritis. *Br. Poult. Sci.*, 58, 418–24.

Kiarie, E. G. and Mills, A. 2019. Role of feed processing on gut health and function in pigs and poultry: conundrum of optimal particle size and hydrothermal regimens. *Front. Vet. Sci.*, 6, 19.

Kim, G. B., Seo, Y. M., Kim, C. H. and Paik, I. K. 2011. Effect of dietary prebiotic supplementation on the performance, intestinal microflora, and immune response of broilers. *Poult. Sci.*, 90, 75–82.

Kim, S. J., Lee, K. W., Kang, C. W. and An, B. K. 2016. Growth performance, relative meat and organ weights, cecal microflora, and blood characteristics in broiler chickens fed diets containing different nutrient density with or without essential oils. *Asian-Australas. J. Anim. Sci.*, 29, 549–54.

Korver, D. R., Zuidhof, M. J. and Lawes, K. R. 2004. Performance characteristics and economic comparison of broiler chickens fed wheat- and triticale-based diets. *Poult. Sci.*, 83, 716–25.

Koynarski, V., Stoev, S., Grozeva, N., Mirtcheva, T., Daskalov, H., Mitev, J. and Mantle, P. 2007. Experimental coccidiosis provoked by *Eimeria acervulina* in chicks simultaneously fed on ochratoxin A contaminated diet. *Res. Vet. Sci.*, 82, 225–31.

Koyuncu, S., Andersson, M. G., Lofstrom, C., Skandamis, P. N., Gounadaki, A., Zentek, J. and Haggblom, P. 2013. Organic acids for control of *Salmonella* in different feed materials. *Bmc. Vet. Res.*, 9, 81.

La Ragione, R. M., Narbad, A., Gasson, M. J. and Woodward, M. J. 2004. *In vivo* characterization of *Lactobacillus johnsonii* Fi9785 for use as a defined competitive exclusion agent against bacterial pathogens in poultry. *Lett. Appl. Microbiol.*, 38, 197–205.

Langhout, D. J., Schutte, J. B., De Jong, J., Sloetjes, H., Verstegen, M. W. and Tamminga, S. 2000. Effect of viscosity on digestion of nutrients in conventional and germ-free chicks. *Br. J. Nutr.*, 83, 533–40.

Leeson, S., Zubair, A. K., Squires, E. J. and Forsberg, C. 1997. Influence of dietary levels of fat, fiber, and copper sulfate and fat rancidity on cecal activity in the growing turkey. *Poult. Sci.*, 76, 59–66.

Lensing, M., Van Der Klis, J. D., Yoon, I. and Moore, D. T. 2012. Efficacy of *Saccharomyces cerevisiae* fermentation product on intestinal health and productivity of coccidian-challenged laying hens. *Poult. Sci.*, 91, 1590–7.

Lepp, D., Roxas, B., Parreira, V. R., Marri, P. R., Rosey, E. L., Gong, J., Songer, J. G., Vedantam, G. and Prescott, J. F. 2010. Identification of novel pathogenicity loci in *Clostridium perfringens* strains that cause avian necrotic enteritis. *PLoS One*, 5, e10795.

Li, J., Wang, W., Xu, S. X., Magarvey, N. A. and Mccormick, J. K. 2011. *Lactobacillus reuteri*-produced cyclic dipeptides quench agr-mediated expression of toxic shock syndrome toxin-1 in staphylococci. *Proc. Natl. Acad. Sci. U. S. A.*, 108, 3360–5.

Lin, Y., Xu, S., Zeng, D., Ni, X., Zhou, M., Zeng, Y., Wang, H., Zhou, Y., Zhu, H., Pan, K. and Li, G. 2017. Disruption in the cecal microbiota of chickens challenged with *Clostridium perfringens* and other factors was alleviated by *Bacillus licheniformis* supplementation. *PLoS One*, 12, e0182426.

Lovland, A. and Kaldhusdal, M. 1999. Liver lesions seen at slaughter as an indicator of necrotic enteritis in broiler flocks. *Fems. Immunol. Med. Microbiol.*, 24, 345–51.

M'Sadeq, S. A., Wu, S. B., Swick, R. A. and Choct, M. 2015a. Towards the control of necrotic enteritis in broiler chickens with in-feed antibiotics phasing-out worldwide. *Anim. Nutr.*, 1, 1–11.

M'Sadeq, S. A., Wu, S. B., Choct, M., Forder, R. and Swick, R. A. 2015b. Use of yeast cell wall extract as a tool to reduce the impact of necrotic enteritis in broilers. *Poult. Sci.*, 94, 898–905.

Maisonnier, S., Gomez, J., Bree, A., Berri, C., Baeza, E. and Carre, B. 2003. Effects of microflora status, dietary bile salts and guar gum on lipid digestibility, intestinal bile salts, and histomorphology in broiler chickens. *Poult. Sci.*, 82, 805–14.

Manafi, M., Hedayati, M., Pirany, N. and Omede, A. A. 2019. Comparison of performance and feed digestibility of the non-antibiotic feed supplement (Novacid) and an antibiotic growth promoter in broiler chickens. *Poult. Sci.*, 98, 904–11.

Manes-Lazaro, R., Van Diemen, P. M., Pin, C., Mayer, M. J., Stevens, M. P. and Narbad, A. 2017. Administration of *Lactobacillus johnsonii* Fi9785 to chickens affects colonisation by *Campylobacter jejuni* and the intestinal microbiota. *Br. Poult. Sci.*, 58, 373–81.

Marotta, F., Garofolo, G., Di Donato, G., Aprea, G., Platone, I., Cianciavicchia, S., Alessiani, A. and Di Giannatale, E. 2015. Population diversity of *Campylobacter jejuni* in poultry and its dynamic of contamination in chicken meat. *Biomed. Res. Int.*, 2015, 859845.

McDougald, L. R. and Fuller, L. 2005. Blackhead disease in turkeys: direct transmission of *Histomonas meleagridis* from bird to bird in a laboratory model. *Avian. Dis.*, 49, 328-31.

McDougald, L. R., Abraham, M. and Beckstead, R. B. 2012. An outbreak of blackhead disease (*Histomonas meleagridis*) in farm-reared bobwhite quail (*Colinus virginianus*). *Avian. Dis.*, 56, 754-6.

Mehde, A. A., Mehdi, W. A., Ozacar, M. and Ozacar, Z. Z. 2018. Evaluation of different saccharides and chitin as eco-friendly additive to improve the magnetic cross-linked enzyme aggregates (CLEAs) activities. *Int. J. Biol. Macromol.*, 118, 2040-50.

Melo, A. D., Amaral, A. F., Schaefer, G., Luciano, F. B., De Andrade, C., Costa, L. B. and Rostagno, M. H. 2015. Antimicrobial effect against different bacterial strains and bacterial adaptation to essential oils used as feed additives. *Can. J. Vet. Res.*, 79, 285-9.

Mettifogo, E., Nunez, L. F., Chacon, J. L., Santander Parra, S. H., Astolfi-Ferreira, C. S., Jerez, J. A., Jones, R. C. and Piantino Ferreira, A. J. 2014. Emergence of enteric viruses in production chickens is a concern for avian health. *Sci. World J.*, 2014, 450423.

Mikkelsen, L. L., Vidanarachchi, J. K., Olnood, C. G., Bao, Y. M., Selle, P. H. and Choct, M. 2009. Effect of potassium diformate on growth performance and gut microbiota in broiler chickens challenged with necrotic enteritis. *Br. Poult. Sci.*, 50, 66-75.

Mith, H., Dure, R., Delcenserie, V., Zhiri, A., Daube, G. and Clinquart, A. 2014. Antimicrobial activities of commercial essential oils and their components against food-borne pathogens and food spoilage bacteria. *Food. Sci. Nutr.*, 2, 403-16.

Mitsch, P., Zitterl-Eglseer, K., Kohler, B., Gabler, C., Losa, R. and Zimpernik, I. 2004. The effect of two different blends of essential oil components on the proliferation of *Clostridium perfringens* in the intestines of broiler chickens. *Poult. Sci.*, 83, 669-75.

Moore, R. J. 2016. Necrotic enteritis predisposing factors in broiler chickens. *Avian Pathol.*, 45, 275-81.

Morrissey, K. L., Widowski, T., Leeson, S., Sandilands, V., Arnone, A. and Torrey, S. 2014. The effect of dietary alterations during rearing on growth, productivity, and behavior in broiler breeder females. *Poult. Sci.*, 93, 285-95.

Mot, D., Timbermont, L., Haesebrouck, F., Ducatelle, R. and Van Immerseel, F. 2014. Progress and problems in vaccination against necrotic enteritis in broiler chickens. *Avian Pathol.*, 43, 290-300.

Muyyarikkandy, M. S. and Amalaradjou, M. A. 2017. *Lactobacillus bulgaricus, Lactobacillus rhamnosus* and *Lactobacillus paracasei* attenuate *Salmonella* Enteritidis, *Salmonella* Heidelberg and *Salmonella* Typhimurium colonization and virulence gene expression *in vitro*. *Int. J. Mol. Sci.*, 18.

Najarian, A., Sharif, S. and Griffiths, M. W. 2019. Evaluation of protective effect of *Lactobacillus acidophilus* La-5 on toxicity and colonization of *Clostridium difficile* in human epithelial cells *in vitro*. *Anaerobe*, 55, 142-51.

Nakphaichit, M., Thanomwongwattana, S., Phraephaisarn, C., Sakamoto, N., Keawsompong, S., Nakayama, J. and Nitisinprasert, S. 2011. The effect of including *Lactobacillus reuteri* Kub-Ac5 during post-hatch feeding on the growth and ileum microbiota of broiler chickens. *Poult. Sci.*, 90, 2753-65.

Namkung, H., Li, M., Gong, J., Yu, H., Cottrill, M. and De Lange, C. F. M. 2004. Impact of feeding blends of organic acids and herbal extracts on growth performance, gut microbiota and digestive function in newly weaned pigs. *Can J Anim Sci*, 84, 697–704.

Oviedo-Rondon, E. O., Hume, M. E., Hernandez, C. and Clemente-Hernandez, S. 2006. Intestinal microbial ecology of broilers vaccinated and challenged with mixed *Eimeria* species, and supplemented with essential oil blends. *Poult. Sci.*, 85, 854–60.

Paiva, D. M., Walk, C. L. and Mcelroy, A. P. 2013. Influence of dietary calcium level, calcium source, and phytase on bird performance and mineral digestibility during a natural necrotic enteritis episode. *Poult. Sci.*, 92, 3125–33.

Palliyeguru, M. W., Rose, S. P. and Mackenzie, A. M. 2010. Effect of dietary protein concentrates on the incidence of subclinical necrotic enteritis and growth performance of broiler chickens. *Poult. Sci.*, 89, 34–43.

Palya, V., Nagy, M., Glavits, R., Ivanics, E., Szalay, D., Dan, A., Suveges, T., Markos, B. and Harrach, B. 2007. Investigation of field outbreaks of turkey haemorrhagic enteritis in Hungary. *Acta. Vet. Hung.*, 55, 135–49.

Pan, D. and Yu, Z. 2014. Intestinal microbiome of poultry and its interaction with host and diet. *Gut Microbes*, 5, 108–19.

Pascual, M., Hugas, M., Badiola, J. I., Monfort, J. M. and Garriga, M. 1999. *Lactobacillus salivarius* CTC2197 prevents *Salmonella enteritidis* colonization in chickens. *Appl. Environ. Microbiol.*, 65, 4981-6.

Persia, M. E., Young, E. L., Utterback, P. L. and Parsons, C. M. 2006. Effects of dietary ingredients and *Eimeria acervulina* infection on chick performance, apparent metabolizable energy, and amino acid digestibility. *Poult. Sci.*, 85, 48–55.

Pessoa, L. M., Morais, S. M., Bevilaqua, C. M. and Luciano, J. H. 2002. Anthelmintic activity of essential oil of *Ocimum gratissimum* Linn. and eugenol against *Haemonchus contortus*. *Vet. Parasitol.*, 109, 59–63.

Pourabedin, M. and Zhao, X. 2015. Prebiotics and gut microbiota in chickens. *Fems. Microbiol. Lett.*, 362, fnv122.

Powers, W. and Capelari, M. 2017. Production, management and the environment symposium: measurement and mitigation of reactive nitrogen species from swine and poultry production. *J. Anim. Sci.*, 95, 2236–40.

Prescott, J. F., Parreira, V. R., Mehdizadeh Gohari, I., Lepp, D. and Gong, J. 2016. The pathogenesis of necrotic enteritis in chickens: what we know and what we need to know: a review. *Avian Pathol.*, 45, 288–94.

Ribet, D. and Cossart, P. 2015. How bacterial pathogens colonize their hosts and invade deeper tissues. *Microbes Infect.*, 17, 173–83.

Ricke, S. C. 2018. Impact of prebiotics on poultry production and food safety. *Yale. J. Biol. Med.*, 91, 151-9.

Ritzi, M. M., Abdelrahman, W., Mohnl, M. and Dalloul, R. A. 2014. Effects of probiotics and application methods on performance and response of broiler chickens to an *Eimeria* challenge. *Poult. Sci.*, 93, 2772-8.

Ritzi, M. M., Abdelrahman, W., Van-Heerden, K., Mohnl, M., Barrett, N. W. and Dalloul, R. A. 2016. Combination of probiotics and coccidiosis vaccine enhances protection against an *Eimeria* challenge. *Vet. Res.*, 47, 111.

Rodjan, P., Soisuwan, K., Thongprajukaew, K., Theapparat, Y., Khongthong, S., Jeenkeawpieam, J. and Salaeharae, T. 2018. Effect of organic acids or probiotics alone or in combination on growth performance, nutrient digestibility, enzyme activities, intestinal morphology and gut microflora in broiler chickens. *J. Anim. Physiol. Anim. Nutr (Berl)*, 102, e931–40.

Sakkas, H. and Papadopoulou, C. 2017. Antimicrobial activity of basil, oregano, and thyme essential oils. *J. Microbiol. Biotechnol.*, 27, 429–38.

Salter, A. M. 2017. Improving the sustainability of global meat and milk production. *Proc. Nutr. Soc.*, 76, 22–7.

Shahid, M. Z., Saima, H., Yasmin, A., Nadeem, M. T., Imran, M. and Afzaal, M. 2018. Antioxidant capacity of cinnamon extract for palm oil stability. *Lipids. Health. Dis.*, 17, 116.

Sharma, N., Gupta, A. and Gautam, N. 2014. Characterization of Bacteriocin like inhibitory substance produced by a new Strain *Brevibacillus borstelensis* Ag1 Isolated from 'Marcha'. *Braz. J. Microbiol.*, 45, 1007–15.

Shin, M. S., Han, S. K., Ji, A. R., Kim, K. S. and Lee, W. K. 2008. Isolation and characterization of bacteriocin-producing bacteria from the gastrointestinal tract of broiler chickens for probiotic use. *J. Appl. Microbiol.*, 105, 2203–12.

Shojadoost, B., Vince, A. R. and Prescott, J. F. 2012. The successful experimental induction of necrotic enteritis in chickens by *Clostridium perfringens*: a critical review. *Vet. Res.*, 43, 74.

Sklan, D., Shelly, M., Makovsky, B., Geyra, A., Klipper, E. and Friedman, A. 2003. The effect of chronic feeding of diacetoxyscirpenol and T-2 toxin on performance, health, small intestinal physiology and antibody production in turkey poults. *Br. Poult. Sci.*, 44, 46–52.

Slawinska, A., Plowiec, A., Siwek, M., Jaroszewski, M. and Bednarczyk, M. 2016. Long-term transcriptomic effects of prebiotics and synbiotics delivered *in ovo* in broiler chickens. *PLoS ONE*, 11, e0168899.

Smirnov, A., Perez, R., Amit-Romach, E., Sklan, D. and Uni, Z. 2005. Mucin dynamics and microbial populations in chicken small intestine are changed by dietary probiotic and antibiotic growth promoter supplementation. *J. Nutr.*, 135, 187–92.

Smith-Palmer, A., Stewart, J. and Fyfe, L. 1998. Antimicrobial properties of plant essential oils and essences against five important food-borne pathogens. *Lett. Appl. Microbiol.*, 26, 118–22.

Strompfova, V., Laukova, A., Marcinakova, M. and Vasilkova, Z. 2010. Testing of probiotic and bacteriocin-producing lactic acid bacteria towards *Eimeria* sp. *Pol. J. Vet. Sci.*, 13, 389–91.

Suresh, G., Das, R. K., Kaur Brar, S., Rouissi, T., Avalos Ramirez, A., Chorfi, Y. and Godbout, S. 2018. Alternatives to antibiotics in poultry feed: molecular perspectives. *Crit. Rev. Microbiol.*, 44, 318–35.

Szkotnicki, J. 2013. Antimicrobial therapy of bacterial diseases in broiler chickens – a comment. *Can. Vet. J.*, 54, 201.

Taha-Abdelaziz, K., Hodgins, D. C., Lammers, A., Alkie, T. N. and Sharif, S. 2018. Effects of early feeding and dietary interventions on development of lymphoid organs and immune competence in neonatal chickens: a review. *Vet. Immunol. Immunopathol.*, 201, 1–11.

Teng, P. Y. and Kim, W. K. 2018. Review: roles of prebiotics in intestinal ecosystem of broilers. *Front. Vet. Sci.*, 5, 245.

Teo, A. Y. and Tan, H. M. 2005. Inhibition of Clostridium perfringens by a novel strain of *Bacillus subtilis* isolated from the gastrointestinal tracts of healthy chickens. *Appl. Environ. Microbiol.*, 71, 4185–90.

Tuomola, E., Crittenden, R., Playne, M., Isolauri, E. and Salminen, S. 2001. Quality assurance criteria for probiotic bacteria. *Am. J. Clin. Nutr.*, 73, 393S–8S.

Udompijitkul, P., Paredes-Sabja, D. and Sarker, M. R. 2012. Inhibitory effects of nisin against *Clostridium perfringens* food poisoning and nonfood-borne isolates. *J. Food. Sci.*, 77, M51-6.

Van Coillie, E., Goris, J., Cleenwerck, I., Grijspeerdt, K., Botteldoorn, N., Van Immerseel, F., De Buck, J., Vancanneyt, M., Swings, J., Herman, L. and Heyndrickx, M. 2007. Identification of lactobacilli isolated from the cloaca and vagina of laying hens and characterization for potential use as probiotics to control *Salmonella* Enteritidis. *J. Appl. Microbiol.*, 102, 1095-106.

Van Immerseel, F., Russell, J. B., Flythe, M. D., Gantois, I., Timbermont, L., Pasmans, F., Haesebrouck, F. and Ducatelle, R. 2006. The use of organic acids to combat *Salmonella* in poultry: a mechanistic explanation of the efficacy. *Avian Pathol.*, 35, 182-8.

Van Immerseel, F., Lyhs, U., Pedersen, K. and Prescott, J. F. 2016. Recent breakthroughs have unveiled the many knowledge gaps in *Clostridium perfringens*-associated necrotic enteritis in chickens: the first International Conference on Necrotic Enteritis in Poultry. *Avian Pathol.*, 45, 269-70.

Varel, V. H., Wells, J. E. and Miller, D. N. 2007. Combination of a urease inhibitor and a plant essential oil to control coliform bacteria, odour production and ammonia loss from cattle waste. *J. Appl. Microbiol.*, 102, 472-7.

Wang, S., Peng, Q., Jia, H. M., Zeng, X. F., Zhu, J. L., Hou, C. L., Liu, X. T., Yang, F. J. and Qiao, S. Y. 2017. Prevention of *Escherichia coli* infection in broiler chickens with *Lactobacillus plantarum* B1. *Poult. Sci.*, 96, 2576-86.

Wati, T., Ghosh, T. K., Syed, B. and Haldar, S. 2015. Comparative efficacy of a phytogenic feed additive and an antibiotic growth promoter on production performance, caecal microbial population and humoral immune response of broiler chickens inoculated with enteric pathogens. *Anim. Nutr.*, 1, 213-19.

Weber, D. J., Rutala, W. A., Fischer, W. A., Kanamori, H. and Sickbert-Bennett, E. E. 2016. Emerging infectious diseases: focus on infection control issues for novel coronaviruses (Severe Acute Respiratory Syndrome-CoV and Middle East Respiratory Syndrome-CoV), hemorrhagic fever viruses (Lassa and Ebola), and highly pathogenic avian influenza viruses, A(H5N1) and A(H7N9). *Am. J. Infect. Control.*, 44, e91-100.

Wigley, P., Hulme, S. D., Powers, C., Beal, R. K., Berchieri Jr., A., Smith, A. and Barrow, P. 2005. Infection of the reproductive tract and eggs with *Salmonella enterica* serovar pullorum in the chicken is associated with suppression of cellular immunity at sexual maturity. *Infect. Immun.*, 73, 2986-90.

Williams, R. B. 2005. Intercurrent coccidiosis and necrotic enteritis of chickens: rational, integrated disease management by maintenance of gut integrity. *Avian Pathol.*, 34, 159-80.

Zhai, H., Liu, H., Wang, S., Wu, J. and Kluenter, A. M. 2018. Potential of essential oils for poultry and pigs. *Anim. Nutr.*, 4, 179-86.

Zubair, A. K., Forsberg, C. W. and Leeson, S. 1996. Effect of dietary fat, fiber, and monensin on cecal activity in turkeys. *Poult. Sci.*, 75, 891-9.

Chapter 11

The interaction between gut microbiota and pathogens in poultry

Ruediger Hauck, Auburn University, USA; and Lisa Bielke and Zhongtang Yu, The Ohio State University, USA

1 Introduction

The poultry microbiota is diverse and complex consisting of hundreds or even thousands of different microbes at varying abundance and with different metabolic functions and physiological and ecological traits (Wei et al., 2013; Oakley et al., 2014). Most members of the gut microbiota of poultry are anaerobic bacteria and considered commensal. Pathogens, including bacteria, viruses and parasites (mainly coccidia), however, do colonize the gut of healthy birds, but they are kept at low abundance and do not cause clinical infections. The high intensity of gut microbes creates stressful and constant interactions among different gut microbes. These interactions range from mutualism (a relationship in which two bacteria benefit from each other's metabolic activities) to competition, amensalism (a relationship in which one species of microbes benefits at the expense of the other due to inhibition of the latter by the former) and parasitism which explains the relationship between bacterial phages and their hosts. These interactions and the interactions of gut microbiota with the host and the diet determine the diversity, composition and structure of the gut microbiota as well as its contribution, both beneficial and detrimental, to the host. The occurrence of intestinal microbes in the gut and the health and diseases are also primarily determined by the above

http://dx.doi.org/10.19103/AS.2019.0059.12

interactions. The common pathogens and their interactions with the gut microbiota in poultry will be discussed in this chapter. The diseases caused by the common pathogens are also discussed but briefly.

2 Common intestinal pathogens and the associated diseases

2.1 Escherichia

Escherichia coli is a normal resident bacterium of the poultry intestinal tract, and most *E. coli* strains are commensal bacteria. Although at low abundance in healthy birds, *E. coli* can overgrow when dysbiosis in the gut microbiota occurs. Some strains of *E. coli* can tightly attach to the mucosa of the gut and respiratory tract (Joerger and Ross, 2005). Some of these *E. coli* strains cause intestinal and extra-intestinal diseases in poultry (Matthijs et al., 2017). Enteropathogenic *E. coli* (EPEC) can infect both chickens and turkeys (Fukui et al., 1995; Guy et al., 2000). One example of EPEC is the attaching and effacing *E. coli*, which can cause attaching/effacing (AE) intestinal lesions and depressed growth (Fukui et al., 1995). Extra-intestinal avian pathogenic *E. coli*, for which the poultry gut is an important reservoir (Ewers et al., 2009), can cause severe respiratory diseases (Peng et al., 2018) and colibacillosis (Kemmett et al., 2013, 2014), both are economically important syndromic diseases of poultry. Although not as severe as other diseases, systemic *E. coli* infection increased early mortalities of commercial broiler chickens and turkeys (Kemmett et al., 2014). Although not a pathogen of poultry, *E. coli* O157:H7 can also be found in poultry gut, especially the caecum (Best et al., 2003). It is important to note that viral infection can enhance *E. coli* infection and worsen lesions (Pakpinyo et al., 2003; Li et al., 2018a).

2.2 Salmonella

Salmonella is a genus of Gram-negative, non-spore forming, predominantly motile bacteria, belonging to the family *Enterobacteriaceae*. They are facultative anaerobes and chemoorganotrophs, obtaining their energy from oxidation and reduction of organic sources in their ecological niche. The genus *Salmonella* currently has two recognized species: *S. bongori* and *S. enterica*, the latter of which is further divided into six subspecies, with *S. enterica* subsp. *enterica* being the most common and relevant to infections in birds and mammals (Brenner et al., 2000; Lamas et al., 2018). There are currently over 2500 serovars of *S. enterica*, 1531 of which belong to *S. enterica* subsp. *enterica* and are associated with gastroenteritis in humans (Rivera-Chavez and Bäumler, 2015). The ubiquitous nature of *S. enterica*, due to its ability to survive in a wide range of hosts that serve as reservoirs and vectors, makes it one of the most relevant foodborne pathogens of worldwide public concern. Historically, poultry products have been implicated as the primary source of human *Salmonella* infection, but recent salmonellosis outbreaks indicate that *Salmonella* sources frequently include other foods such as fresh fruits and vegetables, peanut butter, snack food, tree nuts, breakfast cereal, microwaveable meals and tuna (CDC, 2018). The ability to transmit between and within host populations is centrally important in dictating the epidemiology of *Salmonella* infections, and multiple serovars are well known for transmission between poultry and humans, including Typhimurium, Enteritidis, Heidelberg and Kentucky. In poultry, non-typhoidal *Salmonella* strains do not cause overt disease but are associated with mild inflammation and decreased growth performance (Hume et al., 1998; Menconi et al., 2011; Shivaramaiah et al., 2011a; Raehtz et al., 2018).

2.3 Clostridium

Clostridium spp. are Gram-positive, rod-shaped bacteria. They grow anaerobically and can form spores. *C. perfringens*, the causative agent of necrotic enteritis (NE), is the most important enteric pathogen of poultry. It is ubiquitous and found in the intestinal microbiota of healthy birds, especially the distal small intestine and caecum, while only low numbers of *C. perfringens* are present in the proximal small intestine. Development of lesions is usually associated with rapid increase in *C. perfringens* proliferation to more than 10^5 colony-forming units/g ileal digesta under certain predisposing conditions (Si et al., 2007). A common characteristic of diseases caused by *Clostridium* spp. is that their pathogenesis involves potent extracellular toxins. *C. perfringens* can produce up to four major toxins, which are used to define five toxovars. Most *C. perfringens* isolates causing disease in poultry belong to toxovar A, producing only alpha-toxin, a phospholipase lysing cell membranes (Opengart and Songer, 2013). In addition to the major toxin(s), some isolates produce other, so-called minor toxins. Of these, NetB toxin seems to be most important (Keyburn et al., 2008, 2010), but its importance is still controversial (Giovanardi et al., 2016; Lacey et al., 2016). Two further minor toxins, enterotoxin and beta2 toxin, were once thought important toxins for enteric diseases, but epidemiological data have cast doubt on their role in enteric diseases (Giovanardi et al., 2016; Crespo et al., 2007). Under favourable conditions, *C. perfringens* multiplies quickly to high numbers in the small intestine producing its toxins. The toxins cause necrosis of the enteric mucosa and are absorbed into the bloodstream leading to enterotoxaemia and peracute death of the bird. In some cases, other *Clostridium* spp., such as *C. colinum*, *C. sordellii* and *C. difficile*, have been associated with NE-like disease (Uzal et al., 2016).

2.4 Campylobacter

Campylobacter species, particularly *C. jejuni*, are the leading cause of human enteric infection that causes campylobacteriosis, which is manifested as diarrhoea, cramps, fever and vomiting (Chlebicz and Slizewska, 2018). Up to 1% of the European Union population is infected by *Campylobacter* each year (Humphrey et al., 2014). Poultry is the largest reservoir of *Campylobacter*, and undercooked poultry meat is the cause of campylobacteriosis (Skarp et al., 2016). Traditionally, *C. coli* and *C. jejuni* are considered as normal members of intestinal microbiota of poultry and do not cause any clinical disease (Hermans et al., 2012; Jennings et al., 2011). The mechanism by which poultry does not develop clinical disease upon colonization by *C. jejuni* remains unknown, but the occurrence of some taxa of intestinal bacteria seemed to play a role in the colonization resistance against *Campylobacter* species in poultry gut (Kaakoush et al., 2014). However, a recent study showed that some broiler breeds could develop diarrhoea upon colonization by *C. jejuni*, and immune dysregulation in the intestines was probably attributable (Humphrey et al., 2014). One early study also showed that newly hatched chicks infected within 12 h of hatching could develop gastroenteritis (Welkos, 1984). Additionally, *C. hepaticus* can cause spotty liver disease (SLD) in layer hens mostly (Van et al., 2017a,b; Gregory et al., 2018). Although not prevalent, SLD causes multifocal liver lesions, mortality and decreased egg production. New unclassified species of *Campylobacter* can also cause SLD (Crawshaw et al., 2015).

Besides the aforementioned clinical diseases, *Campylobacter* species, especially *C. jejuni*, can cause other damages to the poultry hosts, including impaired gut integrity, increased gut permeability and altered tight junctions and immune response (Awad et al., 2018). These changes can affect the functions of the gut, such as decreasing nutrient (e.g. glucose) adsorption (Awad et al., 2014) and interference in intracellular Ca^{2+} signalling of the enterocytes (Awad et al., 2015). *C. jejuni* was found to promote *E. coli* translocation to extra-intestinal organs, decrease concentrations of a number of amino acids and alter volatile fatty acid (VFA) profiles in the chicken gut (Awad et al., 2016a), probably due to impaired gut integrity and microbial ecology. Although not the direct pathogen, *C. jejuni* was found associated with vibrionic hepatitis in chickens (Jennings et al., 2011). Future research may reveal more detrimental effects of *Campylobacter* on poultry upon colonization.

2.5 Eimeria

Eimeria spp. are parasites with a distinct life cycle. After oral infection, they undergo several asexual rounds and one sexual round of replication within cells of the intestinal wall, causing the death of the infected cells and inflammation. Seven to nine *Eimeria* spp. plus three operational taxonomic units (OTUs) that have not been assigned to a species are recognized to infect chickens, while five to seven *Eimeria* spp. infecting turkeys are recognized. Of these species, five are clinically relevant in chickens and turkeys, each. *Eimeria* spp. differ in the intestinal segments they infect and in their virulence, causing distinct lesions depending on the infection dose. The severity of lesions a species causes depends on the layer of the intestinal wall that their stages infect, the size of their stages and the number of replication rounds. Species whose stages infect cells in the lower layers of the intestinal wall, that is in the submucosa as opposed to epithelial cells, whose stages develop to a bigger size and who go through more rounds of replication, will cause more severe lesions (McDonald and Shirley, 2009). In addition, infection with certain *Eimeria* spp. predisposes birds to NE (Williams, 2003) due to lesions that compromise gut integrity and favour proliferation of *C. perfringens* (Williams, 2005).

2.6 Viruses

Poultry suffer from enteric viral disease syndromes, which have been given several different names such as runting-stunting syndrome, infectious stunting syndrome, malabsorption syndrome or pale bird syndrome in chickens and poult enteritis complex or poult enteritis mortality syndrome in turkeys. A variety of viruses have been implicated as the cause of enteric disease of poultry, and reoviruses, rotaviruses, astroviruses, parvoviruses and picornaviruses are among the common enteric viruses. Sequencing of intestinal viromes has added more viruses to this list. However, turkey coronavirus (TCV) is the only enteric virus that inevitably causes disease after experimental infections (Ismail et al., 2003; Gomes et al., 2010). None of the other viruses is consistently associated with enteric lesions, and defined viral inocula often cannot reproduce the disease (Day and Zsak, 2013; Jindal et al., 2014).

The syndromes caused by enteric viral infections are generally characterized by high morbidity with low mortality. Diseased birds have diarrhoea, and their intestines are distended, pale and thin-walled, while the intestinal content is watery, gaseous or foamy (Saif et al., 2013). Other viruses including avian influenza viruses (AIV), Marek's disease virus (MDV) or avian leukosis virus (ALV) can infect the intestine, but they are not considered intestinal pathogens in poultry.

3 Interactions between gut pathogens and microbiota and the impact on host nutrition and health

3.1 Escherichia

Colonization of *E. coli* starts in young turkeys and chickens (Leitner and Heller, 1992; Kemmett et al., 2013). Immediately after hatch, intestines were found to have a substantial number of coliform bacteria that increased with time (Leitner and Heller, 1992). Stress can increase pathogenic *E. coli* infection, leading to invasion of blood stream and cause morbidity and mortality. Along the intestinal tract, *E. coli* had a greater occurrence in the caecum than in the upper intestinal tract (Gong et al., 2007). However, using cultivation-based analysis, the crop was found to have more *E. coli* in broiler chickens (Collado and Sanz, 2007). It is also intriguing that phylogenetic group A was the most represented in old birds (21–42 days of age), whereas phylogenetic group B1 was the most represented in 1-day-old chicks (Pasquali et al., 2015).

Many studies have demonstrated a negative correlation between the abundance of *E. coli* and other gut microbes. All these studies used supplementation of one or more bacteria that were potentially antagonistic to *E. coli*. *Enterococcus faecium*, when supplemented to broiler chickens, led to a significant decrease in the population of *E. coli* in the ileal and caecal microbiota (Samli et al., 2010). Dehydrated *E. faecium* also significantly lowered *E. coli* population in broiler chickens challenged with *E. coli* K88 (Cao et al., 2013). A recent study also corroborated the antagonism of *E. faecium* against *E. coli* in broiler chickens as demonstrated by a linear decrease of *E. coli* counts in excreta (Lan et al., 2017). It was proposed that bacteriocin produced by certain strains of *E. faecium* might contribute to its antagonism (Laukova et al., 2015). Interestingly, the gut microbiota of broiler chickens infected with *C. jejuni* was altered with a significantly lower abundance of *E. coli* along the intestinal tract (Awad et al., 2016b). However, it is unknown if *C. jejuni* directly suppressed *E. coli* or *E. coli* was inhibited indirectly.

Numerous studies have evaluated different probiotics with their ability to suppress *E. coli*. These include *Bacillus licheniformis* (Song et al., 2014), *B. subtilis* (Song et al., 2014; Guo et al., 2017; Cheng et al., 2018; Manafi et al., 2017), *B. amyloliquefaciens* (Tsukahara et al., 2018a), *Lactobacillus plantarum* (Song et al., 2014; Wang et al., 2017), *L. acidophilus* (Forte et al., 2016) and *L. salivarius* (Shokryazdan et al., 2017). Mixtures of probiotic bacteria containing *Lactobacillus* species and other bacteria also exhibited antagonism towards *E. coli* (Pourakbari et al., 2016; Ceccarelli et al., 2017). A recent meta-analysis also showed that *E. coli* was among the taxa of category 3 that could be significantly decreased by direct-fed microbials (DFM) and probiotics (Heak et al., 2018). Competition or inhibition are among the antagonistic mechanisms.

Species of *Bdellovibrio*, such as *B. bacteriovorus*, can prey upon and kill Gram-negative bacteria, including the zoonotic pathogens *E. coli* and *Salmonella* (Yair et al., 2009). *E. coli* strains are used to grow *Bdellovibrio* spp. in a laboratory (Lambert et al., 2006; Atterbury et al., 2011b). Thus, *Bdellovibrio* should be able to reduce colonization by *E. coli* in the gut of poultry. In a preliminary study, we detected *Bdellovibrio* in the caecal content samples of chicken (unpublished data). However, no study was found in the literature that has demonstrated reduction of *E. coli* by *Bdellovibrio*, except one study that reported decreased *Enterobacteriaceae* and increased *Bdellovibrio* in chickens that received topinambur (Kleessen et al., 2003). Future studies are warranted to determine if *Bdellovibrio*, and species of *Bacteriovorax*, another group of bacteria that can prey on Gram-negative bacteria, can suppress *E. coli* in chickens.

Most studies focused on the antagonism of bacteria against *E. coli*, and only a few studies reported a positive correlation between *E. coli* and other microbes in the intestinal tract. Among bacteria, only *C. jejuni* challenge was shown to increase *E. coli* colonization (Bereswill et al., 2011; Kaakoush et al., 2014; Sakaridis et al., 2018) (see the *Campylobacter*, Section 3.4, for detail). Viruses, however, were shown to increase *E. coli* colonization, including AIV subtype H9N2 (H9N2 AIV) (Li et al., 2018a) and TCV (Pakpinyo et al., 2003). Damage to the host defense caused by viral infection is one likely mechanism by which these viruses predispose chickens to *E. coli* colonization.

Only a few studies have been reported in the literature that examined how colonization of *E. coli*, either commensal or pathogenic strains, affected several select gut bacteria in chickens. In one study, ETEC (*E. coli* K88) challenge at day 7 of age significantly increased total caecal *E. coli* while decreasing total lactobacilli at days 10 and 35 (Emami et al., 2017). In another challenge study, *E. coli* (not reported if commensal or pathogenic) inoculation significantly increased the population of caecal *E. coli*, coliforms and *Salmonella* (Manafi et al., 2017). The effect of *E. coli* colonization on the overall gut microbiota was not analysed in any of the above two studies. Conceivably, pathogenic *E. coli* can alter the gut microbiota to a greater extent due to damage to the gut barrier functions.

3.2 Salmonella

The relationships of *Salmonella* with other members of the gut microbiota have been studied from the perspectives of how they impact each other. Some of the earliest studies on competitive exclusion were in fact related to *Salmonella* colonization in chickens (Nurmi and Rantala, 1973; Rantala and Nurmi, 1973). Those studies showed that early establishment of the gut microbiota was accelerated by inoculating chicks with the intestinal content from healthy adult chickens and conveyed resistance to *Salmonella* Enteritidis infection. Presumably, this resistance to infection was underpinned by colonization resistance that entails both competition for nutrients and mucosal binding sites and production of inhibitors by commensal bacteria. Competitive exclusion served as the basis for the development of numerous probiotics in the poultry industry (Nurmi and Rantala, 1973; Hume et al., 1998; Menconi et al., 2011). Competition for nutrients as a major mechanism of competitive exclusion was later described by Freter et al. (1983) and termed the 'nutrient-niche hypothesis', which has been reviewed in the context of *Salmonella* Typhimurium by Rivera-Chávez and Bäumler (2015). This hypothesis posits that, for a particular microbial species to survive within a microbiota, it must have an advantage in nutrient acquisition that allows it to grow faster than other species. In the case of pathogens that grow to exceedingly high levels during the course of disease, it would be especially true that such a strategy would be essential, and any species or microbiota capable of removing that advantage from the gut environment may effectively limit pathogen establishment therein.

In monogastric animals, such as avian species, the phyla *Firmicutes* and *Bacteroidetes* exist at high levels in the distal gut, where they degrade nutrients and produce VFAs including acetate, propionate and butyrate (Wei et al., 2013). These VFAs can be utilized by the host as a source of carbon and energy and have antimicrobial activity, particularly owing to their ability to lower intestine pH, particularly in the caeca (Kashket, 1987; Van Immerseel et al., 2006). Indeed, VFA production in the caeca of chickens has long been known as a factor that contributes to colonization resistance (Barnes et al., 1979), and exposure of *Salmonella* to butyrate, propionate or various medium-chain fatty acids has been shown to decrease its ability to invade intestinal epithelial cells and downregulation of expression of *hilA*, a regulator of pathogenicity island 1 (Van Immerseel et al., 2004a,b). Similarly, inoculation of chicks at the day of hatch with a competitive exclusion culture increased VFA production in caeca and reduced the recovery of *Salmonella* as early as 4 h post-inoculation (Hume et al., 1998; Martin et al., 2000).

Other bacteria-derived mechanisms that inhibit *Salmonella* colonization in the gut of poultry include inhibitory substances, such as lactic acid and bacteriocins, that can either kill or slow the growth of the pathogen. Lactic acid bacteria (LAB) and *Bacillus* have been well studied for these properties (Menconi et al., 2011; Higgins et al., 2010; Knap et al., 2011; Wolfenden et al., 2011). Bacteriocins are ribosomally synthesized antibacterial peptides/proteins that can disrupt the cell membrane of their targets, and they are commonly produced among many bacteria, presumably as a competitive factor against other species to generate a selective advantage. Although not yet used to protect chicks from *Salmonella* colonization, bacteriocins have been shown to be effective inhibiting *Salmonella* (Portrait et al., 2000; Audisio et al., 2001; Line et al., 2008; Svetoch et al., 2008). As mentioned above, *B. bacteriovorus* can directly kill *Salmonella* (Dwidar et al., 2012; Yair et al., 2009). When orally administered, this predatory bacterium was able to lower the load of *Salmonella* and associated inflammation in the caeca of chicks (Atterbury et al., 2011a). Methods for selection of potential antimicrobial-substance-producing strains of bacteria exist, most of which involve identification of bacteria that reduce *Salmonella* growth *in vitro*, and some of them have been widely employed to develop probiotics in the poultry industry (Shivaramaiah et al., 2011a; Barbosa et al., 2005; Vicente et al., 2008; Bielke et al., 2012; Park and Kim, 2014). These methods will also aid new understanding of the interactions between *Salmonella* with other bacteria in the poultry gut.

Just as resident microbiota in the gut can impact *Salmonella* colonization, *Salmonella* can impact the resident microbiota. *Salmonella* has the ability to obtain and utilize the nutrients that give it a competitive advantage over other bacteria. These include acquisition or iron and sialic acid and ability to catabolize the available sugars and microbiota-derived H_2 (Rivera-Chavez and Bäumler, 2015; Ng et al., 2013; Maier et al., 2013). Low-level infection of chicks with a wild-type S. Typhimurium has been shown to lower the abundance of *Clostridia* compared with non-infected chicks, while a greater abundance of *Bacteroidia*, *Gammaproteobacteria* and *Verrucomicrobiae* were observed in *Salmonella*-infected birds than in non-infected birds (Park et al., 2017). Using principal coordinates analysis of sequences of 16S rRNA gene amplicons, the same study also showed that *Salmonella* infection altered the overall microbiota. Two other studies also demonstrated altered caecal microbiota in chicks infected with *Salmonella* (Juricova et al., 2013; Ballou et al., 2016). It was also shown that *Salmonella* infection decreased *Clostridiales* (Juricova et al., 2013) while increasing *Enterobacteriales* (Ballou et al., 2016). Interestingly, these changes were associated with delayed microbiota development, in which microbiota of *Salmonella*-infected chicks, at 16 days of age, clustered with that of younger non-infected birds. It was not determined if this was directly attributed to *Salmonella* or indirectly attributed to the inflammation caused by *Salmonella* infection (Juricova et al., 2013). In contrast, Videnska et al. (2013) did not find significant impact of *Salmonella* infection on the intestinal microbiota. Within the class *Clostridia*, increased *Lachnospiraceae* and decreased *Ruminococcaceae* were measured in *Salmonella*-infected chickens through day 28, and changes in caecal microbiota after administration at day hatch were similar to birds continuously fed probiotic, suggesting that *Salmonella* can have a profound and long-lasting influence on the intestinal microbiota (Ballou et al., 2016). These reports collectively show that *Salmonella* infection of young birds can affect the gut microbiota. As in the case of infection with other pathogens, the research challenge is to distinguish between the direct impact from indirect impact.

Salmonella infection in chickens, which is often asymptomatic, can cause mild enteric inflammation characterized by increased mRNA expression of pro inflammatory cytokines including interleukins (such as IL-6, IL-8, IL-12, IL-1β, lipopolysaccharides (LPS)-induced tumour necrosis alpha factor and interferon gamma) (Higgins et al., 2010; Kaiser et al., 2000; Withanage et al., 2005; Haghighi et al., 2008; Quinteiro-Filho et al., 2010; Setta et al., 2012; Arsenault et al., 2016). *Salmonella* infection in mammals also causes inflammation

(Eckmann and Kagnoff, 2001), but the specific cytokine responses are quite different between mammals and chickens. The type III secretion system (T3SS) of *Salmonella* contains virulence factors that promote host inflammation by invading the epithelium, and the inflammatory response in turn creates a unique nutrient environment that selectively promotes *Salmonella* growth (Rivera-Chavez and Bäumler, 2015; Arsenault et al., 2016; Collier-Hyams et al., 2002). This T3SS-guided invasion of host tissue is characterized by heterophil and lymphocyte infiltration of lamina propria, and granulocytes have been noted on the epithelial lining post-infection, which is consistent with increased iNOS mRNA expression (Van Immerseel et al., 2002; Berndt et al., 2007). Cell surface-expressed proteins, such as O-antigens and LPS, also play important roles in the pathogenesis of *Salmonella* by promoting colonization and adherence to avian epithelial cells, which would potentially promote the pro-inflammatory status of the gut (Guard-Petter et al., 1996). The invasion of epithelial cells is known to increase IL-8, which initiates chemotaxis of granulocytes, especially heterophils (Eckmann and Kagnoff, 2001).

In avian tissues, heterophils recognize components of *Salmonella*, such as LPS and flagellin, which result in the characteristic innate immune response of cytokines and other inflammatory mediators described above. Effects of *Salmonella* LPS on heterophil functional activation, as well as inflammatory gene expression, have been linked to the involvement of LPS-binding protein/CD14/toll-like receptor 4 complexes (LBP/CD14/TLR4), which is also associated with macrophage activation (Kogut et al., 2005). Binding of TLR4, an activator of the NF-κB pathway, by LPS increased oxidative burst, and the lack of recognition of LPS by this receptor may be attributed to increased susceptibility to *Salmonella* infection, due to reduced signalling pathways that regulate activation of heterophils. Heterophil activation by TLR4 binding has also been associated with upregulation and expression of IL-6, likely through activation of NF-κB and AP-1, which is a cytokine associated with innate inflammatory responses such as acute phase response and stimulation of B-cell growth (Kogut et al., 2008; Xie et al., 2000). This is consistent with increased acute phase protein levels in *Salmonella*-infected chickens, including alpha-1 acid glycoprotein, ovotransferrin and ceruloplasmin (Xie et al., 2000; Adler et al., 2001; Holt and Gast, 2002).

Inflammation caused by *Salmonella* can have effects beyond the gut. For example, *Salmonella* infection and the induced mucosal permeability of the gut were recently associated with trabecular bone loss and bone microarchitecture abnormalities in femurs (Raehtz et al., 2018). Other gut bacteria have been linked to bone diseases such as bacterial chondronecrosis with osteomyelitis (Wideman, 2016), enterococcal spondylitis (Borst et al., 2012) and turkey osteomyelitis complex (Huff et al., 2006), all of which begin with increased mucosal permeability in the gut allowing bacteria into the circulatory system. Experimentally, these diseases are initiated with some sort of stress or inflammation-inducing events, which emphasizes the importance of controlling inflammatory conditions within the gut, including *Salmonella* infection. Shivaramaiah et al. (2011b) observed that early infection with *Salmonella* Typhimurium resulted in higher incidence and mortality associated with *Eimeria*-induced NE, presumably as a result of inflammation processes induced by *Salmonella* that predisposed the gut to the overgrowth of *C. perfringens*. These examples illustrate that, while the greatest risk to poultry production with regard to *Salmonella* infection is that of a food safety concern, its impacts on poultry production efficiency and disease should also be considered important.

3.3 Clostridium

Interaction of *C. perfringens* with *Eimeria* spp. is an important factor in its pathogenesis in NE, the most costly disease caused by *Clostridium* spp., usually *C. perfringens*, in poultry. In addition, the influence of feed, feed additives and other management factors on *C. perfringens* colonization of poultry has been extensively investigated. However, only limited information is available in the literature about the interaction between *C. perfringens* and other bacteria in chickens, and much of the information was derived from studies where chickens were challenged with *C. perfringens*.

Infection with *C. perfringens* decreased bacterial diversity in the ileum, while it seemed to increase or not change the diversity in the caeca (Stanley et al., 2014; Li et al., 2017). Across all studies, *Enterobacteriaceae* and *Lactobacillaceae* were among the bacteria most affected by infections with *C. perfringens*. Several studies found an increase in the relative abundance of *Enterobacteriaceae* especially *Escherichia/Shigella*, including *E. coli*, in the ileum as well as in the other segments of the small intestine (McReynolds et al., 2004; Liu et al., 2010; Fasina et al., 2016; Li et al., 2017). Infection with *C. perfringens* also increased translocation of *E. coli* to the liver due to damaged gut barrier function (Liu et al., 2010). The increase of *Enterobacteriaceae* is probably linked to the influx of nutrients of host origin after epithelial lesions created by *C. perfringens*.

The effect on *Lactobacillaceae* varied among studies, gut segments and among lactobacilli species. Liu et al. (2010) found a decrease of the abundance of *Lactobacillus* spp. in the ilea of chickens infected with *C. perfringens*, similar to the results of Feng et al. (2010), who found a decreased abundance of *Lactobacillus* spp., especially *L. aviaries* and *L. salivarius*. In the caeca, there was a general decrease in lactobacilli, but different species exhibited different responses. While *L. johnsonii* and *L. fermentum* decreased in abundance, *L. crispatus*, *L. pontis*, *L. ultunese* and *L. salivarius* increased (Stanley et al., 2014). In contrast, Fasina et al. (2016) detected an increase in the relative abundance of total lactobacilli in the jejunum and ileum. Differences in methodologies, gut microbiota in different gut segments and interaction between individual *Lactobacillus* spp. and *C. perfringens* might be reasons for the above discrepancies. Conversely, supplementation of feed with *L. acidophilus* as a probiotic counteracted the changes in the intestinal microbiota caused by *C. perfringens* infection (Li et al., 2017). One plausible reason might be the production of antimicrobial peptides by LAB (Messaoudi et al., 2013). Given the importance of lactobacilli in the gut health, future research is warranted to examine the interactions between individual lactobacilli species and *C. perfringens*.

Other taxa of gut microbiota also changed in abundance after challenge with *C. perfringens*, including decrease of *Peptostreptococcaceae* in the ileum (Li et al., 2017); increase of *Ruminococcus bromii* and decrease of *Anaerostipes butyraticus*, *R. lactaris* in the caeca (Stanley et al., 2014); increase of *Weissella* spp., *Eubacterium halli* and *R. bromii* and decrease of *Clostridium viridae* and an unclassified order of *Mollicutes* that has been associated with irritable bowel syndrome in humans in the caeca (Stanley et al., 2014); and increase of *Streptococcaceae* and *Enterococcaceae* in the jejunum and ileum (Fasina et al., 2016). The abundance of *Clostridiales* was unaffected, but not surprisingly a shift in population towards *C. perfringens* was observed (Stanley et al., 2012). The microbiological underpinning and implication of the above effect remain to be determined.

Few studies have examined the impact of *C. perfringens* infection on the gut fermentation of poultry. In one study (Stanley et al., 2012), *C. perfringens* infection was accompanied with reduced relative abundance of short-chain fatty acid (SCFA)-producing bacteria including *Weissella confusa*, *E. halli* and other *Eubacterium* spp. Lactate concentration in the ileum was increased after infection with *C. perfringens*, while formate, acetate and propionate concentrations in the ileum and isobutyrate and isovalerate concentrations in the caeca were decreased or tended to decrease (Li et al., 2017). Future research is needed to examine the functional consequence of *C. perfringens* infection of poultry, such as fermentation profile, nutrition digestion in the caecum and microbial metabolome in the gut.

Besides the interactions between *C. perfringens* and the intestinal microbiota, the interaction between other clostridial species and gut microbiome remains to be explored. In the literature, only one study was found that assessed the effect of *C. difficile* on the faecal microbiota in chickens (Skraban et al., 2013). Using denaturing high-performance liquid chromatography, these authors found a significant or nearly significant negative correlation between the presence *Acidaminococcus intestini* or *Moniliella* sp. and *C. difficile* as well as a nearly positive correlation between the presence of *Enterococcus cecorum*, *Lactobacillus galinarum* or *Trichosporon asahii* and *C. difficile*. It should be noted that although *C. difficile* is an important common enteric pathogen in humans, it rarely causes infection in poultry.

3.4 Campylobacter

Campylobacter species can be found in the environment including air, litter and drinking water (Kazwala et al., 1990) and feed (Alves et al., 2017), and *Campylobacter* strains present in the environment gain access to and colonize the intestines. No evidence supports vertical transmission (van de Giessen et al., 1992; Callicott et al., 2006; Battersby et al., 2016; Jacobs-Reitsma, 1995). Caeca are the primary habitat of *Campylobacter* in chickens (Beery et al., 1988). Young chicks were found to be more susceptible to *Campylobacter* colonization than older birds. Indeed, *Campylobacter* inoculation at day 1 of hatch resulted in higher colonization rate and CFU recovery in the caecum of broiler chickens than of chickens infected at day 22 (Han et al., 2016). A smaller minimum *C. jejuni* cell number was also required to colonize 2-day-old chickens than to colonize 14-day-old chickens (5×10^3 vs. 5×10^4 cells) (Ringoir et al., 2007). One study using germ-free chicks or antibiotics-treated chickens also showed that a conventional gut microbiota decreased *Campylobacter* colonization of the ileum and caecum (Han et al., 2017). However, other studies showed that colonization with *Campylobacter* starts at 10–45 days of age (Goddard et al., 2014; Evans and Sayers, 2000). In a recent study (Connerton et al., 2018), broiler chickens infected with *Campylobacter* at 20 days became caecal colonized within 2 days, whereas birds infected at 6 days of age only showed partial colonization until 9 days post-infection. This contradicts the principle of colonization resistance. It remains to be confirmed if *Campylobacter* 'needs to wait' until the intestinal environmental (both microbial and physiological) conditions reach a certain stage before colonization.

Dietary factors can also affect *Campylobacter* colonization, with protein sources being a major one (Visscher et al., 2017). Reduced dietary protein content and supplementation with a mixture of amino acids significantly decreased *Campylobacter* colonization in broilers (Visscher et al., 2018). The reasons are not known, but the differences in proteins and amino acid sources probably change the gut microbiota (both composition and metabolism), which in turn affects *Campylobacter* colonization.

The ability of a given strain to colonize the broiler gut is dependent on the strain characteristics, the immune status and health of the bird and the composition of the gut microbiome (Qu et al., 2008). It was concluded that age and *C. jejuni* inoculation had a significant effect on lymphocyte numbers and cytokine expression levels in caecum as well as on gut microbiota composition (Qu et al., 2008). *Campylobacter* interacts with the host upon colonization of the gut in poultry (Awad et al., 2018). Readers interested in the interaction of *Campylobacter* with poultry are referred to a recent review by Awad et al. (2018). *Campylobacter* carrier state in chicken is characterized by multiple changes in the intestinal ecology (Awad et al., 2016b). The focus of this chapter is the microbial interaction between *Campylobacter* and the other members of the gut microbiota and the effect on the gut function.

One study using mice demonstrated that a diet-induced alteration of the intestinal microbiota, including an increase in the abundance of *E. coli* and a decrease in *Lactobacillus*, was associated with a greater susceptibility to *C. jejuni* infection (Bereswill et al., 2011). The reverse relationship between *Lactobacillus* and *Campylobacter* suggests antagonism of the former against the latter. Another study reported an association between higher *C. jejuni* abundance and lower abundance of *Lactobacillus* (Kaakoush et al., 2014). The same research group also showed that an increase in *E. coli* abundance and a decrease of *Lactobacillus* spp. were associated with a greater susceptibility to *C. jejuni* infection. Isolates of lactobacilli from chickens also suggest antagonism of lactobacilli against *Campylobacter* (Dec et al., 2018). In a recent study, CFU of LAB, *Enterobacteriaceae*, *E. coli* and total aerobic counts were found to be significantly higher in *Campylobacter* negative samples (Sakaridis et al., 2018). *Lactobacillaceae* were found to be lower in abundance when *Campylobacter* load was $>8^{10}$ CFU g^{-1}, and at that load *Enterobacteriaceae* reached its highest abundance. However, all the high-load *Campylobacter* samples ($>9^{10}$ CFU g^{-1}) were from the same farm, and farm-specific factors may not be ruled out (Sakaridis et al., 2018). In addition to *Lactobacillaceae*, the candidate genus *Clostridium* XIVa, which is in the family of *Lachnospiraceae*, also lost OTUs (Connerton et al., 2018).

Some taxa of bacteria were found to be positively correlated with *C. jejuni* colonization, including *Enterobacteriaceae* (Sakaridis et al., 2018) and *Clostridium perfringens* (Skanseng et al., 2006; Thibodeau et al., 2015). *Campylobacter* colonization was also accompanied with an increase in *Streptococcus* and *Blautia* in 56-day-old birds irrespective of farms and production types (Sofka et al., 2015). The positive correlation between *C. jejuni* and *C. perfringens* was speculated as mutualism, in which *C. jejuni* acts as a hydrogen sink, which improves growth of some *Clostridia* through increased fermentation (Kaakoush et al., 2014) and organic acid production, which in turn is used by *C. jejuni* as an energy source. However, further studies are needed to establish causality and to investigate whether the presence of *Campylobacter* favours the growth of those taxa or vice versa. It is possible that a general dysbacteriosis leads to colonization by both *Campylobacter* and *Enterobacteriaceae*. Similarly, infectious bursal disease virus, which is an immunosuppressive virus of chickens and allows secondary pathogens to invade or exacerbates their pathogenesis, increased *Campylobacter* colonization in the caecum of broiler chickens (Li et al., 2018b). Taken together,

Campylobacter colonization is associated with changes in certain members of intestinal microbiota, and those antagonistic interactions can be explored as probiotics or DFM to decrease *Campylobacter* colonization in chickens. Competitive exclusion of *Campylobacter* spp. from poultry using K-bacteria and Broilact (Aho et al., 1992; Schoeni and Wong, 1994; Mead et al., 1996) are examples of the use of antagonism to decrease *Campylobacter* infection, though less effective than against other intestinal pathogens such as *Salmonella*. As such, studies on the interaction between *Campylobacter* and gut microbiota will provide new and effective approaches to fight *Campylobacter* colonization and improve public health (Sakaridis et al., 2018).

The effects of *Campylobacter* infection on the gut microbiota of chickens have been assessed using high-throughput sequencing of 16S rRNA gene amplicons (Awad et al., 2016b; Sakaridis et al., 2018), and only one study used a metagenomic approach that revealed an increase in predominance of *Firmicutes* at the expense of *Bacteroidetes* and other phyla upon *C. jejuni* infection (Qu et al., 2008). However, that observation was based on only two chickens. Overall, *Campylobacter* colonization leads to a change in the intestinal microbiota (Haag et al., 2012; Sofka et al., 2015), especially the caecal microbiota, of broiler chickens (Thibodeau et al., 2015). No significant effect on species richness or evenness was reported as indicated by the Shannon-Weaner diversity index. This was also corroborated by another study (Sakaridis et al., 2018). Strong shifts in the bacterial microbiome were thought to be the leading cause for age-dependent colonization of *Campylobacter* in chickens (Han et al., 2016) and in mice (O'Loughlin et al., 2015). Similarly, Johansen et al. (2006) showed that *C. jejuni* colonization affected the development and the diversity and complexity of the caecal microbiota in chickens until 17 days of age. *Campylobacter* colonization was also found to have different effects of the microbiota in different intestinal segments, with *C. jejuni* infection affecting the development and complexity of the microbiotas of the caeca of broilers, but only transiently affecting the ileal microflora (Johansen et al., 2006). Besides alteration of luminal microbiota composition of both small and large intestines, the composition of mucosal microbiota of both small and large intestines was also altered (Awad et al., 2016b).

The *Campylobacter*-induced shifts of bacterial populations can affect metabolic end products derived from the intestinal microbiota in chickens. For example, Awad et al. (2016a) showed that *Campylobacter* infection lowered the concentration of some VFA, thereby raising the luminal pH of the jejunum and caecum of infected birds by approximately 0.5 pH units to ~6.7 (jejunum) and ~7.0 (caecum), which favours the growth of *Campylobacter* (Keener et al., 2004). As reviewed by Awad et al. (2018), *Campylobacter* infection can cause intestinal dysfunction, including change in gut barrier function, pro-inflammatory response, diarrhoea and decreased growth performance. In a recent study, however, early colonization of chicks (day 6) by *Campylobacter* only resulted in a transient growth rate reduction and pro-inflammatory response, although persistent modification of the caecal microbiota, while late colonization produced pro-inflammatory responses with changes in the caecal microbiota without affecting bird growth (Connerton et al., 2018). By the end of the growth cycle at day 35, no difference in growth or feed conversion was noted. As such, studies on the interaction between *Campylobacter* and gut microbiota and mitigation of *Campylobacter* colonization are primarily directed to benefit public health (Sakaridis et al., 2018).

3.5 Eimeria

Most of the *Eimeria* life cycle is intracellular, thereby the interaction between *Eimeria* and gut bacteria is mostly indirect, with direct interaction being minimal. The indirect interaction primarily results from the effect of *Eimeria* infection on host and gut function. Additionally, because the infectious damage to the epithelia can vary at different stages of the life cycle of *Eimeria* spp., the effect the parasitic infection has on the intestinal microbiota can vary significantly at different time points after the infection.

The most consequential interaction between *Eimeria* spp. and gut bacteria is implicated in the pathogenesis of NE. Infections with the coccidia, especially *E. maxima*, create a favourable environment for the growth of *C. perfringens*, predisposing the host for NE. The current dogma is that increased mucus secretion and leakage of plasma proteins after infection with *Eimeria* spp. provide substrates and nutrients for the growth of *C. perfringens*, which is auxotrophic for several amino acids (Timbermont et al., 2011). It is conceivable that the same mechanisms not only have an influence on the growth of *C. perfringens* but also other bacteria in the intestines.

Indeed, increased secretion of mucus, as measured by a larger goblet cell area and differential expression of mucin genes, has been shown after infection with *E. acervulina* and *E. maxima* (Collier et al., 2008; Forder et al., 2012). However, this was not followed by any increase in mucolytic bacteria, which was only observed after co-infection with the two *Eimeria* spp. and *C. perfringens* (Collier et al., 2008). Besides, a low-level leakage of plasma into the intestinal lumen during the first 5 days after infection with *E. acervulina*, *E. necatrix*, *E. maxima*, *E. brunetti* or *E. tenella* was demonstrated using a radioactively labelled marker that had been intravenously injected (Enigk et al., 1970; Joyner et al., 1975). Similarly, intravenous injection of a blue dye 2-6 days after infection with *E. acervulina* caused more intensive staining of the intestinal mucosa of the infected birds than that of uninfected controls (Preston-Mafham and Sykes, 1967).

Infection with *Eimeria* can impair feed digestion and nutrient absorption in the gut, thereby changing the nutrients and substrates including protein and other nutrients available to the intestinal microbiota, which in turn affects the gut microbiota. In early studies, radioactively labelled proteins were absorbed to a lesser degree 4 days after infections with *E. acervulina* (Joyner et al., 1975), 6 days after infection with *E. mitis* (Ruff and Edgar, 1982) and 14 days after infection with *E. necatrix* (Turk, 1972). In contrast, infection with *E. necatrix* or with *E. acervulina* increased absorption of proteins 14 and 28 days after infection whereas *E. brunetti* infection did not influence protein absorption (Turk, 1972). Glucose absorption was also impaired 6 days after the infection with *E. mitis* (Ruff and Edgar, 1982). More recently, a downregulation of the expression of amino acid transporters associated with the brush border membrane was observed 7 days after infection with *E. maxima*. Intriguingly, the basolateral transporters of amino acids were upregulated in the same birds (Fetterer et al., 2014).

Infections with *Eimeria* spp. can alter the pH, viscosity and passage rate of digesta through the alimentary channel. All these changes can have an influence on the gut microbiota. All published studies showed that infection with *Eimeria* spp., namely *E. tenella*, *E. acervulina*, *E. necatrix* and *E. maxima*, reduced the digesta passage rate (Aylott et al., 1968; McKenzie et al., 1982; Schildt and Herrick, 1955; Shane et al., 1985). However, results regarding the effect of *Eimeria* infections on digesta viscosity are inconclusive. Viscosity in the ileum but not in the jejunum was increased 21 days after infection with attenuated *E. acervulina*, *E. maxima*, *E. mitis* and *E. tenella* and decreased 1-4 days after co-infection with *E. maxima* and *C. perfringens* (Tsiouris et al., 2013). Co-infection with *E. acervulina* and *E. praecox* decreased the viscosity of the ileal digesta 7 days after the infection (Waldenstedt et al., 2000). The effect of infections with *Eimeria* spp. on the intestinal pH was dependent on the intestinal segments. While the pH in duodenum, jejunum and ileum tended to decrease after infection with *E. brunetti*, *E. mivati*, *E. maxima* and *E. necatrix* alone or in combination, the pH in the caeca tended to increase (Shane et al., 1985; Tsiouris et al., 2013; Stephens et al., 1974; Ruff and Reid, 1975; Ruff et al., 1975; Perez et al., 2011). The segment-dependent effect of *Eimeria* infection on digesta viscosity and pH can lead to a different impact on the gut microbiota in respective gut segments. However, no study found in the literature has investigated how *Eimeria*-induced alteration of digesta viscosity and passage rate affect gut microbiota.

Another plausible mechanism by which gut microbiota is affected indirectly by *Eimeria* is the modulation of the immune system upon *Eimeria* infection. However, this has not been demonstrated in chickens infected with *Eimeria* spp., but experiments using mice infected with cryptosporidia indicated the importance of this mechanism (Harp et al., 1992; Hu et al., 2014; Lacroix-Lamandé et al., 2014; Lantier et al., 2014). In chickens, mixed infection with *E. acervulina*, *E. maxima* and *E. brunetti* decreased the numbers of the immune-modulating bacterium *Candidatus* Savagella, while infection with *E. tenella* decreased the abundance of *Rikenellaceae*, which in humans is associated with inflammatory bowel disease (Stanley et al., 2014). Given the importance of the immune system on gut microbiota, future research is needed to elucidate how *Eimeria*-induced alteration of immune system affect gut microbiota.

Earlier studies focussed on the interaction between individual species of *Eimeria* and select bacterial taxa that were perceived important to bird health. Because *E. tenella* is the most virulent *Eimeria* spp. infecting chickens, most of the previous work investigated this species. This coccidia species was found to increase clostridia in the caeca during the first week after infection. Seven days after the infection and at later time points, *E. Tenella*-infected birds had lower clostridial counts in the caeca (Waldenstedt et al., 1998; Arakawa and Oe, 1975; Kimura et al., 1976). The results of these earlier experiments are summarized in Fig. 1. Apart from clostridia, some Gram-negative bacteria, such as *Bacteroidaceae* and *Enterobacteriaceae* including *E. coli*, tended to increase after infection with *E. tenella*, whereas a number of Gram-positive bacteria, including species of *Lactobacillus*, *Bifidobacterium*, *Catenibacterium*, *Peptostreptococcus*, *Streptococcus*, *Enterococcus* and *Staphylococcus*, tended to increase (Arakawa and Oe, 1975; Kimura et al., 1976; Johansson and Sarles, 1948; Bradley and Radhakrishnan, 1973; Stanley et al., 2014). Although only a few bacterial species were tested, studies using gnotobiotic chickens and infection with only one bacterial species corroborated the above observation.

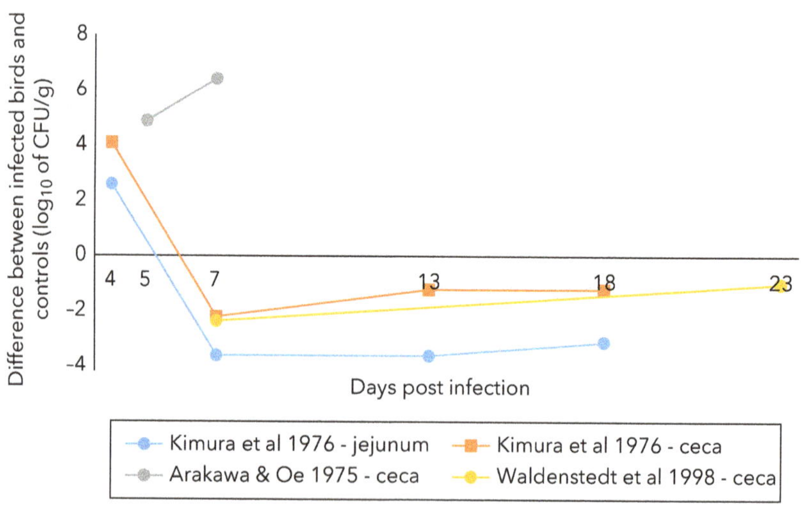

Figure 1 Temporal changes in clostridia counts in the intestines of chickens after *Eimeria tenella* infection according to results by Kimura et al. (1976), Arakawa and Oe (1975) and Waldenstedt et al (1998).

Compared to the non-infected control, clostridial counts in the caeca were eight fold higher in gnotobiotic chickens 6 days after the parasitic infection, while the numbers of *Bacteroides* spp. and *Streptococcus faecalis* decreased 5- to 100-fold (Dykstra and Reid, 1978). The mechanistic nature of these interactions, however, remains to be determined.

Importantly from a food safety point of view, *Salmonella* increased in the caeca of chickens infected with *E. tenella*. This effect could be seen for weeks after the infection and resulted in higher isolation rates of *Salmonella* from liver and eggshell (Arakawa et al., 1981; Takimoto et al., 1984; Qin et al., 1995a,b, 1996). Using gnotobiotic chickens and infection with *Salmonella*, it was found that *E. tenella* infection increased *Salmonella* in the caecum when *Salmonella* infection preceded *Eimeria* infection 6 or 8 days. In contrast, caecal *Salmonella* was decreased by *Eimeria* when the latter infected the gnotobiotic chickens before the bacterial infection (Fukata et al., 1984, 1987). This reflects the exclusion of *Salmonella* by *Eimeria*. Similar to *Salmonella*, *C. jejuni* was increased in the caeca of chickens infected with *E. tenella*. The increase was higher after infection with virulent, in contrast to attenuated *E. tenella* as well as after infection with a higher dose of the coccidia. Interestingly, translocation of *C. jejuni* to liver and spleen were decreased in *E. tenella*-infected chickens (Macdonald et al., 2018). In the case of *Campylobacter* infection, the immune response induced by the coccidia was hypothesized to reduce translocation of the bacteria, while more mucus in the caeca of infected chickens might promote their growth (Macdonald et al., 2018). It remains to be determined how *Eimeria* infection increases or decreases *Salmonella* in chickens in an infection order-dependent manner and if this order-dependent manner also applies to *Campylobacter*.

In regard to other *Eimeria* spp., mixed infection with *E. acervulina* and *E. praecox* decreased *C. perfringens* counts in the caeca 5 days or later after the parasitic infection (Waldenstedt et al., 2000). In contrast, infection with *E. necatrix* increased clostridial counts in duodenum, jejunum, ileum and caecum 3–7 days after the parasitic infection (Baba et al., 1997). *Salmonella* counts in the caecum were higher 14 days after infection with *E. acervulina* and 7 days after infection with *E. maxima*. *Salmonella* numbers in the liver were also increased (Takimoto et al., 1984). Different species of *Eimeria* may have a different effect on gut bacteria.

Recent research gravitates towards the interaction between *Eimeria* spp. and the entire gut microbiota. Typically, gut microbiota is profiled using high-throughput sequencing of 16S rRNA gene amplicons and compared before and after *Eimeria* infection. In one experiment, *Clostridium lacatifermentans* and various *Enterobacteriaceae were* increased in the caecal contents 7 days after a mixed infection with *E. acervulina*, *E. maxima* and *E. tenella*. Interestingly, *Campylobacter* was only detected in birds infected with the coccidia (Hume et al., 2011). A recent study showed decreased bacterial diversity in the caeca 5 and 7 days after infection with *E. tenella*, and the decrease was more pronounced on day 7 (Huang et al., 2018). In a similar study, bacterial diversity in the caeca remained unchanged 4 days after infection with *E. tenella*, but the caecal microbiota composition significantly changed (Macdonald et al., 2017). Other studies using amplicon sequencing of 16S rRNA genes or denaturing gradient gel electrophoresis demonstrated decreased bacterial diversity and homogeneity in the caeca 4 and 14 days after infection with *E. acervulina* alone or mixed infection with *E. acervulina*, *E. maxima* and *E. brunetti* (Stanley et al., 2014; Perez et al., 2011; Wu et al., 2014). Interestingly, the microbiota diversity in the ileum was not changed. This may reflect the intestinal segments that individual *Eimeria* species infect.

These recent studies using amplicon sequencing of 16S rRNA genes confirmed the decrease of Gram-positive, *Firmicutes*, especially *Lactobacillus* spp. and *Faecalibacterium* spp. in the caeca 5 and, to a greater extent, 7 days after infection with *E. tenella* (Macdonald et al., 2017; Huang et al., 2018). Also in line with the early findings, *Clostridium* spp. as well as *Escherichia* spp. and other species of *Enterobacteriaceae* increased in relative abundance upon *E. tenella* infection. It was further noted that although relative abundance of bacteria changed in birds that did not develop lesions, the greatest changes were detected in chickens with severe lesions (Macdonald et al., 2017). This corroborates the above premise that the interaction between *Eimeria* and gut microbiota is largely indirect and the intestinal damage caused by *Eimeria* infection is a primary driver that affects the gut microbiota. Mixed infection with *E. acervulina*, *E. maxima* and *E. brunetti* reduced the relative abundance of *Ruminococcaceae* and *Lachnospiraceae*, while *Clostridium* spp. gained in abundance (Wu et al., 2014). However, another study that used the same triple *Eimeria* infection showed inconsistent results in regard to *Clostridium* spp., with the relative abundance of some species such as *C. aldrichii* and *C. lituseburense* decreasing while that of others including *C. leptum* and *C. celerecrescens* increased (Stanley et al., 2014). Differences in methodologies and gut microbiota between these studies may be among the reasons for the discrepancies.

Germ-free and gnotobiotic chickens were used in the early studies to investigate the influence that select bacteria might have on the outcome of infections with *Eimeria tenella*. The results from comparing germ-free chickens with chickens with normal bacterial microbiota were inconsistent, sometimes showing less severe lesions or a delayed lesion development in germ-free birds than the conventional chickens, while other experiments detected no differences (Bradley and Radhakrishnan, 1973; Clark et al., 1962; Visco and Burns, 1972a). Some of the observed inter-study differences seemed to be due to the use of different chicken lines (Visco and Burns, 1972b). Interestingly, some studies revealed fewer first- and second-generation schizonts in germ-free chickens than conventional chickens, while the number of shed oocysts did not differ (Visco and Burns, 1972a).

The co-infection of germ-free chickens with *E. tenella* and *E. coli*, *C. perfringens*, *Staph. epidermidis*, *Str. faecalis* or one *Lactobacillus* sp. did not produce less severe lesions compared to the birds with normal bacterial microbiota, indicating that severity of intestinal lesion depends only to a minor degree on the interaction between the species of *Eimeria* and bacteria involved (Bradley and Radhakrishnan, 1973; Visco and Burns, 1972c). Similar results were reported in studies that used *E. brunetti* and *E. acervulina* (Hegde et al., 1969; Ruff et al., 1975). In contrast, several probiotics reduced lesions caused by *Eimeria* spp. Examples include *Bacillus* spp. against *E. maxima* (Lee et al., 2010); *Enterococcus faecium*, *Bifidobacterium animalis* and *Lactobacillus salivarius* against a mixed infection with *E. acervulina*, *E. maxima* and *E. tenella* (Giannenas et al., 2014; Ritzi et al., 2014, 2016); as well as *Lactobacillus acidophilus*, *L. fermentum*, *L. plantarum* and *Enterococcus faecium* against *E. tenella* (Chen et al., 2016). *Bacillus amyloliquefaciens* reduced lesions and oocyst counts after infection with *E. tenella* and *E. maxima* (Tsukahara et al., 2018b). It seems that pathogenic bacteria may not aggravate lesions caused by *Eimeria*, while probiotic bacteria can alleviate such lesions. The protective ability of these probiotics against *Eimeria* may be attributed to several factors. In one *in vitro* study, some strains of *E. faecium* and *Lactobacillus* spp. were shown to inhibit invasion of *E. tenella* sporozoites into host epithelial cells (Hessenberger et al., 2016). It was speculated that bacteria formed non-specific barriers by adhering to the epithelial cells, protecting the cells from invasion by the parasite. Intriguingly, some bacteria exhibited protection against invasion of host epithelial cells by *E. tenella* sporozoites even after being killed (Hessenberger et al., 2016). On the other hand, the protective effect of *B. amyloliquefaciens* was correlated with lower counts of *C. perfringens* and *E. coli* and increased lactobacilli, but it remains undetermined if this was the cause or the consequence of reduced lesions and oocyst counts (Tsukahara et al., 2018b). Future research is needed to elucidate the mechanisms by which probiotics protect against *Eimeria* infection.

With most studies focused on how an *Eimeria*-bacteria interaction affects select bacterial species and intestinal lesion, very few studies have examined how the interaction between *Eimeria* and gut microbiota affects gut fermentation. Mixed infection with *E. acervulina*, *E. maxima* and *E. brunetti* increased concentrations of butyric acid, isovaleric acid and isobutyric acid in caecal content while decreasing the concentration of acetic acid 4 days after infection (Stanley et al., 2014; Wu et al., 2014). It is likely that these changes in the concentration of SCFAs are caused by altered microbiota, rather than metabolites produced from *Eimeria*. This is because *Eimeria* performs a low degree of fermentation and respiration compared to the other microbes of the gut microbiota. Indeed, intestinal pH changed in conventional chickens, but not germ-free chickens, after infection with some, although not all, *Eimeria* spp. (Ruff et al., 1975). The functional consequences of changed fermentation profiles after *Eimeria* infection remain to be determined.

3.6 Viruses

A number of studies investigated the epidemiology of enteric viruses in chicken and turkey flocks, looking for correlation of the prevalence of the viruses with disease, lesions or age using PCR (Jindal et al., 2009; Palade et al., 2011; Koo et al., 2013; Mor et al., 2013; Moura-Alvarez et al., 2013, 2014; Domańska-Blicharz et al., 2017; Veen et al., 2017). In other studies, the enteric virome of RNA viruses was determined by metagenomic sequencing (Day and Zsak, 2015; Shah et al., 2016), including virome and bacterial microbiome being investigated in the same birds (Day et al., 2015). While co-infections with more than one enteric virus are frequent, there seems to be a lack of analysis of the correlation between different viruses. Additionally, investigating the bacterial microbiota after experimental infections with enteric viruses and vice versa has been neglected. One series of experiments tested inocula containing various bacteria and/or viruses for their ability to reproduce runting-stunting syndrome, but there was no attempt to isolate or re-isolate the bacteria or viruses from the infected chickens (Montgomery et al., 1997).

Only in recent years, inter-kingdom interactions between viruses, however not enteric viruses, and intestinal microbiota were explored. As in the case of *Eimeria* spp., it is reasonable to assume that the interaction is indirect, primarily mediated by the host response to the infection. Compared to non-infected controls, chickens infected with AIV (H9N2) increased in relative abundance of the phylum *Proteobacteria* and the genera *Vampirovibrio*, *Pseudoflavonifractor*, *Ruminococcus*, *Clostridium* cluster XIVb and *Isobaculum* in the faeces, whereas the relative abundance of the genera *Novosphingobium*, *Sphingomonas*, *Bradyrhizobium* and *Bifidobacterium* was decreased (Yitbarek et al., 2018). There was no difference in the bacterial diversity between infected chickens and controls (Yitbarek et al., 2018). In a similar experiment, infection with AIV (H9N2) increased the relative abundance of *E. coli* and decreased that of bacteria regarded as beneficial including *Lactobacillus* spp. and *Enterococcus* spp. in the ileum (Li et al., 2018a).

Changes in faecal and caecal bacterial microbiota after infection with MDV varied on different days after the infection. In the faeces of infected chickens, differences at the phylum level were only apparent during the first 7 days after infection, when *Proteobacteria* lost relative abundance while *Firmicutes* gained relative abundance over time. Interestingly, the opposite was observed in the non-infected controls. On later days until the end of the experiment 35 days after the infection, differences were only noted at the genus level. Interestingly,

Lactobacillus spp., and also *Haemophilus* spp., were only detected in the faeces of infected chickens, while *Salmonella* had a much higher relative abundance in non-infected birds. Similar differences were seen in caecal samples (Perumbakkam et al., 2014). A follow-up study confirmed the increase in the relative abundance of *Lactobacillus* spp. in the caeca (Perumbakkam et al., 2016). The latter experiment also found changes in the microbiome and predicted metabolic profiles after MDV infection in birds resistant or susceptible to MDV. In the resistant chickens, amino acid metabolism seemed impacted, while in the susceptible birds the lipid metabolism was changed. Differences in the immune response against MDV between these two lines were implicated in the changes of the microbiota (Perumbakkam et al., 2016). Infection with ALV-J and ALV-K increased microbial diversity in the caeca 21 days after the infection compared to non-infected controls. However, the two ALV types influenced the caecal microbiota differently. Bacteria with increased relative abundance in ALV-J-infected chickens included *Escherichia-Shigella*, *Enterococcus* spp. as well as species in the families *Erysipelotrichaceae* and *Helicobacteraceae*. In ALV-K-infected chickens, however, opportunistic pathogens of the phylum *Firmicutes* such as *Staphylococcus* spp. and *Weissella* spp. and some *Bacillales* increased in relative abundance (Ma et al., 2017).

While inter-kingdom interactions between viruses and bacteria have just started to gain attention, interactions between viruses and parasites have been investigated for more than 30 years. Co-infection of *E. acervulina*, *E. mitis* or *E. maxima* with different reovirus isolates brought inconsistent results. In some combinations, the virus increased the frequency of stunting and decreased weight gain compared to chickens only infected with the *Eimeria*, while in other combinations the virus infection did not influence the severity of the disease. Interestingly, one virus reduced the frequency of stunting and increased weight gain compared to chickens only infected with the *E. mitis* or *E. maxima*. However, this effect was observed only in one of two tested chicken lines. Co-infection with the viruses did not influence the severity of the lesions or the shedding of oocysts (Ruff and Rosenberger, 1985a,b). However, co-infection of chickens with *Cryptosporidium baileyi* and two different reovirus isolates seemed to increase shedding of *C. baileyi* as well as of the virus, while birds infected with both pathogens did not develop more severe lesions than with either one alone (Guy et al., 1988). These virus-parasite interactions may be related to competition of the organisms for intra cellular space within the host for reproduction and growth.

4 Summary and future trends

The intestinal pathogens discussed in this chapter can cause costly diseases. They interact with the commensal members of the gut microbiota, resulting in reciprocal effects. The occurrence of a specific pathogen can vary considerably among different birds within the same flock. Such differences can be associated, at least partially, with the variations of the gut environment and microbiota. The overall microbiota and the types and intensity of the interactions among intestinal pathogens and other members of the gut microbiota are important factors that determine the susceptibility or resistance to the colonization of intestinal pathogens. The gut bacteria that exhibit negative correlation can be potentially inhibitory to the pathogens, and they may be further investigated as indicators or biomarkers of 'healthy' gut microbiota, or candidates of probiotics. Although not being demonstrated yet in poultry, in humans, the intestinal colonization with *Campylobacter* seemed to be effectively prevented or decreased by a 'healthy' gut microbiota (Dicksved et al., 2014). Besides the one-to-one correlation analysis currently used in the reported studies, network analysis and mathematical modelling empowered by machine learning, such as random forest and least absolute shrinkage and selection operator (LASSO), can help reveal the potential multidimensional interactions, including commensalism, mutualisms and amensalism, and such high-dimensional relationship can help further understanding the factors that affect the susceptibility of birds to intestinal pathogen infections. Potential biomarkers or predictors of the risk of pathogen infection can also be identified.

Most of the research reported in the literature only examined the occurrence and abundance of pathogens and the gut microbiota with respect to its taxonomic diversity, composition and structure. Because of the functional redundancy of gut microbiota, functional analysis, including metagenome, metatranscriptome and metabolome, will be needed to better characterize the interactions between pathogens and the gut microbiota as a whole. Such functional analyses, together with research on the ecology and physiology of the pathogens of interest, will also allow for mechanistic understanding of the interactions that affect the occurrence and clinical consequence of pathogen colonization and infection, which in turn will help knowledge-based development of new strategies for effective control of the occurrence and spread of the pathogens without the use of antibiotics.

Most of the research reported hitherto used young birds challenged with the pathogen of interest to investigate the interaction between the pathogen and the intestinal microbiota. Although comparative analyses between the controls and the challenged birds provided useful information on the pathogen-microbiota interactions, this approach created an artificial system. Future research that does not use pathogen challenge, but naturally occurring pathogens, shall help reveal the actual interactions that happen in commercial flocks. Such research will provide more pertinent information useful to control intestinal pathogens.

The phasing out of antimicrobial growth promoters calls for non-antibiotic alternatives to control the occurrence and spread of those pathogens. The complexities associated with each pathogen, alone, demonstrates the intricate nature of the gut and importance of maintaining homeostasis to prevent disease conditions. Very minor changes in the gut microbiota or host conditions can shift the environment to favour pathogens that have amazing ability to create advantageous conditions for their survival and growth, which can have long-lasting effects on the gut. While not unreasonable to expect parasitic infections, such as infection with *Eimeria*, to affect the gut microbiota, measuring and mapping these effects become multifaceted as one considers the plethora of downstream effects. Microbiomics empowered by omics technologies, such as metagenomics, metatranscriptomics and metabolomics, will allow research that is mechanistic and insightful into the interactions between gut pathogens and commensal members of the microbiota in poultry.

5 Where to look for further information

Intestinal microbiota research, including the research on poultry gut microbiota and the interactions between gut pathogens and the commensal microbes, advances rapidly, and the most up-to-date scientific information and new learning are published in numerous scientific journals, including microbiology journals and poultry science journals. The following journals are among the journals that publish research in the area discussed in this chapter:

5.1 Scientific journals

Applied and Environmental Microbiology (ASM, ISSN: 0099-2240)
Beneficial Microbes (Wageningen Academic Publishers, 18762883)
British Poultry Science (Taylor & Francis, ISSN: 14661799)
European Poultry Science (Verlag Eugen Ulmer, ISSN: 16129199)
Gut Microbes (Landes BioScience, ISSN: 19)
International Journal of Poultry Science (Asian Network for Scientific Information, ISSN: 16828356)
Journal of Animal Physiology and Animal Nutrition (Wiley, ISSN: 1439-0396)
Journal of Applied Poultry Research (Oxford University Press, ISSN: 15370437)
Journal of Poultry Science (Japan Poultry Science Association, ISSN: 13467395)
Microbiome (BioMed Central, ISSN: 20492618)
PLoS ONE (PLoS ONE, eISSN: 1932-6203)
Poultry Science (Oxford University Press, ISSN: 0032-5791)
World's Poultry Science Journal (Cambridge University Press, ISSN: 00439339)

5.2 Major scientific conferences

Poultry Science Association Annual Meeting, USA
World Veterinary Poultry Association Congress
World's Poultry Congress
International Production & Processing Expo

5.3 Books

The Chicken Health Handbook, 2nd Edition: A Complete Guide to Maximizing Flock Health and Dealing with Disease, Second Edition, by Gail Damerow. ISBN-13: 978-1612124797.

6 References

Adler, K. L., Peng, P. H., Peng, R. K. and Klasing, K. C. 2001. The kinetics of hemopexin and alpha1-acid glycoprotein levels induced by injection of inflammatory agents in chickens. *Avian Dis.* 45(2), 289–96. doi:10.2307/1592967.

Aho, M., Nuotio, L., Nurmi, E. and Kiiskinen, T. 1992. Competitive exclusion of campylobacter from poultry with k-bacteria and broilact. *Int. J. Food Microbiol.* 15(3–4), 265–75. doi:10.1016/0168-1605(92)90057-A.

Alves, M. B., Fonseca, B. B., Melo, R. T., Mendonça, E. P., Nalevaiko, P. C., Girão, L. C., Monteiro, G. P., Silva, P. L. and Rossi, D. A. 2017. Feed can be a source of *Campylobacter jejuni* infection in broilers. *Br. Poult. Sci.* 58(1), 46–9. doi:10.1080/00071668.2016.1258691.

Arakawa, A. and Oe, O. 1975. Reduction of *Clostridium perfringens* by feed additive antibiotics in the ceca of chickens infected with *Eimeria tenella*. *Poult. Sci.* 54(4), 1000–07. doi:10.3382/ps.0541000.

Arakawa, A., Baba, E. and Fukata, T. 1981. *Eimeria tenella* infection enhances *Salmonella typhimurium* infection in chickens. *Poult. Sci.* 60(10), 2203–09. doi:10.3382/ps.0602203.

Arsenault, R. J., Genovese, K. J., He, H., Wu, H., Neish, A. S. and Kogut, M. H. 2016. Wild-type and mutant avra–*Salmonella* induce broadly similar immune pathways in the chicken ceca with key differences in signaling intermediates and inflammation. *Poult. Sci.* 95(2), 354–63. doi:10.3382/ps/pev344.

Atterbury, R. J., Hobley, L., Till, R., Lambert, C., Capeness, M. J., Lerner, T. R., Fenton, A. K., Barrow, P. and Sockett, R. E. 2011a. Effects of orally administered *Bdellovibrio bacteriovorus* on the well-being and *Salmonella* colonization of young chicks. *Appl. Environ. Microbiol.* 77(16), 5794–803. doi:10.1128/AEM.00426-11.

Atterbury, R. J., Hobley, L., Till, R., Lambert, C., Capeness, M. J., Lerner, T. R., Fenton, A. K., Barrow, P. and Sockett, R. E. 2011b. Effects of orally administered *Bdellovibrio bacteriovorus* on the well-being and *Salmonella* colonization of young chicks. *Appl. Environ. Microbiol.* 77(16), 5794–803. doi:10.1128/AEM.00426-11.

Audisio, M. C., Oliver, G. and Apella, M. C. 2001. Effect of different complex carbon sources on growth and bacteriocin synthesis of *Enterococcus faecium*. *Int. J. Food Microbiol.* 63(3), 235–41. doi:10.1016/S0168-1605(00)00429-3.

Awad, W. A., Aschenbach, J. R., Ghareeb, K., Khayal, B., Hess, C. and Hess, M. 2014. *Campylobacter jejuni* influences the expression of nutrient transporter genes in the intestine of chickens. *Vet. Microbiol.* 172(1–2), 195–201. doi:10.1016/j.vetmic.2014.04.001.

Awad, W. A., Smorodchenko, A., Hess, C., Aschenbach, J. R., Molnár, A., Dublecz, K., Khayal, B., Pohl, E. E. and Hess, M. 2015. Increased intracellular calcium level and impaired nutrient absorption are important pathogenicity traits in the chicken intestinal epithelium during *Campylobacter jejuni* colonization. *Appl. Microbiol. Biotechnol.* 99(15), 6431–41. doi:10.1007/s00253-015-6543-z.

Awad, W. A., Dublecz, F., Hess, C., Dublecz, K., Khayal, B., Aschenbach, J. R. and Hess, M. 2016a. *Campylobacter jejuni* colonization promotes the translocation of *Escherichia coli* to extra-intestinal organs and disturbs the short-chain fatty acids profiles in the chicken gut. *Poult. Sci.* 95(10), 2259–65. doi:10.3382/ps/pew151.

Awad, W. A., Mann, E., Dzieciol, M., Hess, C., Schmitz-Esser, S., Wagner, M. and Hess, M. 2016b. Age-related differences in the luminal and mucosa-associated gut microbiome of broiler chickens and shifts associated with *Campylobacter jejuni* infection. *Front. Cell. Infect. Microbiol.* 6, 154. doi:10.3389/fcimb.2016.00154.

Awad, W. A., Hess, C. and Hess, M. 2018. Re-thinking the chicken-*Campylobacter jejuni* interaction: a review. *Avian Pathol.* 47(4), 352–63. doi:10.1080/03079457.2018.1475724.

Aylott, M. V., Vestal, O. H., Stephens, J. F. and Turk, D. E. 1968. Effect of coccidial infection upon passage rates of digestive tract contents of chicks. *Poult. Sci.* 47(3), 900–04. doi:10.3382/ps.0470900.

Baba, E., Ikemoto, T., Fukata, T., Sasai, K., Arakawa, A. and McDougald, L. R. 1997. Clostridial population and the intestinal lesions in chickens infected with *Clostridium perfringens* and *Eimeria necatrix*. *Vet. Microbiol.* 54(3–4), 301–08. doi:10.1016/S0378-1135(96)01289-8.

Ballou, A. L., Ali, R. A., Mendoza, M. A., Ellis, J. C., Hassan, H. M., Croom, W. J. and Koci, M. D. 2016. Development of the chick microbiome: how early exposure influences future microbial diversity. *Front. Vet. Sci.* 3, 2. doi:10.3389/fvets.2016.00002.

Barbosa, T. M., Serra, C. R., La Ragione, R. M., Woodward, M. J. and Henriques, A. O. 2005. Screening for *Bacillus* isolates in the broiler gastrointestinal tract. *Appl. Environ. Microbiol.* 71(2), 968–78. doi:10.1128/AEM.71.2.968-978.2005.

Barnes, E. M., Impey, C. S. and Stevens, B. J. H. 1979. Factors affecting the incidence and anti-*Salmonella* activity of the anaerobic caecal flora of the young chick. *J. Hyg.* 82(2), 263–83. doi:10.1017/S0022172400025687.

Battersby, T., Whyte, P. and Bolton, D. J. 2016. The pattern of *Campylobacter* contamination on broiler farms; external and internal sources. *J. Appl. Microbiol.* 120(4), 1108–18. doi:10.1111/jam.13066.

Beery, J. T., Hugdahl, M. B. and Doyle, M. P. 1988. Colonization of gastrointestinal tracts of chicks by *Campylobacter jejuni*. *Appl. Environ. Microbiol.* 54(10), 2365–70.

Bereswill, S., Plickert, R., Fischer, A., Kühl, A. A., Loddenkemper, C., Batra, A., Siegmund, B., Göbel, U. B. and Heimesaat, M. M. 2011. What you eat is what you get: novel *Campylobacter* models in the quadrangle relationship between nutrition, obesity, microbiota and susceptibility to infection. *Eur. J. Microbiol. Immunol.* 1(3), 237–48. doi:10.1556/EuJMI.1.2011.3.8.

Berndt, A., Wilhelm, A., Jugert, C., Pieper, J., Sachse, K. and Methner, U. 2007. Chicken cecum immune response to *Salmonella enterica* serovars of different levels of invasiveness. *Infect. Immun.* 75(12), 5993–6007. doi:10.1128/IAI.00695-07.

Best, A., La Ragione, R. M., Cooley, W. A., O'Connor, C. D., Velge, P. and Woodward, M. J. 2003. Interaction with avian cells and colonisation of specific pathogen free chicks by shiga-toxin negative *Escherichia coli* o157:H7 (nctc 12900). *Vet. Microbiol.* 93(3), 207–22. doi:10.1016/S0378-1135(03)00031-2.

Bielke, L. R., Tellez, G. and Hargis, B. M. 2012. Successes and failures of bacteriophage treatment of *Enterobacteriaceae* infections in the gastrointestinal tract of domestic animals. In: Kurtboke, I. (Ed.), *Bacteriophages*. InTech, Rijeka, Croatia.

Borst, L. B., Suyemoto, M. M., Robbins, K. M., Lyman, R. L., Martin, M. P. and Barnes, H. J. 2012. Molecular epidemiology of *Enterococcus cecorum* isolates recovered from enterococcal spondylitis outbreaks in the southeastern United States. *Avian Pathol.* 41(5), 479–85. doi:10.1080/03079457.2012.718070.

Bradley, R. E. and Radhakrishnan, C. V. 1973. Coccidiosis in chickens: obligate relationship between *Eimeria tenella* and certain species of cecal microflora in the pathogenesis of the disease. *Avian Dis.* 17(3), 461–76. doi:10.2307/1589145.

Brenner, F. W., Villar, R. G., Angulo, F. J., Tauxe, R. and Swaminathan, B. 2000. *Salmonella* nomenclature. *J. Clin. Microbiol.* 38(7), 2465–67.

Callicott, K. A., Friethriksdottir, V., Reiersen, J., Lowman, R., Bisaillon, J. R., Gunnarsson, E., Berndtson, E., Hiett, K. L., Needleman, D. S. and Stern, N. J. 2006. Lack of evidence for vertical transmission of *Campylobacter* spp. in chickens. *Appl. Environ. Microbiol.* 72(9), 5794–8. doi:10.1128/AEM.02991-05.

Cao, G. T., Zeng, X. F., Chen, A. G., Zhou, L., Zhang, L., Xiao, Y. P. and Yang, C. M. 2013. Effects of a probiotic, *Enterococcus faecium*, on growth performance, intestinal morphology, immune response, and cecal microflora in broiler chickens challenged with *Escherichia coli* k88. *Poult. Sci.* 92(11), 2949–55. doi:10.3382/ps.2013-03366.

CDC. 2018. Outbreaks involving *Salmonella*. Available at: https://www.Cdc.Gov/salmonella/outbreaks.html (accessed on 25 October 2018).

Ceccarelli, D., van Essen-Zandbergen, A., Smid, B., Veldman, K. T., Boender, G. J., Fischer, E. A. J., Mevius, D. J. and van der Goot, J. A. 2017. Competitive exclusion reduces transmission and excretion of extended-spectrum-beta-lactamase-producing *Escherichia coli* in broilers. *Appl. Environ. Microbiol.* 83(11). doi:10.1128/AEM.03439-16.

Chen, C. Y., Chuang, L. T., Chiang, Y. C., Li Lin, C., Lien, Y. Y. and Tsen, H. Y. 2016. Use of a probiotic to ameliorate the growth rate and the inflammation of broiler chickens caused by *Eimeria tenella* infection. *J. Anim. Res. Nutr.* 01(2), 10. doi:10.21767/2572-5459.100010.

Cheng, Y. H., Zhang, N., Han, J. C., Chang, C. W., Hsiao, F. S. and Yu, Y. H. 2018. Optimization of surfactin production from *Bacillus subtilis* in fermentation and its effects on *Clostridium perfringens*-induced necrotic enteritis and growth performance in broilers. *J. Anim. Physiol. Anim. Nutr. (Berl)* 102(5), 1232–44. doi:10.1111/jpn.12937.

Chlebicz, A. and Slizewska, K. 2018. Campylobacteriosis, salmonellosis, yersiniosis, and listeriosis as zoonotic foodborne diseases: a review. *Int. J. Env. Res. Public Health* 15(5). doi:10.3390/ijerph15050863.

Clark, D., Dardas, R. and Smith, C. 1962. Pathological and immunological changes in gnotobiotic chickens due to *Eimeria tenella*. *Poult. Sci.* 41, 1635–36.

Collado, M. C. and Sanz, Y. 2007. Characterization of the gastrointestinal mucosa-associated microbiota of pigs and chickens using culture-based and molecular methodologies. *J. Food Prot.* 70(12), 2799–804. doi:10.4315/0362-028X-70.12.2799.

Collier, C. T., Hofacre, C. L., Payne, A. M., Anderson, D. B., Kaiser, P., Mackie, R. I. and Gaskins, H. R. 2008. Coccidia-induced mucogenesis promotes the onset of necrotic enteritis by supporting *Clostridium perfringens* growth. *Vet. Immunol. Immunopathol.* 122(1-2), 104–15. doi:10.1016/j.vetimm.2007.10.014.

Collier-Hyams, L. S., Zeng, H., Sun, J., Tomlinson, A. D., Bao, Z. Q., Chen, H., Madara, J. L., Orth, K. and Neish, A. S. 2002. Cutting edge: *Salmonella* avra effector inhibits the key proinflammatory, anti-apoptotic NF-kappa B pathway. *J. Immunol.* 169(6), 2846–50. doi:10.4049/jimmunol.169.6.2846.

Connerton, P. L., Richards, P. J., Lafontaine, G. M., O'Kane, P. M., Ghaffar, N., Cummings, N. J., Smith, D. L., Fish, N. M. and Connerton, I. F. 2018. The effect of the timing of exposure to *Campylobacter jejuni* on the gut microbiome and inflammatory responses of broiler chickens. *Microbiome* 6(1), 88. doi:10.1186/s40168-018-0477-5.

Crawshaw, T. R., Chanter, J. I., Young, S. C., Cawthraw, S., Whatmore, A. M., Koylass, M. S., Vidal, A. B., Salguero, F. J. and Irvine, R. M. 2015. Isolation of a novel thermophilic *Campylobacter* from cases of spotty liver disease in laying hens and experimental reproduction of infection and microscopic pathology. *Vet. Microbiol.* 179(3–4), 315–21. doi:10.1016/j.vetmic.2015.06.008.

Crespo, R., Fisher, D. J., Shivaprasad, H. L., Fernández-Miyakawa, M. E. and Uzal, F. A. 2007. Toxinotypes of *Clostridium perfringens* isolated from sick and healthy avian species. *J. Vet. Diagn. Invest.* 19(3), 329–33. doi:10.1177/104063870701900321.

Day, J. M. and Zsak, L. 2013. Recent progress in the characterization of avian enteric viruses. *Avian Dis.* 57(3), 573–80. doi:10.1637/10390-092712-Review.1.

Day, J. M. and Zsak, L. 2015. Investigating turkey enteric picornavirus and its association with enteric disease in poults. *Avian Dis.* 59(1), 138–42. doi:10.1637/10940-092414-RegR.

Day, J. M., Oakley, B. B., Seal, B. S. and Zsak, L. 2015. Comparative analysis of the intestinal bacterial and RNA viral communities from sentinel birds placed on selected broiler chicken farms. *PLOS ONE* 10(1), e0117210. doi:10.1371/journal.pone.0117210.

Dec, M., Nowaczek, A., Urban-Chmiel, R., Stępień-Pyśniak, D. and Wernicki, A. 2018. Probiotic potential of *Lactobacillus* isolates of chicken origin with anti-*Campylobacter* activity. *J. Vet. Med. Sci.* 80(8), 1195–203. doi:10.1292/jvms.18-0092.

Dicksved, J., Ellstrom, P., Engstrand, L. and Rautelin, H. 2014. Susceptibility to *Campylobacter* infection is associated with the species composition of the human fecal microbiota. *MBio* 5(5), e01212-14. doi:10.1128/mBio.01212-14.

Domańska-Blicharz, K., Bocian, Ł., Lisowska, A., Jacukowicz, A., Pikuła, A. and Minta, Z. 2017. Cross-sectional survey of selected enteric viruses in polish turkey flocks between 2008 and 2011. *BMC Vet. Res.* 13(1), 108. doi:10.1186/s12917-017-1013-8.

Dwidar, M., Monnappa, A. K. and Mitchell, R. J. 2012. The dual probiotic and antibiotic nature of *Bdellovibrio bacteriovorus*. *BMB Rep.* 45(2), 71–8. doi:10.5483/BMBRep.2012.45.2.71.

Dykstra, D. D. and Reid, W. M. 1978. Monensin, *Eimeria tenella* infection, and effects on the bacterial populations in the ceca of gnotobiotic chickens. *Poult. Sci.* 57(2), 398–402. doi:10.3382/ps.0570398.

Eckmann, L. and Kagnoff, M. F. 2001. Cytokines in host defense against *Salmonella*. *Microbes Infect.* 3(14–15), 1191–200. doi:10.1016/S1286-4579(01)01479-4.

Emami, N. K., Daneshmand, A., Naeini, S. Z., Graystone, E. N. and Broom, L. J. 2017. Effects of commercial organic acid blends on male broilers challenged with *E. coli* k88: performance, microbiology, intestinal morphology, and immune response. *Poult. Sci.* 96(9), 3254–63. doi:10.3382/ps/pex106.

Enigk, K., Schanzel, H., Scupin, E. and Dey-Hazra, A. 1970. Der enterale plasmaproteinverlust bei der coccidiose Des Huhnes. *Zentralbl. Veterinarmed. B* 17(4), 522–26. doi:10.1111/j.1439-0450.1970.tb01465.x.

Evans, S. J. and Sayers, A. R. 2000. A longitudinal study of *Campylobacter* infection of broiler flocks in Great Britain. *Prev. Vet. Med.* 46(3), 209–23. doi:10.1016/S0167-5877(00)00143-4.

Ewers, C., Antao, E. M., Diehl, I., Philipp, H. C. and Wieler, L. H. 2009. Intestine and environment of the chicken as reservoirs for extraintestinal pathogenic *Escherichia coli* strains with zoonotic potential. *Appl. Environ. Microbiol.* 75(1), 184–92. doi:10.1128/AEM.01324-08.

Fasina, Y. O., Newman, M. M., Stough, J. M. and Liles, M. R. 2016. Effect of *Clostridium perfringens* infection and antibiotic administration on microbiota in the small intestine of broiler chickens. *Poult. Sci.* 95(2), 247–60. doi:10.3382/ps/pev329.

Feng, Y., Gong, J., Yu, H., Jin, Y., Zhu, J. and Han, Y. 2010. Identification of changes in the composition of ileal bacterial microbiota of broiler chickens infected with *Clostridium perfringens*. *Vet. Microbiol.* 140(1–2), 116–21. doi:10.1016/j.vetmic.2009.07.001.

Fetterer, R. H., Miska, K. B., Jenkins, M. C. and Wong, E. A. 2014. Expression of nutrient transporters in duodenum, jejunum, and ileum of *Eimeria maxima*-infected broiler chickens. *Parasitol. Res.* 113(10), 3891–94. doi:10.1007/s00436-014-4114-3.

Forder, R. E. A., Nattrass, G. S., Geier, M. S., Hughes, R. J. and Hynd, P. I. 2012. Quantitative analyses of genes associated with mucin synthesis of broiler chickens with induced necrotic enteritis. *Poult. Sci.* 91(6), 1335–41. doi:10.3382/ps.2011-02062.

Forte, C., Acuti, G., Manuali, E., Casagrande Proietti, P., Pavone, S., Trabalza-Marinucci, M., Moscati, L., Onofri, A., Lorenzetti, C. and Franciosini, M. P. 2016. Effects of two different probiotics on microflora, morphology, and morphometry of gut in organic laying hens. *Poult. Sci.* 95(11), 2528–35. doi:10.3382/ps/pew164.

Freter, R., Brickner, H., Fekete, J., Vickerman, M. M. and Carey, K. E. 1983. Survival and implantation of *Escherichia coli* in the intestinal tract. *Infect. Immun.* 39(2), 686–703.

Fukata, T., Baba, E. and Arakawa, A. 1984. Growth of *Salmonella typhimurium* in the cecum of gnotobiotic chickens with *Eimeria tenella*. *Res. Vet. Sci.* 37(2), 230–33. doi:10.1016/S0034-5288(18)31911-8.

Fukata, T., Baba, E. and Arakawa, A. 1987. Invasion of *Salmonella typhimurium* into the cecal wall of gnotobiotic chickens with *Eimeria tenella*. *Poult. Sci.* 66(4), 760–61. doi:10.3382/ps.0660760.

Fukui, H., Sueyoshi, M., Haritani, M., Nakazawa, M., Naitoh, S., Tani, H. and Uda, Y. 1995. Natural infection with attaching and effacing *Escherichia coli* (O 103:H-) in chicks. *Avian Dis.* 39(4), 912–8. doi:10.2307/1592433.

Giannenas, I., Tsalie, E., Triantafillou, E., Hessenberger, S., Teichmann, K., Mohnl, M. and Tontis, D. 2014. Assessment of probiotics supplementation via feed or water on the growth performance, intestinal morphology and microflora of chickens after experimental infection with *Eimeria acervulina*, *Eimeria maxima* and *Eimeria tenella*. *Avian Pathol.* 43(3), 209–16. doi:10.1080/03079457.2014.899430.

Giovanardi, D., Drigo, I., Vidi, B. D., Agnoletti, F., Viel, L., Capello, K., Berto, G. and Bano, L. 2016. Molecular characterization of *Clostridium perfringens* strains isolated from diseased turkeys in Italy. *Avian Pathol.* 45(3), 376–80. doi:10.1080/03079457.2016.1160364.

Goddard, A. D., Arnold, M. E., Allen, V. M. and Snary, E. L. 2014. Estimating the time at which commercial broiler flocks in Great Britain become infected with *Campylobacter*: a bayesian approach. *Epidemiol. Infect.* 142(9), 1884–92. doi:10.1017/S0950268813002926.

Gomes, D. E., Hirata, K. Y., Saheki, K., Rosa, A. C., Luvizotto, M. C. and Cardoso, T. C. 2010. Pathology and tissue distribution of turkey coronavirus in experimentally infected chicks and turkey poults. *J. Comp. Pathol.* 143(1), 8–13. doi:10.1016/j.jcpa.2009.12.012.

Gong, J., Si, W., Forster, R. J., Huang, R., Yu, H., Yin, Y., Yang, C. and Han, Y. 2007. 16S rRNA gene-based analysis of mucosa-associated bacterial community and phylogeny in the chicken gastrointestinal tracts: from crops to ceca. *FEMS Microbiol. Ecol.* 59(1), 147–57. doi:10.1111/j.1574-6941.2006.00193.x.

Gregory, M., Klein, B., Sahin, O. and Girgis, G. 2018. Isolation and characterization of *Campylobacter hepaticus* from layer chickens with spotty liver disease in the united states. *Avian Dis.* 62(1), 79–85. doi:10.1637/11752-092017-Reg.1.

Guard-Petter, J., Keller, L. H., Rahman, M. M., Carlson, R. W. and Silvers, S. 1996. A novel relationship between o-antigen variation, matrix formation, and invasiveness of *Salmonella enteritidis*. *Epidemiol. Infect.* 117(2), 219–31. doi:10.1017/S0950268800001394.

Guo, J. R., Dong, X. F., Liu, S. and Tong, J. M. 2017. Effects of long-term *Bacillus subtilis* cgmcc 1.921 supplementation on performance, egg quality, and fecal and cecal microbiota of laying hens. *Poult. Sci.* 96(5), 1280–89. doi:10.3382/ps/pew389.

Guy, J. S., Levy, M. G., Ley, D. H., Barnes, H. J. and Gerig, T. M. 1988. Interaction of reovirus and *Cryptosporidium baileyi* in experimentally infected chickens. *Avian Dis.* 32(3), 381–90. doi:10.2307/1590901.

Guy, J. S., Smith, L. G., Breslin, J. J., Vaillancourt, J. P. and Barnes, H. J. 2000. High mortality and growth depression experimentally produced in young turkeys by dual infection with enteropathogenic *Escherichia coli* and turkey coronavirus. *Avian Dis.* 44(1), 105–13. doi:10.2307/1592513.

Haag, L. M., Fischer, A., Otto, B., Plickert, R., Kühl, A. A., Göbel, U. B., Bereswill, S. and Heimesaat, M. M. 2012. Intestinal microbiota shifts towards elevated commensal *Escherichia coli* loads abrogate colonization resistance against *Campylobacter jejuni* in mice. *PLoS ONE* 7(5), e35988. doi:10.1371/journal.pone.0035988.

Haghighi, H. R., Abdul-Careem, M. F., Dara, R. A., Chambers, J. R. and Sharif, S. 2008. Cytokine gene expression in chicken cecal tonsils following treatment with probiotics and *Salmonella* infection. *Vet. Microbiol.* 126(1–3), 225–33. doi:10.1016/j.vetmic.2007.06.026.

Han, Z., Pielsticker, C., Gerzova, L., Rychlik, I. and Rautenschlein, S. 2016. The influence of age on *Campylobacter jejuni* infection in chicken. *Dev. Comp. Immunol.* 62, 58–71. doi:10.1016/j.dci.2016.04.020.

Han, Z., Willer, T., Li, L., Pielsticker, C., Rychlik, I., Velge, P., Kaspers, B. and Rautenschlein, S. 2017. Influence of the gut microbiota composition on *Campylobacter jejuni* colonization in chickens. *Infect. Immun.* 85(11), e00380-17. doi:10.1128/IAI.00380-17.

Harp, J. A., Chen, W. and Harmsen, A. G. 1992. Resistance of severe combined immunodeficient mice to infection with *Cryptosporidium parvum*: the importance of intestinal microflora. *Infect. Immun.* 60(9), 3509–12.

Heak, C., Sukon, P. and Sornplang, P. 2018. Effect of direct-fed microbials on culturable gut microbiotas in broiler chickens: a meta-analysis of controlled trials. *Asian-Australas. J. Anim. Sci.* 31(11), 1781–94. doi:10.5713/ajas.18.0009.

Hegde, K. S., Reid, W. M., Johnson, J. and Womack, H. E. 1969. Pathogenicity of *Eimeria brunetti* in bacteria-free and conventional chickens. *J. Parasitol.* 55(2), 402-05. doi:10.2307/3277422.

Hermans, D., Pasmans, F., Heyndrickx, M., Van Immerseel, F., Martel, A., Van Deun, K. and Haesebrouck, F. 2012. A tolerogenic mucosal immune response leads to persistent *Campylobacter jejuni* colonization in the chicken gut. *Crit. Rev. Microbiol.* 38(1), 17–29. doi:10.3109/1040841X.2011.615298.

Hessenberger, S., Schatzmayr, G. and Teichmann, K. 2016. *In vitro* inhibition of *Eimeria tenella* sporozoite invasion into host cells by probiotics. *Vet. Parasitol.* 229, 93–8. doi:10.1016/j.vetpar.2016.10.001.

Higgins, J. P., Higgins, S. E., Wolfenden, A. D., Henderson, S. N., Torres-Rodriguez, A., Vicente, J. L., Hargis, B. M. and Tellez, G. 2010. Effect of lactic acid bacteria probiotic culture treatment timing on *Salmonella enteritidis* in neonatal broilers. *Poult. Sci.* 89(2), 243–7. doi:10.3382/ps.2009-00436.

Holt, P. S. and Gast, R. K. 2002. Comparison of the effects of infection with *Salmonella enteritidis*, in combination with an induced molt, on serum levels of the acute phase protein, alpha1 acid glycoprotein, in hens. *Poult. Sci.* 81(9), 1295–300. doi:10.1093/ps/81.9.1295.

Hu, G., Feng, Y., O'Hara, S. P. and Chen, X.-M.. 2014. Immunology of *Cryptosporidiosis*. In: Cacciò, S. M., Widmer, G. and Cacciò, S. M., et al. (Eds) *Cryptosporidium: Parasite and Disease*. Springer, Vienna, pp. 423–54.

Huang, G., Tang, X., Bi, F., Hao, Z., Han, Z., Suo, J., Zhang, S., Wang, S., Duan, C., Yu, Z., et al. 2018. *Eimeria tenella* infection perturbs the chicken gut microbiota from the onset of oocyst shedding. *Vet. Parasitol.* 258, 30–7. doi:10.1016/j.vetpar.2018.06.005.

Huff, G., Huff, W., Rath, N., Balog, J., Anthony, N. B. and Nestor, K. 2006. Stress-induced colibacillosis and turkey osteomyelitis complex in turkeys selected for increased body weight. *Poult. Sci.* 85(2), 266–72. doi:10.1093/ps/85.2.266.

Hume, M. E., Corrier, D. E., Nisbet, D. J. and DeLoach, J. R. 1998. Early *Salmonella* challenge time and reduction in chick cecal colonization following treatment with a characterized competitive exclusion culture. *J. Food Prot.* 61(6), 673–6. doi:10.4315/0362-028X-61.6.673.

Hume, M. E., Barbosa, N. A., Dowd, S. E., Sakomura, N. K., Nalian, A. G., Martynova-Van Kley, A. and Oviedo-Rondón, E. O. 2011. Use of pyrosequencing and denaturing gradient gel electrophoresis to examine the effects of probiotics and essential oil blends on digestive microflora in broilers under mixed *Eimeria* infection. *Foodborne Pathog. Dis.* 8(11), 1159–67. doi:10.1089/fpd.2011.0863.

Humphrey, S., Chaloner, G., Kemmett, K., Davidson, N., Williams, N., Kipar, A., Humphrey, T. and Wigley, P. 2014. *Campylobacter jejuni* is not merely a commensal in commercial broiler chickens and affects bird welfare. *MBio* 5(4), e01364–14. doi:10.1128/mBio.01364-14.

Ismail, M. M., Tang, A. Y. and Saif, Y. M. 2003. Pathogenicity of turkey coronavirus in turkeys and chickens. *Avian Dis.* 47(3), 515–22. doi:10.1637/5917.

Jacobs-Reitsma, W. F. 1995. *Campylobacter* bacteria in breeder flocks. *Avian Dis.* 39(2), 355–9. doi:10.2307/1591879.

Jennings, J. L., Sait, L. C., Perrett, C. A., Foster, C., Williams, L. K., Humphrey, T. J. and Cogan, T. A. 2011. *Campylobacter jejuni* is associated with, but not sufficient to cause vibrionic hepatitis in chickens. *Vet. Microbiol.* 149(1–2), 193–9. doi:10.1016/j.vetmic.2010.11.005.

Jindal, N., Patnayak, D. P., Ziegler, A. F., Lago, A. and Goyal, S. M. 2009. A retrospective study on poult enteritis syndrome in Minnesota. *Avian Dis.* 53(2), 268–75. doi:10.1637/8513-110308-Reg.1.

Jindal, N., Mor, S. K. and Goyal, S. M. 2014. Enteric viruses in turkey enteritis. *VirusDisease* 25(2), 173–85. doi:10.1007/s13337-014-0198-8.

Joerger, R. D. and Ross, T. 2005. Genotypic diversity of *Escherichia coli* isolated from cecal content and mucosa of one- to six-week-old broilers. *Poult. Sci.* 84(12), 1902–7. doi:10.1093/ps/84.12.1902.

Johansen, C. H., Bjerrum, L., Finster, K. and Pedersen, K. 2006. Effects of a *Campylobacter jejuni* infection on the development of the intestinal microflora of broiler chickens. *Poult. Sci.* 85(4), 579–87. doi:10.1093/ps/85.4.579.

Johansson, K. R. and Sarles, W. B. 1948. Bacterial population changes in the ceca of young chickens infected with *Eimeria tenella*. *J. Bacteriol.* 56(5), 635–47.

Joyner, L. P., Patterson, D. S. P., Berrett, S., Boarer, C. D., Cheong, F. H. and Norton, C. C. 1975. Amino-acid malabsorption and intestinal leakage of plasma-proteins in young chicks infected with *Eimeria acervulina*. *Avian Pathol.* 4(1), 17–33. doi:10.1080/03079457509353847.

Juricova, H., Videnska, P., Lukac, M., Faldynova, M., Babak, V., Havlickova, H., Sisak, F. and Rychlik, I. 2013. Influence of *Salmonella enterica* serovar Enteritidis infection on the development of the cecum microbiota in newly hatched chicks. *Appl. Environ. Microbiol.* 79(2), 745–7. doi:10.1128/AEM.02628-12.

Kaakoush, N. O., Sodhi, N., Chenu, J. W., Cox, J. M., Riordan, S. M. and Mitchell, H. M. 2014. The interplay between *Campylobacter* and *Helicobacter* species and other gastrointestinal microbiota of commercial broiler chickens. *Gut Pathog.* 6, 18. doi:10.1186/1757-4749-6-18.

Kaiser, P., Rothwell, L., Galyov, E. E., Barrow, P. A., Burnside, J. and Wigley, P. 2000. Differential cytokine expression in avian cells in response to invasion by *Salmonella typhimurium*, *Salmonella enteritidis* and *Salmonella gallinarum*. *Microbiology (Reading, Engl.)* 146(12), 3217–26. doi:10.1099/00221287-146-12-3217.

Kashket, E. R. 1987. Bioenergetics of lactic acid bacteria: cytoplasmic pH and osmotolerance. *FEMS Microbiol. Lett.* 46(3), 233–44. doi:10.1111/j.1574-6968.1987.tb02463.x.

Kazwala, R. R., Collins, J. D., Hannan, J., Crinion, R. A. and O'Mahony, H. 1990. Factors responsible for the introduction and spread of *Campylobacter jejuni* infection in commercial poultry production. *Vet. Rec.* 126(13), 305–6.

Keener, K. M., Bashor, M. P., Curtis, P. A., Sheldon, B. W. and Kathariou, S. 2004. Comprehensive review of *Campylobacter* and poultry processing. *Compr. Rev. Food Sci. Food Saf.* 3(2), 105–16. doi:10.1111/j.1541-4337.2004.tb00060.x.

Kemmett, K., Humphrey, T., Rushton, S., Close, A., Wigley, P. and Williams, N. J. 2013. A longitudinal study simultaneously exploring the carriage of apec virulence associated genes and the molecular epidemiology of faecal and systemic *E. coli* in commercial broiler chickens. *PLoS ONE* 8, e67749.

Kemmett, K., Williams, N. J., Chaloner, G., Humphrey, S., Wigley, P. and Humphrey, T. 2014. The contribution of systemic *Escherichia coli* infection to the early mortalities of commercial broiler chickens. *Avian Pathol.* 43(1), 37–42. doi:10.1080/03079457.2013.866213.

Keyburn, A. L., Boyce, J. D., Vaz, P., Bannam, T. L., Ford, M. E., Parker, D., Di Rubbo, A., Rood, J. I. and Moore, R. J. 2008. NetB, a new toxin that is associated with avian necrotic enteritis caused by *Clostridium perfringens*. *PLoS Pathog.* 4(2), e26. doi:10.1371/journal.ppat.0040026.

Keyburn, A. L., Yan, X. X., Bannam, T. L., Van Immerseel, F., Rood, J. I. and Moore, R. J. 2010. Association between avian necrotic enteritis and *Clostridium perfringens* strains expressing NetB toxin. *Vet. Res.* 41(2), 21. doi:10.1051/vetres/2009069.

Kimura, N., Mimura, F., Nishida, S. and Kobayashi, A. 1976. Studies on the relationship between intestinal flora and cecal coccidiosis in chicken. *Poult. Sci.* 55(4), 1375–83. doi:10.3382/ps.0551375.

Kleessen, B., Elsayed, N. A., Loehren, U., Schroedl, W. and Krueger, M. 2003. Jerusalem artichokes stimulate growth of broiler chickens and protect them against endotoxins and potential cecal pathogens. *J. Food Prot.* 66(11), 2171–5. doi:10.4315/0362-028X-66.11.2171.

Knap, I., Kehlet, A. B., Bennedsen, M., Mathis, G. F., Hofacre, C. L., Lumpkins, B. S., Jensen, M. M., Raun, M. and Lay, A. 2011. *Bacillus subtilis* (DSM17299) significantly reduces *Salmonella* in broilers. *Poult. Sci.* 90(8), 1690–4. doi:10.3382/ps.2010-01056.

Kogut, M. H., He, H. and Kaiser, P. 2005. Lipopolysaccharide binding protein/CD14/TLR4-dependent recognition of *Salmonella* LPS induces the functional activation of chicken heterophils and up-regulation of pro-inflammatory cytokine and chemokine gene expression in these cells. *Anim. Biotechnol.* 16(2), 165–81. doi:10.1080/10495390500264896.

Kogut, M. H., Genovese, K. J., He, H. and Kaiser, P. 2008. Flagellin and lipopolysaccharide up-regulation of IL-6 and CXCLi2 gene expression in chicken heterophils is mediated by ERK1/2-dependent activation of AP-1 and NF-kappaB signaling pathways. *Innate Immun.* 14(4), 213–22. doi:10.1177/1753425908094416.

Koo, B. S., Lee, H. R., Jeon, E. O., Han, M. S., Min, K. C., Lee, S. B. and Mo, I. P. 2013. Molecular survey of enteric viruses in commercial chicken farms in Korea with a history of enteritis. *Poult. Sci.* 92(11), 2876–85. doi:10.3382/ps.2013-03280.

Lacey, J. A., Johanesen, P. A., Lyras, D. and Moore, R. J. 2016. Genomic diversity of necrotic enteritis-associated strains of *Clostridium perfringens*: a review. *Avian Pathol.* 45(3), 302–07. doi:10.1080/03079457.2016.1153799.

Lacroix-Lamandé, S., Guesdon, W., Drouet, F., Potiron, L., Lantier, L. and Laurent, F. 2014. The gut flora is required for the control of intestinal infection by poly(I:C) administration in neonates. *Gut Microbes* 5(4), 533–40. doi:10.4161/gmic.29154.

Lamas, A., Miranda, J. M., Regal, P., Vázquez, B., Franco, C. M. and Cepeda, A. 2018. A comprehensive review of non-enterica subspecies of *Salmonella enterica*. *Microbiol. Res.* 206, 60–73. doi:10.1016/j.micres.2017.09.010.

Lambert, C., Evans, K. J., Till, R., Hobley, L., Capeness, M., Rendulic, S., Schuster, S. C., Aizawa, S. and Sockett, R. E. 2006. Characterizing the flagellar filament and the role of motility in bacterial prey-penetration by *Bdellovibrio bacteriovorus*. *Mol. Microbiol.* 60(2), 274–86. doi:10.1111/j.1365-2958.2006.05081.x.

Lan, R. X., Lee, S. I. and Kim, I. H. 2017. Effects of *Enterococcus faecium* SLB 120 on growth performance, blood parameters, relative organ weight, breast muscle meat quality, excreta microbiota shedding, and noxious gas emission in broilers. *Poult. Sci.* 96(9), 3246-53. doi:10.3382/ps/pex101.

Lantier, L., Drouet, F., Guesdon, W., Mancassola, R., Metton, C., Lo-Man, R., Werts, C., Laurent, F. and Lacroix-Lamandé, S. 2014. Poly(I:C)-induced protection of neonatal mice against intestinal *Cryptosporidium parvum* infection requires an additional TLR5 signal provided by the gut flora. *J. Infect. Dis.* 209(3), 457-67. doi:10.1093/infdis/jit432.

Laukova, A., Kandricakova, A. and Scerbova, J. 2015. Use of bacteriocin-producing, probiotic strain *Enterococcus faecium* AL41 to control intestinal microbiota in farm ostriches. *Lett. Appl. Microbiol.* 60(6), 531-5. doi:10.1111/lam.12409.

Lee, K. W., Lillehoj, H. S., Jang, S. I., Li, G., Lee, S. H., Lillehoj, E. P. and Siragusa, G. R. 2010. Effect of *Bacillus*-based direct-fed microbials on *Eimeria maxima* infection in broiler chickens. *Comp. Immunol. Microbiol. Infect. Dis.* 33(6), e105-10. doi:10.1016/j.cimid.2010.06.001.

Leitner, G. and Heller, E. D. 1992. Colonization of *Escherichia coli* in young turkeys and chickens. *Avian Dis.* 36(2), 211-20. doi:10.2307/1591493.

Li, Z., Wang, W., Liu, D. and Guo, Y. 2017. Effects of *Lactobacillus acidophilus* on gut microbiota composition in broilers challenged with *Clostridium perfringens*. *PLoS ONE* 12(11), e0188634. doi:10.1371/journal.pone.0188634.

Li, H., Liu, X., Chen, F., Zuo, K., Wu, C., Yan, Y., Chen, W., Lin, W. and Xie, Q. 2018a. Avian influenza virus subtype H9N2 affects intestinal microbiota, barrier structure injury, and inflammatory intestinal disease in the chicken ileum. *Viruses* 10(5), 270. doi:10.3390/v10050270.

Li, L., Pielsticker, C., Han, Z., Kubasová, T., Rychlik, I., Kaspers, B. and Rautenschlein, S. 2018b. Infectious bursal disease virus inoculation infection modifies *Campylobacter jejuni*-host interaction in broilers. *Gut Pathog.* 10, 13. doi:10.1186/s13099-018-0241-1.

Line, J. E., Svetoch, E. A., Eruslanov, B. V., Perelygin, V. V., Mitsevich, E. V., Mitsevich, I. P., Levchuk, V. P., Svetoch, O. E., Seal, B. S., Siragusa, G. R., et al. 2008. Isolation and purification of enterocin E-760 with broad antimicrobial activity against gram-positive and gram-negative bacteria. *Antimicrob. Agents Chemother.* 52(3), 1094-100. doi:10.1128/AAC.01569-06.

Liu, D., Guo, Y., Wang, Z. and Yuan, J. 2010. Exogenous lysozyme influences *Clostridium perfringens* colonization and intestinal barrier function in broiler chickens. *Avian Pathol.* 39(1), 17-24. doi:10.1080/03079450903447404.

Ma, X., Wang, Q., Li, H., Xu, C., Cui, N. and Zhao, X. 2017. 16Ss rRNA genes Illumina sequencing revealed differential cecal microbiome in specific pathogen free chickens infected with different subgroup of avian leukosis viruses. *Vet. Microbiol.* 207, 195-204. doi:10.1016/j.vetmic.2017.05.016.

Macdonald, S. E., Nolan, M. J., Harman, K., Boulton, K., Hume, D. A., Tomley, F. M., Stabler, R. A. and Blake, D. P. 2017. Effects of *Eimeria tenella* infection on chicken caecal microbiome diversity, exploring variation associated with severity of pathology. *PLoS ONE* 12(9), e0184890. doi:10.1371/journal.pone.0184890.

Macdonald, S. E., van Diemen, P. M., Martineau, H., Stevens, M. P., Tomley, F. M., Stabler, R. A. and Blake, D. P. 2018. The impact of *Eimeria tenella* co-infection on *Campylobacter jejuni* colonisation of the chicken. *Infect. Immun.* 87(2), e00772-18. doi:10.1128/IAI.00772-18.

Maier, L., Vyas, R., Cordova, C. D., Lindsay, H., Schmidt, T. S., Brugiroux, S., Periaswamy, B., Bauer, R., Sturm, A., Schreiber, F., et al. 2013. Microbiota-derived hydrogen fuels *Salmonella typhimurium* invasion of the gut ecosystem. *Cell Host Microbe* 14(6), 641–51. doi:10.1016/j.chom.2013.11.002.

Manafi, M., Khalaji, S., Hedayati, M. and Pirany, N. 2017. Efficacy of *Bacillus subtilis* and bacitracin methylene disalicylate on growth performance, digestibility, blood metabolites, immunity, and intestinal microbiota after intramuscular inoculation with *Escherichia coli* in broilers. *Poult. Sci.* 96(5), 1174–83. doi:10.3382/ps/pew347.

Martin, C., Dunlap, E., Caldwell, S., Barnhart, E., Keith, N. and Deloach, J. R. 2000. Drinking water delivery of a defined competitive exclusion culture (Rreempt) in 1-day-old broiler chicks. *J. Appl. Poul Res.* 9(1), 88–91. doi:10.1093/japr/9.1.88.

Matthijs, M. G. R., Nieuwenhuis, J. F. and Dwars, R. M. 2017. Signs indicating imminent death in *Escherichia coli*-infected broilers. *Avian Dis.* 61(3), 316–24. doi:10.1637/11509-100316-RegR.

McDonald, V. and Shirley, M. W. 2009. Past and future: vaccination against *Eimeria*. *Parasitology* 136(12), 1477–89. doi:10.1017/S0031182009006349.

McKenzie, M., Colnago, G., Lee, S. and Long, P. 1982. Gut stasis induction by coccidiosis infections in chickens. *Poult. Sci.* 61, 1512–12.

McReynolds, J. L., Byrd, J. A., Anderson, R. C., Moore, R. W., Edrington, T. S., Genovese, K. J., Poole, T. L., Kubena, L. F. and Nisbet, D. J. 2004. Evaluation of immunosuppressants and dietary mechanisms in an experimental disease model for necrotic enteritis. *Poult. Sci.* 83(12), 1948–52. doi:10.1093/ps/83.12.1948.

Mead, G. C., Scott, M. J., Humphrey, T. J. and McAlpine, K. 1996. Observations on the control of *Campylobacter jejuni* infection of poultry by 'competitive exclusion'. *Avian Pathol.* 25(1), 69–79. doi:10.1080/03079459608419121.

Menconi, A., Wolfenden, A. D., Shivaramaiah, S., Terraes, J. C., Urbano, T., Kuttel, J., Kremer, C., Hargis, B. M. and Tellez, G. 2011. Effect of lactic acid bacteria probiotic culture for the treatment of *Salmonella enterica* serovar Heidelberg in neonatal broiler chickens and turkey poults. *Poult. Sci.* 90(3), 561–5. doi:10.3382/ps.2010-01220.

Messaoudi, S., Manai, M., Kergourlay, G., Prévost, H., Connil, N., Chobert, J. M. and Dousset, X. 2013. *Lactobacillus salivarius*: bacteriocin and probiotic activity. *Food Microbiol.* 36(2), 296–304. doi:10.1016/j.fm.2013.05.010.

Montgomery, R. D., Boyle, C. R., Maslin, W. R. and Magee, D. L. 1997. Attempts to reproduce a runting/stunting-type syndrome using infectious agents isolated from affected Mississippi broilers. *Avian Dis.* 41(1), 80–92. doi:10.2307/1592446.

Mor, S. K., Sharafeldin, T. A., Abin, M., Kromm, M., Porter, R. E., Goyal, S. M. and Patnayak, D. P. 2013. The occurrence of enteric viruses in light turkey syndrome. *Avian Pathol.* 42(5), 497–501. doi:10.1080/03079457.2013.832145.

Moura-Alvarez, J., Chacon, J. V., Scanavini, L. S., Nuñez, L. F., Astolfi-Ferreira, C. S., Jones, R. C. and Piantino Ferreira, A. J. 2013. Enteric viruses in Brazilian turkey flocks: single and multiple virus infection frequency according to age and clinical signs of intestinal disease. *Poult. Sci.* 92(4), 945–55. doi:10.3382/ps.2012-02849.

Moura-Alvarez, J., Nuñez, L. F. N., Astolfi-Ferreira, C. S., Knöbl, T., Chacón, J. L., Moreno, A. M., Jones, R. C. and Ferreira, A. J. 2014. Detection of enteric pathogens in turkey flocks affected with severe enteritis, in Brazil. *Trop. Anim. Health Prod.* 46(6), 1051–58. doi:10.1007/s11250-014-0612-7.

Ng, K. M., Ferreyra, J. A., Higginbottom, S. K., Lynch, J. B., Kashyap, P. C., Gopinath, S., Naidu, N., Choudhury, B., Weimer, B. C., Monack, D. M., et al. 2013. Microbiota-liberated host sugars facilitate post-antibiotic expansion of enteric pathogens. *Nature* 502(7469), 96–9. doi:10.1038/nature12503.

Nurmi, E. and Rantala, M. 1973. New aspects of *Salmonella* infection in broiler production. *Nature* 241(5386), 210–11. doi:10.1038/241210a0.

O'Loughlin, J. L., Samuelson, D. R., Braundmeier-Fleming, A. G., White, B. A., Haldorson, G. J., Stone, J. B., Lessmann, J. J., Eucker, T. P. and Konkel, M. E. 2015. The intestinal microbiota influences *Campylobacter jejuni* colonization and extraintestinal dissemination in mice. *Appl. Environ. Microbiol.* 81(14), 4642-50. doi:10.1128/AEM.00281-15.

Oakley, B. B., Lillehoj, H. S., Kogut, M. H., Kim, W. K., Maurer, J. J., Pedroso, A., Lee, M. D., Collett, S. R., Johnson, T. J. and Cox, N. A. 2014. The chicken gastrointestinal microbiome. *FEMS Microbiol. Lett.* 360(2), 100-12. doi:10.1111/1574-6968.12608.

Opengart, K. and Songer, J. G. 2013. Necrotic enteritis. In: Swayne, D. E., Glisson, J. R., McDougald, L. R., Nolan, L. K., Suarez, D. L. and Nair, V. (Eds), *Diseases of Poultry*. Ames, IA, pp. 949-53.

Pakpinyo, S., Ley, D. H., Barnes, H. J., Vaillancourt, J. P. and Guy, J. S. 2003. Enhancement of enteropathogenic *Escherichia coli* pathogenicity in young turkeys by concurrent turkey coronavirus infection. *Avian Dis.* 47(2), 396–405. doi:10.1637/0005-2086(2003)047[0396:EOEECP]2.0.CO;2.

Palade, E. A., Demeter, Z., Hornyák, A., Nemes, C., Kisary, J. and Rusvai, M. 2011. High prevalence of turkey parvovirus in turkey flocks from Hungary experiencing enteric disease syndromes. *Avian Dis.* 55(3), 468-75. doi:10.1637/9688-021711-ResNote.1.

Park, J. H. and Kim, I. H. 2014. Supplemental effect of probiotic *Bacillus subtilis* B2A on productivity, organ weight, intestinal *Salmonella* microflora, and breast meat quality of growing broiler chicks. *Poult. Sci.* 93(8), 2054–9. doi:10.3382/ps.2013-03818.

Park, S. H., Kim, S. A., Rubinelli, P. M., Roto, S. M. and Ricke, S. C. 2017. Microbial compositional changes in broiler chicken cecal contents from birds challenged with different *Salmonella* vaccine candidate strains. *Vaccine* 35(24), 3204–08. doi:10.1016/j.vaccine.2017.04.073.

Pasquali, F., Lucchi, A., Braggio, S., Giovanardi, D., Franchini, A., Stonfer, M. and Manfreda, G. 2015. Genetic diversity of *Escherichia coli* isolates of animal and environmental origins from an integrated poultry production chain. *Vet. Microbiol.* 178(3–4), 230–7. doi:10.1016/j.vetmic.2015.05.007.

Peng, L., Matthijs, M. G. R., Haagsman, H. P. and Veldhuizen, E. J. A. 2018. Avian pathogenic *Escherichia coli*-induced activation of chicken macrophage HD11 cells. *Dev. Comp. Immunol.* 87, 75-83. doi:10.1016/j.dci.2018.05.019.

Perez, V. G., Jacobs, C. M., Barnes, J., Jenkins, M. C., Kuhlenschmidt, M. S., Fahey, G. C., Parsons, C. M. and Pettigrew, J. E. 2011. Effect of corn distillers dried grains with solubles and *Eimeria acervulina* infection on growth performance and the intestinal microbiota of young chicks. *Poult. Sci.* 90(5), 958-64. doi:10.3382/ps.2010-01066.

Perumbakkam, S., Hunt, H. D. and Cheng, H. H. 2014. Marek's disease virus influences the core gut microbiome of the chicken during the early and late phases of viral replication. *FEMS Microbiol. Ecol.* 90(1), 300–12. doi:10.1111/1574-6941.12392.

Perumbakkam, S., Hunt, H. D. and Cheng, H. H. 2016. Differences in CD8αα and cecal microbiome community during proliferation and late cytolytic phases of Marek's disease virus infection are associated with genetic resistance to Marek's disease. *FEMS Microbiol. Ecol.* 92(12). doi:10.1093/femsec/fiw188.

Portrait, V., Cottenceau, G. and Pons, A. M. 2000 A *Fusobacterium mortiferum* strain produces a bacteriocin-like substance(s) inhibiting Salmonella enteritidis. *Lett. Appl. Microbiol.* 31(2), 115–7.

Pourakbari, M., Seidavi, A., Asadpour, L. and Martínez, A. 2016. Probiotic level effects on growth performance, carcass traits, blood parameters, cecal microbiota, and immune response of broilers. *An. Acad. Bras. Ciênc.* 88(2), 1011-21. doi:10.1590/0001-3765201620150071.

Preston-Mafham, R. A. and Sykes, A. H. 1967. Changes in permeability of mucosa during intestinal coccindiosis infections in fowl. *Experientia* 23(11), 972–3.

Qin, Z. R., Arakawa, A., Baba, E., Fukata, T., Miyamoto, T., Sasai, K. and Withanage, G. S. 1995a. *Eimeria tenella* infection induces recrudescence of previous *Salmonella enteritidis* infection in chickens. *Poult. Sci.* 74(11), 1786–92. doi:10.3382/ps.0741786.

Qin, Z. R., Fukata, T., Baba, E. and Arakawa, A. 1995b. Effect of *Eimeria tenella* infection on *Salmonella enteritidis* infection in chickens. *Poult. Sci.* 74(1), 1–7. doi:10.3382/ps.0740001.

Qin, Z., Arakawa, A., Baba, E., Fukata, T. and Sasai, K. 1996. Effect of *Eimeria tenella* infection on the production of *Salmonella enteritidis*-contaminated eggs and susceptibility of laying hens to S. *enteritidis* infection. *Avian Dis.* 40(2), 361-67. doi:10.2307/1592233.

Qu, A., Brulc, J. M., Wilson, M. K., Law, B. F., Theoret, J. R., Joens, L. A., Konkel, M. E., Angly, F., Dinsdale, E. A., Edwards, R. A., et al. 2008. Comparative metagenomics reveals host specific metavirulomes and horizontal gene transfer elements in the chicken cecum microbiome. *PLOS ONE* 3(8), e2945. doi:10.1371/journal.pone.0002945.

Quinteiro-Filho, W. M., Ribeiro, A., Ferraz-de-Paula, V., Pinheiro, M. L., Sakai, M., Sá, L. R., Ferreira, A. J. and Palermo-Neto, J. 2010. Heat stress impairs performance parameters, induces intestinal injury, and decreases macrophage activity in broiler chickens. *Poult. Sci.* 89(9), 1905–14. doi:10.3382/ps.2010-00812.

Raehtz, S., Hargis, B. M., Kuttappan, V. A., Pamukcu, R., Bielke, L. R. and McCabe, L. R. 2018. High molecular weight polymer promotes bone health and prevents bone loss under *Salmonella* challenge in broiler chickens. *Front. Physiol.* 9, 384. doi:10.3389/fphys.2018.00384.

Rantala, M. and Nurmi, E. 1973. Prevention of the growth of *Salmonella infantis* in chicks by the flora of the alimentary tract of chickens. *Br. Poult. Sci.* 14(6), 627–30. doi:10.1080/00071667308416073.

Ringoir, D. D., Szylo, D. and Korolik, V. 2007. Comparison of 2-day-old and 14-day-old chicken colonization models for *Campylobacter jejuni*. *FEMS Immunol. Med. Microbiol.* 49(1), 155–8. doi:10.1111/j.1574-695X.2006.00181.x.

Ritzi, M. M., Abdelrahman, W., Mohnl, M. and Dalloul, R. A. 2014. Effects of probiotics and application methods on performance and response of broiler chickens to an *Eimeria* challenge. *Poult. Sci.* 93(11), 2772–78. doi:10.3382/ps.2014-04207.

Ritzi, M. M., Abdelrahman, W., van-Heerden, K., Mohnl, M., Barrett, N. W. and Dalloul, R. A. 2016 Combination of probiotics and coccidiosis vaccine enhances protection against an *Eimeria* challenge. *Vet. Res.* 47(1), 111. doi:10.1186/s13567-016-0397-y.

Rivera-Chavez, F. and Bäumler, A. J. 2015. The pyromaniac inside you: *Salmonella* metabolism in the host gut. *Annu. Rev. Microbiol.* 69, 31–48. doi:10.1146/annurev-micro-091014-104108.

Ruff, M. D. and Edgar, S. A. 1982. Reduced intestinal absorption in broilers during *Eimeria mitis* infection. *Am. J. Vet. Res.* 43(3), 507–09.

Ruff, M. D. and Reid, W. M. 1975. Coccidiosis and intestinal pH in chickens. *Avian Dis.* 19(1), 52–8.

Ruff, M. D. and Rosenberger, J. K. 1985a. Concurrent infections with reoviruses and coccidia in broilers. *Avian Dis.* 29, 465–78.

Ruff, M. D. and Rosenberger, J. K. 1985b. Interaction of low-pathogenicity reoviruses and low levels of infection with several coccidial species. *Avian Dis.* 29, 1057–65.

Ruff, M. D., Dykstra, D. D., Johnson, J. K. and Reid, W. M. 1975. Effects of *Eimeria brunetti* on intestinal pH in conventional and gnotobiotic chickens. *Avian Pathol.* 4(1), 73–81. doi:10.1080/03079457509353852.

Saif, Y. M., Guy, J. S., Day, J. M., Schultz-Cherry, S. L., Hayhow, C. S. and Zsak, L. 2013. Viral enteric infections. In: Swayne, D. E., Glisson, J. R., McDougald, L. R., Nolan, L. K., Suarez, D. L. and Nair, V. (Eds), *Diseases of Poultry*. Ames, IA, pp. 375–416.

Sakaridis, I., Ellis, R. J., Cawthraw, S. A., van Vliet, A. H. M., Stekel, D. J., Penell, J., Chambers, M., La Ragione, R. M. and Cook, A. J. 2018. Investigating the association between the caecal microbiomes of broilers and *Campylobacter* burden. *Front. Microbiol.* 9, 927. doi:10.3389/fmicb.2018.00927.

Samli, H. E., Dezcan, S., Koc, F., Ozduven, M. L., Okur, A. A. and Senkoylu, N. 2010. Effects of *Enterococcus faecium* supplementation and floor type on performance, morphology of erythrocytes and intestinal microbiota in broiler chickens. *Br. Poult. Sci.* 51(4), 564–8. doi:10.1080/00071668.2010.507241.

Schildt, C. and Herrick, C. 1955. The effect of cecal coccidiosis on the motility of the digestive tract. *J. Parasitol.* 41, 18–19.

Schoeni, J. L. and Wong, A. C. 1994. Inhibition of *Campylobacter jejuni* colonization in chicks by defined competitive exclusion bacteria. *Appl. Environ. Microbiol.* 60(4), 1191–7.

Setta, A. M., Barrow, P. A., Kaiser, P. and Jones, M. A. 2012. Early immune dynamics following infection with *Salmonella enterica* serovars enteritidis, infantis, pullorum and gallinarum: cytokine and chemokine gene expression profile and cellular changes of chicken cecal tonsils. *Comp. Immunol. Microbiol. Infect. Dis.* 35(5), 397–410. doi:10.1016/j.cimid.2012.03.004.

Shah, J. D., Desai, P. T., Zhang, Y., Scharber, S. K., Baller, J., Xing, Z. S. and Cardona, C. J. 2016. Development of the intestinal RNA virus community of healthy broiler chickens. *PLoS ONE* 11(2), e0150094. doi:10.1371/journal.pone.0150094.

Shane, S. M., Gyimah, J. E., Harrington, K. S. and Snider, T. G. 1985. Etiology and pathogenesis of necrotic enteritis. *Vet. Res. Commun.* 9(4), 269–87.

Shivaramaiah, S., Pumford, N. R., Morgan, M. J., Wolfenden, R. E., Wolfenden, A. D., Torres-Rodríguez, A., Hargis, B. M. and Téllez, G. 2011a. Evaluation of *Bacillus* species as potential candidates for direct-fed microbials in commercial poultry. *Poult. Sci.* 90(7), 1574–80. doi:10.3382/ps.2010-00745.

Shivaramaiah, S., Wolfenden, R. E., Barta, J. R., Morgan, M. J., Wolfenden, A. D., Hargis, B. M., Tellez, G. 2011b. The role of an early *Salmonella typhimurium* infection as a predisposing factor for necrotic enteritis in a laboratory challenge model. *Avian Dis.* 55, 319–23.

Shokryazdan, P., Faseleh Jahromi, M., Liang, J. B., Ramasamy, K., Sieo, C. C. and Ho, Y. W. 2017. Effects of a *Lactobacillus salivarius* mixture on performance, intestinal health and serum lipids of broiler chickens. *PLoS ONE* 12(5), e0175959. doi:10.1371/journal.pone.0175959.

Si, W., Gong, J., Han, Y., Yu, H., Brennan, J., Zhou, H. and Chen, S. 2007. Quantification of cell proliferation and alpha-toxin gene expression of *Clostridium perfringens* in the development of necrotic enteritis in broiler chickens. *Appl. Environ. Microbiol.* 73(21), 7110–13. doi:10.1128/AEM.01108-07.

Skanseng, B., Kaldhusdal, M. and Rudi, K. 2006. Comparison of chicken gut colonisation by the pathogens *Campylobacter jejuni* and *Clostridium perfringens* by real-time quantitative PCR. *Mol. Cell. Probes* 20(5), 269–79. doi:10.1016/j.mcp.2006.02.001.

Skarp, C. P. A., Hanninen, M. L. and Rautelin, H. I. K. 2016. Campylobacteriosis: the role of poultry meat. *Clin. Microbiol. Infect.* 22(2), 103–09. doi:10.1016/j.cmi.2015.11.019.

Skraban, J., Dzeroski, S., Zenko, B., Tusar, L. and Rupnik, M. 2013. Changes of poultry faecal microbiota associated with *Clostridium difficile* colonisation. *Vet. Microbiol.* 165(3–4), 416–24. doi:10.1016/j.vetmic.2013.04.014.

Sofka, D., Pfeifer, A., Gleiss, B., Paulsen, P. and Hilbert, F. 2015. Changes within the intestinal flora of broilers by colonisation with *Campylobacter jejuni*. *Berl. Munch. Tierarztl. Wochenschr.* 128(3–4), 104–10.

Song, J., Xiao, K., Ke, Y. L., Jiao, L. F., Hu, C. H., Diao, Q. Y., Shi, B. and Zou, X. T. 2014. Effect of a probiotic mixture on intestinal microflora, morphology, and barrier integrity of broilers subjected to heat stress. *Poult. Sci.* 93(3), 581–8. doi:10.3382/ps.2013-03455.

Stanley, D., Keyburn, A. L., Denman, S. E. and Moore, R. J. 2012. Changes in the caecal microflora of chickens following *Clostridium perfringens* challenge to induce necrotic enteritis. *Vet. Microbiol.* 159(1–2), 155–62. doi:10.1016/j.vetmic.2012.03.032.

Stanley, D., Wu, S. B., Rodgers, N., Swick, R. A. and Moore, R. J. 2014. Differential responses of cecal microbiota to fishmeal, *Eimeria* and *Clostridium perfringens* in a necrotic enteritis challenge model in chickens. *PLoS ONE* 9(8), e104739. doi:10.1371/journal.pone.0104739.

Stephens, J. F., Borst, W. J. and Barnett, B. D. 1974. Some physiological effects of *Eimeria acervulina*, *E. brunetti*, and *E. mivati* infections in young chickens. *Poult. Sci.* 53(5), 1735–42. doi:10.3382/ps.0531735.

Svetoch, E. A., Eruslanov, B. V., Perelygin, V. V., Mitsevich, E. V., Mitsevich, I. P., Borzenkov, V. N., Levchuk, V. P., Svetoch, O. E., Kovalev, Y. N., Stepanshin, Y. G., et al. 2008. Diverse antimicrobial killing by *Enterococcus faecium* E 50-52 bacteriocin. *J. Agric. Food Chem.* 56(6), 1942–8. doi:10.1021/jf073284g.

Takimoto, H., Baba, E., Fukata, T. and Arakawa, A. 1984. Effects of infection of *Eimeria tenella*, *E. acervulina*, and *E. maxima* upon *Salmonella* typhimurium infection in chickens. *Poult. Sci.* 63(3), 478–84. doi:10.3382/ps.0630478.

Thibodeau, A., Fravalo, P., Yergeau, É., Arsenault, J., Lahaye, L. and Letellier, A. 2015. Chicken caecal microbiome modifications induced by *Campylobacter jejuni* colonization and by a non-antibiotic feed additive. *PLOS ONE* 10(7), e0131978. doi:10.1371/journal.pone.0131978.

Timbermont, L., Haesebrouck, F., Ducatelle, R. and Van Immerseel, F. 2011. Necrotic enteritis in broilers: an updated review on the pathogenesis. *Avian Pathol.* 40(4), 341–47. doi:10.1080/03079457.2011.590967.

Tsiouris, V., Georgopoulou, I., Batzios, C., Pappaioannou, N., Diakou, A., Petridou, E., Ducatelle, R. and Fortomaris, P. 2013. The role of an attenuated anticoccidial vaccine on the intestinal ecosystem and on the pathogenesis of experimental necrotic enteritis in broiler chickens. *Avian Pathol.* 42(2), 163–70. doi:10.1080/03079457.2013.776161.

Tsukahara, T., Inoue, R., Nakayama, K. and Inatomi, T. 2018a. Inclusion of *Bacillus amyloliquefaciens* strain TOA5001 in the diet of broilers suppresses the symptoms of coccidiosis by modulating intestinal microbiota. *Anim. Sci. J.* 89(4), 679-87. doi:10.1111/asj.12980.

Tsukahara, T., Inoue, R., Nakayama, K. and Inatomi, T. 2018b. Inclusion of *Bacillus amyloliquefaciens* strain toa5001 in the diet of broilers suppresses the symptoms of coccidiosis by modulating intestinal microbiota. *Anim. Sci. J.* 89(4), 679-87. doi:10.1111/asj.12980.

Turk, D. E. 1972. Protozoan parasitic infections of the chick intestine and protein digestion and absorption. *J. Nutr.* 102(9), 1217–21. doi:10.1093/jn/102.9.1217.

Uzal, F. A., Sentíes-Cué, C. G., Rimoldi, G. and Shivaprasad, H. L. 2016. Non-Clostridium *perfringens* infectious agents producing necrotic enteritis-like lesions in poultry. *Avian Pathol.* 45(3), 326–33. doi:10.1080/03079457.2016.1159282.

Van, T. T. H., Elshagmani, E., Gor, M. C., Anwar, A., Scott, P. C. and Moore, R. J. 2017a. Induction of spotty liver disease in layer hens by infection with *Campylobacter hepaticus*. *Vet. Microbiol.* 199, 85-90. doi:10.1016/j.vetmic.2016.12.033.

Van, T. T. H., Gor, M. C., Anwar, A., Scott, P. C. and Moore, R. J. 2017b. *Campylobacter hepaticus*, the cause of spotty liver disease in chickens, is present throughout the small intestine and caeca of infected birds. *Vet. Microbiol.* 207, 226-30. doi:10.1016/j.vetmic.2017.06.022.

van de Giessen, A., Mazurier, S. I., Jacobs-Reitsma, W., Jansen, W., Berkers, P., Ritmeester, W. and Wernars, K. 1992. Study on the epidemiology and control of *Campylobacter jejuni* in poultry broiler flocks. *Appl. Environ. Microbiol.* 58(6), 1913–7.

Van Immerseel, F., De Buck, J., De Smet, I., Mast, J., Haesebrouck, F. and Ducatelle, R. 2002. Dynamics of immune cell infiltration in the caecal lamina propria of chickens after neonatal infection with a *Salmonella enteritidis* strain. *Dev. Comp. Immunol.* 26(4), 355-64. doi:10.1016/S0145-305X(01)00084-2.

Van Immerseel, F., De Buck, J., Boyen, F., Bohez, L., Pasmans, F., Volf, J., Sevcik, M., Rychlik, I., Haesebrouck, F. and Ducatelle, R. 2004a. Medium-chain fatty acids decrease colonization and invasion through hila suppression shortly after infection of chickens with *Salmonella enterica* serovar enteritidis. *Appl. Environ. Microbiol.* 70(6), 3582–7. doi:10.1128/AEM.70.6.3582-3587.2004.

Van Immerseel, F., De Buck, J., De Smet, I., Pasmans, F., Haesebrouck, F. and Ducatelle, R. 2004b. Interactions of butyric acid- and acetic acid-treated *Salmonella* with chicken primary cecal epithelial cells *in vitro*. *Avian Dis.* 48(2), 384–91. doi:10.1637/7094.

Van Immerseel, F., Russell, J. B., Flythe, M. D., Gantois, I., Timbermont, L., Pasmans, F., Haesebrouck, F. and Ducatelle, R. 2006. The use of organic acids to combat *Salmonella* in poultry: A mechanistic explanation of the efficacy. *Avian Pathol.* 35(3), 182-8. doi:10.1080/03079450600711045.

Veen, T. C., de Bruijn, N. D., Dijkman, R. and de Wit, J. J. 2017, Prevalence of histopathological intestinal lesions and enteric pathogens in dutch commercial broilers with time. *Avian Pathol.* 46, 95–105.

Vicente, J. L., Torres-Rodriguez, A., Higgins, S. E., Pixley, C., Tellez, G., Donoghue, A. M. and Hargis, B. M. 2008. Effect of a selected *Lactobacillus* spp.-based probiotic on *Salmonella enterica* serovar enteritidis-infected broiler chicks. *Avian Dis.* 52(1), 143-6. doi:10.1637/7847-011107-ResNote.

Videnska, P., Sisak, F., Havlickova, H., Faldynova, M. and Rychlik, I. 2013. Influence of *Salmonella enterica* serovar Enteritidis infection on the composition of chicken cecal microbiota. *BMC Vet. Res.* 9, 140. doi:10.1186/1746-6148-9-140.

Visco, R. J. and Burns, W. C. 1972a. *Eimeria tenella* in bacteria-free and conventionalized chicks. *J. Parasitol.* 58(2), 323-31.

Visco, R. J. and Burns, W. C. 1972b. *Eimeria tenella* in bacteria-free chicks of relatively susceptible strains. *J. Parasitol.* 58(3), 586-88.

Visco, R. J. and Burns, W. C. 1972c. *Eimeria tenella* in monoflora and diflora chicks. *J. Parasitol.* 58(3), 576-85.

Visscher, C. F., Abd El-Wahab, A., Ahmed, M. F. E., Hankel, J., Taube, V. and Kamphues, J. 2017. Influence of different protein sources in the broiler diet on the presence of *Campylobacter* spp. in excreta and caecal content. *J. Anim. Physiol. Anim. Nutr. (Berl)* 101 (Suppl. 1), 95-104.

Visscher, C., Klingenberg, L., Hankel, J., Brehm, R., Langeheine, M. and Helmbrecht, A. 2018. Influence of a specific amino acid pattern in the diet on the course of an experimental *Campylobacter jejuni* infection in broilers. *Poult. Sci.* 97(11), 4020-30. doi:10.3382/ps/pey276.

Waldenstedt, L., Elwinger, K., Hooshmand-Rad, P., Thebo, P. and Uggla, A. 1998. Comparison between effects of standard feed and whole wheat supplemented diet on experimental *Eimeria tenella* and *Eimeria maxima* infections in broiler chickens. *Acta Vet. Scand.* 39(4), 461-71.

Waldenstedt, L., Elwinger, K., Lunden, A., Thebo, P., Bedford, M. R. and Uggla, A. 2000. Intestinal digesta viscosity decreases during coccidial infection in broilers. *Br. Poult. Sci.* 41(4), 459-64. doi:10.1080/713654959.

Wang, S., Peng, Q., Jia, H. M., Zeng, X. F., Zhu, J. L., Hou, C. L., Liu, X. T., Yang, F. J. and Qiao, S. Y. 2017. Prevention of *Escherichia coli* infection in broiler chickens with *Lactobacillus plantarum* B1. *Poult. Sci.* 96(8), 2576-86. doi:10.3382/ps/pex061.

Wei, S., Morrison, M. and Yu, Z. 2013. Bacterial census of poultry intestinal microbiome. *Poult. Sci.* 92(3), 671-83. doi:10.3382/ps.2012-02822.

Welkos, S. L. 1984. Experimental gastroenteritis in newly-hatched chicks infected with *Campylobacter jejuni*. *J. Med. Microbiol.* 18(2), 233-48. doi:10.1099/00222615-18-2-233.

Wideman Jr., R. F. 2016. Bacterial chondronecrosis with osteomyelitis and lameness in broilers: a review. *Poult. Sci.* 95(2), 325-44. doi:10.3382/ps/pev320.

Williams, R. B. 2003. Coccidial and clostridial interactions in broilers vaccinated against coccidiosis. *World Poult.* 19, 26-8.

Williams, R. B. 2005. Intercurrent coccidiosis and necrotic enteritis of chickens: rational, integrated disease management by maintenance of gut integrity. *Avian Pathol.* 34(3), 159-80. doi:10.1080/03079450500112195.

Withanage, G. S., Mastroeni, P., Brooks, H. J., Maskell, D. J. and McConnell, I. 2005. Oxidative and nitrosative responses of the chicken macrophage cell line MQ-NCSU to experimental *Salmonella* infection. *Br. Poult. Sci.* 46(3), 261-7. doi:10.1080/00071660500098608.

Wolfenden, R. E., Pumford, N. R., Morgan, M. J., Shivaramaiah, S., Wolfenden, A. D., Pixley, C. M., Green, J., Tellez, G. and Hargis, B. M. 2011. Evaluation of selected direct-fed microbial candidates on live performance and *Salmonella* reduction in commercial turkey brooding houses. *Poult. Sci.* 90(11), 2627-31. doi:10.3382/ps.2011-01360.

Wu, S. B., Stanley, D., Rodgers, N., Swick, R. A. and Moore, R. J. 2014. Two necrotic enteritis predisposing factors, dietary fishmeal and *Eimeria* infection, induce large changes in the caecal microbiota of broiler chickens. *Vet. Microbiol.* 169(3-4), 188-97. doi:10.1016/j.vetmic.2014.01.007.

Xie, H., Rath, N. C., Huff, G. R., Huff, W. E. and Balog, J. M. 2000. Effects of *Salmonella typhimurium* lipopolysaccharide on broiler chickens. *Poult. Sci.* 79(1), 33-40. doi:10.1093/ps/79.1.33.

Yair, S., Yaacov, D., Susan, K. and Jurkevitch, E. 2009. Small eats big: ecology and diversity of *Bdellovibrio* and like organisms, and their dynamics in predator-prey interactions. In: Lichtfouse, E., Navarrete, M., Debaeke, P., et al. (Eds), *Sustainable Agriculture*. Springer Netherlands, Dordrecht, pp. 275-84.

Yitbarek, A., Weese, J. S., Alkie, T. N., Parkinson, J. and Sharif, S. 2018. Influenza A virus subtype H9N2 infection disrupts the composition of intestinal microbiota of chickens. *FEMS Microbiol. Ecol.* 94(1). doi:10.1093/femsec/fix165.

Chapter 12

Microbial ecology and function of the gastrointestinal tract in layer hens

Steven C. Ricke, University of Arkansas, USA

1 Introduction

The commercial egg industry in the United States has undergone extensive changes in bird housing management, food safety issues, and nutrition over the past few decades. The housing management transition from layer hens being confined to cages to cage free systems has not only resulted in new economic challenges, but also for food safety, bird welfare, and performance (Holt et al., 2011; Lay Jr. et al., 2011; Mench et al., 2011; Thompson et al., 2011). When maintained in cage free systems, layer hens are exposed to a different type of environment which impacts not only their behavior but potentially the environmental microbial populations they are exposed to during egg production. This interface with these types of environments would also likely contribute to the level of contact with foodborne pathogens and the microbial contamination of eggs produced under these conditions. Microbial contamination of eggs could influence the types of microorganisms present on the surface of the eggs and subsequent egg quality depending on the ability of specific microorganisms to penetrate the defenses of the eggs. Strategies to limit microbial contamination on eggs have been the subject of numerous

http://dx.doi.org/10.19103/AS.2019.0059.13

research studies over the years, particularly for removal of foodborne pathogens. However, foodborne pathogen colonization in the gastrointestinal tract (GIT) of the hen is considered a critical element for overall prevention of foodborne pathogen establishment in poultry flocks.

From a food safety standpoint, the primary focus over the years in egg production has been directed toward species of the foodborne pathogen *Salmonella*. *Salmonella enterica* subspecies *enterica* serovar Enteritidis (*S.* Enteritidis) has been the species most closely identified with egg laying hens and egg contamination but other species such as *S.* Heidelberg have on occasion also been associated with egg production (Foley et al., 2011; Howard et al. 2012; Kaldhone et al., 2016). *S.* Enteritidis possesses the ability to not only colonize the GIT of the layer hen but, in turn, can invade and infect the reproductive tissues of the hen, contaminating the eggs being produced by the infected hen (Holt, 2003; Ricke 2003a; Gantois et al., 2009; Howard et al., 2012; Ricke, 2017). The threat of *S.* Enteritidis colonization in layer hens occurs at a young age with young layer chicks being at high risk due to limited establishment of a mature GIT microbial population at that stage of their development (Ricke, 2017). However, there is evidence that disturbances of the GIT microbial population in adult hens can also lead to subsequent vulnerability to *S.* Enteritidis colonization as well (Ricke, 2003a, 2017). This has led to a plethora of interventions to attempt to short circuit *S.* Enteritidis colonization of the layer hen GIT (Ricke and Gast, 2016). This includes vaccines, chemical feed amendments such as organic acids and botanicals, and biological agents including probiotics and bacteriophage (Ricke, 2003b, 2018; Galiş et al., 2013; Ricke et al., 2013; Ricke and Gast, 2016).

While some interventions such as vaccines primarily engage the immune system, most of the feed amendments target the layer hen GIT microbial populations either by changing the composition or their metabolic activities. As molecular technologies have advanced, it has now become possible to develop a much higher resolution for profiling the GIT microbiota and the corresponding host responses (Oakley et al., 2014; Stanley et al., 2014; Ricke et al., 2017). As a result, a better understanding of the avian GIT microbiome is becoming a reality and key factors that influence its compositional development as birds become older, fed different diets, or administered feed amendments such as antibiotics that elicit specific antimicrobial properties. This understanding of the avian GIT microbiome is anticipated to help design intervention strategies by identifying the most effective feed additives and the best means for optimal delivery to the GIT. This is certainly true for the layer hen GIT microbiome, but challenges remain to determine how that microbiome impacts layer hen physiology, bird health, and egg laying performance. In this chapter the current knowledge of the layer hen microbiome and its development over the lifetime of the bird, the interaction of the GIT microbial population with foodborne pathogens such as

Salmonella, and the impact of feed additives on the layer hen microbiome will be discussed.

2 Layer hen gastrointestinal tract (GIT) structure and function

Most of what is known about the chicken GIT structure and function has not been differentiated between layers and broilers. Presumably, most activities would be similar with differences more likely related to age and other factors. Therefore, this section will be presented as a general description of the different GIT compartments. The avian GIT gross anatomy has been characterized in a wide range of studies based on microscopic examination dating from over a century ago, but connecting structure with function has been a more recent development (Turk, 1982; Svihus, 2014). The avian GIT begins with the mouth which consists of the beak, tongue, salivary glands, and pharynx (Turk, 1982). Leasure and Link (1940) collected saliva fluids from mature hens and characterized quantity, color, odor, pH, mucin content, and enzyme activities among other properties. They concluded that the pH ranged from 5.92 to 7.15 and that amylase activity was always present with only limited levels of lipase. This is consistent with evidence based on carbon 14 labeling, that some sugars such as glucose are removed in the mouth cavity (Soedarmo et al., 1961).

From the mouth food enters the esophagus, a thin walled flexible tubular structure which allows solid feed to reach the gastric regions of the avian GIT (Zaher et al., 2012). Once traversing the esophagus, feed can either enter the crop or pass directly into the proventriculus and gizzard (Svihus, 2014). The esophagus leads into the crop lined with stratified squamous epithelium which varies in surface texture between the proximal and distal region in male and female birds (Bayer et al., 1975; Turk, 1982). While primarily thought to function as a storage organ, early work indicated that extensive starch hydrolysis occurs with glucose apparently being absorbed in the crop and also being fermented by the crop microbial population forming short chain fatty acids (SCFA) and lactic acid as end products (Soedarmo et al., 1961; Bolton, 1965; Pritchard, 1972; Svihus, 2014). It also appears that fiber can be fermented by the crop microbiota into lactic acid and acetate and depending on retention time can result in considerable fermentation activity (Bayer et al., 1978; Svihus, 2014). Svihus (2014) concluded that the crop made minimal nutritional contributions other than transient storage and extensive moisturization of the feed. The storage role of the crop is important since the proventriculus and gizzard are limited in their storage capacity (Svihus, 2014). Early work indicated that passage rate in the crop was related to feed intake and with larger amounts of feed consumed, larger quantities and proportions were stored in the crop over a period of time (Heuser, 1945). However, this may also depend on the type of feeding regime as Svihus (2014) noted that continuous fed birds may in fact minimize involvement of the crop and bypass most ingested feed into the proventriculus and gizzard versus meal-fed birds which use the crop for storage. Layer hen crop composition has also been shown to respond to oviposition. Mongin (1976) observed that the crops of laying hens were full after 10 h of oviposition but almost empty at 2, 4, and 22 h, while pH declined after oviposition, and was lowest when the crop was full. Total crop calcium was low at 2, 6, and 22 h after oviposition, but was high during shell formation.

The proventriculus is the entry point for the posterior esophagus and is a glandular GIT organ that secretes hydrochloric acid and pepsinogen into the incoming feed to initiate digestion (Turk, 1982). The thick mucosal proventriculus surface contains numerous secretory glands that possess ducts which open into the proventriculus papillae (Turk, 1982). The combination of the proventriculus and gizzard (ventriculus) serve as the functional 'stomach' organs for birds by combining the secretions from the proventriculus with the grinding action of the gizzard (Svihus, 2014). The proventriculus secretions utilize the combination of low pH and pepsin to breach the plant cell walls prevalent in cereal grains by breaking down the fiber components such as hemicelluloses and hydrolyzing the proteins in these cell walls that provide structural integrity (Moran Jr., 2016). In summarizing several research studies, Svihus (2011) concluded that for broilers the proventriculus/gizzard pH ranged between 1.9 and 4.5 while for layers the inclusion of dietary calcium carbonate resulted in gizzard pH levels being more in the range of 4 and 5. Along these lines, Mongin (1976) concluded that the level of HCl excreted in the proventriculus along with the fermentation activity in the layer hen crop impacts calcium solubilization and subsequent intestinal availability and that the amount of HCl secreted is related to egg shell deposition. In summary, the combination of proventriculus activity with the grinding action of the gizzard and subsequent reflux back into the proventriculus results in a more homogenous digesta for entry into the intestine (Svihus, 2014; Moran Jr., 2016).

The gizzard or muscular component of the avian stomach consists of a strong circular muscle and less prominent longitudinal muscle that are responsible for the mixing and grinding associated with the GIT organ (Turk, 1982; Zaher et al., 2012). The combination of two opposing thick lateral muscles and two thin anterior and posterior muscles contained within the long anatomical frame of the gizzard results in what Svihus (2011) has described as a simultaneous 'rotary' and crushing action when the gizzard contracts. The grinding action is necessary to reduce feed particle size and increase availability of the feed entering into the intestinal tract of the bird. Moreover, since the gizzard acts as a sieve, the resulting digesta must be reduced to a certain size before leaving through the gizzard and most digesta particles entering the duodenum are less than 0.1 mm (Svihus, 2011). Both incoming particle size of diet and compositional changes such as increased fiber components can alter the retention in the proventriculus/gizzard and increase the size of the gizzard as a consequence of the need for particle size reduction and the corresponding gizzard size can just as rapidly decrease when these variables are removed from the diet (Svihus, 2011). While most of what is known about changes in gizzard size is based on observations on broilers, this relationship appears to be true for layers as well. For example, Hetland et al. (2005) noted that layers housed on litter floor fed wheat diets and allowed to consume the wood shavings, exhibited gizzard weights 60% higher compared to layers with no access to wood shavings. They concluded that not only did the gizzard possess extensive ability to retain and reduce coarse insoluble fiber particles, but the presence of these structural fibers could actually improve feed utilization in birds receiving high concentration-low fiber diets by increasing gizzard activity.

The small intestine is multilayered with serosal, circular muscular, submucosal, and mucosal layers, respectively, and the tract is composed of three segments, albeit not physically distinguishable from each other, and include a proximal region, the duodenum, followed by the jejunum, and finally the anterior region, the ileum which is in proximity of the ceca (Turk, 1982). Most of the host digestion and subsequent absorption of incoming dietary material occurs in the small intestine and this is initiated in the duodenum with the release of pancreatic digestive enzymes via bile ducts (Turk, 1982). This occurs along with the introduction of bile and pancreatic fluids which result in an increase of the pH of the acidic gizzard contents entering the duodenum to a pH level above 6 and initiating the enzymatic digestion process (Svihus, 2014). The small intestinal lining contains two major cell types, epithelial absorptive cells and mucopolysaccharide producing goblet cells (Turk, 1982). The cell surface is highly folded in the form of villi which contain on their surfaces specialized absorptive columnar epithelial cells referred to as microvilli that project into the lumen while underneath this cell layer is the lamina propria where capillary blood vessels are located (Turk, 1982). The shape of the villi varies with each small intestinal segment and they are functionally distinct with most of the nutrients being hydrolyzed and absorbed in the jejunum, whereas the ileum appears to be more involved in mineral and water absorption (Turk, 1982; Svihus, 2014).

It is not clear whether there would be detectable differences between broiler and layer hen small intestinal function and structure other than the fact that layer hens are held in production much longer than broilers and potential genetic line differences may exist. It might be expected that certain nutrients such as calcium would be more efficiently sequestered in the small intestine of laying hens during egg production given the high demand for calcium in egg shell formation (Hurwitz et al., 1973; Etches, 1987). This fits with the early observations by Hurwitz and Bar (1965) and Hurwitz et al. (1973) using yttrium-91 as a non-absorbed marker to demonstrate that calcium absorption was greater during shell deposition, particularly in the duodenum. Likewise, Waddington et al. (1989) detected an increase in calcium absorption in response to shell formation compared to birds producing soft shelled eggs. Using an *in vivo* perfusion approach, Nys and Mongin (1980) did not see a change in jejunal calcium permeability between egg laying hens before and during egg-shell calcification and concluded that net absorption during egg shell formation was due to an increase in soluble calcium carbonate in the upper intestinal tract and not modulation of passive calcium translocation. In summarizing several research studies, Adedokun and Adeola (2013) concluded that for most animals, calcium absorption occurs throughout the small intestine with active transporters in the duodenum and upper jejunum via the transcellular pathway while the nonsaturable paracellular pathway occurs in all of the small intestinal segments. However, due to their longer tract lengths and increased surface areas coupled with their respective transit times, Adedokun and Adeola (2013) pointed out that the distal jejunum and ileum may be the more major contributors to calcium absorption than previously thought.

At the posterior of the ileum is the ileo-cecal-colic junction and at the terminal ileum end is a muscular ring which serves as a sphincter to the cecal openings that are located posterior to this ring (Turk, 1982). The paired blind ended ceca are attached on either side of the colon and are distal to the muscular ring separating the ileum from the colon (Clench and Mathias, 1995; Svihus et al., 2013). Chicken ceca are not sacculated, but can be distinguished by three regions starting with thicker muscular walls and mucosa at the juncture, followed by thinner walls at the mid-ceca containing longitudinal epithelial ridges on the surface, and finally relatively smooth surfaces at the terminal end of the ceca (Turk, 1982; Svihus et al., 2013). The proximal ceca have been shown to contain villi, lymphoid, and goblet cells with the number of goblet cells declining from the proximal to the terminal end of the ceca (Turk, 1982; Svihus et al., 2013). Circular and longitudinal contractions mix digesta in the ceca, but digesta retention in the ceca is relatively long (potentially over 24 h) with emptying occurring infrequently and expelling of content is by synchronized high-amplitude peristaltic actions of both ceca and the colon (Duke, 1989; Svihus et al., 2013; Svihus, 2014). Attempts to measure passage rates in GIT compartments including the ceca of layer hens fed different diets using the rare earth hafnium have been conducted, but the results were inconclusive presumably because of inconsistent marker retention on the dietary material fed to the layer hens (Dunkley et al., 2008).

Functionality of the ceca is dependent to some extent on the composition of the digesta that is introduced into the ceca and the resulting microbial activities (Svihus et al., 2013). Digesta entering the ceca include finely small and soluble ground particles and potentially low-molecular-weight non-viscous molecules such as short chain oligosaccharides and diets high in fiber require an active gizzard to achieve sufficient particle size reduction (Svihus et al., 2013). It is conceivable, given the fermentation activity that occurs in the ceca, that some prebiotic compounds and soluble fibers escaping microbial breakdown in the small intestine would qualify as candidates for entry into the ceca and this has been reported to occur in layer hen cecal contents (Józefiak et al., 2004; Ricke et al., 2013; Ricke, 2018; Svihus et al., 2013). Other cecal microbial activities that contribute to cecal functionality include generation of SCFA, uric acid breakdown, and subsequent nitrogen recycling with all of the metabolic activities generating end products for potential utilization by the bird (Annison et al., 1968; Goldstein, 1989; Karasawa, 1989; Józefiak et al., 2004; Stanley et al., 2014; Pineda-Quiroga et al., 2019). While the quantitative importance of SCFA and renal nitrogen recycling to the host are unclear, the ceca do serve as the major GIT site for water and electrolyte absorption and this has been described in detail previously thus will not be discussed in the current review (Björnhag, 1989; Braun and Campbell, 1989; Goldstein, 1989; Karasawa, 1989; Svihus et al., 2013; Svihus, 2014).

At the ileo-cecal-colic junction is the avian colon also referred to as the large intestine which extends to the cloaca, but overall length is relatively short in birds compared to mammals (Turk, 1982). The colon surface is populated by goblet cells in the mucosa, lymph nodes, and a few villi (Turk, 1982). The epithelial cells are columnar (Turk, 1982). The hen colon may possess some transport capacity of sugars and amino acids based on early preliminary work characterizing the response of isolated hen mucosa in response to changes in sodium load (Munck et al., 1979). In a follow-up study on hens fed either low or high sodium levels, Lind et al. (1980a) compared short-circuit current responses to sugars and amino acids along with electrolyte transport of isolated intestinal tissue from the hens fed the different diets mounted on glass plates with the mucosal surface facing upward. Short-circuit current responses of colonic tissue from high sodium fed layer hens were stimulated by galactose, glucose, leucine, and lysine but considerably less for galactose, glucose, and lysine in tissue from hens fed low sodium diets leading the authors to suggest co-transport with sodium. In a second study, Lind et al. (1980b) compared the transport kinetics of lysine, leucine, and galactose of colonic tissues from the same layer hen dietary groups and demonstrated that for the high sodium diet, all three nutrients were actively transported across the epithelium reaching steady-state concentrations in the mucosal tissues several fold greater than the surrounding culture media. More recently, secondary active transport of SCFA has also been reported for the hen colon (Holtug, 1989; Holtug et al., 1992). In summary, it remains to be determined how much these colonic functions contribute to laying hen metabolism and what impact the colonic GIT microbiota have on these activities both directly and indirectly.

3 Layer hen gastrointestinal tract (GIT) microbial ecology

While most of the early studies characterizing chicken GIT microbial populations were based on broilers, some work had been conducted on layer chicks and mature hens as well (Ricke, 2017). In examining fecal droppings of pullets, Johansson et al. (1948) detected increases in coliform microbial populations of birds fed certain carbohydrate sources, but specific types of carbohydrates such as lactose also enhanced lactic acid bacteria proliferation in the intestine. In general, they noted that GIT microbial populations increased from the duodenum to the ceca with the ceca possessing the greatest numbers of organisms. Although such trends are consistent with what is now known about the poultry GIT microbial ecology, their reliance on conventional microbiological culture techniques likely missed many of the more strict anaerobes present in the GIT. To gain a better understanding and accurate assessment of the cecal microbial populations, microbiological methods that are more representative of the cecal microbial ecosystem were required.

Once it became clear, particularly in the ceca, that many of the microorganisms in residence were strict anaerobes and would likely not be recovered by conventional microbiological cultivation approaches, cultivation methods involving more strict anaerobic techniques became necessary (Coates and Jayne-Williams, 1966; Barnes et al., 1972; Barnes, 1977, 1979; Ricke and Pillai, 1999; Ricke, 2015). This need for strict anaerobic methodology was further supported by the detection of cecal methane (Shrimpton, 1966), a product characteristic of the strictly anaerobic methanogens whose presence have been confirmed more recently in adult layer hens (Saengkerdsub et al., 2007). Consequently, adaptation of strict anaerobic methodology for isolating rumen microorganisms to isolate and enumerate avian GIT microorganisms proved to be successful in initially characterizing the avian GIT anaerobic populations (Hungate, 1950; Caldwell and Bryant, 1966; Barnes et al., 1972). Based on this approach, Barnes et al. (1972) were able to recover over 20% of the total cecal bacterial as anaerobic bacteria and demonstrated that Gram negative anaerobes appeared in weeks 4 through 6 in male and female chicks.

The role and metabolic ecology of GIT cecal microorganisms in laying hens has been suggested indirectly by nutritional studies. Parsons (1984) used cecectomized 50-week-old Leghorn hens to determine the impact on amino acid excretion in hens fed carbohydrate-based diets. Hens were fasted for 24 h then force-fed nitrogen-free diets containing either nitrogen-free cellulose or an uncooked potato starch-citrus pectin combination with both diets containing basal levels of corn starch and glucose. Excreta were collected from the birds 36 h after feeding and analyzed for amino acids. For both diets, cecectomized birds excreted greater amounts of most amino acids compared to intact birds, leading the author to conclude that substantial proteolysis and deamination occurred in the ceca of these birds due to the limited amounts of fermentable carbohydrate reaching the ceca under these dietary fasting conditions. Amino acid excretion was greater for both intact and cecectomized hens fed the potato starch-citrus pectin combination versus cellulose-fed birds. Parsons (1984) hypothesized that this could be a combination of more sloughing off of endogenous protein from the intestinal lining as well as an increase in microbial protein because of the availability of a more fermentable carbohydrate in the form of starch and pectin versus the less fermentable cellulose. In support of potential cecal microbial deamination activity, Marounek and Rada (1998) detected increases in cecal *in vitro* ammonia production in the second month samples collected from female birds sampled over a 4-month period (collections at 1,2, 3, and 4 months). Other nitrogen compounds such as trimethylamine have also been detected in cecal contents and it has been suggested that they may arise from dietary choline metabolized by layer hen cecal microorganisms (March and MacMillan, 1979). However, as the authors point out it is unclear whether trimethylamine generated in the ceca contributes to sensory issues associated with egg 'taint'.

There is evidence that cecal microorganisms from adult layer hens possess fairly broad carbohydrate fermenting capacity. For example, Dunkley et al. (2007a) used an *in vitro* anaerobic cecal incubation system to screen the capabilities of cecal contents from layer hens to ferment high-fiber dietary sources including alfalfa, soybean hulls, beet pulp, and wheat middlings. Based on SCFA profiles, they concluded that production of SCFA was dependent on the combination of cecal inocula and fiber source as cecal contents alone yielded minimal SCFA. They also noted that acetate was the predominant SCFA produced followed by propionate and butyrate. Saengkerdsub et al. (2006) examined alfalfa in the presence of layer hen cecal inocula incubated anaerobically in crimped serum tubes containing a 50:50 hydrogen-carbon dioxide headspace and recovered predominantly acetate as a fermentation product with much lesser quantities of propionate and butyrate. In addition, they detected methane production which could be inhibited in the presence of nitrocompound inhibitors. Similar decreases in methane have been observed for *in vitro* layer hen cecal contents incubated in the presence of the ionophore, monensin and the methanogen inhibitor 2-bromoethanesulphonic acid (Marounek et al., 1996).

In vivo studies also support this cecal profile for laying hens. Savory and Knox (1991) had reported that SCFA from cecal contents of 12- to 14-week-old female medium hybrid birds fed either grass or cellulose-based diets consisted approximately of two-thirds acetate, followed by smaller quantities of propionate, and butyrate acids, and much smaller amounts of iso-butyrate, valerate, and isovalerate acids. They also observed that increasing the dried grass content led to decreases in acetate and increases in isovalerate, while powdered cellulose-fed hens had increased cecal acetate and grass-fed hens also supplemented with a polysacharidase enzyme exhibited an increase in cecal valerate. The SCFA profile with acetate being the primary SCFA produced has also been observed in laying hen trials when cecal contents were examined from birds fed alfalfa as the sole ingredient for molting (Woodward et al., 2005) or in combinations with layer rations containing anywhere from 100% to 70% alfalfa meal (Dunkley et al., 2007c). Likewise, inherent chicken genetic line differences may also be a factor. Walugembe et al. (2015) compared SCFA profiles between broiler and layer-hen chicks fed a high-fiber diet dried distillers grains with solubles and wheat bran over a 21-day period and detected higher concentrations of cecal acetate and propionate in the broiler chicks versus the layer chicks.

While the microbial activity in the ceca of laying hens has been identified as being potentially important to the host, it is less clear in other GIT compartments. Based on culture identification and enumeration, the upper GIT, namely the crop, gizzard, and small intestine at least in broilers, harbor dominant populations of lactic acid bacteria (Barnes et al., 1972; Salanitro et al., 1978; Yusrizal and Chen 2003; Rehman et al., 2007). The crop of the layer hen also appears to contain fairly predominant populations of lactic acid bacteria (Durant et al., 1999). The layer hen crop GIT populations of lactic acid bacteria and their concomitant production of lactic acid were identified as being a potential barrier to *S.* Enteritidis establishment in molted laying hens as well as being possibly involved in prebiotic metabolism (Durant et al., 1999; Ricke, 2003a, 2018). However, the impact of the microbial inhabitants in the small intestine is less clear. Apajalahti and Vienola (2016) have suggested that the predominant lactic acid bacteria in the small intestine may compete for amino acids with the avian host, based on the fact that most lactic acid bacteria are essentially amino acid auxotrophs, and thus are likely to rely on external amino acid profiles similar to those used by the avian host. Apajalahti and Vienola (2016) have estimated that small intestinal lactobacilli may be capable of assimilating anywhere from 3% to 6% of the total dietary protein. Consequently, Apajalahti and Vienola (2016) concluded that the more rapid assimilation of amino acids by the avian host in the proximal small intestine, the less available amino acids are for lactobacilli in the lower sections of the small intestine and in turn avoid increasing protein flow from the terminal ileum to the ceca where deleterious protein fermentation can occur. Whether this occurs to any extent in the older more mature laying hens remains to be determined, but potential excess protein fermentation into compounds such as amines could negatively influence egg quality and sensory characteristics.

4 Layer hen gastrointestinal tract (GIT) molecular characterization

Despite the introduction of strict anaerobic culture methods, much of the avian GIT microbiota still could not be cultivated and/or if cultivated, was difficult to identify (Ricke and Pillai, 1999; Gabriel et al., 2006; Stanley et al., 2014; Ricke, 2015). Consequently, sole reliance on cultivation techniques introduced opportunities for selective recovery of GIT microorganisms that were not always representative of the microbial population in the original GIT sample despite attempts to use ecologically representative media and cultivation conditions (Ricke, 2015). However, the introduction of molecular identification approaches based on advancements in microbial genetics offered the opportunity for microbial population detection and identification that could be somewhat, if not entirely, independent of cultivation.

One of the earliest molecular applications for GIT microbial characterization involved denatured gradient gel electrophoresis (DGGE), coupled with polymerase chain reaction (PCR) amplification of the 16S rRNA gene resulting in amplified products from unique sequences and subsequent gradient gel separation to create banding patterns representative of microorganisms (Hanning and Ricke, 2011). Hume et al. (2003) used DGGE to compare ileal, jejunal, cecal, and colonic microbial populations in developing Leghorn chicks and cecal samples from molted versus nonmolted adult layer hens. Chicks were administered a competitive exclusion probiotic and compared with a control group not given the probiotic over a 32 grow out period. Based on DGGE dendrogram banding patterns, Hume et al. (2003) detected changes in ileal, jejuna, and colonic predominant microbiota as the chicks became older indicating development of these respective GIT compartment microbial populations with some similarity in adjoining compartments. For cecal contents, they observed more complex banding patterns (more visual bands detected) as the chicks aged. When chicks given the CE culture were compared to the control group chicks, there was a 21% similar relatedness between the respective microbial populations at all ages. In the adult Leghorn layer study, Hume et al. (2003) compared cecal contents of birds that were more than 52 weeks old and had been provided various molting diets, namely no feed (feed withdrawal), low calcium, low calcium/low zinc combination, or alfalfa with a control group of birds that were fully fed and thus not molted. Cecal populations from hens subjected to feed withdrawal were the least similar to the other treatments while the low calcium, low calcium/low zinc, and control birds grouped together versus the alfalfa fed birds. These similarities in hen GIT populations to zinc containing diets have been detected in other studies with molted laying hens. For example, when comparing zinc acetate and zinc propionate molted hens, Ricke et al. (2004) reported over 80% similarity between cecal populations from the two treatment groups and over 60% similarity in crop microbial populations.

Dunkley et al. (2007a) used DGGE to assess the impact of inclusion of fiber into laying hen diets as potential candidates for molting dietary sources in an *in vitro* incubation study with layer hen cecal contents. Based on DGGE banding patterns, they concluded that in general, high-fiber inclusion did lead to detectable differences in cecal populations and that incubations containing similar dietary components such as the different alfalfa/layer ration combinations tended to cluster together. In a follow-up *in vivo* study focused on alfalfa as the molt diet, Dunkley et al. (2007c) collected fecal samples during a 9-day molting period and when the birds were euthanized at the end of the trial, cecal samples were obtained as well. Laying hens received a layer ration (full fed, no molt) or were molted by either complete feed withdrawal or provided a 100% alfalfa crumbles diet. Based on the DGGE dendrograms, Dunkley et al. (2007c) concluded that the full-fed nonmolted and alfalfa-fed molted bird fecal and cecal microbial populations clustered more closely with each other compared to the feed withdrawal birds which paralleled the generally higher SCFA levels detected in these treatment-fed birds. They also noted that the later stages of fecal sample DGGE banding patterns matched up with the cecal DGGE profiles. Janczyk et al. (2009) took this a step farther by excising dominant bands from the DGGE gels and identifying the corresponding microorganisms from 23-week-old laying hens fed the green microalga *Chlorella vulgaris* over a 9-day period. In the crop samples, most of the bands were identified as *Lactobacillus* spp. while in the ceca most of the recovered bands aligned with *Ruminococcaceae*, *Lachnospiraceae*, and lactobacilli. Based on the DGGE dendrograms, Janczyk et al. (2009) concluded that adding *C. vulgaris* increased lactobacilli diversity in the crop and also increased overall cecal microbial consortia diversity in the ceca.

The environment that layer hens are housed and fed may also impact the GIT microbial populations. Cui et al. (2017) also used DGGE and subsequent sequencing of specific gel bands to compare free-range laying hens with caged laying hens GIT microbiota as a function of different feeding environments. They selected 8 week and 30-week-old hens from the respective feed systems for sampling small intestine (duodenum, jejunum, and ileum) and cecal contents to conduct DGGE analyses. Based on comparisons of the gel dendrogram patterns and sequencing of the individual bands, they concluded that both age and feeding regime impacted the laying hen GIT microbial populations. Substantial differences were detected between the older hens and the young hens in both free-range and caged hens. The GIT microbial populations of free-range fed birds were also distinguishable from the hens housed in cages. It would be of interest to compare hens in both systems for longer periods of time and also sample more frequently during the time period to determine whether the GIT microbial populations eventually reach a plateau in terms of diversity and if environment influences when that occurs.

5 Layer hen: next-generation sequencing and gastrointestinal tract (GIT) microbiome analysis

While DGGE proved its utility for detecting shifts in GIT microbial populations when birds are fed substantially different diets or other fairly extensive differences such as bird age or environmental conditions, more detailed resolution of the GIT microbial populations was not possible (Ricke et al., 2015; Lee et al., 2019). The introduction of next-generation sequencing offered an entirely new dimension to not only characterize GIT microbial populations, but also apply molecular taxonomic identification to individual members of the GIT microbial community (Ricke et al., 2015, 2017; Lee et al., 2019). Microbial community analyses based on 16S rRNA gene sequencing provided a more comprehensive assessment of the genetically distinguishable bacteria within that microbial community (Ricke et al., 2015, 2017; Lee et al., 2019). As sequencer technology has advanced, the use of sequencing has become more routine and has become a mainstay for comparing GIT microbial populations in food animals (Hanning and Diaz-Sanchez, 2015; Ricke et al., 2017). In tandem with improved sequencing technology, advancements in bioinformatic tools and availability of more sophisticated computer pipelines have provided the analytical and statistical resolution to differentiate changes in GIT microbial populations and detect shifts in GIT populations in response to any number of factors such as diet, environmental differences, or age progression (Awany et al., 2019; Kers et al., 2018; Read and Holmes, 2017; Ricke et al., 2017; Siegwald et al., 2017; Taboada et al., 2017). Thus far, most GIT microbiome sequencing has been done on broilers but some studies have also examined layer chicks and adult hens as well (Videnska et al., 2013; Stanley et al., 2014; Ricke, 2017).

Callaway et al. (2009) employed bacterial tag-encoded FLX amplicon pyrosequencing to compare cecal sample microbial populations from three treatment groups, control birds fed full layer ration and not molted, hens molted by feed withdrawal, and a third group molted with a 100% alfalfa crumble diet similar to the one used by Dunkley et al. (2007b,c). After 12 days of being on the corresponding treatment, hens were euthanized and cecal samples removed for sequencing. Complete removal of feed during the molt period appeared to diminish cecal microbial community diversity as only 59 genera were detected from their cecal samples versus 64 genera from the full-fed birds and 78 genera from the alfalfa molted birds. When taxonomic analyses of the pyrosequencing data were conducted, the genera *Bacteroides*, *Prevotella*, *Clostridium*, *Phascolarctobacterium*, and *Ruminococcus* were detected in cecal samples across all treatments but not necessarily in all birds per treatment. This decrease in diversity was evident in that the proportional levels of several genera including populations of *Bacteroidales*, *Eubacterium*, *Enterococcus*, *Phascolarctobacterium*, *Firmicutes*, and *Veillonella* were less than half the proportions seen in the full-fed nonmolted and the alfalfa molted birds. In a follow-up study by this research group, Escarcha et al. (2012) used this same pyrosequencing approach to track cecal microbial populations in growing layer chicks fed either 0%, 25%, or 50% alfalfa meal diets from day 7 to day 14 after hatching. They concluded that inclusion of alfalfa in the diets of growing layer chicks increased cecal microbial diversity as indicated by the trend of increased number of individual genera and species detected in the alfalfa-fed birds. However, they also noted that fewer genera were identified from these young layer chicks fed alfalfa than the total genera observed in their previous study with adult laying hens (Callaway et al., 2009). As they suggested, this was not a surprise since these were immature chicks at early stages of their life span compared to fully mature layer hens examined in the previous study.

Other dietary factors such as exposure to antibiotics can also influence the development and characteristics of the young layer chick GIT. Videnska et al. (2013) followed the fecal microbiota from 15- and 46-week-old Lohmann Brown laying hens before and after tetracycline or streptomycin administration. In the initial study, they adapted 12-week-old layers for 3 weeks then introduced antibiotics into the drinking water for 7 days and collected fecal material just before the birds were exposed to the respective antibiotics and continued subsequent daily sampling until the end of the trial. They used pyrosequencing to identify the bacteria from the collected fecal samples. Fecal microbial population responses were detectable within 2 days of exposure to the respective antibiotic. Based on real-time taxon-specific PCR analyses, both antibiotics resulted in increased *Enterobacteriales* and decreased *Bifidobacteriales*. *Lactobacillales* increased after termination of antibiotic exposure, while only tetracycline decreased *Clostridiales*.

In a second experiment, Videnska et al. (2013) examined changes in fecal microbiota of 46-week-old hens exposed to repeated cycle of exposure to the same antibiotics used in the single-cycle experiment, except birds were only given the antibiotics for 2 days since responses of the fecal microbiota in the single-cycle experiment were already detectable within that time frame. After subjecting the birds to antibiotics for 2 days, they were removed and birds recovered for 12 days prior to another round of antibiotic administration. Fecal samples were collected throughout this time period and GIT compartments (crop, gizzard, stomach, duodenum, jejunum, ileum, ceca, and colon) were collected at the end of the study, DNA extracted and subjected to taxon-specific real-time PCR, followed by pyrosequencing of hen GIT samples that yielded median PCR values for the representative microbial taxa. Based on the PCR results for the multiple-cycle hens, *Enterobacteriales* were increased initially by tetracycline but not immediately in the second round of tetracycline, while antibiotics reduced both *Bifidobacteriales* and *Clostridiales* immediately after exposure, with variable recovery of both occurring after antibiotic removal. When microbiome composition based on pyrosequencing of the V3/V4 regions of the 16S rRNA genes was examined, the complexity of the microbiota was reduced by the antibiotics 2 days after exposure in both the single-cycle (12-week-old birds) and multiple-cycle (46-week-old birds) experiments as evidenced by the decrease in the total number of operational taxonomic units (OTUs). They observed that at the genera level, *Escherichia* and *Enterococcus* prevalence increased after antibiotic exposure. Beta diversity analyses revealed a detectable separation between antibiotic-treated birds and nontreated birds. Microbiome GIT comparisons, albeit from limited 46-week-old hen samples, revealed that the crop, gizzard, stomach, and small intestine harbored relatively similar microbial populations, but were distinct from the cecal and colonic microbial populations.

In a follow-up study, Videnska et al. (2014a) screened adult layer hen and young broiler fecal samples from commercial layer farms in Central Europe to determine the prevalence of antibiotic resistance genes in fecal microbial populations and the impact of different production systems. Real-time PCR was used to quantitate selected antibiotic genes previously identified to commonly occur in microbiological communities from poultry and pyrosequencing was used to identify individual bacterial members of the fecal microbiome. While there were some differences among countries, the overall prevalence of antibiotic resistance genes was low with *strA* being the most frequent at 1 in 10000 fecal bacteria and no differences were detected when broiler fecal bacteria were compared with laying hen fecal bacteria. The authors considered this reflective of the fact that these farms did not have a recent history of antibiotic administration. When the broiler fecal microbiota sequences were compared with the layer hen fecal microbiota sequences, with a few exceptions, the microbiomes from the layer hens were generally more complex. Videnska et al. (2014a) were unable to identify OTUs that could be considered unique either to the young broilers or the adult hens, but OTUs associated with young broilers typically were members of the phylum *Firmicutes* while those affiliated with layer hens were members of the phylum *Bacteroidetes*. This is consistent with the observation by Kim et al. (2017) that *Firmicutes* was the predominant phylum identified on broiler carcass surfaces during poultry meat processing. Taking the results from both of these studies together, this may infer that the presence of *Firmicutes* is indicative of fecal contamination of broilers entering the processing plant. Fecal microbiome differences were noted for layer hens from different countries, but the authors attributed this to age differences (30 vs. 61-weeks-old) in hens among the commercial flocks. While fecal samples may be reflective of relative differences and similarities, Stanley et al. (2015)

have expressed some concerns with how quantitatively representative fecal samples are of the cecal microbiota.

Videnska et al. (2014b) followed the development of hen cecal populations from day of hatch until hens reached 60 weeks of age. Lohmann Brown Light layer chicks were placed in a commercial egg layer farm and raised without antibiotics. Subsets of layers were sacrificed weekly (three hens per week) from hatch to the termination of the study and cecal contents removed for pyrosequencing of the V3/V4 variable regions of the 16S rRNA genes. While ten different phyla were detected at least once over the course of the study, the primary phyla identified across all ages of birds were *Proteobacteria*, *Firmicutes*, and *Bacteroidetes*. Further analyses of the microbiome sequence data led the authors to conclude that there were four development phases of cecal microbial community compositional changes over the 60-week period. The first week or phase one was characterized by the phylum *Proteobacteria*, family *Enterobacteriaceae*, and genus *Escherichia*, with *Proteobacteria* dramatically decreasing in the second phase of 2–4 weeks to be replaced by the phylum *Firmicutes* and families *Lachnospiraceae* and *Ruminococcaceae*. This taxonomic transition is consistent with the much earlier reports based on cultural isolation and characterization that indicated that the chicken GIT initial microbial population members that colonized the GIT were facultative, but were replaced over time by more anaerobic fermentative microorganisms. During the third phase from 2 to 6 weeks, Bacteroidetes superseded Firmicutes and by the fourth stage (after 7 weeks) the two phyla were each approximately 50% of the microbiome. It would be of interest to extend laying hen cecal microbiome characterization into even older birds as well as birds entering combinations of molting followed by additional egg laying cycles to determine if these microbial populations remain constant or whether further microbiome transitions occur at later stages during and after molting is induced.

Environmental stresses may also influence the GIT microbial population in the laying hen. For example, heat stress has been identified as a negative impact factor in both broiler and laying hen production that requires sufficient ventilation for mitigation (Xin et al., 2011; Lara and Rostagno, 2013). In layer hens, heat stress can lead to a wide array of responses including decreased feed intake, followed by decreased body weight, feed efficiency, egg production/quality, reduced digestibility, decreased plasma protein and calcium levels, and diminished reproductive function (Mahmoud et al., 1996; Mashaly et al., 2004; Rozenboim et al., 2007; Bozkurt et al., 2012; Deng et al., 2012; Lara and Rostagno, 2013). Zhu et al. (2019) used high-throughput metagenomics to determine if a relationship existed between heat stress in layer hens and their GIT microbiota. To evaluate heat stress impact, fecal samples were collected from layer hens housed in the top cages of a commercial facility from May to July with house temperatures reaching as high as 34°C. Library construction of extracted DNA was conducted using an Illumina HiSeq platform and functional contributions determined via metagenomic assembly and gene prediction bioinformatic programs. They noted that *Firmicutes*, *Bacteroidetes*, and *Proteobacteria* were the predominant phyla but after exposure to heat stress, *Firmicutes* decreased while *Bacteroidetes* were increased to levels higher than the control birds. Relative abundance of specific members of genera belonging to *Firmicutes* were also decreased in birds exposed to heat stress including *Clostridium*, *Ruminococcus*, *Lactobacillus*, and *Turicibacter*. When metabolic pathways were projected based on KEGG analyses, methionine and cysteine pathways and benzoate degradation were elevated in heat stressed birds while retinol and phenylpropanoid biosynthesis were decreased under these same conditions. As more becomes known in terms of the layer hen microbiome response to factors such as heat stress, it will be critical to establish host-microbiome interactions by examining not only microbial composition and functionality, but conduct metabolomic analyses to determine if changes in microbial function are related to changes in GIT metabolite profiles and conduct GIT host tissue transcriptomics to detect host responses to any changes that occur. This would offer potential explanations for why certain layer hen physiological and production characteristics respond accordingly to heat stress. In addition, it would be of interest to examine these responses in the different GIT compartments of the laying hen during the entire life cycle of the hen from hatch to egg laying to determine if certain microbial functions influence intestinal functions such as absorption and competition for nutrients in the GIT.

6 Modulation of the laying hen gastrointestinal tract (GIT) microbiome

As antibiotics have been phased out of poultry production, alternative feed amendments have become more commonly used. Consequently, several strategies have been developed over the years that involve either compounds that directly inhibit pathogenic microorganisms already colonized in the GIT or prevent initial colonization. Among the feed amendments that have been tested and/or implemented in poultry are botanical compounds, organic acids, bacteriophage, probiotics, and prebiotics. Historically, interest in modulating poultry GIT with these types of compounds has been to limit colonization foodborne pathogens such as *Salmonella*. The concept behind most of these feed amendments is to either, directly interfere with *Salmonella* metabolism/survival in the GIT, prevent attachment to the intestinal cells, or support GIT microorganisms that are antagonistic to *Salmonella* establishment. However, there is interest in properties that are beyond pathogen prevention and control that have received more attention in recent years. Improvement in performance, bird health, sensory properties are among those attributes now being considered as part of the screening process for identifying compounds and biological entities that are potentially commercially viable. Most of the research has been done with broilers due to their shorter life span and more rapid responses to beneficial improvements (Joerger, 2003; Wernicki et al., 2017; Dittoe et al., 2018; Ricke, 2018; Teng and Kim, 2018).

While most of the implementation of feed amendments has been focused on broilers, research has also been conducted on laying hen applications. The primary focus has been on *Salmonella* control with emphasis on the serovar *S.* Enteritidis, which historically has been the serovar most closely linked with laying hen egg contamination and salmonellosis outbreaks, although other serovars have also been associated with egg layers (St. Louis et al., 1988; Guard-Petter, 2001; Mollenhorst et al., 2005; Foley et al., 2011, 2013: Howard et al., 2012; Martelli and Davies, 2012; Kaldhone et al., 2016; Ricke, 2017). For laying hens, much of the efforts has been made toward developing optimal vaccines, administration of organic acids and botanicals, and screening prebiotic and probiotic candidates that are effective in limiting *Salmonella* establishment in the laying hen (De Cort et al., 2016). A number of *Salmonella* vaccines have been developed over the years for application on laying hens (Aehle and Curtiss III, 2016; De Cort et al., 2016) but are outside the scope of the current review. Much of the early research on dietary amendments for laying hens has been described in detail in previous reviews (Galiş et al., 2013; Callaway et al., 2016; De Cort et al., 2016; Kollanoor Johny and Venkitanarayanan, 2016; Ricke et al., 2013; Ricke, 2016; Wernicki et al., 2017) and will not be discussed in the current review. Instead, the focus will be on efforts to assess the impact of some of these additives on laying hen GIT ecology and layer hen host responses.

Most organic acid applications have been focused on broilers (Dittoe et al., 2018), but some attention for potential application in laying hens has been explored as well. Wang et al. (2015) collected cecal contents from Hyline Grey laying hens after 28 days of feeding for *in vitro* incubations to examine the ability of a coated sodium butyrate product to reduce ammonia production. The *in vitro* incubations were set up for ammonia gas collection from the headspace and supernatant samples were collected after 12 h to be analyzed for ammonium, nitrate-nitrogen, uricase, urease, uric acid, urea, SCFA, and microbial relative abundance. They reported that addition of sodium butyrate to the *in vitro* incubations reduced ammonia gas generation as well as ammonium in the supernatant along with concomitant decreases in urease and uricase activities in the supernatant and accumulation of urea and uric acid. They concluded that the shift in nitrogen metabolism could be attributed to the impact of sodium butyrate on the cecal microbiota as acetate, butyrate, and isovalerate along with total SCFA were increased, but not propionate or valerate. Likewise, shifts in the cecal microbial composition were noted with relative abundance increases in the phylum *Bacteroidetes* and decreases in *Firmicutes* and *Proteobacteria*. At the genera level, sodium butyrate increased relative abundance of *Bacteroides* and *Faecalibacterium* and decreased *Desulfovibrio*, *Helicobacter*, and *Campylobacter*. The authors concluded that reduction in ammonia could be attributed to reduction in ammonia production, but as they also noted, cecal bacterial ammonia can be derived by multiple pathways including hydrolysis of urea and uric acid as well as via deamination of amino acids. To determine the contribution of these pathways to overall ammonia production will require metagenomic analyses to profile the microbial functional diversity on the layer hen ceca similar to what has been done in broilers by Sergeant et al. (2014) to characterize carbohydrate utilization and SCFA formation pathways.

Several studies have involved the administration of live cultures of probiotic organisms to layer hens. When administered to laying hens, there is evidence that probiotic cultures can influence the GIT microbial responses resulting in potential benefits to the layer hen by modulating the GIT microbiota composition and function, limiting *Salmonella*, establishment, improving egg production, enhancing egg quality, and altering egg chemical composition (Salma et al., 2007; Zhang et al., 2012; Lokhande et al., 2013; Ribeiro Jr. et al., 2014; Tang et al., 2015; Park et al., 2016; Oh et al., 2017; Peralta-Sánchez et al., 2019; Pineda-Quiroga et al., 2019). Probiotic cultures may also interact with the GIT microbiota in minimizing the negative impacts of environmental stresses experienced by laying hens during production. For example, the alleviation of heat stress on the hen GIT has been demonstrated with certain probiotics that were able to mitigate the negative impacts of elevated temperatures on egg production, gut morphology, intestinal mucosal immunity, and the intestinal barrier (Deng et al., 2012; Zhang et al., 2017). Zhang et al. (2017) demonstrated that the presence of a combined *Enterococcus faecium/Bacillus subtilis* probiotic statistically increased the ileal levels of *Lactobacilli* compared to heat stressed layers with no probiotic at 10 and 20 days, and statistically decreased cecal *E. coli* in 20-day old birds versus the no probiotic-treated heat stressed birds.

Peralta-Sánchez et al. (2019) examined the impact of the probiotic bacterium, *Enterococcus faecalis* when administered to 16-week-old laying hens as a feed supplement until termination of the study at 76 weeks. The quantity of eggs produced was recorded daily and eggs were sized and classified 1 day per week. For the GIT microbiome assessment, fecal samples were collected on days 7, 15, 40, and 76, and ileal and cecal samples were collected from subsets of birds euthanized on days 40 and 76, respectively, and GIT samples sequenced using an Illumina MiSeq system. The authors noted that the hens receiving the probiotic retained their egg production levels during the second phase of the study (days 40 to 76) versus the drop in egg production observed in the control birds not supplemented with the probiotic. *Enterococcus* spp.

were recovered from fecal material in high numbers but the probiotic strain appeared to increase in the probiotic-fed birds. When GIT microbiomes were compared, not surprisingly the cecal microbial populations were more diverse than the ileal microbial populations and presence of the probiotic increased complexity of both the ileal and cecal microbial populations. There was also more similarity in microbiome composition between GIT microbial populations from the same diets. It would be interesting to conduct follow-up studies to determine if microbiome compositional changes occur elsewhere in the laying hen such as in the crop and gizzard in the presence of the *Enterococcus* probiotic, as well as determining where the highest concentrations of the probiotic are occurring in the different compartments of the GIT.

The classical definition of a prebiotic as originally described by Gibson and Roberfroid (1995) is essentially some form of food or feed additive that is nondigestible by the host animal, but can be utilized by select GIT bacteria resulting in enhancement of either their growth and/or metabolic activities. In practice, for food animals such as poultry, most prebiotics are some version of nondigestible carbohydrates that have the ability to enter into the GIT and not be hydrolyzed by the host digestive enzymes, hence resulting in intact substrates for the GIT microbial population to ferment (Roto et al., 2015; Ricke, 2016, 2018). The presence of prebiotics in the poultry GIT conceptually favors and subsequently selects for certain GIT bacteria which presumably elicit activities that are beneficial to the host (Ricke, 2016, 2018). There have been studies with layer hens that support this. When Rada et al. (2001) fed 5% inulin, they detected increased bifidobacteria in cecal contents versus birds not fed inulin.

More complex sources of potential prebiotics have been explored for limiting *S.* Enteritidis colonization in layer hens. For example, Kulshreshtha et al. (2017) examined red seaweeds (*Chondrus crispus* CC, and *Sarcodiotheca gaudichaudii,* SG) at 2% and 4% of the diet as potential sources of polysaccharides to limit *S.* Enteritidis establishment in laying hens over a 7-day trial. In a challenge study where 4-week-old birds were orally gavaged with *S.* Enteritidis, they collected and monitored fecal samples, egg production data, cecal microbiota, SCFA, and serum IgA. They observed deceases in fecal *S.* Enteritidis on days 5 and 7 post inoculation, reduction in cecal *S.* Enteritidis in birds supplemented with the 4% CC, and increased serum IgA levels. Real-time PCR estimates of selected cecal bacteria indicated that CC4 increased relative abundance of *L. acidophilus* and cecal SCFA analyses revealed higher levels of propionic acid for CC4 versus the control diet-fed birds.

Other prebiotics such as mannan-oligosaccharides (MOS) are believed to mechanistically behave somewhat differently by interfering with pathogen attachment to GIT epithelial cells via binding with bacterial fimbriae (Hooge, 2004). This is supported by MOS fed to 30-week-old hens in a study by

Fernandez et al. (2002) where they observed decreases in cecal populations of Gram-negative Enterobacteria accompanied by increases in Gram-positive *Enterococcus* spp. The mechanism of MOS may be fairly specific for particular microorganisms in the GIT of the layer hen. When Jahanaian and Ashnagar (2015) fed MOS to layer hens challenged with *Escherichia coli,* they did not detect differences in ileal *E. coli* or total GIT bacteria, but noted decreased *Salmonella* and increased *Lactobacillus* levels. MOS impacts on hen egg performance are less clear. For birds fed MOS over the first 35 days, Jahanaian and Ashnagar (2015) observed increased egg production percentage and egg mass and a decreased feed conversion ratio. In heat stressed layers, Bozkurt et al. (2012) concluded that when MOS and an essential oil mixture was fed, no improvements in efficiency of egg production or humoral immune responses were detected, but amelioration by both MOS and essential oils on eggshell characteristics did occur.

Since several of the GIT bacteria that are supported by prebiotics also happen to be probiotic candidates as well, it is not surprising that combinations of prebiotics and probiotics (referred to as synbiotics) have been explored as a means to select for the probiotic once entering the GIT by providing specific substrates (prebiotics) that can be utilized by them (Patterson and Burkholder, 2003). More is known in broilers on prebiotics and their symbiotic counterparts as opposed to layer hens and egg production (Ricke, 2016, 2018). When a probiotic was used in conjunction with the prebiotic isomaltooligosaccharide (IMO), shell egg Haugh units, color of yolk, specific gravity, and thickness of shell were not impacted (Tang et al., 2015). However, both IMO, alone, and when combined with the probiotic decreased egg yolk cholesterol and total saturated fatty acids, but increased total unsaturated fatty acids, total omega 6, and polyunsaturated fatty acids including linoleic and alpha-linoleic acids.

While dietary intervention strategies at the early stages of young chick development have been a primary focus for introducing feed amendments, there is evidence that they can be influential on cecal microbial ecology when introduced to more mature laying hens. Pineda-Quiroga et al. (2019) conducted a comprehensive examination of the cecal microbial and functional profiles of laying hens fed either a dry whey prebiotic, a *Pediococcus acidilactici* probiotic, or the combination of both as a symbiotic to 57-week-old layer hens. Cecal samples were collected at the end of the 70-day experiment and sequenced using an Illumina HiSeq sequencing platform. The authors noted that the main phylum across all treatments was *Bacteroidetes*, with less *Firmictures*, and much less *Actinobacteria* and *Proteobacteria*. Microbial relative abundances were identified and metagenomic analyses conducted. When treatments were compared, the cecal microbiota populations were different for prebiotic- and symbiotic-fed hens, but the probiotic group was not different from the control. When metagenomic analyses were done, all treatments resulted in increased gene functions related to starch, sucrose, pyruvate, and glycerophospholipid metabolism. When fermentation pathway genes were examined, genes associated with butanoate and propionate metabolism were increased in the prebiotic-fed birds in conjunction with an increase in galactose metabolism. As the authors noted, this is not a surprise since whey contains lactose which when hydrolyzed would yield galactose. The symbiotic supplemented diet promoted pathways associated with starch and sucrose metabolism to greater levels than any of the other treatments. In general, all treatments reduced β-lactam resistant associated genes leading the authors to speculate that the feed additives interfere with the presence of antibiotic resistant bacteria and somehow inhibit the transfer of antibiotic resistant genes from this pool. The authors concluded that the introduction of prebiotic, probiotics, or synbiotics, while influencing specific microbial functions in the ceca as well as microbial compositional differences, the overall cecal ecology, was not extensively disturbed.

There are circumstances where the cecal ecology of hens can be considerably disturbed with deleterious consequences. Of course, when severe pathogen infections such as S. Gallinarum acute fowl typhoid occur, major disruptions can occur in the cecal microbiota with substantial reduction of organisms such as *Lactobacillus* spp. and Bifidobacteria accompanied by considerable increases in *E. coli*, *Staphylococcus* spp., hydrogen sulfide producing Clostridia, and fecal streptococci (Kokosharov, 2001). Even if GIT recovery occurs, Simon et al. (2016) demonstrated that early disruption of the layer chick microbial population can lead to long-term impact on the hen's immune system even after the GIT microbial population has been restored. Dramatic changes in dietary regimes can also be disruptive in fully mature adult hens. For example, adult hens are highly vulnerable to *S.* Enteritidis colonization of the GIT and subsequent infection of the reproductive organs, while undergoing feed withdrawal induced molting to initiate a second egg laying cycle (Holt, 1999; Park et al., 2004; Golden et al., 2008; Ricke, 2003b; Ricke et al., 2010, 2013). Withdrawing feed over several days resulted in alterations in the crop and cecal GIT populations and an increase in *S.* Enteritidis virulence gene expression that led to systemic invasion of laying hen reproductive tissues (Durant et al., 1999; Dunkley et al., 2007a; Ricke, 2003b). Based on microbial population DGGE patterns and SCFA profiles, it became obvious that the crop and cecal microbiota had undergone dramatic changes with the complete removal of feed.

However, there are also means to alleviate some of these GIT disruptions. For example, it became clear that alternative approaches to molting needed to be developed that would retain layer hen GIT microbial populations and their corresponding fermentation activities, but still nutritionally limit the laying hen sufficiently to invoke molting and the cessation of egg production (Ricke, 2003a; Ricke et al., 2013). Several attempts with different layer ration molt formulations were explored and high-fiber diets containing alfalfa or wheat middlings appeared to meet the necessary criteria and prevent *S.* Enteritidis colonization, but were still nutritionally limited to induce molt (Seo et al., 2001; Dunkley et al., 2007a, Yousaf and Chaudhry, 2008; Ricke et al., 2010, 2013). Alfalfa as a molt dietary ingredient was also demonstrated to alter cecal microbial composition, as well as being fermentable by cecal bacteria, with the SCFA being produced corresponding to decreases in *Salmonella* based on *in vitro* studies as well as *in vivo* work with molted layer hens, while retaining optimal egg production in the second egg laying cycle (Donalson et al., 2005; Woodward et al., 2005; Saengkerdsub et al., 2006; Dunkley et al., 2007a,c; Callaway et al., 2009).

Alfalfa molt diets have also been combined with the prebiotic fructooligosaccharide (FOS) in an attempt to further augment the ability of the cecal microbial population to limit *Salmonella* colonization. Based on *in vitro* incubations, Donalson et al. (2007) determined that cecal bacteria from laying hens were capable of fermenting FOS combined with feed and could decrease *S.* Typhimurium growth. In an additional *in vitro* cecal study, Donalson et al. (2008b) concluded that the combining FOS with either alfalfa and the layer rations would for the most part result in higher propionate, butyrate, total SCFA, and lactic acid concentrations. As a follow-up study, Donalson et al. (2008a) conducted challenge studies with *S.* Enteritidis in layer hens being fed a molt diet containing FOS combined with a 90% alfalfa and 10% layer ration in a series of independent bird trials. The results were variable as the addition of FOS on crop and cecal fermentation was detectable in some trials but not others, and there were limited differences between alfalfa alone and in combination with FOS in limiting *S.* Enteritidis GIT colonization and systemic invasion of the laying hens undergoing molt. The inconsistency encountered in these trials may be due to a multitude of reasons and Donalson et al. (2008b) suggested that variable feed intake and/or potentially limited numbers of FOS utilizing GIT microorganisms could be responsible. With the current availability of microbiome and metabolomic analytical tools, these inconsistencies can be explored in more depth and more comprehensive profiling may not only support these explanations, but offer additional factors that may account for some of these types of results.

7 Conclusion and future trends

As more molecular tools become available, much more is becoming known about the GIT microbiota taxa identification and the functions associated with those organisms in the laying hen GIT. Likewise, microbial community genetic diversity comparisons among different dietary regimes or other treatments can be conducted to assess the similarities and differences between the respective microbial populations. Based on this, it does appear that certain dietary supplements such as prebiotics can influence not only the composition of the laying hen microbial consortia but the functionality as well. There are opportunities to further screen and optimize prebiotic choices using the GIT microbial responses in parallel with host health and egg production responses to delineate the best matches for optimizing laying hen performance. This will require more analyses of host GIT tissue functions and in turn, relating that to overall host metabolism and physiology, particularly functions related to optimal egg production.

There is also evidence that the GIT microbiota in the hen may not only restrict establishment of pathogens, but may have impacts on host digestibility, the immune system, and absorption of nutrients. While the GIT microbial composition of broilers and layers no doubt share similarities, there are likely differences associated with genetic lines and simply by the fact that commercial broilers have a much shorter life span. In addition, the relationship between GIT compartments needs to be studied in more depth to determine how much influence each compartment has on the adjoining GIT compartments in microbial compositional development as the layer chick matures. The other question that intuitively accompanies this issue of microbial compositional differences among GIT compartments is how the respective microbial communities influence functionality of the GIT microbial communities in the other GIT compartments. The introduction of omics approaches that include metabolomics should provide the means to sort out and delineate some of these relationships. Finally, the influence of the presence of pathogens such as foodborne *Salmonella* and their subsequent interactions during infection with the indigenous GIT microbiota in the layer hen may also be a factor to consider as suggested by Mon et al. (2015). There could very well be a three-way host-pathogen-microbiome interaction occurring in the laying hen GIT and this in turn could influence host responses such as the immune system to the introduction of live *Salmonella* vaccines to the GIT (Ballou et al., 2016; Park et al., 2017). The future of laying hen GIT research will require a combination of increasingly complex large data sets and the appropriate bioinformatic tools to sort out the key factors that are the most critical to layer hen performance.

8 Where to look for further information

Galiş, A. M., Marcq, C., Marlier, D., Portetelle, D., Van, I., Beckers, Y., and Théwis, A. (2013), Control of *Salmonella* contamination of shell eggs - preharvest and postharvest methods: A review. *Compr. Rev. Food Sci. Food Safety* 12, 155-82.

Gantois, I., Ducatelle, R., Pasmans, F., Haesebrouck, F., Gast, R., Humphrey, T. J., and Van Immerseel, F. (2009), Mechanisms of egg contamination by *Salmonella* Enteritidis. *FEMS Microbiol. Rev.* 33, 718-38.

Hester, P. Y. (Ed.) (2017), *Egg Innovations and Strategies for Improvements*. Academic Press, Elsevier, Inc., London, UK.

Kers, J. G., Velkers, F. C., Fischer, E. A. J., Hermes, G. D. A., Stegeman, J. A., and Smidt, H. (2018), Host and environmental factors affecting the intestinal microbiota in chickens. *Front. Microbiol.* 9, 235. doi: 10.3389/fmicb.2018.00235.

Nys, Y. Bain, M., and Van Immerseel, F. (Eds) (2011), *Improving the Safety and Quality of Eggs and Egg Products. Volume 1: Egg Chemistry, Production and Consumption.* Woodhead Publishing Series in Food Science, Technology and Nutrition: Number 214, Cambridge, UK.

Ricke, S. C. (2017), Insights and challenges of *Salmonella* infections in laying hens. *Curr. Opin. Food Sci.* 18, 43-9.

Ricke, S. C., and Gast, R. K. (Eds) (2016), *Producing Safe Eggs-The Microbial Ecology of Salmonella*. Academic Press, Elsevier, Inc., San Diego, CA.

Svihus, B. (2014), Function of the digestive system. *J. Appl. Poult. Res.* 23, 306-14.

Van Immerseel, F., Nys, Y., and Bain, M. (Eds) (2011), *Improving the Safety and Quality of Eggs and Egg Products. Volume 2: Egg Safety and Nutritional Quality*. Woodhead Publishing Series in Food Science, Technology and Nutrition: Number 214, Cambridge, UK.

9 References

Adedokun, S. A., and Adeola, O. (2013), Calcium and phosphorus digestibility: Metabolic limits. *J. Appl. Poult. Res.* 22, 600-8.

Aehle, S., and Curtiss III, R. (2016), Chapter 14. Current and future perspectives on development of *Salmonella* vaccine technologies. In: S. C. Ricke and R. K. Gast (Eds), *Producing Safe Eggs-The Microbial Ecology of Salmonella*. Academic Press, Elsevier, Inc., San Diego, CA, pp. 281-99.

Annison, K. F., Hill, K. J., and Kenworthy, R. (1968), Volatile fatty acids in the digestive tract of the fowl. *Br. J. Nutr.* 22, 207-16.

Apajalahti, J., and Vienola, K. (2016), Interaction between intestinal microbiota and protein digestion. *Anim. Feed Sci. Technol.* 221, 323-30.

Awany, D., Allali, I., Dalvie, S., Hemmings, S., Mwaikono, K. S., Thomford, N. E., Gomez, A., Mulder, N., and Chimusa, E. R. (2019), Host and microbiome genome-wide association studies: Current state and challenges. *Front. Genet.* 9, 637. doi: 10.3389/fgene.2018.00637.

Ballou, A. L., Ali, R. A., Mendoza, M. A., Ellis, J. C., Hassan, H. M., Croom, W. J., and Koci, M. D. (2016), Development of the chick microbiome: How early exposure influences future microbial diversity. *Front. Vet. Sci.* 3, 2. doi: 10.3389/fvets.2016.00002.

Barnes, E. M. (1977), Ecological concepts of the anaerobic flora in the avian intestine. *Amer. J. Clin. Nutr.* 30, 1793-8.

Barnes, E. M. (1979), The intestinal microflora of poultry and game birds during life and after storage. *J. Appl. Bacteriol.* 46, 407-19.

Barnes, E. M., Mead, G. C., and Barnum, D. A. (1972), The intestinal flora of the chicken in the period 2 to 6 weeks of age, with reference to the anaerobic bacteria. *Br. Poultry Sci.* 13, 311-26.

Bayer, R. C., Chawan, C. B., and Bird, F. H. (1975), Scanning electron microscopy of the chicken crop - The avian rumen? *Poult. Sci.* 54, 703-7.

Bayer, R. C., Hoover, W. H., and Muir, F. V. (1978), Dietary fiber and meal feeding influence on broiler growth and crop fermentation. *Poult. Sci.* 57, 1456-9.

Björnhag, G. (1989), Transport of water and food particles through the avian ceca and colon. *J. Exp. Zool. (Suppl.)* 3, 32–7.

Bolton, W. (1965), Digestion in the crop of the fowl. *Br. Poult. Sci.* 6, 97–102.

Bozkurt, M., Küçükyilmaz, K., Çatli, A. U., Çinar, M., Bintaş, E., and Çöven, F. (2012), Performance, egg quality, and immune response of laying hens fed diets supplemented with mannan-oligosaccharide or an essential oil mixture under moderate and hot environmental conditions. *Poult. Sci.* 91, 1379–86.

Braun, E. J., and Campbell, C. E. (1989), Uric acid decomposition in the lower gastrointestinal tract. *J. Exp. Zool. (Suppl.)* 3, 70–4.

Caldwell, D. R., and Bryant, M. P. (1966), Medium without rumen fluid for nonselective enumeration and isolation of rumen bacteria. *Appl. Microbiol.* 14, 794–801.

Callaway, T. R., Dowd, S. E., Wolcott, R. D., Sun, Y., McReynolds, J. L., Edrington, T. S., Byrd, J. A., Anderson, R. C., Krueger, N., and Nisbet, D. J. (2009), Evaluation of the bacterial diversity in cecal contents of laying hens fed various molting diets by using bacterial tag-encoded FLX amplicon pyrosequencing. *Poult. Sci.* 88, 298–302.

Callaway, T. R., Edrington, T. S., Byrd, J. A., Nisbet, D. J., and Ricke, S. C. (2016), Chapter 15. Use of direct-fed microbials in layer hen production-Performance response and *Salmonella* control. In: S. C. Ricke and R. K. Gast (Eds), *Producing Safe Eggs-The Microbial Ecology of Salmonella*. Academic Press, Elsevier, Inc., San Diego, CA, pp. 301–22.

Clench, M. H., and Mathias, J. R. (1995), The avian cecum: A review. *Wilson Bull.* 107(1), 93–121.

Coates, M. E., and Jayne-Williams, D. J. (1966), Chapter 20. Current views on the role of the gut flora in nutrition of the chick. In: C. Horton-Smith and E. C. Amoroso (Eds), *Physiology of the Domestic Fowl*. Oliver and Boyd, Edinburgh, pp. 181–8.

Cui, Y., Wang, Q., Liu, S., Sun, R., Zhou, Y., and Li, Y. (2017), Age-related variations in intestinal microflora of free-range and caged hens. *Front. Microbiol.* 8, 1310. doi: 10.3389/fmicb.2017.01310.

De Cort, W., Ducatelle, R., and Van Immerseel, F. (2016), Chapter 13. Preharvest measures to improve the safety of eggs. In: S. C. Ricke and R. K. Gast (Eds), *Eggs-The Microbial Ecology of Salmonella*. Academic Press, Elsevier, Inc., San Diego, CA, pp. 259–80.

Deng, W., Dong, X. F., Tong, J. M., and Zhang, Q. (2012), The probiotic *Bacillus licheniformis* ameliorates heat stress-induced impairment of egg production, gut morphology, and intestinal mucosal immunity in laying hens. *Poult. Sci.* 91, 575–82.

Dittoe, D. K., Ricke, S. C., and Kiess, A. S. (2018), Organic acids and potential for modifying the avian gastrointestinal tract and reducing pathogens and disease. *Front. Vet. Sci.* 5, 216. doi: 10.3389/fvets.2018.00216.

Donalson, L. M., Kim, W. K., Woodward, C. L., Hererra, P., Kubena, L. F., Nisbet, D. J., and Ricke, S. C. (2005), Utilizing different ratios of alfalfa and layer ration for molt induction and performance in commercial laying hens. *Poult. Sci.* 84, 362–9.

Donalson, L. M., Kim, W. K., Chalova, V. I., Herrera, P., Woodward, C. L., McReynolds, J. L., Kubena, L. F., Nisbet, D. J., and Ricke, S. C. (2007), *In vitro* anaerobic incubation of *Salmonella enterica* serotype Typhimurum and laying hen cecal bacteria in poultry feed substrates and a fructooligosacharide prebiotic. *Anaerobe* 13, 208–14.

Donalson, L. M., McReynolds, J. L., Kim, W. K., Chalova, V. I., Woodward, C. L., Kubena, L. F., Nisbet, D. J., and Ricke, S. C. (2008a), The influence of a fructooligosaccharide prebiotic combined with alfalfa molt diets on the gastrointestinal tract fermentation,

Salmonella Enteritidis infection and intestinal shedding in laying hens. *Poult. Sci.* 87, 1253–62.

Donalson, L. M., Kim, W. K., Chalova, V. I., Herrera, P., McReynolds, J. L., Gotcheva, V. G., Vidanović, D., Woodward, C. L., Kubena, L. F., Nisbet, D. J., and Ricke, S. C. (2008b), *In vitro* fermentation response of laying hen cecal bacteria to combinations of fructooligosaccharide (FOS) prebiotic with alfalfa or a layer ration. *Poult. Sci.* 87, 1263–75.

Duke, G. E. (1989), Relationship of cecal and colonic motility to diet, habitat, and cecal anatomy in several avian species. *J. Exp. Zool. (Suppl.)* 3, 38–47.

Dunkley, K. D., Dunkley, C. S., Njongmeta, N. L., Callaway, T. R., Hume, M. E., Kubena, L. F., Nisbet, D. J., and Ricke, S. C. (2007a), Comparison of *in vitro* fermentation and molecular microbial profiles of high-fiber feed substrates (HFFS) incubated with chicken cecal inocula. *Poult. Sci.* 86, 801–10.

Dunkley, K. D., McReynolds, J. L., Hume, M. E., Dunkley, C. S., Callaway, T. R., Kubena, L. F., Nisbet, D. J., and Ricke, S. C. (2007b), Molting in *Salmonella* Enteritidis-challenged laying hens fed alfalfa crumbles. I. *Salmonella* Enteritidis colonization and virulence gene *hilA* response. *Poult. Sci.* 86, 1633–9.

Dunkley, K. D., McReynolds, J. L., Hume, M. E., Dunkley, C. S., Callaway, T. R., Kubena, L. F., Nisbet, D. J., and Ricke, S. C. (2007c), Molting in *Salmonella* Enteritidis challenged laying hens fed alfalfa crumbles II. Fermentation and microbial ecology response. *Poult. Sci.* 86, 2101–9.

Dunkley, C. S., Kim, W.-K., James, W. D., Ellis, W. C., McReynolds, J. L., Kubena, L. F., Nisbet, D. J., and Ricke, S. C. (2008), Passage rates in poultry digestion using stable isotope tracer markers. *J. Radioanal. Nucl. Chem.* 276, 35–9.

Durant, J. A., Corrier, D. E., Byrd, J. A., Stanker, L. H., and Ricke, S. C. (1999), Feed deprivation affects crop environment and modulates *Salmonella* Enteritidis colonization and invasion of leghorn hens. *Appl. Environ. Microbiol.* 65, 1919–3.

Escarcha, J. F., Callaway, T. R., Byrd, J. A., Miller, D. N., Edrington, T. S., Anderson, R. C., and Nisbet, D. J. (2012), Effects of dietary alfalfa inclusion on *Salmonella* Typhimurium populations in growing layer chicks. *Foodborne Path. Dis.* 9, 945–1.

Etches, R. J. (1987), Calcium logistics in the laying hen. *J. Nutr.* 117, 619–28.

Fernandez, F., Hinton, M., and Van Gils, B. (2002), Dietary mannan-oligosaccharides and their effect on chicken caecal microflora in relation to *Salmonella* Enteritidis colonization. *Avian Pathol.* 31, 49–58.

Foley, S. L., Nayak, R., Hanning, I. B., Johnson, T. J., Han, J., and Ricke, S. C. (2011), Population dynamics of *Salmonella enterica* serotypes in commercial egg and poultry production. *Appl. Environ. Microbiol.* 77, 4273–9.

Foley, S. L., Johnson, T. J., Ricke, S. C., R. Nayak, R., and Danzeisen, J. (2013), *Salmonella* pathogenicity and host adaptation in chicken-associated serovars. *Microbiol. Mol. Biology Revs.* 77, 582–607.

Gabriel, I., Lessire, M., Mallet, S., and Guillot, J. F. (2006), Microflora of the digestive tract: Critical factors and consequences for poultry. *World's Poult. Sci. J.* 62, 499–511.

Galiş, A. M., Marcq, C., Marlier, D., Portetelle, D., Van, I., Beckers, Y., and Théwis, A. (2013), Control of *Salmonella* contamination of shell eggs - preharvest and postharvest methods: A review. *Comp. Revs. Food Sci. Food Safety* 12, 155–82.

Gantois, I., Ducatelle, R., Pasmans, F., Haesebrouck, F., Gast, R., Humphrey, T. J., and Van Immerseel, F. (2009), Mechanisms of egg contamination by *Salmonella* Enteritidis. *FEMS Microbiol. Rev.* 33, 718–38.

Gibson, G. R., and Roberfroid, M. B. (1995), Dietary modulation of the human colonic microflora: Introducing the concept of prebiotics. *J. Nutr.* 125, 1401–12.

Golden, N. J., Marks, H. H., Coleman, M. E., Schroeder, C. M., Bauer Jr., N. E., and Schlosser, W. D. (2008), Review of induced molting by feed removal and contamination of eggs with *Salmonella enterica* serovar Enteritidis. *Vet. Microbiol.* 131, 215–28.

Goldstein, D. L. (1989), Absorption by the cecum of wild birds: Is there interspecific variation? *J. Exp. Zool. (Suppl.)* 3, 103–10.

Guard-Petter, J. (2001), The chicken, the egg and *Salmonella enteritidis*. *Environ. Microbiol.* 3, 421–30.

Hanning, I., and Diaz-Sanchez, S. (2015), The functionality of the gastrointestinal microbiome in non-human animals. *Microbiome* 3, 51. doi: 10.1186/s40168-015-0113-6.

Hanning, I., and Ricke, S. C. (2011), Prescreening methods of microbial populations for the assessment of sequencing potential. In: Y. M. Kwon and S. C. Ricke (Eds), *Methods in Molecular Microbiology 733 - High-Throughput Next Generation Sequencing: Methods and Applications*. Springer Protocols, Humana Press, New York, NY, pp. 159–70.

Hetland, H., Svihus, B., and Choct, M. (2005), Role of insoluble fiber on gizzard activity in layers. *J. Appl. Poult. Res.* 14, 38–46.

Heuser, G. F. (1945), The rate of passage of feed from the crop of the hen. *Poult. Sci.* 24, 20–4.

Holt, P. S. (1999), Impact of induced molting on immunity and *Salmonella enterica* serovar Enteritidis infection in laying hens. In: A. M. Saeed, R. K. Gast, M. E. Potter and P. G. Wall (Eds), *Salmonella enterica serovar Enteritidis in Humans and Animals - Epidemiology, Pathogenesis, and Control*. Iowa State Press, Ames, IA, pp. 367–75.

Holt, P. S. (2003), Molting and *Salmonella enterica* serovar Enteritidis infection: The problem and some solutions. *Poult. Sci.* 82, 1008–10.

Holt, P. S., Davies, R. H., Dewulf, J., Gast, R. K., Huwe, J. K., Jones, D. R., Waltman, D., and Willian, K. R. (2011), The impact of different housing systems on egg safety and quality. *Poult. Sci.* 90, 251–62.

Holtug, K. (1989), Mechanisms of absorption of short-chain fatty acids - coupling to intracellular regulation. *Acta Vet. Scand. Suppl.* 86, 126–33.

Holtug, K., McEwan, G. T. A., and Skadhauge, E. (1992), Effects of propionate on the aid microclimate of hen (*Gallus domesticus*) colonic mucosa. *Comp. Biochem. Physiol.* 103A, 649–52.

Hooge, D. M. (2004), Meta-analysis of broiler chicken pen trials evaluating dietary mannan oligosaccharide, 1993–2003. *Int. J. Poult. Sci.* 3, 163–74.

Howard, Z. R., O'Bryan, C. A., Crandall, P. G., and Ricke, S. C. (2012), *Salmonella* Enteritidis in shell eggs: Current issues and prospects for control. *Food Res. Int.* 45(2), 755–64.

Hume, M. E., Kubena, L. F., Edrington, T. S., Donskey, C. J., Moore, R. W., Ricke, S. C., and Nisbet, D. J. (2003), Poultry digestive microflora diversity as indicated by denaturing gradient gel electrophoresis. *Poult. Sci.* 82, 1100–7.

Hungate, R. E. (1950), The anaerobic mesophilic cellulolytic bacteria. *Bacteriol. Revs.* 14, 1–49.

Hurwitz, S., and Bar, A. (1965), Absorption of calcium and phosphorus along the gastrointestinal tract of the laying fowl as influenced by dietary calcium and egg shell formation. *J. Nutr.* 86, 433–8.

Hurwitz, S., Bar, A., and Cohen, I. (1973), Regulation of calcium absorption by fowl intestine. *Amer. J. Physiol.* 225, 150-4.

Jahanian, R., and Ashnagar, M. (2015), Effect of dietary supplementation of mannan-oligosaccharides on performance, blood metabolites, ileal digestibility, and gut microflora in Escherichia coli-challenged laying hens. *Poult. Sci.* 94, 2165-72.

Janczyk, P., Halle, B., and Souffrant, W. B. (2009), Microbial community composition of the crop and ceca contents of laying hens fed diets supplemented with *Chlorella vulgaris*. *Poult. Sci.* 88, 2324-32.

Joerger, R. D. (2003), Alternatives to antibiotics: Bacteriocins, antimicrobial peptides and bacteriophages. *Poult. Sci.* 82, 640-7.

Johansson, K. R., Sarles, W. B., and Shapiro, S. K. (1948), The intestinal microflora of hens as influenced by various carbohydrates in a biotin-deficient ration. *Poult. Sci.* 56, 619-34.

Józefiak, D., Rutkowski, A., and Martin, S. A. (2004), Carbohydrate fermentation in the ceca: A review. *Anim. Feed Sci. Technol.* 113, 1-15.

Kaldhone, P. R., Foley, S. L., and Ricke, S. C. (2016), Chapter 12. *Salmonella* Heidelberg in layer hens and egg production: Incidence and potential issues. In: S. C. Ricke and R. K. Gast (Eds), *Producing Safe Eggs-The Microbial Ecology of Salmonella*. Academic Press, Elsevier, Inc., San Diego, CA, pp. 235-56.

Karasawa, Y. (1989), Ammonia production from uric acid, urea, and amino acids and its absorption from the ceca of the cockerel. *J. Exp. Zool. (Suppl.)*, 3, 75-80.

Kers, J. G., Velkers, F. C., Fischer, E. A. J., Hermes, G. D. A., Stegeman, J. A., and Smidt, H. (2018), Host and environmental factors affecting the intestinal microbiota in chickens. *Front. Microbiol.* 9, 235. doi: 10.3389/fmicb.2018.00235.

Kim, S. A., Park, S. H., Lee, S. I., Owens, C. M., and Ricke, S. C. (2017), Assessment of chicken carcass microbiome responses during processing in the presence of commercial antimicrobials using a next generation sequencing approach. *Sci. Rep.* 7, 43354. doi: 10.1038/srep43354.

Kokosharov, T. (2001), Some observations on the caecal microflora of the chickens during experimental acute fowl typhoid. *Revue Méd. Vét.* 152, 531-4.

Kollanoor Johny, A. and Venkitanarayanan, K. (2016), Chapter 17. Preharvest food safety-Potential use of plant-derived compounds in layer chickens. In: S. C. Ricke and R. K. Gast (Eds), *Producing Safe Eggs-The Microbial Ecology of Salmonella*. Elsevier, Inc., San Diego, CA, pp. 347-72.

Kulshreshtha, G., Rathgeber, B., MacIsaac, J., Boulianne, M., Brigitte, L., Stratton, G., Thomas, N. A., Critchley, A. T., Hafting, J. and Prithiviraj, B. (2017), Feed supplementation with red seaweeds, *Chondrus crispus* and *Sarcodiotheca gaudichaudii*, reduce *Salmonella* Enteritidis in laying hens. *Front. Microbiol.* 8, 567. doi: 10.3389/fmicb.2017.00567.

Lara, L. J., and Rostagno, M. H. (2013), Impact of heat stress on poultry production. *Animals* 3, 356-69. doi: 10.3390/ani3020356.

Lay Jr., D. C., Fulton, R. M., Hester, P. Y., Karcher, D. M., Kjaer, J. B., Mench, J. A., Mulens, B. A., Newberry, R. C., Nicol, C. J., O'Sullivan, N. P., and Porter, R. E. (2011), Hen welfare in different housing systems. *Poult. Sci.* 90, 278-94.

Leasure, E. E., and Link, R. P. (1940), Studies on the saliva of the hen. *Poult. Sci.* 19, 131-4.

Lee, S. I., Kim, S. A., Park, S. H., and Ricke, S. C. (2019), Molecular and new generation techniques for rapid detection of foodborne pathogens and characterization of microbial communities in poultry meat. In: K. Venkitanarayanan, S. Thakur, and S. C.

Ricke (Eds), *Food Safety in Poultry Meat Production*. Springer Science, New York, NY, pp. 235-60.

Lind, J., Munck, B. G., Olsen, O., and Skadhauge, E. (1980a), Effects of sugars, amino acids and inhibitors on electrolyte transport across hen colon at different sodium chloride intakes. *J. Physiol.* 365, 315-25.

Lind, J., Munck, B. G., and Olsen, O. (1980b), Effects of dietary intake of sodium chloride on sugar and amino acid transport across isolated hen colon. *J. Physiol.* 365, 327-36.

Lokhande, A., Ingale, S. L., Lee, S. H., Kim, J. S., Lohakare, J. D., Chae, B. J., and Kwon, I. K. (2013), The effects of *Rhodobacter capsulatus* KCTC-2583 on cholesterol metabolism, egg production and quality parameters during the late laying periods in hens. *Asian Austral. J. Anim.* 26, 831-7.

Mahmoud, K. Z., Beck, M. M., Scheideler, S. E., Forman, M. F., Anderson, K. P., and Kachman, S. D. (1996), Acute high environmental temperature and calcium-estrogen relationship in the hen. *Poult. Sci.* 75, 1555-62.

March, B. E., and MacMillan, C. (1979), Trimethylamine production in the caeca and small intestine as a cause of fishy taints in eggs. *Poult. Sci.* 53, 93-8.

Marounek, M., and Rada, V. (1998), Age effect on in vitro fermentation pattern and methane production in the caeca of chickens. *Physiol. Res.* 47, 259-63.

Marounek, M., Rada, V., and Benda, V. (1996), Effect of ionophores and 2-bromoethanesulphonic acid in hen caecal methanogenic cultures. *J. Anim. Feed Sci.* 5, 425-31.

Martelli, F., and Davies, R. H. (2012), *Salmonella* serovars from table eggs: An overview. *Food Res. Int.* 45, 745-54.

Mashaly, M. M., Hendricks III, G. L., Kalama, M. A., Gehad, A. E., Abbas, A. O., and Patterson, P. H. (2004), Effect of heat stress on production parameters and immune responses of commercial laying hens. *Poult. Sci.* 83, 889-94.

Mench, J. A., Sumner, D. A., and Rosen-Molina, J. T. (2011), Sustainability of egg production in the United States - The policy and market context. *Poult. Sci.* 90, 229-40.

Mollenhorst, H., van Woudenbergh, C. J., Bokkers, E. G. M., and de Boer, I. J. M. (2005), Risk factors for *Salmonella* Enteritidis infections in laying hens. *Poult. Sci.* 84, 1308-13.

Mon, K. K. Z., Saelao, P., Halstead, M. M., Chanthavixay, G., Chang, H.-C., Garas, L., Maga, E. A., and Zhou, H. (2015), *Salmonella enterica* serovar Enteritidis infection alters the indigenous microbiota diversity in young layer chicks. *Front. Vet. Sci.* 2, 61. doi: 10.3389/fvets.2015.00061.

Mongin, P. (1976), Composition of crop and gizzard contents in the laying hen. *Br. Poult. Sci.* 17, 499-507.

Moran Jr., E. T. (2016), Gastric digestion of protein through pancreozyme action optimizes intestinal forms for absorption, mucin formation and villus integrity. *Anim. Feed Sci. Technol.* 221, 284-303.

Munck, B. G., Lind, J., and Olsen, O. (1979), Uphill intestinal transport of sugars and amino acids eliminated from the chicken colon by feeding a low sodium diet. *Gastroenterol. Clin. Biol.* 3, 173-4.

Nys, Y., and Mongin, P. (1980), Jejunal calcium permeability in laying hens during egg formation. *Reprod. Nutr. Dévelop.* 20(1A), 155-61.

Oakley, B. B., Lillehoj, H. S., Kogut, M. H., Kim, W. K., Maurer, J. J., Pedroso, A., Lee, M. D., Collett, S. R., Johnson, T. J., and Cox, N. A. (2014), The chicken gastrointestinal microbiome. *FEMS Microbiol. Letts.* 360, 100-12.

Oh, J. K., Pajarillo, E. A. B., Chae, J. P., Kim, I. H., and Kang, D.-K. (2017), Protective effects of *Bacillus subtilis* against *Salmonella* infection in the microbiome of Hy-Line Brown layers. *Asian Austral. J. Anim.* 30, 1332–9.

Park, S. Y., Kim, W. K., Birkhold, S. G., Kubena, L. F., Nisbet, D. J., and Ricke, S. C. (2004), Induced moulting issues and alternative dietary strategies for the egg industry in the United States. *World's Poult. Sci. J.* 60, 196–209.

Park, J. W., Jeong, J. S., Lee, S. I., and Kim, I. H. (2016), Effect of dietary supplementation with a probiotic (*Enterococcus faecium*) on production performance, excreta microflora, ammonia emission, and nutrient utilization in ISA brown laying hens. *Poult. Sci.* 95, 2829–35.

Park, S. H., Kim, S. A., Rubinelli, P. M., Roto, S. M., and Ricke, S. C. (2017), Microbial compositional changes in broiler chicken cecal contents from birds challenged with different *Salmonella* vaccine candidate strains. *Vaccine* 35, 3204–8.

Parsons, C. M. (1984), Influence of caecectomy and source of dietary fibre or starch on excretion of endogenous amino acids by laying hens. *Br. J. Nutr.* 51, 541–8.

Patterson, J. A., and Burkholder, K. (2003), Application of prebiotics and probiotics in poultry production. *Poult. Sci.* 82, 627–31.

Peralta-Sánchez, J. M., Martín-Platero, A. M., Ariza-Romero, J. J., Rabelo-Ruiz, M., Zurita-González, M. J., Baños, A., Rodríguez-Ruano, S. M., Maqueda, M., Valdivia, E., and Martínez-Bueno, M. (2019), Egg production in poultry farming is improved by probiotic bacteria. *Front. Microbiol.* 10, 1042. doi: 10.3389/fmicb.2019.01042.

Pineda-Quiroga, C., Borda-Molina, D., Chaves-Moreno, D., Ruiz, R., Atxaerandio, R., Camarinha-Silva, A., and García-Rodriguez, A. (2019), Microbial and functional profile of ceca from laying hens affected by feeding prebiotics, probiotics, and synbiotics. *Microorganisms* 7, 123. doi.10.3390/microorganisms7050123.

Pritchard, P. J. (1972), Digestion of sugars in the crop. *Comp. Biochem. Physiol.* 43A, 195–205.

Rada, V., Duškova, D., Marounek, M., and Petr, J. (2001), Enrichment of bifidobacteria in the hen caeca by dietary inulin. *Folia Microbiol.* 46, 73–5.

Read, M. N., and Holmes, A. J. (2017), Towards an integrative understanding of diet-host-gut microbiome interactions. *Front. Immunol.* 8, 538. doi: 10.3389/fimmu.2017.00538.

Rehman, H. U., Vahjen, W., Awad, W. A., and Zentek, J. (2007), Indigenous bacteria and bacterial metabolic products in the gastrointestinal tract of broiler chickens. *Arch. Anim. Nutr.* 61, 319–35.

Ribeiro Jr., V., Albino, L. F. T., Rostagno, H. S., Barreto, S. L. T., Hannas, M. I., Harrington, D., de Araujo, F. A., Ferreira Jr., H. C., and Ferriera, M. A. (2014), Effects of the dietary supplementation of *Bacillus subtilis* levels on performance, egg quality and excreta moisture of layers. *Anim. Feed Sci. Tech.* 195, 142–6.

Ricke, S. C. (2003a), The gastrointestinal tract ecology of *Salmonella* Enteritidis colonization in molting hens. *Poult. Sci.* 82, 1003–7.

Ricke, S. C. (2003b), Perspectives on the use of organic acids and short chain fatty acids as antimicrobials. *Poult. Sci.* 82, 632–9.

Ricke, S. C. (2015), Chapter 19. Application of anaerobic microbiology laboratory training and writing comprehension to food safety education. In: S. C. Ricke, J. R. Donaldson, and C. A. Phillips (Eds), *Food Safety: Emerging Issues, Technologies and Systems*. Elsevier, Oxford, UK, pp. 395–419.

Ricke, S. C. (2016), Chapter 16. Gastrointestinal ecology of *Salmonella* Enteritidis in laying hens and intervention by prebiotic and non-digestible carbohydrate dietary supplementation. In: S. C. Ricke and R. K. Gast (Eds), *Producing Safe Eggs-The Microbial Ecology of Salmonella*. Academic Press, Elsevier, Inc., San Diego, CA, pp. 323–45.

Ricke, S.C. (2017) Insights and challenges of Salmonella infections in laying hens. *Curr. Opin. Food Sci.* 18: 43–49.

Ricke, S. C. (2018), Impact of prebiotics on poultry production and food safety. *Yale J. Biol. Med.* 91, 151–9.

Ricke, S. C., and Gast, R. K. (Eds) (2016), *Producing Safe Eggs-The Microbial Ecology of Salmonella*. Elsevier, Inc., San Diego, CA, 436pp.

Ricke, S. C., and Pillai, S. D. (1999), Conventional and molecular methods for understanding probiotic bacteria functionality in gastrointestinal tracts. *Crit. Rev. Microbiol.* 25, 19–38.

Ricke, S. C., Hume, M. E., Park, S. Y., Moore, R. W., Birkhold, S. G., Kubena, L. F., and Nisbet, D. J. (2004), Denaturing gradient gel electrophoresis (DGGE) as a rapid method for assessing gastrointestinal tract microflora responses in laying hens fed similar zinc molt induction diets. *J. Rapid Meth. Aut. Microbiol.* 12, 69–81.

Ricke, S. C., Dunkley, C. S., McReynolds, J. L., Dunkley, K. D., and Nisbet, D. J. (2010), Molting in laying hens and *Salmonella* infection. In: P. J. van der Aar and J. Doppenberg (Eds), *Dynamics in Animal Nutrition*. Wageningen Academic Publishers, Wageningen, The Netherlands, pp. 135–46.

Ricke, S. C., Dunkley, C. S., and Durant, J. A. (2013), A review on development of novel strategies for controlling *Salmonella* Enteritidis colonization in laying hens: Fiber-based molt diets. *Poult. Sci.* 92(2), 502–25.

Ricke, S. C., Dawoud, T. M., and Kwon, Y. M. (2015), Chapter 4. Application of molecular methods for traceability of foodborne pathogens in food safety systems. In: S. C. Ricke, J. R. Donaldson, and C. A. Phillips (Eds), *Food Safety: Emerging Issues, Technologies and Systems*. Elsevier, Oxford, UK, pp. 37–63.

Ricke, S. C., Hacker, J., Yearkey, K., Shi, Z., Park, S. H., and Rainwater, C. (2017), Chapter 19. Unravelling food production microbiomes: Concepts and future directions. In: S. C. Ricke, G. G. Atungulu, S. H. Park, and C. E. Rainwater (Eds), *Food and Feed Safety Systems and Analysis*. Elsevier Inc., San Diego, CA, pp. 347–74.

Roto, S. M., Rubinelli, P. M., and Ricke, S. C. (2015), An introduction to the avian gut microbiota and the effects of yeast-based prebiotic compounds as potential feed additives. *Front. Vet. Sci.* 2, 28, 1–18. doi: 10.3389/fvets.2015.00028.

Rozenboim, I., Tako, E., Gal-Garber, O., Proudman, J. A., and Uni, Z. (2007), The effect of heat stress on ovarian function of laying hens. *Poult. Sci.* 86, 1760–5.

Saengkerdsub, S., Kim, W.-K., Anderson, R. C., Woodward, C. L., Nisbet, D. J., and Ricke, S. C. (2006), Effects of nitrocompounds and feedstuffs on *in vitro* methane production in chicken cecal contents and rumen fluid. *Anaerobe* 12, 85–92.

Saengkerdsub, S., Anderson, R. C., Wilkinson, H. H., Kim, W.-K., Nisbet, D. J., and Ricke, S. C. (2007), Identification and quantification of methanogenic archaea in adult chicken ceca. *Appl. Environ. Microbiol.* 73, 353–6.

Salanitro, J. P., Blake, I. G., Muirhead, P. A., Maglio, M., and Goodman, J. R. (1978), Bacteria isolated from the duodenum, ileum, and cecum of young chicks. *Appl. Environ. Microbiol.* 35, 782–90.

Salma, U., Miah, A. G., Tareq, K. M. A., Maki, T., and Tsujii, H. (2007), Effect of dietary *Rhodobacter capsulatus* on egg-yolk cholesterol and laying hen performance. *Poult. Sci.* 86, 714–19.

Savory, C. J., and Knox, A. J. (1991), Chemical composition of caecal contents in the fowl in relation to dietary fibre level and time of day. *Comp. Biochem. Physiol.* 100A, 739–43.

Seo, K. H., Holt, P. S., and Gast, R. K. (2001), Comparison of *Salmonella* Enteritidis infection in hens molted via long-term feed withdrawal versus full-fed wheat middling. *J. Food Prot.* 64, 1917–21.

Sergeant, M. J., Constantinidou, C., Cogan, T. A., Bedford, M. R., Penn, C. W., and Pallen, M. J. (2014), Extensive microbial and functional diversity within the chicken cecal microbiome. *PLoS ONE* 9(3), e91941. doi:10.1371/journal.pone.0091941.

Shrimpton, D. H. (1966), Metabolism of the intestinal microflora in birds and its possible influence on the composition of flavour precursors in their muscles. *J. Appl. Bacteriol.* 29, 222–30.

Siegwald, L., Touzet, H., Lemoine, Y., Hot, D., Audebert, C., and Caboche, S. (2017), Assessment of common and emerging bioinformatics pipelines for targeted metagenomics. *PLoS ONE* 12(1), e0169563. doi: 10.1371/journal.pone.0169563.

Simon, K., Verwoolde, M. B., Zhang, J., Smidt, H., de Vries Reilingh, G., Kemp, and Lammers, A. A. (2016), Long-term effects of early life microbiota disturbance on adaptive immunity in laying hens. *Poult. Sci.* 95, 1543–54.

Soedarmo, D., Kare, M. R., and Wasserman, R. H. (1961), Observations on the removal of sugar from the mouth and the crop of the chicken. *Poult. Sci.* 40, 123–8.

Stanley, D., Hughes, R. J., and Moore, R. J. (2014), Microbiota of the chicken gastrointestinal tract: Influence of health, productivity and disease. *Appl. Microbiol. Biotechnol.* 98, 4301–10.

Stanley, D., Geier, M. S., Chen, H., Hughes, R. J., and Moore, R. J. (2015), Comparison of fecal and cecal microbiotas reveals qualitative similarities but quantitative differences. *BMC Microbiol.* 15, 51 doi 10.1186/s12866-015-0388-6.

St. Louis, M. E., Morse, D. L., Potter, M. E., DeMelfi, T. M., Guzewich, J. J., Tauxe, R. V., Blake, and the *Salmonella enteritidis* working group (1988), The emergence of grade A eggs as a major source of Salmonella enteritidis infections: New implications for the control of salmonellosis. *J. Amer. Med. Assoc.* 259, 2103–7.

Svihus, B. (2011), The gizzard: Function, influence of diet structure and effects on nutrient availability. *World's Poult. Sci. J.* 67, 207–23.

Svihus, B. (2014), Function of the digestive system. *J. Appl. Poult. Res.* 23, 306–14.

Svihus, B., Choct, M., and Classen, H. L. (2013), Function and nutritional roles of the avian caeca: A review. *World's Poult. Sci. J.* 69, 249–63.

Taboada, E. N., Graham, M. R., Carriço, J. A., and Van Domselaar, G. (2017), Food safety in the age of next generation sequencing, bioinformatics, and open data access. *Front. Microbiol.* 8, 909. doi: 10.3389/fmicb.2017.00909.

Tang, S. G. H., Sieo, C. C., Kalavathy, R., Saad, W. Z., Yong, S. T., and Wong, H. K. (2015), Chemical compositions of egg yolks and egg quality of laying hens fed prebiotic, probiotic, and symbiotic diets. *J. Food Sci.* 80, C1686–C95.

Teng, P.-Y., and Kim, W. K. (2018), Review: Roles of prebiotics in intestinal ecosystem of broilers. *Front. Vet. Sci.* 5, 245. doi: 10.3389/fvets.2018.00245.

Thompson, P. B., Appleby, M., Busch, L., Kalof, L., Miele, M., Norwood, B. F., and Pajor, E. (2011), Values and public acceptability dimensions of sustainable egg production. *Poult. Sci.* 90, 2097–109.

Turk, D. E. (1982), The anatomy of the avian digestive tract as related to feed utilization. *Poult. Sci.* 61, 1225–44.

Videnska, P., Faldynova, M., Juricova, H., Babak, V., Sisak, F., Havlickova, H., and Rychlik, I. (2013), Chicken faecal microbiota and disturbances induced by single or repeated therapy with tetracycline and streptomycin. *BMC Vet. Res.* 9, 30. http://www.biomedcentral.com/1746-6148/9/30.

Videnska, P., Rahman, M. M., Marcela Faldynova, M., Babak, V., Matulova, M. E., Prukner-Radovcic, E.-P., Krizek, I., Smole-Mozina, S., Jasna Kovac, J., Szmolka, A., Nagy, B., Sedlar, K., Cejkova, D., and Rychlik, I. (2014a), Characterization of egg laying hen and broiler fecal microbiota in poultry farms in Croatia, Czech Republic, Hungary and Slovenia. *PLoS ONE* 9(10), e110076. doi: 10.1371/journal.pone.0110076.

Videnska, P., Sedlar, K., Lukac, M., Faldynova, M., Gerzova, L., Cejkova, D., Sisak, F., and Rychlik, I. (2014b), Succession and replacement of bacterial populations in the caecum of egg laying hens over their whole life. *PLoS ONE* 9(12), e115142. doi:10.1371/journal.pone.0115142.

Waddington, D., Peddie, J., Dewar, W. A., and Gilbert, A. B. (1989), Regulation of net intestinal calcium uptake in hens laying obligatory soft-shelled eggs. *Br. Poult. Sci.* 30, 341–51.

Walugembe, M., Hsieh, J. C. F., Koszewski, N. J., Lamont. S. J., Persia, M. E., and Rothschild, M. F. (2015), Effects of dietary fiber on cecal short-chain fatty acid and cecal microbiota of broiler and laying-hen chicks. *Poult. Sci.* 94, 2351–9.

Wang, A., Wang, Y., Liao, X. D., Wu, Y., Liang, J. B., Laudadio, V., and Tufarelli, V. (2015), Sodium butyrate mitigates in vitro ammonia generation in cecal content of laying hens. *Environ. Sci. Pollut. Res.* 23(16), 16272–9. doi: 10.1007/s11356-016-6777-z.

Wernicki, A., Nowaczek, A., and Urban-Chmiel, R. (2017), Bacterial therapy to combat bacterial infections in poultry. *Virol. J.* 14, 179. doi: 10.1186/s12985-017-0849-7.

Woodward, C. L., Kwon, Y. M., Kubena, L. F., Byrd, J. A., Moore, R. W., Nisbet, D. J., and Ricke, S. C. (2005), Reduction of *Salmonella enterica* serovar Enteritidis colonization and invasion by an alfalfa diet during molt in Leghorn hens. *Poult. Sci.* 84, 185–93.

Xin, H., Gates, R. S., Green, A. R., Mitloehner, F. M., Moore Jr., P. A., and Wathes, C. M. (2011), Environmental impacts and sustainability of egg production systems. *Poult. Sci.* 90, 263–77.

Yousaf, M., and Chaudhry, A. S. (2008), History, changing scenarios and future strategies to induce moulting in laying hens. *World's Poult. Sci. J.* 64, 65–75.

Yusrizal, Y., and Chen, T. C. (2003), Effect of adding chicory fructans in feed on faecal and intestinal microflora and excretory volatile ammonia. *Int .J. Poult. Sci.* 2, 188–94.

Zaher, M., El-Ghareeb, A.-W., Hamida Hamdi, H., and AbuAmod, F. (2012), Anatomical, histological and histochemical adaptations of the avian alimentary canal to their food habits: *I-Coturnix coturni*. *Life Sci. J.* 9, 253–75.

Zhang, J. L., Xie, Q. M., Ji, J., Yang, W. H., Wu, Y. B., Li, C., Ma, J. Y., and Bi, Y. Z. (2012), Different combinations of probiotics improve the production performance, egg quality, and immune response of layer hens. *Poult. Sci.* 91, 2755–60.

Zhang, P., Yan, T., Wang, X., Kuang, S., Xiao,Y., Lu, W., and Bi, D. (2017), Probiotic mixture ameliorates heat stress of laying hens by enhancing intestinal barrier function and improving gut microbiota, *Italian J. Anim. Sci.* 16, 2, 292–300. doi: 10.1080/1828051X.2016.1264261.

Zhu, L., Liao, R., Wu, N., Gensheng Zhu, G., and Yang, C. (2019), Heat stress mediates changes in fecal microbiome and functional pathways of laying hens. *Appl. Microbiol. Biotechnol.* 103, 461–72.

Part 3

Feed additives and gut health modulation

Chapter 13

Controlling pathogens in the poultry gut

Osman Yasir Koyun and Todd R. Callaway, University of Georgia, USA

1 Introduction

Poultry production is currently the most efficient (based upon pounds of feed/pound of gain) animal production system and constitutes the base of worldwide affordable protein production for human consumption. Over the last several decades, the poultry industry has produced genetic lines of chickens and turkeys that efficiently utilize feed to build muscle mass that meets the demand for high-quality protein in the United States and around the world. According to the National Chicken Council (2018), per capita consumption of poultry in 2010 was 82.3 pounds, and it is estimated that each American will consume approximately 92 pounds in 2018 in the United States. The most recent Census of Agriculture (USDA, 2015) reported that there were 233 770 poultry farms in the United States in 2012. In 2014, the US poultry industry produced 8.54 billion broilers, 99.8 billion eggs and 238 million turkeys (USDA, 2015). The combined value of production from broilers, eggs, turkeys and the value of sales from chickens in 2014 was $48.3 billion, up 9% from $44.4 billion in 2013 (USDA, 2015). Obviously, chicken production in the United States represents a vital contribution to the gross domestic product (GDP) which is expected to increase in the future as well. Therefore, the future profitability of the poultry industry depends on maintaining (or improving) the current production levels

http://dx.doi.org/10.19103/AS.2019.0059.16

to meet growing demand and expand new markets. In addition to meeting the commercial and global protein needs, poultry products must also be safe and free of pathogenic microorganisms. The presence of food-borne pathogenic bacteria in poultry products, coupled with large-scale food-borne illness outbreaks are serious concerns to both consumers and the industry.

Antibiotics have been widely utilized in human and veterinary medicine for many decades for various purposes. Some of them are used for therapeutic reasons to improve animal health and welfare, while the majority of them were administered for prophylactic purposes since the US Food and Drug Administration (FDA) approved the use of antibiotics in animal feeds in 1951 (Jones and Ricke, 2003). It is well known that despite the dearth of knowledge of how the gastrointestinal tract (GIT) microbiome functions, the use of antibiotics in feed (i.e. antibiotic growth promoters, or AGPs) has reduced the incidence of production diseases, improved animal growth rate and feed conversion efficiency, decreased pathogenic bacterial populations in the GIT, thus reducing substrate/nutrient competition in favour of beneficial bacteria (Economou and Gousia, 2015).

The discovery of antibiotics has been one of the tremendous breakthroughs of the twentieth century in the world of medicine; however, overuse and/or misuse of antibiotics predisposes man to opportunistic poultry-associated pathogens and leads to the emergence of a current concern, 'antibiotic resistance'. Therefore, the European Commission (EC) aimed to remove the antibiotic use in feed, and then totally banned the use of all AGPs as of 1 January 2006 (Castanon, 2007). In the United States, consumers have started to coerce the poultry industry into rearing animals without AGPs (Dibner and Richards, 2005); therefore, they have taken initiatives to phase out the AGPs in animal production systems. However, producers have been experiencing challenges due to the ban of AGPs. Removal of AGPs has resulted in an increase in the prevalence of antibiotic resistance of *Campylobacter* and *Salmonella* to varied antibiotics in EU and the United States (Endtz et al., 1991; EFSA, 2014; Economou and Gousia, 2015). Additionally, as a consequence of banning or phasing out of AGPs, producers have encountered an increase in a condition often referred to as 'dysbacteriosis' (Dittoe et al., 2018), which is also called 'wet litter syndrome', 'bacterial overgrowth in the small intestine' or 'feed passage syndrome' (Huyghebaert et al., 2011).

The GIT has been studied for years, but we still lack sufficient knowledge of its internal workings (i.e. the gut has largely been a 'black box'); however, a broad understanding of the ecology of the GIT and the native microbiota therein has grown in recent years (Oakley et al., 2014; Pan and Yu, 2014; Sergeant et al., 2014; Stanley et al., 2014; Waite and Taylor, 2014; Han et al., 2016; Clavijo and Flórez, 2017; Borda-Molina et al., 2018). Recent studies have demonstrated that the GIT is a complicated microbial environment in which the host and the microbiota are dependent on each other in a symbiotic relationship, and is occupied by a complex microbial consortium that forms an interconnected ecosystem that dramatically impacts its environment (e.g. the host bird) (Vispo and Karasov, 1997; Hooper and Gordon, 2001; Oakley et al., 2014; Pan and Yu, 2014). In addition to the competitive nature of the microbes against pathogens, the beneficial/commensal microorganisms in the GIT have a vital role in the health and modulation of the immune system to protect the host from pathogens (Clavijo and Flórez, 2017). The ability of poultry-associated pathogens to colonize in the GIT means that strategies to control these pathogens must include methods to reduce or eliminate the pathogens from live birds. In order to best target foodborne pathogens in the gut, we must understand the environment in which they live. Therefore, a deeper and more nuanced understanding of the biochemical, physiological and microbiological functions of the GIT microbiota is crucial to the sustainability of the poultry industry to prevent disease and food-borne illnesses and to improve growth as well as performance of broiler chickens.

2 The gastrointestinal microbiota

The microbiota of the GIT has undergone a renaissance in research interest in the past 20 years that rivals the 'golden age' of gastrointestinal microbiology of the 1950s and 1960s (Hungate, 1950; Bryant and Burkey, 1953; Salanitro et al., 1974). In general, our understanding of the function of the normal intestinal microbiota in poultry was limited, although the role of pathogens (both animal and human) and the diseases that they lead to were well documented (Oakley et al., 2014; Pan and Yu, 2014; Sergeant et al., 2014; Stanley et al., 2014; Waite and Taylor, 2014; Clavijo and Flórez, 2017).

The GIT of poultry is composed of the oesophagus, crop, proventriculus, ventriculus or gizzard, small intestine (duodenum, jejunum and ileum), caeca, colon and cloaca. Since the poultry GIT is shorter than that of mammalian animals, the digesta moves through the GIT faster in poultry than in mammals (Hughes, 2008; Rougière and Carré, 2010; Borda-Molina et al., 2018). Therefore, shorter retention time allows bacteria to attach to the mucosal epithelial layer and grow; however, the caeca, which are two blind pouches between the ileum, allow this compartment to have a relatively slower passage rate and are a favourable environment to provide a home for a diversified microbiome (Pan and Yu, 2014).

As feed/digesta moves through the GIT, it encounters specific microbial communities localized within each compartment that can ferment different substrates at different rates, via a variety of pathways, and produce a variety of end products. Physical disruption of the diet begins in the crop, along with microbial colonization of feeds. Microbial populations of the crop are predominantly *Lactobacillus* spp. that are present at cell densities up to 10^9 g^{-1} (Wielen et al., 2002; Stanley et al., 2014), and different species of the *Clostridiaceae* family (Clavijo and Flórez, 2017). Unfortunately, the crop is often colonized by *Campylobacter* species (Byrd et al., 1998; Musgrove et al., 2001; Hinton Jr et al., 2002) and can be colonized with *Salmonella* spp. as well (Hinton et al., 2002; Vaughn et al., 2008). The proventriculus is a thick-walled muscular expanded pouch of the oesophagus, and ventriculus (gizzard) are also dominated by *Lactobacilli* (Oakley et al., 2014). The gizzard has been known for its grinding functions, and most of the chemical breakdown of feed is performed in this part of the GIT (Rehman et al., 2007). The gizzard has a low pH due to gastric juices (pepsin and hydrochloric acid), which results in a relatively lower number of cells (well below 10^8 g^{-1}), and acts as a barrier to the passage of pathogens to the lower GIT (Gabriel et al., 2006; Oakley et al., 2014). The small intestine harbours the highest concentration of bacterial cells (up to 10^9–10^{11} g^{-1}), and is dominated by *Lactobacillus*, *Enterococcus* and *Clostridiaceae* (Han et al., 2016). The caecum has the longest retention time of digesta (12–20 h), and is a site for water absorption and urea recycling, as well as home to a secondary carbohydrate fermentation (Clavijo and Flórez, 2017). All these characteristics make the caecum a convenient environment for microorganisms, resulting in having the largest taxonomic diversity in the GIT, with *Firmicutes*, *Bacteroides*, *Proteobacteria* and *Clostridiaceae* all being caecal residents (Clench and Mathias, 1995; Sergeant et al., 2014; Waite and Taylor, 2014). The caeca is a home to a number of metabolic functional niches and pathways of various members of the caecal community. The activity of hydrogenase in the caeca is attributed to *Megamonas*, *Helicobacter* and *Campylobacter*, and appears to have a triggering role in the production of short chain fatty acids (SCFAs) (Oakley et al., 2014). A study conducted by Sergeant et al. (2014) has revealed that there are over 200 non-starch polysaccharide (NSP)-degrading enzymes as well as potential pathways linked to SCFA and acetate production in the chicken caeca metagenome (Sergeant et al., 2014). This degree of biochemical diversity underlies the ability of the microbes in the avian GIT and is critical to understand the microbial ecology of the gut, which can impact colonization of the gut by food-borne pathogenic bacteria.

In the chicken gastrointestinal microbiota, the existence of pathogenic bacteria is crucial to both animal and human health. Among these pathogens, *Campylobacter* spp. (particularly, *C. jejuni* and *C. coli*), *Salmonella enterica*, *E. coli* and *Clostridium perfringens* species are capable of causing food-borne illness in humans (Oakley et al., 2014). In the United States, food-borne diseases account for an estimated 48 million illnesses each year (Gould et al., 2013). Infections caused by *Campylobacter* spp. and *Salmonella* spp. are responsible for massive medical costs to the US economy, reaching approximately $1.56 billion and $365 million respectively per year (Grant et al., 2016).

Salmonellosis is caused by ingesting *Salmonella* spp. and symptoms generally develop 12-72 h after the consumption of contaminated foods (Grant et al., 2016) and are classified into two categories: (1) typhoid fever in humans (caused by *S. typhi* and *paratyphi*) and (2) gastroenteritis in humans and animals caused by other *S. enterica* serovars (Park et al., 2013; Kim et al., 2006; Barrow, 2007). While typhoid fever is a systemic disease where the infected humans manifest fever, severe abdominal pain/cramps and fatigue (Park et al., 2013), pathogenic *Salmonella* strains lead to gastroenteritis symptoms such as nausea, headache, diarrhoea and fever (D'Aoust and Maurer, 2007; Park et al., 2009). The infection is generally self-limiting and mostly does not require any antimicrobial treatment and recovers in a couple of days; however, young children, elderly and immune-compromised people, who are under a risk of severe/profuse diarrhoea, need to be hospitalized (Grant et al., 2016; Park et al., 2013).

Contaminated foods, water and faeces are common agents that transmit *Salmonella* infections; however, the infection is rarely transmitted by human-to-human contact (Murray, 2000). Faecal-oral transmission of *Salmonella* is the main route of infection in both humans and animals (Ricke et al., 2005) and carcass contamination at slaughter can lead *Salmonella* to the food chain (Barrow, 2007). A considerable amount of research has been conducted for decades to elaborate the pathogenesis of *Salmonella* since it is linked to infections in humans; however, there is still insufficient knowledge on the mechanisms of colonization and pathogenicity of the microorganism in poultry, especially chickens. Once contaminated food is ingested, the bacteria have to move through the alimentary canal and survive the gastric acidity via 'acid tolerance response' (Foster, 1993; Audia et al., 2001; Bearson et al., 2006). Subsequently, *Salmonella* organisms that overcome hurdles along the alimentary canal invade GIT organs and tissues, including small intestine and caecum which is the primary site of colonization in poultry (Swaggerty et al., 2017; Berndt et al., 2007; Humphrey, 2004). Surviving microorganisms compete with the GIT microbiota to adhere to enterocytes or M cells to initiate colonization (Ruby et al., 2012). Flagella and fimbriae present on the bacterial surface assists the progress of adhesion to the GIT epithelium (Van Asten and Van Dijk, 2005).

Campylobacter spp. are ubiquitous zoonotic food-borne pathogens that are some of the most common causes of food-borne illness in the United States (Scallan et al., 2011), more specifically *C. jejuni* and *C. coli* species, infecting approximately 2.4 million humans per year (Grant et al., 2016; Park et al., 2013). Since *Campylobacter* spp. are present in the GIT of poultry as commensal organisms, up to 10^7 cfu g^{-1} (Oakley et al., 2014), poultry meat can be contaminated at processing plants in case of rupture of intestinal organs (Ringoir et al., 2007; Horrocks et al., 2009; Smith et al., 2005). After exposure to contaminated food, *Campylobacter* spp. invades epithelial cells and colonizes ileum, jejunum and colon without showing symptoms; however, in symptomatic cases, infected humans can manifest symptoms such as nausea, abdominal pain, fever and bloody diarrhoea that can last for 1 week (Park et al., 2013; Dasti et al., 2010; Park, 2002). If immunocompromised humans are not treated, infection can pass through the blood stream and result in serious consequences (Grant et al., 2016; Mead et al., 1999).

Worldwide consumer pressure and concerns about the effects of antibiotic use in poultry production have led researchers to focus on novel alternatives to antibiotics. The goals of these alternatives should have similar characteristics with AGPs, such as reducing mortality rates, decreasing the number of incidences of infection and maintaining animal productivity while being environment- and consumer-friendly. There are varied non-antibiotic strategies and/or alternatives studied by the researchers to substitute AGPs in animal production. The most popular are probiotics, prebiotics, organic acids and bacteriophages. None of the alternative strategies elaborated in the following sections compensate comprehensively for the removal of AGPs, but recent studies have had promising results to reduce global concerns over the use of AGPs in the poultry industry.

3 Probiotics and competitive exclusion cultures

One of the most well-established methods to mitigate enteric disease is via the administration of live bacteria, known as probiotics, or when fed to animals, 'Direct Fed Microbials' or DFM (Jayne-Williams and Fuller, 1971; Fuller, 1989). Probiotics are defined as 'a live microbial feed supplement which beneficially affects the host animal by improving its intestinal balance' (Fuller, 1991). The main purpose of the probiotic administration is to improve or establish the growth of groups of bacteria that are competitive with, or antagonistic to, pathogenic bacteria (food-borne pathogens and/or animal health pathogens) (Fuller, 1989, 1991; Buntyn et al., 2016). The term 'probiotic' is an umbrella term that includes a variety of microbial ecology-modifying techniques such as (1) introducing a 'normal' microbe or microbial population to colonize the GIT or (2) supplying a limiting nutrient (often termed a 'prebiotic') that is not digested by the host animal and which enables the native commensal microbial population to expand its role in the GIT. Collectively, the goal of these microbial ecology modification methodologies is to occupy all microbial ecological niches in the GIT and thereby hinder the establishment of an opportunistic pathogenic bacterial population (Callaway et al., 2003a, 2017).

The use of probiotic approaches in order to control pathogens in poultry is a common application in the poultry production. Pathogens can be inhibited by probiotics establishing suboptimal environmental conditions for pathogen growth or colonization in a variety of methods. Beginning in the early 1970s, research focused on why the administration of adult chicken microbiota to newly hatched chicks inhibited the ability of *Salmonella* to colonize newly hatched chicks' intestinal tracts (Nurmi and Rantala, 1973; Rantala and Nurmi, 1973; Lloyd et al., 1974; Nurmi et al., 1992). These types of cultures became widely accepted as a specific subset of probiotic treatments known as competitive exclusion (CE) cultures, reduced the colonization of the chicken intestine by pathogenic bacteria (e.g. *Salmonella* and *Campylobacter*) and mitigated the ability of pathogens to attach to the intestine by occupying binding sites (Nurmi et al., 1992; Schneitz, 2005). CE as a technology now is defined as the addition of a normal gastrointestinal microbiota from a healthy adult bird that can be introduced into a naïve (sterile) GIT of a newly hatched chick (Weinack et al., 1982; Wierup et al., 1988; CVM, 1997). The early establishment of a complicated, mature microbial consortium in the gut means that the normal succession of microbes occupying the gut is 'jumpstarted', preventing pathogenic bacteria from occupying an unoccupied (or under-occupied) environmental niche (Nurmi et al., 1992; Corrier et al., 1993; Nisbet et al., 1993a; Stavric and D'Aoust, 1993; Wray and Davies, 2000). To paraphrase military strategists, CE allows us to 'get there first, with the most' with the bacteria that can prevent pathogens from establishing a foothold and colonizing the gut of newly hatched chickens (Nurmi et al., 1992; Mead, 2000; Schneitz, 2005; Mountzouris et al., 2009).

It has been demonstrated that probiotics not only mitigate the ability of pathogens to attach to the surface of the intestine (Kim et al., 2008), but also reduce pathogen colonization by increasing competition for nutrients, as well as the production of antibacterial substances such as volatile fatty acids (VFAs) (Corrier et al., 1991), which are the end products of the gastrointestinal microbial fermentation of carbohydrates or proteins and can be lethal to some species of bacteria, including *E. coli* O157:H7 and *Salmonella* (Wolin, 1969; Barnes et al., 1979; Callaway et al., 2003a). Some bacteria (both native and those found in probiotic preparations) produce antimicrobial compounds such as antibiotics, bacteriocins or colicins to eliminate competitive bacteria in the same or similar environmental niche (Jack et al., 1995). Antibiotics will be discussed elsewhere, but the protein-based antimicrobials (e.g. bacteriocins/colicins) can be used to eliminate food-borne pathogenic bacteria (Stahl et al., 2004; Cutler et al., 2007; Schulz et al., 2015). Colicins and bacteriocins are peptide antimicrobials that disrupt the structure of membranes, and subsequently the proton motive force that bacteria utilize to take up nutrients (Jack et al., 1995; Montville et al., 1995; Joerger, 2003). The use of colicins/bacteriocins includes the use of organisms that produce these compounds as probiotic components as well, thus colonizing the gut with a competitive organism that produces a toxin to its competitors (Schamberger and Diez-Gonzalez, 2002, 2005; Schamberger et al., 2004; Lee et al., 2008).

Probiotics that have been used in the poultry industry usually fall into three categories: (1) pure cultures of a single microbe, (2) mixed defined cultures and (3) mixed undefined cultures (Corrier et al., 1993; Nisbet, 2002). A wide variety of microbial species have been successfully used as probiotics in humans and livestock, including species of *Bacillus*, *Bifidobacterium*, *Enterococcus*, *E. coli*, *Lactobacillus*, *Lactococcus*, *Streptococcus*, many different yeast species, as well as undefined mixed cultures (Callaway et al., 2006, 2008a; Higgins et al., 2010; Shivaramaiah et al., 2011; Tellez et al., 2012). Much of the poultry probiotic research has focused on using *Lactobacillus* and *Bifidobacterium* spp. bacteria in poultry due to the fact that they are normal beneficial inhabitants of the chicken intestinal tract and play a vital role in resistance to pathogen colonization/infections (Patterson and Burkholder, 2003). In addition to these prokaryotic organisms, it has been demonstrated that eukaryotic organisms such as *Saccharomyces boulardii* and *S. cerevisiae* have the ability to bind *Salmonella* spp., thereby preventing or limiting the attachment of the pathogen to the intestinal wall (Mountzouris et al., 2009; Chen et al., 2012; Sanchez et al., 2013; Broadway et al., 2015). For example, the *in vivo* administration of *Saccharomyces cerevisiae* and *Saccharomyces boulardii* has been reported to improve broiler performance (Onifade et al., 1999) and to reduce pathogen load during transport to slaughter (Line et al., 1997). These pathogen-binding effects have been shown in other monogastric and ruminal species as well (Broadway et al., 2015). In addition to these above-mentioned mechanisms, pathogen inhibition can be the result of a communication between epithelial cells, the host endocrine and immune system and gastrointestinal microorganisms (Hooper and Gordon, 2001; Lyte, 2011; Broadway et al., 2015).

Overall, there have been advantages to the use of probiotics made from a single or multiple species of microorganisms in regard to animal health and food safety in live animals (Higgins et al., 2008; Shivaramaiah et al., 2011), whereas the addition of a complex or adult natural gut microbiota has also been demonstrated to be an effective probiotic approach (Nisbet et al., 1993b, 1996; Droleskey et al., 1995). However, inconsistencies in production of complex probiotics (e.g. quality control issues), governmental regulations, a dearth of understanding the mechanism of action of probiotic cultures all pose significant challenges to the poultry industry. Antibiotics have long been utilized in the poultry industry because they work reliably and consistently, and have rather well known modes of action. Yet in some cases, probiotics can have a more positive impact on food-borne and animal health pathogen carriage than antibiotics do, yet antibiotics are often still fed because they are economically favourable and act predictably. Thus, the differences between the modes of understanding and effects of both antibiotics and probiotics emphasize the need for further research to come to a better understanding of the native (and introduced) gastrointestinal microorganisms and their environment, the avian GIT.

4 Prebiotics

The use of a prebiotic approach to control pathogens in the poultry gut is relatively a newer strategy than the use of probiotics to achieve the same goal (Hume, 2011). Prebiotics are another approach to reducing pathogens in live animals that basically involves an administration of a particular limiting nutrient, which is unavailable to or unused by the host, but is utilized by the intestinal microbial population (Walker and Duffy, 1998). Gibson and Roberfroid (1995) introduced the term 'prebiotic' by defining it as 'a nondigestible food ingredient that beneficially affects the host by selectively stimulating the growth and/or activity of one or a limited number of bacteria in the colon, and thus improves host health' (Gibson and Roberfroid, 1995; Butel et al., 2016). Several criteria have been advanced for a food ingredient to be considered as a prebiotic: (1) be neither degraded nor absorbed in the upper GIT; (2) be selective for beneficial bacteria that are commensal to the colon; (3) be able to modify the colonic flora to form a beneficial composition; and (4) lead to intestinal or systemic changes that are beneficial to the host health (Patterson and Burkholder, 2003; Roberfroid, 2007; Hume, 2011). The Food and Agriculture Organization of the United Nations reported that 'there were no industry-wide guidelines governing the usage of the term prebiotic and that the world prebiotic market offered over 400 prebiotic food products with more than 20 companies producing oligosaccharides and fibers to be used as prebiotics' (Pineiro et al., 2008). Additionally, inulin, fructooligosaccharides (FOS), galactooligosaccharides (GOS), mannanoligosaccharides (MOS), soy oligosaccharides, xylo-oligosaccharides (XOS), pyrodextrins, isomaltooligosaccharides and lactulose have been frequently used as prebiotics in human health, but some of them were not classified by Gibson and Roberfroid (1995) and Roberfroid (2007) as prebiotics since they did not meet the definitional criteria of a prebiotic (Hume, 2011). Recently, the International Scientific Association for Probiotics and Prebiotics (ISAPP) revised and defined prebiotics as 'a substrate that is selectively utilized by host microorganisms conferring a health benefit' (Gibson et al., 2017; Micciche et al., 2018).

Alteration of the resident gut microbial population by using prebiotic supplementation has been the focus of many studies. Bacterial culture-based studies reported that *Lactobacilli* and *Bifidobacterium* species were dominantly present in the gut microbiota of chickens fed diets supplemented by prebiotics (Gaggìa et al., 2010). Additionally, several potential mechanisms have been documented to elucidate the beneficial effects of prebiotic supplementation in the gut microbiota, including competitive exclusion (CE) of pathogenic microorganisms (Callaway et al., 2008a), production of antimicrobial agents (e.g. bacteriocins) (Chen et al., 2007; Munoz et al., 2012) and improving intestinal dynamics and morphology (Chee et al., 2010). This shift in the microbial population can result in an elevated production of vitamins, VFA and antioxidants that further improve the health of the host via intestinal bacterial fermentation (Callaway et al., 2014). Although culture-based techniques have tremendously contributed to our understanding of intestinal microbiota and the interactions within this ecosystem, challenges to cultivate bacterial species still stymie our ability to elucidate the overall ecosystem dynamics. With the advent of using 16S rRNA gene as a taxonomic marker and NGS microbiome-sequencing technology, it has become easier to study changes in complex, real-world microbial populations.

Prebiotics such as FOS and inulin are the most commonly studied prebiotics in humans and animals since they are not hydrolysed by mammalian or avian digestive enzymes and thereby reach the colon undigested, where they can be fermented by the resident intestinal microbiota (Roberfroid et al., 2010). FOS promote the growth of *Lactobacillus* and *Bifidobacterium*, leading to an increase in the production of short chain fatty acids (SCFAs) and lactate, improve the immune system and inhibit *Salmonella* colonization (Cummings and Macfarlane, 2002; Bogusławska-Tryk et al., 2012; Emami et al., 2012; Ricke, 2015). Most strains of *Lactobacillus* and *Bifidobacterium* have the ability to ferment FOS (Ricke, 2015); however, the potential impacts of FOS and inulin are dependent on several factors including the composition of the basal diet, rate of FOS polymerization, the presence of *Bifidobacteria* strains, host characteristics and stress factors (Lee et al., 2010; Bogusławska-Tryk et al., 2012). It has been reported that FOS elevated the population of *Lactobacillus*, and simultaneously inhibited the growth of *C. perfringens* and *E. coli* in broilers (Kim et al., 2011). In an *in vitro* study, the use of FOS inulin resulted in significantly fewer viable intracellular *S.* Enteritidis in prebiotic-treated cells than in untreated cells (Babu et al., 2012). A study conducted on rats showed a decrease in *Salmonella* resistance because of an increase in intestinal permeability (Ten Bruggencate et al., 2005). Additionally, it is thought that SCFAs may cause an elevated expression of *Salmonella* virulence genes despite inhibition of their colonization ability (Durant et al., 2000; Lawhon et al., 2002). It has been reported that inulin altered the caecal microbial metabolic activity without any major impact on the composition of the microbial community of the chicken caecum (Rehman et al., 2008). The supplementation of diets with FOS and inulin in poultry studies yielded unconvincing results; however, it has been demonstrated that FOS combined with probiotics (known as synbiotics) can lead to significant reductions in *Salmonella* (Ricke, 2015).

Mannan oligosaccharides (MOS) are mannose-based oligomers that are commonly found in plants and the wall of the yeast *Saccharomyces cerevisiae*, and linked via β-1,4 glycosidic bonds. This oligosaccharide is thought to reach the lower GIT undigested due to the lack of enzymes in birds to break down the mannan backbone (Pourabedin and Zhao, 2015). Studies have documented that bacterial strains in the genera of *Bacteroides*, *Bacillus* and *Clostridium* produce mannanase (endo-hydrolases) that are capable of cleaving β-1,4 mannopyranosides (Dhawan and Kaur, 2007); however, it is important to consider that other members of the native gut microbiota have roles in the polysaccharide degradation and utilization process (Rakoff-Nahoum et al., 2014). Kim et al. (2011) evaluated the effect of MOS on the ileocaecal microbiota of broilers and reported that MOS decreased *C. perfringens* and *E. coli* populations while it also elevated *Lactobacillus* populations (Kim et al., 2011). In another study, MOS supplementation again altered the caecal microbial composition, and increased the number of species or Operational Taxonomic Units (OTUs) within the phylum *Bacteroidetes* (Corrigan et al., 2011). Brewer's yeast cell walls are a common ration component that are composed of MOS and were reported to modify the microbiome and to decrease *Campylobacter* concentrations (Park et al., 2014, 2017b). Evidence suggests that some MOS-based prebiotics inhibit pathogens that utilize mannose-specific type 1 fimbriae such as *Salmonella* (Newman, 1994; Park et al., 2017a).

When poultry rations were supplemented with yeast extracts (as a MOS source) the number of heterophils was increased, and boosted resistance against disease caused by *E. coli* in challenged and transport-stressed turkey poults (Huff et al., 2010). In another study, the growth of *Lactobacillus* and *Bifidobacterium* species in the caecum was improved and caecal bacterial diversity was increased by MOS supplementation of broiler chickens exposed to stressful suboptimal environmental conditions (Pourabedin et al., 2014). It is believed that mannose-based carbohydrates may bind to pathogen lectins, and thereby stymie the attachment of pathogens to the epithelial surface; consequently, pathogens bound by mannose pass through the GIT without being able to colonize the host (Pourabedin and Zhao, 2015). MOS supplementation resulted in a milder response and an earlier termination of a systemic inflammation in broilers which were challenged with *Salmonella* LPS, compared to subtherapeutic doses of the antibiotic virginiamycin (Baurhoo et al., 2012). In laying hens challenged with the coccidian *Eimeria maxima*, supplementation of *S. cerevisiae* fermentation products also decreased the incidence and severity of lesions (Lensing et al., 2012).

In addition to the above-cited oligosaccharides, research has suggested GOS and soybean meal oligosaccharides (SMO) as potential prebiotic agents for chickens (Lan et al., 2007, Jung et al., 2008). While GOS is naturally present in human milk and it has gained increasing interest as a supplement for infant formulas, only a single study has documented the effect of GOS on faecal microbiota in broilers (Jung et al., 2008). The *in vitro* fermentation of SMO by caecal microbiota of broilers demonstrated an impact on caecal microorganisms, *Bifidobacterium* species were increased as well as VFA concentrations (Lan et al., 2007). In an accompanying *in vivo* trial, SMO increased the population of lactic acid bacteria (Lan et al., 2007). In another study, feed supplementation with a prebiotic mixture containing galactoglucomannan oligosaccharides and arabinoxylan increased caecal *Bifidobacterium* populations and boosted the innate immune response in broiler chickens challenged with *S.* Typhimurium (Faber et al., 2012). Clearly, further research is needed to reveal the potential prebiotic properties of GOS and SMO in chickens to improve animal health and enhance food safety.

5 Bacteriophages

Bacteriophages are viruses that act as intracellular parasites/predators of bacteria that multiply using the bacterium's biosynthetic machinery to produce more phages and have a narrow spectrum, meaning that certain phages may be effective against only a specific strain of bacteria in a mixed population without any negative effect on the normal intestinal microbiota (Callaway et al., 2003a; Clavijo and Flórez, 2017). There are two main kinds of phages characterized by their life cycles as shown in Fig. 1: virulent (or lytic) phages and temperate phages (Clavijo and Flórez, 2017). Virulent phages have a lytic life cycle in which a phage recognizes specific receptors on the outer membrane/surface of bacteria and injects its genetic material into the host bacterium, thereby phage DNA is incorporated into the host chromosome. Subsequently, the phage 'hijacks' the bacterium's metabolic machinery to produce more phages. When intracellular substrates are utilized for phage replication, the host/bacterium lyses and frees new phage particles to reinitiate the cycle. The number of phages rapidly increase as long as targeted bacteria are present in the gut (Callaway et al., 2003a). On the other hand, temperate phages have a lysogenic life cycle in which a phage recognizes the host cell, injects its DNA and incorporates into the bacterium's genome and replicates with it. This DNA can disassemble itself from the genome under certain conditions and begin a lytic life cycle as is stated above for virulent phages. Once it leaves the bacterial genome, the phage's DNA can transfer virulence factors or antibiotic resistance genes to its next host (Clavijo and Flórez, 2017).

Bacteriophages took their place as therapeutic agents to treat infections in humans upon their discovery in 1915 by Frederick Twort and D'Herelle (1919); however their use, also known as 'phage therapy', was disrupted by the large scale production of penicillin during and after the Second World War and phage therapy continued to be practised primarily in the Soviet Union and eastern Europe (Lederberg, 1996; Summers, 2012). Phage therapy as an

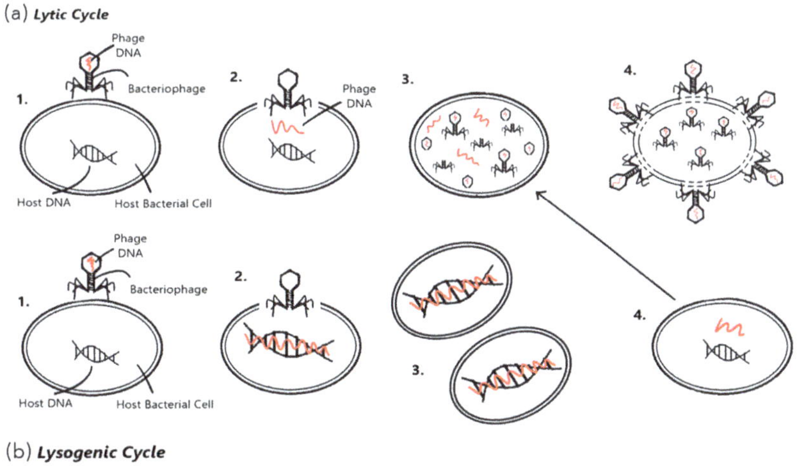

Figure 1 Life cycle of a bacteriophage. (a) Lytic cycle: 1. phage recognizes specific receptors on the outer membrane/surface of bacteria, 2. injects its genetic material into the host bacterium and phage DNA is incorporated into host chromosome, 3. hijacks the bacterium's metabolic activity to produce more phages, 4. host/bacteria lyses and frees new phages. (b) Lysogenic cycle: 1. Phage recognizes specific receptors on the outer membrane/surface of bacteria, 2. injects its genetic material into the host bacterium and phage DNA is incorporated into host chromosome, 3. replication inside the host, cell division and phage DNA is passed on to daughter cells, 4. phage DNA can disassemble itself and begin a lytic cycle.

alternative method to combat against pathogens has advantages, reviewed in exquisite detail by Loc-Carrillo and Abedon (2011). Advantages of using bacteriophages include (1) phages during the killing process have an ability to increase in number where pathogens are located (i.e. *in situ* effect); (2) phages are inherently non-toxic due to their composition (mostly nucleic acids and proteins); (3) phages have a (relatively) high degree of specificity, and therefore, minimally affect the normal intestinal microbiota, and have less potential to induce transmissible/antibiotic resistance mechanisms; (4) phages have an ability to penetrate and eliminate biofilms (Loc-Carrillo and Abedon, 2011).

Pre-harvest use of bacteriophages as antimicrobial biological control strategy in live poultry has not yet been approved by the US Food and Drug Administration (US-FDA); however, research has been conducted around the world to demonstrate the inhibitory effect of bacteriophages in order to control and/or eliminate pathogens (Grant et al., 2016). In recent decades along with the emergence of 'antibiotic-resistant bacteria', the use of bacteriophages has gained a revived interest as an alternative to antibiotics to control pathogenic bacteria in food animals (Smith and Huggins, 1983; Huff et al., 2002; Carrillo et al., 2005, 2008b; Wagenaar et al., 2005; Atterbury et al., 2007). A variety of studies have documented that the colonization of *Campylobacter* (Carrillo et al., 2005; Wagenaar et al., 2005; Scott et al., 2007) and *Salmonella* (Andreatti Filho et al., 2007; Atterbury et al., 2007; Bardina et al., 2012; Wong et al., 2014) has been reduced significantly in poultry by using bacteriophage therapy. Although there has been a success with the use of bacteriophages to reduce pathogens in poultry, several factors should be taken into consideration in order to achieve optimum reduction of the targeted microorganisms. These factors include the concentration ratio of phage compared to the concentration of target bacteria (i.e. multiplicity of infection), selection of the optimum exposure time of target bacteria to phage therapy to obtain maximum reduction, type of phage, the route of phage administration and the composition of phage (single or as a cocktail) (Grant et al., 2016). Any alteration of these factors can affect a phage therapy's efficacy against its target pathogen. Studies have demonstrated that the administration of phages in higher concentrations than the targeted microorganism improved the level of reduction (Bigwood et al., 2009; Bardina et al., 2012), but if the phage:target ratio is too high then a failure to replicate occurs, meaning that the single bacteria is killed but daughter phages are not produced (Callaway et al., 2003b). It has been reported that the use of bacteriophage cocktails are more effective compared to the use of single strain applications (Fiorentin et al., 2005; Andreatti Filho et al., 2007). Additionally, although most studies preferred orally gavaging as an administration method of bacteriophages (Atterbury et al., 2007; Bardina et al., 2012; Wong et al., 2014), Carvalho et al. (2010) supplemented the feed with bacteriophages and obtained better reductions: 1.7 $\log_{10}$ vs. 2 $\log_{10}$ CFU/mL (Carvalho et al., 2010).

Although the use of bacteriophage appears to be effective against bacterial enteric infections and reduces the incidence of food-borne illness, further research is necessary in this area in order for bacteriophages to be considered as a viable and sustainable strategy to control populations of food-borne pathogenic bacteria in food animals.

6 Organic acids

Organic acids (which can include such disparate compounds as VFAs, fatty acids, carboxylic acids or weak acids) are composed of carbon compounds with acidic characteristics, and naturally produced in the GIT of food animals by the microbial fermentation of carbohydrates and/or proteins (Ricke, 2003; Dittoe et al., 2018). Due to their antimicrobial activity, organic acids have largely been utilized for decades in foodstuffs in order to control microorganisms involved with food spoilage and shortened shelf life of food products (Ricke, 2003). The use of organic acids in livestock animals has lately gained increased interest due to (1) the demand to decrease enteropathogens, particularly *Salmonella* in poultry and (2) removal of AGPs from animal production operations around the world (Broom, 2015).

Organic acid toxicity to pathogens have been attributed to a variety of mechanisms, including decreasing the buffering capacity of feed, reducing the pH of drinking water, controlling pathogen microorganisms in digestive tract by reducing pH levels in the gut, improving the availability of nutrients in the feed and their absorption and digestion and promoting immune responses in poultry, which result in a great contribution to the profitability in poultry production and safe poultry products (Thompson and Hinton, 1997; Van Immerseel et al., 2006; Dittoe et al., 2018; Nguyen et al., 2018). Because the use of organic acids does not lead to residues in meat, their use is considered safe from a human nutrition and food safety perspective due to their long history of inclusion in foods around the world (Sterzo et al., 2007). Additionally, organic acids are classified by the FDA as generally recognized as safe (GRAS) for meat products after harvest as well (Mani-Lopez et al., 2012).

Antibacterial mechanisms of organic acids are not fully understood; however, it has been thought that undissociated forms of organic acids can readily penetrate the lipid membrane of the bacterial cell and after the internalization into the neutral pH of the cell cytoplasm, organic acids dissociate into anions and protons, resulting in challenges for bacteria that must maintain a neutral pH cytoplasm. Exportation of excess protons requires consumption of cellular adenosine triphosphate (ATP) and may result in depletion of cellular energy (Russell, 1992; Ricke, 2003; Van Immerseel et al., 2006). The efficacy of organic acids are dependent upon certain factors including acid type, acidity level and inclusion rate of acids, buffering capacity and composition of the diet, amount of intestinal production of acids in GIT, attachment site or receptor on the intestinal epithelium for bacteria, hygienic conditions and animal production system (Thompson and Hinton, 1997; Papatsiros et al., 2013).

Various organic acids have been used as drinking water supplements or as feed additives, often as sodium, potassium or calcium salts, making them easier to include in feed processing due to their less volatile nature (Huyghebaert et al., 2011). Bacterial species that have been documented to be sensitive (i.e. reduced growth and ability of colonization and/or invasion of epithelial cells) to a variety of organic acids include pathogens of concern to both human and animals, such as *Salmonella*, *E. coli*, *Campylobacter* and *Clostridium perfringens* (Wolin, 1969; Tellez et al., 1993; Chaveerach et al., 2002; Van Deun et al., 2008; Huyghebaert et al., 2011; Wales et al., 2013; Nguyen et al., 2018). Studies investigating the effect of organic acids on intestinal pathogen populations have utilized acetic acid, butyric acid, citric acid, capric acid, caproic acid, caprylic acid, formic acid, fumaric acid, gluconic acid, lactic acid, lauric acid, malic acid, propionic acid and sorbic acid (Tellez et al., 1993; Thompson and Hinton, 1997; Van Immerseel et al., 2004; Biggs and Parsons, 2008; Wales et al.,

2013; Lee et al., 2015; Bourassa et al., 2017; Hankel et al., 2018; Nguyen et al., 2018). Thus with this breadth of acids utilized, it is difficult to draw specific conclusions about organic acid inclusion. However, it is evident from the studies that organic acids as alternatives to AGPs in poultry has resulted in promising results to control and/or inhibit pathogens of both human and animal concern and led to the development of numerous commercial products; however, further research is needed to improve their effectiveness.

7 Sodium chlorate

Sodium chlorate is a chemical that can profoundly impact the microbial physiology of pathogenic bacteria that are equipped with the enzyme nitrate reductase. This intracellular enzyme allows certain intestinal bacteria (e.g. *E. coli* and *Salmonella*) to anaerobically reduce nitrate to nitrite as well as converting chlorate to chlorite, which is a cytotoxic end product (Ingledew and Poole, 1984; Stewart, 1988). The accumulation of toxic concentrations of chlorite intracellularly results in death of the pathogens that have the nitrate reductase enzyme (Stewart, 1988). Utilization of chlorate as a pre-harvest intervention strategy has been documented to reduce populations of enteric pathogens in food animals such as chickens (Byrd et al., 2003; McReynolds et al., 2004), turkeys (Moore et al., 2006), sheep (Edrington et al., 2003), cattle (Callaway et al., 2002) and pigs (Anderson et al., 2001). In a study, the effectiveness of an experimental chlorate product (ECP) on a CE culture from a chicken, and the efficacy of ECP in the reduction of *Salmonella* in broilers was investigated (McReynolds et al., 2004). As a result, the combination of the ECP and CE reduced food-borne pathogens that have nitrate reductase enzyme activity (McReynolds et al., 2004). Another study resulted in a reduction of *Salmonella* in turkeys when an ECP administered prior to feed and food withdrawal (Moore et al., 2006). To reduce the initial microbial level of pathogens in poultry before arriving at processing plants is one of the fundamental goals of the poultry industry, and it has been documented that chlorate supplementation is one of the promising intervention strategies. However, further research is needed to determine specific time points to administer this chemical via feed or drinking water to reach the maximum killing effect.

8 Conclusion

Antibiotics used as prophylaxis to prevent illness, or to promote growth have become less favourably viewed by consumers and regulatory authorities and are now illegal in some parts of the world. Therefore antibiotic alternatives have been sought to maintain the production efficiency of the poultry industry. These approaches have included organic acids, bacteriophage, sodium chlorate, as well as pro- and prebiotic approaches. As we develop a deeper understanding of the microbial population of the gut and what constitutes a 'good' and 'bad' microbiome from a production and food safety/animal health perspective, then we will be able to specifically tailor approaches to reduce these pathogens in the live animal and prevent their entry to the gastrointestinal microbial ecosystem. This may include development of novel approaches or simply application of a logical and integrated multiple hurdle scheme.

9 Where to look for further information

For additional information on strategies to control pathogens in the poultry gut, the chicken gastrointestinal microbiota and alternatives to antibiotics in animal production systems, an interested reader is referred to:

- Clavijo, V. and Flórez, M. 2017. The gastrointestinal microbiome and its association with the control of pathogens in broiler chicken production: a review. *Poult. Sci.* 97(3), 1006–21.
- Hume, M. E. 2011. Historic perspective: prebiotics, probiotics, and other alternatives to antibiotics. *Poult. Sci.* 90(11), 2663–9. doi:10.3382/ps.2010-01030.
- Micciche, A. C., Foley, S. L., Pavlidis, H. O., McIntyre, D. R. and Ricke, S. C. 2018. A review of prebiotics against *Salmonella* in poultry: current and future potential for microbiome research applications. *Front. Vet. Sci.* 5, 191. doi:10.3389/fvets.2018.00191.

An interested reader can gain more insight into the concept of probiotics and competitive exclusion strategy by reading the following foundational articles:

- Nurmi, E. and Rantala, M. J. N. 1973. New aspects of *Salmonella* infection in broiler production. *Nature* 241(5386), 210–1. doi:10.1038/241210a0.
- Fuller, R. 1989. Probiotics in man and animals. *J. Appl. Bacteriol.* 66(5), 365–78. doi:10.1111/j.1365-2672.1989.tb05105.x.

Below are a couple of beneficial websites for an interested reader to improve their knowledge on poultry production systems and industry-wide standards in the poultry industry:

- https://www.nationalchickencouncil.org/
- https://www.nal.usda.gov/poultry-and-poultry-production

10 References

Abedon, S. T., Kuhl, S. J., Blasdel, B. G. and Kutter, E. M. 2011. Phage treatment of human infections. *Bacteriophage* 1(2), 66–85. doi:10.4161/bact.1.2.15845.

Anderson, R. C., Buckley, S. A., Callaway, T. R., Genovese, K. J., Kubena, L. F., Harvey, R. B. and Nisbet, D. J. 2001. Effect of sodium chlorate on *Salmonella* Typhimurium concentrations in the weaned pig gut. *J. Food Prot.* 64(2), 255–8. doi:10.4315/0362-028X-64.2.255.

Andreatti Filho, R. L., Higgins, J. P., Higgins, S. E., Gaona, G., Wolfenden, A. D., Tellez, G. and Hargis, B. M. 2007. Ability of bacteriophages isolated from different sources to reduce *Salmonella enterica* serovar Enteritidis *in vitro* and *in vivo*. *Poult. Sci.* 86(9), 1904–9. doi:10.1093/ps/86.9.1904.

Atterbury, R. J., Van Bergen, M. A., Ortiz, F., Lovell, M. A., Harris, J. A., De Boer, A., Wagenaar, J. A., Allen, V. M. and Barrow, P. A. 2007. Bacteriophage therapy to reduce *Salmonella* colonization of broiler chickens. *Appl. Environ. Microbiol.* 73(14), 4543–9. doi:10.1128/AEM.00049-07.

Audia, J. P., Webb, C. C. and Foster, J. W. 2001. Breaking through the acid barrier: an orchestrated response to proton stress by enteric bacteria. *Int. J. Med. Microbiol.* 291(2), 97–106. doi:10.1078/1438-4221-00106.

Babu, U. S., Sommers, K., Harrison, L. M. and Balan, K. 2012. Effects of fructooligosaccharide-inulin on *Salmonella*-killing and inflammatory gene expression in chicken macrophages. *Vet. Immunol. Immunopathol.* 149(1-2), 92–6. doi:10.1016/j.vetimm.2012.05.003.

Bardina, C., Spricigo, D. A., Cortés, P. and Llagostera, M. 2012. Significance of the bacteriophage treatment schedule in reducing *Salmonella* in poultry. *Appl. Environ. Microbiol.* 78(18), 6600–7. doi:10.1128/AEM.01257-12.

Barnes, E. M., Impey, C. S. and Stevens, B. 1979. Factors affecting the incidence and anti-*Salmonella* activity of the anaerobic caecal flora of the young chick. *J. Hyg.* 82(2), 263–83. doi:10.1017/S0022172400025687.

Barrow, P. A. 2007. *Salmonella* infections: immune and non-immune protection with vaccines. *Avian Pathol.* 36(1), 1–13. doi:10.1080/03079450601113167.

Baurhoo, B., Ferket, P., Ashwell, C. M., de Oliviera, J. and Zhao, X. 2012. Cell walls of *Saccharomyces cerevisiae* differentially modulated innate immunity and glucose metabolism during late systemic inflammation. *PLoS ONE* 7(1), e30323. doi:10.1371/journal.pone.0030323.

Bearson, S. M., Bearson, B. L. and Rasmussen, M. A. 2006. Identification of *Salmonella enterica* serovar Typhimurium genes important for survival in the swine gastric environment. *Appl. Environ. Microbiol.* 72(4), 2829–36. doi:10.1128/AEM.72.4.2829-2836.2006.

Berndt, A., Wilhelm, A., Jugert, C., Pieper, J., Sachse, K. and Methner, U. 2007. Chicken cecum immune response to *Salmonella enterica* serovars of different levels of invasiveness. *Infect. Immun.* 75(12), 5993–6007. doi:10.1128/IAI.00695-07.

Biggs, P. and Parsons, C. M. 2008. The effects of several organic acids on growth performance, nutrient digestibilities, and cecal microbial populations in young chicks. *Poult. Sci.* 87(12), 2581–9. doi:10.3382/ps.2008-00080.

Bigwood, T., Hudson, J. A. and Billington, C. 2009. Influence of host and bacteriophage concentrations on the inactivation of food-borne pathogenic bacteria by two phages. *FEMS Microbiol. Lett.* 291(1), 59–64. doi:10.1111/j.1574-6968.2008.01435.x.

Bogusławska-Tryk, M., Piotrowska, A. and Burlikowska, A. 2012. Dietary fructans and their potential beneficial influence on health and performance parameters in broiler chickens. *J. Cent. Eur. Agric.* 13(2), 270–88.

Borda-Molina, D., Seifert, J. and Camarinha-Silva, A. 2018. Current perspectives of the chicken gastrointestinal tract and its microbiome. *Comput. Struct. Biotechnol. J.* 16, 131–9. doi:10.1016/j.csbj.2018.03.002.

Bourassa, D. V., Wilson, K. M., Ritz, C. R., Kiepper, B. K. and Buhr, R. 2017. Evaluation of the addition of organic acids in the feed and/or water for broilers and the subsequent recovery of *Salmonella* Typhimurium from litter and ceca. *Poult. Sci.* 97(1), 64–73. doi:10.3382/ps/pex289.

Broadway, P. R., Carroll, J. A. and Sanchez, N. C. 2015. Live yeast and yeast cell wall supplements enhance immune function and performance in food-producing livestock: a review. *Microorganisms* 3(3), 417–27. doi:10.3390/microorganisms3030417.

Broom, L. J. 2015. Organic acids for improving intestinal health of poultry. *World's Poult. Sci. J.* 71(4), 630–42. doi:10.1017/S0043933915002391.

Bryant, M. P. and Burkey, L. A. 1953. Cultural methods and some characteristics of some of the more numerous groups of bacteria in the bovine rumen. *J. Dairy Sci.* 36(3), 205–17. doi:10.3168/jds.S0022-0302(53)91482-9.

Buntyn, J. O., Schmidt, T. B., Nisbet, D. J. and Callaway, T. R. 2016. The role of direct-fed microbials in conventional livestock production. *Annu. Rev. Anim. Biosci.* 4(4), 335–55. doi:10.1146/annurev-animal-022114-111123.

Butel, M. J. and Waligora-Dupriet, A. J. 2016. Probiotics and prebiotics: what are they and what can they do for us? In: Henderson, B. and Nibali, L. (Eds), *The Human Microbiota and Chronic Disease: Dysbiosis as a Cause of Human Pathology*. John Wiley & Sons, Hoboken, NJ, pp. 467–78..

Byrd, J. A., Corrier, D. E., Hume, M. E., Bailey, R. H., Stanker, L. H. and Hargis, B. M. 1998. Incidence of *Campylobacter* in crops of preharvest market-age broiler chickens. *Poult. Sci.* 77(9), 1303–5. doi:10.1093/ps/77.9.1303.

Byrd, J. A., Anderson, R. C., Callaway, T. R., Moore, R. W., Knape, K. D., Kubena, L. F., Ziprin, R. L. and Nisbet, D. J. 2003. Effect of experimental chlorate product administration in the drinking water on *Salmonella* Typhimurium contamination of broilers. *Poult. Sci.* 82(9), 1403–6. doi:10.1093/ps/82.9.1403.

Callaway, T. R. and Martin, S. A. 2006. Use of competitive exclusion cultures and oligosaccharides. In: *Feedstuffs Direct-fed Microbial, Enzyme and Forage Additive Compendium* (8th edn.). Miller Publishing, Minnetonka, MN, pp. 34–9.

Callaway, T. R., Anderson, R. C., Genovese, K. J., Poole, T. L., Anderson, T. J., Byrd, J. A., Kubena, L. F. and Nisbet, D. J. 2002. Sodium chlorate supplementation reduces *E. coli* O157: H7 populations in cattle. *J. Anim. Sci.* 80(6), 1683–9. doi:10.2527/2002.8061683x.

Callaway, T., Anderson, R., Edrington, T., Elder, R., Genovese, K., Bischoff, K., Poole, T., Jung, Y. S., Harvey, R. and Nisbet, D. J. 2003a. Preslaughter intervention strategies to reduce food-borne pathogens in food animals 1,2. *J. Anim. Sci.* 81(14_suppl_2), E17–23.

Callaway, T., Edrington, T., Anderson, R., Jung, Y. S., Genovese, K., Elder, R. and Nisbet, D. 2003b. Isolation of naturally-occurring bacteriophage from sheep that reduce populations of *E. coli* O157: H7 *in vitro* and *in vivo*. In: *Proc. 5th Int. Symp. on Shiga Toxin-Producing Escherichia coli Infections*, Edinburgh, UK, p. 25.

Callaway, T. R., Edrington, T. S., Anderson, R. C., Harvey, R. B., Genovese, K. J., Kennedy, C. N., Venn, D. W. and Nisbet, D. J. 2008a. Probiotics, prebiotics and competitive exclusion for prophylaxis against bacterial disease. *Anim. Health Res. Rev.* 9(2), 217–25. doi:10.1017/S1466252308001540.

Callaway, T. R., Edrington, T. S., Brabban, A. D., Anderson, R. C., Rossman, M. L., Engler, M. J., Carr, M. A., Genovese, K. J., Keen, J. E., Looper, M. L., et al. 2008b. Bacteriophage isolated from feedlot cattle can reduce *Escherichia coli* O157: H7 populations in ruminant gastrointestinal tracts. *Foodborne Pathog. Dis.* 5(2), 183–91. doi:10.1089/fpd.2007.0057.

Callaway, T. R., Edrington, T. S. and Nisbet, D. J. 2014. Meat science and muscle biology symposium: ecological and dietary impactors of foodborne pathogens and methods to reduce fecal shedding in cattle. *J. Anim. Sci.* 92(4(4)), 1356–65. doi:10.2527/jas.2013-7308.

Callaway, T., Edrington, T., Byrd, J., Nisbet, D., Ricke, S. and Gast, R. 2017. Use of direct-fed microbials in layer hen production-performance response and *Salmonella* control. In: Ricke, S. C. and Gast, R. K. (Eds), *Producing Safe Eggs: Microbial Ecology of Salmonella*. Academic Press, pp. 301–22.

Carrillo, C. L., Atterbury, R. J., El-Shibiny, A., Connerton, P. L., Dillon, E., Scott, A. and Connerton, I. F. 2005. Bacteriophage therapy to reduce *Campylobacter jejuni* colonization of broiler chickens. *Appl. Environ. Microbiol.* 71(11), 6554–63. doi:10.1128/AEM.71.11.6554-6563.2005.

Carvalho, C. M., Gannon, B. W., Halfhide, D. E., Santos, S. B., Hayes, C. M., Roe, J. M. and Azeredo, J. 2010. The *in vivo* efficacy of two administration routes of a phage cocktail to reduce numbers of *Campylobacter coli* and *Campylobacter jejuni* in chickens. *BMC Microbiol.* 10(1), 232. doi:10.1186/1471-2180-10-232.

Castanon, J. I. 2007. History of the use of antibiotic as growth promoters in European poultry feeds. *Poult. Sci.* 86(11), 2466–71. doi:10.3382/ps.2007-00249.

Chaveerach, P., Keuzenkamp, D. A., Urlings, H. A., Lipman, L. J. and Van Knapen, F. 2002. *In vitro* study on the effect of organic acids on *Campylobacter jejuni/coli* populations in mixtures of water and feed. *Poult. Sci.* 81(5), 621–8. doi:10.1093/ps/81.5.621.

Chee, S. H., Iji, P. A., Choct, M., Mikkelsen, L. L. and Kocher, A. 2010. Functional interactions of manno-oligosaccharides with dietary threonine in chicken gastrointestinal tract. I. Growth performance and mucin dynamics. *Br. Poult. Sci.* 51(5), 658–66. doi:10.1080/00071668.2010.517251.

Chen, Y. S., Srionnual, S., Onda, T. and Yanagida, F. 2007. Effects of prebiotic oligosaccharides and trehalose on growth and production of bacteriocins by lactic acid bacteria. *Lett. Appl. Microbiol.* 45(2), 190–3. doi:10.1111/j.1472-765X.2007.02167.x.

Chen, C. Y., Tsen, H. Y., Lin, C. L., Yu, B. and Chen, C. S. 2012. Oral administration of a combination of select lactic acid bacteria strains to reduce the *Salmonella* invasion and inflammation of broiler chicks. *Poult. Sci.* 91(9), 2139–47. doi:10.3382/ps.2012-02237.

Clavijo, V. and Flórez, M. 2017. The gastrointestinal microbiome and its association with the control of pathogens in broiler chicken production: a review. *Poult. Sci.* 97(3), 1006–21.

Clench, M. H. and Mathias, J. R. 1995. The avian cecum: a review. *Wilson Bull.* 107, 93–121.

Corrier, D. E., Hargis, B., Hinton Jr, A., Lindsey, D., Caldwell, D., Manning, J. and DeLoach, J. 1991. Effect of anaerobic cecal microflora and dietary lactose on colonization resistance of layer chicks to invasive *Salmonella enteritidis*. *Avian Dis.* 35(2), 337–43.

Corrier, D. E., Nisbet, D. J., Hollister, A. G., Scanlan, C. M., Hargis, B. M. and DeLoach, J. R. 1993. Development of defined cultures of indigenous cecal bacteria to control salmonellosis in broiler chicks. *Poult. Sci.* 72(6), 1164–8. doi:10.3382/ps.0721164.

Corrigan, A., Horgan, K., Clipson, N. and Murphy, R. A. 2011. Effect of dietary supplementation with a *Saccharomyces cerevisiae* mannan oligosaccharide on the bacterial community structure of broiler cecal contents. *Appl. Environ. Microbiol.* 77(18), 6653–62. doi:10.1128/AEM.05028-11.

Cummings, J. H. and Macfarlane, G. T. 2002. Gastrointestinal effects of prebiotics. *Br. J. Nutr.* 87(S2), S145–51. doi:10.1079/BJNBJN/2002530.

Cutler, S. A., Lonergan, S. M., Cornick, N., Johnson, A. K. and Stahl, C. H. 2007. Dietary inclusion of colicin E1 is effective in preventing postweaning diarrhea caused by F18-positive *Escherichia coli* in pigs. *Antimicrob. Agents Chemother.* 51(11), 3830–5. doi:10.1128/AAC.00360-07.

CVM. 1997. *VM Policy on Competitive Exclusion Products*. FDA, Center for Veterinary Medicine, Office of Management and Communications, Rockville, MD.

D'Aoust, J. Y. and Maurer, J. 2007. *Salmonella* species. In: Doyle, M. P. and Beuchat, L. R. (Eds), *Food Microbiology: Fundamentals and Frontiers* (3rd edn.). American Society of Microbiology, Washington DC, pp. 187–236.

D'Herelle, F. 1919. Sur le role du microbe bacteriophage dans la typhose aviare. *Comptes rendus Acad. Sci. Paris* 169, 932–4.

Dasti, J. I., Tareen, A. M., Lugert, R., Zautner, A. E. and Groß, U. 2010. *Campylobacter jejuni*: a brief overview on pathogenicity-associated factors and disease-mediating mechanisms. *Int. J. Med. Microbiol.* 300(4), 205–11. doi:10.1016/j.ijmm.2009.07.002.

Dhawan, S. and Kaur, J. 2007. Microbial mannanases: an overview of production and applications. *Crit. Rev. Biotechnol.* 27(4), 197–216. doi:10.1080/07388550701775919.

Dibner, J. J. and Richards, J. D. 2005. Antibiotic growth promoters in agriculture: history and mode of action. *Poult. Sci.* 84(4), 634–43. doi:10.1093/ps/84.4.634.

Dittoe, D. K., Ricke, S. C., Kiess, A. S. 2018. Organic acids and potential for modifying the avian gastrointestinal tract and reducing pathogens and disease. *Front. Vet. Sci.* 5. doi:10.3389/fvets.2018.00216.

Droleskey, R. E., Corrier, D. E., Nisbet, D. J. and DeLoach, J. R. 1995. Colonization of mucosal epithelium in chicks treated with a continuous flow culture of 29 characterized bacteria: confirmation by scanning electron microscopy. *J. Food Prot.* 58(8), 837–42. doi:10.4315/0362-028X-58.8.837.

Durant, J. A., Corrier, D. E. and Ricke, S. C. 2000. Short-chain volatile fatty acids modulate the expression of the hilA and invF genes of *Salmonella typhimurium*. *J. Food Prot.* 63(5), 573–8.

Economou, V. and Gousia, P. J. I. 2015. Agriculture and food animals as a source of antimicrobial-resistant bacteria. *Infect. Drug Resist.* 8, 49–61. doi:10.2147/IDR.S55778.

Edrington, T. S., Callaway, T. R., Anderson, R. C., Genovese, K. J., Jung, Y. S., McReynolds, J. L., Bischoff, K. M. and Nisbet, D. J. 2003. Reduction of *E. coli* O157: H7 populations in sheep by supplementation of an experimental sodium chlorate product. *Small Rumin. Res.* 49(2), 173–81. doi:10.1016/S0921-4488(03)00099-3.

EFSA. 2014. The European Union Summary Report on antimicrobial resistance in zoonotic and indicator bacteria from humans, animals and food in 2012. *EFSA Journal* 12(3), 3590.

Emami, N. K., Samie, A., Rahmani, H. and Ruiz-Feria, C. A. 2012. The effect of peppermint essential oil and fructooligosaccharides, as alternatives to virginiamycin, on growth performance, digestibility, gut morphology and immune response of male broilers. *Anim. Feed Sci. Technol.* 175(1–2), 57–64.

Endtz, H. P., Ruijs, G. J., van Klingeren, B., Jansen, W. H., van der Reyden, T. and Mouton, R. P. 1991. Quinolone resistance in *Campylobacter* isolated from man and poultry following the introduction of fluoroquinolones in veterinary medicine. *J. Antimicrob. Chemother.* 27(2), 199–208. doi:10.1093/jac/27.2.199.

Faber, T. A., Dilger, R. N., Iakiviak, M., Hopkins, A. C., Price, N. P. and Fahey, G. C. Jr. 2012. Ingestion of a novel galactoglucomannan oligosaccharide-arabinoxylan (GGMO-AX) complex affected growth performance and fermentative and immunological characteristics of broiler chicks challenged with *Salmonella typhimurium*. *Poult. Sci.* 91(9), 2241–54. doi:10.3382/ps.2012-02189.

Fiorentin, L., Vieira, N. D. and Barioni, W. Jr. 2005. Oral treatment with bacteriophages reduces the concentration of *Salmonella enteritidis* PT4 in caecal contents of broilers. *Avian Pathol.* 34(3), 258–63. doi:10.1080/01445340500112157.

Foster, J. W. 1993. The acid tolerance response of *Salmonella typhimurium* involves transient synthesis of key acid shock proteins. *J. Bacteriol.* 175(7), 1981–7. doi:10.1128/jb.175.7.1981-1987.1993.

Fuller, R. 1989. Probiotics in man and animals. *J. Appl. Bacteriol.* 66(5), 365–78. doi:10.1111/j.1365-2672.1989.tb05105.x.

Fuller, R. 1991. Probiotics in human medicine. *Gut* 32(4), 439–42. doi:10.1136/gut.32.4.439.

Gabriel, I., Lessire, M., Mallet, S., Guillot, J. F. and Guillot, J. 2006. Microflora of the digestive tract: critical factors and consequences for poultry. *Worlds Poult. Sci. J.* 62(3), 499–511. doi:10.1079/WPS2006111.

Gaggìa, F., Mattarelli, P. and Biavati, B. 2010. Probiotics and prebiotics in animal feeding for safe food production. *Int. J. Food Microbiol.* 141, S15–28. doi:10.1016/j.ijfoodmicro.2010.02.031.

Gibson, G. R. and Roberfroid, M. B. 1995. Dietary modulation of the human colonic microbiota: introducing the concept of prebiotics. *J. Nutr.* 125(6), 1401–12. doi:10.1093/jn/125.6.1401.

Gibson, G. R., Hutkins, R., Sanders, M. E., Prescott, S. L., Reimer, R. A., Salminen, S. J., Scott, K., Stanton, C., Swanson, K. S., Cani, J. N. R. G., et al. 2017. Expert consensus document: the International Scientific Association for Probiotics and Prebiotics (ISAPP) consensus statement on the definition and scope of prebiotics. *Nat. Rev. Gastroenterol. Hepatol.* 14(8), 491–502. doi:10.1038/nrgastro.2017.75.

Gould, L. H., Walsh, K. A., Vieira, A. R., Herman, K., Williams, I. T., Hall, A. J., Cole, D. and Centers for Disease Control and Prevention. 2013. Surveillance for foodborne disease outbreaks – United States, 1998–2008. *MMWR Surveill. Summ.* 62(2), 1–34.

Grant, A., Hashem, F. and Parveen, S. 2016. *Salmonella* and *Campylobacter*: antimicrobial resistance and bacteriophage control in poultry. *Food Microbiol.* 53(B), 104–9. doi:10.1016/j.fm.2015.09.008.

Han, G. G., Kim, E. B., Lee, J., Lee, J. Y., Jin, G., Park, J., Huh, C. S., Kwon, I. K., Kil, D. Y., Choi, Y.-J. J. S., et al. 2016. Relationship between the microbiota in different sections of the gastrointestinal tract, and the body weight of broiler chickens. *SpringerPlus* 5(1), 911. doi:10.1186/s40064-016-2604-8.

Hankel, J., Popp, J., Meemken, D., Zeiger, K., Beyerbach, M., Taube, V., Klein, G. and Visscher, C. 2018. Influence of lauric acid on the susceptibility of chickens to an experimental *Campylobacter jejuni* colonisation. *PLoS ONE* 13(9), e0204483. doi:10.1371/journal.pone.0204483.

Higgins, S. E., Higgins, J. P., Wolfenden, A. D., Henderson, S. N., Torres-Rodriguez, A., Tellez, G. and Hargis, B. 2008. Evaluation of a *Lactobacillus*-based probiotic culture for the reduction of *Salmonella enteritidis* in neonatal broiler chicks. *Poult. Sci.* 87(1), 27-31. doi:10.3382/ps.2007-00210.

Higgins, J. P., Higgins, S. E., Wolfenden, A. D., Henderson, S. N., Torres-Rodriguez, A., Vicente, J. L., Hargis, B. M. and Tellez, G. 2010. Effect of lactic acid bacteria probiotic culture treatment timing on *Salmonella enteritidis* in neonatal broilers. *Poult. Sci.* 89(2), 243-7. doi:10.3382/ps.2009-00436.

Hinton, A., Buhr, R. J. and Ingram, K. D. 2002. Carbohydrate-based cocktails that decrease the population of *Salmonella* and *Campylobacter* in the crop of broiler chickens subjected to feed withdrawal. *Poult. Sci.* 81(6), 780-4. doi:10.1093/ps/81.6.780.

Hooper, L. V. and Gordon, J. I. J. S. 2001. Commensal host-bacterial relationships in the gut. *Science* 292(5519), 1115-8. doi:10.1126/science.1058709.

Horrocks, S. M., Anderson, R. C., Nisbet, D. J. and Ricke, S. C. 2009. Incidence and ecology of *Campylobacter jejuni* and *coli* in animals. *Anaerobe* 15(1-2), 18-25. doi:10.1016/j.anaerobe.2008.09.001.

Huff, W. E., Huff, G. R., Rath, N. C., Balog, J. M., Xie, H., Moore, P. A. Jr. and Donoghue, A. M. 2002. Prevention of *Escherichia coli* respiratory infection in broiler chickens with bacteriophage (SPR02). *Poult. Sci.* 81(4), 437-41. doi:10.1093/ps/81.4.437.

Huff, G. R., Huff, W. E., Farnell, M. B., Rath, N. C., Solis de Los Santos, F. and Donoghue, A. M. 2010. Bacterial clearance, heterophil function, and hematological parameters of transport-stressed turkey poults supplemented with dietary yeast extract. *Poult. Sci.* 89(3), 447-56. doi:10.3382/ps.2009-00328.

Hughes, R. 2008. Relationship between digesta transit time and apparent metabolisable energy value of wheat in chickens. *Br. Poult. Sci.* 49(6), 716-20. doi:10.1080/00071660802449145.

Hume, M. E. 2011. Historic perspective: prebiotics, probiotics, and other alternatives to antibiotics. *Poult. Sci.* 90(11), 2663-9. doi:10.3382/ps.2010-01030.

Humphrey, T. 2004. *Salmonella*, stress responses and food safety. *Nat. Rev. Microbiol.* 2(6), 504-9. doi:10.1038/nrmicro907.

Hungate, R. E. 1950. The anaerobic mesophilic cellulolytic bacteria. *Bacteriol. Rev.* 14(1), 1-49.

Huyghebaert, G., Ducatelle, R. and Van Immerseel, F. 2011. An update on alternatives to antimicrobial growth promoters for broilers. *Vet. J.* 187(2), 182-8. doi:10.1016/j.tvjl.2010.03.003.

Ingledew, W. and Poole, R. 1984. The respiratory chains of *Escherichia coli*. *Microbiol. Rev.* 48(3), 222.

Jack, R. W., Tagg, J. R. and Ray, B. 1995. Bacteriocins of gram-positive bacteria. *Microbiol. Rev.* 59(2), 171-200.

Jayne-Williams, D. J. and Fuller, R. 1971. The influence of the intestinal microflora on nutrition. In: Bell, D. J. and Freeman, B. M. (Eds), *Physiology and Biochemistry of the Domestic Food*. Academic Press, London, UK, pp. 74-92.

Joerger, R. D. 2003. Alternatives to antibiotics: bacteriocins, antimicrobial peptides and bacteriophages. *Poult. Sci.* 82(4), 640-7. doi:10.1093/ps/82.4.640.

Jones, F. T. and Ricke, S. C. 2003. Observations on the history of the development of antimicrobials and their use in poultry feeds. *Poult. Sci.* 82(4), 613-7. doi:10.1093/ps/82.4.613.

Jung, S. J., Houde, R., Baurhoo, B., Zhao, X. and Lee, B. H. 2008. Effects of galacto-oligosaccharides and a bifidobacteria lactis-based probiotic strain on the growth performance and fecal microflora of broiler chickens. *Poultry Science* 87(9), 1694–9. doi:10.3382/ps.2007-00489.

Kim, H. J., Park, S. H., Lee, T. H., Nahm, B. H., Chung, Y. H., Seo, K. H. and Kim, H. Y. 2006. Identification of *Salmonella enterica* serovar Typhimurium using specific PCR primers obtained by comparative genomics in *Salmonella* serovars. *J. Food Prot.* 69(7), 1653–61. doi:10.4315/0362-028X-69.7.1653.

Kim, Y., Kim, S. H., Whang, K. Y., Kim, Y. J. and Oh, S. 2008. Inhibition of *Escherichia coli* O157:H7 attachment by interactions between lactic acid bacteria and intestinal epithelial cells. *J. Microbiol. Biotechnol.* 18(7), 1278–85.

Kim, G. B., Seo, Y. M., Kim, C. H. and Paik, I. K. 2011. Effect of dietary prebiotic supplementation on the performance, intestinal microflora, and immune response of broilers. *Poult. Sci.* 90(1), 75–82. doi:10.3382/ps.2010-00732.

Lan, Y., Williams, B. A., Verstegen, M. W. A., Patterson, R. and Tamminga, S. 2007. Soy oligosaccharides *in vitro* fermentation characteristics and its effect on caecal microorganisms of young broiler chickens. *Anim. Feed Sci. Technol.* 133(3-4), 286–97. doi:10.1016/j.anifeedsci.2006.04.011.

Lawhon, S. D., Maurer, R., Suyemoto, M. and Altier, C. 2002. Intestinal short-chain fatty acids alter *Salmonella typhimurium* invasion gene expression and virulence through BarA/SirA. *Mol. Microbiol.* 46(5), 1451–64. doi:10.1046/j.1365-2958.2002.03268.x.

Lederberg, J. 1996. Smaller fleas . . . ad infinitum: therapeutic bacteriophage redux. *Proc. Natl. Acad. Sci. U.S.A.* 93(8), 3167–8. doi:10.1073/pnas.93.8.3167.

Lee, J.-H. and O'Sullivan, D. J. 2010. Genomic insights into bifidobacteria. *Microbiol. Mol. Biol. Rev.* 74(3), 378–416.

Lee, N. K., Lee, J. Y., Kwak, H. G. and Paik, H. D. 2008. Perspectives for the industrial use of bacteriocin in dairy and meat industry. *Korean J. Food Sci. Anim. Resour.* 28(1), 1–8. doi:10.5851/kosfa.2008.28.1.1.

Lee, S. I., Kim, H. S. and Kim, I. 2015. Microencapsulated organic acid blend with MCFAs can be used as an alternative to antibiotics for laying hens. *Turk. J. Vet. Anim. Sci.* 39(5), 520–7. doi:10.3906/vet-1505-36.

Lensing, M., Van der Klis, J. D., Yoon, I. and Moore, D. T. 2012. Efficacy of *Saccharomyces cerevisiae* fermentation product on intestinal health and productivity of coccidian-challenged laying hens. *Poult. Sci.* 91(7), 1590–7. doi:10.3382/ps.2011-01508.

Line, J. E., Bailey, J. S., Cox, N. A. and Stern, N. J. 1997. Yeast treatment to reduce *Salmonella* and *Campylobacter* populations associated with broiler chickens subjected to transport stress. *Poult. Sci.* 76(9), 1227–31. doi:10.1093/ps/76.9.1227.

Lloyd, A. B., Cumming, R. B. and Kent, R. D. 1974. Competitive exclusion as exemplified by *Salmonella typhimurium*. In: *Proc. Australasian Poult. Sci. Conv.*, World Poult. Sci. Assoc. Austral. Br. 155.

Loc-Carrillo, C. and Abedon, S. 2011. Pros and cons of phage therapy. *Bacteriophage* 1(2), 111–4. doi:10.4161/bact.1.2.14590.

Lyte, M. 2011. Probiotics function mechanistically as delivery vehicles for neuroactive compounds: microbial endocrinology in the design and use of probiotics. *BioEssays* 33(8), 574–81. doi:10.1002/bies.201100024.

Mani-Lopez, E., García, H. S. and López-Malo, A. 2012. Organic acids as antimicrobials to control *Salmonella* in meat and poultry products. *Food Res. Int.* 45(2), 713–21. doi:10.1016/j.foodres.2011.04.043.

McReynolds, J. L., Byrd, J. A., Moore, R. W., Anderson, R. C., Poole, T. L., Edrington, T. S., Kubena, L. F. and Nisbet, D. J. 2004. Utilization of the nitrate reductase enzymatic pathway to reduce enteric pathogens in chickens. *Poult. Sci.* 83(11), 1857–60. doi:10.1093/ps/83.11.1857.

Mead, G. C. 2000. Prospects for 'competitive exclusion' treatment to control *Salmonellas* and other foodborne pathogens in poultry. *Vet. J.* 159(2), 111–23. doi:10.1053/tvjl.1999.0423.

Mead, P. S., Slutsker, L., Dietz, V., McCaig, L. F., Bresee, J. S., Shapiro, C., Griffin, P. M. and Tauxe, R. V. 1999. Food-related illness and death in the United States. *Emerg. Infect. Dis.* 5(5), 607–25. doi:10.3201/eid0505.990502.

Micciche, A. C., Foley, S. L., Pavlidis, H. O., McIntyre, D. R. and Ricke, S. C. 2018. A review of prebiotics against *Salmonella* in poultry: current and future potential for microbiome research applications. *Front. Vet. Sci.* 5, 191. doi:10.3389/fvets.2018.00191.

Montville, T. J., Winkowski, K. and Ludescher, R. D. 1995. Models and mechanisms for bacteriocin action and application. *Int. Dairy J.* 5(8), 797–814. doi:10.1016/0958-6946(95)00034-8.

Moore, R. W., Byrd, J. A., Knape, K. D., Anderson, R. C., Callaway, T. R., Edrington, T., Kubena, L. F. and Nisbet, D. J. 2006. The effect of an experimental chlorate product on *Salmonella* recovery of turkeys when administered prior to feed and water withdrawal. *Poult. Sci.* 85(12), 2101–5. doi:10.1093/ps/85.12.2101.

Mountzouris, K. C., Balaskas, C., Xanthakos, I., Tzivinikou, A. and Fegeros, K. J. 2009. Effects of a multi-species probiotic on biomarkers of competitive exclusion efficacy in broilers challenged with *Salmonella enteritidis*. *Br. Poult. Sci.* 50(4), 467–78. doi:10.1080/00071660903110935.

Munoz, M., Mosquera, A., Almeciga-Diaz, C. J., Melendez, A. P. and Sanchez, O. F. 2012. Fructooligosaccharides metabolism and effect on bacteriocin production in *Lactobacillus* strains isolated from ensiled corn and molasses. *Anaerobe* 18(3), 321–30. doi:10.1016/j.anaerobe.2012.01.007.

Murray, C. J. 2000. Environmental aspects of *Salmonella*. In: Barrow, P. A. and Methner, U. (Eds), *Salmonella in Domestic Animals*. CABI, Wallingford, UK, pp. 265–83.

Musgrove, M. T., Berrang, M. E., Byrd, J. A., Stern, N. J. and Cox, N. A. 2001. Detection of *Campylobacter* spp. in ceca and crops with and without enrichment. *Poult. Sci.* 80(6), 825–8. doi:10.1093/ps/80.6.825.

National Chicken Council. 2018. Per capita consumption of poultry and livestock, 1965 to estimated 2019, in pounds. Available at: https://www.nationalchickencouncil.org/about-the-industry/statistics/per-capita-consumption-of-poultry-and-livestock-1965-to-estimated-2012-in-pounds/.

Newman, K. J. 1994. Mannan-oligosaccharides: natural polymers with significant impact on the gastrointestinal microflora and the immune system. In: Lyons, T. P. and Jacques, K. A. (Eds), *Biotechnology in the Feed Industry*. Proceeding of Alltech's Tenth Annual Symposium, Nottingham University Press, Nottingham, pp. 167–75.

Nguyen, D. H., Lee, K. Y., Mohammadigheisar, M. and Kim, I. H. 2018. Evaluation of the blend of organic acids and medium-chain fatty acids in matrix coating as antibiotic growth promoter alternative on growth performance, nutrient digestibility, blood profiles, excreta microflora, and carcass quality in broilers. *Poult. Sci.* 97(12), 4351–8. doi:10.3382/ps/pey339.

Nisbet, D. 2002. Defined competitive exclusion cultures in the prevention of enteropathogen colonisation in poultry and swine. *Antonie Van Leeuwenhoek* 81(1-4), 481–6.

Nisbet, D. J., Corrier, D. E. and DeLoach, J. R. 1993a. Effect of mixed cecal microflora maintained in continuous culture and of dietary lactose on *Salmonella typhimurium* colonization in broiler chicks. *Avian Dis.* 37(2), 528–35. doi:10.2307/1591682.

Nisbet, D. J., Corrier, D. E., Scanlan, C. M., Hollister, A. G., Beier, R. C. and DeLoach, J. R. 1993b. Effect of a defined continuous flow derived bacterial culture and dietary lactose on *Salmonella* colonization in broiler chicks. *Avian Dis.* 37(4), 1017-25.

Nisbet, D. J., Corrier, D. E., Ricke, S. C., Hume, M. E., Byrd, J. A. and DeLoach, J. R. 1996. Maintenance of the biological efficacy in chicks of a cecal competitive-exclusion culture against *Salmonella* by continuous-flow fermentation. *J. Food Prot.* 59(12), 1279-83. doi:10.4315/0362-028X-59.12.1279.

Nurmi, E. and Rantala, M. J. N. 1973. New aspects of Salmonella infection in broiler production. *Nature* 241(5386), 210-1. doi:10.1038/241210a0.

Nurmi, E., Nuotio, L. and Schneitz, C. 1992. The competitive exclusion concept: development and future. *Int. J. Food Microbiol.* 15, 237-40.

Oakley, B. B., Lillehoj, H. S., Kogut, M. H., Kim, W. K., Maurer, J. J., Pedroso, A., Lee, M. D., Collett, S. R., Johnson, T. J. and Cox, N. A. 2014. The chicken gastrointestinal microbiome. *FEMS Microbiol. Lett.* 360(2), 100-12. doi:10.1111/1574-6968.12608.

Onifade, A. A., Odunsi, A. A., Babatunde, G. M., Olorede, B. R. and Muma, E. J. 1999. Comparison of the supplemental effects of *Saccharomyces cerevisiae* and antibiotics in low-protein and high-fibre diets fed to broiler chickens. *Arch. Tierernahr.* 52(1), 29-39.

Pan, D. and Yu, Z. 2014. Intestinal microbiome of poultry and its interaction with host and diet. *Gut Microbes.* 5(1), 108-19. doi:10.4161/gmic.26945.

Papatsiros, V., Katsoulos, P., Koutoulis, K., Karatzia, M., Dedousi, A. and Christodoulopoulos, G. 2013. Alternatives to antibiotics for farm animals. *CAB Rev.* 8(32), 1-15. doi:10.1079/PAVSNNR20138032.

Park, S. F. 2002. The physiology of *Campylobacter* species and its relevance to their role as foodborne pathogens. *Int. J. Food Microbiol.* 74(3), 177-88. doi:10.1016/S0168-1605(01)00678-X.

Park, S. H., Kim, H. J., Cho, W. H., Kim, J. H., Oh, M. H., Kim, S. H., Lee, B. K., Ricke, S. C. and Kim, H. Y. 2009. Identification of *Salmonella enterica* subspecies I, *Salmonella* enterica serovars *Typhimurium*, *Enteritidis* and *Typhi* using multiplex PCR. *FEMS Microbiol. Lett.* 301(1), 137-46. doi:10.1111/j.1574-6968.2009.01809.x.

Park, S. H., Hanning, I., Perrota, A., Bench, B. J., Alm, E. and Ricke, S. C. 2013. Modifying the gastrointestinal ecology in alternatively raised poultry and the potential for molecular and metabolomic assessment. *Poult. Sci.* 92(2), 546-61. doi:10.3382/ps.2012-02734.

Park, S., Gibson, K., Almeida, G. and Ricke, S. C. 2014. Assessment of gastrointestinal microflora in pasture raised chickens fed two commercial prebiotics. *J. Prob. Health* 2(1), 122.

Park, S. H., Perrotta, A., Hanning, I., Diaz-Sanchez, S., Pendleton, S., Alm, E. and Ricke, S. C. 2017a. Pasture flock chicken cecal microbiome responses to prebiotics and plum fiber feed amendments. *Poult. Sci.* 96(6), 1820–30. doi:10.3382/ps/pew441.

Park, S. H., Kim, S. A., Lee, S. I., Rubinelli, P. M., Roto, S. M., Pavlidis, H. O., McIntyre, D. R. and Ricke, S. C. 2017b. Original XPCTM effect on *Salmonella* Typhimurium and cecal microbiota from three different ages of broiler chickens when incubated in an anaerobic *in vitro* culture system. *Front. Microbiol.* 8, 1070. doi:10.3389/fmicb.2017.01070.

Patterson, J. A. and Burkholder, K. M. 2003. Application of prebiotics and probiotics in poultry production. *Poult. Sci.* 82(4), 627-31. doi:10.1093/ps/82.4.627.

Pineiro, M., Asp, N. G., Reid, G., Macfarlane, S., Morelli, L., Brunser, O. and Tuohy, K. J. 2008. FAO technical meeting on prebiotics. *J. Clin. Gastroenterol.* 42, S156-9. doi:10.1097/MCG.0b013e31817f184e.

Pourabedin, M. and Zhao, X. J. 2015. Prebiotics and gut microbiota in chickens. *FEMS Microbiol. Lett.* 362(15), fnv122. doi:10.1093/femsle/fnv122.

Pourabedin, M., Xu, Z., Baurhoo, B., Chevaux, E. and Zhao, X. J. C. 2014. Effects of mannan oligosaccharide and virginiamycin on the cecal microbial community and intestinal morphology of chickens raised under suboptimal conditions. *Can. J. Microbiol.* 60(5), 255-66. doi:10.1139/cjm-2013-0899.

Rakoff-Nahoum, S., Coyne, M. J. and Comstock, L. E. 2014. An ecological network of polysaccharide utilization among human intestinal symbionts. *Curr. Biol.* 24(1), 40-9. doi:10.1016/j.cub.2013.10.077.

Rantala, M. and Nurmi, E. 1973. Prevention of the growth of *Salmonella infantis* in chicks by the flora of the alimentary tract of chickens. *Br. Poult. Sci.* 14(6), 627-30. doi:10.1080/00071667308416073.

Rehman, H. U., Vahjen, W., Awad, W. A. and Zentek, J. 2007. Indigenous bacteria and bacterial metabolic products in the gastrointestinal tract of broiler chickens. *Arch Anim Nutr* 61(5), 319-35. doi:10.1080/17450390701556817.

Rehman, H., Hellweg, P., Taras, D. and Zentek, J. J. P. S. 2008. Effects of dietary inulin on the intestinal short chain fatty acids and microbial ecology in broiler chickens as revealed by denaturing gradient gel electrophoresis. *Poult. Sci.* 87(4), 783-9. doi:10.3382/ps.2007-00271.

Ricke, S. C. 2003. Perspectives on the use of organic acids and short chain fatty acids as antimicrobials. *Poult. Sci.* 82(4), 632-9. doi:10.1093/ps/82.4.632.

Ricke, S. C. 2015. Potential of fructooligosaccharide prebiotics in alternative and nonconventional poultry production systems. *Poult. Sci.* 94(6), 1411-8. doi:10.3382/ps/pev049.

Ricke, S. C., Kundinger, M. M., Miller, D. R. and Keeton, J. T. 2005. Alternatives to antibiotics: chemical and physical antimicrobial interventions and foodborne pathogen response. *Poult. Sci.* 84(4), 667-75. doi:10.1093/ps/84.4.667.

Ringoir, D. D., Szylo, D. and Korolik, V. 2007. Comparison of 2-day-old and 14-day-old chicken colonization models for *Campylobacter jejuni*. *FEMS Immunol. Medical Microbiol.* 49(1), 155-8. doi:10.1111/j.1574-695X.2006.00181.x.

Roberfroid, M. 2007. Prebiotics: the concept revisited. *J. Nutr.* 137(3 Suppl. 2), 830S-7S. doi:10.1093/jn/137.3.830S.

Roberfroid, M., Gibson, G. R., Hoyles, L., McCartney, A. L., Rastall, R., Rowland, I., Wolvers, D., Watzl, B., Szajewska, H., Stahl, B., et al. 2010. Prebiotic effects: metabolic and health benefits. *Br. J. Nutr.* 104(S2), S1-S63. doi:10.1017/S0007114510003363.

Rougière, N. and Carré, B. J. A. 2010. Comparison of gastrointestinal transit times between chickens from D+ and D− genetic lines selected for divergent digestion efficiency. *Animal* 4(11), 1861-72. doi:10.1017/S1751731110001266.

Ruby, T., McLaughlin, L., Gopinath, S. and Monack, D. 2012. *Salmonella*'s long-term relationship with its host. *FEMS Microbiol. Rev.* 36(3), 600-15. doi:10.1111/j.1574-6976.2012.00332.x.

Russell, J. B. 1992. Another explanation for the toxicity of fermentation acids at low pH: anion accumulation versus uncoupling. *J. Appl. Bacteriol.* 73(5), 363–70. doi:10.1111/j.1365-2672.1992.tb04990.x.

Salanitro, J. P., Fairchilds, I. G. and Zgornicki, Y. D. 1974. Isolation, culture, characteristics and identification of anaerobic bacteria from the chicken cecum. *Appl. Microbiol.* 27(4), 678–87.

Sanchez, N. C. B., Young, T. R., Carroll, J. A., Corley, J. R., Rathmann, R. J. and Johnson, B.. 2013. Yeast cell wall supplementation alters aspects of the physiological and acute phase responses of crossbred heifers to an endotoxin challenge. *Innate Immun.* 19(4), 411–9. doi:10.1177/1753425912469673.

Scallan, E., Hoekstra, R. M., Angulo, F. J., Tauxe, R. V., Widdowson, M. A., Roy, S. L., Jones, J. L. and Griffin, P. M. 2011. Foodborne illness acquired in the United States—major pathogens. *Emerg. Infect. Dis.* 17(1), 7–15. doi:10.3201/eid1701.091101p1.

Schamberger, G. P. and Diez-Gonzalez, F. 2002. Selection of recently isolated colicinogenic *Escherichia coli* strains inhibitory to *Escherichia coli* O157:H7. *J. Food Prot.* 65(9), 1381–7. doi:10.4315/0362-028X-65.9.1381.

Schamberger, G. P. and Diez-Gonzalez, F. 2005. Assessment of resistance to colicinogenic *Escherichia coli* by E. coli O157:H7 strains. *J. Appl. Microbiol.* 98(1), 245–52. doi:10.1111/j.1365-2672.2004.02452.x.

Schamberger, G. P., Phillips, R. L., Jacobs, J. L. and Diez-Gonzalez, F. 2004. Reduction of *Escherichia coli* O157:H7 populations in cattle by addition of colicin E7-producing E. coli to feed. *Appl. Environ. Microbiol.* 70(10), 6053–60. doi:10.1128/AEM.70.10.6053-6060.2004.

Schneitz, C. 2005. Competitive exclusion in poultry - 30 years of research. *Food Cont.* 16(8), 657–67. doi:10.1016/j.foodcont.2004.06.002.

Schulz, S., Stephan, A., Hahn, S., Bortesi, L., Jarczowski, F., Bettmann, U., Paschke, A. K., Tuse, D., Stahl, C. H., Giritch, A., et al. 2015. Broad and efficient control of major foodborne pathogenic strains of Escherichia coli by mixtures of plant-produced colicins. *Proc. Nat. Acad. Sci. U.S.A.* 112, E5454–60.

Scott, A. E., Timms, A. R., Connerton, P. L., El-Shibiny, A. and Connerton, I. F. 2007. Bacteriophage influence *Campylobacter jejuni* types populating broiler chickens. *Environ. Microbiol.* 9(9), 2341–53. doi:10.1111/j.1462-2920.2007.01351.x.

Sergeant, M. J., Constantinidou, C., Cogan, T. A., Bedford, M. R., Penn, C. W. and Pallen, M. 2014. Extensive microbial and functional diversity within the chicken cecal microbiome. *PLoS ONE* 9(3), e91941. doi:10.1371/journal.pone.0091941.

Shivaramaiah, S., Pumford, N. R., Morgan, M. J., Wolfenden, R. E., Wolfenden, A. D., Torres-Rodriguez, A., Hargis, B. M. and Téllez, G. 2011. Evaluation of *Bacillus* species as potential candidates for direct-fed microbials in commercial poultry. *Poult. Sci.* 90(7), 1574–80. doi:10.3382/ps.2010-00745.

Smith, H. W. and Huggins, M. B. 1983. Effectiveness of phages in treating experimental *Escherichia coli* diarrhoea in calves, piglets and lambs. *J. Gen. Microbiol.* 129(8), 2659–75. doi:10.1099/00221287-129-8-2659.

Smith, C. K., Kaiser, P., Rothwell, L., Humphrey, T., Barrow, P. A. and Jones, M. A. 2005. *Campylobacter jejuni*-induced cytokine responses in avian cells. *Infection and Immunity* 73(4), 2094–100. doi:10.1128/IAI.73.4.2094-2100.2005.

Stahl, C. H., Callaway, T. R., Lincoln, L. M., Lonergan, S. M. and Genovese, K. J. 2004. Inhibitory activities of colicins against *Escherichia coli* strains responsible for postweaning diarrhea and edema disease in swine. *Antimicrob. Agents Chemother.* 48(8), 3119–21. doi:10.1128/AAC.48.8.3119-3121.2004.

Stanley, D., Hughes, R. J. and Moore, R. 2014. Microbiota of the chicken gastrointestinal tract: influence on health, productivity and disease. *Appl. Microbiol. Biotechnol.* 98(10), 4301–10. doi:10.1007/s00253-014-5646-2.

Stavric, S. and D'Aoust, J.-Y. 1993. Undefined and defined bacterial preparations for competitive exclusion of *Salmonella* in poultry. *J. Food Prot.* 56(2), 173–80. doi:10.4315/0362-028X-56.2.173.

Sterzo, E., Paiva, J., Mesquita, A., Freitas Neto, O. and Berchieri Jr., A. 2007. Organic acids and/or compound with defined microorganisms to control *Salmonella enterica* serovar Enteritidis experimental infection in chickens. *Rev. Bras. Cienc. Avic.* 9(1), 69–73. doi:10.1590/S1516-635X2007000100010.

Stewart, V. 1988. Nitrate respiration in relation to facultative metabolism in enterobacteria. *Microbiol. Rev.* 52(2), 190–232.

Summers, W. 2012. The strange history of phage therapy. *Bacteriophage* 2(2), 130–3. doi:10.4161/bact.20757.

Swaggerty, C. L., Kogut, M. H., He, H., Genovese, K. J., Johnson, C. and Arsenault, R. J. 2017. Differential levels of cecal colonization by *Salmonella enteritidis* in chickens triggers distinct immune kinome profiles. *Frontiers in Veterinary Science* 4, 214. doi:10.3389/fvets.2017.00214.

Tellez, G., Dean, C. E., Corrier, D. E., Deloach, J. R., Jaeger, L. and Hargis, B. M. 1993. Effect of dietary lactose on cecal morphology, pH, organic acids, and *Salmonella enteritidis* organ invasion in Leghorn chicks. *Poult. Sci.* 72(4), 636–42. doi:10.3382/ps.0720636.

Tellez, G., Pixley, C., Wolfenden, R. E., Layton, S. L. and Hargis, B. M. 2012. Probiotics/direct fed microbials for *Salmonella* control in poultry. *Food Res. Int.* 45(2), 628–33. doi:10.1016/j.foodres.2011.03.047.

Ten Bruggencate, S. J. M., Bovee-Oudenhoven, I. M. J., Lettink-Wissink, M. L. G. and Van der Meer, R. J. 2005. Dietary fructooligosaccharides increase intestinal permeability in rats. *J. Nutr.* 135(4), 837–42. doi:10.1093/jn/135.4.837.

Thompson, J. L. and Hinton, M. 1997. Antibacterial activity of formic and propionic acids in the diet of hens on *Salmonellas* in the crop. *Br. Poult. Sci.* 38(1), 59–65. doi:10.1080/00071669708417941.

Twort, F. W. 1915. An investigation on the nature of ultramicroscopic viruses. *The Lancet*; 186, 1241–3; doi:10.1016/S0140-6736(01)20383-3.

USDA. 2015. United States Department of Agriculture National Agricultural Statistics Service poultry statistics fact sheet. Available at: https://www.usda.gov/sites/default/files/documents/nass-poultry-stats-factsheet.pdf.

Van Asten, A. J. and Van Dijk, J. E. 2005. Distribution of 'classic' virulence factors among Salmonella spp. *FEMS Immunology and Medical Microbiology* 44(3), 251–9. doi:10.1016/j.femsim.2005.02.002.

Van Deun, K., Haesebrouck, F., Van Immerseel, F., Ducatelle, R. and Pasmans, F. 2008. Short-chain fatty acids and L-lactate as feed additives to control *Campylobacter jejuni* infections in broilers. *Avian Pathol.* 37(4), 379–83. doi:10.1080/03079450802216603.

Van Immerseel, F., De Buck, J., Boyen, F., Bohez, L., Pasmans, F., Volf, J., Sevcik, M., Rychlik, I., Haesebrouck, F. and Ducatelle, R. 2004. Medium-chain fatty acids decrease colonization and invasion through hilA suppression shortly after infection of chickens with *Salmonella enterica* serovar Enteritidis. *Appl. Environ. Microbiol.* 70(6), 3582-7. doi:10.1128/AEM.70.6.3582-3587.2004.

Van Immerseel, F., Russell, J. B., Flythe, M. D., Gantois, I., Timbermont, L., Pasmans, F., Haesebrouck, F. and Ducatelle, R. 2006. The use of organic acids to combat *Salmonella* in poultry: a mechanistic explanation of the efficacy. *Avian Pathol.* 35(3), 182-8. doi:10.1080/03079450600711045.

Vaughn, L. E., Holt, P. S. and Gast, R. K. 2008. Cellular assessment of crop lymphoid tissue from specific-pathogen-free white leghorn chickens after *Salmonella enteritidis* challenge. *Avian Dis.* 52(4), 657-64. doi:10.1637/8369-052308-Reg.1.

Vispo, C. and Karasov, W. H. 1997. The interaction of avian gut microbes and their host: an elusive symbiosis. In: Mackie, R. I. and White, B. A. (Eds), *Gastrointestinal Microbiology*. Springer, Boston, MA, pp. 116-55.

Wagenaar, J. A., Van Bergen, M. A., Mueller, M. A., Wassenaar, T. M. and Carlton, R. 2005. Phage therapy reduces *Campylobacter jejuni* colonization in broilers. *Vet. Microbiol.* 109(3-4), 275-83. doi:10.1016/j.vetmic.2005.06.002.

Waite, D. W. and Taylor, M. W. 2014. Characterizing the avian gut microbiota: membership, driving influences, and potential function. *Front. Microbiol.* 5, 223. doi:10.3389/fmicb.2014.00223.

Wales, A., McLaren, I., Rabie, A., Gosling, R. J., Martelli, F., Sayers, R. and Davies, R. 2013. Assessment of the anti-*Salmonella* activity of commercial formulations of organic acid products. *Avian Pathol.* 42(3), 268-75. doi:10.1080/03079457.2013.782097.

Walker, W. A., Duffy, L. C. 1998. Diet and bacterial colonization: role of probiotics and prebiotics. *J. Nutr. Biochem.* 9(12), 668-75. doi:10.1016/S0955-2863(98)00058-8.

Weinack, O. M., Snoeyenbos, G. H., Smyser, C. F. and Soerjadi, A. S. 1982. Reciprocal competitive exclusion of *Salmonella* and *Escherichia coli* by native intestinal microflora of the chicken and turkey. *Avian Dis.* 26(3), 585-95. doi:10.2307/1589905.

Wielen, P., Keuzenkamp, D., Lipman, L. J., van Knapen, F. and Biesterveld, S. 2002. Spatial and temporal variation of the intestinal bacterial community in commercially raised broiler chickens during growth. *Microb. Ecol.* 44(3), 286-93.

Wierup, M., Wold-Troell, M., Nurmi, E. and Hakkinen, M. 1988. Epidemiological evaluation of the *Salmonella*-controlling effect of a nationwide use of a competitive exclusion culture in poultry. *Poult. Sci.* 67(7), 1026-33. doi:10.3382/ps.0671026.

Wolin, M. J. 1969. Volatile fatty acids and the inhibition of *Escherichia coli* growth by rumen fluid. *Appl. Microbiol.* 17(1), 83-7.

Wong, C. L., Sieo, C. C., Tan, W. S., Abdullah, N., Hair-Bejo, M., Abu, J. and Ho, Y. W. 2014. Evaluation of a lytic bacteriophage, Φ st1, for biocontrol of *Salmonella enterica* serovar Typhimurium in chickens. *Int. J. Food Microbiol.* 172, 92-101. doi:10.1016/j.ijfoodmicro.2013.11.034.

Wray, C. and Davies, R. H. 2000. Competitive exclusion - an alternative to antibiotics. *Vet. J.* 159(2), 107-8. doi:10.1053/tvjl.1999.0442.

Chapter 14

The role of probiotics in optimizing gut function in poultry

Guillermo Tellez and Juan D. Latorre, University of Arkansas, USA; Margarita A. Arreguin-Nava, Eco-Bio LLC, USA; and Billy M. Hargis, University of Arkansas, USA

1 Introduction

A common human propensity is to regard all microorganisms as 'harmful', in particular, equating bacteria to pathogenic germs. Nothing could be further from the truth. The number of beneficial bacterial species far exceeds the number of pathogenic species and many of the known bacteria are in fact useful or even indispensable for the continued existence of life on Earth. Prokaryotic microorganisms are widespread in all environments on Earth, establishing diverse interactions with many eukaryotic taxa (Bronstein et al., 2006). The cooperative interactions between species (mutualism) have had a central role in the generation and maintenance of life on earth (Kikuchi et al., 2009). Prokaryotes and eukaryotes are involved in diverse forms of mutualism (Saridaki and Bourtzis, 2010). Adaptive diversification is a process intrinsically tied to species interactions (Xie et al., 2010). Yet, the influence of most types of interspecific interactions on adaptive evolutionary diversification remains poorly understood. The endosymbiotic theory states that several key organelles of eukaryotes originated as symbioses between separate single-celled organisms (Degli Esposti et al., 2014). According to this theory, mitochondria and plastids (e.g. chloroplasts), and possibly other organelles, represent formerly free-living bacteria that were taken inside another cell as an endosymbiont, around

http://dx.doi.org/10.19103/AS.2019.0059.17

1.5 billion years ago (Gibson and Hunter, 2010). Molecular and biochemical evidence suggests that the mitochondrion developed from *Proteobacteria* and the chloroplast from *Cyanobacteria* (Mackiewicz et al., 2012). Numerous facultative heritable endosymbionts are reproductive manipulators (Saridaki and Bourtzis, 2010). Nevertheless, many do not manipulate reproduction, so they are expected to confer fitness benefits to their hosts, as has been shown in several studies that report defence against natural enemies, tolerance to environmental stress and increased fecundity (Xie et al., 2010). One example of such a beneficial group of microorganisms is the incredibly complex and abundant ensemble of microbes that harbour in the gastrointestinal tract (GIT) of metazoans. The GIT is more densely populated with microorganisms than any other organ and is an interface where the microbiota may have a pronounced impact on animal biology (Maslowski and Mackay, 2011). More than 50 genera and at least 500–1000 different species are distributed along the length of the GIT in most metazoans (Fraune and Bosch, 2010). The bacterial population of the human caecum and colon is numerically ~10^{13} cfu/g (Neish, 2009), comprising about 40–55% of solid stool matter and weights ~1 kg (Blaser, 2006). Presumably, the assembly of the gut microflora is regulated by the elaborate and combinatorial host-microbial and microbial-microbial interactions predicated on principles refined over the course of evolution (Xu and Gordon, 2003). Comparison of rodents raised without exposure to any microorganisms to those colonized with an assembly of microbiota revealed a wide range of host functions affected by indigenous microbial communities. For example, the microbiota directs the assembly of the gut-associated lymphoid tissue (Martin et al., 2010), helps educate the immune system (McFall-Ngai, 2007), affects the integrity of the intestinal mucosal barrier (Duerkop et al., 2009), modulates proliferation and differentiation of its epithelial lineages (Moran, 2007), regulates angiogenesis (Sekirov et al., 2010), modifies the activity of the enteric nervous system (Tlaskalová-Hogenová et al., 2011) and plays a key role in extracting and processing nutrients consumed in the diet (Walter et al., 2011). The microflora can metabolize proteins and protein degradation products, sulphur-containing compounds, and endogenous and exogenous glycoproteins (Qiu et al., 2012). Some organisms grow on intermediate products of fermentation such as H_2, lactate, succinate, formate and ethanol, converting these compounds to end products including short-chain fatty acids, a process which has direct benefits on digestive physiology (Dass et al., 2007). In particular with diet composition, one must conclude that metazoans literally 'become what we eat'. So, any disorders in this fragile microbial ecosystem (dysbacteriosis) may predispose the host to a whole range of chronic diseases and infections, thereby affecting the production of food animals. On the other hand, over millions of years, animals have developed various means for supporting complex and dynamic consortia of microorganisms during their life cycle (Dale and Moran, 2006). As with most complex ecosystems, it appears that the majority of these microbial species cannot be cultured when removed from the niches in their host animals (Moran, 2007). The fragile composition of the gut microflora can be affected by various factors such as age, diet, environment, stress and medication (Bäckhed, 2011). Furthermore, many factors are involved in shaping gut microflora from infancy such as mode of delivery, type of infant feeding, hospitalization, prematurity, antibiotic use and dietary nutrient composition (Choct, 2009). Dietary ingredients have a profound effect on the composition of the gut microflora, which in turn, regulates the physiology of all animals (Fraune and Bosch, 2010). As such, nutritional components of the diet are of critical importance not only for meeting the nutrient requirements of the host but also shaping the profile of the microbiome, which in turn, will determine the balance between health and disease. As an example, several studies have shown the effect of diet composition on promoting insulin sensitivity, diabetes, cancer and other metabolic disorders (Cani and Delzenne, 2009). Some researchers believe that the alarming increase in autoimmune diseases in the West may be due to a disruption in the ancient relationship between our bodies and a healthy microbiome (Salzman, 2011). Thus, colonization of microbiomes in metazoans begins at birth and is followed by progressive assembly of a complex and dynamic microbial society maintaining a perfect harmony or homeostasis (Di Mauro et al., 2013). However, little is known about how they influence the normal development and physiology of hosts. A transcendent view of vertebrate biology, therefore, requires an understanding of the contributions of these indigenous microbial communities to host development and adult physiology.

2 Experiences of probiotics in poultry

The alarming spread of antibiotic resistance genes has created public concerns leading to new laws that restrict the use of antibiotics as growth promoters in domestic and meat-producing animals in several countries around the world. Such policies have obligated scientists to evaluate different alternatives to antibiotics. Furthermore, the satisfaction of consumer preferences has also become an important strategy for countries looking to export their animal products (Parker, 1990; Dahiya et al., 2006; You and Silbergeld, 2014). Hence, the use of probiotics as alternative tools to antibiotic growth promoters has been increasing and their efficacy has been demonstrated by many investigators around the world (Hammes and Hertel, 2002; Tellez et al., 2013). Nevertheless, some of the mechanisms by which the probiotics improve health and performance parameters in domestic animals are not well understood (Musso et al., 2010). Even though the use of probiotics is not new, it is only in the last two decades that scientists have demonstrated the potential of probiotics to improve health and performance parameters of meat-producing animals (Isolauri et al., 2002; Salminen and Isolauri, 2006; Dominguez-Bello and Blaser, 2008). Studies have shown how probiotic bacteria regulate production of pro-inflammatory and anti-inflammatory cytokines (Borchers et al., 2009; Lyte, 2011), exert antioxidant properties (Farnell et al., 2006; Tao et al., 2006; Zareie et al., 2006; Segawa et al., 2011; Howarth and Wang, 2013) and enhance barrier integrity (Yu et al., 2012). In addition, several investigators have demonstrated the benefits of probiotics on innate immunity (Alvarez-Olmos and Oberhelman, 2001; Vanderpool et al., 2008; Molinaro et al., 2012) as well as on humoral immunity (Arvola et al., 1999; Haghighi et al., 2006; Howarth and Wang, 2013).

Our laboratory has worked to identify probiotic candidates for use in poultry such as a defined lactic acid bacteria (LAB)-based probiotic that was confirmed to increase the resistance of poultry to *Salmonella* spp. infections (Farnell et al., 2006; Higgins et al., 2007; Menconi et al., 2011; Tellez et al., 2012; Biloni et al., 2013) and reduce transit diarrhoea in turkeys (Higgins et al., 2005). In other commercial trials, administration of this probiotic mixture to turkeys and chickens increased performance and reduced costs of production (Vicente et al., 2007, 2008; Torres-Rodriguez et al., 2007). More recently, microarray analysis of caeca in chickens infected with *Salmonella* serovar Enteritidis (SE) and then treated with this probiotic showed significant differences in the expression of some inflammatory genes associated with the NFκβ complex (Higgins et al., 2011).

3 Probiotics and inflammation

Salmonellosis remains one of the most comprehensive food-borne diseases that can be transmitted to humans through animal and plant products (Hernández-Reyes and Schikora, 2013; Zheng et al., 2013; Schleker et al., 2015). Chickens infected with SE show a significant increase of heterophils-to-lymphocyte ratio (Al-Murrani et al., 2002; Shini et al., 2008). In a recent study, we have also observed a marked heterophilia and lymphopenia in chickens challenged with SE. However, these haematological changes were prevented in chickens that received a LAB probiotic 1 h after SE challenge. Furthermore, we observed that the reduction in intestinal colonization by SE was associated with a reduction in intestinal permeability of fluorescein isothiocyanate-dextran (FITC-d) in chickens (Prado-Rebolledo et al., 2017). Due to its molecular size (4kDa), FITC-d is a molecule which does not cross the intact GIT barrier. However, when conditions disturb tight junctions (TJ) between epithelial cells, FITC-d can enter the systemic circulation as demonstrated by an increase in paracellular permeability after oral administration (Yan et al., 2009; Kuttappan et al., 2015; Vicuña et al., 2015a,b). TJ act as intercellular cement between epithelial cells and regulate the permeability and dissemination of microorganisms and antigens (Ulluwishewa et al., 2011). Hence, any damage to this fragile epithelium may lead to chronic intestinal and systemic inflammation (Steed et al., 2010). Uptake and distribution within the host of salmonellae infections are associated with disruption of the TJ complex, loss of barrier function, bacterial translocation and initiation of polymorphonuclear (PMN) cells migration across the intestinal barrier (Köhler et al., 2007). Therefore, appropriate gut barrier function is indispensable to maintain optimal health (Groschwitz and Hogan, 2009; Sharma et al., 2010; Jeon et al., 2013; Pastorelli et al., 2013). Bacterial endotoxin has been shown to activate aldose reductase (AR), a member of the aldo-keto reductase (AKR) superfamily and nuclear factor (Ramana and Srivastava, 2006, 2010). Activation of NF-κB results in the expression of several inflammatory cytokines (Ozinsky et al., 2000; Overman et al., 2012). It is well known that oxidative stress-induced inflammation is a major contributor to several diseases. AR catalytic activity plays a crucial role in some inflammatory diseases associated with disruption of TJ between epithelial cells (Srivastava et al., 2011; Yadav et al., 2011; Pastel et al., 2012). Interestingly, microarray analysis with our LAB probiotic in broiler chickens challenged with SE showed a significant reduction in intestinal gene expression associated with the NF-κB complex and AR (Higgins et al., 2011). Hence, our studies suggest that the probiotic preserved the integrity of the intestinal epithelial cells (IEC), which represent the second layer of innate defence mechanisms of the GIT (Sakamoto et al., 2000; Johansson et al., 2010; Kim and Ho, 2010). These results are in agreement with numerous studies demonstrating that probiotics prevent *Salmonella* translocation, suppress the oxidant-induced intestinal permeability and improve intestinal barrier function (Madsen et al., 2001; Ewaschuk et al., 2007; Mennigen and Bruewer, 2009; Segawa et al., 2011; Hsieh et al., 2015).

Researchers have investigated the effects of probiotics on gene regulation associated with immune modulation, enteropathogens control and homeostasis, using *in vitro* and *in vivo* disease challenge models (Tellez et al., 1993; Timbermont et al., 2010; Kiarie et al., 2013; Sherryll Lynn et al., 2013). Metchnikoff founded the research field of beneficial microorganisms for animals and humans (probiotics), aimed at modulating the intestinal microflora (Metchnikoff and Metchnikoff, 1907). At the moment, new molecular techniques are helping us to understand how the anti-inflammatory, cell integrity and antioxidant properties of probiotics can improve gut and barrier integrity. Given the recent international legislation and domestic consumer pressure to withdraw growth-promoting antibiotics and limit antibiotics available for treatment of bacterial infections, probiotics and direct-fed microbials (DFM) can offer clear alternative options.

4 Risks of overuse of antibiotics

Fluoroquinolones are the third generation of quinolone progress. Nalidixic acid and pipemidic acid are examples of the first generation and currently have limited activity against Gram-negative bacteria. Fluorinated 4-quinolones were introduced to the market in the 1980s and were the top of the line antibiotics, offering a broad spectrum of activity and high efficacy in a wide range of infections both orally and parenterally (Piddock et al., 1990). Nevertheless, history has demonstrated that the extensive use of new antibiotics is eventually shadowed by the appearance of resistance to those chemicals which has become a major global problem. This was demonstrated by the higher incidence of salmonellae and *Campylobacter* infections worldwide, and several reports of fluoroquinolone resistance in clinical isolates for these and other enteric pathogens (Murray, 1986; Uwaydah et al., 1991; Griggs et al., 1994). Hence, the World Health Organization (WHO) published a list of antibiotics that should be reserved for human use only (Couper, 1997), and fluoroquinolones were among them, due to the alarming evidence of quinolone-resistant zoonotic pathogens. Soon after the publication of the WHO report, several countries banned the use of fluoroquinolones in animal production (Rodrigue et al., 1990; Randall et al., 2006). With growing consumer and scientific pressure, the European Union went one step further, creating new legislation banning the use of all antibiotics as growth promoters as of January 2006 (Castanon, 2007). However, in many countries, the indiscriminate use and misuse of antibiotics including fluoroquinolones are still a sad reality. Especially in countries where there is no legislation regulating the use of fluoroquinolones in animal agriculture and where there is an abundance of generic fluoroquinolones at a low cost. Typical management practices in those countries are to treat or dose healthy neonatal chickens and turkey poults with five times the recommended dose of enrofloxacin for five consecutive days in the drinking water (Tellez, 2015, pers. comm.). Interestingly, in those countries, the incidence of *Salmonella* spp. and *Campylobacter* spp. infection rates in both humans and agriculture are also high (Piddock et al., 1990; Acar and Goldstein, 1997; Piddock, 2002; Sierra-Arguello et al., 2016). In recent years, several investigators have shown that the use of certain antibiotics increases enteric colonization of antibiotic-resistant strains of selective enteric pathogens in domestic animals (Smith and Tucker, 1975; Manning et al., 1992), because some of these selective pathogens are extremely resistant to many antibiotics and are capable of rapidly developing resistance when exposed (Piddock and Wise, 1989; Acar and Goldstein, 1997). Antibiotic prophylaxis or treatment has been reported to actually increase the occurrence and severity of these infections in commercial poultry (Seuna and Nurmi, 1979; Niewold, 2007). In addition to the lack of effect of these antibiotics in resistant enteropathogens, some researchers have shown that antibiotics can actually cause disruption in the microbiome (Bartlett, 2002), accompanied with reduction of short-chain fatty acids (Van Der Wielen et al., 2000) and increased luminal pH in the distal GIT (Corrier et al., 1990). To evaluate the common management practice of using five times the recommended dose of enrofloxacin, we recently published a study where neonatal chickens and turkey poults received different doses of this fluoroquinolone for five consecutive days after placement and looking at their susceptibility to salmonellae infections. It was remarkable to observe how this practice makes chickens and turkeys more susceptible to enteric infections (Morales-Barrera et al., 2016). In our study, microbiome analysis of the caecal content revealed that turkey poults treated with enrofloxacin had a lower proportion of *Firmicutes* and *Bacteroidetes* suggesting that the broad spectrum of enrofloxacin had a profound impact upon the microbiome. Interestingly, these poults had the highest proportion of *Proteobacteria*. Such a high dose of antibiotic also had a significant increase in Gammaproteobacteria. Changes in the proportion of phylum and class were associated with higher *Salmonella* Heidelberg intestinal colonization since *Salmonella* belongs to phylum *Proteobacteria*, class Gammaproteobacteria. Furthermore, poults treated with enrofloxacin had lower proportions of clostridia and bacilli when compared with control or probiotic experimental groups. In contrast, poults that received the probiotic had the highest proportion of *Firmicutes* and *Bacteroidetes*, but the lowest amount of *Proteobacteria*. These birds also showed a significant reduction in Gammaproteobacteria, and a higher proportion of clostridia and bacilli. These results suggest that five times the recommended dose of enrofloxacin, a broad-spectrum antibiotic, can have a negative effect on the microbiome that may be responsible for an enhancement of salmonellae colonization, which has been previously demonstrated with other selective enteropathogens (Uwaydah et al., 1991; Manning et al., 1992). Acquisition of resistance to fluoroquinolones has been reported to be a multifaceted process, which includes spontaneous point mutations that result in amino acid substitutions within the topoisomerase subunits GyrA, GyrB, ParC or ParE, as well as reduced expression of outer membrane porins, overexpression of multidrug efflux pumps and/or plasmid-mediated quinolone resistance (Angulo et al., 2000; Engberg et al., 2001). It is remarkable to contemplate that the alarming incidence of certain selective enteric pathogens is associated with the indiscriminate use of some antibiotics in animal agriculture in some countries (Hofer and Reis, 1994; Irino et al., 1996; Borsoi et al., 2009; De Moura et al., 2013). Since poultry products have been identified as important reservoirs of human infections, this is a growing public health concern. Given that fluoroquinolones and other antibiotics are overused in animal production, any effort to diminish the risk of resistance is crucial. The previous investigations involving antibiotics and other enteropathogens suggest that prophylactic utilization of some antibiotics in poultry increases the susceptibility to salmonellae colonization and organ invasion. Therefore, antibiotics should be limited to infections of specific bacteria with known antibiotic sensitivity. In addition, our findings also confirm previous studies suggesting that the use of alternatives such as probiotics can be an effective tool in controlling salmonellae infections.

4.1 *In ovo strategies*

Under commercial conditions, millions of chickens and turkeys hatch in a hostile environment and are exposed for several hours to heat stress and potentially pathogenic bacteria in the hatcheries. Increased stress along with the potential abundance of pathogens in the hatching cabinet leads to ideal conditions for pathogen colonization. It is generally accepted that the natural route of transmission of zoonotic pathogens, such as *Salmonella,* is faecal-oral (White et al., 1997; Galanis et al., 2006). However, published studies have also suggested that airborne transmission of *Salmonella* in poultry is possible (Wathes et al., 1988). Understanding the anatomical and immunological defences of the avian respiratory tract helps to clarify this issue. The architecture of the avian respiratory tract is an important component for susceptibility and resistance against infectious agents. In day-old chickens and turkeys, no or very few infiltrating lymphocytes are seen in the primary bronchi region (Fagerland and Arp, 1990; Śmiałek et al., 2011), and it is not until 3–4 weeks of age that the lymphoid nodules are developed at these locations (Fagerland and Arp, 1993; Drolet et al., 2010). During the following week, the number of IgG-, IgA- or IgM-producing cells continues to increase; however, the bronchial-associated lymphoid tissue (BALT) is not mature until chickens are 6–8 weeks old (Bienenstock, 1980; Bienenstock and McDermott, 2005; De Geus et al., 2012). Hence, commercial neonatal poultry are extremely susceptible to airborne pathogens, regardless of whether or not they are respiratory or enteric bacteria (Arshad et al., 1998). In support of these findings, our laboratory has recently shown that transmission by the faecal-respiratory route is a viable portal of entry for *Salmonella* (Kallapura et al., 2014a,b,c). This mode of infection could explain some clinical expression of relatively low-dose infectivity under field conditions in relation to the high oral dose challenge that is typically required for infection through the oral route in laboratory studies. This also supports previous studies demonstrating the fan-driven spread of *Salmonella* within the hatching cabinet and hatchery incubators (Hashemzadeh et al., 2010).

Recently, we evaluated the *in ovo* application as a practical and reliable way of delivering a probiotic mixed with the diluent of the Marek's disease (MD) vaccine (Teague et al., 2017). Although we previously reported the benefits of spray application of our LAB probiotic in the hatcheries (Wolfenden et al., 2007), this was the first report of *in ovo* application of this defined probiotic, mixed with MD vaccine. Interestingly, embryos, which received the probiotic before hatch, had a significant reduction in SE infection, lactose-positive Gram-negative bacteria without affecting the hatchability when compared with saline-treated chickens (Teague et al., 2017). Several studies conducted in our laboratory suggest that this particular probiotic is able to control salmonellae infections in poultry in both laboratory and commercial conditions (Farnell et al., 2006; Higgins et al., 2007; Vicente et al., 2007; Biloni et al., 2013;

Menconi et al., 2013). This current study further validated the probiotics efficacy via *in ovo* administration by reducing the recovery of SE when chickens were challenged on the day of hatch and cultured 24 h later. These results are in agreement with the work of De Oliveira et al. (2014) who demonstrated that *in ovo* colonization with probiotic could reduce *Salmonella* and other intestinal bacterial infections in poultry. In our study, we also observed that the higher body weight in the probiotic-treated group was due to the increase in villus height, leading to larger villus surface area, thus resulting in better nutrient absorption. These results are meaningful in context with the rapid early growth of broiler chickens. A newly hatched modern broiler chicken increases its body weight by 25% overnight and by 5000% in 5 weeks (Choct, 2009). Similarly, it is also important to consider the productive life of broiler chickens. The full genetic potential of modern chickens starts at conception and the first 21 days of embryo development. During this period, variables such as temperature and oxygen are important and any problem related to them could cause a big impact later in life. Hence, the 21 days of embryogenesis plus the first 7 days of the life of a chicken could potentially represent between 50% and 74% of the life of a commercial broiler chicken, depending on the time they are slaughtered (56 or 77 days) (Cherian, 2011). Therefore, earlier administration of probiotics to embryos can have a profound impact on the growth and overall health of the birds.

5 The use of direct-fed microbials

The use of selected strains from different beneficial microorganisms from the genus *Bacillus* and *Lactobacillus* have shown to be a suitable option for the poultry industry (Tellez et al., 2012). *Bacillus* spp. are a Gram-positive, facultative aerobe, endospore-forming, rod-shaped bacterium normally found in soil and water sources, as well as in the GIT of animals and humans (Hong et al., 2009). Its multiple flagella allow it to move quickly in liquids. *Bacillus* spp. are the most investigated Gram-positive bacteria and a model organism to study bacterial chromosome replication and cell differentiation, and together with other beneficial microbes have been extensively used as a source of industrial enzymes and antibiotics by biotechnology companies (Kunst et al., 1997). When environmental conditions are not favourable for growth and replication of bacteria from the genus *Bacillus*, dramatic metabolic changes occur, such as the induction of chemotaxis, cannibalism, production of macromolecular hydrolases (proteases and carbohydrases) as well as the formation of endospores (González-Pastor et al., 2003; Hong et al., 2005; López et al., 2009). Due to the capacity of bacterial spores to resist harsh environmental conditions and long storage periods, endospores from selected *Bacillus* strains have been used as reliable DFM in animal production (Tellez et al., 2013). Additionally,

Bacillus-DFM have previously been shown to prevent GIT disorders and impart numerous nutritional benefits for animals and humans (Hong et al., 2005; Duc et al., 2004; Cartman et al., 2007; Sen et al., 2012). Recent studies published by our laboratory have shown that approximately 90% of *Bacillus subtilis* spores germinate within 60 min in the presence of feed *in vitro* and *in vivo* in different segments of the GIT (Latorre et al., 2014a). After spore germination into vegetative cells, *Bacillus* spp. become metabolically active to produce chemical compounds that are beneficial to the host and the intestinal microflora (Jadamus et al., 2001; Leser et al., 2008).

In most of the United States and in other countries, including Brazil, broiler feed is based primarily on corn and soybean meal, which supplies the majority of energy and protein in the diet. Utilization of the nutrients contained in corn by broilers is generally considered to be high. Nevertheless, at times it is difficult to formulate least-cost diets using corn and unconventional grains with variable concentrations of anti-nutritional factors are used. Rye (*Secale cereale*) is a cereal member of the wheat tribe (*Triticeae*) and has been reported to contain 152 g of total non-starch polysaccharides (NSP) per kilogram of dry matter (Campbell et al., 1983). When chickens are fed alternative cereal grains such as rye that are high in soluble NSP, high digesta viscosity, poor nutrient digestibility and reduced bone mineralization have been reported before, resulting in decreased growth performance and reduced litter quality conditions caused by sticky droppings (Fengler and Marquardt, 1988). However, different studies have shown that the inclusion of carbohydrases such as xylanase in rye-based diets significantly improved all these negative factors reducing the impact of the anti-nutritional components present in the rye grain (Bedford and Classen, 1993; Silva and Smithard, 2002). Previously, we have evaluated the inclusion of selected *Bacillus*-DFM candidates that produce a different set of extracellular enzymes using different poultry diets *in vitro* (rye, wheat, barley and oat-based diets), resulting in a significant reduction in both digesta viscosity and *Clostridium perfringens* proliferation between control diets and *Bacillus*-DFM supplemented diets (Latorre et al., 2015). Rye has an elevated concentration of highly branched arabinoxylans in comparison to other cereals like wheat or corn (Bach Knudsen, 1997). The high concentration of soluble NSP in rye-based diets also have an impact on the intestinal bacterial population, probably as a consequence of the increased digesta viscosity and prolonged feed passage time (Choct et al., 2009; Bedford and Schulze, 1998). Furthermore, utilization of rye in poultry diets has also been related to malabsorption of lipids, deterioration of bone mineralization and reduced leg soundness (Kiarie et al., 2013) This negative effect on bone quality could also be related to an elevated digesta viscosity, therefore, enhancing the deconjugation of bile acids by the overgrowth intestinal microflora, resulting in a reduction of micelle formation, affecting fat solubilization and absorption of fat-soluble vitamins and minerals (Esteve-Garcia et al., 1997). Since monogastric animals do not have endogenous enzymes capable of hydrolysing the β-linkages present in soluble NSP, exogenous carbohydrases (xylanase, β-glucanase, β-mannanase, α-galactosidase and pectinase) have been used in poultry diets as feed additives in an attempt to reduce the adverse impact of these anti-nutritional factors (Bedford and Schulze, 1998). It has been well documented that inclusion of xylanase in rye-based diets significantly improved viscosity of digesta supernatant, accelerated feed passage time through the GIT and enhanced digestibility of dietary protein and fat sources resulting in an improvement in growth performance (Langhout et al., 1997; Lázaro et al., 2004). The results of the *Bacillus*-DFM study from our laboratory support previous findings in turkey poults fed with rye-based diets (Latorre et al., 2014b). Further studies confirmed that our multiple enzyme-producing *Bacillus*-based DFM improved growth performance, digesta viscosity, bacterial translocation, microbiota composition and bone mineralization in broiler chickens fed with a rye-based diet (Latorre et al., 2015). In our studies with broilers and turkeys, the increase in digesta viscosity observed in the control group was also associated with elevated bacterial translocation to the liver and overgrowth of Gram-negative and anaerobic bacteria in the duodenal section when compared with animals that consumed the *Bacillus*-DFM diet. These differences could be due to fewer substrates available for bacterial growth, generating lower intestinal inflammation and translocation of bacteria when the intestinal viscosity was reduced by the inclusion of the DFM candidate, suggesting more absorption of nutrients by the intestinal brush border of supplemented groups. It has been previously reported that alterations in gut permeability are connected with bacterial translocation in the portal and/or systemic circulation during several types of 'leaky gut' syndromes leading to bacterial septicaemia (Ilan, 2012; Seki and Schnabl, 2012). Furthermore, the significant improvements in performance observed in animals consuming the *Bacillus*-DFM supplemented diet when compared to the unsupplemented control group suggests that the production of enzymes from the combined *Bacillus* spp. strains used as DFM could increase the absorption of nutrients promoting growth performance and a more efficient feed conversion ratio in addition to enhancing the physical and bacteriological conditions of the intestinal content. We also observed that the significant reduction in bone strength and mineralization generated by consumption of rye-based diets confirmed previous research from different authors that have shown that the inclusion of rye in poultry diets is associated with malabsorption of minerals and fat-soluble vitamins (Campbell et al., 1983; Wideman and Prisby, 2012). In our studies, the reduction of digesta viscosity together with the production of phytase by the *Bacillus*-DFM candidate could enhance the absorption of nutrients including minerals, hence improving bone strength and bone mineralization (Latorre et al., 2014a, 2015; Tellez et al.,

2014b). Our studies have shown that chickens and turkeys fed with rye-based diets have an increase in digesta viscosity and bacterial translocation associated with overgrowth of gut microflora, low performance and decreased bone mineralization. However, these adverse effects caused by the utilization of rye in poultry diets can be minimized by the inclusion of a selected *Bacillus*-DFM candidate, thereby enhancing intestinal integrity and absorption of nutrients resulting in an improvement of production performance.

6 Conclusion

Bacterial antimicrobial resistance in both the medical and agricultural fields has become a serious problem worldwide. During the last 15 years, our laboratories have worked towards the identification of probiotic candidates for poultry which can actually displace *Salmonella* and other enteric pathogens which have colonized the GIT of chickens and turkeys, indicating that selection of therapeutically efficacious probiotic cultures with marked performance benefits in poultry is possible, and that defined cultures can sometimes provide an attractive alternative to conventional antimicrobial therapy. Our studies have been focused on specific pathogen reduction, performance under commercial conditions and effects on both idiopathic and defined enteritis. We have also confirmed that selected heat-resistant spore-forming *Bacillus* species can markedly reduce *Salmonella* and *Clostridium* when administered in very high numbers, and we have developed a novel and simple technique for obtaining cultured *Bacillus* spore counts, providing a cost-effective, feed-stable inclusion in commercial poultry diets. In order to select even more effective isolates, we are still currently focused on the mechanistic action of the *Lactobacillus* probiotic previously developed as well as new *Bacillus* candidates. Current indications are that the mechanism of action involves rapid activation of innate host immune mechanisms, providing an exciting possibility for identification of vastly superior and more potent probiotics. In this chapter, we summarized the safety and efficacy of individual monocultures for prophylactic and/or therapeutic efficacy against *Salmonella* infections under both laboratory and field conditions as well as the development of a novel, cost-effective, feed-stable, direct-fed microbials (DFM) with potential for widespread utilization and improved production, delivery and clinical efficacy for animal use.

7 Where to look for further information

During the last decade, the increasing interest in renewable energy sources changed the distribution of corn utilization from human and animal consumption to biofuel production, leading to a continuous rise in feed costs of livestock diets. Therefore, alternative feed ingredients such as distillers dried grains with solubles (DDGS), as well as cereals like wheat, barley and sorghum have become part of the feed matrix to maintain or reduce production costs. However, these raw materials often contain a higher concentration of anti-nutritional factors in comparison to corn, including non-starch polysaccharides which increase digesta viscosity and reduce nutrient absorption in monogastric animals. As a result, the addition of exogenous enzymes in poultry feed has steadily increased to maximize nutrient utilization and maintain performance parameters with diets containing less digestible ingredients. On the other hand, the poultry industry is also facing social concerns regarding the use of antibiotic growth promoters and the development of antibiotic-resistant microorganisms. One alternative among others is the *in ovo* utilization of probiotics candidates based on enzyme production profiles to improve nutrient absorption and intestinal integrity, as well as to maintain a healthy microflora balance in poultry-consuming commercial and alternative diets. Currently, our laboratories are also working on evaluating and selecting different *Bacillus* spp. strains for *in ovo* delivery.

8 References

Acar, J. F. and Goldstein, F. W. 1997. Trends in bacterial resistance to fluoroquinolones. *Clinical Infectious Diseases: an Official Publication of the Infectious Diseases Society of America* 24(Suppl. 1), S67–73. doi:10.1093/clinids/24.Supplement_1.S67.

Al-Murrani, W. K., Al-Rawi, I. K. and Raof, N. M. 2002. Genetic resistance to *Salmonella typhimurium* in two lines of chickens selected as resistant and sensitive on the basis of heterophil/lymphocyte ratio. *British Poultry Science* 43(4), 501–7. doi:10.1080/00 07166022000004408.

Alvarez-Olmos, M. I. and Oberhelman, R. A. 2001. Probiotic agents and infectious diseases: a modern perspective on a traditional therapy. *Clinical Infectious Diseases: an Official Publication of the Infectious Diseases Society of America* 32(11), 1567–76. doi:10.1086/320518.

Angulo, F. J., Johnson, K. R., Tauxe, R. V. and Cohen, M. L. 2000. Origins and consequences of antimicrobial-resistant nontyphoidal *Salmonella*: implications for the use of fluoroquinolones in food animals. *Microbial Drug Resistance* 6(1), 77–83. doi:10.1089/mdr.2000.6.77.

Arshad, M. J., Siddique, M., Rehman, S. and Aslam, M. S. 1998. A comparative study of respiratory phagocytic cell activities in layer chicks. *Medical Journal of Islamic Academy of Sciences* 11(3), 107–10.

Arvola, T., Laiho, K., Torkkeli, S., Mykkänen, H., Salminen, S., Maunula, L. and Isolauri, E. 1999. Prophylactic *Lactobacillus GG* reduces antibiotic-associated diarrhea in children with respiratory infections: a randomized study. *Pediatrics* 104(5), e64. doi:10.1542/peds.104.5.e64.

Bäckhed, F. 2011. Programming of host metabolism by the gut microbiota. *Annals of Nutrition and Metabolism* 58(Suppl. 2), 44–52. doi:10.1159/000328042.

Bartlett, J. G. 2002. Clinical practice. Antibiotic-associated diarrhea. *The New England Journal of Medicine* 346(5), 334–9. doi:10.1056/NEJMcp011603.

Bedford, M. and Classen, H. 1993. An *in vitro* assay for prediction of broiler intestinal viscosity and growth when fed rye-based diets in the presence of exogenous enzymes. *Poultry Science* 72, 137–43.

Bedford, M. R. and Schulze, H. 1998. Exogenous enzymes for pigs and poultry. *Nutrition Research Reviews* 11(1), 91–114. doi:10.1079/NRR19980007.

Bienenstock, J. 1980. Bronchus-associated lymphoid tissue and the source of immunoglobulin-containing cells in the mucosa. *Environmental Health Perspectives* 35, 39–42. doi:10.1289/ehp.803539.

Bienenstock, J. and McDermott, M. R. 2005. Bronchus- and nasal-associated lymphoid tissues. *Immunological Reviews* 206, 22–31. doi:10.1111/j.0105-2896.2005.00299.x.

Biloni, A., Quintana, C. F., Menconi, A., Kallapura, G., Latorre, J., Pixley, C., Layton, S., Dalmagro, M., Hernandez-Velasco, X., Wolfenden, A., et al. 2013. Evaluation of effects of EarlyBird associated with FloraMax-B11 on *Salmonella* Enteritidis, intestinal morphology, and performance of broiler chickens. *Poultry Science* 92(9), 2337–46. doi:10.3382/ps.2013-03279.

Blaser, M. J. 2006. Who are we? Indigenous microbes and the ecology of human diseases. *EMBO Reports* 7(10), 956–60. doi:10.1038/sj.embor.7400812.

Borchers, A. T., Selmi, C., Meyers, F. J., Keen, C. L. and Gershwin, M. E. 2009. Probiotics and immunity. *Journal of Gastroenterology* 44(1), 26–46. doi:10.1007/s00535-008-2296-0.

Borsoi, A., Santin, E., Santos, L. R., Salle, C. T., Moraes, H. L. and Nascimento, V. P. 2009. Inoculation of newly hatched broiler chicks with two Brazilian isolates of *Salmonella* Heidelberg strains with different virulence gene profiles, antimicrobial resistance, and pulsed field gel electrophoresis patterns to intestinal changes evaluation. *Poultry Science* 88(4), 750–8. doi:10.3382/ps.2008-00466.

Bronstein, J. L., Alarcón, R. and Geber, M. 2006. The evolution of plant-insect mutualisms. *The New Phytologist* 172(3), 412–28. doi:10.1111/j.1469-8137.2006.01864.x.

Campbell, G., Classen, H., Reichert, R. and Campbell, L. 1983. Improvement of the nutritive value of rye for broiler chickens by gamma irradiation-induced viscosity reduction. *British Poultry Science* 24, 205–12.

Cani, P. D. and Delzenne, N. M. 2009. The role of the gut microbiota in energy metabolism and metabolic disease. *Current Pharmaceutical Design* 15(13), 1546–58. doi:10.2174/138161209788168164.

Cartman, S. T., La Ragione, R. M. and Woodward, M. J. 2007. Bacterial spore formers as probiotics for poultry. *Food Science and Technology Bulletin: Functional Foods* 4, 21–30.

Castanon, J. I. 2007. History of the use of antibiotic as growth promoters in European poultry feeds. *Poultry Science* 86(11), 2466–71. doi:10.3382/ps.2007-00249.

Cherian, G. 2011. Essential fatty acids and early life programming in meat-type birds. *World's Poultry Science Journal* 67(4), 599–614. doi:10.1017/S0043933911000705.

Choct, M. 2009. Managing gut health through nutrition. *British Poultry Science* 50(1), 9–15. doi:10.1080/00071660802538632.

Corrier, D. E., Hinton Jr., A., Ziprin, R. L., Beier, R. C. and DeLoach, J. R. 1990. Effect of dietary lactose on cecal pH, bacteriostatic volatile fatty acids, and *Salmonella typhimurium* colonization of broiler chicks. *Avian Diseases* 34(3), 617–25. doi:10.2307/1591254.

Couper, M. R. 1997. Strategies for the rational use of antimicrobials. *Clinical Infectious Diseases: an Official Publication of the Infectious Diseases Society of America* 24(Suppl. 1), S154–6. doi:10.1093/clinids/24.Supplement_1.S154.

Dahiya, J. P., Wilkie, D. C., Van Kessel, A. G. and Drew, M. D. 2006. Potential strategies for controlling necrotic enteritis in broiler chickens in post-antibiotic era. *Animal Feed Science and Technology* 129(1-2), 60-88. doi:10.1016/j.anifeedsci.2005.12.003.

Dale, C. and Moran, N. A. 2006. Molecular interactions between bacterial symbionts and their hosts. *Cell* 126(3), 453-65. doi:10.1016/j.cell.2006.07.014.

Dass, N. B., John, A. K., Bassil, A. K., Crumbley, C. W., Shehee, W. R., Maurio, F. P., Moore, G. B., Taylor, C. M. and Sanger, G. J. 2007. The relationship between the effects of short-chain fatty acids on intestinal motility in vitro and GPR43 receptor activation. *Neurogastroenterology and Motility: the Official Journal of the European Gastrointestinal Motility Society* 19(1), 66-74. doi:10.1111/j.1365-2982.2006.00853.x.

De Geus, E. 2012. Respiratory immune responses in the chicken; towards development of mucosal avian influenza virus vaccines. PhD Dissertation. Utrecht University, Utrecht, the Netherlands.

De Moura, H. M., Silva, P. R., da Silva, P. H. C., Souza, N. R., Racanicci, A. M. C. and Santana, A. P. 2013. Antimicrobial resistance of *Campylobacter jejuni* isolated from chicken carcasses in the Federal District, Brazil. *Journal of Food Protection* 76(4), 691-3. doi:10.4315/0362-028X.JFP-12-485.

De Oliveira, J., van der Hoeven-Hangoor, E., van de Linde, I., Montijn, R. and van der Vossen, J. 2014. *In ovo* inoculation of chicken embryos with probiotic bacteria and its effect on posthatch *Salmonella* susceptibility. *Poultry Science* 93, 818-29.

Degli Esposti, M., Chouaia, B., Comandatore, F., Crotti, E., Sassera, D., Lievens, P. M.-J., Daffonchio, D. and Bandi, C. 2014. Evolution of mitochondria reconstructed from the energy metabolism of living bacteria. *PLoS One* 9(5), e96566. doi:10.1371/journal.pone.0096566.

Di Mauro, A., Neu, J., Riezzo, G., Raimondi, F., Martinelli, D., Francavilla, R. and Indrio, F. 2013. Gastrointestinal function development and microbiota. *Italian Journal of Pediatrics* 39(15), 15. doi:10.1186/1824-7288-39-15.

Dominguez-Bello, M. G. and Blaser, M. J. 2008. Do you have a probiotic in your future? *Microbes and Infection* 10(9), 1072-6.

Drolet, J.-P., Frangie, H., Guay, J., Hajoui, O., Hamid, Q. and Mazer, B. D. 2010. B lymphocytes in inflammatory airway diseases. *Clinical and Experimental Allergy* 40(6), 841-9. doi:10.1111/j.1365-2222.2010.03512.x.

Duc, L. H., Hong, H. A., Barbosa, T. M., Henriques, A. O. and Cutting, S. M. 2004. Characterization of *Bacillus* probiotics available for human use. *Applied and Environmental Microbiology* 70, 2161-71.

Duerkop, B. A., Vaishnava, S. and Hooper, L. V. 2009. Immune responses to the microbiota at the intestinal mucosal surface. *Immunity* 31(3), 368-76. doi:10.1016/j.immuni.2009.08.009.

Engberg, J., Aarestrup, F. M., Taylor, D. E., Gerner-Smidt, P. and Nachamkin, I. 2001. Quinolone and macrolide resistance in *Campylobacter jejuni* and *C. coli*: resistance mechanisms and trends in human isolates. *Emerging Infectious Diseases* 7(1), 24-34. doi:10.3201/eid0701.700024.

Esteve-Garcia, E., Brufau, J., Perez-Vendrell, A., Miquel, A. and Duven, K. 1997. Bioefficacy of enzyme preparations containing beta-glucanase and xylanase activities in broiler diets based on barley or wheat, in combination with flavomycin. *Poultry Science* 76(12), 1728-37. doi:10.1093/ps/76.12.1728.

Ewaschuk, J., Endersby, R., Thiel, D., Diaz, H., Backer, J., Ma, M., Churchill, T. and Madsen, K. 2007. Probiotic bacteria prevent hepatic damage and maintain colonic barrier function in a mouse model of sepsis. *Hepatology (Baltimore, MD)* 46(3), 841–50. doi:10.1002/hep.21750.

Fagerland, J. A. and Arp, L. H. 1990. A morphologic study of bronchus-associated lymphoid tissue in turkeys. *The American Journal of Anatomy* 189(1), 24–34. doi:10.1002/aja.1001890104.

Fagerland, J. A. and Arp, L. H. 1993. Structure and development of bronchus-associated lymphoid tissue in conventionally reared broiler chickens. *Avian Diseases* 37(1), 10–8. doi:10.2307/1591451.

Farnell, M. B., Donoghue, A. M., De Los Santos, F. S., Blore, P. J., Hargis, B. M., Tellez, G. and Donoghue, D. J. 2006. Upregulation of oxidative burst and degranulation in chicken heterophils stimulated with probiotic bacteria. *Poultry Science* 85(11), 1900–6. doi:10.1093/ps/85.11.1900.

Fengler, A. I. and Marquardt, R. R. 1988. Water-soluble pentosans from rye: II. Effects on rate of dialysis and on the retention of nutrients by the chick. *Cereal Chemistry* 65(4), 298–302.

Fraune, S. and Bosch, T. C. 2010. Why bacteria matter in animal development and evolution. *BioEssays: News and Reviews in Molecular, Cellular and Developmental Biology* 32(7), 571–80. doi:10.1002/bies.200900192.

Galanis, E., Lo Fo Wong, D. M., Patrick, M. E., Binsztein, N., Cieslik, A., Chalermchikit, T., Aidara-Kane, A., Ellis, A., Angulo, F. J., Wegener, H. C., et al. 2006. Web-based surveillance and global *Salmonella* distribution, 2000–2002. *Emerging Infectious Diseases* 12(3), 381–8. doi:10.3201/eid1205.050854.

Gibson, C. M. and Hunter, M. S. 2010. Extraordinarily widespread and fantastically complex: comparative biology of endosymbiotic bacterial and fungal mutualists of insects. *Ecology Letters* 13(2), 223–34. doi:10.1111/j.1461-0248.2009.01416.x.

González-Pastor, J. E., Hobbs, E. C. and Losick, R. 2003. Cannibalism by sporulating bacteria. *Science* 301(5632), 510–3. doi:10.1126/science.1086462.

Griggs, D. J., Hall, M. C., Jin, Y. F. and Piddock, L. J. 1994. Quinolone resistance in veterinary isolates of *Salmonella*. *The Journal of Antimicrobial Chemotherapy* 33(6), 1173–89. doi:10.1093/jac/33.6.1173.

Groschwitz, K. R. and Hogan, S. P. 2009. Intestinal barrier function: molecular regulation and disease pathogenesis. *The Journal of Allergy and Clinical Immunology* 124(1), 3–20; quiz 21. doi:10.1016/j.jaci.2009.05.038.

Haghighi, H. R., Gong, J., Gyles, C. L., Hayes, M. A., Zhou, H., Sanei, B., Chambers, J. R. and Sharif, S. 2006. Probiotics stimulate production of natural antibodies in chickens. *Clinical and Vaccine Immunology: CVI* 13(9), 975–80. doi:10.1128/CVI.00161-06.

Hammes, W. P. and Hertel, C. 2002. Research approaches for pre- and probiotics: challenges and outlook. *Food Research International* 35, 165–70.

Hashemzadeh, Z., Torshizi, K., Rahimi, S., Razban, V. and Zahraei Salehi, T. 2010. Prevention of *Salmonella* colonization in neonatal broiler chicks by using different routes of probiotic administration in hatchery evaluated by culture and PCR techniques. *Journal of Agricultural Science and Technology* 12, 425–32.

Hernández-Reyes, C. and Schikora, A. 2013. Salmonella, a cross-kingdom pathogen infecting humans and plants. *FEMS Microbiology Letters* 343(1), 1–7. doi:10.1111/1574-6968.12127.

Higgins, S. E., Torres-Rodriguez, A., Vicente, J. L., Sartor, C. D., Pixley, C. M., Nava, G. M., Tellez, G., Barton, J. T. and Hargis, B. M. 2005. Evaluation of intervention strategies for idiopathic diarrhea in commercial turkey brooding houses. *The Journal of Applied Poultry Research* 14(2), 345–8. doi:10.1093/japr/14.2.345.

Higgins, J. P., Higgins, S. E., Vicente, J. L., Wolfenden, A. D., Tellez, G. and Hargis, B. M. 2007. Temporal effects of lactic acid bacteria probiotic culture on Salmonella in neonatal broilers. *Poultry Science* 86(8), 1662–6. doi:10.1093/ps/86.8.1662.

Higgins, S. E., Wolfenden, A. D., Tellez, G., Hargis, B. M. and Porter, T. E. 2011. Transcriptional profiling of cecal gene expression in probiotic-and *Salmonella*-challenged neonatal chicks. *Poultry Science* 90(4), 901–13. doi:10.3382/ps.2010-00907.

Hofer, E. and Reis, E. M. F. dos 1994. *Salmonella* serovars in food poisoning episodes recorded in Brazil from 1982 to 1991. *Revista do Instituto de Medicina Tropical de São Paulo* 36(1), 7–9. doi:10.1590/S0036-46651994000100002.

Hong, H. A., Duc, le H. and Cutting, S. M. 2005. The use of bacterial spore formers as probiotics. *FEMS Microbiology Reviews* 29(4), 813–35. doi:10.1016/j.femsre.2004.12.001.

Hong, H. A., Khaneja, R., Tam, N. M., Cazzato, A., Tan, S., Urdaci, M., Brisson, A., Gasbarrini, A., Barnes, I. and Cutting, S. M. 2009. *Bacillus subtilis* isolated from the human gastrointestinal tract. *Research in Microbiology* 160(2), 134–43. doi:10.1016/j.resmic.2008.11.002.

Howarth, G. S. and Wang, H. 2013. Role of endogenous microbiota, probiotics and their biological products in human health. *Nutrients* 5(1), 58–81. doi:10.3390/nu5010058.

Hsieh, C. Y., Osaka, T., Moriyama, E., Date, Y., Kikuchi, J. and Tsuneda, S. 2015. Strengthening of the intestinal epithelial tight junction by *Bifidobacterium bifidum*. *Physiological Reports* 3(3), e12327. doi:10.14814/phy2.12327.

Ilan, Y. 2012. Leaky gut and the liver: a role for bacterial translocation in nonalcoholic steatohepatitis. *World Journal of Gastroenterology: WJG* 18, 2609.

Irino, K., Fernandes, S. A., Tavechio, A. T., Neves, B. C. and Dias, A. M. 1996. Progression of *Salmonella* Enteritidis phage Type 4 strains in São Paulo State, Brazil. *Revista do Instituto de Medicina Tropical de São Paulo* 38(3), 193–6. doi:10.1590/S0036-46651996000300005.

Isolauri, E., Kirjavainen, P. V. and Salminen, S. 2002. Probiotics: a role in the treatment of intestinal infection and inflammation? *Gut* 50(Suppl. 3), III54–9. doi:10.1136/gut.50.suppl_3.iii54.

Jadamus, A., Vahjen, W. and Simon, O. 2001. Growth behaviour of a spore forming probiotic strain in the gastrointestinal tract of broiler chicken and piglets. *Archives of Animal Nutrition* 54, 1–17.

Jeon, M. K., Klaus, C., Kaemmerer, E. and Gassler, N. 2013. Intestinal barrier: molecular pathways and modifiers. *World Journal of Gastrointestinal Pathophysiology* 4(4), 94–9. doi:10.4291/wjgp.v4.i4.94.

Johansson, M. E., Gustafsson, J. K., Sjöberg, K. E., Petersson, J., Holm, L., Sjövall, H. and Hansson, G. C. 2010. Bacteria penetrate the inner mucus layer before inflammation in the dextran sulfate colitis model. *PLoS One* 5, e12238.

Kallapura, G., Kogut, M., Morgan, M., Pumford, N., Bielke, L., Wolfenden, A., Faulkner, O., Latorre, J., Menconi, A., Hernandez-Velasco, X., et al. 2014a. Fate of *Salmonella* Senftenberg in broiler chickens evaluated by challenge experiments. *Avian Pathology* 43, 305–9.

Kallapura, G., Morgan, M., Pumford, N., Bielke, L., Wolfenden, A., Faulkner, O., Latorre, J., Menconi, A., Hernandez-Velasco, X., Kuttappan, V., et al. 2014b. Evaluation of the respiratory route as a viable portal of entry for Salmonella in poultry via intratracheal challenge of *Salmonella* Enteritidis and *Salmonella* Typhimurium. *Poultry Science* 93, 340–6.

Kallapura, G., Botero, A., Layton, S., Bielke, L. R., Latorre, J. D., Menconi, A., Hernández-Velasco, X., Bueno, D. J., Hargis, B. M. and Téllez, G. 2014c. Evaluation of recovery of *Salmonella* from trachea and ceca in commercial poultry. *Journal of Applied Poultry Research* 23(1), 132–6.

Kiarie, E., Romero, L. F. and Nyachoti, C. M. 2013. The role of added feed enzymes in promoting gut health in swine and poultry. *Nutrition Research Reviews* 26(1), 71–88. doi:10.1017/S0954422413000048.

Kikuchi, Y., Hosokawa, T., Nikoh, N., Meng, X. Y., Kamagata, Y. and Fukatsu, T. 2009. Host-symbiont co-speciation and reductive genome evolution in gut symbiotic bacteria of acanthosomatid stinkbugs. *BMC Biology* 7(1), 2. doi:10.1186/1741-7007-7-2.

Kim, Y. S. and Ho, S. B. 2010. Intestinal goblet cells and mucins in health and disease: recent insights and progress. *Current Gastroenterology Reports* 12, 319–30.

Köhler, H., Sakaguchi, T., Hurley, B. P., Kase, B. A., Reinecker, H. C. and McCormick, B. A. 2007. *Salmonella enterica* serovar Typhimurium regulates intercellular junction proteins and facilitates transepithelial neutrophil and bacterial passage. *American Journal of Physiology. Gastrointestinal and Liver Physiology* 293(1), G178–87. doi:10.1152/ajpgi.00535.2006.

Kunst, F., Ogasawara, N., Moszer, I., Albertini, A. M., Alloni, G., Azevedo, V., Bertero, M. G., Bessieres, P., Bolotin, A., Borchert, S., et al. 1997. The complete genome sequence of the Gram-positive bacterium *Bacillus subtilis*. *Nature* 390(6657), 249–56. doi:10.1038/36786.

Kuttappan, V., Berghman, L., Vicuña, E., Latorre, J., Menconi, A., Wolchok, J., Wolfenden, A., Faulkner, O., Tellez, G., Hargis, B., et al. 2015. Poultry enteric inflammation model with dextran sodium sulfate mediated chemical induction and feed restriction in broilers. *Poultry Science* 94, 1220–6.

Langhout, D., Schutte, J., Geerse, C., Kies, A., De Jong, J. and Verstegen, M. 1997. Effects on chick performance and nutrient digestibility of an endo-xylanase added to a wheat-and rye-based diet in relation to fat source. *British Poultry Science* 38, 557–63.

Latorre, J. D., Hernandez-Velasco, X., Kallapura, G., Menconi, A., Pumford, N. R., Morgan, M. J., Layton, S. L., Bielke, L. R., Hargis, B. M. and Téllez, G. 2014a. Evaluation of germination, distribution, and persistence of *Bacillus subtilis* spores through the gastrointestinal tract of chickens. *Poultry Science* 93(7), 1793–800. doi:10.3382/ps.2013-03809.

Latorre, J. D., Hernandez-Velasco, X., Kogut, M. H., Vicente, J. L., Wolfenden, R., Wolfenden, A., Hargis, B. M., Kuttappan, V. A. and Tellez, G. 2014b. Role of a *Bacillus subtilis* direct-fed microbial on digesta viscosity, bacterial translocation, and bone mineralization in turkey poults fed with a rye-based diet. *Frontiers in Veterinary Science* 1, 26. doi:10.3389/fvets.2014.00026.

Latorre, J. D., Hernandez-Velasco, X., Bielke, L. R., Vicente, J. L., Wolfenden, R., Menconi, A., Hargis, B. M. and Tellez, G. 2015. Evaluation of a *Bacillus* direct-fed microbial candidate on digesta viscosity, bacterial translocation, microbiota composition and bone mineralisation in broiler chickens fed on a rye-based diet. *British Poultry Science* 56(6), 723–32. doi:10.1080/00071668.2015.1101053.

Lázaro, R., Latorre, M., Medel, P., Gracia, M. and Mateos, G. 2004. Feeding regimen and enzyme supplementation to rye-based diets for broilers. *Poultry Science* 83, 152–60.

Leser, T., Knarreborg, A. and Worm, J. 2008. Germination and outgrowth of *Bacillus subtilis* and *Bacillus licheniformis* spores in the gastrointestinal tract of pigs. *Journal of Applied Microbiology* 104, 1025–33.

López, D., Vlamakis, H., Losick, R. and Kolter, R. 2009. Cannibalism enhances biofilm development in *Bacillus subtilis*. *Molecular Microbiology* 74(3), 609–18. doi:10.1111/j.1365-2958.2009.06882.x.

Lyte, M. 2011. Probiotics function mechanistically as delivery vehicles for neuroactive compounds: microbial endocrinology in the design and use of probiotics. *BioEssays: News and Reviews in Molecular, Cellular and Developmental Biology* 33(8), 574–81. doi:10.1002/bies.201100024.

Mackiewicz, P., Bodył, A. and Gagat, P. 2012 Possible import routes of proteins into the cyanobacterial endosymbionts/plastids of *Paulinella chromatophora*. *Theory in Biosciences = Theorie in den Biowissenschaften* 131(1), 1–18. doi:10.1007/s12064-011-0147-7.

Madsen, K., Cornish, A., Soper, P., McKaigney, C., Jijon, H., Yachimec, C., Doyle, J., Jewell, L. and De Simone, C. 2001. Probiotic bacteria enhance murine and human intestinal epithelial barrier function. *Gastroenterology* 121(3), 580–91. doi:10.1053/gast.2001.27224.

Manning, J. G., Hargis, B. M., Hinton Jr., A., Corrier, D. E., DeLoach, J. R. and Creger, C. R. 1992. Effect of nitrofurazone or novobiocin on *Salmonella* Enteritidis cecal colonization and organ invasion in leghorn hens. *Avian Diseases* 36(2), 334–40. doi:10.2307/1591508.

Martin, R., Nauta, A. J., Ben Amor, K., Knippels, L. M., Knol, J. and Garssen, J. 2010. Early life: gut microbiota and immune development in infancy. *Beneficial Microbes* 1(4), 367–82. doi:10.3920/BM2010.0027.

Maslowski, K. M. and Mackay, C. R. 2011. Diet, gut microbiota and immune responses. *Nature Immunology* 12(1), 5–9. doi:10.1038/ni0111-5.

McFall-Ngai, M. 2007. Adaptive immunity: care for the community. *Nature* 445(7124), 153-. doi:10.1038/445153a.

Menconi, A., Wolfenden, A. D., Shivaramaiah, S., Terraes, J. C., Urbano, T., Kuttel, J., Kremer, C., Hargis, B. M. and Tellez, G. 2011. Effect of lactic acid bacteria probiotic culture for the treatment of *Salmonella enterica* serovar Heidelberg in neonatal broiler chickens and turkey poults. *Poultry Science* 90(3), 561–5. doi:10.3382/ps.2010-01220.

Menconi, A., Hernandez-Velasco, X., Latorre, J. D., Kallapura, G., Pumford, N. R., Morgan, M. J., Hargis, B. and Tellez, G. 2013. Effect of chitosan as a biological sanitizer for *Salmonella Typhimurium* and aerobic Gram-negative spoilage bacteria present on chicken skin. *International Journal of Poultry Science* 12(6), 318–21.

Mennigen, R. and Bruewer, M. 2009. Effect of probiotics on intestinal barrier function. *Annals of the New York Academy of Sciences* 1165, 183–9. doi:10.1111/j.1749-6632.2009.04059.x.

Metchnikoff, E. and Metchnikoff, I. 1907. *The Prolongation of Life*. GP Putnam and Sons.

Molinaro, F., Paschetta, E., Cassader, M., Gambino, R. and Musso, G. 2012. Probiotics, prebiotics, energy balance, and obesity: mechanistic insights and therapeutic implications. *Gastroenterology Clinics of North America* 41(4), 843–54. doi:10.1016/j.gtc.2012.08.009.

Morales-Barrera, E., Calhoun, N., Lobato-Tapia, J. L., Lucca, V., Prado-Rebolledo, O., Hernandez-Velasco, X., Merino-Guzman, R., Petrone-García, V. M., Latorre, J. D., Mahaffey, B. D., et al. 2016. Risks involved in the use of Enrofloxacin for *Salmonella*

Enteritidis or *Salmonella* Heidelberg in commercial poultry. *Frontiers in Veterinary Science* 3, 72.

Moran, N. A. 2007. Symbiosis as an adaptive process and source of phenotypic complexity. *Proceedings of the National Academy of Sciences of the United States of America* 104(Suppl. 1), 8627–33. doi:10.1073/pnas.0611659104.

Murray, B. E. 1986. Resistance of *Shigella*, *Salmonella*, and other selected enteric pathogens to antimicrobial agents. *Reviews of Infectious Diseases* 8(Suppl. 2), S172–81. doi:10.1093/clinids/8.Supplement_2.S172.

Musso, G., Gambino, R. and Cassader, M. 2010. Obesity, diabetes, and gut microbiota the hygiene hypothesis expanded? *Diabetes Care* 33(10), 2277–84. doi:10.2337/dc10-0556.

Neish, A. S. 2009. Microbes in gastrointestinal health and disease. *Gastroenterology* 136(1), 65–80. doi:10.1053/j.gastro.2008.10.080.

Niewold, T. A. 2007. The nonantibiotic anti-inflammatory effect of antimicrobial growth promoters, the real mode of action? A hypothesis. *Poultry Science* 86(4), 605–9. doi:10.1093/ps/86.4.605.

Overman, E. L., Rivier, J. E. and Moeser, A. J. 2012. CRF induces intestinal epithelial barrier injury via the release of mast cell proteases and TNF-α. *PLoS One* 7(6), e39935. doi:10.1371/journal.pone.0039935.

Ozinsky, A., Underhill, D. M., Fontenot, J. D., Hajjar, A. M., Smith, K. D., Wilson, C. B., Schroeder, L. and Aderem, A. 2000. The repertoire for pattern recognition of pathogens by the innate immune system is defined by cooperation between toll-like receptors. *Proceedings of the National Academy of Sciences of the United States of America* 97(25), 13766–71. doi:10.1073/pnas.250476497.

Parker, D. S. 1990. Manipulation of the functional activity of the gut by dietary and other means (antibiotics/probiotics) in ruminants. *The Journal of Nutrition* 120(6), 639–48. doi:10.1093/jn/120.6.639.

Pastel, E., Pointud, J. C., Volat, F., Martinez, A. and Lefrançois-Martinez, A. M. 2012. Aldo-keto reductases 1B in endocrinology and metabolism. *Frontiers in Pharmacology* 3, 148. doi:10.3389/fphar.2012.00148.

Pastorelli, L., De Salvo, C., Mercado, J. R., Vecchi, M. and Pizarro, T. T. 2013. Central role of the gut epithelial barrier in the pathogenesis of chronic intestinal inflammation: lessons learned from animal models and human genetics. *Frontiers in Immunology* 4, 280. doi:10.3389/fimmu.2013.00280.

Piddock, L. J. 2002. Fluoroquinolone resistance in *Salmonella* serovars isolated from humans and food animals. *FEMS Microbiology Reviews* 26(1), 3–16. doi:10.1111/j.1574-6976.2002.tb00596.x.

Piddock, L. J. and Wise, R. 1989. Mechanisms of resistance to quinolones and clinical perspectives. *The Journal of Antimicrobial Chemotherapy* 23(4), 475–80. doi:10.1093/jac/23.4.475.

Piddock, L. J., Whale, K. and Wise, R. 1990. Quinolone resistance in *Salmonella*: clinical experience. *The Lancet* 335(8703), 1459. doi:10.1016/0140-6736(90)91484-R.

Prado-Rebolledo, O. F., Delgado-Machuca, J. J., Macedo-Barragan, R. J., Garcia-Márquez, L. J., Morales-Barrera, J. E., Latorre, J. D., Hernandez-Velasco, X. and Tellez, G. 2017. Evaluation of a selected lactic acid bacteria-based probiotic on *Salmonella enterica* serovar Enteritidis colonization and intestinal permeability in broiler chickens. *Avian Pathology: Journal of the W.V.P.A.* 46(1), 90–4. doi:10.1080/03079457.2016.1222808.

Qiu, R., Croom, J., Ali, R. A., Ballou, A. L., Smith, C. D., Ashwell, C. M., Hassan, H. M., Chiang, C. C. and Koci, M. D. 2012. Direct fed microbial supplementation repartitions host energy to the immune system. *Journal of Animal Science* 90(8), 2639–51. doi:10.2527/jas.2011-4611.

Ramana, K. V. and Srivastava, S. K. 2006. Mediation of aldose reductase in lipopolysaccharide-induced inflammatory signals in mouse peritoneal macrophages. *Cytokine* 36(3–4), 115–22. doi:10.1016/j.cyto.2006.11.003.

Ramana, K. V. and Srivastava, S. K. 2010. Aldose reductase: A novel therapeutic target for inflammatory pathologies. *The International Journal of Biochemistry and Cell Biology* 42(1), 17–20. doi:10.1016/j.biocel.2009.09.009.

Randall, L. P., Cooles, S. W., Coldham, N. C., Stapleton, K. S., Piddock, L. J. and Woodward, M. J. 2006. Modification of enrofloxacin treatment regimens for poultry experimentally infected with *Salmonella enterica* serovar Typhimurium DT104 to minimize selection of resistance. *Antimicrobial Agents and Chemotherapy* 50(12), 4030–7. doi:10.1128/AAC.00525-06.

Rodrigue, D. C., Tauxe, R. V. and Rowe, B. 1990. International increase in *Salmonella* Enteritidis: A new pandemic? *Epidemiology and Infection* 105(1), 21–7. doi:10.1017/S0950268800047609.

Salminen, S. and Isolauri, E. 2006. Intestinal colonization, microbiota, and probiotics. *The Journal of Pediatrics* 149(5), S115–20. doi:10.1016/j.jpeds.2006.06.062.

Salzman, N. H. 2011. Microbiota-immune system interaction: an uneasy alliance. *Current Opinion in Microbiology* 14(1), 99–105. doi:10.1016/j.mib.2010.09.018.

Sakamoto, K., Hirose, H., Onizuka, A., Hayashi, M., Futamura, N., Kawamura, Y. and Ezaki, T. 2000. Quantitative study of changes in intestinal morphology and mucus gel on total parenteral nutrition in rats. *Journal of Surgical Research* 94, 99–106.

Saridaki, A. and Bourtzis, K. 2010. Wolbachia: more than just a bug in insects genitals. *Current Opinion in Microbiology* 13(1), 67–72. doi:10.1016/j.mib.2009.11.005.

Schleker, S., Kshirsagar, M. and Klein-Seetharaman, J. 2015. Comparing human-*Salmonella* with plant-*Salmonella* protein-protein interaction predictions. *Frontiers in Microbiology* 6, 45. doi:10.3389/fmicb.2015.00045.

Segawa, S., Fujiya, M., Konishi, H., Ueno, N., Kobayashi, N., Shigyo, T. and Kohgo, Y. 2011. Probiotic-derived polyphosphate enhances the epithelial barrier function and maintains intestinal homeostasis through integrin-p38 MAPK pathway. *PLoS One* 6(8), e23278. doi:10.1371/journal.pone.0023278.

Seki, E. and Schnabl, B. 2012. Role of innate immunity and the microbiota in liver fibrosis: crosstalk between the liver and gut. *The Journal of Physiology* 590, 447–58.

Sekirov, I., Russell, S. L., Antunes, L. C. M. and Finlay, B. B. 2010. Gut microbiota in health and disease. *Physiological Reviews* 90(3), 859–904. doi:10.1152/physrev.00045.2009.

Sen, S., Ingale, S. L., Kim, Y. W., Kim, J. S., Kim, K. H., Lohakare, J. D., Kim, E. K., Kim, H. S., Ryu, M. H., Kwon, I. K., et al. 2012. Effect of supplementation of *Bacillus subtilis* LS 1-2 to broiler diets on growth performance, nutrient retention, caecal microbiology and small intestinal morphology. *Research in Veterinary Science* 93(1), 264–8. doi:10.1016/j.rvsc.2011.05.021.

Seuna, E. and Nurmi, E. 1979. Therapeutical trials with antimicrobial agents and cultured cecal microflora in *Salmonella infantis* infections in chickens. *Poultry Science* 58(5), 1171–4. doi:10.3382/ps.0581171.

Sharma, R., Young, C. and Neu, J. 2010. Molecular modulation of intestinal epithelial barrier: contribution of microbiota. *Journal of Biomedicine and Biotechnology* 2010, 305879. doi:10.1155/2010/305879.

Sherryll Lynn, L., Xochitl, H.-V., Shivaramaiah, C., Jorge, X., Anita, M., Juan David, L., Gopala, K., Vivek Ayamchirakkunnel, K., Ross Elderon, W., Andreatti, R. L., et al. 2013. The effect of a *Lactobacillus*-based probiotic for the control of necrotic enteritis in broilers. *Food and Nutrition Sciences* 4, 1-7.

Shini, S., Kaiser, P., Shini, A. and Bryden, W. L. 2008. Differential alterations in ultrastructural morphology of chicken heterophils and lymphocytes induced by corticosterone and lipopolysaccharide. *Veterinary Immunology and Immunopathology* 122(1-2), 83-93. doi:10.1016/j.vetimm.2007.10.009.

Sierra-Arguello, Y. M., Faulkner, O., Tellez, G., Hargis, B. M. and do Nascimento, V. P. 2016. The use of FTA cards for transport and detection of gyrA mutation of *Campylobacter jejuni* from poultry. *Poultry Science* 95(4), 798-801. doi:10.3382/ps/pev384.

Śmiałek, M., Tykałowski, B., Stenzel, T. and Koncicki, A. 2011. Local immunity of the respiratory mucosal system in chickens and turkeys. *Polish Journal of Veterinary Sciences* 14(2), 291-7. doi:10.2478/v10181-011-0047-2.

Smith, H. W. and Tucker, J. F. 1975. The effect of antibiotic therapy on the faecal excretion of *Salmonella typhimurium* by experimentally infected chickens. *The Journal of Hygiene* 75(2), 275-92. doi:10.1017/S0022172400047306.

Srivastava, S. K., Yadav, U. C., Reddy, A. B., Saxena, A., Tammali, R., Shoeb, M., Ansari, N. H., Bhatnagar, A., Petrash, M. J., Srivastava, S., et al. 2011. Aldose reductase inhibition suppresses oxidative stress-induced inflammatory disorders. *Chemico-Biological Interactions* 191(1-3), 330-8. doi:10.1016/j.cbi.2011.02.023.

Steed, E., Balda, M. S. and Matter, K. 2010. Dynamics and functions of tight junctions. *Trends in Cell Biology* 20(3), 142-9. doi:10.1016/j.tcb.2009.12.002.

Tao, Y., Drabik, K. A., Waypa, T. S., Musch, M. W., Alverdy, J. C., Schneewind, O., Chang, E. B. and Petrof, E. O. 2006. Soluble factors from *Lactobacillus GG* activate MAPKs and induce cytoprotective heat shock proteins in intestinal epithelial cells. *American Journal of Physiology. Cell Physiology* 290(4), C1018-30. doi:10.1152/ajpcell.00131.2005.

Teague, K. D., Graham, L. E., Dunn, J. R., Cheng, H. H., Anthony, N., Latorre, J. D., Menconi, A., Wolfenden, R. E., Wolfenden, A. D., Mahaffey, B. D., et al. 2017. *In ovo* evaluation of FloraMax®-B11 on Marek's disease HVT vaccine protective efficacy, hatchability, microbiota composition, morphometric analysis, and *Salmonella* Enteritidis infection in broiler chickens. *Poultry Science* 96(7), 2074-82. doi:10.3382/ps/pew494.

Tellez, G. I., Jaeger, L., Dean, C. E., Corrier, D. E., DeLoach, J. R., Williams, J. D. and Hargis, B. M. 1993. Effect of prolonged administration of dietary capsaicin on *Salmonella* Enteritidis infection in leghorn chicks. *Avian Diseases* 37(1), 143-8. doi:10.2307/1591467.

Tellez, G., Pixley, C., Wolfenden, R., Layton, S. and Hargis, B. 2012. Probiotics/direct fed microbials for *Salmonella* control in poultry. *Food Research International* 45, 628-33.

Tellez, G., Rodriguez-Fragoso, L., Kuttappan, V., Kallapura, G., Velasco, X., Menconi, A., Latorre, J., Wolfenden, A., Hargis, B. and J. Reyes-Esparza. 2013. Probiotics for human and poultry use in the control of gastrointestinal disease: a review of real-world experiences. *Alternative and Integrative Medicine* 2, 118.

Tellez, G., Latorre, J. D., Kuttappan, V. A., Kogut, M. H., Wolfenden, A., Hernandez-Velasco, X., Hargis, B. M., Bottje, W. G., Bielke, L. R. and Faulkner, O. B. 2014. Utilization of rye as energy source affects bacterial translocation, intestinal viscosity, microbiota composition, and bone mineralization in broiler chickens. *Frontiers in Genetics* 5, 339. doi:10.3389/fgene.2014.00339.

Timbermont, L., Lanckriet, A., Dewulf, J., Nollet, N., Schwarzer, K., Haesebrouck, F., Ducatelle, R. and Van Immerseel, F. 2010. Control of *Clostridium perfringens*-induced necrotic enteritis in broilers by target-released butyric acid, fatty acids, and essential oils. *Avian Pathology: Journal of the W.V.P.A.* 39(2), 117–21. doi:10.1080/03079451003610586.

Tlaskalová-Hogenová, H., Stěpánková, R., Kozáková, H., Hudcovic, T., Vannucci, L., Tučková, L., Rossmann, P., Hrnčíř, T., Kverka, M., Zákostelská, Z., et al. 2011. The role of gut microbiota (commensal bacteria) and the mucosal barrier in the pathogenesis of inflammatory and autoimmune diseases and cancer: contribution of germ-free and gnotobiotic animal models of human diseases. *Cellular and Molecular Immunology* 8(2), 110–20.

Torres-Rodriguez, A., Donoghue, A. M., Donoghue, D. J., Barton, J. T., Tellez, G. and Hargis, B. M. 2007. Performance and condemnation rate analysis of commercial turkey flocks treated with a *Lactobacillus* spp.-based probiotic. *Poultry Science* 86(3), 444–6. doi:10.1093/ps/86.3.444.

Ulluwishewa, D., Anderson, R. C., McNabb, W. C., Moughan, P. J., Wells, J. M. and Roy, N. C. 2011. Regulation of tight junction permeability by intestinal bacteria and dietary components. *The Journal of Nutrition* 141(5), 769–76. doi:10.3945/jn.110.135657.

Uwaydah, A. K., Matar, I., Chacko, K. C. and Davidson, J. C. 1991. The emergence of antimicrobial resistant *Salmonella typhi* in Qatar: epidemiology and therapeutic implications. *Transactions of the Royal Society of Tropical Medicine and Hygiene* 85(6), 790–2. doi:10.1016/0035-9203(91)90457-A.

Van Der Wielen, P. W., Biesterveld, S., Notermans, S., Hofstra, H., Urlings, B. A. and van Knapen, F. 2000. Role of volatile fatty acids in development of the cecal microflora in broiler chickens during growth. *Applied and Environmental Microbiology* 66(6), 2536–40. doi:10.1128/AEM.66.6.2536-2540.2000.

Vanderpool, C., Yan, F. and Polk, D. B. 2008. Mechanisms of probiotic action: implications for therapeutic applications in inflammatory bowel diseases. *Inflammatory Bowel Diseases* 14(11), 1585–96. doi:10.1002/ibd.20525.

Vicente, J., Higgins, S., Bielke, L., Tellez, G., Donoghue, D., Donoghue, A. and Hargis, B. 2007. Effect of probiotic culture candidates on *Salmonella* prevalence in commercial turkey houses. *The Journal of Applied Poultry Research* 16(3), 471–6. doi:10.1093/japr/16.3.471.

Vicente, J. L., Torres-Rodriguez, A., Higgins, S. E., Pixley, C., Tellez, G., Donoghue, A. M. and Hargis, B. M. 2008. Effect of a selected *Lactobacillus* spp.-based probiotic on *Salmonella enterica* serovar enteritidis infected broiler chicks. *Avian Diseases* 52(1), 143–6. doi:10.1637/7847-011107-ResNote.

Vicuña, E. A., Kuttappan, V. A., Galarza-Seeber, R., Latorre, J. D., Faulkner, O. B., Hargis, B. M., Tellez, G. and Bielke, L. R. 2015a. Effect of dexamethasone in feed on intestinal permeability, differential white blood cell counts, and immune organs in broiler chicks. *Poultry Science* 94(9), 2075–80. doi:10.3382/ps/pev211.

Vicuña, E. A., Kuttappan, V. A., Tellez, G., Hernandez-Velasco, X., Seeber-Galarza, R., Latorre, J. D., Faulkner, O. B., Wolfenden, A. D., Hargis, B. M. and Bielke, L. R. 2015b. Dose titration of FITC-D for optimal measurement of enteric inflammation in broiler chicks. *Poultry Science* 94(6), 1353–9. doi:10.3382/ps/pev111.

Walter, J., Britton, R. A. and Roos, S. 2011. Host-microbial symbiosis in the vertebrate gastrointestinal tract and the *Lactobacillus reuteri* paradigm. *Proceedings of the National Academy of Sciences of the United States of America* 108(Suppl. 1), 4645–52. doi:10.1073/pnas.1000099107.

Wathes, C. M., Zaidan, W. A., Pearson, G. R., Hinton, M. and Todd, N. 1988. Aerosol infection of calves and mice with *Salmonella typhimurium*. *The Veterinary Record* 123(23), 590–4.

White, P. L., Baker, A. R. and James, W. O. 1997. Strategies to control *Salmonella* and *Campylobacter* in raw poultry products. *Revue scientifique et technique (International Office of Epizootics)* 16(2), 525–41. doi:10.20506/rst.16.2.1046.

Wideman, R. F. and Prisby, R. D. 2012. Bone circulatory disturbances in the development of spontaneous bacterial chondronecrosis with osteomyelitis: a translational model for the pathogenesis of femoral head necrosis. *Frontiers in Endocrinology* 3, 183-. doi:10.3389/fendo.2012.00183.

Wolfenden, A. D., Pixley, C. M., Higgins, J. P., Higgins, S. E., Hargis, B. M., Tellez, G., Vicente, J. L. and Torres-Rodriguez, A. 2007. Evaluation of spray application of a *Lactobacillus*-based probiotic on *Salmonella* Enteritidis colonization in broiler chickens. *International Journal of Poultry Science* 6(7), 493–6. doi:10.3923/ijps.2007.493.496.

Xie, J., Vilchez, I. and Mateos, M. 2010. Spiroplasma bacteria enhance survival of *Drosophila Hydei* attacked by the parasitic wasp *Leptopilina Heterotoma*. *PLoS One* 5(8), e12149. doi:10.1371/journal.pone.0012149.

Xu, J. and Gordon, J. I. 2003. Honor thy symbionts. *Proceedings of the National Academy of Sciences of the United States of America* 100(18), 10452–9. doi:10.1073/pnas.1734063100.

Yadav, U. C., Ramana, K. V. and Srivastava, S. K. 2011. Aldose reductase inhibition suppresses airway inflammation. *Chemico-Biological Interactions* 191(1–3), 339–45. doi:10.1016/j.cbi.2011.02.014.

Yan, Y., Kolachala, V., Dalmasso, G., Nguyen, H., Laroui, H., Sitaraman, S. V. and Merlin, D. 2009. Temporal and spatial analysis of clinical and molecular parameters in dextran sodium sulfate-induced colitis. *PLoS One* 4(6), e6073. doi:10.1371/journal.pone.0006073.

You, Y. and Silbergeld, E. K. 2014. Learning from agriculture: understanding low-dose antimicrobials as drivers of resistome expansion. *Frontiers in Microbiology* 5, 284. doi:10.3389/fmicb.2014.00284.

Yu, Q., Zhu, L., Wang, Z., Li, P. and Yang, Q. 2012. *Lactobacillus delbrueckii* ssp. Lactis R4 prevents *Salmonella typhimurium* SL1344-induced damage to tight junctions and adherens junctions. *Journal of Microbiology* 50(4), 613–7. doi:10.1007/s12275-012-1596-5.

Zareie, M., Johnson-Henry, K., Jury, J., Yang, P. C., Ngan, B. Y., McKay, D. M., Soderholm, J. D., Perdue, M. H. and Sherman, P. M. 2006. Probiotics prevent bacterial translocation and improve intestinal barrier function in rats following chronic psychological stress. *Gut* 55(11), 1553–60. doi:10.1136/gut.2005.080739.

Zheng, J., Allard, S., Reynolds, S., Millner, P., Arce, G., Blodgett, R. J. and Brown, E. W. 2013. Colonization and internalization of *Salmonella enterica* in tomato plants. *Applied and Environmental Microbiology* 79(8), 2494–502. doi:10.1128/AEM.03704-12.

Chapter 15

Role of prebiotics in poultry gastrointestinal tract health, function, and microbiome composition

Steven C. Ricke, University of Arkansas, USA

1 Introduction

It is becoming apparent that gastrointestinal tract (GIT) function and the microbial inhabitants are important contributors not only for the prevention of pathogen establishment but also for general host health as well. Shifts or changes in the GIT microbiota are becoming more discernible with the emergence of new sequencing technologies that are more definitive as well as metabolomics, and transcriptomic data that can be generated to accompany the taxonomic information. Consequently, factors such as dietary changes, immune response, supplementation of feed amendments, housing, breed, age, and environmental conditions associated with stocking density can all exhibit detectable impacts on not only the host but also the GIT microbiota and function (Korver, 2006; Yegani and Korver, 2008; Choct, 2009; Guardia et al., 2011; Kers et al., 2018). Modifying the GIT microbiota and in turn, GIT function has always been of interest but has become more extensively researched in the past few years. There are several reasons for this. Certainly, the need to generate a GIT more resistant to foodborne pathogens and poultry disease-causing

http://dx.doi.org/10.19103/AS.2019.0059.18

organisms has historically been a primary driver for the development of a wide range of feed amendments that limit pathogen establishment in the GIT. Likewise, GIT modification to produce a generally healthier GIT has been a focal point. Improvement of nutritional efficiency, immune response, and other less defined GIT functions that ultimately result in enhanced bird performance is an emerging goal of dietary additives (Pan and Yu, 2014; Stanley et al., 2014; Ricke, 2018; Swaggerty et al., 2019).

Other factors also impact the search for different types of feed additives. One of the primary motivators has been a general movement away from general antibiotic use for the improvement of bird performance and hence the interest in alternative compounds (Jones and Ricke, 2003; Singer and Hofacre, 2006). In addition, the development of antimicrobials to limit foodborne pathogens such as *Salmonella* and *Campylobacter* spp. occurring in broiler and layer flocks has been an ongoing issue in the poultry industry. Once foodborne pathogens become established in flocks, they can eventually contaminate the retail poultry meat products and table eggs that may in turn become sources of foodborne diseases and general public health concerns. While there are numerous intervention strategies for decontaminating poultry carcasses during processing, reduction at the farm remains an important component for achieving further decreases of foodborne pathogens in overall poultry production (Keener et al., 2004; Ricke et al., 2005; Cox et al., 2011; Sofos et al., 2013).

Several feed additives have been developed and commercially introduced to the poultry industry over the years. Preharvest intervention strategies can be broadly categorized as those that either eliminate the already GIT-colonized foodborne pathogens or prevent their initial GIT colonization. Preharvest food safety control measures that reduce or eliminate foodborne pathogens already established in the poultry GIT include among others, bacteriocins, botanical compounds, bacteriophage, and organic acids (Joerger, 2003; Ricke, 2003a; Ricke et al., 2012a; O'Bryan et al., 2015; Calo et al., 2015; Dittoe et al., 2018; Moye et al., 2018; Wernicki et al., 2017). Feed additives that prevent foodborne pathogen colonization generally interact with either the host or the GIT microbial community or both to alter the GIT ecology and tissue to create a GIT environment hostile to incoming foodborne pathogens. Some approaches such as introducing virulence-attenuated foodborne pathogens trigger the immune system of the bird to generate a much more rapid and specific immune response to inhibit foodborne pathogen infection (Revolledo et al., 2006). Others seek to modify the composition of the GIT microbiota by introducing microbial species into the GIT microbial community that are antagonistic to the corresponding foodborne pathogen, and these live cultures are referred to as probiotics, competitive exclusion cultures, or direct-fed microbials (Fuller, 1989; Nisbet, 2002; Patterson and Burkholder, 2003; Hume, 2011; Callaway and Ricke, 2012; Siragusa and Ricke, 2012). A somewhat different strategy involves adding feed compounds that can serve as substrates for specific members of the GIT microbial population that favor either their metabolic activity and/or growth to alter the GIT ecology by increased end products, for example, short-chain fatty acids (SCFAs) or other compounds such as bacteriocins that in turn, inhibit foodborne pathogen colonization (Patterson and Burkholder, 2003; Hume, 2011; Ricke, 2015, 2018; Micciche et al., 2018; Kim et al., 2019). Over time, the understanding of prebiotics has evolved and the definition of what constitutes a compound that elicits prebiotic properties has been refined accordingly (Hutkins et al., 2016; Gibson et al., 2017; Ricke, 2018). In this chapter, less emphasis will be placed on describing individual classes of prebiotics and their application for limiting foodborne pathogens in poultry as these have been extensively reviewed elsewhere (Pourabedin and Zhao, 2015; Micciche et al., 2018; Ricke, 2015, 2018; Teng and Kim, 2018; Kim et al., 2019). Instead, prebiotics will be discussed in terms of general definitions and mechanisms followed by an assessment of their impact on the poultry GIT microbial consortia along with host GIT functions, and finally future directions for further applications of prebiotics.

2 Prebiotics: definition

Historically compounds that qualified as prebiotics were based on their characteristic of not being digestible in the GIT of the host animal or human consuming them but still considered beneficial to the host once ingested due to their ability to specifically enhance growth and/or metabolic activity of certain GIT bacteria (Gibson and Roberfroid, 1995; Wang, 2009; Bird et al., 2010; Hutkins et al., 2016). Traditionally prebiotics included a select group of oligosaccharides such as fructooligosaccharides (FOS), galactooligosaccharides (GOS) and mannans that were resistant to digestive enzymes and could be hydrolyzed and subsequently fermented by certain lactic acid bacteria and bifidobacteria (Flickinger et al., 2003; Ricke, 2015). Fermentation products generated by these GIT bacteria include lactic acid, SCFAs, acetate, propionate, and butyrate. Increased presence of these fermentation products have been linked to various functions including being antagonistic to foodborne pathogens and serving as sources of energy to the host.

Over the years since the initial definition, the concept of what functionally constitutes a prebiotic has evolved and become more inclusive of a wider range of compounds (Bird et al., 2010; Hutkins et al., 2016; Gibson et al., 2017; Ricke, 2018). Much of why this has occurred is due to the more comprehensive understanding of the GIT microbial community composition achieved from the next-generation sequencing (NGS) of the GIT microbiome (Hanning and Diaz-Sanchez, 2015; Ricke et al., 2017). Using NGS of the variable region of the genomic 16S rDNA, individual bacterial members of GIT microbial consortia can now be identified and the taxa can be assigned at least in some cases to the species level (Ricke et al., 2017). Once individual microorganisms can be identified, it becomes possible to detect the overall shifts in the GIT microbiota in response to a number of variables including the age of the host, dietary changes, and inclusion of specific dietary feed additives such as antibiotics and other antimicrobial types of compounds.

Not surprisingly, NGS analysis of the GIT microbiome has also revealed the broader impact of prebiotics on the GIT microbial community. It is now becoming apparent that the more traditional prebiotic compounds may be influencing more than just a few select GIT bacteria when they are introduced into the diet. In addition, NGS of the GIT microbial community has indicated that other feed components may also be eliciting prebiotic type influences on the GIT microbiota and therefore the concept of prebiotic may need to be considered more expansive that originally conceived. Consequently certain dietary fibers, resistant starch and other related dietary components may possess at least some prebiotic properties when introduced to the GIT microbial community (Bird et al., 2010; Ricke et al., 2013; Ricke, 2018). As more becomes known through NGS taxa identification as well as other approaches such as metabolomics, it is anticipated that more understanding of not only which GIT microorganisms are involved but also the metabolic and fermentation activities associated with these organisms. Even when overall GIT microbial composition does not change in response to prebiotic administration, there may still be a detectable impact on the GIT microbial population. Since GIT bacteria often possess more than one fermentation pathway, metabolomics may reveal shifts in metabolism occurring in the presence of prebiotics that is not reflected in changes in GIT taxa.

Historically, a variety of prebiotics have been examined as potential feed additives for improving poultry health and food safety (Alloui et al., 2013; Pourabedin and Zhao, 2015; Ricke, 2015, 2018; Roto et al., 2015; Teng and Kim, 2018). Initial interest evolved from the potential for modulating the GIT microbiota to limit foodborne pathogen colonization and establishment (Micciche et al., 2018; Kim et al., 2019). However, other benefits to the poultry host beyond prevention of foodborne pathogen establishment have come to light in recent years (Ricke, 2018; Teng and Kim, 2018). More recently, the focus on the impact of feed additives such as prebiotics has been expanded to potential interaction with other aspects of the bird GIT functions such as the immune system of the bird, nutritional and metabolic responses, and overall bird performance including either broiler growth and feed efficiency or egg production in layer hens. Part of this shift in focus is the result of advancement in methodologies to better understand GIT functional mechanisms and relate them to overall bird physiology. In addition, the emphasis on bird performance is a reflection of the ongoing demand for retaining and/or improving the economics of commercial poultry production in the face of rising costs for feed and other production expenses along with changes in management strategies to accommodate animal welfare demands. In the following sections, the impact of prebiotics on different aspects of poultry GIT function will be discussed.

3 The avian upper GIT: potential impact of prebiotics

The avian GIT consists of several compartments, namely, the crop, proventriculus, gizzard, small intestine, large intestine, and ceca. Each compartment possesses distinctive anatomical and functional characteristics that contribute to avian digestion of feed. Some compartments have been much more extensively studied than others, and more is understood about their contribution to the overall avian digestive process. After the initial hatch, there appears to be a similarity in the microbiota composition of the different GIT compartments whether determined by cultural methods and enumerating GIT organisms or molecular profiling (Barnes et al., 1980; van der Wielen et al., 2002). However, as expected, the microbial populations harbored in each compartment vary and become more distinct as the bird ages and the individual compartments become more mature (Barnes et al., 1980; van der Wielen et al., 2002; Stanley et al., 2014). These differences have not been typically considered when examining the impact of feed additives such as prebiotics on the GIT of birds but may be more influential than realized since each compartment leads to the next one. This may be particularly important for feed additives such as prebiotics which are known to shift GIT microbial populations. Depending on how early in the bird's life they are fed, they could influence the subsequent development and composition of the respective microbial populations in the different avian GIT compartments.

The crop, proventriculus (proximal stomach) and gizzard (distal stomach) comprise the upper GIT of the bird (Duke, 2002; Stanley et al., 2014). Less is known about the microbial ecology in the proventriculus and gizzard compared to the crop. The proventriculus in addition to secreting mucus also produces hydrochloric acid and pepsin which contribute to protein digestion (Duke, 2002). The gizzard is a muscular GIT organ that is responsible for grinding incoming digesta and also the starting point for chemical digestion (Duke, 2002; Moore, 1999). The gizzard plays a role in reducing the particle size of structural components of diets with retention being dependent on feed particles becoming reduced to a certain size before passage to the lower GIT (Hetland et al., 2005). It has also been suggested that the gizzard is important in the digestive processing of nonsoluble fibrous components of avian diets, and the inclusion of insoluble fibers may improve the overall digestion of the otherwise non-fiber concentrate diets via the enhancement of gizzard activity (Hetland et al., 2005; Mateos et al., 2012). The presence of grit in the gizzard contributes to the breakdown of digesta entering the gizzard and is dependent on the quantity of grit contained in the gizzard (Moore, 1998).

Based on culture studies as well as more recent microbiome profiling, the gizzard appears to be populated predominantly by the phylum Firmicutes with lactobacilli species and Clostrideaediaceae being the most commonly identified (Rehman et al., 2007; Sekelja et al., 2012; Yeoman et al., 2012; Stanley et al., 2014). The contribution of the indigenous bacteria in this region of the avian GIT to prebiotic metabolism is not known; but based on fermentation profiling, it appears that the contribution may be minimal as SCFA production and pH are reduced in the gizzard (Rehman et al., 2007). There is some evidence that prebiotics may impact the gizzard microbiota. Yusrizal and Chen (2003) observed that oligofructose supplementation of broiler diets did lead to increased levels of lactobacilli counts in the gizzards of male and female broilers but not inulin. In addition, the grinding function of the gizzard could conceivably contribute by making structural components of feeds that possess prebiotic properties more available for GIT microbial hydrolysis and fermentation in the lower GIT. Certainly, the interaction between dietary fiber and the apparent ability of the gizzard to regulate GIT passage appears to impact intestinal nutrient digestion (Hetland et al., 2004). This, in turn, would likely influence the substrate availability to intestinal microbial populations as well.

More is known on the microbial ecology of the crop. Early work indicated that the crop served as a storage site for ingested feed, but passage rate varies considerably depending on the type of diet, lighting of the house, feed intake, texture of the feed, and methodology used to quantify passage rate (Heuser, 1945; Sibbald, 1979, 1980; May et al., 1990). The crop harbors primarily lactobacilli with a wide range of species that vary as the bird ages (Rehman et al., 2007; Sekelja et al., 2012). The crop pH generally ranges approximately from 4 to 5 and the main fermentation products may also vary as both lactic acid and SCFA have been detected with lactic acid and acetate being predominant (Bayer et al., 1978; Durant et al., 1999, 2000; Rehman et al., 2007). Several factors may influence the microbial community composition and fermentation profiles in the crop ranging from diet and passage rate to environmental factors such as bird stocking density (Bayer et al., 1978; Durant et al., 1999; Guardia et al., 2011). Certainly, the mucosal surface of the crop is not uniform as Bayer et al. (1975) used scanning electron microscopy to demonstrate that the region closer to the esophagus was slightly folded and densely populated by bacteria versus the more distal regions which were smoother surfaces and minimally populated by bacteria. Specific lactobacilli species present may also be important as some species are probably homofermentative producing only lactic acid, while others are likely heterofermentative generating mixed fermentation acid profile of either lactic acid or SCFA depending on growth conditions.

Diet composition also appears to have a variable impact on the crop population. Bayer et al. (1978) compared low- and high-fiber diets and detected minimal differences in crop production of acetate or lactate. However, the passage of feed through the crop and emptying of the crop appear to be important. Durant et al. (1999) demonstrated that a 9-day feed removal for inducing molt in adult laying hens dramatically lowered the lactobacilli populations enumerated on Rogosa agar and decreased crop lactic acid and SCFA, but increased crop pH. Similar responses were also observed in broilers after feed withdrawal for 12 to 24 hours (Hinton et al., 2000a,b). These reductions in crop fermentation activity and lactobacilli populations accompanied by a rise in pH support a crop colonization increase in *Salmonella* Enteritidis in laying hens and *S.* Typhimurium in broilers (Durant et al., 1999; Hinton et al., 2000a,b). This has ramifications for egg contamination in layers due to an increase in *Salmonella* virulence and organ invasion during feed withdrawal molt (Durant et al., 1999).

The potential impact of prebiotics on the crop microbial populations and the presence of foodborne pathogens are difficult to define. Presumably, the dominance of lactobacilli, which can utilize prebiotics, would suggest that there are prebiotic fermentation capabilities in the crop. There is evidence of sugar production in the crop from diets containing carbohydrates (Pritchard, 1972). Subsequent absorption of sugars into the bloodstream of the bird has also been demonstrated with carbon-14-labeled sugars (Soedarmo et al., 1961). Durant et al. (2000) detected much lower glucose concentrations in crops isolated from feed withdrawal birds compared to full fed birds, which corresponded with the decreased lactate concentrations also observed in feed withdrawal bird crops. Hinton et al. (2000a) observed that providing a glucose-supplemented cocktail to broilers undergoing feed withdrawal generally increased crop lactobacilli and decreased pH and *S.* Typhimurium colonization in glucose-supplemented birds compared to the non-supplemented birds. Type of sugar may be important as Hinton et al. (2002) demonstrated that sucrose supplementation to broilers artificially infected with *S.* Typhimurium and undergoing 12-hour feed withdrawal exhibited more reduced populations of *Campylobacter* and *S.* Typhimurium in the crop compared to those supplemented with glucose. However, minimal changes in crop pH or lactic acid bacteria were detected. In short, the role of the crop microbial inhabitants in the GIT processing of prebiotics remains unclear. Certainly, the crop is a potential barrier to further colonization of *Salmonella* and can be an important contributor to *Salmonella* dissemination in the poultry processing plant when compromised (Hargis et al., 1995). As microbiome characterization progresses, the opportunity to combine this with complete metabolic profiling of the various sections of the avian GIT may help to more precisely quantify the contribution of the crop to prebiotic digestion (Ricke, 2018).

4 The avian intestinal microbiome, function, and prebiotics

In the avian intestinal tract, most of the GIT microbial community characterization has been conducted on the small intestine compared to the large intestine. Both are generally represented by members of the Firmicutes and Proteobacteria phyla with individual taxonomic groups such as *Lactobacillus* and *Escherichia* being identified in both intestinal sections (Yeoman et al., 2012). Anaerobic isolation and culture-based identification from roll tube methods and media developed for rumen microorganisms revealed that facultative microorganisms (*Streptococcus, Staphylococcus, Lactobacillus* and *Escherichia coli*) tended to be more prevalent in the ileum and duodenum, but in some cases, at least a third of the population isolates were classified as anaerobes and included *Eubacterium, Propionibacterium, Clostridium, Gemmiger,* and *Fusobacterium* (Salanitro et al., 1978). Earlier work by Barnes et al. (1972) indicated that lactobacilli were isolated in highest numbers from the intestinal tract of chicks 2 to 6 weeks old. Since then, the application of molecular identification approaches has improved the resolution of the intestinal populations. For example, Lu et al. (2003) using partial 16S rRNA gene sequencing concluded that almost 70% of the ileal population comprised *Lactobacillus* followed by *Clostridiaceae*, *Streptococcus*, and *Enterococcus*, respectively. When GIT populations were monitored as broilers matured, Lu et al. (2003) concluded that the ileal and cecal profiles did not differ at day 3 and over the entire first 14 days, cecal microbial populations could still be linked to the ileal populations with each GIT compartment becoming more distinct after this time period. Based on 16S rDNA amplification and cloning followed by sequencing, Bjerrum et al. (2006) found that the ileum contents of 40-day-old broilers were dominated by lactobacilli.

The role of intestinal microbial consortia in the fermentation of prebiotics has not been well characterized. Certainly, the prevalence of lactic acid bacteria such as lactobacilli in the ileum and duodenum would represent a population capable of fermenting some prebiotics such as fructooligosaccharides as well as other prebiotic oligosaccharides (Kaplan and Hutkins, 2000; Saminathan et al., 2011; Ricke, 2015). However, as Rehman et al. (2007) has pointed out, the lower SCFA concentrations observed in the small intestine compared to the ceca probably reflects the relatively rapid transit time in the small intestine precluding extensive fermentation activity in this segment of the avian GIT with only soluble carbohydrates and shorter chain components likely fermented. In summarizing studies on the influence of diet on intestinal fermentation Rehman et al. (2007) concluded that certain grain diets did appear to alter fermentation and production of SCFA, but acetate was always predominant. Ricke et al. (1982) did detect increases in total SCFA in the large intestine of chicks fed diets supplemented with xylose or gum arabic compared to either control basal diets or diets containing lignin. For individual SCFA products, acetate was proportionally the predominant SCFA with considerably less butyrate and barely detectable levels of propionate (Ricke et al., 1982).

The interaction between prebiotics and the intestinal microbial consortia has only been minimally documented. Presumably, based on the lower fermentation activity in the intestinal tract, indigestible prebiotics are unlikely to undergo hydrolysis and fermentation by the resident intestinal population. While some changes in intestinal microbial populations have been observed with prebiotic supplementation, the results have been mixed. Xu et al. (2003) fed 4.0 g/kg FOS to male broilers and detected an increased growth of *Bifidobacterium* and *Lactobacillus*, along with inhibition of *Escherichia coli* populations in small intestinal contents. Yusrizal and Chen (2003) observed that oligofructose supplementation of broiler diets led to increased levels of lactobacilli counts in the small intestines of male and female broilers but only the large intestines of female broilers. They also observed a reduction of *Campylobacter* in the presence of oligofructose in the large intestines of both male and female broilers but not in the small intestine.

Not all prebiotics mediate their activity against foodborne pathogen colonization by altering intestinal microbial composition and fermentation. Mannan-oligosaccharides (MOS) such as those derived from yeast cell walls may interact specifically with the intestinal mucosa and prevent binding to epithelial cells by specific bacteria with type-1 fimbriae to the epithelial gut cell wall (Ofek et al., 1977; Ofek and Beachy, 1978; Spring et al., 2000; Hooge, 2004; Baurhoo et al., 2009; Yang et al., 2009; Roto et al., 2015; Teng and Kim, 2018). However, Santovito et al. (2018) has emphasized that the exact antimicrobial mechanism for the yeast cell wall is complex and will require more genetic and biochemical-based research to elucidate the adhesion properties. Improvements in pathogen reduction by yeast cell products and extracts have been observed both *in vitro* and *in vivo* (Roto et al., 2015; Santovito et al., 2018). For example, when Jahanian and Ashnagar (2015) fed different levels of supplementation of MOS (0.05 to 0.2 % of diet), they observed substantial decreases in ileal *Salmonella* and increases at the higher levels of MOS (0.1 to 0.2%) but no overall changes in total ileal bacterial populations. There is also increasing evidence that yeast products may not only directly interact with pathogen attachment but also modulate the intestinal microbial community and the immune system responses including cell-mediated responses and mucosal IgA secretions in chickens (Gómez-Verduzco et al., 2009; Bortoluzzii et al., 2018).

There are indications that prebiotics can influence intestinal function. Histology evidence presented by Hanning et al. (2012) suggests that supplementation with prebiotics can retain intestinal villi length and crypt depth in certain breeds of birds undergoing feed withdrawal, but intestinal microbial populations were not examined in this study. The intestinal segment may be important. De Maesschalck et al. (2015) reported an increase in ileal villi length in broiler chicks fed wheat-rye-based diets supplemented with xylo-oligosaccharides as well as an increase in the colonic lactobacilli population. De Maesschalck et al. (2015) also noted that these responses were accompanied by an improvement in feed conversion ratio. Solis de los Santos et al. (2007) reported that a yeast extract supplement enhanced ileal villus height, surface area, lamina propria thickness and crypt depth in turkey poults but elicited an inconsistent or minimal impact on jejunal and duodenal villus architecture.

Other factors may predispose the impact of prebiotics on intestinal tissue architecture. For example, the age of bird and maturity of intestinal tract development may be a factor in determining intestinal structural responses to prebiotics. Fasina and Olowo (2013) observed inconsistent effects of a yeast product on different segments of the intestine including the ileum in 10-day broiler chicks. In 21-day broiler chicks fed different levels of MOS, Iji et al. (2001) detected an increase in jejunal villi height for the highest level of MOS supplementation compared to control birds not receiving MOS, but jejunal villi surface area or crypt depth and ileal villi height, crypt depth and surface area did not change. Using alkaline phosphatase activity as an indicator, Fasina and Thanissery (2011) detected an improvement in small intestinal ileal maturation in broiler chicks fed diets supplemented with a yeast product. Some of the influence that prebiotics exercise over intestinal development may also be related to shifts in intestinal microbial populations and/or fermentation pathways as the indigenous intestinal microbial community adapts to the continuous exposure of prebiotic-supplemented feed.

Colonization by pathogens can also impact intestinal integrity and supplementation and prebiotics offers an opportunity to potentially ameliorate the negative impact of the presence of pathogens in the intestinal tract. Santos et al. (2013) collected ileal, jejunal, and duodenal histology samples from young chicks infected early with *S.* Enteritidis and quantified villi structure. When they compared infected birds with MOS-supplemented infected birds, MOS decreased ileal and jejunal villi height and crypts depth but increased villi-to-crypt ratio while MOS increased duodenal crypt depth and villi-to-crypt ratios. As more in-depth intestinal tissue analyses are conducted, it would not be surprising that intestinal structural changes would also lead to detectable differences in dietary digestion and nutrient absorption throughout the entire intestinal tract.

Changes in intestinal villi structure would suggest that there are discernible interactions among prebiotics, intestinal microbial communities, and host metabolic functions. Host lipid metabolism is one example of this. Fructooligosaccharides have been linked in animals with a reduction in hepatic triglyceride and very-low-density lipoprotein (VLDL) synthesis (Taylor and Williams, 1998). This is consistent with studies on rats fed oligofructose that led to inhibition of lipogenic enzymes and reduction of fatty acid synthesis in the liver (Delzenne and Kok, 2001). More recently, Sevane et al. (2014) using microarray-based transcriptomic profiling demonstrated that inulin supplementation modulated liver metabolic processes associated with immune system processes and fatty acid metabolism that could be linked to chick performance and growth. Intuitively, systemic metabolic responses to prebiotics would be expected to reflect the differences at the intestinal cell level. Iji et al. (2001) quantitated the protein, DNA, and RNA content of ileal and mucosal homogenates and compared their corresponding ratios (protein/DNA = cell size; protein/RNA = rate of protein synthesis; RNA/DNA = cell metabolic efficiency) to assess intestinal cell size and metabolic activity. While there were minimal differences in jejunal cell metabolic activities, ileal cell RNA content, and protein/RNA and RNA/DNA ratios were greater in the cell from birds fed higher concentrations of MOS. These interactions may in turn also influence bird immunal and nutritional responses (Ricke, 2018; Teng and Kim, 2018; Swaggerty et al., 2019). The general aspects of the influence of dietary modulation on immune function and prebiotics are well summarized in previous reviews and will not be discussed in the current review (Teng and Kim, 2018; Swaggerty et al., 2019). Impact of prebiotic supplementation on the metabolism of other nutrients has been less extensively examined. Supplementation of MOS in laying hen diets resulted in improved ileal digestibility coefficients of dry matter, ether extract and crude protein (Jahanian and Ashnagar, 2015). However, some components of protein digestion such as amino acid digestion responded with mixed results in broilers fed yeast fractions (Biggs and Parsons, 2008). In more specific tissue-based examinations, Iji et al. (2001) detected increases in specific activities of digestive enzymes of maltase, leucine aminopeptidase and alkaline phosphatase from the jejunal brush border in birds fed MOS, whereas no effects were observed in ileal cells. They also observed increased tryptophan uptake in jejunal brush border membrane vesicles for the higher levels of MOS, but no differences in ileal vesicles. More in-depth characterization of the intestinal level responses to prebiotics and linking with host systemic metabolic responses will require further molecular analyses of both the intestinal tissue and the intestinal microbiome along with host tissue responses.

Mineral bioavailability is known to be generally influenced by the presence of prebiotics and has been linked to the production of SCFA from colonic bacteria (Hara, 2002; Morohashi, 2002; Bongers and van den Heuvel, 2003; Yeung et al., 2005). These changes in mineral bioavailability can influence host physiology. For example, Morohashi (2002) concluded that increases in the absorption of minerals such as calcium and magnesium in rats supplemented with fructooligosaccharides led to increased bone surface volume and mineral content. Yeung et al. (2005) has suggested that colonic microbial fermentation of prebiotics may lower colonic pH and enhance iron absorption by changing the valency of dietary iron, increasing absorption surface area, and stimulating epithelial cell transport proteins. Interestingly, the host sequesters iron to limit access by invading pathogens, who in turn have developed high iron chelators known as siderophores to compete with host systems (Payne, 1988; Litwin and Calderwood, 1993; Wooldridge and Williams, 1993). Given the competition for iron by pathogens, the microbial ecosystem dynamics on iron acquisition may be a critical consideration for optimizing both choice and level of supplementation of prebiotics. The type of prebiotics may influence the intestinal microbial population, its relationship with dietary iron, and potential competition with pathogens. To elucidate this will require a more complete metagenomic profile of the intestinal microbiota that reveals which iron metabolic pathways are linked directly to the form and availability of dietary iron.

Interactions between prebiotic and prebiotic-like compounds with mineral metabolism have been examined in poultry. In growing chicks, Van der Aar et al. (1983) reported that brans from corn, wheat, and oats significantly lowered serum zinc levels and kidney magnesium levels. Mineral requirements can be elevated in adult birds subjected to sudden physiological changes. For example, laying hens undergoing feed withdrawal molt to cease egg production experience significant bone integrity changes and strength loss and can be challenged to regain bone integrity after molting because of the elevated requirements for calcium in shell egg formation after molting (Garlich et al., 1984; Newman and Leeson, 1999; Mazzuco and Hester, 2005; Kim et al., 2006a). Kim et al. (2006a) examined laying hens being fed different levels of alfalfa-FOS-supplemented diets during molt using dual-energy X-ray absorptiometry and conventional bone assays and concluded that FOS might help to maintain bone strength during molting. There are undoubtedly multiple interactions occurring between prebiotics, intestinal microbial communities, and the minerals contained in feed matrices entering the intestinal tract. As more in-depth microbial community characterization is connected to fermentation and metabolomic profiles, these complex interactions will become better understood both at the GIT level as well as the impacts on poultry metabolism and physiology.

5 Cecal composition and functional characteristics

Chickens possess two cecal appendages which are visually recognized as blind pouches each consisting of a proximal region near the ileorectum junction and medial-distal section (McNab, 1973; Moretó and Planas, 1989). Functionally the two regions differ with the proximal region exhibiting sugar and amino acid transport capabilities and the presence of villi, lymphoid and goblet cells, comparable to the intestine, while the distal region possesses only mounds and ridges with minimal transport capabilities (Moretó and Planas, 1989; Svihus et al., 2013). Movement of digesta into and out of the ceca is controlled by peristaltic and antiperistaltic contractions (Duke, 1989; Svihus et al., 2013). The passage rate of the digest in the overall poultry GIT can occur within a few hours with the ceca being comparatively much slower (Pan and Yu, 2014). However, precise passage rates are difficult to identify due to a variety of factors including diet composition and particle size among others. Also, Vergara et al. (1989) has pointed out that estimates of passage rates through the avian intestinal tract and ceca may depend on the type of passage rate marker used.

Cecal compositional analyses have been conducted in several bird trials over the years. Svius et al. (2013) concluded, based on the literature published on cecal composition, that the materials entering the ceca were soluble, low-molecular-weight viscous materials that could be in the form of finely ground particles as opposed to the previous assumption of simply being nutrients escaping the small intestine. Cecal pH is generally around 7, but variation does occur (McNab, 1973). In an extensive review of published research, Rehman et al. (2007) summarized cecal pH levels to be anywhere from a low of 5 to as high as 7 based on a wide range of bird maturities, diets, and SCFA concentrations. Shermer et al. (1998) detected glutamate and alanine as the predominant free amino acids in the cecal contents of broilers with glutamate accounting for over 26 molar % of the total amino acids. Citrulline and ornithine, which are considered metabolites of urea cycling, were also detected in broiler cecal contents by Shermer et al. (1998). In a much more recent study, Rubinelli et al. (2017) confirmed the presence of cecal glutamate and ornithine based on metabolomic analyses of *in vitro* cecal incubations. They also detected allantoic acid, methionine, and pantothenic acid. The presence of allantoic acid is an indication of the active breakdown pathway of uric acid (Bacharach, 1957; Carlile, 1984; Kim and Patterson, 2003; Kim et al., 2009). In birds fed high levels of cellulose or grass fiber in combination with a composite polysaccharidase enzyme, Savory and Knox (1991) detected predominant molar levels of free glucose over three-fold greater compared to a combination of all the other sugars pooled together (galactose, mannose, xylose, arabinose, fucose and rhamnose). The cecal carbohydrate content will no doubt be somewhat variable depending on the diets and the cecal microbial hydrolysis capabilities. For example, Rubinelli et al. (2017) detected a decrease in maltose over time in cecal *in vitro* incubations. Similar metabolomic approaches to examine cecal profiles from birds at different ages, fed different diets or different stages of foodborne pathogen colonization would be helpful to delineate cecal microbial populations even when cecal microbial populations remain relatively unchanged.

The ceca are not generally considered essential to the bird based on the numerous nutritional studies using cecetomized birds to determine feed digestibilities (Chaplin, 1989). Removal of ceca is implemented in poultry nutritional studies to avoid confounding the bias resulting from endogenous microbial protein synthesis occurring in the ceca (Parsons, 1996; Angkanaporn et al., 1997). However, certain cecal functional activities can be associated with host benefits that potentially contribute to the bird's overall metabolism. Ceca have been identified as the primary site for water and electrolyte resorption (Chaplin, 1989; Svihus et al., 2013). Certainly, the ability to transport sugars and amino acids in the proximal region would presumably lead to absorption albeit limited to the times that concur with direct contact of the cecal surface and intestinal digesta entering the ceca (Moretó and Planas, 1989). Ammonia production occurs for the most part in the ceca, and there are several cecal microorganisms that contribute by degrading various ammonia-containing compounds. This has been previously reviewed in detail by Karasawa (1989) and will only be briefly discussed here. Amino acids from dietary as well as endogenous sources that are not absorbed in the upper GIT can be deaminated to generate ammonia in the ceca (Karasawa, 1989). Cecal ammonia is also generated from urea contained in dietary material entering the ceca as well as from urine originating in the cloaca (Karasawa, 1989; Svihus et al., 2013). Uric acid coming in as a major end product of nitrogen metabolism that is present in the urine is also converted to ammonia by uric acid-degrading microorganisms (Barnes, 1972; Barnes and Impey, 1974; Mead and Adams, 1975; Mead, 1989; Karasawa, 1989). Ammonia generated from these sources is absorbed in the ceca, and it has been suggested that it could serve as a substrate for biosynthesis of non-essential amino acids (Karasawa, 1989). Strategies to inhibit these organisms have been suggested as a means to reduce ammonia concentration in poultry manure that can lead to reduced air quality (Carlisle, 1984; Kim and Patterson, 2003; Kim et al., 2006b, 2009, 2013; Chalova et al., 2016a,b). Yurizal and Chen (2003) observed reductions in fecal ammonia levels from 2 to 4 weeks in male and female broilers fed oligofructose compared to control birds. This suggests that either oligofructose selects against ammonia-forming fecal bacteria and/or more ammonia is utilized by fecal microorganisms for bacteria protein synthesis. In some GIT ecosystems such as the rumen in ruminant animals, rumen bacteria are capable of assimilating luminal external free ammonia-nitrogen into amino acids and eventually microbial protein (Smith and Bryant, 1979). It is not known how much of this occurs in the cecal microbial population but further characterizations of individual cecal microbial isolates and metagenomic analyses of the cecal microbiome may shed light on the relative contributions of this microbial activity to total cecal ammonia pools.

The high level of fermentation activity and the corresponding presence of substantial levels of SCFAs in the chicken and turkey ceca likely lead to their absorption and metabolism by the bird (Annison et al., 1968; McNab, 1973; Sudo and Duke, 1980; Goldstein, 1989). However, this is only true for SCFA produced locally in the ceca versus those produced elsewhere in the GIT or introduced as dietary amendments. Organic acids are often added as a supplement to poultry feeds to limit foodborne pathogens in the GIT of the birds (Ricke, 2003; Dittoe et al., 2018). Early work by Bolton (1965) indicated that externally administered SCFA rapidly disappeared from the ingesta of adult birds before reaching Merkel's diverticulum and were virtually undetectable in the small intestine. Likewise, using C-14-labeled propionate, Hume et al. (1993) detected minimal amounts of propionate in the lower part of the GIT in broiler chicks. The cecal SCFA profiles based on numerous *in vivo* poultry trials and *in vitro* incubation of cecal contents consist of mostly acetate with lesser amounts of butyrate and propionate (Dunkley et al., 2007a,b; Svihus et al., 2013). Absorbed SCFAs are believed to contribute to the energy requirements of the bird but quantitative estimates of actual utilization as a proportion of total energy have been reported as varying anywhere from 3 to 11% (Annison et al., 1968; Svihus et al., 2013). Additional research needs to be conducted to more precisely determine individual SCFA contributions to bird metabolism and physiology and what GIT factors influence the levels they are utilized.

6 Cecal microbiome: general characteristics

As indicated by the levels of SCFA production, the chicken ceca contain a major population of fermentative microorganisms and are considered a more stable GIT environment compared to the other compartments of the avian GIT (McNab, 1973; Mead, 1989; Stanley et al., 2014; Kers et al., 2018). The cecal contents harbor a relatively diverse microbial population that exhibit a wide range of metabolic activities resulting in the production of SCFA, carbon dioxide and methane as detectable end products. The ceca are considered the primary fermentation compartment of the poultry GIT-containing microorganisms capable of hydrolyzing and fermenting the range of substrates entering the ceca (Józefiak et al., 2004). Presumably, prebiotics entering the ceca that had not already been utilized by upper GIT and intestinal microbial populations would be fermented in the ceca. Depending on the age of the birds of when prebiotics are introduced into the diets suggests that their appearance would impact the direction of microbial composition and development of the cecal microbial population. However, any changes in the microbial composition being detectable may depend to some extent on the diversity of the cecal population at that point in time.

Early work by Huhtanen and Pensack (1965) using anaerobic culture incubation in Brewers jars followed by isolation and enumeration indicated that anaerobic bacteria were present in chick ceca on day 1 and became nearly half of the population by day 6 and the dominant population by day 24. Using anaerobic roll tubes supplemented with rumen fluid Salanitro et al. (1978) enumerated and identified the cecal microbial populations in 14-day-old chicks and estimated recovery based on the ratio of total colony-forming units (CFU) on roll tubes versus total microscopic counts. Based on recovery assessments, anaerobic bacteria in the ceca on average were 45% of the total estimated population and were the dominant part of the cecal population. These strict anaerobes were further identified as anaerobic gram-positive cocci, *Eubacterium*, *Clostridium*, *Gemmiger*, *Fusobacterium*, and *Bacteroides*, while facultative anaerobic lactobacilli were a relatively minor member of the cecal population. Barnes et al. (1972) also used anaerobic roll tubes to follow cecal populations in chicks 2 to 6 weeks old and were able to approximately recover over 20% of the population (10^{11} per gram). Further characterization revealed that the anaerobic streptococci dominated the cecal population at 2 weeks of age but greatly diminished as the birds reached 6 weeks of age. Assessment of cecal populations isolated on roll tubes from 5-week-old broilers by Salanitro et al. (1974) estimated that the majority (77%) were strict anaerobes with the largest percentages identified as *Propionibacterium acnes*, *Eubacterium* sp., *Bacteroides clostridiiformis*, *B. hypermegas* and *B. fragilis*, and spore-forming *Clostridium* sp. with lesser percentages of pleomorphic cocci and *Peptostreptococcus*.

While the initial work with cecal populations revealed that the onset of strictly anaerobic bacteria appears relatively early in the life of the bird, the functional properties of these populations were less well defined or understood. Certainly their involvement in degrading uric acid was identified as a consistent metabolic role for certain members of the cecal population (Barnes, 1972; Barnes and Impey, 1974; Mead and Adams, 1975; Mead, 1989; Karasawa, 1989). Fan et al. (1995) developed a rumen fluid-based carbohydrate differential media for plating and enumerating cecal bacteria in an anaerobic glove box. They based this approach on earlier work with differential carbohydrate media for selecting and enumerating different carbohydrate rumen microbial populations (Leedle and Hespell, 1980) and a previous study using this approach for recovery of total anaerobes from the ceca of young chicks (Nisbet et al., 1994). The agar media used by Fan et al. (1995) contained differing amounts of rumen fluid clarified by centrifugation to remove particulate matter and selective carbohydrates of glucose, galactose, or lactose. They assessed this media by comparing the recovery of 11 cecal bacterial pure cultures isolated from a continuous flow culture inoculated with poultry cecal contents and originally selected for lactose fermentation. Recovery of viable cecal bacteria based on total microscopic counts versus CFUs on the plates was best with the media with the highest level of rumen fluid (16%) and complete carbohydrate containing media followed by selective media containing either lactose or galactose. The higher recoveries on lactose and galactose were not surprising since the cecal bacteria had originated from lactose-based continuous flow cultures, but the results also revealed that such bacteria could utilize multiple carbohydrate sources to some extent as well. This overlap in carbohydrate preference is consistent with the report by Shermer et al. (1998) where they did not detect the differences in glucose, starch, or maltose subgroups of cecal bacteria enumerated from 16-week-old birds.

More recently, Yeh et al. (2019) used commercial 96-well microtiter plates (EcoPlates™, Biolog, Inc., Hayward, CA) containing 31 unique carbon sources to profile carbon source preferences of broiler cecal microbial populations. The determination was made after 120 hours of aerobic incubation, and changes in cecal inocula on individual carbon sources were detected by shifts in tetrazolium production. Based on these results, they noted that 11 carbon substrates could not be used at all and that pyruvic acid methyl ester, glycogen, glucose 1-phosphate, and N-acetyl-D-glucoaamine were most frequently used. Such detailed carbon profiling has the potential for identifying cecal nutritional subgroups of bacteria but as Yeh et al. (2019) points out, incubation with modifications such as a metabolic dye that would be detectable under anaerobic conditions which more closely resembled cecal environments and the addition of more carbon sources is needed. Along these lines, it would be interesting to develop a similar approach that allowed for cecal metabolic screening of prebiotic carbohydrate polymers as well as their hydrolysates.

Attempts to achieve more specific nutritional profiling studies of cecal bacteria have been conducted in combination with molecular characterization. Dunkley et al. (2007a) demonstrated that cecal bacteria from layer hens incubated in *in vitro* batch culture could ferment several high-fiber sources in combination with layer rations producing SCFA profiles similar to those observed from *in vivo* studies. To further differentiate cecal groups, they used denaturing gradient gel electrophoresis (DGGE) banding patterns and observed a 69 to 71% similarity among subgroups which supports the concept of cecal bacteria being capable of fermenting multiple carbohydrates. Using a similar approach, Saengkerdsub et al. (2006) demonstrated that alfalfa and layer rations would support *in vitro* production of methane from layer hen inocula suggesting that methanogens were present in the ceca. In follow-up studies, Saengkerdsub et al. (2007a) were able to quantify methanogens using a most probable number estimate based on methane production along with an independent quantitative PCR 16S rRNA copy number estimation. Methanogen populations ranged from 6.38 and 8.23 cells/g (wet weight) based on cultivation, and for 16S rRNA estimates were between 5.50 and 7.19 $\log_{10}$/g (wet weight), respectively. Individual isolates were identified as being most similar to *Methanobrevibacter woesei* and could be detected in fecal material from chicks 3 to 12 days of age (Saengkerdsub et al., 2007a,b).

As sequencing technologies evolved, it became possible to achieve more comprehensive microbiome characterization in food production systems that could be conducted independently of culture methods (Ricke et al., 2017). Early work using DGGE indicated that changes in layer chick cecal contents occurred as birds matured were fed different diets such as alfalfa-based feeds or provided a probiotic and could be detected by banding patterns in the respective gel profiles (Hume et al., 2003). In follow-up studies, Callaway et al. (2009) and Escarcha et al. (2012) also detected cecal microbial population changes in both alfalfa-fed laying hens undergoing feed withdrawal and layer chicks being fed alfalfa-supplemented diets using bacterial tag-encoded FLX amplicon pyrosequencing based on a partial ribosomal amplification followed by pyrosequencing to identify cecal microorganisms. Zhu et al. (2002) obtained 16S rDNA gene sequences from luminal and mucosal isolated cecal DNA and amplified these sequences with universal primers followed by either cloning or separation via temporal temperature gradient gel electrophoresis (TTGE). Respective TTGE bands and 16S rDNA clone libraries were sequenced using an ABI sequencer. Differences in sequences between the TGGE and random cloning for taxonomic identification was noted by Zhu et al. (2002), but both methods identified common phylogenetic subgroups in all birds even though they encountered difficulties in detecting differences in the cecal wall and cecal lumen microbiota.

To overcome some of the limitations associated with these approaches, Zhu and Joerger (2003) employed a fluorescent *in situ* hybridization (FISH) approach to quantify abundance and distribution of microorganisms from cecal luminal and mucosal samples in an attempt to differentiate the respective microbial communities. They concluded that for the most part, the phylogenetic taxa were distributed fairly similar in the two cecal sites except for enteric bacteria which were higher in the mucosal samples. Gong et al. (2007) compared 16S rRNA gene clone libraries of mucosal microbial populations from the crop to the cecum of 5-week-old broilers and detected a much more diverse population in the cecal mucosal wall (51 Operational Taxonomic Units, OTU) versus the other segments of the GIT (9 to 14 OTUs) and was predominated by clostridia-related sequences followed by other groups including *E. coli*. The predominance of clostridia-related microorganisms is consistent with earlier sequencing studies by Lu et al. (2003) using similar approaches to examine luminal contents of broilers from different time points throughout the grow out period of 49 days. The other major groups detected in the ceca but in lesser abundance included *Fusobacterium*, *Lactobacillus*, and *Bacteroides*. While clostridia were abundant in the ceca at all ages of birds, Lu et al. (2003) noted that cecal populations differed at each of the weekly time intervals (3 to 7, 14 to 28, and 49 days) during the 7 weeks and evolved with increasing complexity as the birds matured. Likewise, the other GIT compartments became more distinct from each other as the birds aged leading Lu et al. (2003) to suggest that the respective microbial populations represented unique physiological functions identified with each GIT compartmental epithelial cell surface and the polysaccharides present in mucin as well as the glycolipids and glycoproteins can serve as nutrient sources for these microorganisms.

As NGS technology has progressed, more comprehensive profiling of cecal and GIT microbial populations based on 16S rRNA gene sequencing became possible. Stanley et al. (2013) examined the cecal microbiota from 207 individual broilers sampled on day 25 across three trials fed the same diets and grown under similar conditions. When cecal microbiomes were compared between trials, not only was variability in feed conversion ratios noted, but cecal microbiomes differed both among birds within a trial as well as across trials. Based on this variability, Stanley et al. (2013) suggested that removal of chicks from maternal laying hen contact in combination with the relatively clean rearing environments resulted in more variability in initial colonization by microorganisms. Such variability would suggest that feed additives introduced early in the life cycle of the chick would potentially influence cecal microbial community development over time leading to cecal populations selected or favored by that feed additive and in turn, perhaps more uniformity in cecal populations when the birds reach maturity. Ballou et al. (2016) followed cecal microbial development over a time period of 28 days in birds either vaccinated for *Salmonella* Typhimurium or fed a probiotic cocktail. While the age of bird was the primary factor influencing microbiome development versus treatment, comparisons within each time point of cecal collection (days 1, 7, 14, and 28) revealed that by day 7 cecal microbiomes could be differentiated among the treatments but functional diversity was less pronounced. In a later study, Park et al. (2017a) demonstrated that even different genetic derivatives of the same *Salmonella* vaccine parent strain could lead to differences in cecal microbiome composition.

Other factors can contribute to cecal microbiome responses to feed additives. Certainly the type of feed-additive treatment may matter. For example, Oakley et al. (2014) did not detect the differences in cecal microbiome response in birds fed various organic acids over a 42-day period. This would indicate that mechanism(s) elicited by feed additives is an important consideration and the extent to which the presence of the particular additive is interactive with the cecal microbiota may dictate cecal microbiome responses. The method used for microbiome sampling may have an impact as well. While direct cecal sampling is always preferred, practical issues, particularly in commercial field studies, preclude the inability to sacrifice birds during a study to retrieve cecal samples. Instead, fecal samples have often been considered as substitutes that avoid sacrificing the bird and as Stanley et al. (2015) point out offers the potential to repeatedly sample the same bird. Stanley et al. (2015) tested the utility of fecal samples as a substitute by comparing cloacal swab microbiomes with cecal samples in birds over three trials. When comparing the respective microbiomes, most of the OTUs could be identified in both sample sources but the cecal samples possessed more diversity (alpha diversity) than fecal samples, and there were significant differences in diversity (beta diversity) between the two types of samples. Based on the correlation comparisons of OTU abundances of the two sample sources Stanley et al. (2015) concluded that there were qualitative similarities but quantitative differences when comparing fecal versus cecal microbiomes. Consequently, they recommended that either type of sample would still allow for the detection of potential treatment differences but interpretation still might depend on sample type. How fecal samples are collected may be important as well. Oakley et al. (2013) collected fresh fecal droppings and noted similarities in microbiome profiles between the fecal and litter samples supporting their observation on potential litter contamination of the fecal samples.

To further examine fecal versus cecal profiles, Oakley and Kogut (2016) collected fecal samples of birds and then collected cecal samples after euthanasia at three different time points over a 42-day time period. They also compared the microbiome taxa responses to quantitative PCR assessment of cytokine gene expression. While fecal and cecal samples exhibited comparable richness and diversity over time, taxonomic analyses revealed bias in the respective sets of samples with genera *Gallibacterium* and *Lactobacillus* were over-represented in fecal droppings versus *Bacteroides* in the cecal samples of these same birds after 1-week post hatch. At 6 weeks post hatch, *Lactobacillus* remained over-representative in the fecal droppings, but now *Clostridium* and *Caloramator* had increased in the cecum. Pro-inflammatory cytokines were positively correlated with phylum Proteobacteria and negatively correlated with phylum Firmicutes. As the authors point out, while these phyla differences would suggest an interaction between the immune system and GIT microbiota, individual members of particular phyla can react differently than the overall phyla. This places further emphasis on the importance of sequence resolution to identify specific genera and species.

Defining the cecal microbial community has evolved tremendously over the years, and it has now become possible to collect a relatively complex data set on the identity of individual members of the cecal microbial consortia. However, much research remains to further improve the resolution on the taxa being identified. In addition, it is becoming more apparent that identification alone is not enough to achieve an in-depth appraisal of cecal microbial function and metabolism. Even though some members of the cecal microbial community may be less numerically prominent, they may still elicit functionally important contributions to the cecal ecosystem and help to support the stability of the overall cecal microbial community. Application of metagenomic and metabolomics should begin to bridge some of these gaps in understanding the interactions among these complex relationships that occur in the ceca.

7 Cecal microbiome and prebiotics: current perspectives and future prospects

Given the extensive fermentation activity associated with the complex microbial population harbored in the avian ceca, it would be expected that the GIT compartment will be most likely to be impacted by the presence of prebiotics. Since the cecum is also the primary site for colonization of foodborne pathogens such as *Salmonella*, it is also the section of the GIT which is usually examined for prebiotic efficacy to prevent pathogen colonization (Dunkley et al., 2009). Numerous studies have been conducted over the years on the effectiveness of prebiotics and *Salmonella* in poultry, and these have been recently reviewed by Micciche et al. (2018) and will not be discussed in detail in the current review except for some summary observations. In general, several mechanisms have been proposed for prebiotic prevention of *Salmonella* colonization including increases in SCFAs antagonistic to *Salmonella*, competition for binding sites on the GIT wall, and support of GIT bacteria that can metabolically out-compete *Salmonella* in the GIT environment (Alloui et al., 2013). However, identifying the exact mechanisms of *Salmonella* antagonism attributable to prebiotics remains elusive to some extent. Part of this may be due to the type of prebiotic and the level of metabolism it undergoes in the presence of GIT and cecal bacteria and its direct impact on the host (Alloui et al., 2013; Ricke, 2018). Some evidence indicates that there are linkages between prebiotics, GIT indigenous bacteria, and inhibition of *Salmonella*. For example, De Maesschalck et al. (2015) detected increases in gene copy number of the butyryl-coenzyme A (butyryl-CoA): acetate CoA transferase enzyme (involved in butyrate production), in the ceca of chickens fed xylo-oligosaccharides. *In vitro* work with cecal cultures and prebiotic-like compounds does indicate that cecal bacteria are certainly involved as the greatest reductions were generally observed with cecal bacteria pre-adapted to the feed additive before introducing *Salmonella* (Rubinelli et al., 2016). Development of the cecal microbiota also appears to be important as cecal inocula from older birds were more inhibitory than inocula from younger birds (Park et al., 2017b; Kim et al., 2018). Identifying core cecal microbial populations via microbiome sequencing that can be directly associated with *Salmonella* inhibition is a potential next step in optimizing choice and dosage of specific prebiotics. This will require *Salmonella* challenge bird studies as well identifying any host factors that may influence the interaction of the cecal microbiota with *Salmonella*.

Campylobacter can be isolated in fairly high numbers in the colon and ceca but can also extensively colonize the small and large intestine (Awad et al., 2016; Umara et al., 2017). Resident *Campylobacter* represent more of a challenge for control by prebiotics in the poultry lower GIT and ceca (Kim et al., 2019). Part of the issue is their relationship to the host and the indigenous GIT microbial community and the potential interdependence that occurs (Indikova et al., 2015). Wigley (2015) has concluded that although historically *Campylobacter* was considered a commensal organism, more recent evidence suggests that under certain circumstances it may, in fact, behave as a pathogen in poultry. As Wigley (2015) points out, some of this may be a consequence of an unbalanced GIT and, in turn, impact the GIT microbial composition and host immune response. Awad et al. (2016) observed definitive changes in the intestinal microbial composition resulting from *C. jejuni* infection and noted that this could potentially lead to intestinal dysfunction. Regardless of its status in the bird GIT, its association with the resident microbial population makes feed-additive strategies to limit its prevalence overly complicated as there would appear to be both direct and indirect mechanism(s) in play. Certainly, some interventions such as *Campylobacter* bacteriophage would intuitively be expected to be specific to its host *Campylobacter* with minimal impact on the non-*Campylobacter* cecal population. Richards et al. (2019) confirmed this with a *Campylobacter jejuni* broiler challenge study where the administered *Campylobacter* phages successfully reduced *Campylobacter* cecal populations but did not alter the diversity comparisons of the non-*Campylobacter* members of the cecal microbiome. Broad-spectrum feed additives such as prebiotics are much more likely to result in a nonspecific collateral influence on the non-*Campylobacter* population with the *Campylobacter* impact being more of an indirect consequence of shifts of the cecal population and fermentation pathways along with alterations in co-dependent metabolic relationships with members of the cecal microbiota.

Results from poultry trials examining the potential for prebiotic-mediated reduction of *Campylobacter* have been somewhat mixed in both conventional and nonconventional poultry production operations and have been summarized by Kim et al. (2019) and will not be discussed in detail in the current review. Kim et al. (2019) concluded that part of this variation in bird trial responses could be attributed to the methodology employed for detection and enumeration of *Campylobacter* particularly culture media. They noted that the selective culture media for *Campylobacter* varied in discriminatory capabilities resulting in non-*Campylobacter* colonies also appearing on the plates depending on the type of media. More emphasis on molecular methods is needed, and advancements are being made to improve molecular-based approaches on detecting and quantifying *Campylobacter* in poultry (Kim et al., 2019; Ricke et al., 2018). Improving the sensitivity and the ability to differentiate among *Campylobacter* species and strains will be important for assessing the prebiotic impact on *Campylobacter* in the cecal tract. There have been some relatively recent applications of microbiome sequencing to assess prebiotic responses and determine the relationship with *Campylobacter*. When examining commercial flocks, Park et al. (2017c) noted that a significantly higher level of *Campylobacter* in the cecal microbiome of commercial broilers fed an antibiotic containing diet than either the control birds or birds fed a yeast prebiotic. Park et al. (2017d) compared different prebiotics in pasture flock chickens and noted that OTUs identified as *Alistipes* and *Lactobacillus intestinalis* may potentially be predictive of *Campylobacter* in the ceca, but further work would need to be done to determine if any metabolic linkages existed.

Identifying associations among prebiotics, microbial composition, and metabolism occurring in the ceca requires functional analyses. Sergeant et al. (2014) generated a comprehensive metagenomic profile from the cecum of 42-day-old broiler-fed wheat-based diets and ionophores, but no antibiotics along with phylogenetic profiling of microbiome sequences from 20 birds. Based on the metagenomic analyses, over 200 non-starch polysaccharide-degrading enzymes could be identified in the chicken cecum with a large fraction of these being associated with oligosaccharide degradation. More specifically, Sergeant et al. (2014) identified polysaccharide utilization systems in cecal *Bacteroidetes*, which they suggested could be involved in non-starch polysaccharide hydrolysis and utilization. Given the predominance of these enzyme systems, it would appear that the cecal microbial community would possess a fairly versatile array of polysaccharide degradative capabilities to ferment most of the common prebiotics entering the ceca. Genes for all three SCFA (acetate, propionate, and butyrate) could be identified representing a wide spread distribution in the metagenome with over 30 acetate kinase/phosphotransferase sequences alone. As more metagenomic data are generated along with metabolomic profiling and transcriptomics, a comprehensive view of the relationship should emerge between incoming prebiotic substrates and the cecal microbiota response of individual microbial members as well as the interactions and cross-feeding that occur within the microbial community.

Location within the cecum of where it is sampled may also need to be considered when evaluating intestinal and cecal responses to prebiotics. Adhikari ad Kwon (2017) collected ileal and cecal samples from 3-week-old broilers and separated these into mucosal and luminal subsets for plating on MRS agar plates to recover the *Lactobacillus* populations. Bacterial pellets were recovered from each MRS plate followed by DNA extraction and sequencing using the V1–V3 region of the 16S rRNA gene. Not all bacteria isolated on the MRS plates proved to be *Lactobacillus*. Taxonomic identification indicated several non-*Lactobacillus* genera were present on the MRS plates including

Bacillus and *Citrobacter* emphasizing that selective plating cannot be assumed to be entirely selective and is in agreement for reports on selective media for other bacteria (Kim et al., 2017). Sequencing of the MRS groups revealed significant differences in beta diversity of the respective GIT sites with higher levels of *Firmicutes* occurring in ileal mucosal and cecal luminal samples versus cecal mucosa. Cecal mucosal MRS samples conversely yielded more *Proteobacteria*. Individual *Lactobacillus* species also varied among GIT regions. *Lactobacillus salivarius* and *L. johnsonii* were more prevalent in ileal mucosa compared to cecal mucosa and lumen, while the cecal lumen contained more *L. crispatus* than the ileal mucosa. More *L. gasseri* appeared in the ileal mucosa compared to the cecal mucosa.

These potential location differences have implications for assessing the impact of feed additives such as prebiotics on cecal microbial populations. Certainly how much the resident mucosal surface microbial population versus the microorganisms in the cecal lumen contributes to the fecal microbiota would need to be factored if fecal sampling is being considered as being representative. More importantly, the overall cecal microbial community based on microbiome sequencing may be somewhat heterogeneous depending on the composition of the mucosal population versus the lumen population. Differences could be reflected not only on how representative samples are from the ceca but their relative exposure to the incoming prebiotic. Since certain genera such as *Lactobacillus* are involved in the direct fermentation of prebiotics the differences in species at different regions of the ceca may also be an important determinant in the relative efficacy of particular prebiotics. For example, *L. salivarius* identified by Adhikari and Kwon (2017) in their cecal studies has been shown to grow relatively well on most of the commercial prebiotics but exhibited preferences for certain oligosaccharides (Saminathan et al., 2011).

Given the potential differences in *Lactobacillus* spp. and niches they occupy in the ceca, it would be beneficial to screen these isolates for their ability to utilize the various prebiotics. This would be particularly useful in the context of Adhikari and Kwon (2017) proposed advocacy for these strains as probiotic candidates for poultry. It is conceivable that further development in combination with assessing the growth capability of prebiotics could lead to optimal symbiotic approaches that combine both the probiotic candidate and its preferred prebiotic substrate. For example, De Maesschalck et al. (2015) not only observed increases in *L. crispatus* in the colons from birds fed xylo-oligosaccharides but detected cross-feeding occurring between *L. crispatus* and butyrate-producing *Anaerostipes butyraticus* when incubated with xylo-oligosaccharides in an *in vitro* system. It is quite likely that other cross-feeding events occurring among the lower GIT, and cecal inhabitants are enhanced in the presence of prebiotics.

8 Summary and conclusions

Prebiotics have great utility as feed additives for poultry. It is anticipated that potential sources of prebiotics and substances that elicit prebiotic-like benefits will continue to be identified. Commercial application will depend on a balance between economics and efficacy. Efficacy toward the detectable and consistent reduction of foodborne pathogens and/or improved health certainly has value but retaining poultry performance is also critical and improved performance obviously would be ideal. Consistency remains an ongoing concern and will be difficult to resolve given the complexity of bird physiology and the GIT ecosystem. Several different strategies are needed to address this concern. Certainly better understanding of the chemical composition of sources of prebiotics and identifying the components that elicit prebiotic properties would be helpful. Screening tools will be needed to identify the bioactive components, and in some cases, there may be evidence of synergism between different components making the composite source of prebiotic more efficacious. Likewise, processing of sources of prebiotics such as cereal grains will need to be examined for any impact that the physical nature, for example, milling might have on the recovery of bioactive components. Feed milling and thermal processes such as pelleting will need to be examined to assess whether the prebiotic properties are lost during heating thus rendering the prebiotic inactive.

Factors associated with the complexity of the GIT system will also need to be addressed. It is becoming more apparent that all compartments of the poultry GIT need to be considered when evaluating the interaction of the GIT with the incoming prebiotic. While most of the focus has been understandably placed on the ceca given its considerable population of fermentative microorganisms, bacteria are also present and presumably active toward prebiotics as they traverse through all of the upper and lower GIT compartments. In addition, host GIT tissue needs to be examined. In certain GIT sections such as the small intestine, there is a dynamic interface between the immune and metabolic cells of the small intestine and the microbial community inhabiting the surfaces of the small intestine. As these surfaces are exposed to the fermentation activities associated with utilization of prebiotics, the concomitant shifting of microbial inhabitants and fermentation patterns could elicit detectable host GIT functional responses.

The cecal microbiota is highly complex and the microbial community response to feeding additives is difficult to interpret. However, the application of 16S-based microbiome sequencing and metagenomics has offered new insights to how specific members of that community are impacted. As sequencing resolution continues to improve and bioinformatic tools become more user-friendly, the amount of available data will expand. As more data become available, it will be possible to construct metabolic networks that identify groups of microorganisms that collectively degrade and utilize complex feed components. This will lead to a better understanding of which microorganisms are cross-feeding and the overall contributions to cecal function during digestion and fermentation of prebiotics. Combining this information with activities occurring in the other GIT compartments offers an opportunity to take a systems approach to evaluate GIT response to a particular prebiotic and potentially begin to predict the overall bird performance responses.

9 Where to look for further information

Gibson, G. R., Hutkins, R. W., Sanders, M. E., Prescott, S. L., Reimer, R. A., Salminen, S. J., Scott, K., Stanton, C., Swanson, K. S., Cani, P. D., Verbeke, K. and Reid, G. 2017. The International Scientific Association for Probiotics and Prebiotics (ISAPP) consensus statement on the definition and scope of prebiotics. *Nat. Rev. Gastroenterol. Hepatol.* 14, 491-502. doi:10.1038/nrgastro.2017.75.

Hutkins, R. W., Krumbeck, J. A., Bindels, L. B., Cani, P. D., Fahey Jr., G. C., Goh, Y. J., Hamaker, B., Martens, E. C., Mills, D. A., Rastal, R. A., Vaughan, E. and Sanders, M. E. 2016. Prebiotics: why definitions matter. *Curr. Opin. Biotech.* 37, 1-7.

Kim, S. A., Jang, M. J., Kim, S. Y., Yang, Y., Pavlidis, H. O. and Ricke, S. C. 2019. Potential for prebiotics as feed additives to limit foodborne *Campylobacter* establishment in the poultry gastrointestinal tract. *Front. Microbiol.* 10, 91. doi:10.3389/fmicb.2019.00091.

Micciche, A. C., Foley, S. L., Pavlidis, H. O. McIntyre, D. R. and Ricke, S. C. 2018. A review of prebiotics against *Salmonella* in poultry: current and future potential for microbiome research application. *Front. Vet. Sci.* 5, 191. doi:10.3389/fvets.2018.00191.

Ricke, S. C. 2018. Impact of prebiotics on poultry production and food safety. *Yale J. Biol. Med.* 91, 151-9.

Ricke, S. C., Hacker, J., Yearkey, K., Shi, Z., Park, S. H. and Rainwater, C. 2017. Unravelling food production microbiomes: concepts and future directions. In: Ricke, S. C., Atungulu, G. G., Park, S. H. and Rainwater, C. E. (Eds), *Food and Feed Safety Systems and Analysis*. Elsevier Inc., San Diego, CA, pp. 347-74. Chapter 19.

Stanley, D., Hughes, R. J. and Moore, R. J. 2014. Microbiota of the chicken gastrointestinal tract: Influence on health, productivity and disease. *App. Microbiol. Biotechnol.* 98, 4301-10.

Teng, P.-Y. and Kim, W. K. 2018. Review: roles of prebiotics in intestinal ecosystem of broilers. *Front. Vet. Sci.* 5, 245. doi:10.3389/fvets.2018.00245.

10 References

Adhikari, B. and Kwon, Y. M. 2017. Characterization of the culturable subpopulations of *Lactobacillus* in the chicken intestinal tract as a resource for probiotic development. *Front. Microbiol.* 8, 1389. doi:10.3389/fmicb.2017.01389.

Alloui, M. N., Szczurek, W. and Świątkiewicz, S. 2013. The usefulness of prebiotics and probiotics in modern poultry nutrition: a review. *Ann. Anim. Sci.* 13, 17-32.

Angkanaporn, K., Ravindran, V. and Bryden, W. L. 1997. Influence of caecectomy and dietary protein concentration on apparent excreta amino acid digestibility in adult cockerals. *Br. Poult. Sci.* 38(3), 270-6. doi:10.1080/00071669708417985.

Annison, E. F., Hill, K. J. and Kenworthy, R. 1968. Volatile fatty acids in the digestive tract of the fowl. *Br. J. Nutr.* 22(2), 207-16. doi:10.1079/BJN19680026.

Awad, W. A., Mann, E., Dzieciol, M., Hess, C., Schmitz-Esser, S., Wagner, M. and Hess, M. 2016. Age-related differences in the luminal and mucosa-associated gut microbiome of broiler chickens and shifts associated with *Campylobacter jejuni* infection. *Front. Cell. Infect. Microbiol.* 6, 154. doi:10.3389/fcimb.2016.00154.

Bachrach, U. 1957. The aerobic breakdown of uric acid by certain pseudomonads. *J. Gen. Microbiol.* 17(1), 1-11. doi:10.1099/00221287-17-1-1.

Ballou, A. L., Ali, R. A., Mendoza, M. A., Ellis, J. C., Hassan, H. M., Croom, W. J. and Koci, M. D. 2016. Development of the chick microbiome: how early exposure influences future microbial diversity. *Front. Vet. Sci.* 3, 2, doi:10.3389/fvets.2016.00002.

Barnes, E. M. 1972. The avian intestinal flora with particular reference to the possible ecological significance of the cecal anaerobic bacteria. *Am. J. Clin. Nutr.* 25(12), 1475-9. doi:10.1093/ajcn/25.12.1475.

Barnes, E. M. and Impey, C. S. 1974. The occurrence and properties of uric acid decomposing anaerobic bacteria in the avian caecum. *J. Appl. Bacteriol.* 37(3), 393-409. doi:10.1111/j.1365-2672.1974.tb00455.x.

Barnes, E. M., Mead, G. C., Barnum, D. A. and Harry, E. G. 1972. The intestinal flora of the chicken in the period 2 to 6 weeks of age, with particular reference to the anaerobic bacteria. *Br. Poult. Sci.* 13(3), 311-26. doi:10.1080/00071667208415953.

Barnes, E. M., Impey, C. S. and Cooper, D. M. 1980. Manipulation of the crop and intestinal flora of the newly hatched chick. *Am. J. Clin. Nutr.* 33(11 Suppl.), 2426-33. doi:10.1093/ajcn/33.11.2426.

Baurhoo, B., Goldflus, F. and Zhao, X. 2009. Purified cell wall of *Saccharomyces cerevisiae* increases protection against intestinal pathogens in broiler chickens. *Int. J. Poult. Sci.* 8(2), 133-7. doi:10.3923/ijps.2009.133.137.

Bayer, R. C., Chawan, C. B. and Bird, F. H. 1975. Scanning electron microscopy of the chicken crop - the avian rumen? *Poult. Sci.* 54(3), 703-7. doi:10.3382/ps.0540703.

Bayer, R. C., Hoover, W. H. and Muir, F. V. 1978. Dietary fiber and meal feeding influence on broiler growth and crop fermentation. *Poult. Sci.* 57(5), 1456-9. doi:10.3382/ps.0571456.

Biggs, P. and Parsons, C. M. 2008. The effects of Grobiotic-P on growth performance, nutrient digestibilities, and cecal microbial populations in young chicks. *Poult. Sci.* 87(9), 1796-803. doi:10.3382/ps.2007-00450.

Bird, A. R., Conlon, M. A., Christopher, C. T. and Topping, D. L. 2010. Resistant starch, large bowel fermentation and a broader perspective of prebiotics and probiotics. *Benef. Microbes* 1(4), 423-31. doi:10.3920/BM2010.0041.

Bjerrum, L., Engberg, R. M., Leser, T. D., Jensen, B. B., Finster, K. and Pedersen, K. 2006. Microbial community composition of the ileum and cecum of broiler chickens as revealed by molecular and culture-based techniques. *Poult. Sci.* 85(7), 1151-64. doi:10.1093/ps/85.7.1151.

Bolton, W. and Dewar, W. A. 1965. The digestibility of acetic, propionic and butyric acids by fowl. *Br. Poult. Sci.* 6(2), 103-5. doi:10.1080/00071666508415562.

Bongers, A. and van den Heuvel, E. G. H. M. 2003. Prebiotics and the bioavailability of minerals and trace elements. *Food Rev. Int.* 19(4), 397–422. doi:10.1081/FRI-120025482.

Bortoluzzi, C., Barbosa, J. G. M., Pereira, R., Fagundes, N. S., Rafael, J. M. and Menten, J. F. M. 2018. Autolyzed yeast (*Saccharomyces cerevisiae*) supplementation improves performance while modulating the intestinal immune-system and microbiology of broiler chickens. *Front. Sustain. Food Syst.* 2, 85, doi:10.3389/fsufs.2018.00085.

Callaway, T. R. and Ricke, S. C. (Eds). 2012. *Direct Fed Microbials/Prebiotics for Animals: Science and Mechanisms of Action*. Springer Science, New York, NY.

Callaway, T. R., Dowd , S. E., Wolcott, R. D., Sun, Y., McReynolds, J. L., Edrington, T. S., Byrd, J. A., Anderson, R. C., Krueger, N. and Nisbet, D. J. 2009. Evaluation of the bacterial diversity in cecal contents of laying hens fed various molting diets by using bacterial tag-encoded FLX amplicon pyrosequencing. *Poult. Sci.* 88(2), 298–302. doi:10.3382/ps.2008-00222.

Calo, J. R., Crandall, P. G., O'Bryan, C. A. and Ricke, S. C. 2015. Essential oils as antimicrobials in food systems – a review. *Food Control* 54, 111–9. doi:10.1016/j.foodcont.2014.12.040.

Carlile, F. S. 1984. Ammonia in poultry houses: a literature review. *Worlds Poult. Sci. J.* 40(2), 99–113. doi:10.1079/WPS19840008.

Chalova, V. I., Kim, J. H., Patterson, P. H., Ricke, S. C. and Kim, W. K. 2016a. Reduction of nitrogen excretion and emissions from poultry: a review for conventional poultry. *Worlds Poult. Sci. J.* 72(3), 509–20. doi:10.1017/S0043933916000477.

Chalova, V. I., Kim, J. H., Patterson, P. H., Ricke, S. C. and Kim, W. K. 2016b. Reduction of nitrogen excretion and emission in poultry: a review for organic poultry. *J. Environ. Sci. Health B* 51(4), 230–5. doi:10.1080/03601234.2015.1120616.

Chaplin, S. B. 1989. Effect of cecectomy on water and nutrient absorption of birds. *J. Exp. Zool. (Suppl.)* 3, 81–6. doi:10.1002/jez.1402520514.

Choct, M. 2009. Managing gut health through nutrition. *Br. Poult. Sci.* 50(1), 9–15. doi:10.1080/00071660802538632.

Cox, N. A., Cason, J. A. and Richardson, L. J. 2011. Minimization of *Salmonella* contamination on raw poultry. *Ann. Rev. Food Sci. Technol.* 2, 75–95. doi:10.1146/annurev-food-022510-133715.

Delzenne, N. M. and Kok, N. 2001. Effects of fructans-type prebiotics on lipid metabolism. *Am. J. Clin. Nutr.* 73(Suppl.), 456S–8S. doi:10.1093/ajcn/73.2.456s.

De Maesschalck, C., Eeckhaut, V., Maertens, L., De Lange, L., Marchal, L., Nezer, C., De Baere, S., Croubels, S., Daube, G., Dewulf, J., Haesebrouck, F., Ducatelle, R., Taminau, B. and Van Immerseel, F. 2015. Effects of xylo-oligosaccharides on broiler chicken performance and microbiota. *Appl. Environ. Microbiol.* 81(17), 5880–8. doi:10.1128/AEM.01616-15.

Dittoe, D. K., Ricke, S. C. and Kiess, A. S. 2018. Organic acids and potential for modifying the avian gastrointestinal tract and reducing pathogens and disease. *Front. Vet. Sci.* 5, 216. doi:10.3389/fvets.2018.00216.

Duke, G. E. 1989. Relationship of cecal and colonic motility to diet, habitat, and cecal anatomy in several avian species. *J. Exp. Zool. (Suppl.)* 3, 38–47.

Duke, G. E. 2002. Digestive processes in poultry from a physiological viewpoint. In: McNab, J. M. and Boorman, K. N. (Eds), *Poultry Feedstuffs: Supply, Composition and Nutritive Value*. CABI Publishing, Oxon, UK, pp. 111–4. Chapter 7.

Dunkley, K. D., Dunkley, C. S., Njongmeta, N. L., Callaway, T. R., Hume, M. E., Kubena, L. F., Nisbet, D. J. and Ricke, S. C. 2007a. Comparison of *in vitro* fermentation and molecular microbial profiles of high-fiber feed substrates (HFFS) incubated with chicken cecal inocula. *Poult. Sci.* 86, 801-10.

Dunkley, K. D., McReynolds, J. L., Hume, M. E., Dunkley, C. S., Callaway, T. R., Kubena, L. F., Nisbet, D. J. and Ricke, S. C. 2007b. Molting in *Salmonella* Enteritidis-challenged laying hens fed alfalfa crumbles.II. Fermentation and microbial ecology response. *Poult. Sci.* 86(10), 2101-9. doi:10.1093/ps/86.10.2101.

Dunkley, K. D., Callaway, T. R., Chalova, V. I., McReynolds, J. L., Hume, M. E., Dunkley, C. S., Kubena, L. F., Nisbet, D. J. and Ricke, S. C. 2009. Foodborne *Salmonella* ecology in the avian gastrointestinal tract. *Anaerobe* 15(1-2), 26-35. doi:10.1016/j.anaerobe.2008.05.007.

Durant, J. A., Corrier, D. E., Byrd, J. A., Stanker, L. H. and Ricke, S. C. 1999. Feed deprivation affects crop environment and modulates *Salmonella enteritidis* colonization and invasion of leghorn hens. *Appl. Environ. Microbiol.* 65(5), 1919-23.

Durant, J. A., Corrier, D. E., Stanker, L. H. and Ricke, S. C. 2000. Expression of the *hilA Salmonella typhimurium* gene in a poultry *Salm. enteritidis* isolate in response to lactate and nutrients. *J. Appl. Microbiol.* 89(1), 63-9. doi:10.1046/j.1365-2672.2000.01089.x.

Escarcha, J. F., Callaway, T. R., Byrd, J. A., Miller, D. N., Edrington, T. S., Anderson, R. C. and Nisbet, D. J. 2012. Effects of dietary alfalfa inclusion on *Salmonella* Typhimurium populations in growing layer chicks. *Foodborne Pathog. Dis.* 9(10), 945-51. doi:10.1089/fpd.2012.1251.

Fan, Y. Y., Ricke, S. C., Scanlan, C. M., Nisbet, D. J., Vargas-Moskola, A. A., Corrier, D. E. and DeLoach, J. R. 1995. Use of differential rumen fluid-based carbohydrate agar media for culturing lactose-selected cecal bacteria from chickens. *J. Food Prot.* 58(4), 361-7. doi:10.4315/0362-028X-58.4.361.

Fasina, Y. O. and Olowo, Y. L. 2013. Effect of a commercial yeast-based product (Mexigen®) on intestinal villi morphology and growth performance of broiler chickens. *Int. J. Poult. Sci.* 12(1), 9-14. doi:10.3923/ijps.2013.9.14.

Fasina, Y. O. and Thanissery, R. R. 2011. Comparative efficacy of a yeast product and bacitracin methylene disalicylate in enhancing early growth and intestinal maturation in broiler chicks from breeder hens of different ages. *Poult. Sci.* 90(5), 1067-73. doi:10.3382/ps.2010-01033.

Flickinger, E. A., Van Loo, J. and Fahey Jr., G. C. 2003. Nutritional responses to the presence of inulin and oligofructose in the diets of domesticated animals: a review. *Crit. Rev. Food Sci. Nutr.* 43(1), 19-60. doi:10.1080/10408690390826446.

Fuller, R. 1989. Probiotics in man and animals. *J. Appl. Bacteriol.* 66(5), 365-78. doi:10.1111/j.1365-2672.1989.tb05105.x.

Garlich, J., Brake, J., Parkhurst, C. R., Thaxton, J. P. and Morgan, G. W. 1984. Physiological profile of caged layers during one production year, molt, and postmolt: egg production, egg shell quality, liver, femur, and blood parameters. *Poult. Sci.* 63(2), 339-43. doi:10.3382/ps.0630339.

Gibson, G. R. and Roberfroid, M. B. 1995. Dietary modulation of the human colonic microflora: introducing the concept of prebiotics. *J. Nutr.* 125(6), 1401-12. doi:10.1093/jn/125.6.1401.

Gibson, G. R., Hutkins, R., Sanders, M. E., Prescott, S. L., Reimer, R. A., Salminen, S. J., Scott, K., Stanton, C., Swanson, K. S., Cani, P. D., Verbeke, K. and Reid, G. 2017. Expert consensus document: The International Scientific Association for Probiotics and Prebiotics (ISAPP) consensus statement on the definition and scope of prebiotics. *Nat. Rev. Gastroenterol. Hepatol.* 14(8), 491–502. doi:10.1038/nrgastro.2017.75.

Goldstein, D. L. 1989. Absorption by the cecum of wild birds: is there interspecific variation? *J. Exp. Zool. (Suppl.)* 3, 103–10. doi:10.1002/jez.1402520517.

Gómez-Verduzco, G., Cortes-Cuevas, A., López-Coello, C., Ávila-González, E. and Nava, G. M. 2009. Dietary supplementation of mannan-oligosaccharide enhances neonatal immune responses in chickens during natural exposure to *Eimeria* spp. *Acta Vet. Scand.* 51, 11. doi:10.1186/1751-0147-51-11.

Gong, J., Si, W., Forster, R. J., Huang, R., Yu, H., Yin, Y., Yang, C. and Han, Y. 2007. 16S rRNA gene-based analysis of mucosa-associated bacterial community and phylogeny in the chicken gastrointestinal tracts: from crops to ceca. *FEMS Microbiol. Ecol.* 59(1), 147–57. doi:10.1111/j.1574-6941.2006.00193.x.

Guardia, S., Konsak, B., Combes, S., Levenez, F., Cauquil, L., Guillot, J. F., Moreau-Vauzelle, C., Lessire, M., Juin, H. and Gabriel, I. 2011. Effects of stocking density on the growth performance and digestive microbiota of broiler chickens. *Poult. Sci.* 90(9), 1878–89. doi:10.3382/ps.2010-01311.

Hanning, I. and Diaz-Sanchez, S. 2015. The functionality of the gastrointestinal microbiome in non-human animals. *Microbiome* 3, 51 doi:. doi:10.1186/s40168-015-0113-6.

Hanning, I., Clement, A., Owens, C., Park, S. H., Pendleton, S., Scott, E. E., Almeida, G., Gonzalez Gil, F. and Ricke, S. C. 2012. Assessment of production performance in two breeds of broilers fed prebiotics as feed additives. *Poult. Sci.* 91(12), 3295–9. doi:10.3382/ps.2012-02557.

Hara, H. 2002. Physiological effects of short-chain fatty acid reduced from prebiotics in the colon. *Biosci. Microflora* 21(1), 35–42. doi:10.12938/bifidus1996.21.35.

Hargis, B. M., Caldwell, D. J., Brewer, R. L., Corrier, D. E. and DeLoach, J. R. 1995. Evaluation of the chicken crop as a source of *Salmonella* contamination for broiler carcasses. *Poult. Sci.* 74(9), 1548–52. doi:10.3382/ps.0741548.

Hetland, H., Choct, M. and Svihus, B. 2004. Role of insoluble non-starch polysaccharides in poultry nutrition. *Worlds Poult. Sci. J.* 60(4), 415–22. doi:10.1079/WPS200325.

Hetland, H., Svihus, B. and Choct, M. 2005. Role of insoluble fiber on gizzard activity in layers. *J. Appl. Poult. Res.* 14(1), 38–46. doi:10.1093/japr/14.1.38.

Heuser, G. F. 1945. The rate of passage of feed from the crop of the hen. *Poult. Sci.* 24(1), 20–4. doi:10.3382/ps.0240020.

Hinton Jr., A., Buhr, R. J. and Ingram, K. D. 2000a. Reduction of *Salmonella* in the crop of broiler chickens subjected to feed withdrawal. *Poult. Sci.* 79(11), 1566–70. doi:10.1093/ps/79.11.1566.

Hinton Jr., A., Buhr, R. J. and Ingram, K. D. 2000b. Physical, chemical, and microbiological changes in the crop of broiler chickens subjected to incremental feed withdrawal. *Poult. Sci.* 79(2), 212–8. doi:10.1093/ps/79.2.212.

Hinton Jr., A., Buhr, R. J. and Ingram, K. D. 2002. Carbohydrate-based cocktails that decrease the population of *Salmonella* and *Campylobacter* in the crop of broiler chickens subjected to feed withdrawal. *Poult. Sci.* 81(6), 780–4. doi:10.1093/ps/81.6.780.

Hooge, D. M. 2004. Meta-analysis of broiler chicken pen trials evaluating dietary mannan oligosaccharide, 1993–2003. *Int. J. Poult. Sci.* 3, 163–74.

Huhtanen, C. N. and Pensack, J. M. 1965. The development of the intestinal flora of the young chick. *Poult. Sci.* 44, 825–30. doi:10.3382/ps.0440825.

Hume, M. E. 2011. Historic perspective: prebiotics, probiotics, and other alternatives to antibiotics. *Poult. Sci.* 90(11), 2663-9. doi:10.3382/ps.2010-01030.

Hume, M. E., Corrier, D. E., Ivie, G. W. and DeLoach, J. R. 1993. Metabolism of [^{14}C] propionic acid in broiler chicks. *Poult. Sci.* 72(5), 786-93. doi:10.3382/ps.0720786.

Hume, M. E., Kubena, L. F., Edrington, T. S., Donskey, C. J., Moore, R. W., Ricke, S. C. and Nisbet, D. J. 2003. Poultry digestive microflora biodiversity as indicated by denaturing gradient gel electrophoresis. *Poult. Sci.* 82(7), 1100-7. doi:10.1093/ps/82.7.1100.

Hutkins, R. W., Krumbeck, J. A., Bindels, L. B., Cani, P. D., Fahey Jr., G. C., Goh, Y. J., Hamaker, B., Martens, E. C., Mills, D. A., Rastal, R. A., Vaughan, E. and Sanders, M. E. 2016. Prebiotics: why definitions matter. *Curr. Opin. Biotechnol.* 37, 1-7. doi:10.1016/j.copbio.2015.09.001.

Iji, P. A., Saki, A. A. and Tivey, D. R. 2001. Intestinal structure and function of broiler chickens on diets supplemented with a mannan oligosaccharide. *J. Sci. Food Agric.* 81(12), 1186-92. doi:10.1002/jsfa.925.

Indikova, I., Humphrey, T. J. and Hilbert, F. 2015. Survival with a helping hand: *Campylobacter* and microbiota. *Front. Microbiol.* 6 (Article 1266), 1-6. doi:10.3389/fmicb.2015.01266.

Jahanian, R. and Ashnagar, M. 2015. Effect of dietary supplementation of mannan-oligosaccharides on performance, blood metabolites, ileal nutrient digestibility, and gut microflora in *Escherichia coli*-challenged laying hens. *Poult. Sci.* 94(9), 2165-72. doi:10.3382/ps/pev180.

Joerger, R. D. 2003. Alternatives to antibiotics: bacteriocins, antimicrobial peptides and bacteriophages. *Poult. Sci.* 82(4), 640-7. doi:10.1093/ps/82.4.640.

Jones, F. T. and Ricke, S. C. 2003. Observations on the history of the development of antimicrobials and their use in poultry feeds. *Poult. Sci.* 82(4), 613-17. doi:10.1093/ps/82.4.613.

Józefiak, D., Rutkowski, A. and Martin, S. A. 2004. Carbohydrate fermentation in the ceca: a review. *Anim. Feed Sci. Technol.* 113(1-4), 1-15. doi:10.1016/j.anifeedsci.2003.09.007.

Kaplan, H. and Hutkins, R. W. 2000. Fermentation of fructooligosaccharides by lactic acid bacteria and bifidobacteria. *Appl. Environ. Microbiol.* 66(6), 2682-4. doi:10.1128/aem.66.6.2682-2684.2000.

Karasawa, Y. 1989. Ammonia production from uric acid, urea, and amino acids and its absorption from the ceca of the cockerel. *J. Exp. Zool. (Suppl.)* 3, 75-80. doi:10.1002/jez.1402520513.

Keener, K. M., Bashor, M. P., Curtis, P. A., Sheldon, B. W. and Kathariou, S. 2004. Comprehensive review of *Campylobacter* and poultry processing. *Compr. Rev. Food Sci. Food Saf.* 3(2), 105-16. doi:10.1111/j.1541-4337.2004.tb00060.x.

Kers, J. G., Velkers, F. C., Fischer, E. A. J., Hermes, G. D. A., Stegeman, J. A. and Smidt, H. 2018. Host and environmental factors affecting the intestinal microbiota in chickens. *Front. Microbiol.* 9, 235. doi:10.3389/fmicb.2018.00235.

Kim, W. K. and Patterson, P. H. 2003. Effect of minerals on activity of microbial uricase to reduce ammonia volatilization in poultry manure. *Poult. Sci.* 82(2), 223-31. doi:10.1093/ps/82.2.223.

Kim, W. K., Donalson, L. M., Mitchell, A. D., Kubena, L. F., Nisbet, D. J. and Ricke, S. C. 2006a. Effects of alfalfa and fructooligosaccharide on molting parameters and bone qualities using dual energy X-ray absorptiometry and conventional bone assays. *Poult. Sci.* 85(1), 15–20. doi:10.1093/ps/85.1.15.

Kim, W. K., Anderson, R. C., Ratliff, A. L., Nisbet, D. J. and Ricke, S. C. 2006b. Growth inhibition by nitrocompounds of selected uric acid–utilizing microorganisms isolated from poultry manure. *J. Environ. Sci. Health B* 41(1), 97–107. doi:10.1080/03601230500234950.

Kim, W. K., Weeks, L. J., Anderson, R. C., Nisbet, D. J., Dunkley, K. and Ricke, S. C. 2009. Effects of nitrocompounds on uric acid-utilizing microorganisms, nitrogen retention, and microbial community in laying hen manure. *J. Environ. Sci. Health B* 44(4), 403–6. doi:10.1080/03601230902801133.

Kim, W. K., Patterson, P. H., Rodriguez-LeCompte, J. C. and Ricke, S. C. 2013. The potential to reduce poultry nitrogen emissions with specific uricase egg yolk feed grade antibodies. *Worlds Poult. Sci. J.* 69(1), 45–56. doi:10.1017/S0043933913000056.

Kim, S. A., Hong Park, S., In Lee, S., Owens, C. M. and Ricke, S. C. 2017. Assessment of chicken carcass microbiome responses during processing in the presence of commercial antimicrobials using a next generation sequencing approach. *Sci. Rep.* 7, 43354. doi:10.1038/srep43354.

Kim, S. A., Rubinelli, P. M., Park, S. H. and Ricke, S. C. 2018. Ability of Arkansas LaKast and LaKast hybrid rice bran to reduce *Salmonella* Typhimurium in chicken cecal incubations and effects on cecal microbiota. *Front. Microbiol.* 9, 134. doi:10.3389/fmicb.2018.00134.

Kim, S. A., Jang, M. J., Kim, S. Y., Yang, Y., Pavlidis, H. O. and Ricke, S. C. 2019. Potential for prebiotics as feed additives to limit foodborne *Campylobacter* establishment in the poultry gastrointestinal tract. *Front. Microbiol.* 10, 91. doi:10.3389/fmicb.2019.00091.

Korver, D. R. 2006. Overview of the immune dynamics of the digestive system. *J. Appl. Poult. Res.* 15(1), 123–35. doi:10.1093/japr/15.1.123.

Leedle, J. A. Z. and Hespell, R. B. 1980 Differential carbohydrate media and anaerobic replica plating techniques in delineating carbohydrate-utilizing subgroups in rumen bacterial populations. *Appl. Environ. Microbiol.* 39(4), 709–19.

Litwin, C. M. and Calderwood, S. B. 1993. Role of iron in regulation of virulence genes. *Clin. Microbiol. Rev.* 6(2), 137–49. doi:10.1128/cmr.6.2.137.

Lu, J., Idris, U., Harmon, B., Hofacre, C., Maurer,J. J. and Lee, M. D. 2003. Diversity and succession of the intestinal bacterial community of the maturing broiler chicken. *Appl. Environ. Microbiol.* 69(11), 6816–24. doi:10.1128/aem.69.11.6816-6824.2003.

Mateos, G. G., Jiménez-Moreno, E., Serrano, M. P. and Lazaro, R. P. 2012. Poultry responses to high levels of dietary fiber sources varying in physical and chemical characteristics. *J. Appl. Poult. Res.* 21(1), 156–74. doi:10.3382/japr.2011-00477.

May, J. D., Lott, B. D. and Deaton, J. W. 1990. The effect of light and environmental temperature on broiler digestive tract contents after feed withdrawal. *Poult. Sci.* 69(10), 1681–4. doi:10.3382/ps.0691681.

Mazzuco, H. and Hester, P. Y. 2005. The effect of an induced molt and a second cycle of lay on skeletal integrity of white leghorns. *Poult. Sci.* 84(5), 771–81. doi:10.1093/ps/84.5.771.

McNab, J. M. 1973. The avian caeca: a review. *Worlds Poult. Sci. J.* 29(3), 251–63. doi:10.1079/WPS19730014.

Mead, G. C. 1989. Microbes of the avian cecum: types present and substrates utilized. *J. Exp. Zool. (Suppl.)* 3, 48–54. doi:10.1002/jez.1402520508.

Mead, G. C. and Adams, B. W. 1975. Some observations on the caecal microflora of the chick during the first two weeks of life. *Br. Poult. Sci.* 16(2), 169–76. doi:10.1080/00071667508416174.

Micciche, A. C., Foley, S. L., Pavlidis, H. O., McIntyre, D. R. and Ricke, S. C. 2018. A review of prebiotics against *Salmonella* in poultry: current and future potential for microbiome research applications. *Front. Vet. Sci.* 5, 191. doi:10.3389/fvets.2018.00191.

Moore, S. J. 1998. Use of an artificial gizzard to investigate the effect of grit on the breakdown of grass. *J. Zool. Lond.* 246(1), 119–24. doi:10.1111/j.1469-7998.1998.tb00140.x.

Moore, S. J. 1999. Food breakdown in an avian herbivore: who needs teeth? *Aust. J. Zool.* 47(6), 625–32. doi:10.1071/ZO99051.

Moretó, M. and Planas, J. M. 1989. Sugar and amino acid transport properties of the chicken ceca. *J. Exp. Zool. (Suppl.)* 3, 111–6. doi:10.1002/jez.1402520518.

Morohashi, T. 2002. The effect on bone of stimulated intestinal mineral absorption following fructooligosaccharide consumption in rats. *Biosci. Microflora* 21(1), 21–5. doi:10.12938/bifidus1996.21.21.

Moye, Z. D., Woolston, J. and Sulakvelidze, A. 2018. Bacteriophage applications for food production and processing. *Viruses* 10(4), 205, doi:10.3390/v10040205.

Newman, S. and Leeson, S. 1999. The effect of feed deprivation and subsequent refeeding on the bone characteristics of aged hens. *Poult. Sci.* 78(12), 1658–63. doi:10.1093/ps/78.12.1658.

Nisbet, D. 2002. Defined competitive exclusion cultures in the prevention of enteropathogen colonisation in poultry and swine. *Antonie Leeuwenhoek* 81(1–4), 481–6. doi:10.1023/A:1020541603877.

Nisbet, D. J., Ricke, S. C., Scanlan, C. M., Corrier, D. E., Hollister, A. G. and DeLoach, J. R. 1994. Inoculation of broiler chicks with a continuous-flow derived bacterial culture facilitates early cecal bacterial colonization and increases resistance to *Salmonella typhimurium*. *J. Food Prot.* 57(1), 12–5. doi:10.4315/0362-028X-57.1.12.

Oakley, B. B. and Kogut, M. H. 2016. Spatial and temporal changes in the broiler chicken cecal and fecal microbiomes and correlations of bacterial taxa with cytokine gene expression. *Front. Vet. Sci.* 3, 11. doi:10.3389/fvets.2016.00011.

Oakley, B. B., Morales, C. A., Line, J., Berrang, M. E., Meinersmann, R. J., Tillman, G. E., Wise, M. G., Siragusa, G. R., Hiett, K. L. and Seal, B. S. 2013. The poultry-associated microbiome: network analysis and farm-to-fork characterizations. *PLoS ONE* 8(2), e57190. doi:10.1371/journal.pone.0057190.

Oakley, B. B., Buhr, R. J., Ritz, C. W., Kiepper, B. H., Berrang, M. E., Seal, B. S. and Cox, N. A. 2014. Successional changes in the chicken cecal microbiome during 42 days of growth are independent of organic acid feed additives. *BMC Vet. Res.* 10, 282. doi:10.1186/s12917-014-0282-8.

O'Bryan, C. A., Pendleton, S. J., Crandall, P. G. and Ricke, S. C. 2015. Potential of plant essential oils and their components in animal agriculture - *in vitro* studies on antibacterial mode of action. *Front. Vet. Sci.* 2(Article 35), 1–8. doi:10.3389/fvets.2015.00035.

Ofek, I. and Beachy, E. H. 1978. Mannose binding and epithelial cell adherence of *Escherichia coli*. *Infect. Immun.* 22(1), 247–54.

Ofek, I., Mirelman, D. and Sharon, N. 1977. Adherence of *Escherichia coli* to human mucosal cells mediated by mannose receptors. *Nature* 265(5595), 623–5. doi:10.1038/265623a0.

Pan, D. and Yu, Z. 2014. Intestinal microbiome of poultry and its interaction with host and diet. *Gut Microbes* 5(1), 108–19. doi:10.4161/gmic.26945.

Park, S. H., Kim, S. A., Rubinelli, P. M., Roto, S. M. and Ricke, S. C. 2017a. Microbial compositional changes in broiler chicken cecal contents from birds challenged with different *Salmonella* vaccine candidate strains. *Vaccine* 35(24), 3204–8. doi:10.1016/j.vaccine.2017.04.073.

Park, S. H., Kim, S. A., Lee, S. I., Rubinelli, P. M., Roto, S. M., Pavlidis, H. O., McIntyre, D. R. and Ricke, S. C. 2017b. Original XPC™ effect on *Salmonella* Typhimurium and cecal microbiota from three different ages of broiler chickens when incubated in an anaerobic in vitro culture system. *Front. Microbiol.* 8, 1070. doi:10.3389/fmicb.2017.01070.

Park, S. H., Lee, S. I., Kim, S. A., Christensen, K. and Ricke, S. C. 2017c. Comparison of antibiotic supplementation versus a yeast-based prebiotic on the cecal microbiome of commercial broilers. *PLoS ONE* 12(8), e0182805. doi:10.1371/journal.pone.0182805.

Park, S. H., Perrotta, A., Hanning, I., Diaz-Sanchez, S., Pendleton, S., Alm, E. and Ricke, S. C. 2017d. Pasture flock chicken cecal microbiome responses to prebiotics and plum fiber feed amendments. *Poult. Sci.* 96(6), 1820–30. doi:10.3382/ps/pew441.

Parsons, C. M. 1996. Digestible amino acids for poultry and swine. *Anim. Feed Sci. Technol.* 59(1–3), 147–53. doi:10.1016/0377-8401(95)00895-0.

Patterson, J. A. and Burkholder, K. M. 2003. Application of prebiotics and probiotics in poultry production. *Poult. Sci.* 82(4), 627–31. doi:10.1093/ps/82.4.627.

Payne, S. M. 1988. Iron and virulence in the family *Enterobacteriaceae*. *Crit. Rev. Microbiol.* 16(2), 81–111. doi:10.3109/10408418809104468.

Pourabedin, M. and Zhao, X. 2015. Prebiotics and gut microbiota in chickens. *FEMS Microbiol. Lett.* 362(15), fnv122. doi:10.1093/femsle/fnv122.

Pritchard, P. J. 1972. Digestion of sugars in the crop. *Comp. Biochem. Physiol.* 43A, 195–205.

Rehman, H. U., Vahjen, W., Awad, W. A. and Zentek, J. 2007. Indigenous bacteria and bacterial metabolic products in the gastrointestinal tract of broiler chickens. *Arch. Anim. Nutr.* 61(5), 319–35. doi:10.1080/17450390701556817.

Revolledo, L., Ferreira, A. J. P. and Mead, G. C. 2006. Prospects in *Salmonella* control: competitive exclusion, probiotics, and enhancement of avian intestinal immunity. *J. Appl. Poult. Res.* 15(2), 341–51. doi:10.1093/japr/15.2.341.

Richards, P. J., Connerton, P. L. and Connerton, I. F. 2019. Phage biocontrol of *Campylobacter jejuni* in chickens does not produce collateral effects on the gut microbiota. *Front. Microbiol.* 10, 476. doi:10.3389/fmicb.2019.00476.

Ricke, S. C. 2003. Perspectives on the use of organic acids and short chain fatty acids as antimicrobials. *Poult. Sci.* 82(4), 632–9. doi:10.1093/ps/82.4.632.

Ricke, S. C. 2015. Potential of fructooligosaccharide prebiotics in alternative and nonconventional poultry production systems. *Poult. Sci.* 94(6), 1411–8. doi:10.3382/ps/pev049.

Ricke, S. C. 2018. Impact of prebiotics on poultry production and food safety. *Yale J. Biol. Med.* 91(2), 151–9.

Ricke, S. C., van der Aar, P. J., Fahey Jr., G. C. and Berger, L. L. 1982. Influence of dietary fibers on performance and fermentation characteristics of gut contents from growing chicks. *Poult. Sci.* 61(7), 1335–43. doi:10.3382/ps.0611335.

Ricke, S. C., Kundinger, M. M., Miller, D. R. and Keeton, J. T. 2005. Alternatives to antibiotics: chemical and physical antimicrobial interventions and foodborne pathogen response. *Poult. Sci.* 84(4), 667–75. doi:10.1093/ps/84.4.667.

Ricke, S. C., Hererra, P. and Biswas, D. 2012. Bacteriophages for potential food safety applications in organic meat production. In: Ricke, S. C., Van Loo, E. J., Johnson, M. G. and O'Bryan, C. A. (Eds), *Organic Meat Production and Processing*. Wiley Scientific/IFT, New York, NY, pp. 407–24. Chapter 23.

Ricke, S. C., Dunkley, C. S. and Durant, J. A. 2013. A review on development of novel strategies for controlling *Salmonella* Enteritidis colonization in layer hens: fiber-based molt diets. *Poult. Sci.* 92(2), 502–25. doi:10.3382/ps.2012-02763.

Ricke, S. C., Hacker, J., Yearkey, K., Shi, Z., Park, S. H. and Rainwater, C. 2017. Unravelling food production microbiomes: concepts and future directions. In: Ricke, S. C., Atungulu, G. G., Park, S. H. and Rainwater, C. E. (Eds), *Food and Feed Safety Systems and Analysis*. Elsevier Inc., San Diego, CA, pp. 347–74. Chapter 19.

Ricke, S. C., Feye, K. M., Chaney, W. E., Shi, Z., Pavlidis, H. and Yang, Y. 2018. Developments in rapid detection methods for the detection of foodborne *Campylobacter* in the United States. *Front. Microbiol.* 9, 3280. doi:10.3389/fmicb.2018.03280.

Roto, S. M., Rubinelli, P. M. and Ricke, S. C. 2015. An introduction to the avian gut microbiota and the effects of yeast-based prebiotic compounds as potential feed additives. *Front. Vet. Sci.* 2(Article 28), 1–18. doi:10.3389/fvets.2015.00028.

Rubinelli, P. M., Roto, S., Kim, S. A., Park, S. H., Pavlidis, H. O., McIntyre, D. and Ricke, S. C. 2016. Reduction of *Salmonella* Typhimurium by fermentation metabolites of Diamond V Original XPC in an *in vitro* anaerobic mixed chicken cecal culture. *Front. Vet. Sci.* 3, 83. doi:10.3389/fvets.2016.00083.

Rubinelli, P. M., Kim, S. A., Park, S. H., Roto, S. M., Nealon, N. J., Ryan, E. P. and Ricke, S. C. 2017. Differential effects of rice bran cultivars to limit *Salmonella* Typhimurium in chicken cecal *in vitro* incubations and impact on the cecal microbiome and metabolome. *PLoS ONE* 12(9), e0185002. doi:10.1371/journal.pone.0185002.

Saengkerdsub, S., Kim, W. K., Anderson, R. C., Nisbet, D. J. and Ricke, S. C. 2006. Effects of nitrocompounds and feedstuffs on *in vitro* methane production in chicken cecal contents and rumen fluid. *Anaerobe* 12(2), 85–92. doi:10.1016/j.anaerobe.2005.11.006.

Saengkerdsub, S., Anderson, R. C., Wilkinson, H. H., Kim, W. K., Nisbet, D. J. and Ricke, S. C. 2007a. Identification and quantification of methanogenic archaea in adult chicken ceca. *Appl. Environ. Microbiol.* 73(1), 353–6. doi:10.1128/AEM.01931-06.

Saengkerdsub, S., Herrera, P., Woodward, C. L., Anderson, R. C., Nisbet, D. J. and Ricke, S. C. 2007b. Detection of methane and quantification of methanogenic archaea in faeces from young broiler chickens using real-time PCR. *Lett. Appl. Microbiol.* 45(6), 629–34. doi:10.1111/j.1472-765X.2007.02243.x.

Salanitro, J. P., Blake, I. G. and Muirhead, P. A. 1974. Studies on the cecal microflora of commercial broiler chickens. *Appl. Microbiol.* 28(3), 439–47.

Salanitro, J. P., Blake, I. G., Muirhead, P. A., Maglio, M. and Goodman, J. R. 1978. Bacteria isolated from the duodenum, ileum, and cecum of young chicks. *Appl. Environ. Microbiol.* 35(4), 782–90.

Saminathan, M., Sieo, C. C., Kalavathy, R., Abdullah, N. and Ho, Y. W. 2011. Effect of prebiotic oligosaccharides on growth of *Lactobacillus* strains used as a probiotic for chickens. *Afr. J. Microbiol. Res.* 5(1), 57–64.

Santos, E. G., Costa, F. G. P., Silva, J. H. V., Martins, T. D. D., Figueiredo-Lima, D. F., Macari, M., Oliveira, C. J. B. and Givisiez., P. E. N. 2013. Protective effect of mannan oligosaccharides against early colonization by *Salmonella* Enteritidis in chicks is improved by higher dietary threonine levels. *J. Appl. Microbiol.* 114(4), 1158–65. doi:10.1111/jam.12108.

Santovito, E., Greco, D., Logrieco, A. F. and Avantaggiato, G. 2018. Eubiotics for food security at farm level: yeast cell wall products and their antimicrobial potential against pathogenic bacteria. *Foodborne Pathog. Dis.* 15(9), 531–7. doi:10.1089/fpd.2018.2430.

Savory, C. J. and Knox, A. I. 1991. Chemical composition of caecal contents in the fowl in realtion to dietary fibre level and time of day. *Comp. Biochem. Physiol.* 100A, 739–43.

Sekelja, M., Rud, I., Knutsen, S. H., Denstadl, V., Westereng, B., Naes, T. and Rudi, K. 2012. Abrupt temporal fluctuations in the chicken fecal microbiota are explained by its gastrointestinal origin. *Appl. Environ. Microbiol.* 78(8), 2941–8. doi:10.1128/AEM.05391-11.

Sergeant, M. J., Constantinidou, C., Cogan, T. A., Bedford, M. R., Penn, C. W. and Pallen, M. J. 2014. Extensive microbial and functional diversity within the chicken cecal microbiome. *PLoS ONE* 9(3), e91941. doi:10.1371/journal.pone.0091941.

Sevane, N., Bialade, F., Velasco, S., Rebolé, A., Rodríguez, M. L., Ortiz, L. T., Cañón, J. and Dunner, S. 2014. Dietary inulin supplementation modifies significantly the liver transcriptomic profile of broiler chickens. *PLoS ONE* 9(6), e98942. doi:10.1371/journal.pone.0098942.

Shermer, C. L., Maciorowski, K. G., Bailey, C. A., Byers, F. M. and Ricke, S. C. 1998. Caecal metabolites and microbial populations in chickens consuming diets containing a mined humate compound. *J. Sci. Food Agric.* 77(4), 479–86. doi:10.1002/(SICI)1097-0010(199808)77:4<479::AID-JSFA607>3.0.CO;2-L.

Sibbald, I. R. 1979. Passage of feed through the adult rooster. *Poult. Sci.* 58(2), 446–59. doi:10.3382/ps.0580446.

Sibbald, I. R. 1980. The passage of oat and other feed residues through the adult cockerel. *Poult. Sci.* 59(9), 2136–44. doi:10.3382/ps.0592136.

Singer, R. S. and Hofacre, C. L. 2006. Potential impacts of antibiotic use in poultry production. *Avian Dis.* 50(2), 161–72. doi:10.1637/7569-033106R.1.

Siragusa, G. R. and Ricke, S. C. 2012. Probiotics as pathogen control agents for organic meat production. In: Ricke, S. C., Van Loo, E. J., Johnson, M. G. and O'Bryan, C. A. (Eds), *Organic Meat Production and Processing*. Wiley Scientific/IFT, New York, NY, pp. 331–49. Chapter 20.

Smith, C. J. and Bryant, M. P. 1979. Introduction to metabolic activities of intestinal bacteria. *Am. J. Clin. Nutr.* 32(1), 149–57. doi:10.1093/ajcn/32.1.149.

Soedarmo, D., Kare, M. R. and Wasserman, R. H. 1961. Observations on the removal of sugar from the mouth and the crop of the chicken. *Poult. Sci.* 40(1), 123–8. doi:10.3382/ps.0400123.

Sofos, J. N., Flick, G., Nychas, G.-J., O'Bryan, C. A., Ricke, S. C. and Crandall, P. G. 2013. Meat, poultry, and seafood. In: Doyle, M. P. and Buchanan, R. L. (Eds), *Food Microbiology-Fundamentals and Frontiers* (4th edn.). American Society for Microbiology, Washington DC, pp. 111–67. Chapter 6.

Solis de los Santos, F. S., Dnoghue, A. M., Farnell, M. B., Huff, G. R., Huff, W. E. and Donoghue, D. J. 2007. Gastrintestinal maturation is accelerated in turkey poults supplemented with a mannan-oligosaccharide yeast extract (Alphamune). *Poult. Sci.* 86(5), 921–30. doi:10.1093/ps/86.5.921.

Spring, P., Wenk, C., Dawson, K. A. and Newman, K. E. 2000. The effects of dietary mannanoligosaccharides on cecal parameters and the concentrations of enteric bacteria in the ceca of *Salmonella*-challenged broiler chicks. *Poult. Sci.* 79(2), 205–11. doi:10.1093/ps/79.2.205.

Stanley, D., Geier, M. S., Hughes, R. J., Denman, S. E. and Moore, R. J. 2013. Highly variable microbiota development in the chicken gastrointestinal tract. *PLoS ONE* 8(12), e84290. doi:10.1371/journal.pone.0084290.

Stanley, D., Hughes, R. J. and Moore, R. J. 2014. Microbiota of the chicken gastrointestinal tract: influence on health, productivity and disease. *Appl. Microbiol. Biotechnol.* 98(10), 4301–10. doi:10.1007/s00253-014-5646-2.

Stanley, D., Geier, M. S., Chen, H., Hughes, R. J. and Moore, R. J. 2015. Comparison of fecal and cecal microbiotas reveals qualitative similarities but quantitative differences. *BMC Microbiol.* 15, 51 doi:10.1186/s12866-015-0388-6.

Sudo, S. Z. and Duke, G. E. 1980. Kinetics of absorption of volatile fatty acids from the ceca of domestic turkeys. *Comp. Biochem. Physiol.* 67A, 231–7.

Svihus, B., Choct, M. and Classen, H. L. 2013. Function and nutritional roles of the avian caeca: a review. *Worlds Poult. Sci. J.* 69(2), 249–64. doi:10.1017/S0043933913000287.

Swaggerty, C. L., Callaway, T. R., Kogut, M. H., Piva, A. and Grilli, E. 2019. Modulation of the immune response to improve health and reduce foodborne pathogens in poultry. *Microorganisms* 7(3), 65, doi:10.3390/microorganisms7030065.

Taylor, G. R. J. and Williams, C. M. 1998. Effects of probiotics and prebiotics on blood lipids. *Br. J. Nutr.* 80 (Suppl. 2), S225–30. doi:10.1017/S0007114500006073.

Teng, P. Y. and Kim, W. K. 2018. Review: roles of prebiotics in intestinal ecosystem of broilers. *Front. Vet. Sci.* 5, 245. doi:10.3389/fvets.2018.00245.

Umaraw, P., Prajapati, A., Verma, A. K., Pathak, V. and Singh, V. P. 2017. Control of *Campylobacter* in poultry industry from farm to poultry processing unit: a review. *Crit. Rev. Food Sci. Nutr.* 57(4), 659–65. doi:10.1080/10408398.2014.935847.

Van der Aar, P. J., Fahey Jr., G. C., Ricke, S. C., Allen, S. E. and Berger, L. L. 1983. Effects of dietary fibers on mineral status of chicks. *J. Nutr.* 113(3), 653–61. doi:10.1093/jn/113.3.653.

Van der Wielen, P. W. J. J., Keuzenkamp, D. A., Lipman, L. J. A., van Knapen, F. and Biesterveld, S. 2002. Spatial and temporal variation of the intestinal bacterial community in commercially raised broiler chickens during growth. *Microb. Ecol.* 44(3), 286–93. doi:10.1007/s00248-002-2015-y.

Vergara, P., Ferrando, C., Jiménez, M., Fernández, E. and Goñalons, E. 1989. Factors determining gastrointestinal transit time of several markers in the domestic fowl. *Q. J. Exp. Physiol.* 74(6), 867–74. doi:10.1113/expphysiol.1989.sp003357.

Wang, Y. 2009. Prebiotics: present and future science and technology. *Food Res. Int.* 42(1), 8–12. doi:10.1016/j.foodres.2008.09.001.

Wernicki, A., Nowaczek, A. and Urban-Chmiel, R. 2017. Bacteriophage therapy to combat bacterial infections in poultry. *Virol. J.* 14(1), 179. doi:10.1186/s12985-017-0849-7.

Wigley, P. 2015. Blurred lines: pathogens, commensals, and the healthy gut. *Front. Vet. Sci.* 2, 40. doi:10.3389/fvets.2015.00040.

Wooldridge, K. G. and Williams, P. H. 1993. Iron uptake mechanisms of pathogenic bacteria. *FEMS Microbiol. Rev.* 12(4), 325–48. doi:10.1111/j.1574-6976.1993.tb00026.x.

Xu, Z. R., Hu, C. H., Xia, M. S., Zhan, X. A. and Wang, M. Q. 2003. Effects of dietary fructooligosaccharide on digestive enzyme activities, intestinal microflora and morphology of male broilers. *Poult. Sci.* 82(6), 1030–6. doi:10.1093/ps/82.6.1030.

Yang, Y., Iji, P. A. and Choct, M. 2009. Dietary modulation of gut microflora in broiler chickens: a review of the role of six kinds of alternatives to in-feed antibiotics. *Worlds Poult. Sci. J.* 65(1), 97–114. doi:10.1017/S0043933909000087.

Yegani, M. and Korver, D. R. 2008. Factors affecting intestinal health in poultry. *Poult. Sci.* 87(10), 2052–63. doi:10.3382/ps.2008-00091.

Yeh, H. Y., Line, J. E. and Hinton Jr., A. 2019. Community-level physiological profiling for microbial community function in broiler ceca. *Curr. Microbiol.* 76(2), 173–7. doi:10.1007/s00284-018-1602-1.

Yeoman, C. J., Chia, N., Jeraldo, P., Sipos, M., Goldenfield, N. D. and White, B. A. 2012. The microbiome of the chicken gastrointestinal tract. *Anim. Health Res. Rev.* 13(1), 89–99. doi:10.1017/S1466252312000138.

Yeung, C. K., Glahn, R. E., Welch, R. M. and Miller, D. D. 2005. Prebiotics and iron bioavailability – is there a connection? *J. Food Sci.* 70(5), R88–92. doi:10.1111/j.1365-2621.2005.tb09984.x.

Yusrizal and Chen, T. C. 2003. Effect of adding chicory fructans in feed on fecal and intestinal microflora and excreta volatile ammonia. *Int. J. Poult. Sci.* 2, 188–94.

Zhu, X. Y. and Joerger, R. D. 2003. Composition of microbiota in content and mucus from cecae of broiler chickens as measured by fluorescent in situ hybridization with group-specific, 16S rRNA-targeted oligonucleotide probes. *Poult. Sci.* 82(8), 1242–9. doi:10.1093/ps/82.8.1242.

Zhu, X. Y., Zhong, T., Pandya, Y. and Joerger, R. D. 2002. 16S rRNA-based analysis of microbiota from the cecum of broiler chickens. *Appl. Environ. Microbiol.* 68(1), 124–37. doi:10.1128/aem.68.1.124-137.2002.

Chapter 16

The role of synbiotics in optimizing gut function in poultry

Guillermo Tellez and Juan D. Latorre, University of Arkansas, USA; Margarita A. Arreguin-Nava, Eco-Bio LLC, USA; and Billy M. Hargis, University of Arkansas, USA

1 Introduction

There is a tendency to regard all microorganisms as harmful; to equate bacteria with germs. Nothing could be further from the truth. The number of non-pathogenic species far exceeds the number of pathogenic species, and many of the known bacteria are useful, even essential for the continued existence of life on earth (Fraune and Bosch, 2010). One example is microorganisms which inhabit the gastrointestinal tract (GIT) of animals. The GIT harbours an incredibly complex and abundant ensemble of microbes. The intestine is in contact with components of this microflora from birth, yet little is known about their influence on healthy development and physiology. The GIT is more densely populated with microorganisms than any other organ and is an interface where the microflora may have a pronounced impact on animal biology (López-Garcia et al., 2017). Throughout millions of years of evolution, animals have developed the means for supporting complex and dynamic consortia of microorganisms during their life cycle (Wren, 2000). An excellent view of vertebrate biology, therefore, requires an understanding of the contributions of these indigenous microbial communities to host development and adult physiology (McFall-Ngai, 2001). The fragile composition of the gut microflora can be affected by various factors such as age, diet, environment, stress and medication (Xu et al., 2007). As with most complex ecosystems, it appears that most species cannot be cultured

http://dx.doi.org/10.19103/AS.2019.0059.19

when removed from their niches. Biodiversity awaits systematic application of molecular enumeration techniques, such as genotyping DNA or encoding 16S rRNA genes. Colonization begins at birth and is followed by mounting assembly of a complex and dynamic microbial society (Kikuchi et al., 2009). Assembly is presumably regulated by elaborate and combinatorial microbial-microbial and host-microbial interactions predicated on principles refined throughout animal evolution (Tellez, 2014). Comparisons of rodents raised without exposure to any microorganisms to animals that have assembled a microbiota since birth, or those that have been colonized with components of the microbiota during or after completion of postnatal development, have revealed a range of host functions affected by indigenous microbial communities (Blaser, 2006). For example, the microbiota directs the assembly of the gut-associated lymphoid tissue, helps educate the immune system, affects the integrity of the intestinal mucosal barrier, modulates proliferation and differentiation of its epithelial lineages, regulates angiogenesis, modifies the activity of the enteric nervous system and plays a crucial role in extracting and processing nutrients consumed in the diet (Kau et al., 2011; Hadrich, 2018). The microflora can metabolize proteins and protein degradation products, sulphur-containing compounds, and endogenous and exogenous glycoproteins (O'Hara and Shanahan, 2006). Some organisms grow on intermediate products of fermentation such as H_2, lactate, succinate, formate and ethanol and convert these to end products including short-chain fatty acids (SCFA), a process which has a direct impact on digestive physiology (Van Der Wielen et al., 2000). These and other mechanisms of bacteria biology remain virtually unknown. Researchers in this area are focusing on elucidating these mechanisms as well as manipulating the bacteria and the gastrointestinal environment towards achieving optimal health a number of foods or food components, that provide beneficial roles (for growth and health) beyond ordinary nutrition, leading to the development of the concept of nutraceuticals (Subbiah, 2007). In general, nutraceuticals can be defined as food or food components that have a role in modifying and maintaining normal physiological functions that support the healthy host (Sugiharto, 2016). These nutraceuticals also help in protecting the host against infectious diseases (Hailu et al., 2009). Nutraceuticals may range from isolated nutrients (vitamin, mineral, amino acids, fatty acids), herbal products (polyphenols, herbs, spices), dietary supplements (probiotics, prebiotics, synbiotics, organic acids, antioxidants, enzymes) to genetically modified foods. In this chapter, we will only focus on probiotics, prebiotics and synbiotics.

The use of lactic acid bacteria (LAB) as feed supplements goes back to pre-Christian times when humans consumed fermented milk. It was not until the last century that Eli Metchnikoff, working at the Pasteur Institute in Paris, evaluated the subject from a scientific basis. Metchnikoff documented a direct link between human longevity and the necessity of maintaining a healthy balance of the beneficial and pathological microorganisms residing in the human gut. Metchnikoff was awarded the Nobel Prize in Physiology in 1908 for his discovery of phagocytes and other immune system components, but his accurate description of vital elements in the body's intestinal flora is equally notable (Metchnikoff, 1907). He developed and prescribed to his patients bacteriotherapy, that is the use of LAB in dietary regimens. In support of this, he cited the observation that Bulgarian peasants consumed large quantities of soured milk and lived long lives (Metchnikoff, 1907). He did not doubt the causal relationship, and subsequent events have, in part, confirmed his thesis. He isolated what he called the 'Bulgarian bacillus' from soured milk and used this in subsequent trials. This organism was probably what became known as *Lactobacillus bulgaricus* and is now called *L. delbrueckii* subsp. *bulgaricus* which is one of the organisms used to ferment milk and produce yogurt. After Metchnikoff's death in 1916 the centre of activity moved to the United States. In the late 1940s interest in the gut microflora was stimulated by two research developments. First, the finding that antibiotics included in the feed of farm animals promoted their growth (Dhama et al., 2014). A desire to discover the mechanism of this effect led to increased study on the composition of the gut microflora and the way in which it might be affecting the host animal. Secondly, the increased ready availability of germ-free animals provided a technique for testing the effect that the newly discovered intestinal inhabitants were having on the host (Tlaskalová-Hogenová et al., 2011). This increased knowledge also showed that *L. acidophilus* was not the only *Lactobacillus* in the intestine and a wide range of different organisms came to be studied and later used in probiotic preparations.

2 Probiotics

Probiotics are a single or mixed culture of living microorganisms which when administrated in adequate numbers exerts health benefits for the host by improving the host intestinal microbial balance, enhancing of colonization resistance against pathogens and improving the immune responses (Schrezenmeir and de Vrese, 2001). The species of microorganisms currently being used in probiotic preparations are varied, and LAB, that is *Lactobacillus* spp., *Streptococcus thermophilus, Enterococcus faecium, Enterococcus faecalis* and *Bifidobacterium* spp., are the most common type of bacteria used as probiotics. The definite mechanism through which probiotics may improve the defence and performance of chickens remains unclear, but some possible modes of action have been proposed: (1) maintaining a healthy balance of bacteria in the gut by competitive exclusion (the process by which beneficial bacteria exclude potential pathogenic bacteria through competition for attachment site in the intestine and nutrients) and antagonism (inhibit the growth of pathogenic bacteria by producing, for example lactic acids); (2) promoting gut maturation and integrity; (3) modulating the immune system and preventing inflammation; (4) improving the metabolism by increasing digestive enzyme activity and decreasing bacterial enzyme activity and ammonia production; (5) improving feed intake and digestion (as a result from the improved microbial balance in the gut); and (6) neutralizing enterotoxins and stimulating the immune system (Yurong et al., 2005; Howarth and Wang, 2013). Many probiotic effects are mediated through immune regulation, mainly through balance control of pro-inflammatory and anti-inflammatory cytokines (Vanderpool et al., 2008). However, other studies have shown that some probiotics also exert antioxidant properties and enhance barrier integrity (Prado-Rebolledo et al., 2017). Also, several investigators have demonstrated the benefits of probiotics on innate immunity (Molinaro et al., 2012) as well as on humoral immunity (Howarth and Wang, 2013).

Our laboratory has worked to identify probiotic candidates for use in poultry. FloraMax-B11® is a defined LAB-based probiotic that was confirmed to increase the resistance of poultry to *Salmonella* spp. infections (Farnell et al., 2006; Higgins et al., 2007). Extensive laboratory and field research conducted with this defined LAB culture has demonstrated accelerated development of healthy microflora in chickens and turkeys, providing increased resistance to *Salmonella* sp. infections (Vicente et al., 2007; Higgins et al., 2007; Menconi et al., 2011). Published experimental and commercial studies have shown that these selected probiotic organisms can reduce idiopathic diarrhoea in commercial turkey brooding houses (Higgins et al., 2005). Large-scale commercial trials have indicated that proper administration of this probiotic mixture to turkeys and chickens increased performance and reduced costs of production (Torres-Rodriguez et al., 2007; Vicente et al., 2007). More recently, microarray analysis of gut mRNA expression showed differences in birds treated with this probiotic in genes associated with the NFκB complex (Higgins et al., 2011). These data have demonstrated that the selection of therapeutically efficacious probiotic cultures with marked performance benefits in poultry is possible, and that defined cultures can sometimes provide an attractive alternative to conventional antimicrobial therapy.

2.1 A **Bacillus** *spore-based probiotic for* **Salmonella** *control and performance enhancement in poultry*

Despite the success shown by the development of the LAB probiotic for use in commercial poultry (above), there is still an urgent need for commercial probiotics that are shelf-stable, cost-effective and feed-stable (tolerance to heat pelletization process) to increase compliance and widespread utilization. Among a large number of probiotic products in use today some are bacterial spore formers, mostly of the genus *Bacillus*. Used primarily in their spore form, some (though not all) have been shown to prevent selected gastrointestinal disorders and the diversity of species used, and their applications are astonishing. While not all *Bacillus* spores are highly heat tolerant, some specific isolates are the most robust life form known on earth and can be used under extreme heat conditions (Vreeland et al., 2000). At present, our laboratory's aim is to develop a novel, cost-effective, feed-stable probiotic with widespread utilization and improve probiotic production, delivery and clinical efficacy for human and animal use. We have demonstrated that one *Bacillus subtilis* spore isolate was as effective as FloraMax-B11® for *Salmonella* reduction (Shivaramaiah et al., 2011; Wolfenden et al., 2011). Other isolates or combinations of isolates with increased potency and efficacy may be identified with continued research. Some of these environmental *Bacillus* isolates have been evaluated *in vitro* for antimicrobial activity against selected bacterial pathogens, heat stability and the ability to grow to high numbers. Unpublished experimental evaluations have confirmed improved body weight gain as well as *Salmonella* sp. or *Clostridium perfringens* reduction in commercial turkey and broiler operations when compared with medicated (nitarsone) or control non-medicated diets, respectively. Our preliminary data suggest that these isolates could be an effective alternative to antibiotic growth promoters for commercial poultry. Importantly, improved efficiency of amplification and sporulation is essential to gain widespread industry acceptance of a feed-based probiotic for ante-mortem food-borne pathogen intervention, as well as cost-effectiveness. Recently, both vegetative growth and sporulation rate have been optimized in our laboratory, which may lead to new efficiencies for commercial amplification and manufacture of a cost-effective product at very high spore counts (Wolfenden et al., 2010).

3 Prebiotics

The concept of prebiotics is relatively new; it developed in response to the notion that non-digestible food ingredients (e.g. non-digestible oligosaccharides) are selectively fermented by one or more bacteria known to have positive effects on gut physiology (Schrezenmeir and de Vrese, 2001). Bacteria fed by a preferential food substrate have a proliferative advantage over other bacteria. Some prebiotics have shown to selectively stimulate the growth of endogenous LAB and bifidobacteria in the gut to improve the health of the host (Pourabedin and Zhao, 2015).

Prebiotics may provide energy for the growth of endogenous favourable bacteria in the gut, such as bifidobacteria and lactobacilli, thus improving the host microbial balance. In this notion, prebiotics may have more benefits compared with probiotics, in that prebiotics stimulate the bacteria (commensal bacteria) which have adapted to the environment of the GIT (Liu et al., 2015). Prebiotics have been reported to enhance the host defence and reduce mortality of bird caused by the invasion of gut pathogens (Ducatelle et al., 2015). The mechanism by which prebiotics exert this feature remains less elucidated, but it is likely that the capacity of prebiotics to increase the number of LAB in the gut may aid the competitive exclusion of pathogens from the GIT of birds (Pourabedin and Zhao, 2015). The increased production of SCFAs with the administration of prebiotics resulting in increased intestinal acidity may also contribute to the suppression of pathogens in the gut of chicken. Prebiotics have also been reported to enhance the immune response of chicken, resulting in rapid clearance of pathogens from the gut (Ajuwon, 2016).

With regard to the immune-enhancing effect of prebiotics, this may in part be due to direct interaction between prebiotics and gut immune cells as well as due to an indirect action of prebiotics via preferential colonization of beneficial microbes and microbial products that interact with immune cells (Collins and Gibson, 1999). Overall, prebiotics may have a similar mechanism as probiotics in supporting the gut health of chicken. The most common prebiotics used in poultry are oligosaccharides, including inulin, fructooligosaccharides (FOS), mannan-oligosaccharides (MOS), galactooligosaccharides (GOS), soya oligosaccharides (SOS), xylo-oligosaccharides (XOS), pyrodextrins, isomaltooligosaccharides (IMO) and lactulose (Molinaro et al., 2012; Pandey et al., 2015).

Prebiotic research on poultry has been performed since 1990 and, as a result, an extensive database of research is accessible in this area. Prebiotics in broiler diets have been shown to increase lactobacilli counts in the GIT. Also, increased bifidobacteria and decreased clostridia have been reported in some studies that investigated the microbial effects of prebiotic supplementation (Van den Broek et al., 2008). Some authors reported decreased *Salmonella* and coliforms (Dhama et al., 2008; Janssens et al., 2004). Some other pathogenic bacteria like streptococci, staphylococci, bacilli and yeast have also been reported to decrease with prebiotic supplementation (Parracho et al., 2007). Regarding intestinal morphology, increased intestinal villus height was reported when prebiotics were included in the broiler diet. Other changes of intestinal characteristics have been observed, including increased gut length (Hamilton-Miller, 2004).

Bacteria fed by a preferential food substrate have a proliferative advantage over other bacteria. Prebiotics selectively modify the colonic microflora and can potentially influence gut metabolism (Hedin et al., 2007). The presence of healthy gut microflora may improve the metabolism of host birds in various ways, including absorptive capacity, protein metabolism, energy metabolism and fibre digestion and gut maturation (Everard et al., 2011). A healthy population of these beneficial bacteria in the digestive tract enhances the digestion and absorption of nutrients, detoxification and elimination processes, and helps boost the immune system (Teitelbaum and Walker, 2002). Some studies have shown that prebiotics enhance the performance of egg-laying birds and positively affect mineral utilization and improve eggshell and bone quality.

3.1 Prebiotic properties of Aspergillus niger to control food-borne pathogens improve performance and bone mineralization in poultry

The commercially available mycelium product of *Aspergillus niger*, Fermacto®, referred to as *Aspergillus* meal (AM), has no live cells or spores and is proven to enhance the digestive efficiency of the GI tract. AM contains 16% protein and 45% fibre, and may be used with low levels of protein and amino acid diets to improve performance in commercial poultry (Harms and Miles, 1988; Torres-Rodriguez et al., 2005). Even though the exact mechanisms of action for prebiotics have not been defined, it may be speculated that the effect is due to changing intestinal flora that promote the growth of beneficial bacteria. This product has also been shown to benefit poultry through stimulation of growth, most probably by increasing absorption of feed ingredients and improving digestibility. Additionally, *Aspergillus* fibre contains beta-glucans, FOS, chitosan and mannan-oligosaccharides (MOS) (Uchima et al., 2011; Hernandez-Patlan et al., 2018). Beta-glucan is a powerful immune-enhancing nutritional supplement (Jonker et al., 2010). This unique compound affects the intestinal villi and primes the innate immune system to help the body defend itself against viral and bacterial invaders (Teitelbaum and Walker, 2002; Hooge et al., 2003). MOS protects the GIT from invading toxins by binding the active toxin sites. FOS and chitosan refer to a class of host non-digestible carbohydrates that are readily fermented by the beneficial bacteria in the intestine (Kim et al., 2006). A healthy population of these beneficial bacteria in the digestive tract enhances the digestion and absorption of nutrients, detoxification and elimination processes, and helps boost the immune system.

In previous work, we have shown that dietary AM induces significant changes on intestinal morphometry in turkey poults. Increased number of acid mucin cells in the duodenum, neutral mucin cells in the ileum and sulphomucin cells in the duodenum and ileum, as well as increased villi height and villi surface area of both duodenum and ileum when compared to control, suggest that AM prebiotic has an impact on the mucosal architecture and goblet cell proliferation in the duodenum and ileum of neonate poults (Tellez et al., 2010). In another study, dietary AM prebiotic supplemented for 30 days, significantly increased the body weight of neonate poults and improved feed conversion when compared with poults that received the basal control diet. Interestingly, energy and protein content in the ileum was significantly lower in poults that received dietary AM prebiotic compared with control poults, suggesting better digestibility and absorption of those nutrients, which are in agreement with the morphometric changes observed previously (Reginatto et al., 2011). Furthermore, we also observed significant increases in tibia weight, diameter, breaking strength, ash, calcium and phosphorus in poults that received dietary AM when compared with neonatal poults that received the basal control diet (Reginatto et al., 2011). FOS have been shown to stimulate calcium (Ca) and magnesium (Mg) absorption in the intestine and increase bone mineral concentrations (Scholz-Ahrens et al., 2007).

Several studies have demonstrated that feeding probiotics can achieve prevention of *Salmonella* colonization in chickens (Janssens et al., 2004; Van Immerseel et al., 2002; Burkholder et al., 2008). Finally, chitosan is a modified, natural biopolymer derived by deacetylation of chitin, the main component of the cell walls of fungi and exoskeletons of arthropods. As mentioned before, chitosan exhibits numerous beneficial effects, including strong anti-microbial and antioxidative activities (Filipkowska et al., 2014). Its application in agriculture, horticulture, environmental science, industry, microbiology and medicine are well reported (Hernandez-Patlan et al., 2018). There have been numerous studies that report the use of chitosan as a mucosal adjuvant, by enhancing IgA levels (Ravi Kumar, 2000). The commercial prebiotic supplement derived from *Aspergillus niger* mycelium is unique because it contains all of the above mentioned prebiotic ingredients.

In another study conducted in our laboratory, we evaluated the effect of 0.2% dietary AM against horizontal transmission of *Salmonella* spp. in turkeys and chickens (Londero et al., 2011). The results of this study showed that dietary supplementation with 0.2% AM was able to reduce *Salmonella* Enteritidis horizontal transmission in turkeys, and *Salmonella* Typhimurium horizontal transmission in broiler chickens, by reducing the overall colonization levels in birds. Although the mechanism of action is not entirely understood, the reduction in *Salmonella* colonization may be related to a synergistic effect between beta-glucan, MOS, chitosan and FOS present in the *Aspergillus niger* mycelium. The GIT serves as the interface between diet and the metabolic events that sustain life. Intestinal villi, which play a crucial role in digestion and absorption of nutrients, are underdeveloped at hatch and maximum absorption capacity is attained by 10 days of age. Understanding and optimizing the maturation and development of the intestine in poultry will improve feed efficiency, growth and overall health of the bird (Pourabedin and Zhao, 2015).

4 Synbiotics

Both probiotics and prebiotics have been shown to provide beneficial effects on the gut of birds. When probiotics and prebiotics are combined, they form synbiotics (Schrezenmeir and de Vrese, 2001). This combination could improve the survival and persistence of the health-promoting organism in the gut of birds because its specific substrate is available for fermentation. Several studies have shown the potential benefits of synbiotics on the intestinal microbial ecosystem and immune functions of chicken (Pandey et al., 2015; Téllez et al., 2015).

Regarding performance, Synbiotics have have been found to be more effective than prebiotics in improving growth of broilers when administered *in ovo* (Madej et al., 2015). The improvement of intestinal morphology and nutrient absorption due to feeding synbiotics seems to contribute to the enhanced performance of broiler chicken (Dunislawska et al., 2017). It has been suggested that synbiotics are much more efficient when used in combination than singly (Scholz-Ahrens et al., 2007). The balance of healthy gut microflora may improve the metabolism of the host bird in various ways, including absorptive capacity, protein metabolism, energy metabolism and fibre digestion, energy conversion and gut maturation (Sugiharto, 2016). The consumption of synbiotics thus contributes to immunostimulation and a beneficial balance of microbiota in the gut (Awad et al., 2009).

Probiotic numbers have been enhanced by prebiotics that selectively stimulate the growth and activity of one or a limited number of bacterial species already resident in the large intestine, and, thus, improves host health. In this way, prebiotics selectively modify the colonic microflora and can potentially influence gut metabolism. However, the bacterial nutrient package will not be advantageous without the presence of the targeted, beneficial bacteria and likewise, the live microbial product will not succeed if the environment into which it is introduced is unfavourable (Liu et al., 2015; Dhama et al., 2008).

4.1 Role of synbiotics in digestive physiology: SCFA production

SCFA increases from undetectable levels in the caeca of day-of-hatch chicks to the highest concentration at day 15 of age as the enteric microflora become established (Yang et al., 2011). The primary fermentative reaction in the human colon or chicken caecum is similar to that in obligate herbivores: hydrolysis of polysaccharides, oligosaccharides and disaccharides to their constituent sugars, which are then fermented resulting in increased biomass. Carbohydrate hydrolysis is promoted by hydrolases secreted by bacterial cells that are able to digest a range of carbohydrates which monogastric animals would not otherwise be able to digest (Xu et al., 2003). Fermentation yields metabolizable energy for microbial growth and maintenance and also metabolic end products. Nitrogen for protein synthesis can come from either urea, undigested dietary protein or endogenous secretions (Van den Broek et al., 2008). The principal products are SCFA together with gases (CO_2, CH_4

and H_2) and some heat. Carbohydrates entering the large intestine can alter gut physiology in two ways: physical presence and fermentation. Effects of SCFA can be divided into those occurring in the lumen and those arising from their uptake and metabolism by the cells of the large bowel wall. SCFA are the principal luminal anions. They are relatively weak acids with pKa values of 4.8, and raising their concentrations through fermentation lowers digesta pH. SCFA also serves as an indispensable source of energy for the gut wall, providing up to 50% of the daily energy requirements of colonocytes (Dimitrov, 2011). Fermentable carbohydrates can alter the microbial ecology significantly by acting as substrates or supplying SCFA. Much attention has been directed towards the study of specific beneficial LAB, rather than the flora as a whole. However, the SCFA have diverse functions with regard to host and microbial physiology (Tellez et al., 2006).

4.1.1 Blood flow and muscular activity

In vitro studies have shown that incubation with SCFA at concentrations as low as 3 mM dilate precontracted colonic resistance arterioles in separate human colonic segments (Molinaro et al., 2012). Greater colonic blood flow has been observed with the infusion of acetate, propionate or butyrate (Plöger et al., 2012). The mechanism of action of SCFA on blood flow does not involve either prostaglandins or α- or β-adrenoreceptor-linked pathways. The mechanisms of action may involve local neural networks as well as chemoreceptors together with direct effects on smooth muscle cells (Van Der Wielen et al., 2000). SCFA produced in the colon and entering the portal circulation seem to influence the upper gut musculature. These actions are essential for the maintenance of the function of the whole gastrointestinal system, not just the colon. It is expected that greater blood flow enhances tissue oxygenation and transport of absorbed nutrients (Braniste et al., 2014).

4.1.2 Enterocyte proliferation

In rats, SCFA stimulates the growth of colorectal and ileal mucosal cells when they are delivered colorectal or intraperitoneally. In addition to promoting growth, the major SCFA (especially butyrate) appears to lower the risk of malignant transformation in the colon (Shen et al., 2013). Secondary bile acids are cytotoxic, and in rats fed deoxycholate plus cholesterol, cell proliferation as measured by incorporation of [3*H*]thymidine was increased (Begley et al., 2005). Some of the effects of SCFA may be due to low intra-colorectal pH rather than any specific SCFA. At pH 6, bile acids are mostly protonated and insoluble and so would not be taken up by colonocytes. Additionally, lower pH inhibits the bacterial conversion of primary to secondary bile acids and therefore lowers their carcinogenic potential (Hofmann, 1999).

4.1.3 Mucin production

Evidence has been presented that mucus production and release is stimulated locally by endogenous production of SCFA by gut microflora. Additionally, some studies have been completed evaluating the influence of specific beneficial or probiotic organisms on mucin production (Plöger et al., 2012). *In vitro* studies with *Lactobacillus plantarum* 299v suggest that the ability of organisms to inhibit adherence of attaching and effacing organisms to intestinal epithelial cells is mediated through their ability to increase expression of MUC2 and MUC3 intestinal mucins (Montagne et al., 2004). The benefits of probiotics mediated through intestinal mucin upregulation may have broader applicability than enteropathogen intervention in poultry. Several investigators have shown that the increase in mucin production following probiotic administration inhibited replication, disease symptoms and shedding of rotavirus. In the proximal colon, an increase in the butyrate concentration altered crypt depth and the number of mucus-containing cells; the increase in butyrate was highly correlated with the number of neutral mucin-containing cells (Van Immerseel et al., 2002; Schippa and Conte, 2014).

4.2 Role of synbiotics in poultry production

The GIT serves as the interface between diet and the metabolic events that sustain life. In poultry, intestinal villi, which play a crucial role in digestion and absorption of nutrients, are underdeveloped at hatch and maximum absorption capacity is attained by 10 days of age. Understanding and optimizing the maturation and development of the intestine in poultry will improve feed efficiency, growth and overall health of the bird. In the immediate post-hatch period birds must undergo the transition from energy supplied by the endogenous nutrients of the yolk to exogenous carbohydrate-rich feed (Awad et al., 2009). During that critical time, dramatic changes occur both in the intestinal size and morphology. Maturational changes also affect the epithelial cell membranes, a primary mechanical interface between the internal environment of the host and the luminal contents. Studies on nutrition and metabolism during the early phase of growth in chicks may help in optimizing nutritional management for maximum growth. By dietary means, it is possible to affect the development of the gut and the competitiveness of both beneficial and harmful bacteria, which can alter not only gut dynamics but also many physiological processes due to the end products metabolized by symbiotic gut microflora (Maiorano et al., 2012). Additives such as probiotics and prebiotics are now extensively used throughout the world. The chemical nature of these additives are well understood, but the manner by which they benefit the animal is not (Sugiharto, 2016).

4.3 Synbiotics as an alternative to antibiotics for control of bacterial pathogens and improved performance in poultry

Bacterial antimicrobial resistance in both the medical and agricultural fields has become a serious problem worldwide. Antibiotic-resistant strains of bacteria are an increasing threat to animal and human health, with resistance mechanisms having been identified and described for all known antimicrobials currently available for clinical use (Kiser, 1976). There is currently increased public and scientific interest regarding the administration of therapeutic and sub-therapeutic antimicrobials to animals, due primarily to the emergence and dissemination of multiple antibiotic-resistant zoonotic bacterial pathogens (Shea, 2003). Social pressures have led to the creation of regulations to restrict antibiotic use in poultry and livestock production (Niewold, 2007). There is a need to evaluate potential antibiotic alternatives to improve disease resistance in high-intensity food animal production. Nutritional approaches to counteract the debilitating effects of stress and infection may provide producers with useful alternatives to antibiotics (Joerger, 2003). Improving the disease resistance of animals grown without antibiotics will not only benefit the animals' health, welfare and production efficiency but is also a key strategy in the effort to improve the microbiological safety of poultry products. Most of the experiments conducted with pro-, pre- and synbiotics have focused on improving the microbial health, performance and decreasing carcass contamination of young meat birds (Teillant and Laxminarayan, 2015; Ajuwon, 2016).

5 Conclusion and future trends

Overall in this chapter pro-, pre- and synbiotics have been discussed concerning the systemic effects they exert on the host's health, metabolism and immune system. Probiotics and synbiotics have systemic effects on the host's healthy metabolism and immune system. Utilization of prebiotics by probiotics should be a prerequisite for symbiotic selection, in order to maintain a good synergy between the two and maximize the beneficial effects. By establishing the underlying mechanisms of probiotics, prebiotics and their combination (synbiotics), scientists would be able to design enhanced functional foods to improve host health. The ability to regulate the composition of the microbiota by symbiotic products is an exciting approach in the control and treatment of some major diseases and to increase performance. The recent advances in technology have enabled the deep sequencing and analysis of the beautiful diversity of the microorganisms in the GIT, and should be able to prevent diseases and lead to maintain better health.

6 Where to look for further information

During the last decade, the increasing interest in renewable energy sources changed the distribution of corn utilization from human and animal consumption to biofuel production, leading to a continuous rise in feed costs of livestock diets. Therefore, alternative feed ingredients such as distillers dried grains with solubles (DDGS) as well as cereals like wheat, barley and sorghum have become part of the feed matrix to maintain or reduce production costs. However, these raw materials often contain a higher concentration of anti-nutritional factors in comparison to corn, including non-starch polysaccharides which increase digesta viscosity and reduce nutrient absorption in monogastric animals. As a result, the addition of exogenous enzymes in poultry feed has steadily increased to maximize nutrient utilization and maintain performance parameters with diets containing less digestible ingredients. On the other hand, the poultry industry is also facing social concerns regarding the use of antibiotic growth promoters and the development of antibiotic-resistant microorganisms. One alternative among others is the utilization of direct-fed microbials as substitutes for antibiotics growth promoters and also as a prophylactic practice to reduce the incidence of bacterial gastrointestinal diseases. Currently, our laboratories are also working on evaluating and selecting different *Bacillus* spp. strains as DFM candidates based on enzyme production profiles to improve nutrient absorption and intestinal integrity, as well as to maintain a healthy microflora balance in poultry-consuming commercial and alternative diets.

7 References

Ajuwon, K. M. 2016. Toward a better understanding of mechanisms of probiotics and prebiotics action in poultry species. *The Journal of Applied Poultry Research* 25(2), 277–83. doi:10.3382/japr/pfv074.

Awad, W. A., Ghareeb, K., Abdel-Raheem, S. and Böhm, J. 2009. Effects of dietary inclusion of probiotic and synbiotic on growth performance, organ weights, and intestinal histomorphology of broiler chickens. *Poultry Science* 88(1), 49–56. doi:10.3382/ps.2008-00244.

Begley, M., Gahan, C. G. and Hill, C. 2005. The interaction between bacteria and bile. *FEMS Microbiology Reviews* 29(4), 625–51. doi:10.1016/j.femsre.2004.09.003.

Blaser, M. J. 2006. Who are we? Indigenous microbes and the ecology of human diseases. *EMBO Reports* 7(10), 956–60. doi:10.1038/sj.embor.7400812.

Braniste, V., Al-Asmakh, M., Kowal, C., Anuar, F., Abbaspour, A., Tóth, M., Korecka, A., Bakocevic, N., Ng, L. G., Kundu, P., et al. 2014. The gut microbiota influences blood-brain barrier permeability in mice. *Science Translational Medicine* 6(263), 263ra158. doi:10.1126/scitranslmed.3009759.

Burkholder, K. M., Thompson, K. L., Einstein, M. E., Applegate, T. J. and Patterson, J. A. 2008. Influence of stressors on normal intestinal microbiota, intestinal morphology, and susceptibility to *Salmonella* Enteritidis colonization in broilers. *Poultry Science* 87(9), 1734–41. doi:10.3382/ps.2008-00107.

Collins, M. D. and Gibson, G. R. 1999. Probiotics, prebiotics, and synbiotics: approaches for modulating the microbial ecology of the gut. *The American Journal of Clinical Nutrition* 69(5), 1052S–7S. doi:10.1093/ajcn/69.5.1052s.

Dhama, K., Mahendran, M., Tomar, S., Chauhan, R. 2008. Beneficial effects of probiotics and prebiotics in livestock and poultry: the current perspectives. *Intas Polivet* 9, 1-12.

Dhama, K., Tiwari, R., Khan, R. U., Chakrabort, S., Gopi, M., Karthik, K., Saminathan, M., Desingu, P. A. and Sunkara, L. T. 2014. Growth promoters and novel feed additives improving poultry production and health, bioactive principles and beneficial applications: the trends and advances–a review. *International Journal of Pharmacology* 10, 129–59. doi:10.3923/ijp.2014.129.159.

Dimitrov, D. V. 2011. The human gutome: nutrigenomics of the host-microbiome interactions. *Omics: a Journal of Integrative Biology* 15(7–8), 419–30. doi:10.1089/omi.2010.0109.

Ducatelle, R., Eeckhaut, V., Haesebrouck, F. and Van Immerseel, F. 2015. A review on prebiotics and probiotics for the control of dysbiosis: present status and future perspectives. *Animal: an International Journal of Animal Bioscience* 9(1), 43–8. doi:10.1017/S1751731114002584.

Dunislawska, A., Slawinska, A., Stadnicka, K., Bednarczyk, M., Gulewicz, P., Jozefiak, D. and Siwek, M. 2017. Synbiotics for broiler chickens–*in vitro* design and evaluation of the influence on host and selected microbiota populations following *in ovo* delivery. *PLoS One* 12(1), e0168587. doi:10.1371/journal.pone.0168587.

Everard, A., Lazarevic, V., Derrien, M., Girard, M., Muccioli, G. G., Neyrinck, A. M., Possemiers, S., Van Holle, A., François, P., de Vos, W. M., et al. 2011. Responses of gut microbiota and glucose and lipid metabolism to prebiotics in genetic obese and diet-induced leptin-resistant mice. *Diabetes* 60(11), 2775–86. doi:10.2337/db11-0227.

Farnell, M. B., Donoghue, A. M., De Los Santos, F. S., Blore, P. J., Hargis, B. M., Tellez, G. and Donoghue, D. J. 2006. Upregulation of oxidative burst and degranulation in chicken heterophils stimulated with probiotic bacteria. *Poultry Science* 85(11), 1900–6. doi:10.1093/ps/85.11.1900.

Filipkowska, U., Jóźwiak, T. and Szymczyk, P. 2014. Application of cross-linked chitosan for phosphate removal from aqueous solutions. *Progress on the Chemistry and Application of Chitin and Its Derivatives* 19, 5–14.

Fraune, S. and Bosch, T. C. 2010. Why bacteria matter in animal development and evolution. *BioEssays: News and Reviews in Molecular, Cellular and Developmental Biology* 32(7), 571–80. doi:10.1002/bies.200900192.

Hadrich, D. 2018. Microbiome research is becoming the key to better understanding health and nutrition. *Frontiers in Genetics* 9, 212. doi:10.3389/fgene.2018.00212.

Hailu, G., Boecker, A., Henson, S. and Cranfield, J. 2009. Consumer valuation of functional foods and nutraceuticals in Canada. A conjoint study using probiotics. *Appetite* 52(2), 257–65. doi:10.1016/j.appet.2008.10.002.

Hamilton-Miller, J. M. 2004. Probiotics and prebiotics in the elderly. *Postgraduate Medical Journal* 80(946), 447–51. doi:10.1136/pgmj.2003.015339.

Harms, R. H. and Miles, R. D. 1988. Research note: influence of Fermacto on the performance of laying hens when fed diets with different levels of methionine. *Poultry Science* 67(5), 842–4. doi:10.3382/ps.0670842.

Hedin, C., Whelan, K. and Lindsay, J. O. 2007. Evidence for the use of probiotics and prebiotics in inflammatory bowel disease: a review of clinical trials. *The Proceedings of the Nutrition Society* 66(3), 307–15. doi:10.1017/S0029665107005563.

Hernandez-Patlan, D., Solis-Cruz, B., Hargis, B. M. and Tellez, G. 2018. Chitoneous materials for control of foodborne pathogens and mycotoxins in poultry. *Chitin-Chitosan-Myriad Functionalities in Science and Technology*. IntechOpen. doi:10.5772/intechopen.76041.

Higgins, S. E., Torres-Rodriguez, A., Vicente, J. L., Sartor, C. D., Pixley, C. M., Nava, G. M., Tellez, G., Barton, J. T. and Hargis, B. M. 2005. Evaluation of intervention strategies for idiopathic diarrhea in commercial turkey brooding houses. *The Journal of Applied Poultry Research* 14(2), 345–8. doi:10.1093/japr/14.2.345.

Higgins, J. P., Higgins, S. E., Vicente, J. L., Wolfenden, A. D., Tellez, G. and Hargis, B. M. 2007. Temporal effects of lactic acid bacteria probiotic culture on Salmonella in neonatal broilers. *Poultry Science* 86(8), 1662–6. doi:10.1093/ps/86.8.1662.

Higgins, S. E., Wolfenden, A. D., Tellez, G., Hargis, B. M. and Porter, T. E. 2011. Transcriptional profiling of cecal gene expression in probiotic-and *Salmonella*-challenged neonatal chicks. *Poultry Science* 90(4), 901–13. doi:10.3382/ps.2010-00907.

Hofmann, A. F. 1999. Bile acids: the good, the bad, and the ugly. *News in Physiological Sciences: an International Journal of Physiology Produced Jointly by the International Union of Physiological Sciences and the American Physiological Society* 14, 24–9. doi:10.1152/physiologyonline.1999.14.1.24.

Hooge, D. M., Sims, M. D., Sefton, A. E., Connolly, A. and Spring, P. 2003. Effect of dietary mannan oligosaccharide, with or without bacitracin or virginiamycin, on live performance of broiler chickens at relatively high stocking density on new litter. *The Journal of Applied Poultry Research* 12(4), 461–7. doi:10.1093/japr/12.4.461.

Howarth, G. S. and Wang, H. 2013. Role of endogenous microbiota, probiotics and their biological products in human health. *Nutrients* 5(1), 58–81. doi:10.3390/nu5010058.

Janssens, G. P., Millet, S., Van Immerseel, F., De Buck, J. and Hesta, M. 2004. The impact of prebiotics and salmonellosis on apparent nutrient digestibility and *Salmonella typhimurium* var. Copenhagen excretion in adult pigeons (*Columba livia domestica*). *Poultry Science* 83(11), 1884–90. doi:10.1093/ps/83.11.1884.

Joerger, R. D. 2003. Alternatives to antibiotics: bacteriocins, antimicrobial peptides and bacteriophages. *Poultry Science* 82(4), 640–7. doi:10.1093/ps/82.4.640.

Jonker, D., Kuper, C. F., Maquet, V., Nollevaux, G. and Gautier, S. 2010. Subchronic (13-week) oral toxicity study in rats with fungal chitin-glucan from Aspergillus niger. *Food and Chemical Toxicology: an International Journal Published for the British Industrial Biological Research Association* 48(10), 2695–701. doi:10.1016/j.fct.2010.06.042.

Kau, A. L., Ahern, P. P., Griffin, N. W., Goodman, A. L. and Gordon, J. I. 2011. Human nutrition, the gut microbiome and the immune system. *Nature* 474(7351), 327–36. doi:10.1038/nature10213.

Kikuchi, Y., Hosokawa, T., Nikoh, N., Meng, X. Y., Kamagata, Y. and Fukatsu, T. 2009. Host-symbiont co-speciation and reductive genome evolution in gut symbiotic bacteria of acanthosomatid stinkbugs. *BMC Biology* 7, 2. doi:10.1186/1741-7007-7-2.

Kim, W. K., Donalson, L. M., Mitchell, A. D., Kubena, L. F., Nisbet, D. J. and Ricke, S. C. 2006. Effects of alfalfa and fructooligosaccharide on molting parameters and bone qualities using dual energy X-ray absorptiometry and conventional bone assays. *Poultry Science* 85(1), 15–20. doi:10.1093/ps/85.1.15.

Kiser, J. S. 1976. A perspective on the use of antibiotics in animal feeds. *Journal of Animal Science* 42(4), 1058–72. doi:10.2527/jas1976.4241058x.

Liu, X., Cao, S. and Zhang, X. 2015. Modulation of gut microbiota-brain axis by probiotics, prebiotics, and diet. *Journal of Agricultural and Food Chemistry* 63(36), 7885–95. doi:10.1021/acs.jafc.5b02404.

Londero, A., Menconi, A., Reginatto, A. R., Bacocina, I., Wolfenden, A., Shivaramai, S., Hargis, B. M. and Tellez, G. 2011. Effect of an aspergillus meal prebiotic on *Salmonella* infection in turkeys and broiler chickens. *International Journal of Poultry Science* 10(12), 946–51. doi:10.3923/ijps.2011.946.951.

López-Garcia, P., Eme, L. and Moreira, D. 2017. Symbiosis in eukaryotic evolution. *Journal of Theoretical Biology* 434, 20–33. doi:10.1016/j.jtbi.2017.02.031.

Madej, J. P., Stefaniak, T. and Bednarczyk, M. 2015. Effect of *in ovo*-delivered prebiotics and synbiotics on lymphoid-organs' morphology in chickens. *Poultry Science* 94(6), 1209–19. doi:10.3382/ps/pev076.

Maiorano, G., Sobolewska, A., Cianciullo, D., Walasik, K., Elminowska-Wenda, G., Slawinska, A., Tavaniello, S., Zylinska, J., Bardowski, J. and Bednarczyk, M. 2012. Influence of in ovo prebiotic and synbiotic administration on meat quality of broiler chickens. *Poultry Science* 91(11), 2963–9. doi:10.3382/ps.2012-02208.

McFall-Ngai, M. J. 2001. Identifying 'prime suspects': symbioses and the evolution of multicellularity. *Comparative Biochemistry and Physiology. Part B, Biochemistry and Molecular Biology* 129(4), 711–23. doi:10.1016/S1096-4959(01)00406-7.

Menconi, A., Wolfenden, A. D., Shivaramaiah, S., Terraes, J. C., Urbano, T., Kuttel, J., Kremer, C., Hargis, B. M. and Tellez, G. 2011. Effect of lactic acid bacteria probiotic culture for the treatment of *Salmonella enterica* serovar Heidelberg in neonatal broiler chickens and turkey poults. *Poultry Science* 90(3), 561–5. doi:10.3382/ps.2010-01220.

Metchnikoff, E. 1907. Essais optimistes. Paris. The prolongation of life. Optimistic studies. Translated and edited by P. Chalmers Mitchell.

Molinaro, F., Paschetta, E., Cassader, M., Gambino, R. and Musso, G. 2012. Probiotics, prebiotics, energy balance, and obesity: mechanistic insights and therapeutic implications. *Gastroenterology Clinics of North America* 41(4), 843–54. doi:10.1016/j.gtc.2012.08.009.

Montagne, L., Piel, C. and Lalles, J. P. 2004. Effect of diet on mucin kinetics and composition: nutrition and health implications. *Nutrition Reviews* 62(3), 105–14. doi:10.1111/j.1753-4887.2004.tb00031.x.

Niewold, T. A. 2007. The nonantibiotic anti-inflammatory effect of antimicrobial growth promoters, the real mode of action? A hypothesis. *Poultry Science* 86(4), 605–9. doi:10.1093/ps/86.4.605.

O'Hara, A. M. and Shanahan, F. 2006. The gut flora as a forgotten organ. *EMBO Reports* 7(7), 688–93. doi:10.1038/sj.embor.7400731.

Pandey, K. R., Naik, S. R. and Vakil, B. V. 2015. Probiotics, prebiotics and synbiotics – a review. *Journal of Food Science and Technology* 52(12), 7577–87. doi:10.1007/s13197-015-1921-1.

Parracho, H., McCartney, A. L. and Gibson, G. R. 2007. Probiotics and prebiotics in infant nutrition. *The Proceedings of the Nutrition Society* 66(3), 405–11. doi:10.1017/S0029665107005678.

Plöger, S., Stumpff, F., Penner, G. B., Schulzke, J. D., Gäbel, G., Martens, H., Shen, Z., Günzel, D. and Aschenbach, J. R. 2012. Microbial butyrate and its role for barrier function in the gastrointestinal tract. *Annals of the New York Academy of Sciences* 1258, 52–9. doi:10.1111/j.1749-6632.2012.06553.x.

Pourabedin, M. and Zhao, X. 2015. Prebiotics and gut microbiota in chickens. *FEMS Microbiology Letters* 362(15), fnv122. doi:10.1093/femsle/fnv122.

Prado-Rebolledo, O. F., Delgado-Machuca, J. J., Macedo-Barragan, R. J., Garcia-Márquez, L. J., Morales-Barrera, J. E., Latorre, J. D., Hernandez-Velasco, X. and Tellez, G. 2017. Evaluation of a selected lactic acid bacteria-based probiotic on *Salmonella enterica* serovar Enteritidis colonization and intestinal permeability in broiler chickens. *Avian Pathology: Journal of the W.V.P.A.* 46(1), 90–4. doi:10.1080/03079457.2016.1222808.

Ravi Kumar, M. N. V. 2000. A review of chitin and chitosan applications. *Reactive and Functional Polymers* 46(1), 1-27. doi:10.1016/S1381-5148(00)00038-9.

Reginatto, A. R., Menconi, A., Londero, A., Lovato, M., Rosa, A. P., Shivaramaiah, S., Wolfenden, A. D., Huff, W. E., Huff, G. R., Rath, N. C., et al. 2011. Effects of dietary *Aspergillus* meal prebiotic on turkey poults production parameters and bone qualities. *International Journal of Poultry Science* 10(7), 496-9. doi:10.3923/ijps.2011.496.499.

Schippa, S. and Conte, M. P. 2014. Dysbiotic events in gut microbiota: impact on human health. *Nutrients* 6(12), 5786–805. doi:10.3390/nu6125786.

Scholz-Ahrens, K. E., Ade, P., Marten, B., Weber, P., Timm, W., Avarsigmail, Y., Glüer, C. C. and Schrezenmeir, J. 2007. Prebiotics, probiotics, and Synbiotics affect mineral absorption, bone mineral content, and bone structure. *The Journal of Nutrition* 137(3 Suppl. 2), 838S–46S. doi:10.1093/jn/137.3.838S.

Schrezenmeir, J. and de Vrese, M. 2001. Probiotics, prebiotics, and synbiotics – approaching a definition. *The American Journal of Clinical Nutrition* 73(2 Suppl.), 361S–4S. doi:10.1093/ajcn/73.2.361s.

Shea, K. M. 2003. Antibiotic resistance: what is the impact of agricultural uses of antibiotics on children's health? *Pediatrics* 112(1 Pt 2), 253–8.

Shen, J., Obin, M. S. and Zhao, L. 2013. The gut microbiota, obesity and insulin resistance. *Molecular Aspects of Medicine* 34(1), 39–58. doi:10.1016/j.mam.2012.11.001.

Shivaramaiah, S., Pumford, N. R., Morgan, M. J., Wolfenden, R. E., Wolfenden, A. D., Torres-Rodr, A., Hargis, B. M. and Téllez, G. 2011. Evaluation of *Bacillus* species as potential candidates for direct-fed microbials in commercial poultry. *Poultry Science* 90(7), 1574–80. doi:10.3382/ps.2010-00745.

Subbiah, M. T. 2007. Nutrigenetics and nutraceuticals: the next wave riding on personalized medicine. *Translational Research: the Journal of Laboratory and Clinical Medicine* 149(2), 55–61. doi:10.1016/j.trsl.2006.09.003.

Sugiharto, S. 2016. Role of nutraceuticals in gut health and growth performance of poultry. *Journal of the Saudi Society of Agricultural Sciences* 15(2), 99–111. doi:10.1016/j.jssas.2014.06.001

Teillant, A. and Laxminarayan, R. 2015. Economics of antibiotic use in U.S. swine and poultry production. *Choices* 30(1).

Teitelbaum, J. E. and Walker, W. A. 2002. Nutritional impact of pre-and probiotics as protective gastrointestinal organisms. *Annual Review of Nutrition* 22, 107–38. doi:10.1146/annurev.nutr.22.110901.145412.

Tellez, G. 2014. Prokaryotes versus eukaryotes: who is hosting whom? *Frontiers in Veterinary Science* 1, 3. doi:10.3389/fvets.2014.00003.

Tellez, G., Higgins, S. E., Donoghue, A. M. and Hargis, B. M. 2006. Digestive physiology and the role of microorganisms. *The Journal of Applied Poultry Research* 15(1), 136–44. doi:10.1093/japr/15.1.136.

Tellez, G., Nava, G. M., Vicente, J. L., De Franceschi, M., Morales, E. J., Prado, O., Terraes, J. C. and Hargis, B. M. 2010. Evaluation of dietary aspergillus meal on intestinal morphometry in turkey poults. *International Journal of Poultry Science* 9(9), 875–8. doi:10.3923/ijps.2010.875.878.

Téllez, G., Lauková, A., Latorre, J. D., Hernandez-Velasco, X., Hargis, B. M. and Callaway, T. 2015. Food-producing animals and their health in relation to human health. *Microbial Ecology in Health and Disease* 26(1), 25876. doi:10.3402/mehd.v26.25876.

Tlaskalová-Hogenová, H., Štěpánková, R., Kozáková, H., Hudcovic, T., Vannucci, L., Tučková, L., Rossmann, P., Hrnčíř, T., Kverka, M., Zákostelská, Z., et al. 2011. The role of gut microbiota (commensal bacteria) and the mucosal barrier in the pathogenesis of inflammatory and autoimmune diseases and cancer: contribution of germ-free and gnotobiotic animal models of human diseases. *Cellular and Molecular Immunology* 8(2), 110–20. doi:10.1038/cmi.2010.67.

Torres-Rodriguez, A., Sartor, C., Higgins, S. E., Wolfenden, A. D., Bielke, L. R., Pixley, C. M., Sutton, L., Tellez, G. and Hargis, B. M. 2005. Effect of aspergillus meal prebiotic (Fermacto) on performance of broiler chickens in the starter phase and fed low protein diets. *The Journal of Applied Poultry Research* 14(4), 665–9. doi:10.1093/japr/14.4.665.

Torres-Rodriguez, A., Higgins, S. E., Vicente, J. L. S., Wolfenden, A. D., Gaona-Ramirez, G., Barton, J. T., Tellez, G., Donoghue, A. M. and Hargis, B. M. 2007. Effect of lactose as a prebiotic on turkey body weight under commercial conditions. *The Journal of Applied Poultry Research* 16(4), 635–41. doi:10.3382/japr.2006-00127.

Uchima, C. A., Tokuda, G., Watanabe, H., Kitamoto, K. and Arioka, M. 2011. Heterologous expression and characterization of a glucose-stimulated β-glucosidase from the termite Neotermes koshunensis in *Aspergillus oryzae*. *Applied Microbiology and Biotechnology* 89(6), 1761–71. doi:10.1007/s00253-010-2963-y.

Van den Broek, L. A., Hinz, S. W., Beldman, G., Vincken, J.-P. and Voragen, A. G. 2008. Bifidobacterium carbohydrases-their role in breakdown and synthesis of (potential) prebiotics. *Molecular Nutrition and Food Research* 52, 146–63.

Van Der Wielen, P. W., Biesterveld, S., Notermans, S., Hofstra, H., Urlings, B. A. and van Knapen, F. 2000. Role of volatile fatty acids in development of the cecal microflora in broiler chickens during growth. *Applied and Environmental Microbiology* 66(6), 2536–40. doi:10.1128/AEM.66.6.2536-2540.2000.

Van Immerseel, F., De Buck, J., De Smet, I., Mast, J., Haesebrouck, F. and Ducatelle, R. 2002. Dynamics of immune cell infiltration in the caecal lamina propria of chickens after neonatal infection with a *Salmonella* Enteritidis strain. *Developmental and Comparative Immunology* 26(4), 355–64.

Vanderpool, C., Yan, F. and Polk, D. B. 2008. Mechanisms of probiotic action: implications for therapeutic applications in inflammatory bowel diseases. *Inflammatory Bowel Diseases* 14(11), 1585–96. doi:10.1002/ibd.20525.

Vicente, J., Wolfenden, A., Torres-Rodriguez, A., Higgins, S., Tellez, G. and Hargis, B. 2007. Effect of a *Lactobacillus* species-based probiotic and dietary lactose prebiotic on turkey poult performance with or without *Salmonella* Enteritidis challenge. *The Journal of Applied Poultry Research* 16(3), 361–4. doi:10.1093/japr/16.3.361.

Vreeland, R. H., Rosenzweig, W. D. and Powers, D. W. 2000. Isolation of a 250 million-year-old halotolerant bacterium from a primary salt crystal. *Nature* 407(6806), 897–900. doi:10.1038/35038060.

Wolfenden, R. E., Pumford, N. R., Morgan, M. J., Shivaramaiah, S., Wolfenden, A. D., Tellez, G. and Hargis, B. M. 2010. Evaluation of a screening and selection method for Bacillus isolates for use as effective direct-fed microbials in commercial poultry. *International Journal of Poultry Science* 9(4), 317-23. doi:10.3923/ijps.2010.317.323.

Wolfenden, R. E., Pumford, N. R., Morgan, M. J., Shivaramaiah, S., Wolfenden, A. D., Pixley, C. M., Green, J., Tellez, G. and Hargis, B. M. 2011. Evaluation of selected direct-fed microbial candidates on live performance and Salmonella reduction in commercial turkey brooding houses. *Poultry Science* 90(11), 2627-31. doi:10.3382/ps.2011-01360.

Wren, B. W. 2000. Microbial genome analysis: insights into virulence, host adaptation and evolution. *Nature Reviews. Genetics* 1(1), 30–9. doi:10.1038/35049551.

Xu, Z. R., Hu, C. H., Xia, M. S., Zhan, X. A. and Wang, M. Q. 2003. Effects of dietary fructooligosaccharide on digestive enzyme activities, intestinal microflora and morphology of male broilers. *Poultry Science* 82(6), 1030–6. doi:10.1093/ps/82.6.1030.

Xu, J., Mahowald, M. A., Ley, R. E., Lozupone, C. A., Hamady, M., Martens, E. C., Henrissat, B., Coutinho, P. M., Minx, P., Latreille, P., et al. 2007. Evolution of symbiotic bacteria in the distal human intestine. *PLoS Biology* 5(7), e156. doi:10.1371/journal.pbio.0050156.

Yang, X. J., Li, W. L., Feng, Y. and Yao, J. H. 2011. Effects of immune stress on growth performance, immunity, and cecal microflora in chickens. *Poultry Science* 90(12), 2740-6. doi:10.3382/ps.2011-01591.

Yurong, Y., Ruiping, S., Shimin, Z. and Yibao, J. 2005. Effect of probiotics on intestinal mucosal immunity and ultrastructure of cecal tonsils of chickens. *Archives of Animal Nutrition* 59(4), 237-46. doi:10.1080/17450390500216928.

Chapter 17

Short chain organic acids: microbial ecology and antimicrobial activity in the poultry gastrointestinal tract

Steven C. Ricke, University of Arkansas, USA

1 Introduction

The demand continues to increase for feed additives that achieve improvements in food safety and reduction of foodborne pathogens in poultry production. Numerous feed additive strategies have been examined or are currently being commercially implemented to either prevent establishment of foodborne pathogens in the gastrointestinal tract (GIT) of poultry or reduce the levels of pathogens already colonized in the GIT. Feed additives vary in their specificity with some being more broad-spectrum than others in terms of targeting microorganisms as antimicrobials. While most of the focus is on antimicrobial activity in the poultry GIT, efficacy in the feed prior to feeding is also a consideration for reducing foodborne pathogens in feeds being stored. Activity in the feed depends upon several variables including water activity of the feed, ability to withstand thermal processing such as what occurs during pelleting, and uniform distribution in the feed to ensure direct contact with

http://dx.doi.org/10.19103/AS.2019.0059.20

foodborne pathogen contaminants. Other factors may play a role as well, such as dosage, feed matrix, and the composition of individual ingredients. While not examined extensively, overall microbial community composition could also be a contributing factor.

Organic acids have been commonly used as feed additives for several decades. These acids are comprised of lactate, short chain fatty acids (SCFA) that include acetate, propionate, butyrate, and formate and medium (MCFA) and long-chain fatty acids (Wales et al., 2010; Dittoe et al., 2018). Most of the research has been conducted on SCFA with more recent emphasis on the longer chain organic acids. Interest in SCFA has centered around applications in animal and poultry feeds to serve as feed additives for inhibiting and reducing *Salmonella* in contaminated feed (Cherrington et al., 1991a,b; Ricke, 2003, 2017). However, there are also substantial quantities of SCFA and lactate produced from fermentation by GIT in the poultry and the types and levels vary depending on the GIT compartment. Production of these short chain organic acids certainly can serve as a barrier to pathogen establishment in the poultry GIT, but they may also have other GIT functions as well.

The overall goal of this chapter is to assess the impact of short chain organic acids on prevention of foodborne pathogens in poultry production and feed sources. Most of the emphasis will be on SCFA as the longer chain acids have been reviewed elsewhere and are beyond the scope of the current chapter (Wales et al., 2010). Likewise, the protected, encapsulated, and modified organic acids will not be discussed as these have been described in several recent reviews (Broom, 2015; Moquet et al., 2016; Bedford and Gong, 2018). The generation of SCFA by the indigenous microorganisms in the poultry GIT will be discussed and examined for their potential influence on foodborne pathogen colonization and functionality in the poultry GIT. Potential impact of the presence of these SCFA on poultry GIT microbial communities whether internally produced or externally introduced will also be considered. Finally, potential issues that impact the efficacy of SCFA as feed amendments will be discussed.

2 Short chain organic acid production in the upper poultry gastrointestinal tract

The poultry GIT harbors a complex array of microorganisms in the different compartments from the crop to the ceca. Regardless of the GIT compartment, the microorganisms at each site are universally characterized by their ability to ferment and generate short chain organic acids as end products. The primary fermentation end products include lactate along with the SCFA, namely, acetate, propionate, and butyrate. The amount of each being produced is dependent on several factors. Certainly, the types of microorganisms prevalent are a critical factor, but age of bird, dietary composition, interactions among GIT microorganisms, presence of pathogens, and other factors, some of which have yet to be identified, play a role as well (Gabriel et al., 2006; Choct, 2009; Ricke et al., 2013). Specific alterations in diets such as the inclusion of antibiotics or other antimicrobials, biological feed amendments including prebiotic or probiotic supplements may also have direct or at the very least an indirect impact on either the microbial composition of the GIT or the metabolism of particular members of the GIT microbial populations (Roto et al., 2015; Ricke, 2018). Historically these factors were difficult to identify with specific changes in the poultry GIT microbial community, but the introduction of next-generation sequencing of the 16S rDNA region of microbial populations has provided greater resolution of taxa identification and microbial community diversity comparisons (Ricke et al., 2017). This information coupled with metabolomic analyses and application of metabolic networks have offered opportunities to better integrate the interactions among the host, the GIT microbial population, and dietary differences (Read and Holmes, 2017). Individual poultry GIT compartments and their characteristic short chain organic acid production profiles will be discussed as follows.

The crop has been examined anatomically as well as from a microbial composition standpoint for a number of years. Initially viewed as simply a storage GIT organ, it has received more attention in recent years as both a site for fermentative microorganisms and a critically important source of foodborne pathogen contamination in poultry processing plants (Bayer et al., 1978; Hargis et al., 1995). Lactobacilli are predominant in the crop with several species occurring over time as the bird matures (Rehman et al., 2007). The gizzard also is predominated by lactobacilli, but they appear to generate less acetate and lactate than what is observed in the crop (Rehman et al., 2007). While all of the SCFA have been detected in the crop, acetate generally occurs in the highest concentration (Rehman et al., 2007). Lactate is also detected in the crop and may vary depending on the level of feed intake and can decrease considerably when feed is removed for extended periods of time (Rehman et al., 2007; Durant et al., 1999). This decrease has been associated with a detectable decrease in numbers of lactobacilli in laying hens undergoing a feed withdrawal molt period of several days (Durant et al., 1999b). This combination of decreases in crop lactobacilli and lactate and an increase in pH appears to be directly related to increases in *S.* Enteritidis crop populations when these changes occur (Durant et al., 1999b). Similar trends have been noted by Hinton Jr. et al. (2000a,b) in broilers undergoing much shorter feed withdrawal time periods of hours versus the days of feed withdrawal undergone by the laying hens in the studies conducted by Durant et al. (1999b). This has also been observed in commercial birds undergoing feed withdrawal prior to entering the processing plant (Ramirez et al., 1997). This led Hinton Jr. et al. (2000a,b) to point out that feed withdrawal may impact the crop as a critical reservoir for *Salmonella* contamination in the poultry processing plant. This suggests that the crop microbial population and concomitant fermentation may act as a barrier to establishment of pathogens. Therefore, the retaining and/or enhancing the crop microbial population fermentation should be considered as part of any feed additive strategies such as short chain organic acid supplementation proposed for modulating the chicken GIT to become more hostile to pathogens such as *Salmonella*.

In general, the small intestine does not exhibit extensive fermentation activity compared to the ceca, but SCFA, primarily acetate, and lactate can be detected and do appear in some studies to be dependent on dietary composition (Rehman et al., 2007). Whether differences occur on the intestinal tract mucosal surfaces versus the intestinal lumen and, in turn contribute to differences in SCFA and lactate concentrations remains to be determined. The microbial ecology of intestinal mucosal surfaces may also be critical for the nutrition of the bird as these microorganisms are presumably in direct competition for nutrients being processed by the intestinal cells. Application of transcriptomic and proteomic approaches on intestinal tissues may help to elucidate whether specific mechanisms exist that directly respond to the presence of short chain organic acids generated by the GIT fermentative microbial community.

3 Cecal fermentation and generation of short chain organic acids

The ceca are considered the primary site of GIT microbial fermentation in the chicken and the source of SCFA generated by a highly diverse microbial consortia. Visible as a pair of blind pouches, they are attached to the colon distal to the muscular ring dividing the ileum from the colon (Svihus et al., 2013). The ceca have been associated with a variety of functions over the years, but it is well established that they are the major site of microbial fermentation and harbor a complex microbial consortia capable of producing SCFA and other fermentation products (Józefiak et al., 2004). Based on *in vitro* and *in vivo* studies, fermentation products produced by cecal microbiota generally include ammonia, carbon dioxide, SCFA, and methane as the predominant compounds (Józefiak et al., 2004; Dunkley et al., 2007a,b; Saengkerdsub et al., 2006, 2007a,b). However, there may be a much wider range of products generated by the cecal microbiota, most of which may be produced in fairly low concentrations and would be considered highly variable depending on diet and other factors. For example, using gas chromatography-mass spectrometry metabolomic analyses, Rubinelli et al. (2017) detected over 500 compounds generated from *in vitro* cecal incubations of feed and rice bran, many of which were not identifiable. Of the metabolites identified that were increased in the presence of the rice bran-supplemented *in vitro* cecal incubations, several not surprisingly were associated with SCFA synthesis while other metabolites included nitrogen metabolites such as glutamate and methionine.

It is assumed that nondigestible components of the diet entering from the upper GIT of the bird would be available as substrates for the cecal microbial population. This would include dietary components classified as prebiotic oligosaccharides such as fructooligosaccharides, mannan-oligosaccharides, and other similar nondigestible carbohydrate polymers (Rehman et al., 2009; Kim et al., 2019; Micciche et al., 2018; Ricke, 2018). Certainly, *in vitro* incubation studies utilizing inocula from ceca would indicate that there are microbial members in the ceca capable of fermenting most fiber sources that would be potential dietary constituents. For example, Dunkley et al. (2007a) screened soybean meal, soybean hull, beet pulp, wheat middlings, ground sorghum, cottonseed meal, and alfalfa meal as high-fiber sources in an *in vitro* anaerobic incubation system using cecal inocula from laying hens over 50 weeks old. When they profiled the SCFA production from these incubations, the high-fiber dietary sources clearly supported SCFA production with acetate as the predominant SCFA followed by lesser concentrations of propionate and butyrate and trace amounts of valerate, isovalerate, and isobutyrate. This general pattern of the predominant SCFA being acetate appears to be characteristic of cecal fermentation populations in general as it is consistent with SCFA profiles from *in vitro* broiler cecal studies (Lan et al., 2005; Rubinelli et al., 2016). Likewise, similar profiles have been observed from cecal contents sampled from *in vivo* studies with laying hens (Woodward et al., 2005; Dunkley et al., 2007b) and broilers fed a wide range of diets (Rehman et al., 2007).

While the general patterns for cecal SCFA appear to be relatively consistent, differences do exist due to several factors. Certainly diet composition, in particular the presence and level of nondigestible carbohydrates and insoluble fibers entering the ceca, would presumably impact the ultimate concentrations of individual SCFA. Dunkley et al. (2007a) detected differences in levels of acetate, propionate, and butyrate as a function of fiber source at the end of *in vitro* 24-h cecal incubations of layer hen cecal inocula. They also noted differences when different ratios of alfalfa and layer ration were combined in increments of 100% alfalfa (no layer ration) to 70% alfalfa (30% layer ration) were compared. This is consistent with the differences observed by Lan et al. (2005) for nondigestible carbohydrates incubated with adult broiler cecal contents. Similar trends have also been reported for *in vivo* bird studies. Ricke et al. (1982) observed differences in cecal SCFA patterns in young chicks fed different forms of carbohydrates versus lignin, a non-carbohydrate fiber component. Kareem et al. (2017) combined a postbiotic (probiotic metabolites) and inulin and fed to broiler chicks over a 42-day period to examine cecal SCFA and cecal microbiota responses. They detected increases in cecal acetate production compared to control diets but not in total SCFA propionate, or butyrate. When Walugembe et al. (2015) fed a high-fiber diet containing dried distiller grains and wheat bran to broiler and layer chicks, they did not detect a decrease in either acetate or propionate, but reported a decrease in butyrate levels in the high-fiber-fed chicks. Responses to metabolites have also been observed for *in vitro* studies as well. For example, Rubinelli et al. (2016) detected increases in acetate when incubating broiler cecal contents in the presence of a yeast fermentate, but also detected an increase in butyrate production in an *in vitro* system. In short, both the composition of the diet source and the type of feed additive appear to influence the concentrations of SCFA produced as well as which individual SCFA are increased. Other dietary factors such as particle size may also play a role as changing the size does impact digestibility and bird performance while also influencing gizzard development (Hetland et al., 2004; Amerah et al., 2007).

Non-dietary factors such as poultry genetic lines of birds may also contribute to differences in short chain organic acid levels. For example, Walugembe et al. (2015) observed higher concentrations of acetate and propionate in cecal contents of broiler chicks when compared to layer chicks aged 21 days. This distinction in SCFA patterns is supported by evidence that the development of the respective GIT microbiota does differ between broilers and layers (Walugembe et al., 2015). Age of the bird and development of the cecal microbiota also influence SCFA concentrations. Van der Wielen et al. (2000) followed cecal bacterial populations, pH, SCFA, and lactate levels in commercial broilers from day 1 to nearly 40 days of age. While cecal microbiota populations in the younger birds consisted primarily of *Enterobacteriaceae*, enterococci, and lactobacilli, by day 12 the total numbers of anaerobes were 1 to 1.5 colony forming units (CFU) logs higher than the aerobic bacteria and remained so until the birds reached the end of the study. While pH ranged from 5.5 to 6.0 throughout this time period, organic acids paralleled the cecal microbial shifts with lactate decreasing to levels below detection by day 15 versus acetate already being detected in day 3 chicks and increasing thereafter until stabilizing as the predominant SCFA at day 15. Levels of cecal propionate and butyrate followed similar patterns as acetate and their concentrations albeit less than acetate also reached their respective plateaus in birds at day 15.

Józefiak et al. (2004) have likened the SCFA produced in the ceca to be similar to the SCFA profiles characteristic of the rumen and the human large intestine. Based on this general premise, it is conceivable that similar pathways for SCFA formation and at least some of the cecal microorganisms producing these SCFA may be similar taxonomically or at least functionally to those found in the other two GIT ecosystems. Early work based primarily on anaerobic cultural isolations and enumeration established that a fairly densely populated and diverse group of anaerobes established in the ceca as the chickens become older results in a mature cecal population capable of fermentation and producing SCFA (Mead, 1989). It is apparent that the cecal environment becomes sufficiently anaerobic to support methanogens (Miller et al., 1986; Marounek and Rada, 1998; Marounek et al., 1999; Saengkerdsub et al., 2006). While the methanogens isolated from the ceca are not nearly diverse as those characterized in the rumen, they can be detected in the young bird and their numbers can be correlated quantitatively with methane production in cecal contents (Saengkerdsub and Ricke, 2014). In the rumen, methanogens play an important role in fermentation pathways by serving as a hydrogen sink and supporting more oxidized pathways such as the formation of acetate versus more reduced end products (Wolin and Miller, 1982). Whether they serve a similar function in chicken cecal is not known, but their presence may be a factor in the variation observed in acetate versus other SCFA productions in different studies. However, other hydrogen sinks may exist. Based on 16S rDNA sequencing of cecal samples from 42-day old broilers, Sergeant et al. (2014) did not detect methanogens, sulfate reducers, or acetogens that would be candidates to utilize excess hydrogen. This led them to speculate based on their metagenome analyses that the presence of 12 uptake hydrogenases associated with some of the most abundant genera, namely *Megamonas, Helicobacter,* and *Campylobacter,* were involved in hydrogen utilization.

The application of 16S rDNA sequencing of GIT microbiomes has also paved the way to developing a much more comprehensive identification of the individual members of the respective GIT microbial communities including the ceca and their contributions to SCFA metabolism. Recent reviews (Oakley et al., 2013; Stanley et al., 2014) have summarized the research using these techniques and have drawn some general conclusions. Oakley et al. (2013) reported that at the phyla level cecal populations are dominated by Firmicutes and Bacteroides. In summarizing several earlier studies, Stanley et al. (2014) concluded the occurrence of an abundance of *Clostridium*, *Ruminococcus*, *Eubacterium*, *Faecalibacterium*, and *Lactobacillus* species as well as a number of unknown and uncultured phylotypes. However, as Stanley et al. (2014) have pointed out, the cecal microbiota composition is likely to be quite variable among flocks. The factors that impact that variation remain to be determined, but presumably environment, host immune responses, and substrate availability may be involved to some degree.

Based on cecal SCFA profiles, the fermentation by the cecal microorganisms would seem to be fairly constant regardless of potential compositional changes. In GIT systems such as the rumen, there is considerable redundancy in microbial function both in the initial stages of catabolism of incoming substrates and end product generation during fermentation (Weimer, 2015). For example, there are several organisms capable of hydrolyzing cellulose and numerous rumen bacteria that generate one or more of the primary fermentation products, namely, acetate, propionate, or butyrate (Russell et al., 1992; Weimer et al., 1999). For SCFA generation, presumably there are multiple poultry cecal bacteria capable of producing one or more of these individual SCFA. Applying metagenomic analyses, Sergeant et al. (2014) were able to identify predominant pathways for each of the major SCFA. The dominant nature of acetate as the primary cecal SCFA was evident by the preponderance of over 30 acetate kinase/phosphotransferase sequences. Propionate appeared to be generated mostly via methylmalonyl-CoA decarboxylase and methylmalonyl-CoA epimerase and could be found in microorganisms such as *Bacteroidetes*. The primary series of enzymes associated with butyrate production were identified as 3-hydroxybutyryl-CoA dehydrogenase, phosphotransbutyrylase, and butyrate kinase. Presence of these genes and their encoded pathways should allow for direct functional and quantitative assessment of diet changes and feed amendments that relates microbial compositional responses with specific metabolic pathways.

4 Functions of cecal short chain organic acids: host metabolism

The generation of short chain organic acids such as SCFA in the ceca potentially serves multiple functions. Rumen SCFA are a major energy source for the ruminant animal as well as primary carbon sources (Russell et al., 1992; Józefiak et al., 2004). In birds it is less clear whether SCFA are a major contributor to bird metabolism. As Józefiak et al. (2004) have pointed out, SCFA capture potential metabolizable energy from the fermentation of carbohydrates such as non-starch polysaccharides entering the ceca. In some avian species, dilution of metabolizable energy via increased dietary fiber intake results in increases in gizzard and hindgut sizes and are accompanied by changes in digestibility and potential extraction of energy from these fiber sources via formation of SCFA (Redig, 1989). For example, changes in the microbiota with an increase in organisms capable of hydrolyzing cellulose can be seen when comparing low-fiber versus high-fiber–fed domestic turkeys (Bedbury and Duke, 1983; Duke et al., 1984; Redig, 1989). Certainly, based on cecal *in vitro* incubation studies and *in vivo* studies with fiber-fed laying hens, it would appear that chicken cecal microorganisms are present that possess the ability to ferment fiber components and generate SCFA (Marounek et al., 1999; Lan et al., 2005; Saengkerdsub et al., 2006; Dunkley et al. 2007a,b; Ricke et al., 2013). This is consistent with the observation by Stanley et al. (2013) who concluded that birds exhibiting high-energy efficiency feed extraction capacity were also characterized as having more cecal bacteria that could produce enzymes that degraded cellulose and/or resistant starch.

However, the key question is whether the SCFA produced from these cecal fermentations are absorbed and actually used by the chicken as an energy source. Goldstein (1989) speculated that cecal transport systems for SCFA produced in the ceca could exist. Early work by Annison et al. (1968) compared germ free with conventionally raised birds and measured SCFA produced in the various compartments of the GIT along with the levels in the portal and peripheral blood. They detected all of the SCFA produced by the microorganisms in the portal blood but only acetate and formic acid in the peripheral blood. Based on the comparisons with germ free and conventionally raised birds, they concluded that most of the acetate in the blood was of endogenous origin with approximately 25% exogenously derived from microbial fermentation and that total acetate contributed 11% and 6% of total metabolizable energy in fed and starved mature birds, respectively. In summarizing results from later studies, Svihus et al. (2013) concluded that the cecal fermentation energy contribution to the energy requirements of the bird ranged from 3% to 5%.

Absorption of SCFA may occur in other regions of the chicken GIT as well and this may influence subsequent metabolism. For example, Hume et al. (1993a) administered C-14 labeled propionate by oral gavage to 11-day-old chicks and birds were euthanized at 15 and 60 min after gavaging followed by removal of the various compartments of the GIT (crop, gizzard, proventriculus, small intestine, large intestine, and ceca) along with the liver and blood samples also being taken for extraction and radiolabel analyses. In following the radiolabeled propionate, Hume et at. (1993a) noted that some of the label ended up in other fermentation products such as lactate, indicating that GIT metabolism of propionate into intermediary metabolites may have occurred. They detected most of the extracted propionate in the foregut and serum with minimal amounts of propionate in the lower part of the GIT, indicating that most of the absorption occurred in the foregut of the chick. However, over time much of the radiolabel ended up in carbon dioxide with up to 75% of the original dose by 3 h. This outcome led the authors to conclude that propionate absorbed from the GIT was either rapidly assimilated or degraded. In summary, additional studies need to be done to track individual SCFA produced in the GIT to elucidate metabolic pathways and routes of incorporation versus degradation and how these participate in host energy generating pathways. Once such pathways are more precisely identified, it may be possible to construct metabolic balance assessments of the contributions of these GIT-SCFA. More in-depth characterization of intestinal and cecal tissues using transcriptomic and proteomic approaches would also help to determine whether there are active mechanisms such as specific transporters for SCFA that support their potential contribution to utilization by the host and factors that may influence the level of contribution to bird metabolism.

5 Functions of cecal short chain organic acids: pathogen inhibition

The other GIT functional activity that the GIT microbial generation of short chain organic acids contributes is the inhibition of pathogens in the GIT (Ricke, 2003; Józefiak et al., 2004). By serving as a barrier to pathogen colonization, they exhibit protective properties to the host and limit the dissemination of pathogens in poultry flocks. This is reflective in the inverse relationship between colonization of pathogens and the development of the GIT microbial community. As the GIT microbial community becomes more diverse with increased bird maturity, this more complex GIT microbial ecosystem is more resilient to colonization by pathogens. For example, Van der Wielen et al. (2000) demonstrated a negative correlation between *Enterobacteriaceae* and appearance of acetate, propionate, and butyrate in the broiler ceca when these microorganisms declined in lock step with the increases in SCFA. They further demonstrated this relationship with decreases of isolated pure cultures of *Enterobacteriaceae* when incubated *in vitro* and exposed to SCFA. This is consistent with the earlier observations by McHan and Shotts (1993) who observed at least 50% decreases in *Salmonella* Typhimurium growth when incubated in the presence of a mixture of short chain organic acids at pH 5.

Given the potentially protective relationship of the mature GIT microbial community and prevention of pathogen establishment, it is not surprising that attempts to manipulate or accelerate the development of a highly fermentative GIT microbial community have become commonplace in the poultry industry. Certainly, this trait is often affiliated with probiotics and prebiotics as one of the primary underlying antibacterial mechanisms for these biological feed additives and their relative effectiveness against foodborne pathogens such as *Salmonella* and *Campylobacter* (Nisbet, 2002; Callaway and Ricke, 2012; Siragusa and Ricke, 2012; Micciche et al., 2018; Kim et al., 2019). In general, the basic premise behind these biological additives is to either provide specific substrates that favor GIT bacteria that are considered beneficial to the host and antagonistic to pathogens (prebiotics) or directly introduce such organisms as live cultures to the host animal recipient (probiotics) or as combinations (synbiotics) of both prebiotics and probiotics (Patterson and Burkholder, 2003; Hume, 2011; Callaway and Ricke, 2012). In all cases, the strategy is to enhance early establishment of microorganisms in the GIT that display antagonistic properties such as highly active fermentation capabilities which in turn serve as GIT ecological barriers to incoming pathogens.

The antibacterial mechanisms associated with short chain organic acids have been described extensively in several reviews and will not be discussed in detail here (Cherrington et al., 1991a; Russell, 1992; Ricke, 2003; Van Immerseel et al., 2006). Much of the information on specific mechanisms associated with short chain organic acids and pathogens has been elucidated by pure culture studies under laboratory incubation conditions. However, in the GIT the relationship between short chain organic acids and pathogens may be more complicated due to the complex composition of the GIT microbial population, their relationship with production of fermentation products, and other antagonistic mechanisms that the population as a whole may elicit toward pathogens. Other factors such as competition for specific nutrients between indigenous GIT organisms and pathogens, occupation of pathogen attachment sites, generation of strict anaerobic conditions, and other less defined factors such as the GIT microbiota interaction with the host immune system may also contribute to the GIT hostility toward pathogens.

Conceptionally, the generation of fermentation products lowers the external pH to which some pathogens are more sensitive than fermentative bacteria, but their presence can also be directly antagonistic to particular pathogens (Russell, 1992; Józefiak et al., 2004; Ricke, 2003; Van Immerseel et al., 2006). This may also be a function of particular SCFA and the specific microorganisms that produce them. There are probiotic/competitive exclusion studies with chickens that illustrate this. In a series of studies, a 29 characterized continuous flow propionate producing culture that included facultative and obligate anaerobic cecal bacteria originating from *Salmonella*-free broiler chicks was developed and proven to be effective against *Salmonella* colonization of *Salmonella* marker strain challenged young chicks, as indicated by the increase in propionate and total SCFA in treated birds versus control non-treated birds (Corrier et al., 1995). A key characteristic was that once the mixed continuous flow culture was introduced to the young chicks, cecal propionate markedly increased and cecal *Salmonella* populations decreased accordingly (Corrier et al., 1995; Nisbet et al., 1996a). However, when follow-up studies were conducted with variations on the generation of the continuous flow culture, Nisbet et al. (1996b) demonstrated that any alteration in the composition of the 29 culture, such as removal of individual species, resulted in failure to increase propionate in the chick ceca and did not protect against *Salmonella* colonization. As pointed out by Nisbet et al. (1996b), this emphasizes that *Salmonella* inhibition by competitive exclusion microorganisms is probably a concerted effort that involves not just one organism producing a certain SCFA but the metabolic interactions of the entire GIT microbial community invoking multiple inhibitory mechanisms with propionate production being a biological indicator of successful control of *Salmonella*. This also suggests that there may be other metabolite candidates for indicators of efficacious probiotics. Gaining a better understanding of the cecal populations specifically associated with minimal *Salmonella* colonization using a combination of microbiome characterization and metabolomic profiling may reveal core microbial populations that could serve as potential probiotics and provide a metabolite signature for proof of successful establishment.

The potential interactions observed between GIT microbiota, SCFA, and pathogens may consist of broader ecological mechanisms associated with the GIT ecosystem. Therefore, it may not be a matter of direct inhibition of the pathogen itself by a particular SCFA, but an impact on pathogen activity while residing in the GIT. For example, Rivera-Chávez et al. (2016) demonstrated that decreases in butyrate occurred when streptomycin was used to reduce a commensal butyrate-producing *Clostridia* from the mouse intestinal lumen. This led to increased epithelial oxygenation and aerobic expansion of *Salmonella* Typhimurium. They observed that tributyrin supplementation restored epithelial hypoxia and limited *Salmonella* expansion. Fukuda et al. (2011) concluded that increased acetate production by *Bifidobacterium* protected mice from a lethal dose of enterohemorrhagic *Escherichia coli* O157:H7 by inhibiting shiga toxin translocation from the gut lumen to the blood stream even though GIT *E. coli* O157:H7 numbers were not decreased.

In a series of tissue culture studies by Durant et al. (1999a, 2000a,b), adherence and invasion of HEp-2 mammalian cells by *S.* Typhimurium was influenced by the presence of SCFA. To monitor the impact on virulence gene expression, Durant et al. (2000c) used *S.* Typhimurium chromosomal *lacZY* transcriptional gene fusion strains of two virulence genes, *hilA*, a regulatory gene of *Salmonella* invasion gene that responds to environmental stimuli and *invF*, which promotes the transcription of the *sipBCD* genes. Gene expression was quantified as a β-galactosidase assay after the *S.* Typhimurium strains were exposed to SCFA over time. Expression levels of *invF-lacZY* and *hilA-lacZY* fusions in the presence of all SCFA (acetate, propionate, and butyrate, individually, or the combination all three) were generally higher than the controls at 1 h at pH 6 but only higher for acetate exposed bacterial cells at pH 7. This led Durant et al. (2000c) to conclude that SCFA may influence virulence expression in the GIT, but pH could also be a contributing factor. However, there is also evidence that exposure of *Salmonella* to GIT levels of SCFA at neutral pH could activate acid tolerance mechanisms in *Salmonella* enabling it to resist normally lethal low pH levels (Kwon and Ricke, 1998). The ability of *Salmonella* to resist SCFA to some extent may be related to the fact that it is capable of growing under anaerobic incubation conditions and generating fermentation acids consisting mostly of acetate with lesser amounts of propionate and butyrate (Dunkley et al., 2008, 2009).

In summary, more needs to be done to delineate the multiple factors potentially involved with foodborne pathogens and their responses to SCFA produced during fermentation by indigenous cecal and GIT microbiota. Characterizing the relationships between specific members of the GIT microbial community and the resulting SCFA profiles in conjunction with the rise and fall of pathogen levels may help to sort out some of the complexity of these ecosystems. Furthermore, delineating the regulatory pathways in organisms such as *Salmonella* and how its anaerobic metabolism is related to virulence expression may help to explain its success in the GIT and the environmental cues that dictate the organism's interaction with the host. This may explain why certain probiotics and prebiotics may be more effective than others and/or variable in their efficacy and why *Salmonella* suddenly becomes invasive under certain environmental conditions in the GIT.

6 Feed contamination and feed additives: general concepts

Biological contamination of animal feeds can occur at any point in feed processing and subsequent delivery to animal production operations (Ricke, 2005, 2017). Biological contaminants can be quite varied and can consist of a wide range of fungi, bacteria, and viruses (Ricke, 2005, 2017; Maciorowski et al., 2006b, 2007). Much of this variation is due to the complex nature of environmental sources of potential contaminants, numerous vectors that feeds are exposed to, and the potential for cross contamination during certain processes such as feed milling and animal housing (Ricke, 2005, 2017). Control of the contaminants is equally variable, ranging from physical processes such as application of heat during pelleting of feed to supplementation with a series of chemicals as feed additives that include aldehydes, botanicals such as essential oils, and acids, among others (Williams, 1981c; Wales et al., 2010; Ricke, 2017). Some of these have been more successful than others in controlling particular types of feed contamination. Cost, worker safety, regulatory restrictions, ease of application, animal palatability, and efficacy are all criteria that have to be considered when designing a feed treatment application strategy (Ricke, 2017). In addition, the type of biological contaminant that is being targeted for control must be considered as a fungal organism would presumably respond differently than a bacterium or a virus. Likewise, a sporeformer may require a different intervention strategy than a bacterium that does not form spores. Finally, there are potential differences among different bacterial species and even among strains in their responses to antimicrobials as well as survival capabilities in complex matrices such as feed (Lianou and Koutsoumanis, 2013; Andino et al., 2014).

While there are several biological contaminants that are of either animal disease or food safety interest for limiting in feeds, the foodborne pathogen *Salmonella* has received most of the attention. Occurrence of *Salmonella* in animal feeds has been documented for a number of years, in some countries for well over a half century and continues to remain of interest due to potential linkages with public health (Williams, 1981a; Jones and Richardson, 2004; Maciorowski et al., 2004; Jones, 2011). Much of the early *Salmonella* isolation and characterization studies in poultry feeds were summarized in a comprehensive review by Williams (1981a). Since then, several studies have been conducted to survey *Salmonella* frequencies in feed mill processing operations, poultry farms, feed ingredients, and different poultry production feeds from breeder flocks, and broiler houses resulting in a wide range of estimated contamination levels and identification of different serovars of *Salmonella* (Whyte et al., 2003; Jones and Richardson, 2004; Maciorowski et al., 2004; Jarquin et al., 2009; Wales et al., 2010). In summarizing *Salmonella* feed studies, Jones (2011) concluded that *Salmonella* spp. in feeds present several challenges including their ability to persist for long periods of time and thus achieving representative sampling of the feed directly may be problematic given the randomness and vast quantities of feed being processed at any given time.

The overall strategy for implementing feed additives as control agents for microorganisms such as *Salmonella* has been twofold. Reducing or eliminating *Salmonella* in the feed itself is considered a first step. This represents a challenge as feed additive efficacy will be a function of the physical form and chemical composition of the feed. In addition, the relatively infrequent and somewhat random occurrence of *Salmonella* in feed makes monitoring effectiveness in the large animal and poultry feeds a nearly impossible task (Jones and Ricke, 1994). Given this infrequency, Jones (2011) suggested that sampling of the environment in the feed mill may be a more optimal sampling approach and this in turn would allow for the identification of potential critical control points. The further detection and isolation challenge resides in over 2500 potential serovars of *Salmonella* that have been identified (Finstad et al., 2012; Foley et al., 2011, 2013; Howard et al., 2012). Finally, development of specific detection and quantitation approaches for recovering *Salmonella* from feeds has undergone continued refinement as more is known about the relationship between feed and *Salmonella* contamination.

Historically, *Salmonella* spp. were isolated from feed and depending on the research goal of the particular study, enumerated using combinations of selective enrichment and plating followed by serotyping for further identification (Williams, 1981b; Maciorowski et al., 2006a). Non-culture-based methods have also been used. For example, Alvarez et al. (2003) used microarrays to identify and track *Salmonella* across different feed mills. Maciorowski et al. (2005) have extensively reviewed polymerase chain reaction (PCR) approaches for detection, identification, and quantitation of *Salmonella* in feeds and this will not be discussed in detail here. Regardless of the detection method, several factors must be considered to achieve accurate assessment of *Salmonella* contamination in feeds including both the size of individual samples and the frequency of sampling. Survival of *Salmonella* over time in feed also varies among strains and may influence recovery of certain *Salmonella* strains in stored feeds (Andino et al., 2014).

In addition, variability in physiological responses by different serovars could influence *Salmonella* culture outcomes from feeds containing acids as feed additives. For example, González-Gil et al. (2012) compared virulence response of multiple strains of *Salmonella* Typhimurium, *S.* Enteritidis, *S.* Kentucky along with single strains of *S.* Seftenberg, *S.* Heidelberg, *S.* Mbandanka, *S.* Newport, *S.* Bairely, *S.* Javiana, *S.* Montevideo, and *S.* Infantis. Serovars were exposed to either HCl or acetic acid at pH levels of 7.2, 6.2, or 5.5 and RNA recovered for quantitative reverse transcriptase real-time PCR of the virulence regulatory gene *hilA*. Several fold increases in *hilA* expression levels occurred between 0 h and 2 h for some of the *S.* Typhimurium strains, along with *S.* Senftenberg, *S.* Heidelberg, *S.* Mbandanka, *S.* Montevideo, and *S.* Infantis. Other genetic differences may also exist between serovars. For example, Joerger et al. (2009) observed differences in *S.* Kentucky versus other serovars in sensitivity to acetic acid. In a follow-up study, microarray transcriptional profiles were compared between *S.* Kentucky and *S.* Enteritidis and more genes were up- or downregulated in *S.* Kentucky versus *S.* Enteritidis after exposure to HCL or acetic acid (Joerger et al., 2012). Strain differences in response to acid stress have also been reported. Shah et al. (2012) detected differences in acid resistance among *S.* Enteritidis strains even though genomic hybridization arrays revealed virtually no genetic differences.

These differences are not only critical for evaluating impact of acid treatments against different serovars in feeds, but could also impact recovery of the respective serovar on plating media and the subsequent masking that may occur in poorly buffered recovery media (Carrique-Mas et al., 2007; Cox et al., 2013, 2016). This in turn could influence interpretation regarding serovar prevalence and acid-based control strategies for treating feeds. One potential solution would be to employ molecular methods that avoid some of the need to quantitatively recover cells in a culture media and depending on the method, allow for differentiation of the respective serovars (Koyuncu et al., 2011; Maciorowski et al., 2005; Malorny et al., 2008; Park et al., 2014; Park and Ricke, 2015; Pumfrey and Nelson, 1991; Ricke et al., 1998, 2018; Schelin et al., 2013; Soria et al., 2011). Along these lines, employing a recovery media that resuscitates injured bacterial cells and combining this with a molecular detection approach may help to alleviate this. For example, Kim et al. (2017) combined a most probable number (MPN) microtiter plate assay incorporating nonselective media to recover *Salmonella* combined with qPCR to detect *Salmonella* in the individual wells, and used the MPN profile outcome based on wells with positive qPCR results to quantitate *Salmonella*.

7 Activities of short chain organic acids in the feed

Short chain organic acids have a long history as antimicrobial additives for poultry and other production animals as well as processing of their meat products (Cherrington et al., 1991a; Mani-López et al., 2012; Ricke, 2005; Ricke, 2003, 2017; Van Immerseel et al., 2006; Wales et al., 2010; Bedford and Gong, 2018). Short chain organic acids have also been applied directly to feed as a contaminant control (Williams, 1981c; Cherrington et al., 1991a,b; Maciorowski et al., 2007; Ricke, 2003, 2005, 2017). Much of the emphasis has been directed toward reducing foodborne pathogens in feeds with a particular focus on *Salmonella* (Williams, 1981a; Van Immerseel et al., 2002; Maciorowski et al., 2004, 2006b; Jones, 2011; Ricke, 2017; Wales et al., 2010). Short chain organic acids have been used either singly or in combination as feed additives in a number of feed studies. Impact on *Salmonella* levels in poultry and other animals can be categorized as direct effects on *Salmonella* and indigenous microbial populations in the feed and/or the more indirect effect on *Salmonella* and other pathogens colonizing the bird's GIT after the treated feed has been consumed. The other factors to consider is the form at which the acid is applied and the corresponding concentration. Much of the details of individual factors have been discussed in a comprehensive review by Wales et al. (2010) and therefore only an overview of the key points will be listed here as they relate specifically to short chain organic acids. In general, reduction of feedborne *Salmonella* by acid additives is influenced by the concentration of the acid and final pH with buffered products appearing to be, for the most part, less effective (Ha et al., 1998a; Wales et al., 2010).

Over the years, specific studies have been conducted to determine the direct impact of organic acids on *Salmonella* populations in feeds. Vanderwal (1979) designed studies to decontaminate mixed feeds based on the concept that all ingredients would be present and the opportunity for recontamination would be minimal prior to delivery to the animal facilities. They compared candidate organic acids, formic, acetic, propionic, and lactic acids in doses ranging from 0.5% to 1.2% on *Enterobacteriaceae* and concluded that formic acid resulted in fewest days required to achieve a decimal reduction. When Pumfrey and Nelson (1991) used a combined DNA-MPN approach to assess the effect of different concentrations of a commercial formic-lactic-propionic acid combination formulation, they concluded that effectiveness of the chemical preservative was both bacterial cell concentration and time dependent. It would seem that efforts to assess naturally occurring *Salmonella* in feeds which would presumably be much lower and more infrequent than externally added *Salmonella* would require not only a combination of a buffered recovery media but a quantitative method with increased detection sensitivity.

The concept of treatment of mixed feeds toward the end of feed processing being the better strategy for containment of *Salmonella* is supported by other studies. For example, when Humphrey and Lanning (1988) compared naturally occurring *Salmonella* serovar profiles before and after application of 0.5% formic acid, they noted significant reduction in the breeder feed as well as a reduction in infection of chicks but little impact on broiler feed ingredient contamination levels or subsequent infection of broilers. They attributed this to contaminated imported fish and soy meal. This is in concert with the suggestion by Wales et al. (2010) that higher acid concentrations should be used in individual feed ingredients prior to feed mixing to avoid animal palatability issues and feed mill equipment deterioration. This approach, based on the results from Humphrey and Lanning (1988), would also indicate that it might be prudent to more effectively decontaminate ingredients prior to mixing into the final diet.

Several factors associated with the feed matrix will influence the antimicrobial efficacy of the organic acid applied to the feed. For example, temperature and moisture content of the feed have been stated by Van Immerseel et al. (2006) to impact the corresponding organic acid activity. This fits with early observations by Williams and Benson (1978) that *S.* Typhimurium survived best in poultry feed and litter at lower temperatures. More recently, even without added antimicrobials, Koyuncu et al. (2013) also saw decreased *Salmonella* antimicrobial efficacy of acids applied to feeds when the temperature was reduced. Morita et al. (2005) examined the impact of temperature, relative humidity, and a commercial mixture of formic and propionic acids concentration on *S.* Agona contamination levels in oilmeal from commercial oilmeal manufacturing facilities. Both increased temperature and increased relative humidity caused reductions in *S.* Agona. The serovar was not detected at a 1% w/w for 2 days at 45°C but only minimal reduction was achieved at lower temperatures. This supports the results of Milillo and Ricke (2010) that increased temperature in the presence of organic acids leads to synergistic antagonism of *Salmonella.* Perhaps application of organic acids prior to the heat associated with pelleting could lead to a more consistent and complete reduction of *Salmonella*. However, as Jones (2011) points out, the presence of chemicals, particularly higher concentrations required for killing *Salmonella,* may be corrosive to milling equipment and therefore not practical.

Various feed ingredients such as those of animal protein origin versus plant origin have been attributed with different *Salmonella* isolation frequencies (Crump et al., 2002; Hsieh et al., 2016; Jiang, 2016; Veldman et al., 1995; Wierup and Häggblom, 2010; Williams, 1981a). Therefore, it might be expected that feed composition could influence efficacy of an acid feed additive. For example, Ge et al. (2013) observed that animal by-products yielded a substantially higher recovery of *Salmonella* positive samples compared to plant by-product samples. Furthermore, there is some indication that microenvironments associated with the surfaces of certain food particles can protect enteric pathogens from severe acidic conditions (Waterman and Small, 1998). However, most studies have detected minimal impact on *Salmonella* survival either as a function of protein level or animal versus plant source of protein (Ha et al., 1998b). Koyuncu et al. (2013) compared formic acid alone or different combinations of formic and propionic acids, and sodium formate on survival of several *Salmonella* serovars in various feed matrices. They not only observed different acid tolerances among the *Salmonella* serovars (*S.* Infantis being the most tolerant, then *S.* Patten, *S.* Seftenberg, and *S.* Typhimurium) but also efficacy differences among the feed types with the greatest reductions in pelleted and compound mash feed, then rapeseed meal with minimal reduction occurring in soybean meal. They also noted less effect with lower treatment temperatures. Consequently, they concluded that evaluation of candidate feed acids should take into account not only the respective *Salmonella* serovar, but the corresponding feed matrix and the temperature.

Carrique-Mas et al. (2007) suggested that the application approach of a feed additive could be important as they and others (Rouse et al., 1988) have noted that pretreatment prior to inoculation of *Salmonella* results in a more rapid decline in bacterial populations suggesting that pretreated feeds may be more resistant to subsequent contamination. This has practical significance as the potential for *Salmonella* cross contamination during milling is a concern. For example, in a study by Jones and Richardson (2004), *Salmonella* recontamination originated from dust in the feed mill and they concluded that potential cross contamination between areas of the mill operation could occur. Likewise, Jones (2011) concluded that thermal processes such as pelleting could reduce *Salmonella* levels, but recontamination may occur post-pelleting and suggested that the addition of chemical disinfectants could diminish potential recontamination. Along these lines, Cochrane et al. (2016) examined post-rendering chemical treatment of rendered feed ingredients by comparing a wide range of feed additives including a MCFA blend (caproic, caprylic, and capric acids) with an organic acid blend (lactic, formic, propionic, and benzoic acids), an EO blend (garlic oleoresin, turmeric oleoresin, capsicum oleoresin, rosemary extract, and wild oregano), sodium bisulfate, and a commercial formaldehyde product. They initially treated the rendered protein feed ingredients (feather meal, blood meal, MBM, and poultry by-product meal) with the corresponding feed additive followed by spray inoculation with a *S.* Typhimurium strain. They observed that feed ingredient matrix influenced *Salmonella* persistence as blood meal and MBM yielded similar populations to each other and were higher than the populations recovered from feather meal and poultry by-product meal. Out of all the products examined, they concluded that the MCFA blend and the formaldehyde commercial product were the most effective in preventing *S.* Typhimurium post-processing contamination, but time and feed matrix type were all factors in reducing *S.* Typhimurium levels.

Studies have been conducted over the years to directly compare feed additive organic acid blends with formaldehyde. In their early work, Smyser and Snoeyenbos (1979) compared 12 different compounds as antimicrobials for *Salmonella* when these compounds were added to meat and bone meal

(MBM). Several acids and non-acid antimicrobials were examined including among others, acetic acid, oleic acid, propionate salts, benzoic, sorbic, methylparaben, formalin at 0.05%, 0.1%, 0.12%, and 0.2% (w/w), and some commercial blends. *S.* Infantis that was nalidixic acid resistant was used as the marker strain to inoculate the samples set at a moisture level in the MBM to support *Salmonella* growth. Plate enumerations were conducted starting at 2 to 3 days post-inoculation and continued for anywhere from 1 to 2 weeks thereafter. All compounds except formalin at levels greater than 0.1% failed to prevent *S.* Infantis growth. The authors noted that while initial declines in *S.* Infantis occurred for many of the additives, the pH of the feed mixtures also became alkaline over time with spoilage ensuing.

8 Short chain organic acids: feeding studies

While direct reduction of pathogens in feeds by inclusion of short chain organic acids could be viewed as a critical step, retention of antimicrobial activity once the treated feed enters the GIT would seemingly provide additional benefit. Several studies have examined organic acid treated feeds impact on pathogen establishment in the poultry GIT and the results in general have been mixed. Iba and Berchieri Jr. (1995) tested a commercial formic-propionic acid mixture using an experimental *Salmonella* broiler chicken infection model. Of the four serovars tested, *S.* Enteritidis, *S.* Typhimurium, and *S.* Agona were not detected in cecal contents of birds fed the treated feed but *S.* Infantis could still be detected. This acid combination also appears to reduce the mortality and morbidity associated with fowl typhoid caused by the serovar *S.* Gallinarum (Berchieri Jr. and Barrow, 1996). Similar results were reported for *S.* Pullorum when a 2% formic-propionic acid combination was included in the diet and fed to young layer chicks (Al-Tarazi and Alshawabkeh, 2003).

Al-Natour and Alshawabkeh (2005) compared *S.* Gallinarum response to feed containing different levels of 85% formic acid added at either 0%, 0.5%, 1.0%, or 1.5% v/w of the diet mix. While all levels decreased *S.* Gallinarum in the feed samples, the 0.5% formic acid did not reduce colonization of S. Gallinarum in the infected broiler chicks in any of the segments of the avian GIT (crop, small intestine, large intestine, and ceca), but the higher levels did decrease colonization and this corresponded to a decrease in GIT pH levels in these same segments. More recently, Bourassa et al. (2018) using a seeder broiler pen of birds challenged with a *S.* Typhimurium isolate housed next to non-challenged birds, demonstrated that feeding formic acid decreased cecal *S.* Typhimurium over time but did not significantly impact the recovery of *S.* Typhimurium in the litter or the environment.

Physical form could also influence the effectiveness of an antimicrobial in the GIT. Papenbrock et al. (2005) applied potassium diformate to a cereal grain-based diet subjected to coarse grinding (over 50% of feed particles greater than 1.4 m in diameter) and compared this with pelleted and finely ground diets in pigs experimentally infected with *S.* Derby. Coarse grinding resulted in increased population levels of colonic chime lactobacilli and led to a reduced *Salmonella* excretion rate, shorter shedding periods, and translocation reductions in the infected pigs and thus ultimately decreasing dissemination of *S.* Derby among the group of pigs. This is consistent with earlier work by Mikkelsen et al. (2004) who observed that coarsely ground meal fed to pigs exhibited a greater concentration of undissociated lactic acid in the gastric contents and this in turn correlated with the death rate of *S.* Typhimurium. Given the importance of the avian gizzard and its influence on particle size of incoming feeds, it would be interesting to determine the relationship between size and development of the gizzard and the relative efficacy of an organic acid feed additive of feeds with different particle sizes.

Age of bird may be a factor as well. When Thompson and Hinton (1997) examined the antibacterial activity of a commercial formic (68%)-propionic (20%) acid treatment in 1-year-old laying hens, they did not see a change in GIT pH, although they did detect increased levels of these acids in the crop and gizzard, but a decrease in lactic acid in the crop. They assumed the decrease in lactic acid reflected a negative impact on the crop indigenous lactic acid bacteria. However, the chemical additive was bactericidal to *S.* Enteritidis when tested in an *in vitro* crop simulation. After examining the dynamics of chemical feed additives in the upper part of the avian GIT, they concluded that the upper GIT is probably the right target. This is consistent with the conclusion of Hume et al. (1993b) that most of these acids may not even reach the lower part of the GIT.

This is also in line with the emphasis placed by Moquet et al. (2016) on the release location for impacting the mode of action of butyrate derivatives in the bird's GIT. The form of the acid administered may be important for retaining activity as well. For example, in summarizing previous reports, Dennis and Blanchard (2004) noted that a high percentage of formic acid in the form of potassium diformate passed through the stomach into the small and large intestine of pigs. Hume et al. (1993b) did not see consistent reductions in *S.* Typhimurium when feeding propionate at a level of 30 μmol/g in the feed and concluded that dietary propionic acid was not effective and that this was due to propionate absorption in the upper GIT of the bird (Hume et al., 1993a). Oakley et al. (2014) followed cecal microbiome changes in broilers over 42 days fed feed supplemented with various combinations of organic acids and orally challenged with *S.* Typhimurium. Cecal contents were sampled on days 0, 7, 21, and 42. While cecal microbiome composition changed considerably over time as the birds matured, organic acids had minimal impact on either *Salmonella* levels or the cecal microbiome composition. However, as pointed out earlier, there is some question as to whether organic acids do in fact actually reach the cecal compartment and if not this could at least partially explain the relative lack of response by the cecal microbiome. It would be interesting in future studies to monitor microbial composition changes in the upper GIT of these birds fed organic acids and/or compare with birds fed encapsulated or protected forms of organic acids used in the studies by Van Immerseel et al. (2004, 2005).

Now that GIT bacterial populations can be identified through next-generation-based microbiome sequencing (Roto et al., 2015; Ricke et al., 2017), it is possible to examine GIT microbial compositional changes in these resident populations before, during, and after exposure to short chain organic acid feed additives. In future studies, microbiome populations in the upper part of avian GIT should also be examined in more detail. Not only could overall bacterial population shifts and changes be determined in response to feed chemicals, it would be of interest to conduct extensive omics analyses to see if more 'acid tolerant' resident bacteria are selected as exposure to acid chemical additives are fed to the birds over their lifetime and if their presence and/or metabolic activities creates additional barriers to pathogen colonization.

9 Conclusion

While a fairly wide range of chemical, physical, and biological agents have been examined and in some cases commercially applied over the years as feed additives, organic acids remain one of the more frequently used feed additives from a commercial standpoint. These products are considered generally effective as feed additives, but not only are the short chain organic acids chemically distinct, they also may possess different antimicrobial mechanism(s) against pathogens such as *Salmonella*. Likewise, their activity in the GIT, once consumed by the animal or bird, may be different as well. Organic acids in general appear to behave differently against the various *Salmonella* serovars. Tolerance/resistance mechanisms may be similar, although survivability among serovars in the presence of different organic acids does appear to differ and this seems to be true even within strains of the same serovar. Other genomic responses such as virulence gene levels also seem to be different among strains within a particular serovar. Most of this antimicrobial mechanism research has been done with the three SCFA, namely, acetate, propionic, and butyric acids and much less is known about serovar specificity against formic acid or lactate. This should be a focus of future research as more organic acids are examined as potential feed additives.

There are clearly different impacts of feed additives on *Salmonella* depending on the feed matrix. However, there are multiple factors associated with the feed matrix to be considered when examining the efficacy. Certainly, water activity of the feed matrix is important as it appears that hydration enhances the activity of most of these antimicrobial compounds and they are relatively less effective in dry forms of the same feeds. The water activity of the feed may be important as it sets up the physiological state (growth or stationary phase) of the corresponding *Salmonella* serovar in the feed matrix. The physiological status may also affect detection and recovery strategies when attempting to culture and/or enumerate *Salmonella* originating from feed matrices. Certainly 'masking' by carryover of feed additive acids into recovery media has been identified as a critical issue for organic acids. Temperature of the feed matrix also appears to be a major contributor to *Salmonella* response to feed additives and as it becomes more elevated antimicrobial activity is enhanced (Milillo et al., 2011). Herein are opportunities for potentially creating multiple hurdle combinations of feed additives and heating to create synergistic antimicrobial activity. However, adding antimicrobials prior to a thermal process such as pelleting may have deleterious effects on equipment depending on how caustic the antimicrobial is that has been added and overall antimicrobial impact may be lessened if heating leads to vaporization of the feed antimicrobial compound.

Feed composition appears to matter as well. While there are a range of studies looking at different feeds and antimicrobial combinations, few attempt to do exhaustive comparative assessments. The study by Cochrane et al. (2016) indicates that the source of the rendered protein certainly may matter both in terms of *Salmonella* survival and its corresponding recovery as well as antimicrobial effectiveness of the feed additive. While not well documented, it would be assumed that the interaction of organic acids with the feed matrix could be different. This may be important once these treated feeds are consumed by the animal or bird. The stability of the antimicrobial in the feed as it becomes digesta in the GIT may impact how effective it is against pathogens in the tract. It appears that at least some organic acids such as propionic acid rapidly disappear in the upper part of the GIT and therefore have little impact in the lower regions. It is conceivable that encapsulated products are more stable in the GIT and therefore are more likely to reach the lower parts of the GIT. It would be interesting to compare labeled encapsulated products with their unprotected counterparts similar to the work done by Hume et al. (1993a) with propionate to determine whether that is indeed the case and how much is metabolized by the bird.

While it would be assumed that these antimicrobials would be effective against non-*Salmonella* pathogenic or zoonotic microorganisms that are physiologically similar to *Salmonella*, most of the focus has been on *Salmonella* in feed applications of antimicrobials. Certainly, compounds effective against *Salmonella* would be assumed to be equally effective against other foodborne pathogens such as pathogenic *Escherichia coli* and *Listeria* which have been or could potentially be associated occasionally with feeds. However, sporeformers such as *Clostridia* would be much more of a challenge given their inherent resistance. Given challenges associated with natural occurring, sporeformers such as *Clostridia,* the addition of surrogate sporeformer organisms may be an option to test effectiveness of feed antimicrobials against sporeformers in much the same fashion as they would be for testing the effectiveness of thermal processes. Okelo et al. (2008) suggested that *Bacillus stearothermophilus* spores could be a potential surrogate organism to assess efficacy of feed sterilization during the extrusion step in feed processing. As far as animal viruses are concerned, some preliminary work has been done to test chemical feed additives against PEDv in swine feed and ingredients (Cochrane et al., 2015). More research needs to be done with chemical mitigation of other animal and avian viruses. However, in some cases, depending on the pathogenicity and concerns over their potential environmental dissemination, this may require that any feed additive experiments conducted be done with surrogate nonpathogenic viruses as substitutes.

Short chain organic acids have potential to be effective feed additives. Historically, most of the emphasis has been on pathogen reduction in the feed but now more is being placed on GIT efficacy. Pathogen reduction in the presence of acids has yielded mixed results, but this may be a consequence of GIT location where most of the acid is likely present as some studies indicate that very little reaches the lower GIT of the bird. Derivatization or encapsulation to protect short chain acids offers a potential means to deliver these antimicrobials to the lower sections of the GIT. Given the considerable barrier that fermentation acids generated in the lower GIT present to pathogen establishment, increasing the delivery of external organic acids to these lower GIT sections would appear to have the potential to be highly effective. However, the issue of cost for encapsulation would have to be considered before commercial implementation. Emphasis on the impact of organic acids on GIT function and bird health is relatively new phenomena and it would be anticipated that there would be extensive host functional reactions to the presence of high levels of external organic acids. Bedford and Gong (2018), in summarizing the responses of animal production to butyrate and its derivatives, concluded that in addition to pathogen control there were several positive effects including enhanced reduction of inflammation, improved growth performance as well as improved carcass traits. It will be of interest to determine host responses to the other short chain organic acids as to whether similar host responses occur or if unique host characteristics are prevalent with each individual organic acid. If different, then opportunities to optimize host responses with short chain organic acid mixtures and their corresponding protected versions could be taken advantage for optimization of host GIT health and animal performance.

10 Where to look for further information

Bedford, A. and Gong, J. (2018) Implications of butyrate and its derivatives for gut health and animal production. *Anim. Nutr.* 4, 151-9.

Broom, L. J. (2015) Organic acids for improving intestinal health of poultry. *World's Poult. Sci. J.* 71, 630-42.

Cherrington, C. A., Hinton, M., Mead, G. C. and Chopra, I. (1991a) Organic acids: Chemistry: antimicrobial activity and practical applications. *Adv. Microb. Physiol.* 32, 87-108.

Dittoe, D. K., Ricke, S. C. and Kiess, A. S. (2018) Organic acids and potential for modifying the avian gastrointestinal tract and reducing pathogens and disease. *Front. Vet. Sci.* 5, 216. doi:10.3389/fvets.2018.00216.

Jones, F. T. (2011) A review of practical *Salmonella* control measures in animal feed. *J. Appl. Poult. Res.* 20, 102-113.

Maciorowski, K. G., Pillai, S. D. Jones, F. T. and Ricke, S. C. (2005) Polymerase chain reaction detection of foodborne *Salmonella* spp. in animal feeds. *Crit. Revs. Microbiol.* 31, 45-53.

Moquet, P. C. A., Onrust, L., Van Immerseel, F., Ducatelle, R., Hendriks, W. H. and Kwakkel, R. P. (2016) Importance of release location on the mode of action of butyrate derivatives in the avian gastrointestinal tract. *World's Poult. Sci. J.* 72, 61-80.

Ricke, S. C. (2017) Chapter 8. Feed hygiene. In: J. Dewulf and F. Van Immerseel (Eds), *Biosecurity in Animal Production and Veterinary Medicine*. ACCO, Leuven, Belgium, pp. 144-76.

Ricke, S. C. (2003) Perspectives on the use of organic acids and short chain fatty acids as antimicrobials. *Poult. Sci.* 82, 632-39.

Van Immerseel, F., Russell, J. B., Flythe, M. D., Gantois, I., Timbermont, L., Pasmans, F., Haesebrouck, F. and Ducatelle, R. (2006) The use of organic acids to combat *Salmonella* in poultry: A mechanistic explanation of the efficacy. *Avian Pathol.* 35, 182-8.

Wales, A. D., Allen, V. M. and Davies, R. H. (2010) Chemical treatment of animal feed and water for the control of *Salmonella*. *Foodborne Pathog. Dis.* 7, 1-15.

11 References

Al-Natour, M. Q. and Alshawabkeh, K. M. (2005) Using varying levels of formic acid to limit growth of *Salmonella gallinarum* in contaminated broiler feed. *Asian-Australas. J. Anim. Sci.* 18, 390-5.

Al-Tarazi, Y. H. and Alshawabkeh, K. M. (2003) Effect of dietary formic acid and propionic acids on *Salmonella pullorum* shedding and morality in layer chicks after experimental infection. *J. Vet. Med. B* 50, 112-17.

Alvarez, J., Porwollik, S., Laconcha, I., Gisakis, V., Vivanco, A. B., Gonzalez, I., Echenagusia, S., Zabala, N., Blackmer, F., McClelland, M., Rementaria, A. and Garaizar, J. (2003) Detection of *Salmonella enterica* serovar California strain spreading in Spanish feed mills and genetic characterization with DNA microarrays *Appl. Environ. Microbiol.* 69, 7531-4.

Amerah, A. M., Ravindran, V., Lentle, R. G. and Thomas, D. G. (2007) Feed particle size: Implications on the digestion and performance of poultry. *World's Poult. Sci. J.* 63, 439-55.

Andino, A., Pendleton, S., Zhang, N., Chen, W., Critzer, F. and Hanning, I. (2014) Survival of *Salmonella enterica* in poultry feed is strain dependent. *Poult. Sci.* 93, 441–7.

Annison, K. F., Hill, K. J. and Kenworthy, R. (1968) Volatile fatty acids in the digestive tract of the fowl. *Br. J. Nutr.* 22, 207–16.

Bayer, R. C., Hoover, W. H. and Muir, F. V. (1978) Dietary fiber and meal feeding influence on broiler growth and crop fermentation. *Poult. Sci.* 57, 1456–9.

Bedbury, H. P. and Duke, G. E. (1983) Cecal microflora of turkeys fed low or high fiber diets: Enumeration, identification and determination of cellulolytic activity. *Poult. Sci.* 62, 675–82.

Bedford, A. and Gong, J. (2018) Implications of butyrate and its derivatives for gut health and animal production. *Anim. Nutr.* 4, 151–9.

Berchieri Jr., A. and Barrow, P. A. (1996) Reduction in incidence of experimental fowl typhoid by incorporation of commercial formic acid preparation (Bio-Add™) into poultry feed. *Poult. Sci.*75, 339–41.

Bourassa, D. V., Wilson, K. M., Ritz, C. R., Kiepper, B. K. and Buhr, R. J. (2018) Evaluation of the addition of organic acids in the feed and/or water for broilers and the subsequent recovery of *Salmonella* Typhimurium from litter and ceca. *Poult. Sci.* 97, 64–73.

Broom, L. J. (2015) Organic acids for improving intestinal health of poultry. *World's Poult. Sci. J.* 71, 630–42.

Callaway, T. R. and Ricke, S. C. (Eds) (2012) *Direct Fed Microbials/Prebiotics for Animals: Science and Mechanisms of Action*. Springer Science, New York, NY, 206pp.

Carrique-Mas, J. J., Bedford, S. and Davies, R. H. (2007) Organic acid and formaldehyde treatment of animal feeds to control *Salmonella*: Efficacy and masking during culture. *J. Appl. Microbiol.* 103, 88–96.

Cherrington, C. A., Hinton, M., Mead, G. C. and Chopra, I. (1991a) Organic acids: Chemistry: antimicrobial activity and practical applications. *Adv. Microb. Physiol.* 32, 87–108.

Cherrington, C. A., Chopra, I. and Hinton, M. (1991b) Acid control of feed for the control of *Salmonella* infections in poultry. *Vet. Annu.* 31, 90–5.

Choct, M. (2009) Managing gut health through nutrition. *Br. Poult. Sci.* 50, 9–15.

Cochrane, R. A., Dritz, S. S., Woodworth, J. C. and Jones, C. K. (2015) Evaluating chemical mitigation of PEDv in swine feed and ingredients. *J. Anim. Sci.* 92(E-Suppl. 2) (Abstract).

Cochrane, R. A., Huss, A. R., Aldrich, G. C., Stark, C. R. and Jones, C. A. (2016) Evaluating chemical mitigation of *Salmonella* Typhimurium ATCC 14028 in animal feed ingredients. *J. Food Prot.* 79, 672–6.

Corrier, D. E., Nisbet, D. G., Scanlan, C. M., Hollister, A. G. and DeLoach, J. R. (1995) Control of *Salmonella typhimurium* colonization in broiler chicks with a continuous-flow characterized mixed culture of cecal bacteria. *Poult. Sci.* 74, 916–24.

Cox, N. A., Cason, J. A., Buhr, R. J., Richardson, K. E., Richardson, L. J., Rigsby, L. L., Fedorka-Cray, P. J. (2013) Variation in preenrichment pH of poultry feed and feed ingredients after incubation periods up to 48 hours. *J. Appl. Poult. Res.* 22, 190–5.

Cox, N. A., Richardson, K. F., Cosby, D. E., Berrang, M. F., Cason, J. A., Rigsby, L. L., Holcombe, N. and DeRome, L. (2016) Injury and death of various *Salmonella* serotypes due to acidic conditions. *J. Appl. Poult. Res.* 25, 62–6.

Crump, J. A., Griffin, P. M. and Angulo, F. J. (2002) Bacterial contamination of animal feed and its relationship to human foodborne illness. *Clin. Inf. Dis.* 35, 859–65.

Dennis, I. and Blanchard, P. (2004) Effect of feeding potassium diformate on incidence of *Salmonella* infection on a commercial unit. *Pig J.* 54, 157-60.

Dittoe, D. K., Ricke, S. C. and Kiess, A. S. (2018) Organic acids and potential for modifying the avian gastrointestinal tract and reducing pathogens and disease. *Front. Vet. Sci.* 5, 216. doi:10.3389/fvets.2018.00216.

Duke, G. E., Eccleston, E., Kirkwood, S., Louis, C. F. and Bedbury, H. P. (1984) Cellulose digestion by domestic turkeys fed low of high fiber diets. *J. Nutr.* 114, 95-102.

Dunkley, K. D., Dunkley, C. S., Njongmeta, N. L., Callaway, T. R., Hume, M. E., Kubena, L. F., Nisbet, D. J. and Ricke, S. C. (2007a) Comparison of *in vitro* fermentation and molecular microbial profiles of high-fiber feed substrates (HFFS) incubated with chicken cecal inocula. *Poult. Sci.* 86, 801-10.

Dunkley, K. D., McReynolds, J. L., Hume, M. E., Dunkley, C. S., Callaway, T. R., Kubena, L. F., Nisbet, D. J. and Ricke, S. C. (2007b) Molting in *Salmonella* Enteritidis-challenged laying hens fed alfalfa crumbles. II. Fermentation and microbial ecology response. *Poult. Sci.* 86, 2101-2109.

Dunkley, K. D., Callaway, T. R., Chalova, V. I. Anderson, R. C., Kundinger, M. M., Dunkley, C. S., Nisbet, D. J. and Ricke, S. C. (2008) Growth and genetic responses of *Salmonella* Typhimurium to pH-shifts in an anaerobic continuous culture. *Anaerobe* 14, 35-42.

Dunkley, K. D., Callaway, T. R., O'Bryan, C., Kundinger, M. M., Dunkley, C. S., Anderson, R. C., Nisbet, D. J., Crandall, P. G. and Ricke, S. C. (2009) Cell yields and fermentation responses of a *Salmonella* Typhimurium poultry isolate at different dilution rates in an anaerobic steady state continuous culture. *Antonie van Leeuwenhoek* 96, 537-44.

Durant, J. A., Lowry, V. K., Nisbet, D. J., Stanker, L. H. Corrier, D. E. and Ricke, S. C. (1999a) Short-chain volatile fatty acids affect the adherence and invasion of HEp-2 cells by *Salmonella typhimurium*. *J. Environ. Sci. Health Part B* 34, 1083-99.

Durant, J. A., Corrier, D. E., Byrd, J. A., Stanker, L. H. and Ricke, S. C. (1999b) Feed deprivation affects crop environment and modulates *Salmonella enteritidis* colonization and invasion of leghorn hens. *Appl. Environ. Microbiol.* 65, 1919-23.

Durant, J. A., Lowry, V. K., Nisbet, D. J., Stanker, L. H., Corrier, D. E. and Ricke, S. C. (2000a) Late logarithmic *Salmonella typhimurium* HEp-2 cell-association and invasion response to short chain volatile fatty acid addition. *J. Food Safety* 20, 1-11.

Durant, J. A., Lowry, V. K., Nisbet, D. J., Stanker, L. H., Corrier, D. E. and Ricke, S. C. (2000b) Short-chain fatty acids alter HEp-2 cell association and invasion by stationary growth phase *Salmonella typhimurium*. *J. Food Sci.* 65, 1206-9.

Durant, J. A., Corrier, D. E. and Ricke, S. C. (2000c) Short-chain volatile fatty acids modulate the expression of the *hilA* and *invF* genes of *Salmonella* Typhimurium. *J. Food Prot.* 63, 573-8.

Finstad, S., O'Bryan, C. A., Marcy, J. A., Crandall, P. G. and Ricke, S. C. (2012) *Salmonella* and broiler production in the United States: Relationship to foodborne salmonellosis. *Food Res. Int.* 45, 789-94.

Foley, S., Nayak, R., Hanning, I. B., Johnson, T. J., Han, J. and Ricke, S. C. (2011) Population dynamics of *Salmonella enterica* serotypes in commercial egg and poultry production. *Appl. Environ. Microbiol.* 77, 4273-9.

Foley, S. L., Johnson, T. J., Ricke, S. C., Nayak, R. and Danzeisen, J. (2013) *Salmonella* pathogenicity and host adaptation in chicken-associated serovars. *Microbiol. Mol. Biol. Revs.* 77, 582-607.

Fukuda, S., Toh, H., Hase, K., Oshima, K., Nakanishi, Y., Yoshimura, K., Tobe, T., Clarke, J. M., Topping, D. L., Suzuki, T., Taylor, T. D., Itoh, K., Kikuchi, J., Morita, H., Hattori, M. and

Ohno, H. (2011) Bifidobacteria can protect from enteropathogenic infection through production of acetate. *Nature* 469, 543–7.

Gabriel, I., Lessire, M., Mallet, S. and Guillot, J. F. (2006) Microflora of the digestive tract: Critical factors and consequences for poultry. *Worlds Poult. Sci. J.* 62, 499–511.

Ge, B., LaFon, P. C., Carter, P. J., McDermott, S. D., Abbott, J., Glenn, A., Ayers, S. L., Friedman, S. L., Paige, J. C., Wagner, D. D., Zhao, S., McDermott, P. F. and Rasmussen, M. A. (2013) Retrospective analysis of *Salmonella*, *Campylobacter*, *Escherichia coli*, and *Enterococcus* in animal feed ingredients. *Foodborne Pathog. Dis.* 10, 684–91.

Goldstein, D. L. (1989) Absorption by the cecum of wild birds: Is there interspecific variation? *J. Exp. Zool. (Suppl.)*, 3, 103–10.

González-Gil, F., Le Bolloch, A., Pendleton, S., Zhang, N., Wallis, A. and Hanning, I. (2012) Expression of *hilA* in response to mild acid stress in *Salmonella enterica* is serovar and strain dependent. *J. Food Sci.* 77, M292–M297.

Ha, S. D., Maciorowski, K. G., Jones, F. T., Kwon, Y. M. and Ricke, S. C. (1998a) Survivability of indigenous feed microflora and a *Salmonella typhimurium* marker strain in poultry feed treated with buffered propionic acid. *Anim. Feed Sci. Technol.* 75, 145–55.

Ha, S. D., Maciorowski, K. G., Kwon, Y. M., Jones, F. T. and Ricke, S. C. (1998b) Indigenous feed microflora and *Salmonella typhimurium* marker strain survival in poultry feed with varying levels of protein. *Anim. Feed Sci. Technol.* 76, 23–33.

Hargis, B. M., Caldwell, D. J., Brewer, R. L., Corrier, D. E. and DeLoach, J. R. (1995) Evaluation of the chicken crop as a source of *Salmonella* contamination for broiler carcasses. *Poult. Sci.* 74, 1548–52.

Hetland, H., Choct, M. and Svihus, B. (2004) Role of insoluble non-starch polysaccharides in poultry nutrition. *World's Poult. Sci. J.* 60, 415–22.

Hinton Jr., A., Buhr, R. J. and Ingram, K. D. (2000a) Reduction of *Salmonella* in the crop of broiler chickens subjected to feed withdrawal. *Poult. Sci.* 79, 1566–70.

Hinton Jr., A., Buhr, R. J. and Ingram, K. D. (2000b) Physical, chemical, and microbiological changes in the crop of broiler chickens subjected to incremental feed withdrawal. *Poult. Sci.* 79, 212–18.

Howard, Z. R., O'Bryan, C. A., Crandall, P. G. and Ricke, S. C. (2012) *Salmonella* Enteritidis in shell eggs: Current issues and prospects for control. *Food Res. Int.* 45, 755–64.

Hseih, Y.-C., Poole, T. L., Runyon, M., Hume, M. and Herrman, T. J. (2016) Prevalence of nontyphoidal *Salmonella* and *Salmonella* strains with conjugative antimicrobial-resistant serovars contaminating animal feed in Texas. *J. Food Prot.* 79, 194–204.

Hume, M. E. (2011) Historic perspective: Prebiotics, probiotics, and other alternatives to antibiotics. *Poult. Sci.* 90, 2663–9.

Hume, M. E., Corrier, D. E., Ivie, G. W. and DeLoach, J. R. (1993a) Metabolism of [^{14}C] propionic acid in broiler chicks. *Poult. Sci.* 72, 786–93.

Hume, M. E., Corrier, D. E., Ambrus, S., Hinton Jr., A. and DeLoach, J. R. (1993b) Effectiveness of dietary propionic acid in controlling *Salmonella typhimurium* colonization in broiler chicks. *Avian Dis.* 37, 1051–6.

Humphrey, T. J. and Lanning, D. G. (1988) The vertical transmission of salmonellas and formic acid treatment of chicken feed. *Epidemiol. Infect.* 100, 43–9.

Iba, A. M. and Berchieri Jr., A. (1995) Studies on the use of a formic acid-propionic acid mixture (Bio-add™) to control experimental *Salmonella* infection in broiler chickens. *Avian Path.* 24, 303–11.

Jarquin, R., Hanning, I., Ahn, S. and Ricke, S. C. (2009) Development of rapid detection and genetic characterization of *Salmonella* in poultry breeder feeds. *Sensors* 9, 5308–23.

Jiang, X. (2016) Prevalence and characterization of *Salmonella in* animal meals collected from rendering plants. *J. Food Prot.* 79, 1026–31.

Joerger, R. D., Sartori, C. A. and Kniel, K. E. (2009) Comparison of genetic and physiological properties of *Salmonella enterica* isolates from chickens reveals one major difference between serovar Kentucky and other serovars: Response to acid. *Foodborne Pathog. Dis.* 6, 503–12.

Joerger, R. D., Sartori, C., Frye, J. G., Jennifer B. Turpin, J. B., Schmidt, C., McClelland, M. and Porwollik, S. (2012) Gene expression analysis of *Salmonella enteric* Enteritidis Nal^R and *Salmonella enterica* Kentucky 3795 exposed to HCl and acetic acid in rich medium. *Foodborne Pathog. Dis.* 9, 331–7.

Jones, F. T. (2011) A review of practical *Salmonella* control measures in animal feed. *J. Appl. Poult. Res.* 20, 102–13.

Jones, F. T. and Richardson, K. E. (2004) *Salmonella* in commercially manufactured feeds. *Poult. Sci.* 83, 384–91.

Jones, F. T. and Ricke, S. C. (1994) Researchers propose tentative HACCP plan for feed manufacturers. *Feedstuffs* 66(18), 32, 36–8, 40–2.

Józefiak, D., Rutkowski, A. and Martin, S. A. (2004) Carbohydrate fermentation in the ceca: A review. *Anim. Feed Sci. Technol.* 113, 1–15.

Kareem, K. Y., Loh, T. C., Foo, H. L., Asmara, S. A. and Akit, H. (2017) Influence of postbiotic RG14 and inulin combination on cecal microbiota, organic acid concentration, and cytokine expression in broiler chickens. *Poult. Sci.* 96, 966–75.

Kim, S. A., Park, S. I., Lee, S. I. and Ricke, S. C. (2017) Development of a rapid method to quantify *Salmonella* Typhimurium using a combination of MPN with a qPCR and a shortened time. *Food Microbiol.* 65, 7–18.

Kim, S. A., Jang, M. J., Kim, S. Y., Yang, Y., Pavlidis, H. O. and Ricke, S. C. (2019) Potential for prebiotics as feed additives to limit foodborne *Campylobacter* establishment in the poultry gastrointestinal tract. *Front. Microbiol.* 10, 91. doi:10.3389/fmicb.2019.00091.

Koyuncu, S., Andersson, G., Vos, P. and Häggblom, P. (2011) DNA microarray for tracing *Salmonella* in the feed chain. *Int. J. Food Microbiol.* 145, 518–22.

Koyuncu, S., Andersson, M. G., Löfström, C., Skandamis, P. N., Gounadaki, A., Zentek, J. and Häggblom, P. (2013) Organic acids for control of *Salmonella* in different feed materials. *BMC Vet. Res.* 9(81), 1–9.

Kwon, Y. M. and Ricke, S. C. (1998) Induction of acid resistance of *Salmonella typhimurium* by exposure to short-chain fatty acids. *Appl. Environ. Microbiol.* 64, 3458–63.

Lan, Y., Williams, B. A., Tamminga, S., Boer, H., Akkermans, A., Erdi, G. and Verstegen, M. W. A. (2005) *In vitro* fermentation kinetics of some non-digestible carbohydrates by the caecal microbial community of broilers. *Anim. Feed Sci. Technol.* 123–4, 687–702.

Lianou, A. and Koutsoumanis, K. P. (2013) Strain variability of the behavior of foodborne bacterial pathogens: A review. *Int. J. Food Microbiol.* 167, 310–21.

Maciorowski, K. G., Jones, F. T., Pillai, S. D. and Ricke, S. C. (2004) Incidence, sources, and control of food-borne *Salmonella* spp. in poultry feed. *World's Poult. Sci. J.* 60, 446–57.

Maciorowski, K. G., Pillai, S. D. Jones, F. T. and Ricke, S. C. (2005) Polymerase chain reaction detection of foodborne *Salmonella* spp. in animal feeds. *Crit. Rev. Microbiol.* 31, 45–53.

Maciorowski, K. G., Hererra, P., Jones, F. T., Pillai, S. D. and Ricke, S. C. (2006a) Cultural and immunological detection methods for *Salmonella* spp. in animal feeds - A review. *Vet. Res. Comm.* 30, 127-37.

Maciorowski, K. G., Herrera, P., Kundinger, M. M. and Ricke, S. C. (2006b) Animal feed production and contamination by foodborne *Salmonella*. *J. Consum. Prot. Food Safety* 1, 197-209.

Maciorowski, K. G., Herrera, P., Jones, F. T., Pillai, S. D. and Ricke, S. C. (2007) Effects of poultry and livestock of feed contamination with bacteria and fungi. *Anim. Feed Sci. Technol.* 133, 109-36.

Malorny, B., Löfström, C., Wagner, M., Krämer, N. and Hoorfar, J. (2008) Enumeration of *Salmonella* in food and feed samples by real-time PCR for quantitative microbial risk assessment. *Appl. Environ. Microbiol.* 74, 1299-304.

Mani-López, E., García, H. S. and López-Malo, A. (2012) Organic acids as antimicrobials to control *Salmonella* in meat and poultry products. *Food Res. Int.* 45, 713-21.

Marounek, M. and Rada V. (1998) Age effect on in vitro fermentation pattern and methane production in the caeca of chickens. *Physiol. Res.* 47, 259-63.

Marounek, M., Suchorska, O. and Savka, O. (1999) Effect of substrate and feed antibiotics on in vitro production of volatile fatty acids and methane in caecal contents of chickens. *Anim. Feed Sci. Technol.* 80, 223-30.

McHan, F. and Shotts, E. B. (1993) Effect of short-chain fatty acids on the growth of *Salmonella typhimurium* in an *in vitro* system. *Avian Dis.* 37, 396-98.

Mead, G. C. (1989) Microbes of the avian cecum: Types present and substrates utilized. *J. Exp. Zool. (Suppl.)* 3, 48-54.

Micciche, A. C., Foley, S. L., Pavlidis, H. O., McIntyre, D. R. and Ricke, S. C. (2018) A review of prebiotics against *Salmonella* in poultry: Current and future potential for microbiome research application. *Front. Vet. Sci.* 5, 191. doi:10.3389/fvets.2018.00191.

Mikkelsen, L. L., Naughton, P. J., Hedemann, M. S. and Jensen, B. B. (2004) Effects of physical properties of feed on microbial ecology and survival of *Salmonella enterica* serovar Typhimurium in the pig gastrointestinal tract. *Appl. Environ. Microbiol.* 70, 3485-92.

Milillo, S. R. and Ricke, S. C. (2010) Synergistic reduction of *Salmonella* in a model raw chicken media using a combined thermal and organic acid salt intervention treatment. *J. Food Sci.* 75, M121-5.

Milillo, S. R., Martin, E., Muthaiyan, A. and Ricke, S. C. (2011) Immediate reduction of *Salmonella enterica* serotype Typhimurium following exposure to multiple-hurdle treatments with heated, acidified organic acid salt solutions. *Appl. Environ. Microbiol.* 77, 3765-72.

Miller, T. L., Wolin, M. J. and Kusel., E. A. (1986) Isolation and characterization of methanogens from animal feces. *Syst. Appl. Microbiol.* 8, 234-8.

Moquet, P. C. A., Onrust, L., Van Immerseel, F., Ducatelle, R., Hendriks, W. H. and Kwakkel, R. P. (2016) Importance of release location on the mode of action of butyrate derivatives in the avian gastrointestinal tract. *World's Poult. Sci. J.* 72, 61-80.

Morita, T., Iida, T. and Kamata, S.-I. (2005) Condition of oilmeal and control of *Salmonella* using organic acid in an oilmeal manufacturing plant. *Jpn. J. Anim. Hygiene* 31, 19-24.

Nisbet, D. J. (2002) Defined competitive exclusion cultures in the prevention of enteropathogen colonisation in poultry and swine. *Antonie van Leeuwenhoek* 81, 481-6.

Nisbet, D. J., Corrier, D. E., Ricke, S. C., Hune, M. E., Byrd II, J. A. and DeLoach, J. R. (1996a) Maintenance of the biological efficacy in chicks of a cecal competitive-exclusion against *Salmonella* continuous-flow fermentation. *J. Food Prot.* 58, 1279-83.

Nisbet, D. J., Corrier, D. E., Ricke, S. C., Hume, M. E., Byrd II, J. A. and DeLoach, J. R. (1996b) Cecal propionic acid as a biological indicator of the early establishment of a microbial ecosystem inhibitory to *Salmonella* in chicks. *Anaerobe* 2, 345-50.

Oakley, B. B., Morales, C. A., Line, J., Berrang, M. E., Meinersmann, R. J., Tillman, G. E., Wise, M. G., Siragusa, G. R., Hiett, K. L. and Seal, B. S. (2013) The poultry-associated microbiome: Network analysis and farm-to-fork characterizations. *PLoS ONE* 8(2), e57190. doi:10.1371/journal.pone.0057190.

Oakley, B. B., Buhr, R. J., Ritz, C. W., Kiepper, B. H., Berrang, M. E., Seal, B. S. and Cox, N. A. (2014) Successional changes in the chicken cecal microbiome during 42 days of growth are independent of organic acid feed additives. *BMC Vet. Res.* 10, 282. http://www.biomedcentral.com/1746-6148/10/282.

Okelo, P. O., Joseph, S. W., Wagner, D. D., Wheaton, F. W., Douglass, L. W. and Carr, L. E. (2008) Improvements in reduction of feed contamination: An alternative monitor of bacterial killing during feed extrusion. *J. Appl. Poult. Res.* 17, 219-28.

Papenbrock, S., Stemme, K., Amtsberg, G., Verspohl, J. and Kamphues, J. (2005) Investigations on prophylactic effects of coarse feed structure and/or potassium diformate on the microflora in the digestive tract of weaned piglets experimentally infected with *Salmonella* Derby. *J. Anim. Physiol. Anim. Nutr.* 89, 84-7.

Park, S. H. and Ricke, S. C. (2015) Development of multiplex and quantitative PCR assays for simultaneous detection of *Salmonella* genus, *Salmonella* subspecies I, *S.* Enteritidis, *S.* Heidelberg, and *S.* Typhimurium. *J. Appl. Microbiol.* 118, 152-60.

Park, S. H., Aydin, M., Khatiwara, A., Dolan, M. C., Gilmore, D. F., Bouldin, J. L., Ahn, S. and Ricke, S. C. (2014) Current and emerging technologies for rapid detection and characterization of *Salmonella* in poultry and poultry products. *Food Microbiol.* 38, 250-62.

Patterson, J. A. and Burkholder, K. M. (2003) Application of prebiotics and probiotics in poultry production. *Poult. Sci.* 82, 627-31.

Pumfrey, L. and Nelson, C. E. (1991) Use of a most probable number method modified with a deoxyribonucleic probe to monitor control by food preservatives of natural *Salmonella* contamination in animal meat meals. *Poult. Sci.* 70, 780-4.

Ramirez, G. A., Sarlin, L. L., Caldwell, D. J., Yezak Jr., C. R., Hume, M. E., Corrier, D. E., DeLoach, J. R. and Hargis, B. M. (1997) Effect of feed withdrawal on the incidence of *Salmonella* in the crops and ceca of market age broiler chickens. *Poult. Sci.* 76, 654-6.

Read, M. N. and Holmes, A. J. (2017) Towards an integrative understanding of diet-host-gut microbiome interactions. *Front. Immunol.* 8, 538. doi: 10.3389/fimmu.2017.00538.

Redig, P. T. (1989) The avian ceca: Obligate combustion chambers or facultative afterburners? - The conditioning influence of diet. *J. Exp. Zool. (Suppl.)* 3, 66-9.

Rehman, H. U., Vahjen, W., Awad, W. A. and Zentek, J. (2007) Indigenous bacteria and bacterial metabolic products in the gastrointestinal tract of broiler chickens. *Arch. Anim. Nutr.* 61, 319-35.

Rehman, H. U., Vahjen, W., Kohl-Parisini, A., Ijaz, A. and Zentek, J. (2009) Influence of fermentable carbohydrates on the intestinal bacteria and enteropathogens in broilers. *World's Poult. Sci. J.* 65, 75-89.

Ricke, S. C. (2003) Perspectives on the use of organic acids and short chain fatty acids as antimicrobials. *Poult. Sci.* 82, 632-9.

Ricke, S. C. (2005) Chapter 7. Ensuring the safety of poultry feed. In: G. C. Mead (Ed.), *Food Safety Control in Poultry Industry*. Woodhead Publishing Limited, Cambridge, UK, pp. 174-94.

Ricke, S. C. (2017) Chapter 8. Feed hygiene. In: J. Dewulf and F. Van Immerseel (Eds), *Biosecurity in Animal Production and Veterinary Medicine*. ACCO, Leuven, Belgium, pp. 144-76.

Ricke, S. C. (2018) Impact of prebiotics on poultry production and food safety. *Yale J. Biol. Med.* 91, 151-9.

Ricke, S. C., Pillai, S. D., Norton, R. A., Maciorowski, K. G. and Jones, F. T. (1998) Applicability of rapid methods for detection of *Salmonella* spp. in poultry feeds: A review. *J. Rapid Meths. Auto. Microbiol.* 6, 239-58.

Ricke, S. C., Dunkley, C. S. and Durant, J. A. (2013) A review on development of novel strategies for controlling *Salmonella* Enteritidis colonization in layer hens: Fiber-based molt diets. *Poult. Sci.* 92, 502-25.

Ricke, S. C., Hacker, J., Yearkey, K., Shi, Z., Park, S. H. and Rainwater, C. (2017) Chapter 19. Unravelling food production microbiomes: Concepts and future directions. In: Ricke, S. C., Atungulu, G. G., Park, S. H. and Rainwater, C. E. (Eds), *Food and Feed Safety Systems and Analysis*. Elsevier Inc., San Diego, CA, pp. 347-74.

Ricke, S. C., Kim, S. A. and Park, S. H. (2018) Molecular-based identification and detection of *Salmonella* in food production systems: Current perspectives. *J. Appl. Microbiol.* 125, 313-27.

Rivera-Chávez, F., Zhang, L. F., Faber, F., Lopez, C. A., Byndloss, M. X., Olsan, E. E., Xu, G., Velazquez, E. M., Lebrilla, C. B., Winter, S. E. and Bäumler, A. J. (2016) Depletion of butyrate-producing *Clostridia* from the gut microbiota drives an aerobic luminal expansion of *Salmonella*. *Cell Host Microbe* 19, 443-54.

Roto, S. M., Rubinelli, P. M. and Ricke, S. C. (2015) An introduction to the avian gut microbiota and the effects of yeast-based prebiotic compounds as potential feed additives. *Front. Vet. Sci.* 2, 28, 1-18. doi: 10.3389/fvets.2015.00028.

Rouse, J., Rolow, A. and Nelson, C. E. (1988) Research note: Effect of chemical treatment of poultry feed on survival of *Salmonella*. *Poult. Sci.* 67, 1225-8.

Rubinelli, P., Roto, S., Kim, S. A., Park, S. H., Pavlidis, H. O., McIntyre, D. and Ricke, S. C. (2016) Reduction of *Salmonella* Typhimurium by fermentation metabolites of Diamond V Original XPC in an *in vitro* anaerobic mixed chicken cecal culture. *Front. Vet. Sci.* 3, 83. doi: 10.3389/fvets.2016.00083.

Rubinelli, P. M., Kim, S. A., Park, S. H., Roto, S. M., Nealon, N. J., Ryan, E. P. and Ricke, S. C. (2017) Differential effects of rice bran cultivars to limit *Salmonella* Typhimurium in chicken cecal *in vitro* incubations and impact on the cecal microbiome and metabolome. *PLoS ONE* 12(9), e0185002. doi: 10.1371/journal.pone.0185002.

Russell, J. B. (1992) Another explanation for the toxicity of fermentation acids at low pH: Anion accumulation versus uncoupling. *J. Appl. Bacteriol.* 73, 363-70.

Russell, J. B., O'Connor, J., Fox, D. G., Van Soest, P. J. and Sniffen, C. J. (1992) A net carbohydrate and protein system for evaluating cattle diets: I. Ruminal fermentation. *J. Anim. Sci.* 70, 3551-61.

Saengkerdsub, S. and Ricke, S. C. (2014) Ecology and characteristics of methanogenic archaea in animals and humans. *Crit. Rev. Microbiol.* 40, 97-116.

Saengkerdsub, S., Kim, W.-K., Anderson, R. C., Nisbet, D. J. and Ricke, S. C. (2006) Effects of nitrocompounds and feedstuffs on in vitro methane production in chicken cecal contents and rumen fluid. *Anaerobe* 12, 85–92.

Saengkerdsub, S., Anderson, R. C., Wilkinson, H. H., Kim, W.-K., Nisbet, D. J. and Ricke, S. C. (2007a) Identification and quantification of methanogenic archaea in adult chicken ceca. *Appl. Environ. Microbiol.* 73, 353–6.

Saengkerdsub, S., Herrera, P., Woodward, C. L., Anderson, R. C., Nisbet, D. J. and Ricke, S. C. (2007b) Detection of methane and quantification of methanogenic archaea in faeces from young broiler chickens using real-time PCR. *Lett. Appl. Microbiol.* 45, 629–34.

Schelin, J., Andersson, G., Vigre, H., Norling, B., Häggblom, P. Hoorfar, J., Rädström, P. and Löfström (2013) Evaluation of pre-PCR processing approaches for enumeration of *Salmonella enterica* in naturally contaminated feed. *J. Appl. Microbiol.* 116, 167–78.

Sergeant, M. J., Constantinidou, C., Cogan, T. A., Bedford, M. R., Penn, C. W. and Pallen, M. J. (2014) Extensive microbial and functional diversity within the chicken cecal microbiome. *PLoS ONE* 9(3), e91941. doi: 10.1371/journal.pone.0091941.

Shah, D. H., Casavant, C., Hawley, Q., Addwebi, T., Call, D. R. and Guard, J. (2012) *Salmonella* Enteritidis strains from poultry exhibit differential responses to acid stress, oxidative stress, and survival in the egg albumen. *Foodborne Pathog. Dis.* 9, 258–64.

Smyser, C. F. and Snoeyenbos, G. H. (1979) Evaluation of organic acids and other compounds in *Salmonella* antagonists in meat and bone meal. *Poult. Sci.* 58, 50–4.

Soria, M. C., Soria, M. A., Bueno, D. J. and Colazo, J. L. (2011) A comparative study of culture methods and polymerase chain reaction assay for *Salmonella* detection in poultry feed. *Poult. Sci.* 90, 2606–18.

Stanley, D., Geier, M. S., Denman, S. E., Haring, V. R., Crowley, T. M., Hughes, R. J. and Moore, R. J. (2013) Identification of chicken intestinal microbiota correlated with the efficiency of energy extraction from feed. *Vet. Microbiol.* 164, 85–92.

Svihus, B., Choct, M. and Classen, H. L. (2013) Function and nutritional roles of the avian caeca: A review. *World's Poult. Sci. J.* 69, 249–63.

Thompson, J. L. and Hinton, M. (1997) Antibacterial activity of formic and propionic acids in the diet of hens and *Salmonellas* in the crop. *Br. Poult. Sci.* 38, 59–65.

Vanderwal, P. (1979) Salmonella control of feedstuffs by pelleting or acid treatment. *World's Poult. Sci. J.* 30, 70–9.

Van der Wielen, P. W., Biesterveld, J. J., Notermans, S., Hofstra, H., Urlings, B. A. P. and van Knapen, F. (2000) Role of volatile fatty acids in development of the cecal microflora in broiler chickens during growth. *Appl. Environ. Microbiol.* 66, 2536–40.

Van Immerseel, F., Cauwerts, K., Devriese, L. A., Haesebrouck, F. and Ducatelle, R. (2002) Feed additives to control *Salmonella* in feed. *World's Poult. Sci. J.* 58, 501–13.

Van Immerseel, F., Fievez, V., de Buck, J., Pasmans, F., Martel, A., Haesebrouck, F. and Ducatelle, R. (2004) Microencapsulated short-chain fatty acids in feed modify colonization and invasion early after infection with *Salmonella* Enteritidis in young chickens. *Poult. Sci.* 83, 69–74.

Van Immerseel, F., Boyen, F., Gantois, I., Timbermont, L., Bohez, L., Pasmans, F., Haesebrouck, F. and Ducatelle, R. (2005) Supplementation of coated butyric acid in the feed reduces colonization and shedding of *Salmonella* in poultry. *Poult. Sci.* 84:1851–6.

Van Immerseel, F., Russell, J. B., Flythe, M. D., Gantois, I., Timbermont, L., Pasmans, F., Haesebrouck, F. and Ducatelle, R. (2006) The use of organic acids to combat *Salmonella* in poultry: A mechanistic explanation of the efficacy. *Avian Path.* 35, 182-8.

Veldman, A., Vahl, A., Boggreve, G. J. and Fuller, D. C. (1995) A survey of the incidence of *Salmonella* species and Enterobacteriaceae in poultry feeds and feed components. *Vet. Rec.* 136, 169-72.

Walugembe, M., Hsieh, J. C. F., Koszewski, N. J., Lamont, S. J., Persia, M. E. and Rothschild, M. F. (2015) Effects of dietary fiber on cecal short-chain fatty acid and cecal microbiota of broiler and laying-hen chicks. *Poult. Sci.* 94, 2351-59.

Wales, A. D., Allen, V. M. and Davies, R. H. (2010) Chemical treatment of animal feed and water for the control of *Salmonella*. *Foodborne Pathog. Dis.*7, 1-15.

Waterman, S. R. and Small, P. L. C. (1998) Acid-sensitive enteric pathogens are protected from killing under extremely acidic conditions of pH 2.5 when they are inoculated into certain food sources. *Appl. Environ. Microbiol.* 64, 3882-6.

Weimer, P. J. (2015) Redundancy, resilience, and host specificity of the ruminal microbiota: Implications for engineering improved ruminal fermentations. *Front. Microbiol.* 6, 296. doi: 10.3389/fmicb.2015.00296.

Weimer, P. J., Russell, J. B. and Muck, R. E. (1999) Lessons from the cow: What the ruminant animal can teach us about consolidated bioprocessing of cellulosic biomass. *Bioresour. Technol.* 100, 5323-31.

Whyte, P., McGill, K. and Collins, J. D. (2003) A survey of the prevalence of *Salmonella* and other enteric pathogens in a commercial poultry feedmill. *J. Food Safety* 23, 13-24.

Wierup, M. and Häggblom, P. (2010) An assessment of soybeans and other vegetable proteins as source of salmonella contamination in pig production. *Acta Vet. Scand.* 52(15), 1-9.

Williams, J. E. (1981a) *Salmonellas* in poultry feeds - A worldwide review. Part I. Introduction. *World's Poult. Sci. J.* 37, 6-19.

Williams, J. E. (1981b) *Salmonellas* in poultry feeds - A worldwide review. Part II. Methods in isolation and identification. *World's Poult. Sci. J.* 37, 19-25.

Williams, J. E. (1981c) *Salmonellas* in poultry feeds - A worldwide review. Part III. Methods in control and elimination. *World's Poult. Sci. J.* 37, 97-105.

Williams, J. E. and Benson, S. T. (1978) Survival of *Salmonella typhimurium* in poultry feed and litter at three temperatures. *Avian Dis.* 22, 742-7.

Wolin, M. J. and Miller, T. L. (1982) Interspecies hydrogen transfer: 15 years later. *Amer. Soc. For. Microbiol. News* 48, 561-5.

Woodward, C. L., Kwon, Y. M., Kubena, L. F., Byrd, J. A., Moore, R. W., Nisbet, D. J. and Ricke, S. C. (2005) Reduction of *Salmonella enterica* serovar Enteritidis colonization and invasion by an alfalfa diet during molt in Leghorn hens. *Poult. Sci.* 84, 185-93.

Chapter 18

The role of essential oils and other botanicals in optimizing gut function in poultry

Divek V. T. Nair, Grace Dewi and Anup Kollanoor-Johny, University of Minnesota, USA

1 Introduction

The diet plays a significant role in maintaining the intestinal homeostasis in poultry. A well-balanced diet is also needed for strengthening the defence systems against invading pathogens helping the bird to reach its maximum production potential. Changes in diet may result in variations in gut physiology, allowing invading pathogens to colonize the gut, subsequently reducing the performance in birds (Sugiharto, 2016). In this regard, antibiotic growth promoters (AGPs) have offered a means of improving the animal productivity through the promotion of intestinal health and reduction in the incidence of diseases (Hao et al., 2014).

Utilization of antibiotics in agriculture can be traced back to Britain in 1938 when prontosil, a sulphonamide manufactured by the German pharmaceutical company Bayer, was marketed for use in animals (Lesche, 2007). The growth-promoting properties of antibiotics in food-producing animals was exemplified somewhat serendipitously when researchers at American Cyanamid's Lederle Laboratories fed chicks with crude preparations of *Streptomyces aureofaciens* as an alternative source of vitamin B_{12}. The increased weight gain in birds obtained from feeding the crude preparation was later attributed to the presence of trace amounts of aureomycin (chlortetracycline) and other antibiotics (Goldbloom and Steigmann, 1951).

http://dx.doi.org/10.19103/AS.2019.0059.22

Since March 1950, numerous scientific publications have reported the effects of feeding antibiotics and illustrated its revolutionary role in improving animal production (Jukes and Williams, 1953). The growth-promoting effects of antibiotics were attributed to their antibacterial properties based on evidence that illustrated the lack of their effect in germ-free chicks (Coates et al., 1952; Luckey, 1952). It was established that the suppression of mildly harmful bacteria was what contributed to the antibiotic growth effect (Jukes, 1985). Additionally, the activity of antibiotics in the intestine was found to be imperative to its growth-promoting effects as antibiotics such as chloramphenicol, despite its wide antibacterial potency, produced weak impacts on growth potentially due to its absorption through the gastric wall preventing it from reaching the intestinal tract at low dosages (Jukes and Williams, 1953).

Shortly after the announcement of their benefits in 1949, AGPs were rapidly adopted by farmers, fuelled in part by growing demand for meat during the post-war era. The improved growth rate and efficiency of feed conversion to body weight proved to be so lucrative that Lederle had difficulties keeping up with the demands and the company had to prorate available supply among their customers (Jukes, 1985). The role of antibiotics in agriculture was initially to support producers by controlling disease pressure, increase yields and contain economic risks. However, with the rise of meat consumption and increased market demand for meat, antibiotics became necessary for livestock production systems to meet the market demands. Adoption of antibiotic use became so widespread that by the mid-1950s, streptomycin was used to treat and prevent bacterial plant infections and tetracycline was used in the United States to prevent spoilage of fish, shellfish and poultry (Kirchhelle, 2018).

2 The emergence of regulations to curb antibiotic resistance

The practice of feeding sub-therapeutic doses of antibiotics to farm animals became widespread as several studies conducted at that time could not find any adverse effects on public health. The investigations found no detectable amounts of aureomycin in meat from animals that were fed higher levels of the antibiotic, and it was of no concern as many antibiotics were inactivated during cooking. Furthermore, no reports of resistance towards the antibiotics in pathogenic bacteria had surfaced at the time (Jukes and Williams, 1953). However, in the most recent decades, the emergence of antibiotic resistance in pathogenic bacteria has raised concerns with regard to the use of antibiotics in agriculture.

The overall contribution of antibiotic use in agriculture to the emergence of antimicrobial resistance remains contested as direct causality is difficult to establish due to the nature of antibiotic selection pressure. However, studies have illustrated a close association between the levels of antimicrobial use in animals at a population level to the prevalence of antimicrobial-resistant bacteria in animals and humans (Aarestrup, 2005; Schwarz et al., 2001). Furthermore, studies conducted across seven European countries found a correlation between consumption of antimicrobials and the prevalence of antimicrobial-resistant commensal *Escherichia coli* in food-producing animals (Chantziaras et al., 2014). Studies have suggested that the emergence and spread of antibiotic-resistant bacteria are tied to their exposure to low dosages of antibiotics at which growth-promoting antimicrobials are administered (You and Silbergeld, 2014).

Warnings regarding the hazards of antibiotics have been voiced by experts since the 1940s but were often overshadowed by the immediate benefits of antibiotics in improving agricultural productivity. International concerns over the rise of antimicrobial resistance were soon observed in the decades that followed. The challenge of creating regulations lies in the complexity of the risk-benefit matrix of the inexpensive production of protein and the less tangible considerations of antibiotic stewardship. In response to the growing concern over antimicrobial resistance, several jurisdictions across the world have responded by requiring a veterinary prescription to utilize antibiotics for food-producing animals (Maron et al., 2013).

The ban on the use of antibiotics as feed additives first began in Sweden in 1986, Switzerland in 1999 and Denmark in 2000 (Wenk, 2003). Since January 2006, the European Union has placed a ban on the use of antibiotics for growth promotion purposes in the feed (EC Regulation No. 1831/2003). In the United States, antimicrobials are approved for use in food-animal production systems to treat, control and prevent diseases, as well as for growth promotion purposes. Unlike the contemporary EU restrictions on AGPs, the U.S. Food and Drug Administration (FDA) developed voluntary guidelines to phase out antibiotic growth promotion through changes in the label. In January 2017, the FDA completed the voluntary transitioning of medically important antimicrobials which are used in animal feed or water from 'over-the-counter' status to VFD or Rx marketing status under the Guidance for Industry (GFI) #213 in an effort to promote antimicrobial stewardship in animal production (FDA, 2013, 2017). As the concern continues to grow among consumers, more countries have established increasingly stringent policies regarding antibiotic use and veterinary oversight in animal production systems. Several leading U.S. trade partners are among those jurisdictions which adopted these policies (Maron et al., 2013).

Recent phase out of antibiotics from the U.S. poultry production has provided a vast avenue for the introduction of parasitic and bacterial infections in poultry; most of these invaders colonize and affect the intestinal tract (Van Immerseel et al., 2004; Yegani and Korver, 2008). For example, a coccidiosis infection leads to mucosal degradation and can predispose the birds to other infections such as necrotic enteritis which cause mortality or production loss in survivors (Van Immerseel et al., 2004; Yegani and Korver, 2008b). In addition, the colonization of the gut with pathogens such as *Salmonella* can lead to microbiome changes in the gut (Nair et al., 2018). The alteration in normal microbiota affects the availability of the nutrients to the birds and also affects the ability of the birds to counteract pathogens through competitive exclusion (Mead, 2000). Therefore, the alterations in intestinal morphology, digestion, microbial balance or immunity can affect the health status of the birds, ultimately leading to a lowered production performance. These situations have reignited the exploration for viable alternatives or adjuncts to antibiotics which could help maintain the productivity and economic viability of animal production systems, including poultry.

3 Phytobiotics: an emerging group of alternatives

Along with their provision of food and oxygen, humans have utilized a plethora of plants for a variety of purposes tapping their medicinal value, food preservation and flavour enhancement properties which can be dated back to the time of the ancient Egyptians (Matthews et al., 2017). However, the incorporation of plant extracts in a wide variety of traditional medicines is not unique to Egyptian culture as they are also documented in India and China (Potterat and Hamburger, 2007). Fossil records can date human use of plants for medicinal purposes around 60 000 years ago in the Middle Paleolithic age. A plethora of modern drugs has since been developed based on their initial use in ethnobotanical medicine (Fabricant and Farnsworth, 2001). Interest in the medicinal properties of plants and their derivatives has garnered renewed attention in part due to the need to combat antimicrobial resistance and the growing demand for reduced use of synthetic preservatives (Coates et al., 2011; Perricone et al., 2015). The use of phytobiotics as growth and health promoters has received renewed attention due to the increasing concern over the non-therapeutic use of antibiotics. Plants represent an untapped resource for antimicrobials as there are over 35 000 plant species anecdotally known to serve the medicinal purpose of which only a few have been investigated by scientists (Aguilar, 2001; Yoo et al., 2018).

3.1 Classification

The terms 'phytobiotic' and 'phytogenic' have been interchangeably used in publications since they offer the same meaning (Windisch and Kroismayr, 2007; Puvača et al., 2013). In this chapter, we will use the term 'phytobiotic' whenever a broad meaning is implied and 'plant-derived compounds or molecules' when specific compounds or molecules (e.g. *trans*-cinnamaldehyde) are discussed.

Windisch and Kroismayr (2007) have defined phytobiotics as plant products that are added to feed to improve the performance of food-producing animals. However, these products are complex with varying compositions, biological origins, as well as purity (Yang et al., 2009). In this regard, they broadly classified phytobiotics into four categories by their origin and processing. The four categories are:

1. Essential oils - volatile and lipophilic compounds obtained from plants through hydrodistillation.
2. Botanicals - the entire or processed parts of plants including the bark, roots and leaves.
3. Herbs - flowering, non-woody and non-persistent plants.
4. Oleoresins - the extracts obtained using non-aqueous solvents.

Different phytobiotics share a common feature, in that they often contain a complex blend of bioactive components. A majority of the compounds responsible for the observed effects have been identified as secondary metabolites, which are distinct from the primary metabolites (amino acids, proteins, lipids and polysaccharides), in that they are not essential for metabolic processes of the plant and do not appear to play a direct role in plant physiology (Dixon, 2001; Jones and Dangl, 2006). The presence and concentrations of various compounds in plants are influenced by several factors, including the plant's environment, growth conditions and storage conditions after harvest, in addition to the extraction processes (Huyghebaert et al., 2011). Moreover, the presence of these compounds is influenced by the plant's interaction with their environment and may act as a defence mechanism by which it protects itself from predators, pathogens or physiological stressors. In this chapter, we are focussing on the two major categories, botanicals and essential oils, based on their potential for improving gut function in poultry.

3.1.1 Essential oils

Essential oils are not strictly oils (as they are produced by hydrodistillation), but they are poorly soluble in water and thus are categorized as oils (Calo et al., 2015). They are a class of volatile oils obtained from plants with characteristic odours and properties of the plant that are often used in perfumes, flavours and pharmaceuticals (Wenk, 2003). Paracelsus von Hohenheim has been credited for establishing the term 'essential oil' based on the Latin phrase '*Quinta essentia*' meaning the 'effective component of a drug' (Guenther, 1948).

Although the use of aromatic plants can be dated back to prehistoric times, the practice of distillation to extract the medicinal and fragrant components of the plant only began in the eighteenth century (Lawless, 1995). Essential oils have since been utilized widely in numerous fields with applications in food, cosmetics, agriculture and pharmaceutical industries (Nychas, 1995; Schmidt et al., 2008).

Essential oils often contain a complex mixture of volatile, lipophilic, hydrophobic and aromatic compounds which are the secondary metabolites produced by the plant in response to their environment, and as a defence mechanism against predators and pathogens (Attokaran, 2017; Ballhorn et al., 2011; Chávez-González et al., 2016). As the concentrations of these secondary metabolites are influenced by the environmental and climatic conditions to which the plant was exposed, variations in the composition of essential oils are often observed even among plants of the same species. The compounds could be further classified by the levels at which they are present, with two to three major components present at high concentrations ranging from 20% to 70%, and trace components occurring at lower concentrations (Kollanoor-Johny and Venkitanarayanan, 2017). The major components often dictate the biological properties of essential oils. These compounds could be broadly grouped into the major terpene hydrocarbons (terpenes and terpenoids), and aromatic and aliphatic compounds (Bakkali et al., 2008; Pichersky et al., 2006) and have been discussed previously (Kollanoor-Johny and Venkitanarayanan, 2017).

As essential oils are hydrodistilled extracts from plants, they concentrate specifically the volatile and lipophilic components of the plant. Hence, the essential oil of a plant often varies in composition from the plant itself. For example, oregano plants (*Origanum vulgare* subsp. *hirtum*) contain a variety of glycosidically bound volatile and non-volatile constituents which exhibit biological activity after enzymatic or acid hydrolysis that are not extracted into the essential oil (Milos, 2000). Based on this information, Florou-Paneri et al. (2005b) investigated to examine whether the difference in composition would elicit different biological responses when incorporated into poultry diets. The study found that both oregano essential oil and the herb were effective in delaying lipid oxidation during refrigerated storage, though 200 mg/kg of essential oil was equivalent in efficacy to that of 10 g/kg of the herb.

Among the different classes of phytobiotics, essential oils have been more extensively studied for their antimicrobial, antiparasitic, anti-inflammatory and anti-oxidative properties (Hashemi and Davoodi, 2011). Numerous investigations have been conducted to study the antimicrobial properties of several plant-derived compounds both *in vitro* and, to a lesser extent, in animal models (Borges et al., 2015; Burt, 2004; Holley and Patel, 2005; Newman, 2008; Newman and Cragg, 2012; Osbourn, 1996; Shahid et al., 2009; Tagboto and Townson, 2001; Upadhyaya et al., 2013).

Most of the studies with essential oils in poultry have focussed on a major mechanism of action, and not comprehensively evaluated from a gut health perspective. Selected essential oil ingredients that possess a generally recognized as safe (GRAS) status have been tested in poultry for their antimicrobial effects against pathogens, but have not visualized the 'gut' as a premise for interactive effects (Burt, 2004; Kollanoor-Johny et al., 2008, 2012a,b; Surendran-Nair et al., 2017; Upadhyaya et al., 2017). Some of the essential oils have been found to increase production performance in poultry (Jang et al., 2007; Kalemba and Kunicka, 2003; Zhai et al., 2018). Additionally, some were found to modulate the immune response in poultry and to help the restoration of the gut microbiota after an infection process (Yin et al., 2017). Also, the supplementation of some plant-derived compounds was found to increase the beneficial gut microbiota in poultry. These mechanisms eventually might lead to improved gut health and production performance in poultry (Sugiharto, 2016; Yang et al., 2009; Yegani and Korver, 2008a; Zhai et al., 2018). Research that investigated these aspects of essential oils has been dealt in detail in the following sections of this chapter.

3.1.2 Botanicals

With regard to its purpose in animal feed, botanicals are defined as the entire or processed parts of plants which include the bark, leaves and roots (Windisch and Kroismayr, 2007). However, the standard definition found in the *Merriam-Webster* dictionary is broader and classifies a 'botanical' as a substance obtained or derived from a plant. The National Institutes of Health (NIH) classifies a botanical as a plant or part of a plant with medicinal properties, flavour or scent, and considers herbs as a subset of botanicals (NIH, 2011). Aside from the exploration of their use in animal production systems, botanicals are a lucrative industry that manufactures and markets human herbal supplements. They have long been accepted and prescribed all over the world with about US$7 billion market in Europe, US$3 billion in North America, US$2.4 billion in Japan and US$2.7 billion in the rest of Asia (Glaser, 1999).

Some studies have been conducted to explore the potential applications of botanicals in production systems, primarily in Europe (Barug et al., 2006). Studies have found that the secondary metabolites found in plants are capable of eliciting beneficial effects in food products as well as mammalian metabolism. Rodehutscord and Kluth (2002) concluded that botanicals are capable of improving growth rate through increased feed intake. In another study, supplementation of ginger root powder in poultry feed was found to reduce serum cholesterol levels (Zomrawi et al., 2012). With regard to their antibacterial properties, several Chinese herbs, including Tibet bitterroot, goldhead rhizome and skullcap root, along with garlic reduced Gram-positive and Gram-negative organisms (Wenk, 2003). Research that investigated various aspects of the use of botanicals is illustrated in the following sections of this chapter.

4 Potential role of phytobiotics to improve gut health in poultry

The definition of gut health has yet to unveil its true meaning. However, it covers multiple aspects of the gastrointestinal (GI) tract, including effective digestion and absorption of food, absence of GI illness, healthy and stable intestinal microbiota, and active immune status, all resulting in a state of well-being of the animal (Bischoff, 2011). Any impairment in the GI tract would adversely affect underlying mechanisms which in turn affects the nutrient and fluid uptake, immune tolerance function, defence against infections and maintenance of energy homeostasis via signalling mechanisms in the brain (Bischoff, 2011). Although deciphering mechanisms of action of essential oils and botanicals are its preliminary stages, due to their wide range of activity and potency, it is highly likely that they could play significant roles in optimizing gut health in food animals, including poultry. Better gut health could mean better performance and well-being of the bird.

4.1 Phytobiotics as digestive conditioning agents

A few studies in poultry have indicated the potential of phytobiotics as digestive conditioning agents. The digestive conditioning is the process of enhancing the digestive enzyme activity and improving the nutrient digestibility and bioavailability in the bird (Brenes and Roura, 2010). The essential oils or their blends are reported to stimulate the secretion of digestive enzymes, although the effects could vary according to the age of birds, type and concentration of the phytobiotic compounds present in the blend and the duration of the supplementation. Studies have shown that the phytobiotic compounds increased the activity of trypsin, amylase, maltase, chymotrypsin, lipase and protease at the intestinal or pancreatic level (Basmacioğlu Malayoğlu et al., 2010; Jamroz et al., 2005; Hashemipour et al., 2013; Lee et al., 2003; Jang et al., 2007).

A commonly used essential oil, oregano oil, containing 84% carvacrol and 1.8% thymol, when supplemented through the feed at 500 mg/kg for 21 days resulted in increased chymotrypsin activity in the pancreatic tissues of broilers. Similarly, the supplementation (250 or 500 mg/kg) increased chymotrypsin activity in the intestinal digesta of broilers compared to the control group fed a standard diet (Basmacioğlu Malayoğlu et al., 2010). The supplementation did not cause any change in the pancreatic and intestinal lipase or amylase activity, faecal digestibility coefficients for dry matter and ether extract. However, 500 mg/kg supplementation resulted in an increased digestibility constant for crude protein. Moreover, the supplementation did not alter the intestinal pH, serum triglyceride and HDL cholesterol when compared to the control. Conversely, the essential oil at 500 mg/kg resulted in an increase of serum total cholesterol compared to the broilers fed with control diet (358.45 mg/dL vs. 443.83 mg/dL) (Basmacioğlu Malayoğlu et al., 2010).

The increased activity of total pancreatic trypsin was observed when a commercial feed containing 50 mg/kg essential oil blend (2.9% active ingredients, including thymol) was supplemented from day 3 of hatch to 35 days of growth in broilers. The essential oil blend resulted in increased activity of pancreatic α-amylase and intestinal maltase compared to the control diet (Jang et al., 2007). In another study, a blend of plant extracts containing 100 mg/kg capsaicin, cinnamaldehyde and carvacrol supplemented in feed for 41 days to broilers increased lipase activity in the pancreas and intestine by 38–46%. In younger birds, a similar effect was not observed, and the supplementation resulted in a lower pancreatic α-amylase activity in 21-day-old broilers. However, the blend had no effects on apparent ileal digestibility of nutrients such as crude protein, fibre and amino acids (Jamroz et al., 2005).

The supplementation of an equal mixture of thymol and carvacrol in broilers for 42 days at 200 mg/kg increased digestive enzyme activity, including pancreatic and intestinal trypsin, protease and lipase on day 24. These changes were absent in the broilers at 42 days of age. Moreover, no changes were observed for intestinal or pancreatic amylase activity either at 24 or 42 days of supplementation (Hashemipour et al., 2013). Interestingly, when the herbs [thyme, oregano, rosemary, marjoram and yarrow (10 g/kg of each)] were included in broiler diets from day 7 to day 28, no effects on the major digestibility coefficients were observed (Cross et al., 2007).

A blend of oregano, anise and citrus peel essential oils at 25 ppm supplemented for 42 days to broilers revealed lower serum cholesterol and very low-density lipoprotein (VLDL) levels, whereas the same treatment resulted in higher total polyphenolic compounds and total flavonoids in blood. The supplementation did not affect the pH of the different regions of broiler intestine such as duodenum, jejunum, ileum and caecum. Also, the essential oil supplementation did not change the caecal volatile fatty acid content. However, the ileal ammonia concentration was lower in essential oil-treated groups compared to the control group (Hong et al., 2012). In another study, an aqueous extract of garlic (*Allium sativum*) in drinking water increased serum HDL and decreased total cholesterol, LDL and triglyceride in birds (Rahimi et al., 2011). In addition, a 200 ppm essential oil extract from oregano, cinnamon and pepper, and a 5000-ppm Labiatae (mint family of flowering plants) extract from sage, thyme and rosemary resulted in increased ileal digestibility.

The Labiatae extracts in the starter feed improved the faecal digestibility of dry matter, and the Labiatae and essential oil extract increased the ether extract digestibility. However, the treatments did not improve crude protein digestibility. The starch digestibility at the ileum was also improved by extract supplementation. However, in the finisher diet, the supplementation increased the faecal digestibility of dry matter and crude protein. In another study, a phytobiotic additive containing essential oils of thyme and star anise at three different inclusion levels (150, 750 and 1500 mg/kg) showed an increase in the apparent ileal digestibilities of crude ash, crude protein, crude fat, calcium and phosphorus. The digestibilities of these nutrients were high in all age groups (days 21, 35 and 42) in the essential oil groups (Amad et al., 2011).

4.2 Phytobiotic mechanisms of antibacterial action

Pathogens affect gut health adversely. Antibiotics at their sub-therapeutic concentrations are thought to control pathogens invading the GI tract very effectively. Several studies had detected better growth performance when antibiotics were supplemented to birds (Miles et al., 2006; Thomke and Elwinger, 1998). The high activity of antibiotics against pathogens could be attributed to the specific mechanisms of antimicrobial action (Kapoor et al., 2017). Conversely, essential oils and their constituent compounds have multiple components with active groups on their chemical structure that possess multiple mechanisms of action. Researchers have explored this potential, and their studies are discussed below.

Reports indicate that the antimicrobial efficacy of the phenol group containing essential oils are higher than the others, and the activity decreases in the order of aldehydes > ketones > alcohols > ethers > hydrocarbons. The phenolic compounds (e.g. thymol, carvacrol and eugenol) exhibit a broad spectrum of activity due to the presence of the hydroxyl group which forms a hydrogen bond with active enzyme centres (Kalemba and Kunicka, 2003). It has been noted that a concentration of carvacrol as low as 0.01 mM affected the membrane potential in *Bacillus cereus*, whereas 0.25 mM and 1 mM increased the ion leakage from the cell. A concentration of 2 mM depleted the ATP pool in the bacterial cell (Ultee et al., 1999). The delocalized electrons and hydroxyl group at the *ortho* position to methyl group on phenolic ring in carvacrol cause the release of protons to the cytoplasm, diminishes the pH gradient across the membrane and affects the proton motive force (Ultee et al., 2002). Although the hydroxyl group in thymol is located at the *meta* position to methyl group on phenolic ring, the antimicrobial activity is still comparable to that of carvacrol (Ultee et al., 2002). Eugenol has a methoxyl group on the phenolic ring which affects the smooth release of protons (Ben Arfa et al., 2006). The conjugated double bonds and the long methyl side chain outside the ring structure contribute to the antimicrobial activity of cinnamaldehyde (Chang et al., 2001).

Essential oils are hydrophobic, and their primary targets are the cytoplasmic membranes. The plant-derived compounds such as thymol, carvacrol, limonene, eugenol and cinnamaldehyde alter the cytoplasmic membrane structure of the pathogens such as *E. coli* O157: H7, *Staphylococcus aureus*, *Salmonella enterica* serovar Typhimurium, *Pseudomonas fluorescens* and *Brochothrix thermosphacta*. They act by reducing the composition of unsaturated fatty acids in the cell membranes and increasing the long chain fatty acids with the alterations in the morphology of these pathogens (Di Pasqua et al., 2007). Spanish oregano, Chinese cinnamon and savoury essential oils target the cell membrane and cell walls of pathogens such as *E. coli* O157: H7 and *Listeria monocytogenes* which destroy the cell membrane and cause depletion of intracellular ATP (Oussalah et al., 2006). In addition, oregano (*Origanum compactum*) and clove essential oils and their components such as thymol and eugenol were found to cause cell membrane damage and cellular lysis in *E. coli* and *B. subtilis* (Rhayour et al., 2003). The combinations of oregano and bergamot oils, oregano and perilla oils, basil and bergamot oils, and oregano and basil oils at their minimum inhibitory concentrations caused morphological alterations in the cellular membranes of *B. subtilis*, *Saccharomyces cerevisiae*, *S. aureus* and *E. coli*, respectively (Lv et al., 2011).

Plant-derived compounds such as carvacrol and eugenol affect the toxin production in pathogens. Carvacrol inhibits diarrhoeal toxin production in *B. cereus*, and oregano essential oil affects the enterotoxin production in *S. aureus* (Akthar et al., 2014; Ultee et al., 2000; Ultee and Smid, 2001). Eugenol downregulates the expression of the genes *sea, seb, tst* and *hla* in *S. aureus* which reduces the production of exotoxins by the pathogen (Qiu et al., 2010). The recent finding also suggests that the subinhibitory concentrations of the plant-derived compounds, especially eugenol, downregulates critical genes associated with *S.* Enteritidis virulence mechanism, including its adhesion to and invasion of avian epithelial cells (Upadhyaya et al., 2015; Kollanoor-Johny et al., 2012a). Moreover, eugenol and *trans*-cinnamaldehyde, at their subinhibitory concentrations, downregulate the genes associated with flagellar motility, Pathogenicity Island 1, invasion of intestinal epithelial cells and transcription of membrane proteins in *S.* Enteritidis phage type 8 (Kollanoor-Johny et al., 2017).

In short, the plant-derived compounds target cellular membranes of the pathogens, affect the proton motive force and enzyme systems, cause leakage of the intracellular contents, downregulate virulence mechanisms and inhibit metabolic pathways which eventually lead to the reduced infectivity or death of the pathogens (Burt and Reinders, 2003; Kollanoor-Johny et al., 2017; Venkitanarayanan et al., 2013; Lv et al., 2011; Surendran-Nair et al., 2017; Burt, 2004).

4.3 Antimicrobial effects of phytobiotics on specific poultry pathogens

4.3.1 Eimeria

Avian coccidiosis is one of the most devastating diseases in poultry caused by a protozoan parasite of the genus *Eimeria*. The economic losses are mainly due to mortality and morbidity or poor growth performance in surviving birds (Allen and Fetterer, 2002). Literature indicates that essential oils and botanicals are good choices against coccidial infections in poultry.

It is reported that oregano essential oil (500 ppm) reduced coccidiosis lesion scores in the upper and middle intestine similar to the uninfected broilers compared to the birds challenged with coccidia and without the essential oil treatment. In addition, the supplementation significantly reduced the excretion of oocysts. The essential oil supplemented groups had lower oocysts per gram of faeces, dropping scores and litter scores (Mohiti-Asli and Ghanaatparast-Rashti, 2015). Similarly, oregano essential oil (5% essential oil from *Origanum vulgare* subsp. *hirtum*) at 300 mg/kg resulted in reduced infection of *Eimeria tenella* in broilers infected with 5×10^4 sporulated oocysts at 14 days of age. The essential oil groups showed a lower lesion score after 7 days of challenge. In addition, the oocyst excretion (number of oocysts per gram of excreta) after the challenge was significantly lower in oregano essential oil-treated groups compared to the challenged groups without any supplementation. Interestingly, the supplementation resulted in milder bloody diarrhoea compared to the infected positive control (Giannenas et al., 2003).

In another study, a supplementation of blend of essential oils derived from oregano, laurel leaf and lavender at 50 mg/kg containing carvacrol (24.5%), 1, 8-cineole (20.1%), camphor (12.1%) and thymol (6%) as active ingredients significantly improved the body weight gain and feed conversion ratio in broilers after infection with a mixed *Eimeria* population of 35×10^4 oocysts at 14 days of age. The supplementation also reduced the excretion of oocysts/bird after the infection (Bozkurt et al., 2012a). These studies indicate that essential oils or their blend containing active compounds improve the gut resistance against *Eimeria* infection, and significantly reduce the outcomes of infection, including reduction of the lesion scores and excretion of oocysts. Also, the essential oils improve the performance in recovered birds after the infection, compared to the controls.

4.3.2 Clostridium perfringens

Necrotic enteritis caused by *Clostridium perfringens* is a serious concern in poultry production (Van Immerseel et al., 2004). The acute form of necrotic enteritis occurs in 2- to 5-week-old broilers leading to clinical infections and high mortality. Necrotic enteritis can also occur as a subclinical form in broilers leading to necrosis in the intestine affecting the productivity of the birds (Van Immerseel et al., 2004; Engström et al., 2003; Lovland et al., 2004).

The essential oil blend containing 25% thymol and 25% carvacrol at 240 mg/kg supplemented through feed significantly reduced the macroscopic lesions in the ileum resulting from *C. perfringens* infection in broilers on days 21 and 28. In addition, the supplementation reduced the ileal and caecal *E. coli* populations in the *C. perfringens* infected groups (Du et al., 2015). In another study, the supplementation of fresh *Artemisia annua* leaf extract at 250 mg/kg reduced *C. perfringens* populations in the ileum and caecum and reduced the lesions associated with infection in the small intestine (Engberg et al., 2012). In yet another study, broilers supplemented with a blend of ginger and carvacrol at 1.5 g/kg significantly increased villus length and villus length:crypt depth ratio in *C. perfringens*-infected groups compared to the infected control groups without supplementation. The supplementation significantly decreased the gross pathological and histopathological lesions associated with the infection (Jerzsele et al., 2012). The plant-derived compounds, thymol and cinnamaldehyde, and oil of eucalyptus at 150–200 mg/kg significantly reduced the number of broilers infected with *C. perfringens* inoculated at 4 × 10^8 CFU and resulted in marked decrease in the macroscopic lesions (Timbermont et al., 2010). In another study, the essential oil blend containing thymol, eugenol, curcumin and piperin reduced *C. perfringens* populations in the jejunum, caecum, cloaca and faeces. In addition, the supplementation significantly reduced the percentage of positive samples from these tissue samples (Mitsch et al., 2004).

Yin et al. (2017) reported that a supplementation of essential oil blend containing 25% thymol and 25% carvacrol at 120 mg/kg in the feed was effective against the virulence factors of *C. perfringens* in broilers. The supplementation significantly reduced the virulence factors such as VF 0073-ClpE (ATP-dependent protease), VF0124-LPS (endotoxin) and VF0350-BSH (salt hydrolase) compared to the control birds challenged with *C. perfringens*.

The publications above suggest that essential oils could be an effective intervention against *C. perfringens* and its virulence factors. Additionally, they could improve the gut health by augmenting beneficial microbial populations. As mentioned earlier, because the phytogenic compounds have the potential to control *Eimeria*, they can also be used to better control *C. perfringens* since mucosal damage associated with *Eimeria* is one of the major predisposing causes of *C. perfringens* infection (Van Immerseel et al., 2004).

4.3.3 Campylobacter

The essential oils and their active compounds have been found to be effective against *Campylobacter* in poultry. The dietary supplementation of carvacrol at

120, 200 and 300 mg/kg of diet for 35 days had a significant effect on reducing *Campylobacter* colonization in the broilers. During the starter and grower phase (0–21 days), the relative abundance of the *Lactobacillus* population was increased in the supplemented groups. This effect was correlated with the occurrence of less colonization of *C. jejuni* during these days (Kelly et al., 2017).

The pharmacokinetic and antimicrobial activity of carvacrol was determined in broilers against *C. jejuni* (Allaoua et al., 2018). The pharmacokinetic assay revealed a minimum presence of the carvacrol in the blood. Carvacrol was mostly present in the distal part of the intestine such as caecum. In addition, it was excreted through the faeces as well. The galenic form (a product formulation of carvacrol with solubilizing agents, stabilizing agents and adsorbent carrier to use in feed) preparation (0.5 or 5 g/kg feed) containing 1.65% active ingredient significantly reduced the colonization of *Campylobacter* by 1.2 and 1.5 logs in broilers on day 35 after a challenge with 1.68×10^8 CFU/bird on day 21. In another study, Arsi et al. (2014) reported that supplementation of either 0.25% or 2% thymol, 1% carvacrol or a combination of carvacrol and thymol (0.5% each) significantly reduced the caecal colonization of *C. jejuni* in broilers, and the observed reductions were between 1 and 2 $\log_{10}$ CFU/mL of caecal contents.

4.3.4 Salmonella

The supplementation of *trans*-cinnamaldehyde (0.5% and 0.75%) and eugenol (0.75% and 1.0%) through feed significantly reduced (>3 $\log_{10}$ CFU/g) *S.* Enteritidis colonization in broiler chicks (Kollanoor-Johny et al., 2012a). Similarly, 0.75% *trans*-cinnamaldehyde and 1% eugenol were also efficient in reducing *S.* Enteritidis colonization in market-age broilers (Kollanoor-Johny et al., 2012b). A blend of essential oils containing carvacrol, thymol, eucalyptol and lemon at 0.5% through drinking water significantly reduced *S.* Heidelberg colonization in broiler crops at 42 days of age. However, these supplementations did not affect the colonization of *S.* Heidelberg in the caecum and/or the faecal shedding (Alali et al., 2013a). In addition, a commercial product containing 0.025% essential oil blend significantly reduced the colonization of *S.* Heidelberg in the crops of broilers on day 42 when supplemented through the water (Alali et al., 2013b). A blend of essential oil containing *trans*-cinnamaldehyde and thymol at 100 mg/kg in broiler diet significantly reduced *Salmonella*-positive caecal samples on day 42 (Amerah et al., 2012). *Trans*-cinnamaldehyde was also found to be effective in preventing vertical transmission of the pathogen in layer chickens. An in-feed supplementation of *trans*-cinnamaldehyde at 1.0% and 1.5% for 66 days resulted in decreased transmission of *S.* Enteritidis to the eggshell and yolk in 40- or 25-week-old layers. In addition, the supplementation of *trans*-cinnamaldehyde significantly reduced *Salmonella* colonization in caecum, liver and oviduct (Upadhyaya et al., 2015). Finally, a low concentration of *trans*-cinnamaldehyde (0.03%) when supplemented through drinking water as a therapeutic agent to turkey poults reduced the caecal colonization of the pathogen (Nair et al., 2017).

4.4 Phytobiotics as modulators of gut microbiota

The essential oils and their blends are found to affect the microbial population in the poultry gut. The effects of essential oil treatments on intestinal microorganisms in broilers had been previously reviewed (Zeng et al., 2015); most studies were based on targeted bacterial populations rather than the entire microbial community profile (gut microbiome).

Oregano and garlic essential oils at 300 mg/kg concentration each or their combination (150 mg/kg, each) for 42 days did not alter *Streptococcus, Lactobacillus* spp. and coliform counts in the ileal digesta of broilers (Kırkpınar et al., 2011). However, the study reported lower *Clostridium* counts. In another study, supplementation of a plant extract blend (100 mg/kg) containing capsaicin, cinnamaldehyde and carvacrol for 41 days in broilers increased *Lactobacillus* spp. In addition, the supplementation resulted in reduced *E. coli, C. perfringens* and fungi populations in the intestinal contents (Jamroz et al., 2005). In yet another study, the supplementation of plant extract mixture [thymol (≥10%), eugenol (≥0.5%) and piperine (≥0.05%)] at 30 mg/kg for 56 days in turkey poults resulted in an increase in lactic acid bacteria in the caecum, and significantly reduced the coliform populations. However, these bacterial populations were not affected in the ileum and crop (Giannenas et al., 2014). Similarly, supplementation of 0.1% aqueous extract of thyme (*Thymus vulgaris*), coneflower (*Echinacea purpurea*), garlic (*Allium sativum*) or a blend of these extracts in drinking water promoted the lactic acid bacteria populations in the ileocaecal digesta. As expected, a decrease in *E. coli* population was observed as well (Rahimi et al., 2011). In a more recent study, Yin et al. (2017) reported that the supplementation of essential oil blend containing 25% thymol and 25% carvacrol at 120 mg/kg in broilers significantly increased the relative abundance of Firmicutes and decreased the abundance of Cyanobacteria and Proteobacteria in the ileum compared to the control. At the taxa level, the supplementation increased the relative abundance of Clostridiales and Lactobacillales as well. However, the supplementation reduced Streptophyta and Rickettsiales populations. The blend resulted in an increased abundance of *Lactobacillus crispatus* and *Lactobacillus agilis* populations in the ileum (Yin et al., 2017).

There are also reports indicating that the plant extracts did not affect any of the tested microbial communities. For example, the herbs from thyme, oregano, rosemary, marjoram and yarrow (10 g/kg of each) did not result in changes to the population of coliforms, lactic acid bacteria, *C. perfringens* or total anaerobes in the caecal contents and faeces when the broilers were fed the blend for 29 days (Cross et al., 2007). In another study, it was found that supplementation of a commercial feed containing 25 or 50 mg/kg essential oil blend containing 2.9% active ingredients, including thymol, fed for 35 days did not significantly affect *E. coli* and *Lactobacilli* populations in the ileocaecal digesta (Jang et al., 2007). Also, supplementing a blend of essential oils including oregano, anise and citrus peel at 25 mg/kg for 42 days did not affect the total bacterial counts, including *Lactobacillus*, coliforms, *Salmonella, Enterococci* and *Clostridium* in the ileum (Hong et al., 2012). In another study, it was reported that the supplementation of oregano essential oil in broiler feed (300 and 600 mg/kg), although it reduced the population of *E. coli* in the caecum, showed no effects on the lactic acid bacterial populations (Roofchaee et al., 2011).

4.5 Phytobiotics as immunomodulators

An immunomodulator is an agent that modifies or normalizes the immune system (American Academy of Allergy Asthma and Immunology, 2018). For example, thyme oil in poultry diets (0.5 g/kg) increased IgA concentrations in the duodenum, enhanced phagocytic activity in the blood and improved intestinal barrier integrity (Placha et al., 2014). In another study, the supplementation of essential oil components such as an equal mixture of thymol and carvacrol in broilers at 200 mg/kg for 42 days resulted in an increased hypersensitivity response and IgG titres. In addition, the supplementation decreased heterophil to lymphocyte ratio compared to the control group which is a direct indication of the effect of these essential oil components in reducing stress in broilers. However, the treatments did not change haematological parameters and lymphoid organ weights (bursa, spleen and thymus) (Hashemipour et al., 2013). In yet another study, it was reported that a 0.1% aqueous extract of thyme (*Thymus vulgaris*), coneflower (*Echinacea purpurea*), garlic (*Allium sativum*) or a blend of these extracts in drinking water resulted in an increase in the relative weight of bursa of Fabricius for garlic-fed groups, whereas the treatments did not affect the weight of the spleen. In addition, the cutaneous basophilic hypersensitivity response to phytohemagglutinins and antibody response to sheep RBCs was higher for coneflower-supplemented birds (Rahimi et al., 2011).

However, oregano essential oil at 250 mg/kg or 500 mg/kg did not alter IgG or IgM levels in a study conducted on broilers (Basmacioğlu Malayoğlu et al., 2010). In another study, a blend of essential oils including oregano, anise and citrus peel at 25 mg/kg for 42 days did not increase IgG concentrations (Hong et al., 2012). Moreover, layers fed with essential oils (24 mg/kg) containing oregano oil (*Origanum* sp.), laurel leaf oil (*Laurus nobilis*), sage leaf oil (*Salvia triloba*), myrtle leaf oil (*Myrtus communis*), fennel seed oil (*Foeniculum vulgare*) and citrus peel oil (*Citrus* sp.) from 36–51 weeks had no impact on the humoral immunity of layers (Bozkurt et al., 2012b).

5 Benefits of improving gut health on poultry production

The improvement in gut function could result in increased production performance in poultry (Brenes and Roura, 2010). Although a definite connection between gut health and production was not established, the following studies reveal the effects of phytobiotic supplementation on growth performance, and carcass and egg quality in poultry.

5.1 Effects on growth performance

When oregano essential oil (84% carvacrol and 1.8% thymol) at 250 mg/kg was supplemented in the diet, it resulted in an increase in body weight of the broilers at their first week of age. An average difference of 5 g/bird was observed at 250 mg/kg (Basmacioğlu Malayoğlu et al., 2010). Also, in a different study, when oregano essential oil at a higher concentration of 600 mg/kg was supplemented in broiler diet, it resulted in increased body weight gain in broilers. In addition, the inclusion of 600 mg/kg or 1200 mg/kg oregano essential oil significantly improved the feed conversion ratio (FCR) in the grower period and throughout the experiment in broilers (Roofchaee et al., 2011). Similarly, supplementation of an equal mixture of thymol and carvacrol (200 mg/kg) for 42 days increased the average daily gain and FCR (Hashemipour et al., 2013). Contrastingly, a study conducted by Kırkpınar et al. (2011) revealed that 300 mg/kg oregano essential oil for 42 days had an adverse effect on broiler body weight.

A study conducted by Amad et al. (2011) using essential oils of thyme and star anise in broiler diets revealed an improvement in FCR during the growth phase. In another study, the supplementation of a mixture of plant extracts [thymol (≥10%), eugenol (≥0.5%) and piperine (≥0.05%)] at 30 mg/kg in turkey poults for 56 days resulted in an increased performance, including an increase in the body weight (Giannenas et al., 2014). Another study conducted using a blend of essential oils derived from oregano, anise and citrus peel revealed an increase in feed to gain ratio in broilers fed with essential oil compounds for 42 days. However, the total weight gain (g/bird) and total feed intake (g/bird) was similar to the control groups (Hong et al., 2012).

Similarly, essential oil extracts (200 mg/kg) from oregano, cinnamon and pepper, and 5000 ppm Labiatae extract from sage, thyme and rosemary resulted in no adverse effect in the feed intake or feed conversion. However, during the younger age (14–21 days age) the Labiatae extract resulted in a faster growth rate of broilers than control and essential oil extracts (68.8 vs. 63.9 and 61.6 g/day, respectively) (Hernandez et al., 2004). In layers also, the essential oil blend containing oregano oil (*Origanum* sp.), laurel leaf oil (*Laurus nobilis*), sage leaf oil (*Salvia triloba*), myrtle leaf oil (*Myrtus communis*), fennel seed oil (*Foeniculum vulgare*) and citrus peel oil (*Citrus* sp.) increased the body weight (Bozkurt et al., 2012b). In an earlier study, the plant extract blend (100 mg/kg) containing capsaicin, cinnamaldehyde and carvacrol supplemented for 41 days resulted in an increase in FCR in broilers (Jamroz et al., 2005).

The essential oil blend (25 mg/kg or 50 mg/kg) in broiler diets containing thymol as active component resulted in comparable growth performance, feed intake and feed/gain ratio in the control birds during 35 days of supplementation (Jang et al., 2007). In another study, it was reported that the supplementation of a blend of essential oils containing clove and cinnamon at a dose rate of 100 mg/kg diet resulted in comparable body weights and average daily gain in broilers at 46 days of age (Isabel and Santos, 2009). Similarly, the supplementation of garlic essential oil at 300 mg/kg for 42 days resulted in comparable body weights, average daily gain and FCR in broilers.

5.2 Effects on carcass quality

The supplementation of essential oil compounds results in comparable or improved carcass quality parameters to that of birds fed with a control diet unless the feed intake was significantly affected. For example, the supplementation of clove and cinnamon with a dose rate of 100 mg/kg in broiler diet resulted in an increase in breast weight in broilers at 46 days of age (Isabel and Santos, 2009). The plant extract (100 mg/kg) containing capsaicin, cinnamaldehyde and carvacrol increased the breast muscle proportion in eviscerated carcasses by 1.2% (Jamroz et al., 2005).

The supplementation of a blend of essential oils at 25 mg/kg including oregano, anise and citrus peel in broilers for 42 days did not affect the water-holding capacity, dry matter, fat content and colour of the muscles. In addition, the treatments resulted in comparable dressing percentage and abdominal fat relative weight to that of the control groups. Upon the sensory evaluation, it was revealed that the essential oil treatments caused an increased tenderness and juiciness of the breast and thigh muscles, respectively. The sensory panel did not find any difference in the appearance and flavour for both muscles between the two treatment groups (Hong et al., 2012). In addition, the supplementation of an equal mixture of thymol and carvacrol at 200 mg/kg in broilers for 42 days resulted in increased antioxidant characteristics of the thigh muscles such as increased superoxide dismutase and glutathione peroxidase activity and a decreased malondialdehyde level. The supplementation also resulted in a lowering of total saturated fatty acids and increase of total polyunsaturated fatty acids in serum and thighs. The total monounsaturated fatty acids were also increased in thigh muscles (Hashemipour et al., 2013).

The supplementation of oregano or garlic essential oils at 300 mg/kg or a combination of both (150 mg/kg, each) for 42 days resulted in no change in the carcass yield, relative weights of proventriculus, gizzard, duodenum, jejunum, ileum, colon, caecum, liver, pancreas, spleen, heart and bursa in broilers (Kırkpınar et al., 2011). Additionally, essential oils of thyme and star anise up to 1500 mg/kg diet did not affect average weights of the liver, heart, kidneys and spleen in broilers (Amad et al., 2011). In a different study, the supplementation of 0.1% aqueous extract of thyme (*Thymus vulgaris*), coneflower (*Echinacea purpurea*), garlic (*Allium sativum*) or a blend of these extracts in drinking water for 42 days did not affect carcass yield, relative weight of fat pad, liver, pancreas, proventriculus, gizzard and heart which was comparable to the broilers fed with a control diet. However, in a different study, thyme reduced the relative percentage weight of small intestine compared to the controls (2.84 vs. 2.48) (Rahimi et al., 2011). Moreover, the supplementation of 200 mg/kg essential oil extract from oregano, cinnamon and pepper, and 5000 mg/kg Labiatae extract from sage, thyme and rosemary also resulted in the similar weights for proventriculus, gizzard, liver, pancreas, or large or small intestine as that of the control group (Hernandez et al., 2004).

5.3 Effects on egg quality

The blend of essential oils (24 mg/kg) containing oregano oil (*Origanum* sp.), laurel leaf oil (*Laurus nobilis*), sage leaf oil (*Salvia triloba*), myrtle leaf oil (*Myrtus communis*), fennel seed oil (*Foeniculum vulgare*) and citrus peel oil (*Citrus* sp.) supplemented to layers from 36 to 51 weeks of age resulted in an increased eggshell weight. The essential oil supplementation significantly decreased the albumen weight; however, the yolk weight was not affected by the treatments. The supplementation did not have any impact on the egg production, egg weight, egg mass, feed consumption, FCR and shell-less egg ratio (Bozkurt et al., 2012b). In another study, it was observed that oregano essential oil supplementation at a dose rate of 100 mg/kg reduced the lipid oxidation in egg yolks. Likewise, the supplementation did not affect the egg weight and shape, yolk diameter, height and colour, Haugh units and shell thickness (Florou-Paneri et al., 2005a).

A combination of essential oils was found to positively affect the egg production parameters under cold stress conditions. Thyme and peppermint essential oils (100 mg/kg each) have been found to increase egg production in layers when supplemented to 42-week-old layers for 8 weeks under cold stress. The supplementation increased the egg mass, shell thickness and Haugh units. In addition, the combined essential oil supplementation significantly reduced the serum cholesterol level. However, the supplementation did not affect serum glucose, triglycerides, uric acid and albumin concentrations (Akbari et al., 2016). The egg production and egg mass increased when cinnamon essential oil (40 mg/kg) was supplemented to 42-week-old layers for 8 weeks under cold stress conditions. However, egg index, Haugh unit, shell thickness, shell weight and yolk colour were not affected by the supplementation. In addition, the supplementation did not affect the serum cholesterol, uric acid and albumin concentrations (Torki et al., 2015).

Olgun and Yıldız (2014) determined the effect of essential oil additives from thyme, black cumin, fennel, anise and rosemary on performance, eggshell quality, reproductive traits and mineral excretion in quails. In this study, 91-day quails were supplemented with the essential oils for 60 days. The results revealed that the 50–600 mg/kg supplementation did not have any effect on performance parameters, eggshell weight, fertility and hatchability of fertile and set eggs. With an exception, the supplementation at 50 mg/kg essential oil mixture improved egg-breaking strength. In addition, the lower level of supplementation significantly increased eggshell thickness and decreased the excretion of ash and minerals such as calcium, phosphorus, magnesium, manganese, zinc and cadmium. Additionally, long-term supplementation (66 days) of *trans*-cinnamaldehyde at 1% or 1.5% in 40- or 25-week-old layers did not affect the sensory quality, especially the taste of the cooked eggs (Upadhyaya et al., 2015).

6 Conclusion and future trends

Essential oils, botanicals or their active compounds are being widely investigated for their potential benefits in poultry. In response to the U.S. federal rule to phase out antibiotics from production, more focus has been given to explore their antimicrobial properties in live production. In recent years, gut health – composite normalcy of the gut contributed by various interacting factors – has gained popularity in the scientific world, warranting comprehensive research to understand the benefits and risks of alternatives. In poultry, maintenance or restoration of gut health by phytobiotics could be broadly associated with their digestive conditioning functions, antimicrobial activity, immunomodulatory effects and modulation of beneficial microbiota. Moreover, these compounds could promote growth and performance parameters in poultry, potentially indicating improved gut health.

On a closer look at the published literature, it is clear that the major function of phytobiotics is to render antimicrobial activity against pathogens in poultry. However, more bird studies are warranted to obtain consistency in data. Studies should include investigating potential shifts in the gut microbiome due to the pathogen and/or alternatives. The essential oils and the active compounds have a promising future as digestive conditioning agents. Almost all studies discussed in this chapter indicated positive effects of phytobiotics in augmenting digestive enzymes in the GI tract, resulting in better digestibility and absorption of nutrients from the intestine. Sufficient literature is lacking to judge on their efficacy on immunomodulation.

Due to a lack of coordinated research effort on potential phytobiotics for use in poultry, there is a dearth of reliable information on the inclusion rates of compounds in poultry diets. The variability observed across studies could be dependent on the type and concentration of active components present in the essential oils and botanicals. The composition of the compounds can vary based on the plant part from which the compound is extracted, the climatic and environmental conditions, soil type and the stage of the harvest of plants for oil production. Also, different classes of bioactive molecules such as phenols, aldehydes, ketones, alcohols, esters, ethers or hydrocarbons also vary in their efficacy depending on their interaction with each other (Bassolé and Juliani, 2012). These factors create difficulty in comparing essential oil compounds among themselves and to compare them with other alternatives such as prebiotics, probiotics and synbiotics. In addition, more research is required in areas of residue detection and labelling of these compounds for poultry use.

Research should also focus on the composition of active molecules present in the essential oils for better repeatability and reproducibility. Moreover, the interaction of these compounds with the complex menstruum of feed and their transformation during storage has not yet been studied. In addition, detailed investigations are necessary to elucidate the pharmacokinetics and toxic effects of these molecules in poultry. The option to supplement these compounds through water and its cost-benefit analysis should be a priority as well. Technologies such as microencapsulation, emulsions and spray drying applications could be adapted to increase the miscibility and to reduce the volatility of essential oil compounds in water.

7 References

Aarestrup, F. M. 2005. Veterinary drug usage and antimicrobial resistance in bacteria of animal origin. *Basic Clin. Pharmacol. Toxicol.* 96(4), 271–81. doi:10.1111/j.1742-7843.2005.pto960401.x.

Aguilar, G. 2001. Access to genetic resources and protection of traditional knowledge in the territories of indigenous peoples. *Environ. Sci. Policy* 4(4–5), 241–56. doi:10.1016/S1462-9011(01)00028-4.

Akbari, M., Torki, M. and Kaviani, K. 2016. Single and combined effects of peppermint and thyme essential oils on productive performance, egg quality traits, and blood parameters of laying hens reared under cold stress condition (6.8±3°C). *Int. J. Biometeorol.* 60(3), 447–54. doi:10.1007/s00484-015-1042-6.

Akthar, M. S., Degaga, B. and Azam, T. 2014. Antimicrobial activity of essential oils extracted from medicinal plants against the pathogenic microorganisms: A review. *Issues Biol. Sci. and Pharm. Res.* 2(1), 1-7.

Alali, W. Q., Hofacre, C. L., Mathis, G. F. and Faltys, G. 2013a. Effect of essential oil compound on shedding and colonization of *Salmonella enterica* serovar Heidelberg in broilers. *Poult. Sci.* 92(3), 836-41. doi:10.3382/ps.2012-02783.

Alali, W. Q., Hofacre, C. L., Mathis, G. F., Faltys, G., Ricke, S. C. and Doyle, M. P. 2013b. Effect of non-pharmaceutical compounds on shedding and colonization of *Salmonella enterica* serovar Heidelberg in broilers. *Food Control* 31(1), 125-8. doi:10.1016/j.foodcont.2012.10.001.

Allaoua, M., Etienne, P., Noirot, V., Carayon, J. L., Téné, N., Bonnafé, E. and Treilhou, M. 2018. Pharmacokinetic and antimicrobial activity of a new carvacrol-based product against a human pathogen, *Campylobacter jejuni*. *J. Appl. Microbiol.* 125(4), 1162-74. doi:10.1111/jam.13915.

Allen, P. C. and Fetterer, R. H. 2002. Recent advances in biology and immunobiology of *Eimeria* species and in diagnosis and control of infection with these coccidian parasites of poultry. *Clin. Microbiol. Rev.* 15(1), 58-65. doi:10.1128/CMR.15.1.58-65.2002.

Amad, A. A., Männer, K., Wendler, K. R., Neumann, K. and Zentek, J. 2011. Effects of a phytogenic feed additive on growth performance and ileal nutrient digestibility in broiler chickens. *Poult. Sci.* 90(12), 2811-6. doi:10.3382/ps.2011-01515.

Amerah, A. M., Mathis, G. and Hofacre, C. L. 2012. Effect of xylanase and a blend of essential oils on performance and *Salmonella* colonization of broiler chickens challenged with *Salmonella* Heidelberg. *Poult. Sci.* 91(4), 943-7. doi:10.3382/ps.2011-01922.

American Academy of Allergy Asthma and Immunology. 2018. Immunomodulators-definition. Available at: https://www.aaaai.org/conditions-and-treatments/conditions-dictionary/immunomodulators (accessed on 6 October 2018).

Arsi, K., Donoghue, A. M., Venkitanarayanan, K., Kollanoor-Johny, A., Fanatico, A. C., Blore, P. J. and Donoghue, D. J. 2014. The Efficacy of the natural plant extracts, thymol and carvacrol against *Campylobacter* colonization in broiler chickens. *J. Food Saf.* 34(4), 321-5. doi:10.1111/jfs.12129.

Attokaran, M. 2017. Essential oils. In: Attokaran, M. (Ed.), *Natural Food Flavors and Colorants* (2nd edn.). Chichester, UK: John Wiley & Sons, Ltd, pp. 17-9.

Bakkali, F., Averbeck, S., Averbeck, D. and Idaomar, M. 2008. Biological effects of essential oils - A review. *Food Chem. Toxicol.* 46(2), 446-75. doi:10.1016/j.fct.2007.09.106.

Ballhorn, D. J., Kautz, S., Jensen, M., Schmitt, I., Heil, M. and Hegeman, A. D. 2011. Genetic and environmental interactions determine plant defences against herbivores. *J. Ecol.* 99(1), 313-26. doi:10.1111/j.1365-2745.2010.01747.x.

Barug, D., de Jong, J., Kies, A. K. and Verstegen, M. W. A. (Eds). 2006. *Antimicrobial Growth Promoters*. Wageningen, The Netherlands: Wageningen Academic Publishers.

Basmacioğlu Malayoğlu, H., Baysal, Ş., Misirlioğlu, Z., Polat, M., Yilmaz, H. and Turan, N. 2010. Effects of oregano essential oil with or without feed enzymes on growth performance, digestive enzyme, nutrient digestibility, lipid metabolism and immune response of broilers fed on wheat-soybean meal diets. *Br. Poult. Sci.* 51(1), 67-80. doi:10.1080/00071660903573702.

Bassolé, I. H. N. and Juliani, H. R. 2012. Essential oils in combination and their antimicrobial properties. *Molecules* 17(4), 3989-4006. doi:10.3390/molecules17043989.

Ben Arfa, A., Combes, S., Preziosi-Belloy, L., Gontard, N. and Chalier, P. 2006. Antimicrobial activity of carvacrol related to its chemical structure. *Lett. Appl. Microbiol.* 43(2), 149-54. doi:10.1111/j.1472-765X.2006.01938.x.

Bischoff, S. C. 2011. 'Gut health': A new objective in medicine? *BMC Med.* 9(1), 24.

Borges, A., Saavedra, M. and Simoes, M. 2015. Insights on antimicrobial resistance, biofilms and the use of phytochemicals as new antimicrobial agents. *Curr. Med. Chem.* 22(21), 2590-614. doi:10.2174/0929867322666150530210522.

Bozkurt, M., Selek, N., Küçükyilmaz, K., Eren, H., Güven, E., Çatli, A. U. and Çinar, M. 2012a. Effects of dietary supplementation with a herbal extract on the performance of broilers infected with a mixture of *Eimeria* species. *Br. Poult. Sci.* 53(3), 325-32. doi:10.1080/00071668.2012.699673.

Bozkurt, M., Küçükyilmaz, K., Çatli, A. U., Çınar, M., Bintaş, E. and Çöven, F. 2012b. Performance, egg quality, and immune response of laying hens fed diets supplemented with mannan-oligosaccharide or an essential oil mixture under moderate and hot environmental conditions. *Poult. Sci.* 91(6), 1379-86. doi:10.3382/ps.2011-02023.

Brenes, A. and Roura, E. 2010. Essential oils in poultry nutrition: Main effects and modes of action. *Anim. Feed Sci. Technol.* 158(1-2), 1-14. doi:10.1016/j.anifeedsci.2010.03.007.

Burt, S. 2004. Essential oils: Their antibacterial properties and potential applications in foods–A review. *Int. J. Food Microbiol.* 94(3), 223-53. doi:10.1016/j.ijfoodmicro.2004.03.022.

Burt, S. A. and Reinders, R. D. 2003. Antibacterial activity of selected plant essential oils against *Escherichia coli* O157: H7. *Lett. Appl. Microbiol.* 36(3), 162-7. doi:10.1046/j.1472-765X.2003.01285.x.

Calo, J. R., Crandall, P. G., O'Bryan, C. A. and Ricke, S. C. 2015. Essential oils as antimicrobials in food systems - A review. *Food Control* 54, 111-9. doi:10.1016/j.foodcont.2014.12.040.

Chang, S. T., Chen, P. F. and Chang, S. C. 2001. Antibacterial activity of leaf essential oils and their constituents from *Cinnamomum osmophloeum*. *J. Ethnopharmacol.* 77(1), 123-7. doi:10.1016/S0378-8741(01)00273-2.

Chantziaras, I., Boyen, F., Callens, B. and Dewulf, J. 2014. Correlation between veterinary antimicrobial use and antimicrobial resistance in food-producing animals: A report on seven countries. *J. Antimicrob. Chemother.* 69(3), 827-34. doi:10.1093/jac/dkt443.

Chávez-González, M. L., Rodríguez-Herrera, R. and Aguilar, C. N. 2016. Essential oils: A natural alternative to combat antibiotics resistance. In: Kon, K. and Raj, M. (Eds), *Antibiotic Resistance: Mechanisms and New Antimicrobial Approaches*. San Diego, CA: Academic Press, pp. 227-37.

Coates, M. E., Dickinson, C. D., Harrison, G. F., Kon, S. K., Porter, J. W. G., Cummins, S. H. and Cuthbertson, W. F. J. 1952. A mode of action of antibiotics in chick nutrition. *J. Sci. Food Agric.* 3(1), 43-8. doi:10.1002/jsfa.2740030108.

Coates, A. R., Halls, G. and Hu, Y. 2011. Novel classes of antibiotics or more of the same? *Br. J. Pharmacol.* 163(1), 184-94. doi:10.1111/j.1476-5381.2011.01250.x.

Cross, D. E., McDevitt, R. M., Hillman, K. and Acamovic, T. 2007. The effect of herbs and their associated essential oils on performance, Dietary digestibility and gut microflora in chickens from 7 to 28 days of age. *Br. Poult. Sci.* 48(4), 496-506. doi:10.1080/00071660701463221.

Di Pasqua, R., Betts, G., Hoskins, N., Edwards, M., Ercolini, D. and Mauriello, G. 2007. Membrane toxicity of antimicrobial compounds from essential oils. *J. Agric. Food Chem.* 55(12), 4863–70. doi:10.1021/jf0636465.

Dixon, R. A. 2001. Natural products and plant disease resistance. *Nature* 411(6839), 843–7. doi:10.1038/35081178.

Du, E., Gan, L., Li, Z., Wang, W., Liu, D. and Guo, Y. 2015. *In vitro* antibacterial activity of thymol and carvacrol and their effects on broiler chickens challenged with *Clostridium perfringens*. *J. Anim. Sci. Biotechnol.* 6(1), 58. doi:10.1186/s40104-015-0055-7.

Engberg, R. M., Grevsen, K., Ivarsen, E., Fretté, X., Christensen, L. P., Højberg, O., Jensen, B. B. and Canibe, N. 2012. The effect of *Artemisia annua* on broiler performance, on intestinal microbiota and on the course of a *Clostridium perfringens* infection applying a necrotic enteritis disease model. *Avian Pathol.* 41(4), 369–76. doi:10.1080/03079457.2012.696185.

Engström, B. E., Fermer, C., Lindberg, A., Saarinen, E., Båverud, V. and Gunnarsson, A. 2003. Molecular typing of isolates of *Clostridium perfringens* from healthy and diseased poultry. *Vet. Microbiol.* 94(3), 225–35. doi:10.1016/S0378-1135(03)00106-8.

Fabricant, D. S. and Farnsworth, N. R. 2001. The value of plants used in traditional medicine for drug discovery. *Environ. Health Perspect.* 109 (Suppl. 1), 69–75. doi:10.1289/ehp.01109s169.

FDA. 2013. Guidance for Industry #213: New animal drugs and new animal drug combination products administered in or on medicated feed or drinking water of food-producing animals. Available at: https://www.fda.gov/downloads/AnimalVeterinary/GuidanceComplianceEnforcement/GuidanceforIndustry/UCM299624.pdf (accessed 10 August 2018).

FDA. 2017. Fact sheet: Veterinary feed directive final rule and next steps. Available at: https://www.fda.gov/animalveterinary/developmentapprovalprocess/ucm449019.htm.

Florou-Paneri, P., Nikolakakis, I., Giannenas, I., Koidis, A., Botsoglou, E., Dotas, V. and Mitsopoulos, I. 2005a. Hen performance and egg quality as affected by dietary oregano essential oil and alpha-tocopheryl acetate supplementation. *Int. J. Poult. Sci.* 4(7), 449–54.

Florou-Paneri, P., Palatos, G., Govaris, A., Botsoglou, D., Giannenas, I. and Ambrosiadis, I. 2005b. Oregano herb versus oregano essential oil as feed supplements to increase the oxidative stability of turkey meat. *Int. J. Poult. Sci.* 4(11), 866–71.

Giannenas, I., Florou-Paneri, P., Papazahariadou, M., Christaki, E., Botsoglou, N. A. and Spais, A. B. 2003. Effect of dietary supplementation with oregano essential oil on performance of broilers after experimental infection with *Eimeria tenella*. *Arch. Anim. Nutr.* 57(2), 99–106.

Giannenas, I., Papaneophytou, C. P., Tsalie, E., Pappas, I., Triantafillou, E., Tontis, D. and Kontopidis, G. A. 2014. Dietary supplementation of benzoic acid and essential oil compounds affects buffering capacity of the feeds, performance of turkey poults and their antioxidant status, pH in the digestive tract, intestinal microbiota and morphology. *Asian-Australas. J. Anim. Sci.* 27(2), 225–36. doi:10.5713/ajas.2013.13376.

Glaser, V. 1999. Billion-dollar market blossoms as botanicals take root. *Nat. Biotechnol.* 17(1), 17–8. doi:10.1038/5186.

Goldbloom, R. S. and Steigmann, F. 1951. Aureomycin therapy in hepatic insufficiency. *Gastroenterology* 18(1), 93–9.

Guenther, E. 1948. *The Essential Oils-Volume 1: History-Origin in Plants Production-Analysis*. New York, NY: D Van Nostrand Company.

Hao, H., Cheng, G., Iqbal, Z., Ai, X., Hussain, H. I., Huang, L., Dai, M., Wang, Y., Liu, Z. and Yuan, Z. 2014. Benefits and risks of antimicrobial use in food-producing animals. *Front. Microbiol.* 5, 288. doi:10.3389/fmicb.2014.00288.

Hashemi, S. R. and Davoodi, H. 2011. Herbal plants and their derivatives as growth and health promoters in animal nutrition. *Vet. Res. Commun.* 35(3), 169-80. doi:10.1007/s11259-010-9458-2.

Hashemipour, H., Kermanshahi, H., Golian, A. and Veldkamp, T. 2013. Effect of thymol and carvacrol feed supplementation on performance, antioxidant enzyme activities, fatty acid composition, digestive enzyme activities, and immune response in broiler chickens. *Poult. Sci.* 92(8), 2059-69. doi:10.3382/ps.2012-02685.

Hernandez, F., Madrid, J., Garcia, V., Orengo, J. and Megias, M. D. 2004. Influence of two plant extracts on broilers performance, digestibility, and digestive organ size. *Poult. Sci.* 83(2), 169-74. doi:10.1093/ps/83.2.169.

Holley, R. A. and Patel, D. 2005. Improvement in shelf-life and safety of perishable foods by plant essential oils and smoke antimicrobials. *Food Microbiol.* 22(4), 273-92. doi:10.1016/j.fm.2004.08.006.

Hong, J. C., Steiner, T., Aufy, A. and Lien, T. F. 2012. Effects of supplemental essential oil on growth performance, lipid metabolites and immunity, intestinal characteristics, microbiota and carcass traits in broilers. *Livest. Sci.* 144(3), 253-62. doi:10.1016/j.livsci.2011.12.008.

Huyghebaert, G., Ducatelle, R. and Immerseel, F. V. 2011. An update on alternatives to antimicrobial growth promoters for broilers. *Vet. J.* 187(2), 182-8.

Isabel, B. and Santos, Y. 2009. Effects of dietary organic acids and essential oils on growth performance and carcass characteristics of broiler chickens. *J. Appl. Poult. Res.* 18(3), 472-6. doi:10.3382/japr.2008-00096.

Jamroz, D., Wiliczkiewicz, A., Wertelecki, T., Orda, J. and Skorupińska, J. 2005. Use of active substances of plant origin in chicken diets based on maize and locally grown cereals. *Br. Poult. Sci.* 46(4), 485-93. doi:10.1080/00071660500191056.

Jang, I. S., Ko, Y. H., Kang, S. Y. and Lee, C. Y. 2007. Effect of a commercial essential oil on growth performance, digestive enzyme activity and intestinal microflora population in broiler chickens. *Anim. Feed Sci. Technol.* 134(3-4), 304-15. doi:10.1016/j.anifeedsci.2006.06.009.

Jerzsele, A., Szeker, K., Csizinszky, R., Gere, E., Jakab, C., Mallo, J. J. and Galfi, P. 2012. Efficacy of protected sodium butyrate, a protected blend of essential oils, their combination, and *Bacillus amyloliquefaciens* spore suspension against artificially induced necrotic enteritis in broilers. *Poult. Sci.* 91(4), 837-43. doi:10.3382/ps.2011-01853.

Jones, J. D. G. and Dangl, J. L. 2006. The plant immune system. *Nature* 444(7117), 323-9. doi:10.1038/nature05286.

Jukes, T. H. 1985. Some historical notes on chlortetracycline. *Rev. Infect. Dis.* 7(5), 702-7. doi:10.1093/clinids/7.5.702.

Jukes, T. H. and Williams, W. L. 1953. Nutritional effects of antibiotics. *Pharmacol. Rev.* 5(4), 381-420.

Kalemba, D. A. A. K. and Kunicka, A. 2003. Antibacterial and antifungal properties of essential oils. *Curr. Med. Chem.* 10(10), 813-29. doi:10.2174/0929867033457719.

Kapoor, G., Saigal, S. and Elongavan, A. 2017. Action and resistance mechanisms of antibiotics: A guide for clinicians. *J. Anesthesiol. Clin. Pharmacol.* 33(3), 300-5. doi:10.4103/joacp.JOACP_349_15.

Kelly, C., Gundogdu, O., Pircalabioru, G., Cean, A., Scates, P., Linton, M., Pinkerton, L., Magowan, E., Stef, L., Simiz, E., et al. 2017. The *in vitro* and *in vivo* effect of carvacrol in preventing *campylobacter* infection, colonization and in improving productivity of chicken broilers. *Foodborne Pathog. Dis.* 14(6), 341-9. doi:10.1089/fpd.2016.2265.

Kirchhelle, C. 2018. Pharming animals: A global history of antibiotics in food production (1935-2017). *Palgrave Commun.* 4(1), 96.

Kırkpınar, F., Ünlü, H. B. and Özdemir, G. 2011. Effects of oregano and garlic essential oils on performance, carcase, organ and blood characteristics and intestinal microflora of broilers. *Livest. Sci.* 137(1-3), 219-25.

Kollanoor-Johny, A. and Venkitanarayanan, K. 2017. Preharvest food safety–potential use of plant-derived compounds in layer chickens. In: Ricke, S. C. and Gast, R. K. (Eds), *Producing Safe Eggs: Microbial Ecology of Salmonella*. San Diego, CA: Academic Press, pp. 347-72.

Kollanoor-Johny, A., Darre, M. J., Hoagland, T. A., Schreiber, D. T., Donoghue, A. M., Donoghue, D. J. and Venkitanarayanan, K. 2008. Antibacterial effect of *trans*-cinnamaldehyde on *Salmonella* enteritidis and *Campylobacter jejuni* in chicken drinking water. *J. Appl. Poult. Res.* 17(4), 490-7.

Kollanoor-Johny, A., Mattson, T., Baskaran, S. A., Amalaradjou, M. A., Babapoor, S., March, B., Valipe, S., Darre, M., Hoagland, T., Schreiber, D., et al. 2012a. Reduction of *Salmonella enterica* serovar Enteritidis colonization in 20-day-old broiler chickens by the plant-derived compounds *trans*-cinnamaldehyde and eugenol. *Appl. Environ. Microbiol.* 78(8), 2981-7. doi:10.1128/AEM.07643-11.

Kollanoor-Johny, A., Upadhyay, A., Baskaran, S. A., Upadhyaya, I., Mooyottu, S., Mishra, N., Darre, M. J., Khan, M. I., Donoghue, A. M., Donoghue, D. J., et al. 2012b. Effect of therapeutic supplementation of the plant compounds *trans*-cinnamaldehyde and eugenol on *Salmonella* enterica serovar Enteritidis colonization in market-age broiler chickens. *J. Appl. Poult. Res.* 21(4), 816-22. doi:10.3382/japr.2012-00540.

Kollanoor-Johny, A., Frye, J. G., Donoghue, A., Donoghue, D. J., Porwollik, S., McClelland, M. and Venkitanarayanan, K. 2017. Gene expression response of *Salmonella enterica* serotype Enteritidis phage Type 8 to sub inhibitory concentrations of the plant-derived compounds *trans*-cinnamaldehyde and eugenol. *Front. Microbiol.* 8, 1828. doi:10.3389/fmicb.2017.01828.

Lawless, J. 1995. *The Illustrated Encyclopedia of Essential Oils: The Complete Guide to the Use of Oils in Aromatic and Herbalism*. Shaftesbury: Element Books.

Lee, K. W., Everts, H., Kappert, H. J., Frehner, M., Losa, R. and Beynen, A. C. 2003. Effects of dietary essential oil components on growth performance, digestive enzymes and lipid metabolism in female broiler chickens. *Br. Poult. Sci.* 44(3), 450-7. doi:10.1080/0007166031000085508.

Lesche, J. E. 2007. *The First Miracle Drugs: How the Sulfa Drugs Transformed Medicine*. New York, NY: Oxford University Press.

Lovland, A., Kaldhusdal, M., Redhead, K., Skjerve, E. and Lillehaug, A. 2004. Maternal vaccination against subclinical necrotic enteritis in broilers. *Avian Pathol.* 33(1), 81-90.

Luckey, T. D. 1952. Effect of feeding antibiotics upon the growth rate of germ-free birds. In Lobund Institute Colloquium, Notre Dame, IN.

Lv, F., Liang, H., Yuan, Q. and Li, C. 2011. *In vitro* antimicrobial effects and mechanism of action of selected plant essential oil combinations against four food-related microorganisms. *Food Res. Int.* 44(9), 3057–64. doi:10.1016/j.foodres.2011.07.030.

Maron, D. F., Smith, T. J. and Nachman, K. E. 2013. Restrictions on antimicrobial use in food animal production: An international regulatory and economic survey. *Global. Health* 9(1), 48. doi:10.1186/1744-8603-9-48.

Matthews, K., Kniel, K. E. and Montville, T. J. 2017. *Food Microbiology: An Introduction* (4th edn.). American Society of Microbiology. Washington DC: ASM Press.

Mead, G. C. 2000. Prospects for 'competitive exclusion' treatment to control salmonellas and other foodborne pathogens in poultry. *Vet. J.* 159(2), 111–23.

Miles, R. D., Butcher, G. D., Henry, P. R. and Littell, R. C. 2006. Effect of antibiotic growth promoters on broiler performance, intestinal growth parameters, and quantitative morphology. *Poult. Sci.* 85(3), 476–85. doi:10.1093/ps/85.3.476.

Milos, M. 2000. Chemical composition and antioxidant effect of glycosidically bound volatile compounds from oregano (*Origanum vulgare* L. ssp. *hirtum*). *Food Chem.* 71(1), 79–83. doi:10.1016/S0308-8146(00)00144-8..

Mitsch, P., Zitterl-Eglseer, K., Köhler, B., Gabler, C., Losa, R. and Zimpernik, I. 2004. The effect of two different blends of essential oil components on the proliferation of *Clostridium perfringens* in the intestines of broiler chickens. *Poult. Sci.* 83(4), 669–75. doi:10.1093/ps/83.4.669.

Mohiti-Asli, M. and Ghanaatparast-Rashti, M. 2015. Dietary oregano essential oil alleviates experimentally induced coccidiosis in broilers. *Prev. Vet. Med.* 120(2), 195–202. doi:10.1016/j.prevetmed.2015.03.014.

Nair, D. V. T., Thomas, J. V. and Kollanoor-Johny, A. 2017. Effect of supplementation of *trans*-cinnamaldehyde with or without oxytetracycline on multidrug-resistant *Salmonella* Heidelberg in turkey poults. Poultry Science Association Annual Meeting Abstr. 96 (E-suppl. 1), p. 193. Available at: https://www.poultryscience.org/psa17/abstracts/193.pdf (accessed on 29 September 2018).

Nair, D. V. T., Johnson, T., Noll, S. and Kollanoor-Johny, A. 2018. Effect of supplementation of an allochthonous probiotic bacterium, *Propionibacterium freudenreichii* ssp. *freudenreichii*, on the cecal microbiome of commercial turkeys challenged with multidrug-resistant *Salmonella* Heidelberg. Poultry Science Association Annual Meeting Abstr. 97 (E-suppl. 1), p. 212. Available at: https://www.poultryscience.org/psa18/abstracts/207.pdf (accessed on 29 September 2018).

Newman, D. J. 2008. Natural products as leads to potential drugs: An old process or the new hope for drug discovery? *J. Med. Chem.* 51(9), 2589–99.

Newman, D. J. and Cragg, G. M. 2012. Natural products as sources of new drugs over the 30 years from 1981 to 2010. *J. Nat. Prod.* 75(3), 311–35. doi:10.1021/np200906s.

NIH. 2011. Botanical dietary supplements. Available at: https://ods.od.nih.gov/factsheets/BotanicalBackground-HealthProfessional/ (accessed on 10 August 2018).

Nychas, G. J. E. 1995. Natural antimicrobials from plants. In: Gould, G. W. (Ed.), *New Methods of Food Preservation*. Boston, MA: Springer, pp. 58–89.

Olgun, O. and Yıldız, A. Ö. 2014. Effect of dietary supplementation of essential oils mixture on performance, eggshell quality, hatchability, and mineral excretion in quail breeders. *Environ. Sci. Pollut. Res. Int.* 21(23), 13434–9. doi:10.1007/s11356-014-3285-x.

Osbourn, A. E. 1996. Preformed antimicrobial compounds and plant defense against fungal attack. *Plant Cell* 8(10), 1821–31. doi:10.2307/3870232.

Oussalah, M., Caillet, S. and Lacroix, M. 2006. Mechanism of action of Spanish oregano, Chinese cinnamon, and savory essential oils against cell membranes and walls of *Escherichia coli* O157: H7 and *Listeria monocytogenes*. *J. Food Prot.* 69(5), 1046–55. doi:10.4315/0362-028X-69.5.1046.

Perricone, M., Arace, E., Corbo, M. R., Sinigaglia, M. and Bevilacqua, A. 2015. Bioactivity of essential oils: A review on their interaction with food components. *Front. Microbiol.* 6(Feb), 1–7.

Pichersky, E., Noel, J. P. and Dudareva, N. 2006. Biosynthesis of plant volatiles: Nature's diversity and ingenuity. *Science* 311(5762), 808–11. doi:10.1126/science.1118510.

Placha, I., Takacova, J., Ryzner, M., Cobanova, K., Laukova, A., Strompfova, V., Venglovska, K. and Faix, S. 2014. Effect of thyme essential oil and selenium on intestine integrity and antioxidant status of broilers. *Br. Poult. Sci.* 55(1), 105–14. doi:10.1080/00071668.2013.873772.

Potterat, O. and Hamburger, M. 2007. Drug discovery and development with plant-derived compounds. In: Petersen, F. and Amstutz, R. (Eds), *Natural Compounds as Drugs Volume I*. Basel: Birkhäuser Verlag Basel, pp. 45–118.

Puvača, N., Stanaćev, V., Glamočić, D., Lević, J., Perić, L. and Milić, D. 2013. Beneficial effects of phytoadditives in broiler nutrition. *Worlds Poult. Sci. J.* 69(1), 27–34.

Qiu, J., Feng, H., Lu, J., Xiang, H., Wang, D., Dong, J., Wang, J., Wang, X., Liu, J. and Deng, X. 2010. Eugenol reduces the expression of virulence-related exoproteins in *Staphylococcus aureus*. *Appl. Environ. Microbiol.* 76(17), 5846–51. doi:10.1128/AEM.00704-10.

Rahimi, S., Zadeh, Z. T., Torshizi, M. K., Omidbaigi, R. and Rokni, H. 2011. Immune system, blood factors and intestinal selected bacterial population in broiler chickens. *J. Agric. Sci. Technol.* 13, 527–39.

Rhayour, K., Bouchikhi, T. and Tantaoui-Elaraki, A., Sendide, K. and Remmal, A. 2003. The mechanism of bactericidal action of oregano and clove essential oils and of their phenolic major components on *Escherichia coli* and *Bacillus subtilis*. *J. Essent. Oil Res.* 15(5), 356–62.

Rodehutscord, M. and Kluth., H. 2002. Tierfütterung ohne antibiotisch wirkende Leistungsförderer. *Züchtungskunde* 74(6), 445–52.

Roofchaee, A., Irani, M., Ebrahimzadeh, M. A. and Akbari, M. R. 2011. Effect of dietary oregano (*Origanum vulgare* L.) essential oil on growth performance, cecal microflora and serum antioxidant activity of broiler chickens. *Afr. J. Biotechnol.* 10(32), 6177–83.

Schmidt, B., Ribnicky, D. M., Poulev, A., Logendra, S., Cefalu, W. T. and Raskin, I. 2008. A natural history of botanical therapeutics. *Metabolism* 57 (Suppl. 1), S3–9. doi:10.1016/j.metabol.2008.03.001.

Schwarz, S., Kehrenberg, C. and Walsh, T. R. 2001. Use of antimicrobial agents in veterinary medicine and food animal production. *Int. J. Antimicrob. Agents* 17(6), 431–7.

Shahid, M., Shahzad, A., Sobia, F., Sahai, A., Tripathi, T., Singh, A., Khan, H. M. and Umesh 2009. Plant natural products as a potential source for antibacterial agents: Recent trends. *Antiinfect. Agents Med. Chem.* 8(3), 211–25. doi:10.2174/187152109788680199.

Sugiharto, S. 2016. Role of nutraceuticals in gut health and growth performance of poultry. *J. Saudi Soc. Agri. Sci.* 15(2), 99–111. doi:10.1016/j.jssas.2014.06.001.

Surendran-Nair, M., Upadhyaya, I., Amalaradjou, M. A. R. and Venkitanarayanan, K. 2017. Antimicrobial food additives and disinfectants. In: Singh, O. V. (Ed.), *Foodborne Pathogens and Antibiotic Resistance*. Hoboken, NJ: John Wiley & Sons, Inc., pp. 275–301.

Tagboto, S. and Townson, S. 2001. Antiparasitic properties of medicinal plants and other naturally occurring products. In: Baker, J. R., Muller, R. and Rollinson, D. (Eds), *Advances in Parasitology Volume 50*. New York, NY: Academic Press, pp. 199-295.

Thomke, S. and Elwinger, K. 1998. Growth promotants in feeding pigs and poultry. I. Growth and feed efficiency responses to antibiotic growth promotants. *Ann. Zootech.* 47(2), 85-97. doi:10.1051/animres:19980201.

Timbermont, L., Lanckriet, A., Dewulf, J., Nollet, N., Schwarzer, K., Haesebrouck, F., Ducatelle, R. and Van Immerseel, F. 2010. Control of *Clostridium perfringens*-induced necrotic enteritis in broilers by target-released butyric acid, fatty acids and essential oils. *Avian Pathol.* 39(2), 117-21. doi:10.1080/03079451003610586.

Torki, M., Akbari, M. and Kaviani, K. 2015. Single and combined effects of zinc and cinnamon essential oil in diet on productive performance, egg quality traits, and blood parameters of laying hens reared under cold stress condition. *Int. J. Biometeorol.* 59(9), 1169-77. doi:10.1007/s00484-014-0928-z.

Ultee, A. and Smid, E. J. 2001. Influence of carvacrol on growth and toxin production by *Bacillus cereus*. *Int. J. Food Microbiol.* 64(3), 373-8. doi:10.1016/S0168-1605(00)00480-3.

Ultee, A., Kets, E. P. W. and Smid, E. J. 1999. Mechanisms of action of carvacrol on the foodborne pathogen *Bacillus cereus*. *Appl. Environ. Microbiol.* 65(10), 4606-10.

Ultee, A., Kets, E. P., Alberda, M., Hoekstra, F. A. and Smid, E. J. 2000. Adaptation of the foodborne pathogen Bacillus cereus to carvacrol. *Arch. Microbiol.* 174(4), 233-8.

Ultee, A., Bennik, M. H. J. and Moezelaar, R. 2002. The phenolic hydroxyl group of carvacrol is essential for action against the foodborne pathogen *Bacillus cereus*. *Appl. Environ. Microbiol.* 68(4), 1561-8.

Upadhyaya, I., Upadhyay, A., Kollanoor-Johny, A., Darre, M. J. and Venkitanarayanan, K. 2013. Effect of plant-derived antimicrobials on *Salmonella* Enteritidis adhesion to and invasion of primary chicken oviduct epithelial cells *in vitro* and virulence gene expression. *Int. J. Mol. Sci.* 14(5), 10608-25. doi:10.3390/ijms140510608.

Upadhyaya, I., Upadhyay, A., Kollanoor-Johny, A., Mooyottu, S., Baskaran, S. A., Yin, H. B., Schreiber, D. T., Khan, M. I., Darre, M. J., Curtis, P. A., et al. 2015. In-feed supplementation of trans-cinnamaldehyde reduces layer-chicken egg-borne transmission of *Salmonella enterica* serovar Enteritidis. *Appl. Environ. Microbiol.* 81(9), 2985-94. doi:10.1128/AEM.03809-14.

Upadhyaya, I., Yin, H. B., Surendran-Nair, M. and Venkitanarayanan, K. 2017. Natural approaches for improving postharvest safety of egg and egg products. In: Ricke, S. C. and Gast, R. K. (Eds), *Producing Safe Eggs*. San Diego, CA: Academic Press, pp. 391-420.

Van Immerseel, F. V., Buck, J. D., Pasmans, F., Huyghebaert, G., Haesebrouck, F. and Ducatelle, R. 2004. *Clostridium perfringens* in poultry: An emerging threat for animal and public health. *Avian Pathol.* 33(6), 537-49. doi:10.1080/03079450400013162.

Venkitanarayanan, K., Kollanoor-Johny, A., Darre, M. J., Donoghue, A. M. and Donoghue, D. J. 2013. Use of plant-derived antimicrobials for improving the safety of poultry products. *Poult. Sci.* 92(2), 493-501. doi:10.3382/ps.2012-02764.

Wenk, C. 2003. Herbs and botanicals as feed additives in monogastric animals. *Asian Australas. J. Anim. Sci.* 16(2), 282-9.

Windisch, W. and Kroismayr, A. 2007. The effect of phytobiotics on performance and gut function in monogastrics. *Biomin World Nutr. Forum*. Available at: https://en.engo

rmix.com/MA-feed-machinery/articles/the-effect-phytobiotics-performance-t285/p0.htm (accessed on 10 August 2018).

Yang, Y., Iji, P. A. and Choct, M. 2009. Dietary modulation of gut microflora in broiler chickens: A review of the role of six kinds of alternatives to in-feed antibiotics. *Worlds Poult. Sci. J.* 65(1), 97–114.

Yegani, M. and Korver, D. R. 2008. Factors affecting intestinal health in poultry. *Poult. Sci.* 87(10), 2052–63. doi:10.3382/ps.2008-00091.

Yin, D., Du, E., Yuan, J., Gao, J., Wang, Y., Aggrey, S. E. and Guo, Y. 2017. Supplemental thymol and carvacrol increases ileum *Lactobacillus* population and reduces effect of necrotic enteritis caused by *Clostridium perfringes* in chickens. *Sci. Rep.* 7(1), 7334. doi:10.1038/s41598-017-07420-4.

Yoo, S., Nam, H. and Lee, D. 2018. Phenotype-oriented network analysis for discovering pharmacological effects of natural compounds. *Sci. Rep.* 8(1), 11667. doi:10.1038/s41598-018-30138-w.

You, Y. and Silbergeld, E. K. 2014. Learning from agriculture: Understanding low-dose antimicrobials as drivers of resistome expansion. *Front. Microbiol.* 5, 284. doi:10.3389/fmicb.2014.00284.

Zeng, Z., Zhang, S., Wang, H. and Piao, X. 2015. Essential oil and aromatic plants as feed additives in non-ruminant nutrition: A review. *J. Anim. Sci. Biotechnol.* 6(1), 7. doi:10.1186/s40104-015-0004-5.

Zhai, H., Liu, H., Wang, S., Wu, J. and Kluenter, A. M. 2018. Potential of essential oils for poultry and pigs. *Anim. Nutr.* 4(2), 179–86. doi:10.1016/j.aninu.2018.01.005.

Zomrawi, W. B., Atti, K. A., Dousa, B. M. and Mahala, A. G. 2012. The effect of ginger root powder (*Zingiber officinale*) supplementation on broiler chicks performance, blood and serum constituents. *Online J. Anim. Feed Res.* 2, 457–60.

Chapter 19

The role of specific cereal grain dietary components in poultry gut function

Paul Iji, Fiji National University, Fiji Islands and University of New England, Australia; Apeh Omede, University of New England, Australia and Kogi State University, Nigeria; Medani Abdallh, University of New England, Australia and University of Khartoum, Sudan; and Emmanuel Ahiwe, University of New England, Australia and Federal University of Technology - Owerri, Nigeria

1 Introduction

The digestive tract is in close proximity with feed; as such feed factors have a primary physical interaction with the luminal mucosa prior to any secondary or tertiary effects of nutrients. The nature of the feed factors determines the kind of changes that occur in the intestinal mucosa. Physical interactions may alter the morphometry of the mucosa, including changes to cell proliferation and migration, to alter the size of the villi and also the life of the individual cells. Nutrients and their constituent chemical compounds interact with cellular and molecular processes, to further alter the physical structure and functions of the gut. The changes are also dependent on the nature of the chemical constituent.

http://dx.doi.org/10.19103/AS.2019.0059.23

Cereal grains are included in poultry diets primarily to supply energy, from their huge reserve of starch. However, starch is neither the only carbohydrate nor the only nutrient in cereal grains. Cereals contain appreciable levels of protein, sometimes as high as 20% and some cereals are also rich in oil. However, it is the carbohydrate fractions that tend to affect the structure and functions of the gut more than the other constituents. Starch is easily digested by poultry, although there are differences in digestibility, depending on the proportions of amylose and amylopectin (McDonald et al., 2011; Svihus, 2014a). Cereal grains also contain other carbohydrates, which are not starch. Some carbohydrates are made up of glucose, such as starch, but the linkages between the glucose molecules create different compounds, which may become resistant to animal enzymes, particularly amylase. In addition to these non-starch polysaccharides (NSP) and glucose-based carbohydrates, there are other polysaccharides that are non-starch but are made up of simple sugars with less than five carbon atoms.

Cereal grains also contain other types of carbohydrates, for example, pectin, pentosans, lignin, oligosaccharides and so forth. These, and the more common NSPs such as β-glucans and xylans, have a major impact on the movement of digesta along the tract and may alter the functions as well as the structure of the gastrointestinal tract (GIT). Research has shown direct effects of the dietary fractions on the intestinal mucosa, but often, the presence of these dietary components elicit a reaction from outside the lumen, particularly from the pancreas (Iji et al., 2001b; Mirzaie et al., 2012). Although not linked directly to any particular grain fraction, the feeding of whole grains is another factor that directly affects GIT function from the gizzard to the small intestine (SI), and sometimes, indirectly affect the pancreas too (Amerah et al., 2007; Biggs and Parsons, 2009). In this chapter, we will examine the impact of cereal grain fractions on the structure and function of the GIT function of poultry. A major focus is on the anti-nutritive factors that are present in these ingredients as the interactions between the regular nutrients and body are not considered to be inimical to productivity or health.

2 The poultry gut

Poultry are equipped with an efficient GIT that aids in feed intake, digestion, absorption, utilization of nutrients and defence. These processes help to achieve and maintain adequate growth and health of the birds (Sugiharto, 2016). The GIT is a narrow tube that begins from the beak and ends at the cloaca of all avian species. The GIT is separated into seven parts, namely (1) mouth/beak, (2) oesophagus, (3) crop, (4) proventriculus, (5) gizzard, (6) SI (consisting of the duodenum (with a pancreas located in between the duodenal loop), jejunum and ileum) and (7) large intestine (consisting of the caeca, colon and rectum). These parts are linked into a long, narrow tube as shown in Fig. 1.

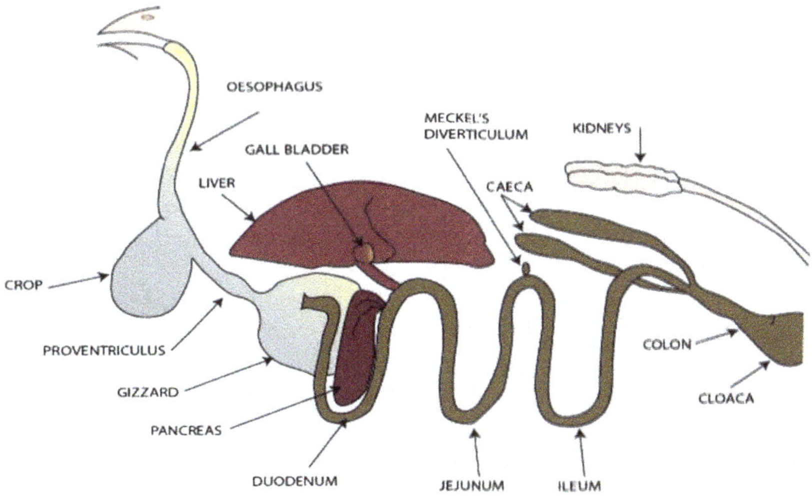

Figure 1 The gastrointestinal tract (GIT) of poultry. Source: adapted from PoultryHub (n.d.).

The action of the gut in ensuring adequate performance and health of poultry may be attributed to the morphological, digestive, microbial and immune response activities that take place in the GIT (Yegani and Korver, 2008). Since a key function of this important part of poultry includes conversion and digestion of ingested feed into less complex components which are readily absorbable and utilizable by the birds, adequate knowledge and understanding of various mechanisms behind the functionality of different parts of the GIT becomes imperative. This will go a long way to aid poultry keepers and researchers to develop improved strategies to further enhance poultry performance and health.

2.1 Gut integrity and factors that influence it

The GIT contributes considerably in maintaining the performance and health integrity of poultry. The GIT has been reported to be the most extensively exposed surface of the body and is constantly exposed to a wide variety of potentially harmful substances. The GIT acts as a selective layer between the tissues of the bird and its luminal environment. The GIT consists of several layers of physical, chemical, immunological and microbiological components (Yegani and Korver, 2008). For poultry to produce or grow at an optimum level, the GIT and its accessories should also be in optimum condition. Any deviation from a normal state or condition of the GIT of poultry caused by external or internal factors may affect its integrity and also go a long way in affecting their performance and health. Poultry gut integrity has been reported to be influenced by several factors, which are worth highlighting.

2.1.1 Degradation of mucus layer

Degradation of mucus layer is usually caused by microorganisms such as bacteria, viruses, fungi, parasites and external and endogenous toxins. The mucus layer is the barrier-like surface that prevents the entrance of unwanted materials into the bloodstream. When the mucus layer is degraded, the integrity of the gut is said to be compromised. The result of these degradations includes epithelial cell destruction, the vascular supply is interrupted and probable distortion of the immune system.

2.1.2 External stressors

Conditions such as overcrowding, temperature variations, transportation, inadequate feeding, improper handling, humidity and other factors that can exert stress on a *bird* can influence the gut integrity.

2.1.3 Feed toxins

Gut integrity can be negatively compromised by feed toxins, resulting in poor performance, health issues and even increased mortality.

2.1.4 Nutritional factors

Nutrient deficiencies, the nature of the diet and feeding of contaminated feed tend to affect gut integrity.

2.1.5 Disease and injuries

Disease conditions, injuries and infections have been reported to affect the nutritional and health status of poultry with a severe effect on their gut health and integrity.

2.1.6 Microorganisms

The gut of poultry serves as a microbial habitat or microbiome. These microbes positively or negatively influence gut health. They assert this effect through causing changes in the morphology of the gut when they feed, produce toxins and activate inflammatory immune response.

3 Functions of the gastrointestinal tract (GIT)

The performance and health of poultry are influenced by the status of the GIT. The GIT of a bird carries out both nutritional and health functions as shown in Table 1. For the GIT to carry out this nutritional and health maintenance function

Table 1 Regions of the poultry GIT and their functions

GIT part	Function
Mouth/beak	First point of entry of feed. Used to pick and peck feed or material and then swallowed through the oesophagus to the crop. Starch hydrolysis to sugar is done by the enzyme, ptyalin, that is contained in the saliva of poultry. However, because of the short time the feed stays in poultry mouth the hydrolysis process is minor.
Oesophagus	This part of the gastrointestinal tract (GIT) aids in the transportation of feed from the mouth to the stomach.
Crop	The crop is simply a small sac/pouch used to store and moisten feed. The ptyalin in the saliva from the mouth continues the hydrolysis of starch to sugar here. The crop serves as a temporary feed storage organ.
Proventriculus	The lining of the proventriculus secretes HCl, pepsin and mucus, which commences and aids actual enzymatic digestion. The pH in this environment is low.
Gizzard	Mechanical grinding of feed takes place in the gizzard. Like in the proventriculus, the gizzard also has a low pH environment. It sets the rate of passage through the GIT.
Pancreas	Although not a direct division of the GIT, the pancreas is an important organ, with connect ducts into the duodenum through which digestive enzymes are secreted. Pancreatic enzymes are typically initiators of digestion, which is then completed by terminal enzymes, most of them membrane-bound to the intestinal mucosa. Research has shown that the pancreas also responds to dietary factors, in terms of structural (Engberg et al., 2002; Agah and Norollahi, 2008) as well as functional changes (Brzęk et al., 2013).
Small intestine (consisting of the duodenum, jejunum and ileum)	Digesta from the gizzard is mixed with bile salts and digestive enzymes in the small intestine. It also serves as a major site of chemical digestion and nutrient absorption. The intestine has two kinds of glands: (a) intestinal glands secrete amylolytic, proteolytic and lipolytic enzymes and (b) glands of the mucous membrane secrete maltase, isomaltase, peptidase, saccharase and palatinase.
Large intestine (consisting of the caeca, colon and cloaca)	The caeca contain large numbers of bacteria which break down indigestible plant material. They empty every 24–48 h and are refilled with contents from the colon. They also aid in water absorption. After digesta leaves the colon, the faecal material passes into the cloaca where it is mixed with uric acid. The faecal sample mixed with uric acid is expelled through the cloaca.

Source: adapted from Yeoman et al. (2012) and Svihus (2014b).

efficiently, its condition or integrity has to be intact. Any deviation from the normal condition of the gut may have a negative impact on the performance and health status of poultry. Several factors such as diet type and quality, stress, toxins, disease, hormonal influence, microbial activities and so forth have been reported to influence the functionality of the GIT of poultry positively or negatively (Muramatsu et al., 1994; Yegani and Korver, 2008). The GIT of poultry is equipped with several means by which it can reduce or eliminate the effect of some of these factors. However, in some cases where its defence mechanism fails, exogenous additives, for instance, antibiotics, probiotics, prebiotics, acidifiers and so forth have been used to improve the condition of the GIT, thereby preventing poor performance and health issues (Cheng et al., 2014; Dhama et al., 2014; Roto et al., 2015).

4 General structure of cereal grains

Grain consists of three major parts: bran which is the protective or outer coating layer that is made up of several layers (10–16%); endosperm which is the main part of the grain kernel (80–85%) is mostly starch and thus considered the energy-rich part; and germ is a very small part (2–3%) lying normally at one end of the seed.

4.1 Bran

Bran is the outer fibrous layer of cereal grain, it is a rich source of insoluble fibre in all types of grains. It also contains some soluble fibres. In addition to it being high in fibre, bran is a good source of essential fatty acids and it contains considerable amounts of starch and protein (Wrigley, 2017). Bran contains most of the B-group vitamins such as thiamin, niacin, riboflavin and folate in the grain, and also contains the majority of cereal minerals (more than 90%) including Ca, Mg, Fe, K and selenium (Anil, 2012). The average fibre content of the bran of two widely used grains, corn and wheat, are 7.5–8.2% and 40–42%, respectively. Wheat bran has the highest fibre content, and oat, barley and rice brans contain around 4%, 10% and 28% crude fibre, respectively. Fat content of the bran of rice (10.3%), oat (7.8%) and sorghum (6.2%) is on the higher side among cereal varieties. Corn, oat and barley brans are rich in starch, while oat, wheat rye and sorghum brans have high protein content (Table 2). Due to its high concentrations of both vitamins and minerals in bran, it gives the nutty flavour to the whole grain. In addition to the nutrients mentioned above, bran contains high levels of NSP; it is also considered as the main source of anti-nutrients in cereals, such as phytic acid, tannins, saponin and enzyme inhibitors (Knudsen, 2014) (Fig. 2).

4.2 Endosperm

The endosperm constitutes the largest part of whole grain (80–85%); it is covered by an aleurone frame which consists of one layer in most cereals, including corn, wheat, sorghum, oat and rye, and three layers in barley and rice (Evers and Millar, 2002). From a nutritional point of view, it is called starchy endosperm because most of the cereals' starch and energy-producing carbohydrates occur

Table 2 Nutrient composition of common cereal grains

Grains	Energy (kcal/kg)	CP (%)	Total fat (%)	Crude fibre (%)	Carb (%)	Ca (mg/kg)	Fe (mg/kg)	K (mg/kg)	P (mg/kg)	Mg (mg/kg)
Corn	3650	9.7	6.43	7.3	74.3	70	27.1	2780	2100	1270
Wheat	3390	13.7	2.5	12.2	75.9	340	35.2	4310	5080	1440
Rice - brown	3620	7.5	2.7	3.4	76.2	330	18	2680	2640	1430
Rice - white	3600	6.5	0.52	2.8	79.3	90	43.6	860	1080	350
Rye	3380	9.4	1.77	14.6	75.9	240	26.3	5100	3320	1100
Oat	3890	16.9	6.90	10.6	66.3	540	47.2	4290	5230	1770
Sorghum	3290	11.1	7.30	6.3	72.1	130	33.6	3630	2890	1650
Barley	3520	9.9	3.39	15.6	73.5	330	36	4520	2640	1330
Millet	3780	11.02	4.22	8.5	72.9	80	30.1	1950	2850	1140
Triticale	3360	13.18	1.81	14.6	72.1	370	25.7	3320	3580	1300

Information from the USDA database https://ndb.nal.usda.gov/ndb/search/list. Values are for whole grain. Wheat values are for durum type, while corn nutrient composition is for the yellow type.

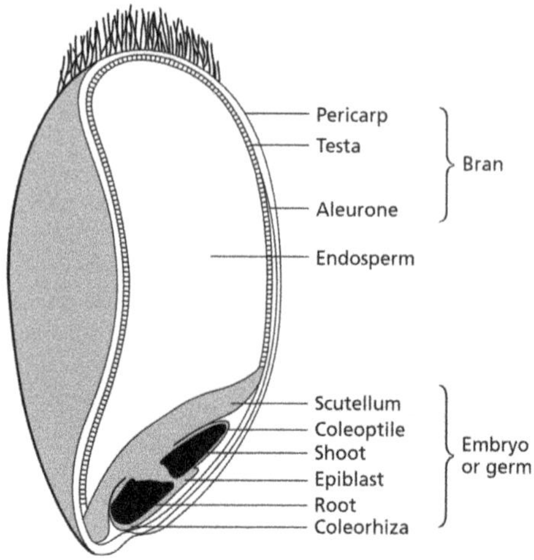

Figure 2 Cross-sectional structure of wheat grain. Source: adapted from McDonald et al. (2011).

in it. This starchy part also contains protein, some vitamins and low amounts of minerals especially in the aleurone layer. The endosperm cell walls consist of polysaccharides (cellulose, arabinoxylans and β-glucans), protein and phenolic acids (Evers et al., 1999). The NSP found in the aleurone layer, which surrounds the endosperm and enzyme inhibitors (especially amylase and protease inhibitors), are the main anti-nutritional factors (ANF) found in endosperm.

4.3 Germ or embryo

Germ, the smallest part of whole grain, represents around 3% of the grain, on average, and reflects the living part of the grain because when seeded it can grow into a new plant. Structurally, germ has a thin wall because it is the grain embryo (Kent and Evers, 1994). Germ is an excellent source of fat, plant sterols, B-group vitamins, vitamin E, minerals and some enzymes. No anti-nutrients are found in germ, although it is high in oil content, so it is better to remove this component, to prevent rancidity, especially during long periods of storage (Koehler and Wieser, 2013).

5 Nutrient composition of cereal grains

Cereal grains are fed mainly to provide energy to poultry; most of the energy in cereal grains comes from carbohydrates due to a large component of starch. The typical nutrient composition of key cereal grains is shown in Table 2. Cereal grains also contain non-starch carbohydrates, which constitute the major anti-nutrients. Corn, wheat, sorghum, barley, rye, oats, triticale and millet are the most commercially used cereal grains. Corn, wheat and rice represent about 87% of total annual grain production worldwide (Ahiwe et al., 2018). Corn is considered the dominant cereal grain grown in the world, followed by wheat, which is recorded as the second most-produced cereal. Cereal grains are cultivated in large quantities and provide more starch in comparison with other types of crops. The nutritive value of cereal grains varies between grain type, location, season, cultivation, harvesting and handling conditions (Wrigley, 2010). The seed in the majority of grain-producing plants is covered by a fibrous shell to protect it from weather changes, pests, water and disease. The whole grain of majority of cereals approximately consists of 12-14% water, 65-75% carbohydrates, 7-12% protein, 2-6% fat and 1-3% minerals (Belitz et al., 2009).

5.1 Carbohydrates

Carbohydrates are the major nutrient in cereal grains. Carbohydrates in cereal grains are present as polysaccharides, mainly as starch (55-70%), NSP (3-15%), cellulose (less than 2.5%), monosaccharides and disaccharides (1-3%) and oligosaccharides (less than 1%) (Zeeman et al., 2010; Wrigley, 2017). Generally, white rice is characterized by a high content of carbohydrates (79.3%), whereas the lowest carbohydrate content is in oats, with an average of 66.3%. The average energy content of cereal grains ranges between 3290 kcal/kg in sorghum and 3890 kcal/kg in oat. Corn, oat, millet and rice have the highest energy content compared to other cereals (Table 2).

5.2 Mono- and disaccharides

Monosaccharides (glucose, galactose and fructose) represent the base and smallest components of carbohydrate. Glucose and fructose are the only two monosaccharides found in cereal grains in significant quantities, ranging from 0.9% to 1.8% in sorghum, 0.4% to 0.5% in wheat and 0.3% to 0.9% in corn (Shelton and Lee, 2000). Sucrose is the only disaccharide found in cereals in significant amounts, with levels of 0.9-1.9% in corn, 0.57-0.8% in wheat, 80-85% in sorghum, 1.7-1.9% in rye and 1.9-2.2% in barley (Shelton and Lee, 2000). Mono- and disaccharides, as simple sugars together, represent around 1-2% of whole grain (McKevith, 2004).

5.3 Oligosaccharides

More than two units of monosaccharide units are attached with glycosidic linkage through condensation reactions to form oligosaccharides. Raffinose, stachyose, verbascose, inulin, fructo- and galacto-oligosaccharides are the main oligosaccharides present in cereal grains (Biesiekierski et al., 2011). Oat contains moderate amounts of raffinose and stachyose with a very low content of fructo-oligosaccharides. Wheat has a high concentration of raffinose, while triticale contains less fructo-oligosaccharides in comparison to the other cereals. Wheat, barley and rye have a substantial concentration (0.5%, 0.3% and 0.2%, respectively) of all types of oligosaccharides (Henry and Saini, 1989). Oligosaccharides are normally hydrolysed to monosaccharides by bacterial fermentation in the non-ruminant intestine, which produces short-chain fatty acids (SCFA) and gases (CO_2, methane and hydrogen) as a result.

5.4 Polysaccharides

Starch is the main polysaccharide in cereal grains, followed by cellulose and glucofructans (Zeeman et al., 2010). Starch is present in the form of amylose in most varieties of cereals and amylopectin in waxy types of cereals such as rice and corn (Baghurst et al., 1996). Depending on the type of cereal grains, starch is present in different sizes and shapes as in cell granules within the starchy endosperm of whole grain. NSP are polysaccharides other than starch, which are considered as dietary fibre and are present in cereals mainly in the form of arabinoxylans and β-glucans (Knudsen, 2014). Arabinoxylans represent the main fraction of cereal NSP (85–90%), while β-glucans are only 9–14%. The average content of NSP in cereals ranges between 2% and 8% for arabinoxylans and 1–7% for β-glucans. Rye has the highest arabinoxylan content (7%) compared to other cereals (Knudsen and Laerke, 2010), while wheat contains only 2% (Courtin and Delcour, 2002). On the other hand, β-glucans are high in barley (>5%) and oat (>4%), with an average of less than 2% in other cereals (Wood, 2001).

5.5 Fibres

Dietary fibres of cereal grains consist of soluble fibres, insoluble fibres, resistant starch and some oligosaccharides. Soluble fibres in cereals are soluble arabinoxylans and β-glucans which are NSP, and pentosans, which are 5-C sugars, while insoluble fibres are lignins, cellulose and hemi-celluloses (Anderson et al., 2009). Resistant starch is the type of starch that cannot be digested in the SI. It is found in the high-amylose varieties of corn and rice (Falomo, 2016). Some oligosaccharides such as fructo-oligosaccharides and inulin are considered a part of dietary fibres (Sampath et al., 2008). The crude fibre content of cereal grains is around 3–16%. Barley, rye, triticale and wheat are the most fibrous (15.6%, 14.6%, 14.6% and 12.2%, respectively) among the cereal grains (Table 2).

6 Anti-nutritive components of cereal grains

Cereal grains contain substantial amounts of ANF, such as phytate, enzyme (mainly amylase and protease) inhibitors, polyphenols and tannins. In addition, the high fibre content (especially insoluble fibre) is considered as anti-nutritive due to its role in reducing the digestibility and availability of some nutrients in cereal grains (Huisman and Tolman, 1992; Pariza, 1996). Some ANF have a negative effect on the utilization of cereal nutrients, for example, through a reduction in mineral bioavailability and digestibility of macronutrients (Evers and Millar, 2002).

6.1 Phytate

Phytic acid (phytate) is the storage form of phosphorus found in cereals, oilseeds and legume seeds. The activity of phytic acid leads to the formation of a mineral-phytate bond, which reduces the bioavailability of minerals, especially calcium, zinc, iron and magnesium (Leeson and Summers, 2001). Phytate also reduces the solubility, functionality, digestibility and absorption of proteins and carbohydrates. Phytate is normally found in all plants as a phosphorus (energy) reservoir for seed germination and acts as a guard against oxidative stress throughout the life of the seed. As a result, phytate-phosphorus is unavailable to non-ruminant animals, including poultry (Liener, 1989). Phytate also reportedly inhibits enzymes such as pepsin, amylase and trypsin (Sandberg and Andersson, 1988). Phytic acid accounts for 50–80% of the total phosphorus in different cereals and represents about 0.50–1.89% of whole grain dry weight. The average content of phytate varies between different varieties of cereal grains, representing 1.05% in corn, 0.96% in rye, 1.37% in oat, 1.19% in barley, 0.96–2.2% in sorghum and 1.0–3% in wheat (Coulibaly et al., 2011).

6.2 Tannins

Tannins are found in cereals, particularly sorghum and millet. Tannins are mainly concentrated in bran, causing tannin-protein complexes, which inactivate the digestive enzymes and reduce protein digestibility by formation of a protein-ionizable iron interaction. The presence of tannins in cereals can cause low feed intake, reduce growth, decrease iron absorption and reduce protein and amino acid digestibility (Salunkhe et al., 1990; Reddy and Pierson, 1994; Haard and Chism 1996; Haard, 1999).

6.3 Enzyme inhibitors

Amylase, protease, trypsin and chymotrypsin inhibitors exist in most cereal grains. Partial inhibition of these important digestive enzymes reduces the digestion of related nutrients and causes metabolic disturbance of the utilization of some nutrients (Shewry, 1999). Wheat, rye, barley, oat, rice and sorghum contain α-amylase inhibitors, while protease inhibitors have been found in most cereal grains (Franco et al., 2002; Habib and Fazili, 2007).

6.4 Lectins

Lectins have anti-nutrient activity towards sugar components. They are carbohydrate-binding proteins that bind glycans of glycoproteins, glycolipids and polysaccharides with high affinity (Chrispeels and Raikhel 1991). Lectins are found in wheat, rye, barley, oat, corn and rice, but have not been reported in sorghum or millet. Wheat germ agglutinin (lectin) has adverse health effects, with the ability to adhere to cell surfaces such as the epithelial layer of the gut. Accordingly, they could impair some digestion and absorption activities, through changes to the bacterial profiles and gut immune function (Brunsgaard, 1998).

6.5 Saponins

Saponins are sterol or triterpene glycosides, found in most cereal grains. These molecules have strong antifungal activity and are well known to be haemolytic. Oat has significant amounts of saponins, but other cereals are low in saponins (Osbourn, 2003). Saponins have been reported to reduce the uptake of certain nutrients, including glucose and cholesterol from the gut and, in turn, cause growth depression (Ujowundu et al., 2008).

7 Role of cereal grain components in poultry gut function

Cereal grains are valuable sources of nutrients for poultry but at the same time contain factors, which reduce digestibility of nutrients. There are not many studies directly comparing the effects of different cereals on digestive function, rather there is greater knowledge on how poultry grow or produce when fed diets based on different cereal grains. Shakouri et al. (2009) did a comparative study on the response of broiler chickens to diets based on barley, maize, sorghum and wheat, although the major focus was on the efficacy of supplemental microbial enzymes that target nutrients or ANF in the key grains used for poultry feeding. However, it is obvious that the various components of these ingredients possess physico-chemical properties that can affect gut function, depending on the level of inclusion in the diet (Jørgensen et al., 1996; Jiménez-Moreno et al., 2010, 2011; Mateos et al., 2012). Some of the aspects that may be affected by these factors include intestinal structure, digestive function, stability of the gut microbiota and possibly, immunocompetence derived from the gut.

7.1 Intestinal structure

A certain amount of cereal dietary components is required for proper development and physiological function of the gut. Studies by Sklan et al. (2003) and Amerah et al. (2009) revealed that increased dietary fibre reduced the length of the SI, contrary to an earlier study (Siri et al., 1992), in which the inclusion of pectin in the diets of layer chickens resulted in increases in the length of the oesophagus and that of the SI, caeca and rectum. Previous studies showed that diets lacking structural components result in dilated proventricular walls, and an enlargement of the organ, which may affect its functionality, while increased dietary fibre in diets decreased the weight of the proventriculus (González-Alvarado et al., 2008; Jiménez-Moreno et al., 2009). Similarly, González-Alvarado et al. (2007) and Svihus (2011) reported that increased dietary fibre resulted in increased gizzard weight and contents. In another study, the inclusion of 10% fine cellulose from oat hulls in a wheat-based diet exerted a positive effect on gizzard development (Hetland et al., 2003), which affected other physiological functions that are linked to the gizzard, for example, the enhancement of digestive secretions, including HCl, bile acid and endogenous enzymes (Mateos et al., 2012). Some of these effects are highlighted in Table 3 (Hetland et al., 2003).

The presence of pectin in these cereals as well as in most soluble dietary fibre sources presents a problem of increased digesta viscosity within the gut, which could impede the absorption of nutrients (Iji et al., 2001a). A study investigating the effects of different cereals (maize, wheat and barley) reported lowered gross responses and increased intestinal viscosity in birds fed wheat and barley in comparison to those fed maize (Mathlouthi et al., 2002), confirming that NSP are accountable for a higher viscosity in the gut. The ratio of villus height to crypt depth is a vital parameter for assessing gut functions in terms of the digestive capacity of the SI (Mateos et al., 2012). Several factors,

Table 3 Gut functions as affected by 10% inclusion of oat hulls in a wheat-based diet for broilers

	Wheat diet	
	No oat hulls	10% oat hulls added
Gizzard weight (g/kg)	20.6	26.0
Pancreas (g)	3.7	4.0
Pancreas (g/kg live weight)	2.1	2.2
Amylase (U/g jejunal DM)	146	255
Amylase (U/g feed, jejunum)	65	137
Bile acid (jejunum mg/g)	11.7	18.0
Bile acid (jejunum mg/g feed)	5	10

Source: adapted from Hetland et al. (2003).

including the viscosity of dietary components from cereal grains, level of inclusion in diets and age of birds, most probably have an influence on the response of the epithelial mucosa to diets containing dietary components from cereal grains. Hence, Iji et al. (2001b) reported deeper crypts in the jejunum and ileum in 28-day-old broilers when the diet was supplemented with a highly viscous dietary component (5% gum xanthan), an effect that was translated into larger villi, at the ileum. This condition may enhance gut function in terms of nutrient absorption and utilization but that would depend on the viscosity of subsequent diets. On the contrary, a lower level of lignin (2.5%) in diets for broilers resulted in reduction in villus height (Baurhoo et al., 2007).

7.2 Digestive function

The effects of dietary components of cereal grains on nutrient digestibility, digestive enzyme secretion and activities are still not clearly understood. When 10% oat hulls were supplemented in wheat-based diets for broilers, Hetland et al. (2003) reported an increase in the relative weight of gizzard. The weight of the pancreas was not affected but amylase activity was considerably increased in birds fed diets supplemented with 10% oat hulls as shown in Table 3 (Hetland et al., 2003). In a previous study by Svihus and Hetland (2001), the inclusion of 10% fine cellulose powder in wheat-based diets for broilers was found to benefit ileal starch digestibility. Kluth and Rodehutscord (2009) also reported that increasing the level of cellulose increased crude protein and amino acid flows at the terminal ileum of broilers.

7.3 Stability of gut microbiota

Poultry gut microbiome is mostly impacted by diet on account of residual dietary components that escape digestion and absorption in the upper gut. These serve as substrates for the growth of intestinal microbes (Pan and Yu, 2014). The most significant impact emanates from feeding wheat-, barley- or rye-based diets to poultry. These diets contain high levels of indigestible, water-soluble NSP, which may enhance the proliferation of *Clostridium perfringens* and predispose young chicks to necrotic enteritis, while corn-based diets, with lower levels of water-soluble NSP may not cause such problems (Annett et al., 2002; Jia et al., 2009). The lengthy passage time of the intestinal contents associated with high levels of viscous NSP in diets creates a conducive environment in the gut that may result in increased population of pathogenic bacteria such as *Escherichia coli* and *C. perfringens* (Józefiak et al., 2006; Hashemipour et al., 2016). Feeding diets rich in NSP has also been reported to result in increased bacterial translocation through the epithelium of the intestinal mucosa into portal circulation (Tellez et al., 2015). Hammons et al. (2010) showed that a small variation in dietary cereal grain composition can affect gut microbiota. They further reported that a standard corn-soybean diet favoured *Lactobacillus agilis* type R5, while a diet high in wheat middlings favoured *L. agilis* type R1. Wheat, barley, rice, rye, oat and sorghum are the major sources of arabinoxylans. As arabinoxylans cannot be degraded by non-ruminant animals, they stimulate the growth of resident bacteria such as *Bacteroides, Bifidobacterium, Clostridium, Lactobacillus* and *Eubacterium* once they reach the colon (Rivière et al., 2014). However, and although less effective in poultry (Józefiak et al., 2004), resistant starch, NSP and oligosaccharides may be fermented by commensal bacteria of the GIT, producing lactic acid and SCFA such as acetate, propionate and butyrate. The SCFA reduce the population of pathogenic bacteria by decreasing the pH of the GIT, which creates unfavourable conditions for their growth, while favouring the growth of commensal anaerobic population (Bederska-Łojewska et al., 2017).

8 Possible mechanisms by which cereal grain dietary components affect gut function

There seems to be a general agreement that gut viscosity is the main mechanism by which cereal grain dietary components influence gut function. An increase in gut viscosity attributed to the presence of water-soluble NSP in diets has been suggested as one mechanism by which nutrient digestibility is reduced (Fengler and Marquardt, 1988; Choct and Annison, 1992). Soluble fibres, such as those present in barley, wheat and rye, form viscous gels that can trap nutrients and slow down the rates of passage and digestion (Englyst, 1989; Bedford et al., 1991; Veldman and Vahl, 1994). Increased digesta viscosity also affects how well nutrients mix with pancreatic enzymes and bile acids, which consequently results in reduced rate of nutrient absorption (Fengler and Marquardt, 1988). There is also evidence suggesting that increased digesta viscosity affects gut microbiota, especially by enhancing bacterial fermentation (Annison, 1993) and thus negatively affects micelle formation (Coates et al., 1981) due to increased deconjugation of bile acids (Cole and Fuller, 1984).

9 Conclusion

Cereal grains remain the most important components of poultry feed and the increased production of these grains has partly been responsible for the rapid growth of the poultry industry. However, there are major structural and chemical differences between the grains, some of which underpin the differences in nutritive value. No cereal grain is completely devoid of deleterious factors, but there are differences in the nature and concentrations of these factors. In this chapter, we have highlighted some of the key grain factors that affect the structural as well as functional development of the GIT in poultry, particularly broiler chickens. Some of the negative effects can be reduced through physical processing, improved diet formulation and inclusion of specific supplements. Such intervention has largely led to sustained or even increased productivity, although there may be lingering subtle changes in the GIT level.

10 Where to look for further information

Further information on the effect of various cereal grains on poultry gut can be found in the book by McDonald et al. (2011).

Some related articles have considered the importance of chicken gut and factors that influence it and they are as follows:

- Shang, Y., Kumar, S., Oakley, B. and Kim, W. K. 2018. Chicken gut microbiota: importance and detection technology. *Front. Vet. Sci.* 5, 254. doi:10.3389/fvets.2018.00254.
- Pan, D. and Yu, Z. 2013. Intestinal microbiome of poultry and its interaction with host and diet. *Gut Microbes* 5(1), 108–19. doi:10.4161/gmic.26945.

11 References

Agah, M. J. and Norollahi, H. 2008. Effect of feed form and duration time in growing period on broilers performance. *Int. J. Poult. Sci.* 7(11), 1074–7. doi:10.3923/ijps.2008.1074.1077.

Ahiwe, E. U., Omede, A. A., Abdallh, M. B. and Iji, P. A. 2018. Managing dietary energy intake by broiler chickens to reduce production costs and improve product quality. In: Banu, Y. and Turgay, T. (Eds), *Animal Husbandry and Nutrition*. IntechOpen, London, UK, pp. 115–45. Chapter 6.

Amerah, A. M., Ravindran, V., Lentle, R. G. and Thomas, D. G. 2007. Feed particle size: implications on the digestion and performance of poultry. *Worlds Poult. Sci. J.* 63(3), 439–55. doi:10.1017/S0043933907001560.

Amerah, A. M., Ravindran, V. and Lentle, R. G. 2009. Influence of insoluble fibre and whole wheat inclusion on the performance, digestive tract development and ileal microbiota profile of broiler chickens. *Br. Poult. Sci.* 50(3), 366–75. doi:10.1080/00071660902865901.

Anderson, J. W., Baird, P., Davis Jr., R. H., Ferreri, S., Knudtson, M., Koraym, A., Waters, V. and Williams, C. L. 2009. Health benefits of dietary fiber. *Nutr. Rev.* 67(4), 188–205. doi:10.1111/j.1753-4887.2009.00189.x.

Anil, M. 2012. Effects of wheat bran, corn bran, rice bran and oat bran supplementation and the properties of pide. *J. Food Process. Preserv.* 36(3), 276–83. doi:10.1111/j.1745-4549.2011.00592.x.

Annett, C. B., Viste, J. R., Chirino-Trejo, M., Classen, H. L., Middleton, D. M. and Simko, E. 2002. Necrotic enteritis: effect of barley, wheat and corn diets on proliferation of *Clostridium perfringens* type A. *Avian Pathol.* 31(5), 598–601. doi:10.1080/0307945021000024544.

Annison, G. 1993. The role of wheat non-starch polysaccharides in broiler nutrition. *Aust. J. Agric. Res.* 44(3), 405–22. doi:10.1071/AR9930405.

Baghurst, P. A., Baghurst, K. I. and Record, S. J. 1996. Dietary fibre, nonstarch polysaccharides and resistant starch - a review. *Food Aust.* 48, S3–5.

Baurhoo, B., Phillip, L. and Ruiz-Feria, C. A. 2007. Effects of purified lignin and mannan oligosaccharides on intestinal integrity and microbial populations in the ceca and litter of broiler chickens. *Poult. Sci.* 86(6), 1070–8. doi:10.1093/ps/86.6.1070.

Bederska-Łojewska, D., Świątkiewicz, S., Arczewska-Włosek, A. and Schwarz, T. 2017. Rye non-starch polysaccharides: their impact on poultry intestinal physiology, nutrients digestibility and performance indices - a review. *Ann. Anim. Sci.* 17(2), 351–69. doi:10.1515/aoas-2016-0090.

Bedford, M. R., Classen, H. L. and Campbell, G. L. 1991. The effect of pelleting, salt and pentosanase on the viscosity of intestinal contents and the performance of broilers fed rye. *Poult. Sci.* 70(7), 1571–7. doi:10.3382/ps.0701571.

Belitz, H.-D., Grosch, W. and Schieberle, P. 2009. Cereals and cereal products. In: Belitz, H.-D., Grosch, W. and Schieberle, P. (Eds), *Food Chemistry* (4th edn.). Springer, Berlin, pp. 670–5.

Biesiekierski, J. R., Rosella, O., Rose, R., Liels, K., Barrett, J. S., Shepherd, S. J., Gibson, P. R. and Muir, J. G. 2011. Quantification of fructans, galacto-oligosaccharides and other short-chain carbohydrates in processed grains and cereals. *J. Hum. Nutr. Diet.* 24(2), 154–76. doi:10.1111/j.1365-277X.2010.01139.x.

Biggs, P. and Parsons, C. M. 2009. The effects of whole grains on nutrient digestibilities, growth performance, and cecal short-chain fatty acid concentrations in young chicks fed ground corn-soybean meal diets. *Poult. Sci.* 88(9), 1893–905, doi:10.3382/ps.2008-00437.

Brunsgaard, G. 1998. Effects of cereal type and feed particle size on morphological characteristics, epithelial cell proliferation, and lectin binding patterns in the large intestine of pigs. *J. Anim. Sci.* 76(11), 2787–98. doi:10.2527/1998.76112787x.

Brzęk, P., Ciminari, M. E., Kohl, K. D., Lessner, K., Karasov, W. H. and Caviedes-Vidal, E. 2013. Effect of age and diet composition on activity of pancreatic enzymes in birds. *J. Comp. Physiol. B, Biochem. Syst. Environ. Physiol.* 183(5), 685–97. doi:10.1007/s00360-012-0731-2.

Cheng, G., Hao, H., Xie, S., Wang, X., Dai, M., Huang, L. and Yuan, Z. 2014. Antibiotic alternatives: the substitution of antibiotics in animal husbandry? *Front. Microbiol.* 5, 217. doi:10.3389/fmicb.2014.00217.

Choct, M. and Annison, G. 1992. The inhibition of nutrient digestion by wheat pentosans. *Br. J. Nutr.* 67(1), 123–32. doi:10.1079/BJN19920014.

Chrispeels, M. J. and Raikhel, N. V. 1991. Lectins, lectin genes, and their role in plant defense. *Plant Cell* 3(1), 1–9. doi:10.1105/tpc.3.1.1.

Coates, M. E., Cole, C. B., Fuller, R., Houghton, S. B. and Yokota, H. 1981. The gut microflora and the uptake of glucose from the small intestine of the chick. *Br. Poult. Sci.* 22(3), 289–94. doi:10.1080/00071688108447888.

Cole, C. B. and Fuller, R. 1984. Bile acid deconjugation and attachment of chicken gut bacteria: their possible role in growth depression. *Br. Poult. Sci.* 25(2), 227–31. doi:10.1080/00071668408454861.

Coulibaly, A., Kouakou, B. and Chen, J. 2011. Phytic acid in cereal grains: structure, healthy or harmful ways to reduce phytic acid in cereal grains and their effects on nutritional quality. *Am. J. Plant Nutr. Fert. Tech.* 1(1), 1–22. doi:10.3923/ajpnft.2011.1.22.

Courtin, C. M. and Delcour, J. A. 2002. Arabinoxylans and endoxylanases in wheat flour breadmaking. *J. Cereal Sci.* 35(3), 225–43.

Dhama, K., Tiwari, R., Ullah, R., Chakraborty, S., Gopi, M., Karthik, K., Saminathan, M., Desingu, P. M. and Sunkara, L. T. 2014. Growth promoters and novel feed additives improving poultry production and health, bioactive principles and beneficial applications: the trends and advances – a review. *Int. J. Pharm.* 10, 129–59.

Engberg, R. M., Hedemann, M. S. and Jensen, B. B. 2002. The influence of grinding and pelleting of feed on the microbial composition and activity in the digestive tract of broiler chickens. *Br. Poult. Sci.* 43(4), 569–79. doi:10.1080/0007166022000004480.

Englyst, H. 1989. Classification and measurement of plant polysaccharides. *Anim. Feed Sci. Tech.* 23(1–3), 27–42. doi:10.1016/0377-8401(89)90087-4.

Evers, T. and Millar, S. 2002. Cereal grain structure and development: some implications for quality. *J. Cereal Sci.* 36(3), 261–84. doi:10.1006/jcrs.2002.0435.

Evers, A. D., O'Brien, L. and Blakeney, A. B. 1999. Cereal structure and composition. *Aust. J. Agric. Res.* 50(5), 629–50. doi:10.1071/AR98158.

Falomo, O. O. 2016. *In vitro* evaluation of resistant starch using corn. MSc thesis. Southern Illinois University at Carbondale.

Fengler, A. I. and Marquardt, R. R. 1988. Water-soluble pentosans from rye: effects on rate of dialysis and on the retention of nutrients by the chick. *Cereal Chem.* 65, 298–302.

Franco, O. L., Rigden, D. J., Melo, F. R. and Grossi-de-Sá, M. F. 2002. Plant α-amylase inhibitors and their interaction with insect α-amylases: structure, function and potential for crop protection. *Eur. J. Biochem.* 269(2), 397–412.

González-Alvarado, J. M., Jiménez-Moreno, E., Lázaro, R. and Mateos, G. G. 2007. Effects of type of cereal, heat processing of the cereal, and inclusion of fiber in the diet on productive performance and digestive traits of broilers. *Poult. Sci.* 86(8), 1705–15. doi:10.1093/ps/86.8.1705.

González-Alvarado, J. M., Jiménez-Moreno, E., Valencia, D. G., Lázaro, R. and Mateos, G. G. 2008. Effects of fiber source and heat processing of the cereal on the development and pH of the gastrointestinal tract of broilers fed diets based on corn or rice. *Poult. Sci.* 87(9), 1779–95. doi:10.3382/ps.2008-00070.

Haard, N. F. 1999. *Fermented Cereals: A Global Perspective (No. 138).* Food and Agriculture Organization, Rome.

Haard, N. M. and Chism, G. W. 1996. Characteristics of edible plant tissues. In: Fennema, O. (Ed.), *Food Chemistry.* Marcel Dekker Inc., New York, p. 963.

Habib, H. and Fazili, K. M. 2007. Plant protease inhibitors: a defense strategy in plants. *Biotech. Mol. Bio. Rev.* 2(3), 68–85.

Hammons, S., Oh, P. L., Martínez, I., Clark, K., Schlegel, V. L., Sitorius, E., Scheideler, S. E. and Walter, J. 2010. A small variation in diet influences the *Lactobacillus* strain composition in the crop of broiler chickens. *Syst. Appl. Microbiol.* 33(5), 275–81. doi:10.1016/j.syapm.2010.04.003.

Hashemipour, H., Khaksar, V., Rubio, L. A., Veldkamp, T. and van Krimpen, M. M. 2016. Effect of feed supplementation with a thymol plus carvacrol mixture, in combination or not with an NSP-degrading enzyme, on productive and physiological parameters of broilers fed on wheat-based diets. *Anim. Feed. Sci. Tech.* 211, 117–31. doi:10.1016/j.anifeedsci.2015.09.023.

Henry, R. J. and Saini, H. S. 1989. Characterization of cereal sugars and oligosaccharides. *Cereal Chem.* 66(5), 362–5.

Hetland, H., Svihus, B. and Krogdahl, A. 2003. Effects of oat hulls and wood shavings on digestion in broilers and layers fed diets based on whole or ground wheat. *Br. Poult. Sci.* 44(2), 275–82. doi:10.1080/0007166031000124595.

Huisman, J. and Tolman, G. H. 1992. Antinutritional factors in the plant proteins of diets for non-ruminants. *Recent Adv. Anim. Nutr.* 68(1), 101–10.

Iji, P. A., Saki, A. A. and Tivey, D. R. 2001a. Intestinal structure and function of broiler chickens on diets supplemented with a mannan oligosaccharide. *J. Sci. Food Agric.* 81(12), 1186–92. doi:10.1002/jsfa.925.

Iji, P. A., Saki, A. A. and Tivey, D. R. 2001b. Intestinal development and body growth of broiler chicks on diets supplemented with non-starch polysaccharides. *Anim. Feed Sci. Tech.* 89(3–4), 175–88. doi:10.1016/S0377-8401(00)00223-6.

Jia, W., Slominski, B. A., Bruce, H. L., Blank, G., Crow, G. and Jones, O. 2009. Effects of diet type and enzyme addition on growth performance and gut health of broiler chickens during subclinical *Clostridium perfringens* challenge. *Poult. Sci.* 88(1), 132–40. doi:10.3382/ps.2008-00204.

Jiménez-Moreno, E., González-Alvarado, J. M., González-Serrano, A., Lázaro, R. and Mateos, G. G. 2009. Effect of dietary fiber and fat on performance and digestive traits of broilers from one to twenty-one days of age. *Poult. Sci.* 88(12), 2562–74. doi:10.3382/ps.2009-00179.

Jiménez-Moreno, E., González-Alvarado, J. M., González-Sánchez, D., Lázaro, R. and Mateos, G. G. 2010. Effects of type and particle size of dietary fiber on growth performance and digestive traits of broilers from 1 to 21 days of age. *Poult. Sci.* 89(10), 2197–212. doi:10.3382/ps.2010-00771.

Jiménez-Moreno, E., Chamorro, S., Frikha, M., Safaa, H. M., Lázaro, R. and Mateos, G. G. 2011. Effects of increasing levels of pea hulls in the diet on productive performance and digestive traits of broilers from one to eighteen days of age. *Anim. Feed Sci. Tech.* 168(1–2), 100–12. doi:10.1016/j.anifeedsci.2011.03.013.

Jørgensen, H., Zhao, X. Q., Knudsen, K. E. B. and Eggum, B. O. 1996. The influence of dietary fibre source and level on the development of the gastrointestinal tract, digestibility and energy metabolism in broiler chickens. *Br. J. Nutr.* 75(3), 379–95. doi:10.1079/BJN19960141.

Józefiak, D., Rutkowskia, A. and Marti, S. A. 2004. Carbohydrate fermentation in the avian ceca: a review. *Anim. Feed. Sci. Technol.* 113(1–4), 1–15. doi:10.1016/j.anifeedsci.2003.09.007.

Józefiak, D., Rutkowski, A., Jensen, B. B. and Engberg, R. M. 2006. The effect of β-glucanase supplementation of barley- and oat-based diets on growth performance and fermentation in broiler chicken gastrointestinal tract. *Br. Poult. Sci.* 47(1), 57–64. doi:10.1080/00071660500475145.

Kent, N. L. and Evers, A. D. 1994. *Kent's Technology of Cereals* (4th edn.). Elsevier, Oxford.

Kluth, H. and Rodehutscord, M. 2009. Effect of inclusion of cellulose in the diet on the inevitable endogenous amino acid losses in the ileum of broiler chickens. *Poult. Sci.* 88(6), 1199–205. doi:10.3382/ps.2008-00385.

Knudsen, K. E. B. 2014. Fiber and nonstarch polysaccharide content and variation in common crops used in broiler diets. *Poult. Sci.* 93(9), 2380–93. doi:10.3382/ps.2014-03902.

Knudsen, K. E. B. and Lærke, H. N. 2010. Rye arabinoxylans: molecular structure, physicochemical properties and physiological effects in the gastrointestinal tract. *Cereal Chem.* 87(4), 353–62. doi:10.1094/CCHEM-87-4-0353.

Koehler, P. and Wieser, H. 2013. Chemistry of cereal grains. In: Gobbetti, M. and Gänzle, M. (Eds), *Handbook on Sourdough Biotechnology*. Springer, Boston, MA, pp. 11–45.

Leeson, S. and Summers, J. D. 2001. Minerals. In: *Nutrition of the Chicken* (4th edn.). University Books, Guelph, Ontario, Canada.

Liener, I. E. 1989. Anti nutritional factors in legume seeds: state of art. In: Huismen, J., van der Poel, T. F. B. and Liener, I. E. (Eds), *Recent Advances of Research on Anti-Nutritional Factors in Legume Seeds*. Pudoc, Wageningen, The Netherlands, pp. 6–13.

Mateos, G. G., Jiménez-Moreno, E., Serrano, M. P. and Lázaro, R. P. 2012. Poultry response to high levels of dietary fiber sources varying in physical and chemical characteristics. *J. Appl. Poult. Res.* 21(1), 156–74, doi:10.3382/japr.2011-00477.

Mathlouthi, N., Mallet, S., Saulnier, L., Quemener, B. and Larbier, M. 2002. Effects of xylanase and β-glucanase addition on performance, nutrient digestibility, and physico-chemical conditions in the small intestine contents and caecal microflora of broiler chickens fed a wheat and barley-based diet. *Anim. Res.* 51(5), 395–406. doi:10.1051/animres:2002034.

McDonald, P., Morgan, C. A., Sinclair, L. A. and Wilkinson, R. G. 2011. *Animal Nutrition* (7th edn.). Prentice Hall, Harlow, UK, pp. 1–714. ISBN-10: 9781408204238..

McKevith, B. 2004. Nutritional aspects of cereals. *Nutr. Bull.* 29(2), 111–42. doi:10.1111/j.1467-3010.2004.00418.x.

Mirzaie, S., Zaghari, M., Aminzadeh, S., Shivazad, M. and Mateos, G. G. 2012. Effect of wheat inclusion and xylanase supplementation of the diet on productive performance, nutrient retention and endogenous intestinal enzyme activity of laying hens. *Poult. Sci.* 91(2), 413–25. doi:10.3382/ps.2011-01686.

Muramatsu, T., Nakajima, S. and Okumura, J. 1994. Modification of energy metabolism by the presence of the gut microflora in the chicken. *Br. J. Nutr.* 71(5), 709–17. doi:10.1079/BJN19940178.

Osbourn, A. E. 2003. Saponins in cereals. *Phytochemistry* 62(1), 1–4. doi:10.1016/S0031-9422(02)00393-X.

Pan, D. and Yu, Z. 2014. Intestinal microbiome of poultry and its interaction with host and diet. *Gut Microbes* 5(1), 108–19. doi:10.4161/gmic.26945.

Pariza, M. W. 1996. Toxic substances. In: Fennema, O. R. (Ed.), *Food Chemistry*. Marcel Dekker, New York, pp. 825–40.

Reddy, N. R. and Pierson, M. D. 1994. Reduction in antinutritional and toxic components in plant foods by fermentation. *Food Res. Inter.* 27(3), 281–90. doi:10.1016/0963-9969(94)90096-5.

Rivière, A., Moens, F., Selak, M., Maes, D., Weckx, S. and De Vuyst, L. 2014. The ability of bifidobacteria to degrade arabinoxylan oligosaccharide constituents and derived oligosaccharides is strain dependent. *Appl. Environ. Microbiol.* 80(1), 204–17. doi:10.1128/AEM.02853-13.

Roto, S. M., Rubinelli, P. M. and Ricke, S. C. 2015. An introduction to the avian gut microbiota and the effects of yeast-based prebiotic-type compounds as potential feed additives. *Front. Vet. Sci.* 2(28), 28. doi:10.3389/fvets.2015.00028.

Salunkhe, D. K., Chavan, J. K. and Kadam, S. S. 1990. *Dietary Tannins: Consequences and Remedies*. CRC Press, Boca Raton, FL.

Sampath, S., Rao, M. T., Reddy, K. K., Arun, K. and Reddy, P. V. M. 2008. Effect of germination on oligosaccharides in cereals and pulses. *J. Food Sci. Tech. Mysore* 45(2), 196–8.

Sandberg, A. S. and Andersson, H. 1988. Effect of dietary phytase on the digestion of phytate in the stomach and small intestines of humans. *J. Nutr.* 118(4), 469–73. doi:10.1093/jn/118.4.469.

Shakouri, M. D., Iji, P. A., Mikkelsen, L. L. and Cowieson, A. J. 2009. Intestinal function and gut microflora of broiler chickens as influenced by cereal grains and microbial enzyme supplementation. *J. Anim. Physiol. Anim. Nutr.* 93(5), 647–58. doi:10.1111/j.1439-0396.2008.00852.x.

Shelton, D. R. and Lee, W. J. 2000. Cereal carbohydrates. In: Kulp, K. and Ponte, J. G., Jr. (Eds), *Handbook of Cereal Science and Technology* (2nd edn.). Marcel Dekker, New York, pp. 385–416.

Shewry, P. R. 1999. Enzyme inhibitors of seeds: types and properties. In: Shewry, P. R. and Casey, R. (Eds), *Seed Proteins*. Springer, Dordrecht, the Netherlands, pp. 1–10.

Siri, S., Tobioka, H. and Tasaki, I. 1992. Effects of dietary fibers on growth performance, development of internal organs, protein and energy utilization, and lipid content of growing chicks. *Jpn. Poult. Sci.* 29(2), 106–14. doi:10.2141/jpsa.29.106.

Sklan, D., Smirnov, A. and Plavnik, I. 2003. The effect of dietary fiber on the small intestines and apparent digestion in the turkey. *Br. Poult. Sci.* 44(5), 735–40. doi:10.1080/00071660310001643750.

Sugiharto, S. 2016. Role of nutraceuticals in gut health and growth performance of poultry. *J. Saudi Soc. Agric. Sci.* 15(2), 99–111. doi:10.1016/j.jssas.2014.06.001.

Svihus, B. 2011. The gizzard: function, influence of diet structure and effects on nutrient availability. *Worlds Poult. Sci. J.* 67(2), 207–24. doi:10.1017/S0043933911000249.

Svihus, B. 2014a. Starch digestion capacity of poultry. *Poult. Sci.* 93(9), 2394–9. doi:10.3382/ps.2014-03905.

Svihus, B. 2014b. Function of the digestive system. *J. Appl. Poult. Res.* 23(2), 306–14, doi:10.3382/japr.2014-00937.

Svihus, B. and Hetland, H. 2001. Ileal starch digestibility in growing broiler chickens fed on a wheat-based diet is improved by mash feeding, dilution with cellulose or whole wheat inclusion. *Br. Poult. Sci.* 42(5), 633–7. doi:10.1080/00071660120088461.

Tellez, G., Latorre, J. D., Kuttappan, V. A., Hargis, B. M. and Hernandez-Velasco, X. 2015. Rye affects bacterial translocation, intestinal viscosity, microbiota composition and bone mineralization in turkey poults. *PLOS ONE* 10(4), e0122390. doi:10.1371/journal.pone.0122390.

Ujowundu, C. O., Igwe, C. U., Enemor, V. H. A., Nwaogu, L. A. and Okafor, O. E. 2008. Nutritive and anti-nutritive properties of *Boerhavia diffusa* and *Commelina nudiflora* leaves. *Pakistan J. Nutr.* 7(1), 90–2. doi:10.3923/pjn.2008.90.92.

Veldman, A. and Vahl, H. A. 1994. Xylanase in broiler diets with differences in characteristics and content of wheat. *Br. Poult. Sci.* 35(4), 537–50. doi:10.1080/00071669408417719.

Wood, P. J. 2001. Cereal β-glucans: structure, properties and health claims. In: McCleary, B. V. and Prosky, L. (Eds), *Advanced Dietary Fibre Technology*. Blackwell Science Ltd., Oxford, UK.

Wrigley, C. 2010. Cereal-grain morphology and composition. In: Wrigley, C. (Ed.), *Cereal Grains: Assessing and Managing Quality*. Woodhead Publishing Series in Food Science, Technology and Nutrition. Elsevier, Duxford, UK, pp. 24–44. doi:10.1533/9781845699529.1.24.

Wrigley, C. 2017. Cereal-grain morphology and composition. In: Wrigley, C. (Ed.), *Cereal Grains: Assessing and Managing Quality* (2nd edn.). Woodhead Publishing Series in Food Science, Technology and Nutrition. Elsevier, Duxford, UK, pp. 55–87. doi:10.1016/B978-0-08-100719-8.00004-8.

Yegani, M. and Korver, D. R. 2008. Factors affecting intestinal health in poultry. *Poult. Sci.* 87(10), 2052–63. doi:10.3382/ps.2008-00091.

Yeoman, C. J., Chia, N., Jeraldo, P., Sipos, M., Goldenfeld, N. D. and White, B. A. 2012. The microbiome of the chicken gastrointestinal tract. *Anim. Health Res. Rev.* 13(1), 89–99. doi:10.1017/S1466252312000138.

Zeeman, S. C., Kossmann, J. and Smith, A. M. 2010. Starch: its metabolism, evolution, and biotechnological modification in plants. *Annu. Rev. Plant Biol.* 61, 209–34. doi:10.1146/annurev-arplant-042809-112301.

Index